Information Theoretic Security and Privacy of Information Systems

Gain a solid understanding of how information theoretic approaches can inform the design of more secure information systems and networks with this authoritative text. With a particular focus on theoretical models and analytical results, leading researchers show how techniques derived from the principles of source and channel coding can provide new ways of addressing issues of data security, embedded security, privacy, and authentication in modern information systems. A wide range of wireless and cyber-physical systems is considered, including 5G cellular networks, the Tactile Internet, biometric identification systems, online data repositories, and smart electricity grids. This is an invaluable guide for both researchers and graduate students working in communications engineering, and industry practitioners and regulators interested in improving security in the next generation of information systems.

Rafael F. Schaefer is an Assistant Professor at the Technische Universität Berlin, having previously worked at Princeton University.

Holger Boche is a Full Professor at the Technische Universität München, a member of the German Academy of Sciences, and a Fellow of the IEEE. He is also a co-editor of *Mechanisms and Games for Dynamic Spectrum Allocation* (Cambridge University Press, 2013).

Ashish Khisti is an Associate Professor and a Canada Research Chair in the Department of Electrical and Computer Engineering at the University of Toronto.

H. Vincent Poor is the Michael Henry Strater University Professor at Princeton University, a member of the US National Academies of Engineering and Sciences, and a Fellow of the IEEE. He has co-authored and co-edited several books, including *Quickest Detection* (Cambridge University Press, 2008) and *Smart Grid Communications and Networking* (Cambridge University Press, 2012).

Information Theoretic Security and Privacy of Information Systems

RAFAEL F. SCHAEFER
Technische Universität Berlin

HOLGER BOCHE
Technische Universität München

ASHISH KHISTI
University of Toronto

H. VINCENT POOR
Princeton University

CAMBRIDGE
UNIVERSITY PRESS

University Printing House, Cambridge CB2 8BS, United Kingdom

One Liberty Plaza, 20th Floor, New York, NY 10006, USA

477 Williamstown Road, Port Melbourne, VIC 3207, Australia

4843/24, 2nd Floor, Ansari Road, Daryaganj, Delhi – 110002, India

79 Anson Road, #06-04/06, Singapore 079906

Cambridge University Press is part of the University of Cambridge.

It furthers the University's mission by disseminating knowledge in the pursuit of education, learning, and research at the highest international levels of excellence.

www.cambridge.org
Information on this title: www.cambridge.org/9781107132269
10.1017/9781316450840

© Cambridge University Press 2017

This publication is in copyright. Subject to statutory exception
and to the provisions of relevant collective licensing agreements,
no reproduction of any part may take place without the written
permission of Cambridge University Press.

First published 2017

Printed in the United Kingdom by Clays, St Ives plc

A catalogue record for this publication is available from the British Library

ISBN 978-1-107-13226-9 Hardback

Cambridge University Press has no responsibility for the persistence or accuracy of URLs for external or third-party Internet Web sites referred to in this publication, and does not guarantee that any content on such Web sites is, or will remain, accurate or appropriate.

Contents

List of Contributors		*page* xv
Preface		xix

Part I Theoretical Foundations — 1

1 Effective Secrecy: Reliability, Confusion, and Stealth — 3
J. Hou, G. Kramer, and M. Bloch

- 1.1 Introduction — 3
- 1.2 Preliminaries — 5
  - 1.2.1 Terminology — 5
  - 1.2.2 Notation — 6
  - 1.2.3 Wiretap Channel — 6
- 1.3 Effective Secrecy Capacity — 7
  - 1.3.1 Examples — 8
  - 1.3.2 Achievability — 9
  - 1.3.3 Converse — 13
  - 1.3.4 Choice of Security Measures — 15
- 1.4 Hypothesis Testing — 16
- 1.5 Conclusion — 18
- *References* — 19

2 Error-Free Perfect Secrecy Systems — 21
S.-W. Ho, T. Chan, A. Grant, and C. Uduwerelle

- 2.1 Introduction — 21
  - 2.1.1 Organization — 24
  - 2.1.2 Notation — 25
- 2.2 Error-Free Perfect Secrecy Systems — 25
- 2.3 Residual Secret Randomness and Expected Key Consumption — 30
  - 2.3.1 The First Justification of $H(R|U,X)$ — 32
  - 2.3.2 The Second Justification of $H(R|U,X)$ — 36
  - 2.3.3 Some Properties of $I(R;U,X)$ and $I(R;X)$ — 37

	2.4	Tradeoff between Key Consumption and Number of Channel Uses	38
		2.4.1 Minimal Expected Key Consumption	39
		2.4.2 Minimal Number of Channel Uses	45
		2.4.3 The Fundamental Tradeoff	46
	2.5	Proof of Theorem 2.8	47
	2.6	Conclusion	48
	References		50

3 Secure Source Coding 51
P. Cuff and C. Schieler

3.1	Introduction	51
3.2	Lossless Compression, Perfect Secrecy	53
	3.2.1 Discrete Memoryless Source	55
3.3	Lossless Compression, Imperfect Secrecy	58
	3.3.1 Rate–Distortion Theory	58
	3.3.2 Henchman	62
	3.3.3 Repeated Guessing	64
	3.3.4 Best Guess	65
3.4	Lossy Compression, Imperfect Secrecy	67
	3.4.1 Rate–Distortion Theory	68
	3.4.2 Henchman	70
3.5	Equivocation	72
References		75

4 Networked Secure Source Coding 77
K. Kittichokechai, T. J. Oechtering, and M. Skoglund

4.1	Introduction	77
	4.1.1 Information Theoretic Measure for Privacy	78
	4.1.2 Fundamental Tradeoff	78
	4.1.3 Models	79
4.2	Lossy Source Coding in the Presence of an Eavesdropper	81
	4.2.1 Problem Setup	81
	4.2.2 Rate–Distortion–Leakage Tradeoff Result	82
	4.2.3 Example	84
4.3	Network with One Helper	85
	4.3.1 Eavesdropper Observes Helper Description Link	85
	4.3.2 Eavesdropper Observes Source Description Link	90
4.4	Network with an Intermediate Node	92
	4.4.1 Models	93
	4.4.2 Problem Setup	93
	4.4.3 Rate–Distortion–Leakage Tradeoff Results	94
4.5	Network with Reconstruction Privacy	96
	4.5.1 End-User Privacy at Eavesdropper	98

		4.5.2 Causal Side Information	100
		4.5.3 Example	101
	4.6	Open Problems and Concluding Remark	102
	References		104

Part II Secure Communication   107

5 Secrecy Rate Maximization in Gaussian MIMO Wiretap Channels   109
S. Loyka and C. D. Charalambous

5.1	Introduction		109
5.2	MIMO Wiretap Channel		111
5.3	Rank-1 Solution		113
5.4	Full-Rank Solution		113
5.5	Weak Eavesdropper		117
5.6	Isotropic Eavesdropper and Capacity Bounds		120
	5.6.1	High SNR Regime	124
	5.6.2	When Is the Eavesdropper Negligible?	125
	5.6.3	Low SNR Regime	126
5.7	Omnidirectional Eavesdropper		126
5.8	Identical Right Singular Vectors		127
5.9	When Is ZF Signaling Optimal?		129
5.10	When Is Standard Water-Filling Optimal?		130
5.11	When Is Isotropic Signaling Optimal?		131
5.12	An Algorithm for Global Maximization of Secrecy Rates		131
5.13	Appendix		135
	5.13.1	Proof of Theorem 5.4	135
	5.13.2	Proof of Theorem 5.5	136
References			137

6 MIMO Wiretap Channels   140
M. Nafea and A. Yener

6.1	Introduction		140
6.2	Secrecy Capacity of the MIMO Wiretap Channel		142
6.3	High-SNR Secrecy Capacity for the MIMO Wiretap Channel		146
6.4	MIMO Wiretap Channel with a Cooperative Jammer		148
	6.4.1	Converse for Theorem 6.3	150
	6.4.2	Achievability for Theorem 6.3	153
	6.4.3	Interpretation of Results	158
6.5	MIMO Wiretap Channel with Unknown Eavesdropper Channel		159
	6.5.1	Proof of Theorem 6.4	161
6.6	MIMO MAC Gaussian Wiretap Channel with Unknown Eavesdropper Channel		175
6.7	Conclusions		178
References			178

7	**MISO Wiretap Channel with Strictly Causal CSI: A Topological Viewpoint**	181
	Z. H. Awan and A. Sezgin	
7.1	Introduction	181
7.2	State of the Art and Preliminaries	181
7.3	Secrecy Capacity Characterization	182
7.4	Secure Degrees of Freedom	184
	7.4.1 Impact of CSIT	184
	7.4.2 Topological Diversity	189
7.5	GSDoF of MISO Wiretap Channel with Delayed CSIT	189
	7.5.1 Upper Bound	191
	7.5.2 Fixed Topology ($\lambda_{11} = 1$)	192
	7.5.3 Fixed Topology ($\lambda_{1\alpha} = 1$)	192
7.6	Discussion and Directions for Future Work	194
	7.6.1 MISO Broadcast Channel with Topology	194
	7.6.2 Synergistic Benefits of CSIT with Topology	194
	7.6.3 Extension to K-User Case	195
7.7	Appendix	195
	7.7.1 Entropy	195
	7.7.2 Degrees of Freedom	196
	7.7.3 Generalized Degrees of Freedom	196
	References	197

8	**Physical-Layer Security with Delayed, Hybrid, and Alternating Channel State Knowledge**	200
	P. Mukherjee, R. Tandon, and S. Ulukus	
8.1	Introduction	200
8.2	The MISO Broadcast Channel	202
	8.2.1 Modeling the Quality of CSIT	203
	8.2.2 Security Requirements	204
	8.2.3 A Degrees-of-Freedom Perspective	204
8.3	The MISO BCCM in Homogeneous CSIT States	205
	8.3.1 Perfect CSIT from Both Users (PP)	205
	8.3.2 No CSIT from Any User (NN)	206
	8.3.3 Delayed CSIT from Both Users (DD)	206
8.4	The MISO BCCM in Hybrid CSIT States	211
	8.4.1 Achievable Schemes for States PN and DN	211
	8.4.2 Converse Tools	213
8.5	The MISO BCCM in Alternating CSIT States	214
	8.5.1 Achievability	220
8.6	Conclusions and Open Problems	225
	References	226

9 Stochastic Orders, Alignments, and Ergodic Secrecy Capacity — 231
P.-H. Lin and E. A. Jorswieck

- 9.1 Introduction — 231
- 9.2 Preliminaries — 233
  - 9.2.1 Properties of Wiretap Channels — 233
  - 9.2.2 Properties of Stochastic Orders — 234
- 9.3 System Model — 235
- 9.4 The Relation between Degradedness and Stochastic Orders — 236
- 9.5 The Fast Fading Wiretap Channel with Statistical CSIT — 240
  - 9.5.1 The Layered Erasure Wiretap Channel — 240
  - 9.5.2 The Fast Fading Gaussian Wiretap Channel with Statistical CSIT — 242
- 9.6 Numerical Results — 246
- 9.7 Multiple-Antenna Fading Wiretap Channel with Statistical CSIT — 247
  - 9.7.1 Multiple Antennas without Channel Enhancement — 247
  - 9.7.2 Multiple Antennas with Channel Enhancement — 254
- 9.8 Conclusion — 255
- *References* — 256

10 The Discrete Memoryless Arbitrarily Varying Wiretap Channel — 258
J. Nötzel, M. Wiese, and H. Boche

- 10.1 Introduction — 258
  - 10.1.1 System Model — 258
  - 10.1.2 Historical Background — 260
  - 10.1.3 New Approaches and New Results — 267
- 10.2 Notation and Definitions — 272
  - 10.2.1 Basic Notation — 272
  - 10.2.2 Models and Operational Definitions — 274
- 10.3 Main Results and Insights — 278
  - 10.3.1 Assisted Capacities: Coding Theorems for $C_{S,\mathrm{ran}}^{\mathrm{mean}}$, $C_{S,\mathrm{ran}}^{\mathrm{max}}$, and C_{key} — 278
  - 10.3.2 The Non-Assisted Capacity — 280
  - 10.3.3 Open Questions — 287
- 10.4 Proofs and Intermediate Technical Results — 287
  - 10.4.1 Technical Definitions, Results, and Facts — 288
  - 10.4.2 Basic Quantities and Estimates — 290
  - 10.4.3 Proofs of Lemmas — 293
  - 10.4.4 Proofs of Theorems — 295
- *References* — 309

11	Super-Activation as a Unique Feature of Secure Communication over Arbitrarily Varying Channels	313

R. F. Schaefer, H. Boche, and H. V. Poor

11.1	Introduction	313
11.2	Problem Motivation	315
11.3	Notation	316
11.4	Arbitrarily Varying Wiretap Channel	316
	11.4.1 System Model	317
	11.4.2 Code Concepts	319
	11.4.3 Capacity Results	321
11.5	Super-Activation of Orthogonal AVWCs	322
11.6	Super-Additivity of Orthogonal AVCs	324
11.7	Discussion	326
	References	328

Part III Secret Key Generation and Authentication   331

12	Multiple Secret Key Generation: Information Theoretic Models and Key Capacity Regions	333

H. Zhang, Y. Liang, L. Lai, and S. Shamai (Shitz)

12.1	Introduction	333
12.2	Hierarchical Model	335
	12.2.1 Model Description	335
	12.2.2 Key Capacity Region	336
12.3	Cellular Model	340
	12.3.1 Two Key Generation over Three Terminals	340
	12.3.2 Two Key Generation Assisted by a Helper	343
12.4	Generating Multiple Keys under the PIN Model	352
	12.4.1 Two Pairs Case	353
	12.4.2 General Case	355
12.5	Discussion and Future Topics	357
	References	359

13	Secret Key Generation for Physical Unclonable Functions	362

M. Pehl, M. Hiller, and G. Sigl

13.1	Introduction	362
13.2	Notation	364
13.3	An Information Theoretical View on Key Storage with PUFs	365
	13.3.1 Source Model	365
	13.3.2 Communication Channel	366
	13.3.3 Key Agreement	366
	13.3.4 Rate and Capacity	367
	13.3.5 Attack Vectors	368
	13.3.6 Summary	369

		13.4	Unified Algebraic Description of Secret Key and Helper Data Generation	370
			13.4.1 Background for Further Analysis	372
			13.4.2 Vulnerability of the Pre- and Postprocessing	372
			13.4.3 Leakage of the Algebraic Core	374
		13.5	Algebraic Core Representations of State-of-the-Art Helper Data Generation	377
			13.5.1 Fuzzy Commitment	377
			13.5.2 Code-Offset Fuzzy Extractor	378
			13.5.3 Fuzzy Extractor with Syndrome Construction	378
			13.5.4 Parity Construction	379
			13.5.5 Systematic Low Leakage Coding	379
			13.5.6 Index-Based Syndrome Coding	380
			13.5.7 Complementary IBS	383
			13.5.8 Summary of State-of-the-Art Syndrome Decoders	385
		13.6	Conclusions	387
		References		387
14	**Wireless Physical-Layer Authentication for the Internet of Things**			390
	G. Caparra et al.			
		14.1	IoT Authentication Overview	390
		14.2	State of the Art	392
			14.2.1 Physical-Layer Authentication	393
		14.3	IoT Channel-Based Authentication	395
			14.3.1 Authentication Protocol	396
			14.3.2 Authentication Protocol Performance	399
		14.4	Centralized Anchor Node Selection	402
			14.4.1 Energy-Efficient Anchor Node Selection	404
			14.4.2 Signaling-Efficient Anchor Selection	405
			14.4.3 A Tradeoff between Energy Efficiency and Signaling Efficiency	407
		14.5	Distributed Anchor Node Selection	410
			14.5.1 Distributed Configuration Selection	410
			14.5.2 Distributed SNR-Based Anchor Node Selection	411
		14.6	Performance Summary and Conclusions	414
			14.6.1 Summary	414
		References		415
Part IV	**Data Systems and Related Applications**			419
15	**Information Theoretic Analysis of the Performance of Biometric Authentication Systems**			421
	T. Ignatenko and F. M. J. Willems			
		15.1	Introduction	421
			15.1.1 Chapter Organization	423

	15.2	Enrollment and Authentication Statistics	423
	15.3	Traditional Authentication Systems	424
		15.3.1 Scenario and Objective	424
		15.3.2 Achievability Definition and Result	424
		15.3.3 Discussion	426
	15.4	Rate-Constrained Authentication Systems	427
		15.4.1 Scenario and Objective	427
		15.4.2 Achievability Definition and Result	427
		15.4.3 Discussion	429
	15.5	Secret-Key-Based Authentication Systems	430
		15.5.1 Scenario and Objective	430
		15.5.2 System Building Blocks: Encoder, Decoder, and Equality Checker; FRR and mFAR	430
		15.5.3 Definition of Achievability and Statement of Result	431
		15.5.4 Relation to Ahlswede–Csiszár Secret Generation	431
		15.5.5 Discussion	434
	15.6	Proof of Theorem 15.3	434
		15.6.1 Achievability Proof for Theorem 15.3	434
		15.6.2 Converse for Theorem 15.3	437
	15.7	Privacy Leakage in Secret-Based Systems	439
	15.8	Proof of Theorem 15.5	439
		15.8.1 Achievability Part for Theorem 15.5	440
		15.8.2 Converse for Theorem 15.5	442
	15.9	Conclusions and Final Remarks	443
	References		443
16	**Joint Privacy and Security of Multiple Biometric Systems**		**445**
	A. Goldberg and S. C. Draper		
	16.1	Introduction	445
		16.1.1 Biometric System Design Requirements	446
		16.1.2 Security and Privacy Leakage	447
		16.1.3 Related Work	450
	16.2	Problem Formulation	451
	16.3	Design Space	455
		16.3.1 Geometric Intuition	457
		16.3.2 Scaling Complexity	460
		16.3.3 Equivalent Designs	461
	16.4	The Fixed-Basis Case	462
		16.4.1 Impact of the Restriction	462
		16.4.2 Optimization of Fixed-Basis Designs	463
		16.4.3 Resulting Privacy/Security Tradeoff	469
		16.4.4 Observed Form of Optimal Solutions	470
	16.5	Conclusions	471
	References		471

17 Information Theoretic Approaches to Privacy-Preserving Information Access and Dissemination 473
G. Fanti and K. Ramchandran

- 17.1 Introduction 473
- 17.2 Information Dissemination 475
  - 17.2.1 Anonymous Broadcast Messaging 476
  - 17.2.2 Anonymous Point-to-Point Messaging 482
- 17.3 Information Access 483
  - 17.3.1 Adversarial Models 484
  - 17.3.2 Private Information Retrieval 485
  - 17.3.3 Private Streaming Search 491
  - 17.3.4 Encrypted Databases 493
- 17.4 Conclusions 493
- *References* 494

18 Privacy in the Smart Grid: Information, Control, and Games 499
H. V. Poor

- 18.1 Introduction 499
- 18.2 Information: A General Formalism 500
- 18.3 Control: Smart Meter Privacy 504
- 18.4 Games: Competitive Privacy 510
- 18.5 Conclusion 515
- *References* 516

19 Security in Distributed Storage Systems 519
S. El Rouayheb, S. Goparaju, and K. Ramchandran

- 19.1 Introduction 519
  - 19.1.1 Related Literature 521
  - 19.1.2 Organization 522
- 19.2 Model and Notation 522
  - 19.2.1 System Model 522
  - 19.2.2 Security and Adversary Model 524
  - 19.2.3 Secrecy Capacity and Notation 524
  - 19.2.4 Flow Graph Representation 526
- 19.3 Secrecy against Passive Eavesdropping 526
  - 19.3.1 Upper Bound on the Secrecy Capacity 527
  - 19.3.2 Achievability of the Secrecy Capacity 528
  - 19.3.3 Secrecy via Separation Schemes 531
  - 19.3.4 Separation Secrecy Capacity of Linear Optimal Repair MDS Codes 533
  - 19.3.5 Universally Secure Optimal Repair MDS Codes 537
- 19.4 Security against an Omniscient Adversary 537
  - 19.4.1 Upper Bound on the Resiliency Capacity 538
  - 19.4.2 Achievability of the Resiliency Capacity Upper Bound 539

19.5	Security against a Limited-Knowledge Adversary		541
	19.5.1	Resiliency Capacity	542
	19.5.2	Secure Scheme Example	542
	19.5.3	Proof of Theorem 19.9	545
19.6	Conclusion and Open Problems		547
References			548

Index   554

Contributors

Zohaib Hassan Awan
Lehrstuhl für Digitale Kommunikationssysteme, Ruhr-Universität Bochum

Matthieu Bloch
School of Electrical and Computer Engineering, Georgia Institute of Technology

Holger Boche
Chair of Theoretical Information Technology, Technische Universität München

Gianluca Caparra
Department of Information Engineering, University of Padova

Marco Centenaro
Department of Information Engineering, University of Padova

Terence Chan
Institute for Telecommunications Research, University of South Australia

Charalambos D. Charalambous
Department of Electrical and Computer Engineering, University of Cyprus

Paul Cuff
Department of Electrical Engineering, Princeton University

Stark C. Draper
Department of Electrical and Computer Engineering, University of Toronto

Salim El Rouayheb
Department of Electrical and Computer Engineering, Illinois Institute of Technology

Giulia Fanti
Department of Electrical Engineering and Computer Science, University of California, Berkeley

Adina Goldberg
Department of Electrical and Computer Engineering, University of Toronto

Sreechakra Goparaju
Qualcomm Institute, University of California, San Diego

Alex Grant
Myriota, Adelaide, Australia

Matthias Hiller
Fraunhofer Institute for Applied and Integrated Security

Siu-Wai Ho
Institute for Telecommunications Research, University of South Australia

Jie Hou
European Patent Office Munich

Tanya Ignatenko
Electrical Engineering Department, Eindhoven University of Technology

Eduard A. Jorswieck
Chair for Communications Theory, Technische Universität Dresden

Kittipong Kittichokechai
Communications and Information Theory Chair, Technische Universität Berlin

Gerhard Kramer
Chair of Communications Engineering, Technische Universität München

Lifeng Lai
Department of Electrical and Computer Engineering, University of California, Davis

Nicola Laurenti
Department of Information Engineering, University of Padova

Yingbin Liang
Department of Electrical Engineering and Computer Science, Syracuse University

Pin-Hsun Lin
Chair for Communications Theory, Technische Universität Dresden

Sergey Loyka
School of Electrical Engineering and Computer Science, University of Ottawa

Pritam Mukherjee
Department of Electrical and Computer Engineering, University of Maryland

Mohamed Nafea
Wireless Communications and Networking Laboratory (WCAN), Electrical Engineering Department, The Pennsylvania State University

Janis Nötzel
Física Teòrica: Informació i Fenòmens Quàntics, Universitat Autònoma de Barcelona

Tobias J. Oechtering
Information Science and Engineering Department, School of Electrical Engineering and ACCESS Linnaeus Center, KTH Royal Institute of Technology

Michael Pehl
Chair of Security in Information Technology, Technische Universität München

H. Vincent Poor
Department of Electrical Engineering, Princeton University

Kannan Ramchandran
Department of Electrical Engineering and Computer Science, University of California, Berkeley

Rafael F. Schaefer
Information Theory and Applications Chair, Technische Universität Berlin

Curt Schieler
Lincoln Laboratory, Massachusetts Institute of Technology

Aydin Sezgin
Lehrstuhl für Digitale Kommunikationssysteme, Ruhr-Universität Bochum

Shlomo Shamai (Shitz)
Department of Electrical Engineering, Technion-Israel Institute of Technology

Georg Sigl
Chair of Security in Information Technology, Technische Universität München
and
Fraunhofer Institute for Applied and Integrated Security

Mikael Skoglund
Information Science and Engineering Department, School of Electrical Engineering and ACCESS Linnaeus Center, KTH Royal Institute of Technology

Ravi Tandon
Department of Electrical and Computer Engineering, University of Arizona

Stefano Tomasin
Department of Information Engineering, University of Padova

Chinthani Uduwerelle
Institute for Telecommunications Research, University of South Australia

Sennur Ulukus
Department of Electrical and Computer Engineering, University of Maryland

Lorenzo Vangelista
Department of Information Engineering, University of Padova

Moritz Wiese
ACCESS Linnaeus Center and Automatic Control Lab, School of Electrical Engineering, KTH Royal Institute of Technology

Frans M. J. Willems
Electrical Engineering Department, Eindhoven University of Technology

Aylin Yener
Wireless Communications and Networking Laboratory (WCAN), Electrical Engineering Department, The Pennsylvania State University

Huishuai Zhang
Department of Electrical Engineering and Computer Science, Syracuse University

Preface

The ubiquity of information technologies such as wireless communications, biometric identification systems, online data repositories, and smart electricity grids has created new challenges in information security and privacy. Traditional approaches based on cryptography are often inadequate in such complex systems and fundamentally new techniques must be developed. Information theory provides fundamental limits that can guide the development of methods for addressing these challenges, and the purpose of this book is to introduce the reader to state-of-the-art developments in this field.

As a prototypical example of a system in which such methods can play an important role, one can consider a communication system. In a typical configuration, there is an architectural separation between data encryption and error correction in such systems. The encryption module is based on cryptographic principles and abstracts out the underlying communication channel as an ideal bit-pipe. The error correction module is typically implemented at the physical layer. It adds redundancy into the source message in order to combat channel impairments or multiuser interference and transforms the noisy communication channel into a reliable bit-pipe. While such a separation-based architecture has long been an obvious solution in most systems, a number of applications have emerged in recent years where encryption mechanisms must be aware of the noise structure in the underlying channel, and likewise the error correction and data compression methods must be aware of the associated secrecy constraints required by the application. Such joint approaches can be studied by developing new mathematical models of communication systems that impose both reliability constraints and secrecy constraints. Similar considerations arise throughout the information and communication technologies, and information theoretic approaches can point the way to fundamentally new solutions for such technologies. We refer to this emerging field of research as *information theoretic approaches to security and privacy* (ITASP). It is notable that this approach leads to guaranteeing information security irrespective of the computational power of the adversary and is a fundamental departure from current computation-based cryptographic solutions. In this book we will highlight among others the following application areas where principles of ITASP have been particularly effective.

Wireless systems: Mobile links are traditionally considered to be the weakest links in network security. The current separation-based architectures allow for a variety of attacks that exploit the broadcast nature of wireless links. Fortunately, when we consider ITASP, a number of new solutions emerge that have been traditionally overlooked by the separation-based architecture. In fact, for ITASP, the broadcast nature of wireless links turns out to be a strength, allowing for new methods of secret key generation, as well as key-less confidential data transmission, to be developed. Physical layer design of wireless systems has dramatically advanced in the last decade to support the growing demands from end users effectively utilizing cooperative relays, multiple-antenna arrays, and channel adaptation strategies, and these advances are making their way into reality. Moreover, emerging concepts such as the Internet of Things introduce network architectural and scaling issues that make traditional methods of providing information security impractical. It is only natural that we now can envision that the physical-layer-based approaches rooted in ITASP can significantly enhance the security and privacy of wireless networks in the near future.

Biometric systems: Biometrics such as fingerprints, iris scan, etc. provide the most compelling means of authentication as they directly use the physical attributes of a user. Surprisingly, practical systems suffer from a serious privacy concern. Many of the commercially available systems store the biometric reading measured during the enrollment phase in the clear. This is because successive biometric readings of the same user are somewhat different due to measurement noise. Thus the one-way hash functions widely used when storing passwords cannot be used in such systems. Interestingly, it has been recently recognized that a class of secure hash functions that are robust to measurement noise can be implemented in such systems. The principles of such robust hash functions are intimately tied to a well-studied problem in ITASP, i.e., the "source model" in secret key generation.

Smart-grid systems: There has been a strong push toward modernizing the electric power grid in most developed countries, using advanced metering infrastructure (AMI) and sensors within the grid. Privacy and security concerns are a major issue in the design and deployment of such systems. For example, AMI systems report real-time measurements of the electricity consumption of each individual household. While such information can be vital for load balancing, it also raises serious privacy concerns, as end-user behavior can be easily inferred from the instantaneous electricity usage. Thus a fundamental tradeoff between utility and privacy exists in such systems, and the ITASP framework is a natural way of characterizing such a tradeoff.

This book provides scientific insights into the emerging field of ITASP. Presenting contributions from prominent researchers in this area, the book not only gives an overview of state-of-the-art research results but also builds a strong foundation for designing future systems incorporating ITASP. The book is organized in four parts: (I) theoretical foundations; (II) secure communication; (III) secret key generation and authentication; and (IV) data systems and related applications. The first part focuses on fundamental concepts of information theoretic security in general. The second part of the book focuses on secure communication. These range from multiple-input multiple-output wiretap channels to different concepts of imperfect channel state information and its implications. The third part discusses secret key generation and authentication over the wireless channel. And finally, the fourth part focuses on data systems and related applications such as biometric authentication, smart grid, and distributed storage systems.

While this treatment primarily adopts an analytical approach, implications of theoretical results and associated insights are discussed explicitly. Often the study of relatively simple information theoretic models can lead to surprising insights that can fundamentally change the way we approach the design of security mechanisms. Thus a significant number of chapters presented in this book are devoted to the study of such models. With a modern presentation style, the book aims to go beyond a dry recapture of published theoretical results and reach a broader audience by providing insights to the consequences of technological and theoretical developments. The reader will also benefit from diverse perspectives on the underlying issues brought to the book by multiple prominent contributing authors.

This treatment is suitable for graduate students and other researchers who wish to gain an entrée into this field, or to expand their existing knowledge of it. It can serve, for example, as the basis for an advanced graduate course in ITASP. It is also of interest to practicing engineers and computer scientists working in communications and information technology who wish to gain an understanding of the possibilities of this emerging field.

Part I

Theoretical Foundations

1 Effective Secrecy: Reliability, Confusion, and Stealth

Jie Hou, Gerhard Kramer, and Matthieu Bloch

A security measure called effective secrecy is defined that includes strong secrecy and stealth communication. Effective secrecy ensures that a message cannot be deciphered and that the presence of meaningful communication is hidden. To measure stealth we use resolvability via informational divergence and relate this to binary hypothesis testing. Results are developed for wiretap channels.

1.1 Introduction

The wiretap channel is depicted in Fig. 1.1 and has a message M that should be decoded reliably at one receiver (Bob) while being kept secret from a second receiver (Eve). Wyner [1] derived the *secrecy capacity* when the channel $P_{YZ|X}$ is *physically degraded*, i.e., X–Y–Z forms a Markov chain. Csiszár and Körner [2] extended the results to broadcast channels with confidential messages. In both [1] and [2], secrecy is measured by a *normalized* mutual information between M and Eve's output string $Z^n = Z_1 Z_2 \ldots Z_n$, i.e., the secrecy requirement is

$$\frac{1}{n} I(M; Z^n) \leq S, \tag{1.1}$$

where we interpret S as a *leakage* rate. An interesting case is to choose S positive and small, in which case the requirement (1.1) is referred to as *weak secrecy*. However, as $n \to \infty$ the eavesdropper can obtain nS bits of M, which grows with n.

Instead, the papers [3, 4] advocated using *strong secrecy* where secrecy is measured by the *unnormalized* mutual information $I(M; Z^n)$ and one requires

$$I(M; Z^n) \leq \xi \tag{1.2}$$

for any $\xi > 0$ and sufficiently large n. We remark that Wyner's random codes already ensured strong secrecy since S in (1.1) can scale inverse-exponentially with n. Also, for *finite n*, whether we use (1.1) with $S = \xi/n$ or (1.2) is obviously immaterial, i.e., the distinction between weak and strong secrecy is of asymptotic nature only.

In related work, Han and Verdú [5] studied *resolvability* based on *variational distance* that addresses the number of bits needed to mimic a marginal distribution of a prescribed joint distribution. Hayashi [6] and Bloch and Laneman [7] used resolvability to prove secrecy, and they extended results in [2] to continuous random variables and channels

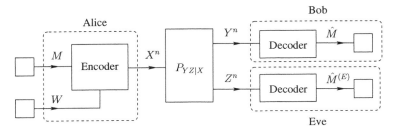

Figure 1.1 A wiretap channel with message M and random symbol W.

with memory. We also use the resolvability-based approach but replace variational distance by informational divergence (or Kullback–Leibler divergence).

The main contribution of this work is to define and justify the usefulness of a security measure that includes not only reliability and strong secrecy but also *stealth*. In particular, we measure secrecy by the informational divergence

$$D(P_{MZ^n} \| P_M Q_{Z^n}), \tag{1.3}$$

where P_{MZ^n} is the joint distribution of MZ^n, P_M is the distribution of M, P_{Z^n} is the distribution of Z^n, and Q_{Z^n} is the distribution that the eavesdropper expects to observe when the source is *not* communicating useful messages. We call this security measure *effective secrecy*. We show that classic random codes achieve effective secrecy by using a recently developed simplified proof [8] of resolvability based on informational divergence (see also [9, Lemma 11]).

It turns out that the effective secrecy measure (1.3) was considered a few months before our work [10] by Han, Endo, and Sasaki [11, 12]. Their motivation for using (1.3) was simply that it gives a secrecy measure that is stronger than strong secrecy. Our motivation was operational: the divergence (1.3) measures secrecy and stealth simultaneously. In particular, one can check that (see (1.7) below)

$$D(P_{MZ^n} \| P_M Q_{Z^n}) = \underbrace{I(M;Z^n)}_{\text{secrecy measure}} + \underbrace{D(P_{Z^n} \| Q_{Z^n})}_{\text{stealth measure}}, \tag{1.4}$$

where we measure secrecy and stealth by using $I(M;Z^n)$ and $D(P_{Z^n} \| Q_{Z^n})$, respectively. We justify the latter measure by using binary hypothesis testing in Section 1.4. Thus, by making $D(P_{MZ^n} \| P_M Q_{Z^n}) \to 0$ we not only keep the message secret from the eavesdropper but also hide the presence of meaningful communication. Of course, one can instead study secrecy and stealth separately rather than using (1.4); see Section 1.3.4 below. We combine these concepts mainly for convenience of the proofs.

The choice of default behavior Q_{Z^n} in (1.3) and (1.4) will depend on the application. For example, if the default behavior is to send a codeword, then $Q_{Z^n} = P_{Z^n}$ and one achieves stealth for free. On the other hand, if the default behavior is $Q_{Z^n} = Q_Z^n$, where Q_Z^n is a product distribution (see Section 1.2.2 below), then code design requires more care. We mostly focus on the case $Q_{Z^n} = Q_Z^n$.

This paper is organized as follows. In Section 1.2, we review terminology on low probability of detection and low probability of intercept communications. We further describe notation and state the problem we study. In Section 1.3 we state and prove the main result. Section 1.4 relates this result to hypothesis testing, and Section 1.5 concludes the paper.

1.2 Preliminaries

1.2.1 Terminology

Many applications require hiding communication or information, and often the same concept is labeled with different words. We therefore begin by reviewing terminology in selected documents, and describe how we use the word *stealth*.

The United States Committee on National Security Systems Glossary [13] has the following definitions:

- Low probability of detection (LPD): Result of measures used to hide or disguise intentional electromagnetic transmissions.
- Low probability of intercept (LPI): Result of measures used to resist attempts by adversaries to analyze the parameters of a transmission to determine if it is a signal of interest.

The document [14, p. 6] has similar but slightly different terminology. There, LPI refers generically to communication methods whose primary purpose

is to prevent an unauthorized listener from determining the presence or location of the transmitter, in order to decrease the possibility of both electronic attack (jamming) and physical attack.

The same document [14, p. 9] refers to [15] as describing

four sequential operations that exploitation systems attempt to perform:

1. Cover the signal, that is, a receiver is tuned to some or all of the frequency intervals being occupied by the signal when the signal is actually being transmitted.
2. Detect the signal, that is, make a decision about whether the power in the intercept bandwidth is a signal plus noise and interference or just noise and interference.
3. Intercept the signal, that is, extract features of the signal to determine if it is a signal of interest or not.
4. Exploit the signal, that is, extract additional signal features as necessary and then demodulate the baseband signal to generate a stream of binary digits.

Secrecy deals with operation (4), i.e., the secrecy constraint prevents exploitation of the signal to generate a stream of (meaningful) binary digits. Our focus in this paper is on either operation (2) or (3), depending on how one interprets the above text. For example, suppose the default behavior Q_{Z^n} has Alice sending a signal whose power is (or more generally whose statistics are) sufficiently similar to interference. In this case, we are interested in operation (2). As a second example, suppose the default behavior Q_{Z^n} has Alice sending either a message-carrying signal or a default signal at irregular intervals with low probability. In this case, it does not matter if Eve detects that a signal was transmitted as long as she cannot determine if it is a message-carrying signal or not. We are thus concerned with operation (3).

We generically refer to scenarios in which message-carrying signals are shaped to resemble an innocent signal as *stealth* communication. The extreme scenario where signals are shaped to hide in noise has been referred to as covert communication [16], deniable communication [17], and undetectable communication [18]. This extreme type of stealth usually requires that X^n is a zero-power string so that no positive communication rate is possible. More precisely, the number of bits that can be communicated reliably over a noisy channel (often) scales as $\sqrt{n}$ [16, 17, 19, 20]. However, as we have seen, stealth communication rates are positive if random strings such as codewords are sent even if no information transmission occurs.

1.2.2 Notation

Random variables are written with upper case letters and their realizations with the corresponding lower case letters. Superscripts denote strings of variables/symbols, e.g., $X^n = X_1 X_2 \ldots X_n$. Subscripts denote the position of a variable/symbol in a string. For instance, X_i denotes the ith variable in X^n. We use X_i^n to denote $X_i, \ldots, X_n$, $1 \leq i \leq n$. A random variable X has probability distribution P_X and the support of P_X is denoted as $\mathrm{supp}(P_X)$. We write probabilities with subscripts $P_X(x)$ but we drop the subscripts if the arguments of the distribution are lower case versions of the random variables. For example, we write $P(x) = P_X(x)$. If the X_i, $i = 1, \ldots, n$, are independent and identically distributed (i.i.d.) according to P_X, then we have $P(x^n) = \prod_{i=1}^n P_X(x_i)$ and we write $P_{X^n} = P_X^n$. We also use Q_X^n to refer to strings of i.i.d. random variables. Calligraphic letters denote sets. The size of a set $\mathcal{S}$ is denoted as $|\mathcal{S}|$ and the complement is denoted as $\mathcal{S}^c$. For X with alphabet $\mathcal{X}$, we denote $P_X(\mathcal{S}) = \sum_{x \in \mathcal{S}} P_X(x)$ for any $\mathcal{S} \subseteq \mathcal{X}$. We use $\mathcal{T}_\epsilon^n(P_X)$ to denote the set of letter-typical strings (or finite sequences) of length n with respect to the probability distribution P_X and the non-negative number ϵ [21, Ch. 3], [22], i.e., we have

$$\mathcal{T}_\epsilon^n(P_X) = \left\{ x^n : \left| \frac{N(a|x^n)}{n} - P_X(a) \right| \leq \epsilon P_X(a), \forall a \in \mathcal{X} \right\},$$

where $N(a|x^n)$ is the number of occurrences of a in x^n.

1.2.3 Wiretap Channel

Consider the wiretap channel depicted in Fig. 1.1. Alice has a message M that is destined for Bob but should be kept secret from Eve. The message M is uniformly distributed over $\{1, \ldots, L\}$, $L = 2^{nR}$, and an encoder $f(\cdot)$ maps M, W to the string

$$X^n = f(M, W) \tag{1.5}$$

with the help of a random variable W that is independent of M and uniformly distributed over $\{1, \ldots, L_1\}$, $L_1 = 2^{nR_1}$. The purpose of W is to confuse Eve so that she learns little about M. X^n is transmitted through a memoryless channel $P_{YZ|X}$. Bob observes the channel output Y^n while Eve observes Z^n. The pair MZ^n has the joint distribution P_{MZ^n}. Bob estimates $\hat{M}$ from Y^n and the average error probability is

$$P_e^{(n)} = \Pr\left[\hat{M} \neq M\right]. \tag{1.6}$$

Eve tries to learn M from Z^n and secrecy is measured by

$$D(P_{MZ^n}\|P_M Q_{Z^n}) = \sum_{\substack{(m,z^n)\\ \in \mathrm{supp}(P_{MZ^n})}} P(m,z^n)\log\left(\frac{P(m,z^n)}{P(m)\cdot Q(z^n)}\cdot\frac{P(z^n)}{P(z^n)}\right)$$

$$= \sum_{\substack{(m,z^n)\\ \in \mathrm{supp}(P_{MZ^n})}} P(m,z^n)\left(\log\frac{P(z^n|m)}{P(z^n)} + \log\frac{P(z^n)}{Q(z^n)}\right)$$

$$= I(M;Z^n) + D(P_{Z^n}\|Q_{Z^n}), \tag{1.7}$$

where P_{Z^n} is the distribution that Eve observes and Q_{Z^n} is the *default* distribution that Eve expects to observe if Alice is *not* sending useful information.

For example, suppose Alice's default behavior is to transmit x^n with memoryless distribution $Q_X^n(x^n)$. The default output distribution is then

$$Q(z^n) = \sum_{x^n \in \mathrm{supp}(Q_X^n)} Q_X^n(x^n) P_{Z|X}^n(z^n|x^n) = P_Z^n(z^n), \tag{1.8}$$

where $P_Z(z) = \sum_x Q_X(x) P_{Z|X}(z|x)$. When Alice sends useful messages, then P_{Z^n} and Q_{Z^n} are different in general. But if we can make $D(P_{MZ^n}\|P_M Q_{Z^n})$ small then both $I(M;Z^n)$ and $D(P_{Z^n}\|Q_{Z^n})$ are small, which in turn implies (as we shall see) that Eve learns little about M and cannot recognize whether Alice is communicating anything meaningful.

We will consider the case $Q(z^n) = Q_Z^n(z^n)$, i.e., the default behavior has a memoryless distribution. Of course, other distributions may also be interesting. We say that a rate R is *achievable* if for any $\xi_1, \xi_2 > 0$ there is a sufficiently large n and an encoder and a decoder such that

$$P_e^{(n)} \leq \xi_1, \tag{1.9}$$

$$D(P_{MZ^n}\|P_M Q_Z^n) \leq \xi_2. \tag{1.10}$$

The *effective secrecy capacity* C_S is the supremum of the set of achievable R. We wish to determine C_S.

1.3 Effective Secrecy Capacity

We prove the following result:

THEOREM 1.1 *C_S is zero if there is no Q_X such that $P_Z = Q_Z$, and otherwise*

$$C_S = \max_{Q_{VX}: P_Z = Q_Z} [I(V;Y) - I(V;Z)], \tag{1.11}$$

where the maximization is over all joint distributions Q_{VX} and we have the Markov chain

$$V\text{--}X\text{--}YZ. \tag{1.12}$$

One may restrict the cardinality of V to $|\mathcal{V}| \leq |\mathcal{X}|$.

REMARK 1.1 Theorem 1.1 applies to random variables with discrete and finite alphabets. However, extensions to real-valued channels such as additive white Gaussian noise (AWGN) channels with power constraints are possible.

REMARK 1.2 If one can choose Q_Z to be the P_Z corresponding to the capacity-achieving output distribution of the wiretap channel, then the effective-secrecy capacity is the same as the wiretap channel capacity with (weak or) strong secrecy.

REMARK 1.3 Consider a physically degraded channel where X–Y–Z forms a Markov chain. We have (see [2, p. 342])

$$\begin{aligned}I(V;Y) - I(V;Z) &= I(V;Y|Z)\\ &\leq I(X;Y|Z)\\ &= I(X;Y) - I(X;Z),\end{aligned} \qquad (1.13)$$

so that choosing $V = X$ achieves capacity.

REMARK 1.4 The capacity (1.11) of a general wiretap channel depends only on the marginals $P(y|x)$ and $P(z|x)$. Hence, the capacity of a (stochastically degraded) channel whose marginals $P(y|x)$ and $P(z|x)$ are the same as those of a physically degraded channel has the same capacity as this physically degraded channel.

1.3.1 Examples

We consider two examples to show how the stealth requirement impacts C_S. These examples show that fixing $P_Z = Q_Z$ can make calculating C_S rather easy.

EXAMPLE 1.1 Consider the binary symmetric channels (BSCs)

$$Y = X \oplus A_1, \quad Z = X \oplus A_2, \qquad (1.14)$$

where the alphabet of all random variables is $\{0,1\}$, the operator $\oplus$ is addition modulo 2, A_1 and A_2 are independent of X, $P_{A_1}(1) = p_1$, and $P_{A_2}(1) = p_2$. Suppose that $0 \leq p_1 \leq p_2 \leq 1/2$. If $p_2 = 1/2$ then $I(X;Z) = 0$ and the only interesting case is uniform Q_Z, for which C_S is the same as the BSC capacity. So consider $p_2 < 1/2$ and suppose the default behavior is $Q_Z(1) = q$, where $p_2 \leq q \leq (1-p_2)$. We compute

$$P_X(1) = \frac{q - p_2}{1 - 2p_2} \qquad (1.15)$$

and, since the channel is stochastically degraded, we have

$$\begin{aligned}C_S = H_b(p_2) - H_b(p_1) - H_b(q)\\ + H_b\left((q - p_2)\frac{1 - 2p_1}{1 - 2p_2} + p_1\right),\end{aligned} \qquad (1.16)$$

where $H_b(\cdot)$ is the binary entropy function. Choosing $q = 1/2$ gives $H(X) = 1$ and we recover the wiretap channel capacity, as expected. But if $q = p_2$ or $q = 1 - p_2$ then $H(X) = 0$ and $C_S = 0$.

EXAMPLE 1.2 Consider next the AWGN channels

$$Y = X + A_1, \quad Z = X + A_2 \qquad (1.17)$$

with the power constraint $E[X^2] \leq P$. The random variables A_1 and A_2 are independent of X, Gaussian, zero-mean, and have variances N_1 and N_2, respectively. We consider $0 \leq N_1 \leq N_2$. Suppose the default Z is a Gaussian random variable with zero mean and variance Q, where $N_2 \leq Q \leq P + N_2$. We thus require that X is zero-mean Gaussian with variance $Q - N_2$. We assume that Theorem 1.1 applies to such channels, and since the channel is stochastically degraded, we compute

$$C_S = \frac{1}{2} \log_2 \left(1 + \frac{Q - N_2}{N_1} \right) - \frac{1}{2} \log_2 \left(\frac{Q}{N_2} \right), \qquad (1.18)$$

where the capacity is measured in bits per channel use. Choosing $Q = P + N_2$ implies $E[X^2] = P$, and we recover the wiretap channel capacity. But if $Q = N_2$ then $E[X^2] = 0$ and $C_S = 0$ (this is the regime of covert communication [16–18], see Section 1.2.1). Furthermore, the power required for the default transmissions increases with C_S.

1.3.2 Achievability

We use random coding and the proof technique of [8]. We assume that there is a Q_X for which $P_Z = Q_Z$.

Random code: Fix a distribution Q_X for which $P_Z = Q_Z$ and generate $L \cdot L_1$ codewords $x^n(m,w)$, $m = 1, \ldots, L$, $w = 1, \ldots, L_1$ using $\prod_{i=1}^{n} Q_X(x_i(m,w))$. This defines the codebook

$$\mathcal{C} = \{x^n(m,w), m = 1, \ldots, L, w = 1, \ldots, L_1\} \qquad (1.19)$$

and we denote the random codebook by

$$\widetilde{\mathcal{C}} = \{X^n(m,w)\}_{(m,w)=(1,1)}^{(L,L_1)}. \qquad (1.20)$$

Encoding: To send a message m, Alice chooses w uniformly from $\{1, \ldots, L_1\}$ and transmits $x^n(m,w)$. Hence, for any $\mathcal{C}$ we have

$$P_{X^n|\widetilde{\mathcal{C}}}(x^n(m,w)|\mathcal{C}) = \frac{1}{L \cdot L_1}. \qquad (1.21)$$

Since (1.21) is not $Q_X^n(x^n(m,w))$, the P_{Z^n} is not the desired Q_Z^n in general, see (1.8). Furthermore, we have

$$P_{Z^n|M\widetilde{\mathcal{C}}}(z^n|m,\mathcal{C}) = \sum_{w=1}^{L_1} \frac{1}{L_1} \cdot P_{Z|X}^n(z^n|x^n(m,w)), \qquad (1.22)$$

$$P_{Z^n|\widetilde{\mathcal{C}}}(z^n|\mathcal{C}) = \sum_{m=1}^{L} \sum_{w=1}^{L_1} \frac{1}{L \cdot L_1} \cdot P_{Z|X}^n(z^n|x^n(m,w)). \qquad (1.23)$$

Bob: Bob knows $\mathcal{C}$ and puts out $(\hat{m}, \hat{w})$ if there is a unique pair $(\hat{m}, \hat{w})$ satisfying the typicality check

$$(x^n(\hat{m},\hat{w}), y^n) \in \mathcal{T}_\epsilon^n(Q_X P_{Y|X}). \tag{1.24}$$

Otherwise he puts out $(\hat{m}, \hat{w}) = (1, 1)$.

Analysis: Define the real-valued random variables

$$E_1 = \Pr\left[(\hat{M}, \hat{W}) \neq (M, W) | \widetilde{\mathcal{C}}\right], \tag{1.25}$$

$$E_2 = D\left(P_{MZ^n | \widetilde{\mathcal{C}}} \,\|\, P_M Q_Z^n\right). \tag{1.26}$$

E_1 is Bob's block decoding error probability, and E_2 represents the security (secrecy and stealth) with respect to Eve's receiver. Using standard arguments (see [22]), $\mathrm{E}[E_1]$ can be made small with large n as long as

$$R + R_1 < I(X;Y). \tag{1.27}$$

To bound $\mathrm{E}[E_2]$, we use the steps in [8, Eq. (9)] to write

$$\begin{aligned}
&\mathrm{E}\left[D\left(P_{MZ^n|\widetilde{\mathcal{C}}} \,\|\, P_M Q_Z^n\right)\right] \\
&\stackrel{(a)}{=} \mathrm{E}\left[\log \frac{\sum_{j=1}^{L_1} P_{Z|X}^n(Z^n | X^n(M,j))}{L_1 \cdot Q_Z^n(Z^n)}\right] \\
&= \sum_{m=1}^{L} \sum_{w=1}^{L_1} \frac{1}{L \cdot L_1} \mathrm{E}\left[\log \frac{\sum_{j=1}^{L_1} P_{Z|X}^n(Z^n | X^n(m,j))}{L_1 \cdot Q_Z^n(Z^n)} \,\bigg|\, M=m, W=w\right] \\
&\stackrel{(b)}{\leq} \sum_{m=1}^{L} \sum_{w=1}^{L_1} \frac{1}{L \cdot L_1} \mathrm{E}\left[\log\left(\frac{P_{Z|X}^n(Z^n | X^n(m,w))}{L_1 \cdot Q_Z^n(Z^n)} + 1\right) \,\bigg|\, M=m, W=w\right] \\
&\stackrel{(c)}{=} \mathrm{E}\left[\log\left(\frac{P_{Z|X}^n(Z^n | X^n)}{L_1 \cdot Q_Z^n(Z^n)} + 1\right)\right],
\end{aligned} \tag{1.28}$$

where

(a) follows from (1.22) and by taking the expectation over M, W, $X^n(1,1), \ldots, X^n(L, L_1)$, Z^n;

(b) follows by applying Jensen's inequality to the expectation over the $X^n(m,j), j \neq w$ for a fixed m, using (1.8), and using $P_Z = Q_Z$;

(c) follows by choosing $X^n Z^n \sim Q_X^n P_{Z|X}^n$.

We may write (1.28) as

$$\mathrm{E}\left[\log\left(\frac{P_{Z|X}^n(Z^n|X^n)}{L_1 \cdot Q_Z^n(Z^n)} + 1\right)\right] = d_1 + d_2, \tag{1.29}$$

where the expectation is split based on typical pairs via

$$d_1 = \sum_{(x^n,z^n)\in\mathcal{T}_\epsilon^n(Q_X P_{Z|X})} Q_X^n(x^n) P_{Z|X}^n(z^n|x^n) \log\left(\frac{P_{Z|X}^n(z^n|x^n)}{L_1 \cdot Q_Z^n(z^n)} + 1\right),$$

$$d_2 = \sum_{\substack{(x^n,z^n)\notin\mathcal{T}_\epsilon^n(Q_X P_{Z|X}) \\ (x^n,z^n)\in\mathrm{supp}(Q_X^n P_{Z|X}^n)}} Q_X^n(x^n) P_{Z|X}^n(z^n|x^n) \log\left(\frac{P_{Z|X}^n(z^n|x^n)}{L_1 \cdot Q_Z^n(z^n)} + 1\right).$$

Using standard inequalities (see [22, Lemmas 18 and 20]) we have

$$d_1 \leq \log\left[\frac{2^{-n(1-\epsilon)H(Z|X)}}{L_1 \cdot 2^{-n(1+\epsilon)[H(Z)+D(P_Z\|Q_Z)]}} + 1\right]$$

$$\stackrel{(a)}{=} \log\left[2^{-n(R_1 - I(X;Z) - \kappa\epsilon)} + 1\right]$$

$$\leq \log(e) \cdot 2^{-n(R_1 - I(X;Z) - \kappa\epsilon)}, \tag{1.30}$$

where (a) follows because we chose Q_X so that $P_Z = Q_Z$, and κ is a constant independent of n. We find that $d_1 \to 0$ if $n \to \infty$ and

$$R_1 > I(X;Z) + \kappa\epsilon. \tag{1.31}$$

Next, consider d_2 and a pair (x^n, z^n) in the support of $Q_X^n P_{Z|X}^n$. We have

$$Q_Z^n(z^n) = P_Z^n(z^n) = \sum_{\tilde{x}^n \in \mathrm{supp}(Q_X^n)} Q_X^n(\tilde{x}^n) P_{Z|X}^n(z^n|\tilde{x}^n), \tag{1.32}$$

so that $Q_Z^n(z^n)$ is positive. We bound (see [22, Lemma 17])

$$d_2 \leq \sum_{\substack{(x^n,z^n)\notin\mathcal{T}_\epsilon^n(Q_X P_{Z|X}) \\ (x^n,z^n)\in\mathrm{supp}(Q_X^n P_{Z|X}^n)}} Q_X^n(x^n) P_{Z|X}^n(z^n|x^n) \log\left[\left(\frac{1}{\mu_Z}\right)^n + 1\right]$$

$$\leq 2|\mathcal{X}| \cdot |\mathcal{Z}| \cdot e^{-n\epsilon^2 \mu_{XZ}/3} \log\left[\left(\frac{1}{\mu_Z}\right)^n + 1\right], \tag{1.33}$$

where

$$\mu_Z = \min_{z \in \mathrm{supp}(Q_Z)} Q(z), \tag{1.34}$$

$$\mu_{XZ} = \min_{(x,z) \in \mathrm{supp}(Q_X P_{Z|X})} Q(x) P(z|x). \tag{1.35}$$

If $\frac{1}{\mu_Z} < 1$, we have

$$d_2 \leq 2|\mathcal{X}| \cdot |\mathcal{Z}| \cdot e^{-n\epsilon^2 \mu_{XZ}/3} \cdot \log 2 \tag{1.36}$$

and $d_2 \to 0$ as $n \to \infty$. If $\frac{1}{\mu_Z} \geq 1$, we have

$$d_2 \leq 2|\mathcal{X}| \cdot |\mathcal{Z}| \cdot e^{-n\epsilon^2 \mu_{XZ}/3} \cdot n \cdot \log\left(\frac{1}{\mu_Z} + 1\right) \tag{1.37}$$

and $d_2 \to 0$ as $n \to \infty$.

For any $\xi_1, \xi_2 > 0$, define the event
$$\mathcal{E} = \{E_1 > \xi_1 \text{ or } E_2 > \xi_2\}. \tag{1.38}$$

Using the union bound and Markov's inequality, we obtain
$$\begin{aligned} \Pr[\mathcal{E}] &\leq \Pr[E_1 > \xi_1] + \Pr[E_2 > \xi_2] \\ &\leq \frac{1}{\xi_1} \mathrm{E}[E_1] + \frac{1}{\xi_2} \mathrm{E}[E_2]. \end{aligned} \tag{1.39}$$

Combining the above, we can make $\Pr[\mathcal{E}] \to 0$ as $n \to \infty$ as long as
$$R + R_1 < I(X;Y), \tag{1.40}$$
$$R_1 > I(X;Z). \tag{1.41}$$

We hence have the achievability of any R satisfying
$$0 \leq R < \max_{Q_X : P_Z = Q_Z} [I(X;Y) - I(X;Z)]. \tag{1.42}$$

Of course, if the right-hand side of (1.42) is non-positive, then we require $R = 0$.

Finally, following [2] we prefix a channel $Q_{X|V}$ to the channel $P_{YZ|X}$ and obtain a new channel $Q_{YZ|V}$ where
$$Q(y,z|v) = \sum_x Q(x|v) P(y,z|x). \tag{1.43}$$

Using a similar analysis as above, we have the achievability of any R satisfying
$$0 \leq R < \max_{Q_{VX} : P_Z = Q_Z} [I(V;Y) - I(V;Z)], \tag{1.44}$$

where the maximization is over all Q_{VX} satisfying (1.12). Again, if the right-hand side of (1.44) is non-positive, then we require $R = 0$. As usual, the purpose of adding the auxiliary variable V is to potentially increase R. Note that $V = X$ recovers (1.42). Hence, the right-hand side of (1.42) is at most the right-hand side of (1.44).

REMARK 1.5 *The average divergence* $\mathrm{E}[D(P_{MZ^n|\widetilde{C}} \| P_M Q_Z^n)]$ *is the sum of* $I(M\widetilde{C};Z^n)$ *and* $D(P_{Z^n} \| Q_Z^n)$ *[6, Section III] (see also [8, Section III-B]). To see this, consider*
$$\begin{aligned} &\mathrm{E}[D(P_{MZ^n|\widetilde{C}} \| P_M Q_Z^n)] \\ &= D(P_{MZ^n|\widetilde{C}} \| P_M Q_Z^n | P_{\widetilde{C}}) \\ &\stackrel{(a)}{=} D(P_{Z^n|M\widetilde{C}} \| Q_Z^n | P_M P_{\widetilde{C}}) \\ &= D(P_{Z^n|M\widetilde{C}} \| P_{Z^n} | P_M P_{\widetilde{C}}) + D(P_{Z^n} \| Q_Z^n) \\ &= I(M\widetilde{C}; Z^n) + D(P_{Z^n} \| Q_Z^n), \end{aligned} \tag{1.45}$$

where (a) follows by the independence of M and the codewords. Therefore, as $\mathrm{E}[D(P_{MZ^n|\widetilde{C}} \| P_M Q_Z^n)] \to 0$ *we have* $I(M\widetilde{C}; Z^n) \to 0$, *which means that $M\widetilde{C}$ and Z^n are (almost) independent. This makes sense, since for effective secrecy the adversary learns little about M and the presence of meaningful transmission.*

REMARK 1.6 *Reference [23] investigates the minimum amount of randomness required for the prefix channel in (1.43).*

REMARK 1.7 *The above steps ensure an* average *effective secrecy and error probability across the messages. A proof for a* maximum *effective secrecy and error probability proceeds in the usual fashion. Consider a code $\tilde{\mathcal{C}}$ with codewords of length n and rate R such that $D(P_{MZ^n|\tilde{\mathcal{C}}} \| P_M Q_Z^n) < \delta$ and $\Pr[(\hat{M}, \hat{W}) \neq (M, W)|\tilde{\mathcal{C}}] < \epsilon$. Define the following per-message metrics:*

$$P_e(m) = \Pr\left[(\hat{M}, \hat{W}) \neq (M, W)|M = m, \tilde{\mathcal{C}}\right], \qquad (1.46)$$

$$S(m) = D\left(P_{Z^n|M=m,\tilde{\mathcal{C}}} \| Q_Z^n\right). \qquad (1.47)$$

Then, we have

$$\Pr[P_e(M) \geq 3\epsilon, S(M) \geq 3\delta]$$
$$\leq \Pr[P_e(M) \geq 3\epsilon] + \Pr[S(M) \geq 3\delta]$$
$$\leq \frac{\Pr\left[(\hat{M}, \hat{W}) \neq (M, W)|\tilde{\mathcal{C}}\right]}{3\epsilon} + \frac{D\left(P_{MZ^n|\tilde{\mathcal{C}}} \| P_M Q_Z^n\right)}{3\delta} \leq \frac{2}{3} < 1. \qquad (1.48)$$

Hence, at least one-third of the messages m in $\tilde{\mathcal{C}}$ satisfy $P_e(m) < 3\epsilon$ and $S(m) < 3\delta$. By removing all codewords corresponding to the other messages from the codebook, we obtain a code with maximum error probability 3ϵ and effective secrecy 3δ for every message. In addition, the resulting message rate is $R - \frac{\log_2 3}{n}$, which can be made arbitrarily close to the original rate R with n large enough.

1.3.3 Converse

If we can choose Q_Z to be the P_Z corresponding to the capacity-achieving output distribution of the (weak or) strong secrecy capacity, then the converse follows from [2, Theorem 1]. An alternative proof uses a *telescoping identity* [24, Section G], see [10]. We repeat these steps below to prove the converse for the more general case we are interested in.

Suppose Q_Z is fixed. We first consider stealth and follow the same steps as in [25, p. 60] to develop (1.10) as

$$\xi_2 \geq D(P_{MZ^n} \| P_M Q_Z^n) = \left[\sum_{z^n} P(z^n) \sum_{i=1}^n \log \frac{1}{Q_Z(z_i)}\right] - H(Z^n|M)$$

$$\overset{(a)}{\geq} \sum_{i=1}^n \left[\sum_z P_{Z_i}(z) \log \frac{1}{Q_Z(z)}\right] - \sum_{i=1}^n H(Z_i) = \sum_{i=1}^n D(P_{Z_i} \| Q_Z)$$

$$\overset{(b)}{\geq} n D(P_{Z_T} \| Q_Z), \qquad (1.49)$$

where

(a) follows by $H(Z^n|M) \leq H(Z^n) \leq \sum_{i=1}^n H(Z_i)$;

(b) follows by the convexity of $D(P\|Q)$ in P, and where the "time-sharing" random variable T is independent of all other random variables and uniformly distributed over $\{1,\ldots,n\}$. In other words, for all z we have

$$P_{Z_T}(z) = \frac{1}{n}\sum_{i=1}^{n} P_{Z_i}(z). \tag{1.50}$$

Letting $\xi_2 \to 0$, the bounds (1.49) imply that we must have $P_Z = Q_Z$. This also implies that if there is no Q_X for which $P_Z = Q_Z$ then the capacity is zero.

Next, we consider reliability and secrecy. Suppose that for some $\xi_1, \xi_2 > 0$ there exists an n, an encoder, and a decoder such that (1.9) and (1.10) are satisfied. We have

$$nR = H(M)$$
$$= I(M;Y^n) + H(M|Y^n)$$
$$\overset{(a)}{\leq} I(M;Y^n) + \left(H_b(P_e^{(n)}) + P_e^{(n)} \cdot nR\right)$$
$$\overset{(b)}{\leq} I(M;Y^n) - I(M;Z^n) + \xi_2 + (1+\xi_1 \cdot nR), \tag{1.51}$$

where (a) follows from Fano's inequality, and (b) follows from the reliability constraint (1.9) and the security constraint (1.10) – see also (1.7). Using the telescoping identity [24, Eqs. (9) and (11)] we have

$$\frac{1}{n}\left[I(M;Y^n) - I(M;Z^n)\right] = \sum_{i=1}^{n}[I(M;Z_{i+1}^n Y^i) - I(M;Z_i^n Y^{i-1})]$$
$$= \frac{1}{n}\sum_{i=1}^{n}[I(M;Y_i|Y^{i-1}Z_{i+1}^n) - I(M;Z_i;|Y^{i-1}Z_{i+1}^n)]$$
$$\overset{(a)}{=} I(M;Y_T|Y^{T-1}Z_{T+1}^n T) - I(M;Z_T|Y^{T-1}Z_{T+1}^n T)$$
$$\overset{(b)}{=} I(V;Y|U) - I(V;Z|U) \overset{(c)}{\leq} \max_{Q_{UVX}} [I(V;Y|U) - I(V;Z|U)]$$
$$= \max_{u} \max_{Q_{VX|U=u}} [I(V;Y|U=u) - I(V;Z|U=u)]$$
$$= \max_{Q_{VX}} [I(V;Y) - I(V;Z)], \tag{1.52}$$

where

(a) follows by choosing T as for (1.49);
(b) follows by defining

$$U = Y^{T-1}Z_{T+1}^n T, \; V = MU,$$
$$X = X_T, \; Y = Y_T, \; Z = Z_T; \tag{1.53}$$

(c) follows by maximizing over all Q_{UVX} so that $D(P_Z\|Q_Z) \leq \delta$ for any choice of $\delta > 0$.

Combining (1.51) and (1.52) we have, for any $\delta > 0$,

$$R \leq \frac{\max_{Q_{VX}} [I(V;Y) - I(V;Z)]}{1 - \xi_1} + \frac{\xi_2 + 1}{(1 - \xi_1)n}, \qquad (1.54)$$

where the maximization is over all Q_{VX} so that $D(P_Z \| Q_Z) \leq \delta$. Letting $\xi_1 \to 0$ and $n \to \infty$, we have

$$R \leq \max_{Q_{VX}} [I(V;Y) - I(V;Z)] \qquad (1.55)$$

where the maximization is over all Q_{VX} satisfying the Markov chain (1.12). The cardinality bound in Theorem 1.1 was derived in [26, Theorem 22.1]. To complete the converse, observe that we have

$$R \leq \max_{Q_{VX}: D(P_Z \| Q_Z) = 0} [I(V;Y) - I(V;Z)] \qquad (1.56)$$

because the function $f : Q_{VX} \mapsto I(V;Y) - I(V;Z)$ is continuous and the set $\{Q_{VX} : D(P_Z \| Q_Z) \leq \delta\}$ is compact.

1.3.4 Choice of Security Measures

Effective secrecy includes both strong secrecy and stealth. One may argue that using only $I(M; Z^n)$ or $D(P_{Z^n} \| Q_Z^n)$ would suffice to measure secrecy. However, it is easy to find examples where one achieves secrecy but not stealth, or where one achieves stealth but not secrecy.

EXAMPLE 1.3 Suppose Alice uses $\widetilde{Q}_X$ rather than Q_X to generate the codebook and (1.31) is satisfied. The new $\widetilde{Q}_X$ causes $\widetilde{Q}_Z^n \neq Q_Z^n$ in general. Hence, as $n \to \infty$ we have $I(M; Z^n) \to 0$ but $D(\widetilde{Q}_Z^n \| Q_Z^n) \to \infty$, so that Eve recognizes that Alice transmits useful information.

EXAMPLE 1.4 We have $\mathbb{E}[D(P_{Z^n} \| Q_Z^n)] \to 0$ as $n \to \infty$ as long as (see [8, Theorem 1])

$$R + R_1 > I(X;Z). \qquad (1.57)$$

If Alice is not careful and chooses R_1 such that (1.31) is violated and (1.57) is satisfied, then $D(P_{Z^n} \| Q_Z^n)$ can be made small, but even for large n we have

$$I(M; Z^n) = I \text{ for some } I > 0. \qquad (1.58)$$

For example, Alice might choose R_1 to be too small. Thus, although the communication makes $D(P_{Z^n} \| Q_Z^n)$ small, Eve learns

$$I(M; Z^n) \approx n[I(X;Z) - R_1] \qquad (1.59)$$

bits about M.

1.4 Hypothesis Testing

We relate $D(P_{Z^n}\|Q_Z^n)$ to hypothesis testing to demonstrate this quantity's operational meaning for stealth. We show that as long as (1.57) is satisfied, the best Eve can do to detect Alice's action is to guess.

For every channel output z^n, Eve considers two hypotheses:

$$H_0 = Q_Z^n, \tag{1.60}$$

$$H_1 = P_{Z^n}. \tag{1.61}$$

If H_0 is accepted, then Eve decides that Alice's transmission is not meaningful, whereas if H_1 is accepted, then Eve decides that Alice sent an information message. We define two kinds of error probabilities:

$$\alpha = \Pr\{H_1 \text{ is accepted} \mid H_0 \text{ is true}\}, \tag{1.62}$$

$$\beta = \Pr\{H_0 \text{ is accepted} \mid H_1 \text{ is true}\}. \tag{1.63}$$

The value α is referred to as *the level of significance* [27] or *false alarm probability*, while β is the *misdetection probability*. In practice, a false alarm can be expensive. Therefore, Eve would like to minimize β for a given tolerance level of α. To this end, for every z^n Eve performs a ratio test,

$$\frac{Q_Z^n(z^n)}{P_{Z^n}(z^n)} = r, \tag{1.64}$$

and makes a decision depending on a threshold F, $F \geq 0$, namely

$$\begin{cases} H_0 \text{ is accepted} & \text{if } r > F \\ H_1 \text{ is accepted} & \text{if } r \leq F. \end{cases} \tag{1.65}$$

Define the set of z^n for which H_0 is accepted as

$$\mathcal{A}_F^n = \left\{ z^n : \frac{Q_Z^n(z^n)}{P_{Z^n}(z^n)} > F \right\}, \tag{1.66}$$

and $(\mathcal{A}_F^n)^c$ is the set of z^n for which H_1 is accepted (see Fig. 1.2). Eve chooses the threshold F and we have

$$\alpha = Q_Z^n((\mathcal{A}_F^n)^c) = 1 - Q_Z^n(\mathcal{A}_F^n),$$
$$\beta = P_{Z^n}(\mathcal{A}_F^n). \tag{1.67}$$

The ratio test (1.65) with an optimized F is the *Neyman–Pearson test*, which is optimal [27, Theorem 3.2.1] in the sense that it minimizes β for a given α. We have the following lemma.

LEMMA 1.1 *If $D(P_{Z^n}\|Q_Z^n) \leq \xi_2$ then we have*

$$1 - g(\xi_2) \leq \alpha + \beta \leq 1 + g(\xi_2), \tag{1.68}$$

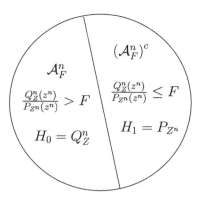

Figure 1.2 Example of the decision regions $\mathcal{A}_F^n$ and $\left(\mathcal{A}_F^n\right)^c$.

where
$$g(\xi_2) = \sqrt{\xi_2 \cdot 2\ln 2}, \tag{1.69}$$

which goes to 0 as $\xi_2 \to 0$.

Proof Since $D(P_{Z^n}||Q_Z^n) \leq \xi_2$, Pinsker's inequality [28, Theorem 11.6.1] gives

$$\|P_{Z^n} - Q_Z^n\|_{\text{TV}} \leq \sqrt{\xi_2 \cdot 2\ln 2} = g(\xi_2), \tag{1.70}$$

where $\|P_X - Q_X\|_{\text{TV}} = \sum_{x \in \mathcal{X}} |P(x) - Q(x)|$ is the variational distance between P_X and Q_X. We further have

$$\|P_{Z^n} - Q_Z^n\|_{\text{TV}} \geq \sum_{z^n \in \mathcal{A}_F^n} \left|P_{Z^n}(z^n) - Q_Z^n(z^n)\right|$$
$$\overset{(a)}{\geq} \left|P_{Z^n}(\mathcal{A}_F^n) - Q_Z^n(\mathcal{A}_F^n)\right|$$
$$= |\beta - (1-\alpha)|, \tag{1.71}$$

where (a) follows by the triangle inequality. Combining (1.70) and (1.71), we have the bounds (1.68).

Figure 1.3 illustrates the optimal tradeoff between α and β for stealth communication, i.e., when (1.57) is satisfied. As $n \to \infty$ and $\xi_2 \to 0$, we have

$$D(P_{Z^n}||Q_Z^n) \to 0, \tag{1.72}$$
$$\alpha + \beta \to 1. \tag{1.73}$$

If Eve allows no false alarm ($\alpha = 0$), then she always ends up with misdetection ($\beta = 1$). If Eve tolerates no misdetection ($\beta = 0$), she pays a high price ($\alpha = 1$). Further, for any given α, the optimal misdetection probability is

$$\beta_{\text{opt}} = 1 - \alpha. \tag{1.74}$$

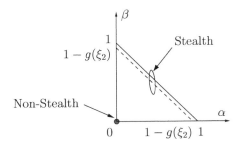

Figure 1.3 Optimal tradeoff between α and β.

But Eve does not need to see Z^n to achieve β_{opt}, she may simply randomly choose some $\mathcal{A}'$ such that

$$Q_Z^n((\mathcal{A}')^c) = \alpha. \tag{1.75}$$

The best strategy is thus to guess. On the other hand, if

$$\lim_{n \to \infty} D(P_{Z^n} \| Q_Z^n) > 0 \tag{1.76}$$

then Eve detects Alice's action and we can have

$$\alpha + \beta = 0. \tag{1.77}$$

We thus operate in one of two regimes in Fig. 1.3, either near $(\alpha, \beta) = (0,0)$ or near the line $\alpha + \beta = 1$.

1.5 Conclusion

We have introduced a new security measure called effective secrecy that includes strong secrecy and stealth. The measure is based on unnormalized informational divergence. The stealth requirement may reduce capacity as compared to the wiretap channel capacity. However, if one can choose the default behavior, then the effective secrecy capacity is the same as the wiretap channel capacity. Finally, in contrast to covert communication, stealth lets one hide the presence of meaningful communication at a positive rate if the default behavior is chosen appropriately.

Acknowledgments

J. Hou thanks Rafael Schaefer for useful discussions.

J. Hou and G. Kramer were supported by an Alexander von Humboldt Professorship endowed by the German Federal Ministry of Education and Research. M. Bloch was supported in part by the National Science Foundation within the Directorate for Computer and Information Science and Engineering under Grant CIF-1320298.

References

[1] A. D. Wyner, "The wire-tap channel," *Bell Syst. Tech. J.*, vol. 54, pp. 1355–1387, Oct. 1975.

[2] I. Csiszár and J. Körner, "Broadcast channels with confidential messages," *IEEE Trans. Inf. Theory*, vol. 24, no. 3, pp. 339–348, May 1978.

[3] U. M. Maurer and S. Wolf, "Information-theoretic key agreement: From weak to strong secrecy for free," *Lecture Notes in Computer Science*, vol. 1807, pp. 351–368, 2000.

[4] I. Csiszár, "Almost independence and secrecy capacity," *Probl. Pered. Inform.*, vol. 32, no. 1, pp. 48–57, 1996.

[5] T. S. Han and S. Verdú, "Approximation theory of output statistics," *IEEE Trans. Inf. Theory*, vol. 39, no. 3, pp. 752–772, May 1993.

[6] M. Hayashi, "General nonasymptotic and asymptotic formulas in channel resolvability and identification capacity and their application to the wiretap channel," *IEEE Trans. Inf. Theory*, vol. 52, no. 4, pp. 1562–1575, Apr. 2006.

[7] M. R. Bloch and J. N. Laneman, "Strong secrecy from channel resolvability," *IEEE Trans. Inf. Theory*, vol. 59, no. 12, pp. 8077–8098, Dec. 2013.

[8] J. Hou and G. Kramer, "Informational divergence approximations to product distributions," in *Proc. Canadian Workshop Inf. Theory*, Toronto, ON, Canada, Jun. 2013, pp. 76–81.

[9] A. Winter, "Secret, public and quantum correlation cost of triples of random variables," in *Proc. IEEE Int. Symp. Inf. Theory*, Adelaide, Australia, Sep. 2005, pp. 2270–2274.

[10] J. Hou and G. Kramer, "Effective secrecy: Reliability, confusion and stealth," Nov. 2013. [Online]. Available: http://arxiv.org/abs/1311.1411

[11] T. S. Han, H. Endo, and M. Sasaki, "Reliability and secrecy functions of the wiretap channel under cost constraint," Jul. 2013. [Online]. Available: http://arxiv.org/abs/1307.0608

[12] T. S. Han, H. Endo, and M. Sasaki, "Reliability and secrecy functions of the wiretap channel under cost constraint," *IEEE Trans. Inf. Theory*, vol. 52, no. 4, pp. 1562–1575, Apr. 2015.

[13] Committee on National Security Systems, "Committee on National Security Systems (CNSS) Glossary," Tech. Rep., Apr. 2015, CNSSI 4009. [Online]. Available: www.cnss.gov

[14] G. E. Prescott, "Performance metrics for low probability of intercept communication systems," Air Force Office of Scientific Research, Bolling AFB, Washington DC, Tech. Rep., Oct. 1993, grant AFOSR-91-0018.

[15] D. L. Nicholson, *Spread Spectrum Signal Design: LPE and AJ Systems*. Rockville, MD: Computer Science Press, 1988.

[16] B. Bash, D. Goeckel, and D. Towsley, "Limits of reliable communication with low probability of detection on AWGN channels," *IEEE J. Sel. Areas Commun.*, vol. 31, no. 9, pp. 1921–1930, Sep. 2013.

[17] P. H. Che, M. Bakshi, and S. Jaggi, "Reliable deniable communcation: Hiding messages in noise," in *Proc. IEEE Int. Symp. Inf. Theory*, Istanbul, Turkey, Jul. 2013, pp. 2945–2949.

[18] S. Lee, R. Baxley, M. Weitnauer, and B. Walkenhorst, "Achieving undetectable communication," *IEEE J. Sel. Topics Signal Proc.*, vol. 9, no. 7, pp. 1195–1205, Oct. 2015.

[19] L. Wang, G. W. Wornell, and L. Zheng, "Fundamental limits of communication with low probability of detection," *IEEE Trans. Inf. Theory*, vol. 62, no. 6, pp. 3493–3503, June 2016.

[20] M. R. Bloch, "Covert communication over noisy channels: A resolvability perspective," *IEEE Trans. Inf. Theory*, vol. 62, no. 5, pp. 2334–2354, May 2016.

[21] J. L. Massey, *Applied Digital Information Theory*, ETH Zurich, Zurich, Switzerland, 1980–1998.

[22] A. Orlitsky and J. R. Roche, "Coding for computing," *IEEE Trans. Inf. Theory*, vol. 47, no. 3, pp. 903–917, Mar. 2001.

[23] S. Watanabe and Y. Oohama, "The optimal use of rate-limited randomness in broadcast channels with confidential messages," *IEEE Trans. Inf. Theory*, vol. 61, no. 2, pp. 983–995, Feb. 2015.

[24] G. Kramer, "Teaching IT: An identity for the Gelfand-Pinsker converse," *IEEE Inf. Theory Society Newsletter*, vol. 61, no. 4, pp. 4–6, Dec. 2011.

[25] J. Hou, "Coding for relay networks and effective secrecy for wire-tap channels," Ph.D. dissertation, Technische Universität München, Munich, Germany, 2014.

[26] A. El Gamal and Y.-H. Kim, *Network Information Theory*. Cambridge: Cambridge University Press, 2011.

[27] E. L. Lehmann and J. P. Romano, *Testing Statistical Hypotheses*, 3rd edn. New York: Springer, 2005.

[28] T. M. Cover and J. A. Thomas, *Elements of Information Theory*, 2nd edn. Chichester: Wiley & Sons, 2006.

2 Error-Free Perfect Secrecy Systems

Siu-Wai Ho, Terence Chan, Alex Grant, and Chinthani Uduwerelle

Shannon's fundamental bound for perfect secrecy says that the source entropy cannot be larger than the entropy of the secret key initially shared by the sender and the legitimate receiver. Massey gave an information theoretic proof of this result, and his proof does not require independence of the key and the source message. By further assuming independence, some stronger results, which govern the probability distributions of the key and the ciphertext, can be shown. These results illustrate that the key entropy is not less than the logarithm of the message sample size in any cipher achieving perfect secrecy, even if the source distribution is fixed. The same bound also applies to the entropy of the ciphertext. These results still hold if the source message has been compressed before encryption.

The above observation leads to different research problems studied in this chapter. When the source distribution is non-uniform, the entropy of the key is required to be strictly greater than the source entropy, and hence some randomness in the key is wasted. To deal with this problem, this chapter investigates cipher systems that contain residual secret randomness after they are used. A collection of such systems can be used to generate a new secret key. The aforementioned entropy bound only gives the minimum size of the pre-shared secret key. A new measure for key consumption, i.e., the entropy difference between the pre-shared secret key and the newly generated key, is proposed and justified in this chapter. Key consumption is shown to be bounded below by the source entropy, and the lower bound can be achieved by the codes proposed in this chapter. Furthermore, the existence of a fundamental tradeoff between the expected key consumption and the number of channel uses for conveying a ciphertext is shown.

2.1 Introduction

Cipher systems with *perfect secrecy* were studied by Shannon in his seminal paper [1] (see also [2]). With reference to Fig. 2.1, a cipher system is defined by three components: a source message U, a ciphertext X, and a key R. Here, R is the collection of secret randomness shared only by the sender and the legitimate receiver.[1] The sender

---

[1] We use R here to denote secret randomness shared between sender and receiver because the symbol K is reserved to denote the key extracted from R in an encryption process.

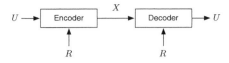

Figure 2.1 A cipher system.

encrypts the message U, together with the key R, into the ciphertext X. This ciphertext will be transmitted to the legitimate receiver via a public channel. A cipher system is *perfectly secure*, or, equivalently, satisfies the perfect secrecy constraint, if the message U and the ciphertext X are statistically independent (i.e., $I(U;X) = 0$). In this case, an adversary who eavesdrops on the public channel and learns X (but does not have R) will not be able to infer any information about the message U. On the other hand, the legitimate receiver can decrypt the message U from the received ciphertext X together with the secret key R. A cipher system is *error free* (i.e., the probability of a decoding error is zero) if $H(U \mid XR) = 0$.

The focus of this chapter is to study the amount of resources required in a cipher system especially for non-uniform source distribution. We will reveal some new concepts about the key and the ciphertext that have not been covered in [1, 2]. The journey starts from revisiting the minimum requirement on the secret randomness R given by Shannon. By considering a *deterministic cipher*, where X is a deterministic function of R and U, Shannon argued that the number of messages is equal to the number of possible ciphertexts, and that the number of different keys is not less than the number of messages [1, p. 681],

$$|\mathcal{X}| = |\mathcal{U}| \leq |\mathcal{R}|,$$

where $\mathcal{X}$, $\mathcal{U}$, and $\mathcal{R}$ are the respective supports of X, U, and R. In order to design a perfectly secure cipher system protecting a source with *unknown source distribution* P_U, Shannon argued that

$$H(R) \geq \log |\mathcal{U}| \geq H(U). \tag{2.1}$$

He also made the important observation [1, p. 682] that

the amount of uncertainty we can introduce into the solution cannot be greater than the key uncertainty.

In other words,

$$H(R) \geq H(U). \tag{2.2}$$

Massey [2] called (2.2) Shannon's fundamental bound for perfect secrecy, and gave an information theoretic proof for this result. It is important to note that Massey's proof [2] does not require U and R to be statistically independent.

Now, suppose U and R are indeed independent (which is common in practice). We will see that (2.2) can be improved. For any source distribution P_U,

$$P_R(r) \leq |\mathcal{U}|^{-1}, \quad \forall r \in \mathcal{R}. \tag{2.3}$$

As a consequence, for any cipher system achieving perfect secrecy, the logarithm of the message sample size cannot be larger than the entropy of the secret key,

$$H(R) \geq \log |\mathcal{U}|. \tag{2.4}$$

Comparing with (2.1), inequalities (2.3) and (2.4) are valid even if the source distribution P_U is fixed and known. If the size of the source alphabet is unbounded $|\mathcal{U}| = \infty$, (2.3) implies that it is impossible to construct a suitable key defined on a countable alphabet (because $P_R(r) \leq 0$ for all r). Note that we do not make any assumption on whether P_R is uniform or defined on a finite alphabet. One of the examples in this chapter (Example 2.5) uses a non-uniform P_R to demonstrate some benefits. So (2.3) gives a necessary condition for the choice of P_R.

This chapter is based on the model in Fig. 2.1. Despite its apparent simplicity, it is the most general encoder possible, covering many interesting special cases. For example, suppose the distribution of U is non-uniform. One may expect that the optimal encoder may first compress and then encrypt the compressed output, as depicted in Fig. 2.2. Roughly speaking, compression converts a message into a sequence of independent and identically distributed (i.i.d.) symbols, and hence it can maximize the adversary's decoding error probability in some systems [3, Theorem 3]. On the other hand, the compressed output also has a smaller expected file size and hence may require less key for encryption. This compression-before-encryption approach was also considered by Shannon [1, p. 682]. In fact, Shannon believed that, after removing redundancy in the source,

a bit of key completely conceals a bit of message information.

The above compression-before-encryption encoder is indeed a special case of our encoder in Fig. 2.1, which can be viewed as a joint compression–encryption coder. We will illustrate that encoders in Fig. 2.2 are not optimal in some cases.

For the size of ciphertext, we will later prove

$$H(X) \geq \log |\mathcal{U}|. \tag{2.5}$$

Note that the source entropy does not appear in both (2.4) and (2.5). The following toy example, however, suggests that we need to rethink the role of source entropy when both perfect secrecy and error-free decoding are required.

EXAMPLE 2.1   *For a secret message U with $P_U(0) = \frac{3}{5}$ and $P_U(1) = \frac{2}{5}$, we define a cipher system as follows. Let $\mathcal{R} = \{0,1,2,3,4\}$ and let R be uniformly distributed in $\mathcal{R}$ and independent of U. If $U = 0$, let the ciphertext $X = (B + R) \mod 5$, where B is a random variable independently generated according to $P_B(0) = P_B(1) = P_B(2) = \frac{1}{3}$. If $U = 1$, let $X = (C + R) \mod 5$, where random variable C is independently generated*

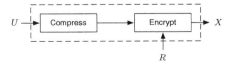

Figure 2.2   Compression before encryption.

according to $P_C(3) = P_C(4) = \frac{1}{2}$. It is easy to verify that $I(U;X) = 0$ so that perfect secrecy is satisfied. Error-free decoding can be achieved because $U = 0$ if $(X - R)$ mod $5 < 3$ and $U = 1$ otherwise. Although this cipher system requires $H(R) = \log 5$, which is strictly larger than the lower bound given in (2.4), an extra secret (i.e., either B or C) is transmitted. The extra secret can be used as a secret key in future encryption of other messages. So a more "fair" way to calculate the expected amount of secret randomness used in this system should be given by

$$\log 5 - \left(\frac{3}{5}\log 3 + \frac{2}{5}\log 2\right) = H(U). \tag{2.6}$$

The quest for the role of source entropy in perfect secrecy systems leads to the major contribution of this chapter, i.e., the introduction of two new concepts – *residual secret randomness* $H(R|UX)$ and *expected key consumption* $I(R;UX)$. Previously in the literature, the amount of key required in a cipher system has been measured by the entropy of the common secret key $H(R)$. We will argue in this chapter that $H(R)$ is only valid for measuring the *initial key requirement*, by which we mean the amount of secret randomness that must be shared between the sender and the legitimate receiver prior to transmission of the ciphertext. As we shall see, the actual key consumption will be measured by $I(R;UX)$, and an amount of $H(R|U,X)$ of secret randomness is left after a cipher system is used each time. In the second part of this chapter, we will design an efficient cipher system, where $I(R;UX)$ will be optimized.

Besides expected key consumption, we also want to minimize the *number of channel uses* required to transmit the ciphertext X from the source to the legitimate receiver. Naturally, we can encode the ciphertext X using a Huffman code [4]. Let $\lambda(X)$ be the codeword length. In this case, the expected codeword length $\mathbf{E}[\lambda(X)]$ satisfies $H(X) \leq \mathbf{E}[\lambda(X)] \leq H(X) + 1$. Note that for two random variables X and X', it is possible that $H(X) < H(X')$, but $\mathbf{E}[\lambda(X)] > \mathbf{E}[\lambda(X')]$. One example is when $P_X = (0.3, 0.23, 0.2, 0.17, 0.1)$ and $P_{X'} = (0.25, 0.25, 0.25, 0.15, 0.1)$. However, we still use $H(X)$ instead of $\mathbf{E}[\lambda(X)]$ as a measure for the number of channel uses required in a cipher system for two reasons: first, $H(X)$ is a lower bound for $\mathbf{E}[\lambda(X)]$, and is in fact a very good estimate for $\mathbf{E}[\lambda(X)]$ when $H(X)$ is large; second, the problem itself is more tractable when using $H(X)$ rather than $\mathbf{E}[\lambda(X)]$.

We will show that there exists a fundamental tradeoff between the expected key consumption and the number of channel uses. In fact, if the source distribution is not uniform, then the minimum expected key consumption and the minimum number of channel uses cannot be simultaneously achieved. We will propose a code design that can achieve the minimum expected key consumption. Simultaneously minimizing the expected key consumption and number of channel uses will be considered. This chapter will explore whether compression before encryption can give us some benefits in these minimization problems.

2.1.1 Organization

In Section 2.2, we will formalize the system model, and give basic results about the bounds on $H(R)$ and $H(X)$. These bounds will lead to our main contribution in

Section 2.3. We will consider the case where a collection of cipher systems has been used. New system parameters including residual secret randomness and expected key consumption will be defined. Two justifications of residual secret randomness will be shown in Sections 2.3.1 and 2.3.2. The justification of expected key consumption will be given in Section 2.3.3. Section 2.4 will focus on two regimes: (1) minimal expected key consumption, and (2) minimal number of channel uses. The existence of a fundamental non-trivial tradeoff will be illustrated.

2.1.2 Notation

Random variables are denoted by capital letters, e.g. X, and their realizations and supports are respectively denoted by small letters x and calligraphic letters $\mathcal{X}$. All random variables are defined on countable alphabets. For simplicity, all logarithms are with base 2.

2.2 Error-Free Perfect Secrecy Systems

DEFINITION 2.1 (Error-free perfect secrecy system) A cipher system (R, U, X) is called an *error-free perfect secrecy* (EPS) system if

$$I(U;X) = 0, \qquad (2.7)$$

$$H(U \mid R, X) = 0, \qquad (2.8)$$

$$I(U;R) = 0. \qquad (2.9)$$

Here, (2.7) ensures perfect secrecy via independence between the ciphertext X and source message U. As such, an eavesdropper learning X can infer no information about the message U. The constraint (2.8) ensures that the receiver can reconstruct U from R and X without error. Finally, (2.9) requires that the shared secret key R is independent of the message U.

The constraints (2.7) and (2.8) were originally used in [2] to prove Shannon's fundamental bound (2.2) for perfect secrecy. In our system, we further require (2.9) for the practical reason that R is usually shared prior to the independent generation of the message U. Furthermore, Definition 2.1 admits the general probabilistic encoder. For the receiver, it is however sufficient to consider deterministic decoding since U is a function of R and X from (2.8). In other words, there exists a decoding function g such that

$$P_{URX}(u,r,x) = P_{RX}(r,x)\mathbf{1}\{u = g(r,x)\}. \qquad (2.10)$$

THEOREM 2.1 (Upper bounds on $P_X(x)$ and $P_R(r)$) *Suppose P_U is known when we construct an error-free perfect secrecy system (R, U, X) satisfying (2.7)–(2.9). Then*

$$\max_{x \in \mathcal{X}} P_X(x) \leq |\mathcal{U}|^{-1}, \qquad (2.11)$$

$$\max_{r \in \mathcal{R}} P_R(r) \leq |\mathcal{U}|^{-1}, \qquad (2.12)$$

where $\mathcal{U}$ is the support of the message U.

Before a formal proof of Theorem 2.1 is shown, we want to illustrate how the independence and functional dependence constraints in (2.7)–(2.9) can be "visualized" in a diagram, which helps to construct the proofs of Theorem 2.1 and Theorem 2.10 in Section 2.4.1. Consider a unit square as shown in Fig. 2.3(a). We divide it into $|\mathcal{U}|$ rows and $|\mathcal{R}|$ columns such that the width of rows and columns are equal to $P_U(u)$ and $P_R(r)$, respectively. The area of the cell with coordinate (u,r) is equal to $P_U(u)P_R(r) = P_{UR}(u,r)$, from (2.9). Now, we partition the cell (u,r) into columns proportional to $P_{X|U=u,R=r}$ and assign values for each partition. For example, if $P_{X|UR}(x|1,1) > 0$ for $x = 1$ or 2, then 1 and 2 are assigned to the cell $(1,1)$ as shown in Fig. 2.3(b). So the area of the partition with label x in the uth row and rth column is equal to $P_{URX}(u,r,x)$. Since the ciphertext x can be generated from the same message u with different key r, the same x can appear more than once in the same row but in different columns. However, the same x cannot appear more than once in the same column but in different rows, due to (2.8). Otherwise, the decoder will be confused. Therefore, if we sum the widths of the partitions with same number, say $x = 1$, the summation is less than the width of the unit square which is equal to 1 (i.e., $\sum_r P_R(r)$). This can be seen in Fig. 2.3(c). Finally, the total width of the partitions with label x in row u is equal to $\frac{P_{XU}(xu)}{P_U(u)} = P_X(x)$ due to (2.7). This means that in Fig. 2.3(c), the total length of the solid lines in each row is equal to $P_X(x)$. Therefore, $P_X(x) \leq |\mathcal{U}|^{-1}$. Formally, we can write the above argument in a compact way as follows.

Proof of Theorem 2.1 For any $x \in \mathcal{X}$, (2.7) gives

$$|\mathcal{U}|P_X(x) = \sum_u P_X(x) \tag{2.13}$$

$$= \sum_u P_{X|U}(x \mid u) \tag{2.14}$$

$$= \sum_u \sum_{r: P_{URX}(u,r,x)>0} \frac{P_{URX}(u,r,x)}{P_U(u)} \tag{2.15}$$

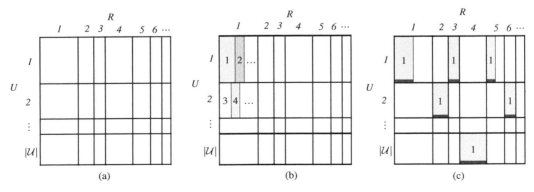

Figure 2.3 Illustration of the relationship between (U,R,X) satisfying (2.7)–(2.9). A solid line is at the bottom of each partition with label 1 in (c).

$$= \sum_u \sum_{r: P_{URX}(u,r,x)>0} \frac{P_{RX}(r,x) \mathbf{1}\{u = g(r,x)\}}{P_U(u)} \tag{2.16}$$

$$= \sum_{r: P_{RX}(r,x)>0} \frac{P_{RX}(r,x)}{P_U(g(r,x))} \tag{2.17}$$

$$= \sum_{r: P_{RX}(r,x)>0} P_{RX}(r,x) \frac{P_{X|UR}(x \mid g(r,x),r) P_R(r)}{P_{URX}(g(r,x),r,x)} \tag{2.18}$$

$$= \sum_{r: P_{RX}(r,x)>0} P_{X|UR}(x \mid g(r,x),r) P_R(r), \tag{2.19}$$

where (2.16) and (2.18) follow from (2.10) and (2.9), respectively. Thus,

$$P_X(x) \le \frac{1}{|\mathcal{U}|} \sum_{r: P_{RX}(r,x)>0} P_R(r) \tag{2.20}$$

$$\le \frac{1}{|\mathcal{U}|}, \tag{2.21}$$

due to (2.10). This establishes (2.11). Finally, due to the symmetric roles of X and R in (2.7)–(2.9), (2.12) is proved by swapping the roles of X and R in the argument from (2.13) to (2.21).

COROLLARY 2.1 (Lower bounds on $H(X)$ and $H(R)$) *Suppose P_U is known when we construct an error-free perfect secrecy system (R, U, X) satisfying (2.7)–(2.9). Then*

$$\log |\mathcal{U}| \le H(X), \tag{2.22}$$

with equality if and only if $P_X(x) = |\mathcal{U}|^{-1}$ for all $x \in \mathcal{X}$. Also,

$$\log |\mathcal{U}| \le H(R), \tag{2.23}$$

with equality if and only if $P_R(r) = |\mathcal{U}|^{-1}$ for all $r \in \mathcal{R}$. If the source distribution is not uniform, $H(X)$ and $H(R)$ are strictly greater than $H(U)$.

Proof Consider

$$H(X) - \log |\mathcal{U}| = \sum_x P_X(x) \log \frac{|\mathcal{U}|^{-1}}{P_X(x)} \ge \sum_x P_X(x) \log 1 = 0, \tag{2.24}$$

where the inequality follows from (2.11) and its equality holds if and only if $P_X(x) = |\mathcal{U}|^{-1}$ for all $x \in \mathcal{X}$. Finally, (2.23) follows from (2.12) due to an argument similar to (2.24) with X replaced by R.

COROLLARY 2.2 *No error-free perfect secrecy system can be constructed if the source message U has a countably infinite support.*

Proof Assume in contradiction that an EPS system exists for a source message $U \sim P_U$ with countably infinite support, $|\mathcal{U}| = \infty$. Note that (2.13)–(2.21) are still valid in this case. However, the conclusion that $|\mathcal{U}| P_X(x) \le 1$ for any $x \in \mathcal{X}$ contradicts that $|\mathcal{U}| = \infty$.

The following four remarks emphasize some of the (perhaps unexpected) consequences of Theorem 2.1.

1. If the source message U is defined on a countably infinite alphabet and R has a countable support, it is impossible to design an EPS system (Corollary 2.2). Therefore, if a cipher system is required for such a source, at least one of the constraints (2.7)–(2.9) must be relaxed. A similar result for secret sharing can be found in [5, Theorem 1] [6, Theorem 5.3]. Note that Corollary 2.2 is directly obtained from (2.12), and it cannot be shown from (2.23).
2. The results in Theorem 2.1 and Corollary 2.1 have led to some constrained non-Shannon type inequalities [7].
3. The bound in (2.12) can be obtained by an alternative proof.[2] Consider (R,U,X) satisfying (2.7)–(2.9), where the encoder and decoder are specified by $P_{XR|U}$ and $P_{U|XR}$, respectively. Note that $P_{XR|U} \equiv P_{X|UR}P_R$ due to (2.9). Now, consider another cipher system (U',R',X') where $P_{R'} = P_R$. The source U' is uniformly distributed in $\mathcal{U}$ and $I(U';R') = 0$. By using the same encoder $P_{X|UR}P_{R'}$ and decoder $P_{U|XR}$, it can be verified that the new system still satisfies (2.7)–(2.9) with (U,R,X) replaced by (U',R',X'). Since $P_{R'} = P_R$, the rest is to prove that $P_R(r) \leq \frac{1}{|\mathcal{U}|}$ for a uniform distribution P_U. The above trick together with [8, Lemma 2] have been used in the proof of [9, Lemma 3.2] for a similar problem in secret sharing. However, our proof of Theorem 2.1 requires fewer steps.
4. Due to (2.2), $H(U)$ is presented as the critical quantity setting a lower bound on $H(R)$. However, Corollary 2.1 shows that $H(R)$ and $H(X)$ can be arbitrarily large, as long as the size of the support of U is also arbitrarily large, *even when $H(U)$ is small*. The following example further illustrates that compared with the entropy bound in (2.23), the bound (2.12) in terms of probability masses gives a stronger requirement on R.

EXAMPLE 2.2   *Suppose $P_U = (0.3, 0.3, 0.3, 0.1)$ so that $H(U) = 1.895$ bits and $\log|\mathcal{U}| = 2$ bits.*

1. *Consider R chosen independently of U according to $P_R = (0.4, 0.2, 0.2, 0.2)$ so that $H(R) = 1.922$ bits and $H(U) < H(R) < \log|\mathcal{U}|$. Although P_R satisfies Shannon's fundamental bound (2.2), Corollary 2.1, in particular (2.23), shows this choice of key R is insufficient to achieve error-free perfect secrecy.*
2. *Consider $P_R = (0.4, 0.15, 0.15, 0.15, 0.15)$ so that $H(R) = 2.171$ bits and $H(U) < \log|\mathcal{U}| < H(R)$. However, this choice of key R is insufficient for error-free perfect secrecy, since from (2.12), $\max_r P_R(r) = 0.4 > 0.25 = |\mathcal{U}|^{-1}$.*

Theorem 2.1 and Corollary 2.1 not only apply to systems of the form shown in Fig. 2.1 (which includes Fig. 2.2 as a special case), but also to multi-letter variations. For example, we can accumulate n secret messages $(M_1, M_2, \ldots, M_n)$ from the source and treat these n messages together as one super-symbol U, i.e., $U = (M_1, M_2, \ldots, M_n)$. It is reasonable to consider finite n because practical systems have (1) only finite resources

---

[2] Private communication with Ueli Maurer.

to store the super-symbol, and (2) delay constraint. Unless the source has some special structure, the distribution of U cannot be uniform for any n if each M_i is not uniform. For example, if the source is stationary and memoryless with generic random variable M, accumulating symbols will only make $H(X)$ and $H(R)$ grow with $n \log |\mathcal{M}|$, where $\mathcal{M}$ is the support of M. On the other hand, one may argue that we just need to encrypt the typical sequences [4] from the source. Then $H(X)$ and $H(R)$ grow with $nH(M)$. However, this approach can only achieve arbitrary small error probability that violates the error-free requirement in EPS systems.

One may argue that the coding rate of X could be less than $H(X)$ because the sender and receiver share the same side information R, and $I(X;R) > 0$ is possible. In other words, a compressor may be appended to the encoder in Fig. 2.1 in order to reduce the size of the ciphertext. This configuration is shown in Fig. 2.4. However, we cannot simply apply the results from source coding with side information here, because the ciphertext still needs to satisfy the security constraint. If the new output Y satisfies the perfect secrecy and zero-error constraints, i.e., $I(U;Y) = H(U|RY) = 0$, then (R, U, Y) in Fig. 2.4 is simply another EPS system, governed by Theorem 2.1 and Corollary 2.1.

We have seen that the encoder in Fig. 2.1 subsumes many special cases like the ones in Figs. 2.2 and 2.4. Another important special case is shown in Fig. 2.5, where a key K is extracted from a bank of shared secret randomness R according to the source message U. In Section 2.3.2, we will consider situation where $H(R)$ is much bigger than $\log |\mathcal{U}|$. Therefore, only part of R, which is K, will be used to encrypt U. In these schemes, the portion of R used to construct K depends on the realization of U.

To complete this section, we show that the lower bounds (2.22) and (2.23) are simultaneously achievable using a one-time pad [10].

DEFINITION 2.2 (One-time pad)   Without lost of generality, let $\mathcal{U} = \{0, \ldots, \mu - 1\}$ be the support of U. Let R be independent of U and uniformly distributed in $\mathcal{U}$, and let X be generated according to the *one-time pad* as $X = (U + R) \mod \mu$. Then U can be recovered via $(X - R) \mod \mu$.

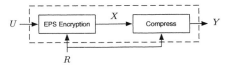

Figure 2.4   Compressing the output of an EPS cipher.

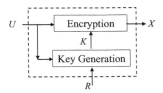

Figure 2.5   A key K is extracted from a bank of shared secret randomness R according to the source message U.

It is easy to verify that (2.7)–(2.9) are satisfied and $H(X) = H(R) = \log \mu$. Therefore, we have proved the following theorem.

THEOREM 2.2 (Achieving the minimum $H(X)$ and $H(R)$) Let $\mathcal{U}$ be the support of U. The one-time pad of Definition 2.2 is an EPS system achieving $H(X) = \log |\mathcal{U}|$ and $H(R) = \log |\mathcal{U}|$.

2.3 Residual Secret Randomness and Expected Key Consumption

The tight lower bounds on $H(X)$ and $H(R)$ depend only on $\log |\mathcal{U}|$ but not $H(U)$ from Theorem 2.1 and Corollary 2.1. Therefore, it is natural to ask: What is the role of $H(U)$ in an EPS system? The investigation into this question leads us to a new notion, "expected key consumption," which cannot be found in the literature. We will show in this section that $H(U)$ is a tight lower bound on the expected key consumption.

Since the one-time pad can simultaneously minimize $H(X)$ and $H(R)$ from Theorem 2.2, it seems to be "optimal" and leaves no room for improvement. However, this conclusion in fact stems from a folk theorem that the "required size of the secret key" is measured by the key entropy. The hidden interpretation behind this folklore is that *minimizing $H(R)$ is equivalent to minimizing the key consumption*. However, this interpretation can be misleading, as we will see in this section. We will illustrate that the expected key consumption should be measured by $I(R;U,X)$, which can be strictly less than $H(R)$. The following example illustrates some of the basic ideas that will be elaborated in this section. In contrast to Example 2.1, this example demonstrates how the shared secret randomness is updated after an EPS system is used.

EXAMPLE 2.3 *Suppose a spy inspects a target in a secret mission. Let U_i be the status of the target at time i for $i \geq 0$. Assume that the U_i are i.i.d. with a generic probability distribution $P_U(0) = 0.5$, $P_U(1) = 0.25$, and $P_U(2) = P_U(3) = 0.125$. The spy reports U_i to the headquarters, which shares a secret key $R = \{B_1, B_2, \ldots, B_{210}\}$ with the spy, where B_i for $1 \leq i \leq 210$ are independent and uniformly distributed over $\{0,1\}$. The mission is over when the spy cannot further transmit U_i by an EPS system. Of course, the spy wants to send as many U_i as possible in the mission. We consider the following two schemes for the spy.*

Scheme I: A one-time pad is used so that exactly 2 bits from R are used to encrypt each U_i. Therefore, $(U_1, U_2, \ldots, U_{105})$ can be successfully reported.

Scheme II: A more sophisticated scheme is used by the spy. After observing U_1, a new random variable U' is defined as

$$U' = \begin{cases} (0, B_{211}, B_{212}), & U_1 = 0 \\ (1, 0, B_{211}), & U_1 = 1 \\ (1, 1, 0), & U_1 = 2 \\ (1, 1, 1), & U_1 = 3 \end{cases} \quad (2.25)$$

where B_{211} and B_{212} are generated from flipping a fair coin by the spy such that all B_i are independent. By letting $K = (B_1, B_2, B_3)$, the spy sends $X = U' \oplus K$ to the headquarters. Upon receiving X, U_1 can be determined from $X \oplus K$, where K is solely a function of R. The headquarters can further retrieve (B_{211}, B_{212}) or only B_{211} for $U_1 = 0$ or 1, respectively. Let

$$R' = \begin{cases} (B_4, B_5, \ldots, B_{210}), & U \in \{2,3\} \\ (B_4, B_5, \ldots, B_{210}, B_{211}), & U = 1 \\ (B_4, B_5, \ldots, B_{210}, B_{211}, B_{212}), & U = 0. \end{cases} \quad (2.26)$$

We refer to R' as the *residual secret randomness* shared by the spy and the headquarters. Note that R' is not deterministic from R. For U_i with $i > 1$, R' will be used and updated according to a similar method as shown in (2.25)–(2.26).

In the worst case that $U_i = 2$ or 3 for all i, the spy can only report $\frac{210}{3} = 70$ messages, i.e., $(U_1, \ldots, U_{70})$. In the best scenario where $U_i = 0$ always, the spy can report $(U_1, \ldots, U_{210})$. This is similar to lossless source coding. In lossless source coding, different messages are encoded into codewords with different lengths according to the rule of thumb that more frequent messages are assigned to shorter codewords; the aim is to minimize the expected codeword length. If we consider the expected key consumption in Scheme II, it is given by

$$P_U(0) \cdot 1 + P_U(1) \cdot 2 + P_U(2) \cdot 3 + P_U(3) \cdot 3 = 1.75. \quad (2.27)$$

Thus the expected number of U_i that can be reported is $\frac{210}{1.75} = 120$ larger than the expected number in Scheme I (one-time pad).

Example 2.3 clearly illustrates that $H(K)$ is not the right measure for key consumption. Note that $H(U) = I(R; U, X) = 1.75$, equal to the expected key consumption in (2.27). It turns out that this is not mere coincidence. In the following example, we illustrate an alternative interpretation of expected key consumption.

EXAMPLE 2.4 *Suppose Alice and Bob generate a secret key R at a rate ω bits per second through certain methods like quantum key distribution [11], key agreement in wireless networks [12, 13], visible light communications [14], etc. Following the definitions of U_i and Schemes I and II in Example 2.3, Alice chooses one of the schemes to build a secret channel for sending U_i to Bob. Using Scheme I, a secret channel with transmission rate ω bits per second can be built. However, a larger rate $\frac{2\omega}{1.75}$ bits per second can be obtained if Scheme II is used instead.*

It is important to clarify how much secret randomness is left after a message has been encrypted and transmitted. Equally, it is important to quantify the expected key consumption. The following three notions will be investigated in this section.

DEFINITION 2.3 The *residual secret randomness* of an error-free perfect secrecy system is $H(R \mid UX)$.

DEFINITION 2.4 The *expected key consumption* of an error-free perfect secrecy system is $I(R; U, X)$.

DEFINITION 2.5 The *excess key consumption* of an error-free perfect secrecy system is $I(R;X)$.

In Sections 2.3.1 and 2.3.2, we will formally provide both converse and achievability results to show that after an EPS system is used, $H(R \mid UX)$ is the expected amount of remaining key that can be used for encryption of the next message. This justifies $H(R|U,X)$ as the right measure of residual secret randomness in a used EPS system. Since the sender and the receiver initially share a quantity $H(R)$ of secret randomness, the key consumption is equal to $H(R) - H(R|U,X) = I(R;U,X)$. We will provide achievable schemes to show that the minimal key consumption is $H(U)$ and, hence, the excess key consumption is $I(R;U,X) - H(U)$, which is equal to $I(R;X)$ in an EPS system.

2.3.1 The First Justification of $H(R|U,X)$

Consider the model depicted in Fig. 2.6. Suppose the sender and the receiver share secret randomness $(R_1, R_2, \ldots, R_n)$ at the beginning. Once the sender knows the message U_i, the ciphertext X_i is generated and sent to the receiver.[3] Assume that (U_i, R_i, X_i) is an EPS system for $1 \leq i \leq n$, where (U_i, R_i, X_i) are i.i.d. with generic distribution P_{URX}. We use (U,R,X) to denote the generic random variables. As we have seen in Examples 2.1 and 2.3, some secret randomness may be left in each used EPS system (U_i, R_i, X_i). Suppose we want to securely send additional messages $(V_1, \ldots, V_m)$. Since the shared R^n has been used, the sender and the receiver need to first establish a new secret key $S^m = (S_1, \ldots, S_m)$, where S_i are i.i.d. with generic distribution P_S. To generate this new key S^m, we assume that the sender can send a secret message A to the receiver. The new secret key S^m will be used to encrypt a second sequence of messages V^m, generating a ciphertext sequence Y^m such that $\{(V_i, S_i, Y_i)\}_{i=1}^m$ is another sequence of EPS systems. The focus here is to investigate the amount of residual secret randomness

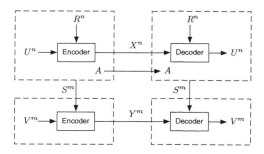

Figure 2.6 Generating a new secret key S^m.

---

[3] In some scenarios, as explained in the paragraph after Example 2.2, the sender may have to flush out X_i immediately for certain reasons: (1) delay constraint or (2) memory constraint if U_i is indeed a super-symbol. Note that the most efficient way to progressively encrypt a data stream with perfect secrecy and zero decoding error is an interesting problem, but it is outside the scope of this chapter.

left in an EPS system (U,R,X) by comparing the sizes of A and S^m. We make the following further assumptions in this model.

Assume

$$I(V^m; S^m, U^n, X^n) = 0, \qquad (2.28)$$

$$I(U^n, X^n; Y^m \mid V^m, S^m) = 0, \qquad (2.29)$$

$$H(S^m \mid R^n, U^n, X^n, A) = 0, \qquad (2.30)$$

$$I(S^m; U^n, X^n) = 0. \qquad (2.31)$$

These assumptions are adopted with the following reasoning. We assume in (2.28) that the new message V^m is generated independently of the previous uses of the EPS systems. Also, (2.29) holds due to (U^n, X^n)–(V^m, S^m)–Y^m forming a Markov chain. The sender and the receiver can agree on S^m without error due to (2.30). The justification of (2.31) is given as follows.

Although $\{(U_i, R_i, X_i)\}_{i=1}^n$ and $\{(V_i, S_i, Y_i)\}_{i=1}^m$ are individually sequences of EPS systems, it is possible that their combination is not secure, $I(X^n, Y^m; U^n, V^m) > 0$. For example, suppose U^n and V^m are i.i.d. with uniform distribution and $m = n$. If $S^m = U^n$, then using the one-time pad, $I(V^m; Y^m) = 0$ but $I(X^n, Y^m; U^n, V^m) \geq H(U^n)$. The following theorem shows that joint EPS systems satisfying (2.28)–(2.31) are still perfectly secure.

THEOREM 2.3 *Consider two sequences of i.i.d. EPS systems $\{(U_i, R_i, X_i)\}_{i=1}^n$ and $\{(V_i, S_i, Y_i)\}_{i=1}^m$ satisfying (2.28)–(2.31). Then the joint EPS system is still perfectly secure,*

$$I(X^n, Y^m; U^n, V^m) = 0. \qquad (2.32)$$

Proof By assumption,

$$I(U^n; X^n) = I(V^m; Y^m) = I(V^m; S^m, U^n, X^n) = I(U^n, X^n; Y^m \mid V^m, S^m)$$

$$= I(S^m; U^n, X^n) = 0. \qquad (2.33)$$

Note that

$$I(S^m; U^n, X^n) + I(V^m; S^m, U^n, X^n) + I(V^m; Y^m) + I(U^n; X^n)$$
$$+ I(U^n, X^n; Y^m \mid V^m, S^m) - I(X^n, Y^m; U^n, V^m)$$
$$= I(U^n, X^n; S^m \mid V^m, Y^m) + I(V^m; S^m) + I(X^n; Y^m) + I(U^n; V^m) \geq 0.$$

Together with (2.33), $I(X^n, Y^m; U^n, V^m) \leq 0$. Since $I(X^n, Y^m; U^n, V^m) \geq 0$, (2.32) is verified.

In order to generate a new key S^m, a secret auxiliary random variable A is sent from the sender to the receiver. Here, A is generated by a probabilistic encoder with $\{(R_i, U_i, X_i)\}_{i=1}^n$ as input. In Scheme II in Example 2.3, suppose we want to restore 210 secret bits after the system is used once. This can be done by letting A be uniformly distributed in a set with size $2, 4, 8, 8$ for $U_1 = 0, 1, 2, 3$, respectively. As shown in

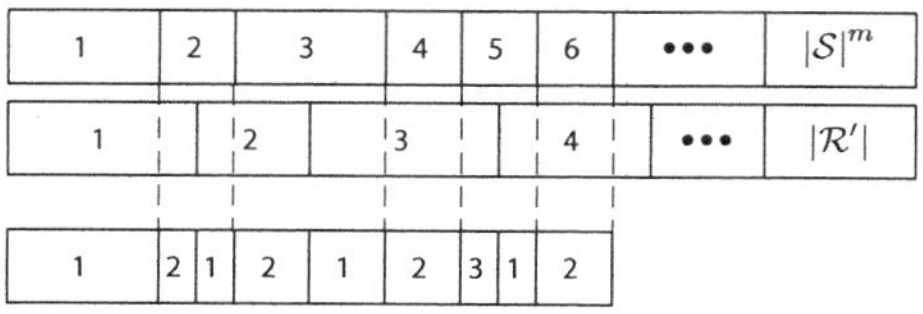

Figure 2.7 The width of the ith cell in the first row is equal to $P_{S^m}(i)$, and the width of the ith cell in the second row is equal to $P_{R^n|U^n,X^n}(i|u^n,x^n)$. An assignment of A is shown in the third row.

Example 2.3, the distribution of A depends on the realization of (R^n, U^n, X^n) so that we measure the expected size of A by $H(A \mid R^n U^n X^n)$. Since we can directly treat A as the new secret key S^m, it is reasonable to expect that $H(S^m) \geq H(A \mid R^n U^n X^n)$. Therefore, it is of interest to know by how much $H(S^m)$ can exceed $H(A \mid R^n, U^n, X^n)$ for a given sequence of EPS systems. The following theorem shows that the additional secret randomness, which can be extracted from $\{(R_i, U_i, X_i)\}_{i=1}^n$ with help from A, is measured by the residual secret randomness $H(R \mid U, X)$.

THEOREM 2.4 (Justification 1) *Consider two sequences of i.i.d. EPS systems* $\{(U_i, R_i, X_i)\}_{i=1}^n$ *and* $\{(V_i, S_i, Y_i)\}_{i=1}^m$, *and any A. If* (2.28)–(2.31) *are satisfied, then*

$$H(S^m) - H(A \mid R^n U^n X^n) \leq nH(R \mid U, X). \qquad (2.34)$$

On the other hand, it is possible to generate S^m such that (2.28)–(2.31) *are satisfied and*

$$H(S^m) - H(A \mid R^n, U^n, X^n) \geq nH(R \mid U, X) - \log 2 \qquad (2.35)$$

for a sufficiently large m such that

$$\max_{s^m} P_{S^m}(s^m) < \min_{r^n, u^n, x^n} P_{R^n|U^n X^n}(r^n \mid u^n, x^n).$$

Proof We first prove (2.34) by showing that

$$H(S^m) = I(S^m; U^n, X^n) + H(S^m \mid A, R^n, U^n, X^n) + I(S^m; A, R^n \mid U^n, X^n)$$
$$= I(S^m; A, R^n \mid U^n, X^n) \qquad (2.36)$$
$$\leq H(A, R^n \mid U^n, X^n) \qquad (2.37)$$
$$= H(A \mid R^n, U^n, X^n) + H(R^n \mid U^n, X^n) \qquad (2.38)$$
$$= H(A \mid R^n, U^n, X^n) + nH(R \mid U, X), \qquad (2.39)$$

where (2.36) follows from (2.30)–(2.31) and (2.39) follows from the fact that $\{(U_i, R_i, X_i)\}_{i=1}^n$ is a sequence of i.i.d. EPS systems.

The proof of the achievability part in (2.35) is via construction. With reference to Fig. 2.7, consider two partitions of the unit interval into disjoint "cells." The width of cell i in the first partition is $P_{S^m}(i)$ for $1 \leq i \leq |\mathcal{S}|^m$, where $\mathcal{S}$ is the support of S_i. Consider $U^n = u^n$ and $X^n = x^n$. The width of cell i in the second partition is $P_{R^n|U^n,X^n}(i \mid u^n, x^n)$ for $1 \leq i \leq |\mathcal{R}|^n$. The distribution of A is constructed to divide the second partition as shown in Fig. 2.7.

To simplify notations, we consider the support of R^n to be a set of consecutive integers $\{1,\ldots,|\mathcal{R}'|\}$ when $U^n = u^n$ and $X^n = x^n$. Suppose $R^n = r$ and let

$$b = \min\left\{ j : \sum_{i=1}^{j} P_S(i) > \sum_{i=1}^{r-1} P_{R^n|U^n,X^n}\left(i \mid u^n, x^n\right) \right\}. \tag{2.40}$$

For $j \geq 1$, A is defined by $P_A(j) = \frac{a(j)}{P_{R^n|U^n,X^n}(r|u^n,x^n)}$, where

$$\sum_{i=1}^{j} a(i) = \min\left\{ \sum_{i=1}^{b+j-1} P_S(i), \sum_{i=1}^{r} P_{R^n|U^n,X^n}\left(i \mid u^n, x^n\right) \right\}$$

$$- \sum_{i=1}^{r-1} P_{R^n|U^n,X^n}\left(i \mid u^n, x^n\right). \tag{2.41}$$

For the example in Fig. 2.7, when $R^n = 1$,

$$P_A(1) = \frac{P_{S^m}(1)}{P_{R^n|U^n,X^n}\left(1 \mid u^n, x^n\right)} = 1 - P_A(2). \tag{2.42}$$

When $R^n = 2$,

$$P_A(1) = \frac{P_{S^m}(1) + P_{S^m}(2) - P_{R^n|U^n,X^n}\left(1 \mid u^n, x^n\right)}{P_{R^n|U^n,X^n}\left(2 \mid u^n, x^n\right)} = 1 - P_A(2). \tag{2.43}$$

By definition, S^m is determined from R^n and A for any fixed $U^n = u^n$ and $X^n = x^n$. On the other hand, A is also determined from S^m and R^n. Therefore,

$$H\left(S^m \mid A, R^n, U^n, X^n\right) = H\left(A \mid S^m, R^n, U^n, X^n\right) = 0. \tag{2.44}$$

By choosing m sufficiently large, such that

$$\max_{s^m} P_{S^m}(s^m) < \min_{r^n, u^n, x^n} P_{R^n|U^n,X^n}\left(r^n \mid u^n, x^n\right), \tag{2.45}$$

R^n can take at most two possible values for any given (S^m, U^n, X^n), and hence

$$H\left(R^n \mid S^m, U^n, X^n\right) \leq \log 2. \tag{2.46}$$

Therefore,

$$H\left(A \mid R^n, U^n, X^n\right) \tag{2.47}$$

$$= I\left(A; S^m \mid R^n, U^n, X^n\right) + H\left(A \mid S^m, R^n, U^n, X^n\right) \tag{2.48}$$

$$= I\left(A; S^m \mid R^n, U^n, X^n\right) + H\left(S^m \mid A, R^n, U^n, X^n\right) \tag{2.49}$$

$$= H\left(S^m \mid R^n, U^n, X^n\right) \tag{2.50}$$

$$= H\left(S^m\right) - H\left(R^n \mid U^n, X^n\right) + H\left(R^n \mid S^m, U^n, X^n\right) - I(S^m; U^n, X^n) \tag{2.51}$$

$$\leq H\left(S^m\right) - H\left(R^n \mid U^n, X^n\right) + H\left(R^n \mid S^m, U^n, X^n\right) \tag{2.52}$$

$$\leq H\left(S^m\right) - H\left(R^n \mid U^n, X^n\right) + \log 2, \tag{2.53}$$

where (2.49) and (2.53) follow from (2.44) and (2.46), respectively. Since $\{(U_i, R_i, X_i)\}_{i=1}^{n}$ is a sequence of i.i.d. EPS systems, (2.35) is verified.

For any (U^n, X^n), the same $P_{S^m|U^n,X^n} \equiv P_{S^m}$ is generated. Therefore, (2.31) is verified. Since S^m is determined by (R^n, U^n, X^n, A), (2.30) is verified, and (2.28) can also be verified as V^m is independent of (R^n, U^n, X^n, A). Finally, (2.29) is due to the fact that $\{(V_i, S_i, Y_i)\}_{i=1}^{m}$ is a sequence of EPS systems.

Roughly speaking, Theorem 2.4 shows that for large n and m, the optimal algorithm with the help of A can extract approximately

$$nH(R \mid U, X)$$

bits of residual secret randomness from $\{(R_i, U_i, X_i)\}_{i=1}^{n}$. In [15], we considered another algorithm generating a new secret key with asymptotic rate $H(R \mid U, X)$ without using an auxiliary secret random variable.

2.3.2 The Second Justification of $H(R|U,X)$

In this subsection, we will see that $H(R|U,X)$ also appears in another model. Consider Fig. 2.8, in which the sender and receiver share a secret key R, and two EPS systems are used sequentially by the sender to securely transmit two (possibly correlated) messages U and V. In the first round, the sender encodes the message U into X, which is transmitted to the receiver as described in Section 2.2. In the second round, the sender further encodes V (or more generally both U and V) into Y, which is then transmitted to the receiver. As before, we require

$$H(U \mid R, X) = H(V \mid R, X, Y) = 0 \tag{2.54}$$

for zero-error decoding, and require

$$I(U, V; X, Y) = 0 \tag{2.55}$$

for perfect secrecy.

THEOREM 2.5 (Justification 2) *Consider the "two-round error-free perfect secrecy system" of Fig. 2.8. If*

$$I(U, V; X, Y) = H(U \mid R, X) = H(V \mid R, X, Y) = 0, \tag{2.56}$$

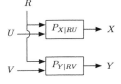

Figure 2.8 Using an error-free perfect secrecy system twice.

then the entropy of the second message V conditioning on the first message U is bounded above by the residual secret randomness,

$$H(V \mid U) \leq H(R \mid U,X). \tag{2.57}$$

Proof Note that

$$H(R \mid U,X) - H(V \mid U) + I(U,V;X,Y) + H(U \mid R,X) + H(V \mid R,X,Y)$$
$$= I(V,Y;U \mid R,X) + I(R,U;Y \mid X) + I(U;X)$$
$$+ H(U \mid R,X,Y) + H(R \mid U,V,X,Y) \geq 0,$$

which can be verified by rewriting all the terms in joint entropies. Together with (2.56), (2.57) is verified.

Theorem 2.5 implies that the maximum amount of information that can be secretly transmitted in the second round is bounded above by the residual secret randomness $H(R \mid UX)$. Hence a bigger $H(R \mid UX)$ means that more secret randomness is left for the encryption of the second message V.

2.3.3 Some Properties of $I(R;U,X)$ and $I(R;X)$

As the sender and receiver initially share $H(R)$ bits of secret randomness, and $H(R \mid U,X)$ bits of residual secret randomness are left, the expected key consumption for each use of the EPS system is

$$H(R) - H(R \mid U,X) = I(R;U,X), \tag{2.58}$$

which is the quantity proposed in Definition 2.4. Next, we exhibit an important property of $I(R;U,X)$.

THEOREM 2.6 *In an error-free perfect secrecy system, the expected key consumption is bounded below by the source entropy,*

$$I(R;U,X) \geq H(U), \tag{2.59}$$

where equality holds if and only if $I(R;X) = 0$.

Proof The information diagram for the random variables U,X,R involved in an error-free perfect secrecy system satisfying (2.7)–(2.9) is shown in Fig. 2.9(a). The theorem follows from

$$I(R;U,X) - H(U) = I(X;R) \geq 0. \tag{2.60}$$

In Section 2.4.1, we will describe EPS coding scheme achieving $I(R;U,X) = H(U)$. Therefore, $I(X;R)$ measures the difference between the expected key consumption of an EPS system and the minimum possible key consumption, justifying the excess key consumption in Definition 2.5. The information diagram for the optimal case $I(X;R) = 0$ is shown in Fig. 2.9(b).

We summarize this section by the following three remarks.

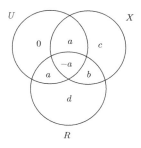 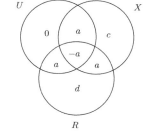

(a) General EPS system. (b) Minimum expected key consumption, achieving equality in (2.59).

Figure 2.9 Information diagrams for (U,R,X) in an EPS system.

1. Theorems 2.4 and 2.5 provide strong justification of $I(R;U,X)$ as the expected key consumption required to achieve error-free perfect secrecy. Theorem 2.6 shows that the expected key consumption cannot be less than the source entropy. Recall that Theorem 2.1 and Corollary 2.1 give the lower bounds on the initial key requirement. Therefore, we have distinguished between two different concepts: (1) expected key consumption in a multi-round system, and (2) the initial key requirement for a one-shot system. In contrast to the bound $H(R) \geq H(U)$ [1, 2], Theorem 2.6 more precisely presents the role of $H(U)$ in an error-free perfect secrecy system.
2. From (2.7)–(2.9) we can show that

$$H(R) = H(U) + I(X;R) + H(R \mid U,X). \qquad (2.61)$$

Thus, the key entropy $H(R)$ can be interpreted as consisting of three parts: the minimum amount of secret randomness used to protect the source, the excess key consumption that has been wasted, and the residual secret randomness that can be used in future.

3. If the source distribution is uniform, it is easy to verify that the one-time pad achieves minimal key consumption, i.e., $I(R;U,X) = H(U)$.

2.4 Tradeoff between Key Consumption and Number of Channel Uses

Although the one-time pad simultaneously achieves the minimal expected key consumption and the minimum number of channel uses for a uniform source, we are going to demonstrate the existence of a fundamental, non-trivial tradeoff between the expected key consumption and the number of channel uses for general non-uniform sources. Our main results, Corollary 2.1 proved earlier and Theorems 2.6–2.12 to be proved below, are summarized in the sketch in Fig. 2.10.

We will consider two important regimes. First, in Section 2.4.1, we will consider the regime in which ciphers minimize the key consumption $I(R;U,X)$. Conversely, in Section 2.4.2 we consider systems that minimize the number of channel uses $H(X)$.

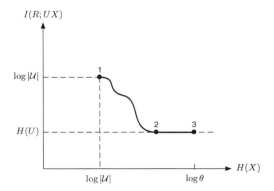

Figure 2.10 A sketch to illustrate the tradeoff between $I(R;U,X)$ and $H(X)$.

Point 1 in Fig. 2.10 is due to Theorem 2.12 in Section 2.4.2, and has the smallest $I(R;U,X)$ among all EPS systems with $H(X) = \log|\mathcal{U}|$. We will show that this point can always be achieved by a one-time pad.

Point 2 in Fig. 2.10 has the smallest $H(X)$ among all the EPS systems with $I(R;U,X) = H(U)$. For this point, Theorem 2.10 in Section 2.4.1 gives the lower bound on $H(X)$, which is strictly greater than $\log|\mathcal{U}|$ if P_U is not uniform.

If P_U has only rational probability masses, Theorem 2.7 in Section 2.4.1 shows that Point 3 in Fig. 2.10 can be achieved by a generalization of the one-time pad, the *partition code* (to be introduced in Definition 2.6).

If all the probability masses in P_U are integer multiples of the smallest probability mass in P_U, then Point 2 in Fig. 2.10 coincides with Point 3 by the partition code shown in Theorem 2.11. Otherwise, Point 3 can differ from Point 2, as will be demonstrated in Example 2.5.

The existence, continuity, and non-increasing in $H(X)$ properties of the curved portion of the tradeoff curve are established in Section 2.4.3.

2.4.1 Minimal Expected Key Consumption

We first consider EPS systems that achieve minimal expected key consumption. From Theorem 2.6, an error-free perfect secrecy system with minimal key consumption satisfies (2.7)–(2.9) and $I(R;U,X) = H(U)$, or, equivalently,

$$I(X;R) = 0. \tag{2.62}$$

Due to (2.62), one may connect the requirement of minimum expected key consumption to homophonic substitution [16], which is now explained. Instead of dealing with a single plaintext U, homophonic substitution is a precoding which converts a *plaintext sequence* U^n into a random sequence, say V^m, such that (1) the mapping is one-to-many, and (2) each V_i in the output sequence V^m is close or equal to uniformity, which is defined as each V_i is equiprobable and independent [17, 18]. Suppose the alphabets of U^n and X^n are the same. If a homophonic code that can achieve uniformity is applied to a source before a one-time pad is used to achieve $I(U^n;X^n) = 0$, then the ciphertext is independent of the key R [17], i.e., $I(X^n;R) = 0$ [cf. (2.62)]. In this case, the key cannot be reconstructed by a ciphertext-only attack regardless of the statistics and the length of the plaintext sequence. Therefore, homophonic coding schemes achieving uniformity have mainly been studied in the literature. Homophonic code is usually applied to protect the key in the scenarios where $H(R) < H(U^n)$. In this case, $I(U^n;X^n) > 0$ is expected but $I(X^n;R) = 0$ is the best that a cipher can achieve. In contrast, $I(X^n;R) > 0$ is allowed in this chapter, but we insist on perfect secrecy, i.e., $I(U^n;X^n) = 0$, throughout the chapter. Another difference is that $H(R) \geq \log|\mathcal{U}^n| \geq H(U^n)$ from Corollary 2.1. In Example 2.5, we will show that variable-length homophonic code [19, 20] together with a one-time pad is not a good solution if we want to minimize the size of ciphertext under the perfect secrecy assumption. Therefore, we are looking for something different from homophonic code. Now, we consider a coding scheme that can achieve minimal key consumption for some source distributions. Part of the coding scheme is similar to homophonic code, but it can generate a new key after it has been used.

DEFINITION 2.6 (Partition code $\mathcal{C}(\Psi)$) Assume that U is a random variable defined on $\{1,\ldots,\ell\}$. Let $\Psi = (\psi_1, \psi_2, \ldots, \psi_\ell)$ and let $\theta = \sum_{i=1}^{\ell} \psi_i$, where ψ_i and θ are positive integers. Let A' be a random variable such that

$$\Pr(A'=j \mid U=i) = \begin{cases} \frac{1}{\psi_i} & \text{if } 1 \leq j \leq \psi_i, \\ 0 & \text{otherwise.} \end{cases}$$

Let $A = \sum_{i=1}^{U-1} \psi_i + A' - 1$, R be uniformly distributed on the set $\{0, 1, \ldots, \theta - 1\}$, and $X = A + R \mod \theta$. The cipher system (R, U, X) thus defined is called the *partition code* $\mathcal{C}(\Psi)$.

The one-time pad can be seen as a special case of a partition code with $\Psi = (1, 1, \ldots, 1)$. Also, the cipher system in Example 2.1 is indeed a partition code with $\Psi = (3, 2)$. We now explain the rationale behind partition codes. Due to Theorem 2.1 and Corollary 2.1, the secret randomness required to encrypt U is at least the logarithm of the support size of U. If P_U is not uniform, we need to use secret randomness more than $H(U)$. By combining the message U and an extra message A', a new message $A = (U, A')$ is created where A' is chosen according to the realization of U such that P_A is equal (or close) to uniform. After that, secret randomness equal (or close) to $H(A)$ is used to encrypt A. Although $H(A) > H(U)$, the extra secret randomness used to protect A' can be extracted back by creating a new secret key from the received A'. The overall consumption of secret randomness is thus reduced. So partition code is different from homophonic code in two aspects: (1) partition code is not applied to protect the used key, and (2) new secret randomness is generated in partition code. Now, some properties of a partition code are shown in the following lemma.

LEMMA 2.1 *A partition code $\mathcal{C}(\Psi)$ satisfies (2.7)–(2.9) and hence is an EPS system. Furthermore,*

$$H(X) = H(R) = \log \theta, \qquad (2.63)$$

$$I(R;U,X) = H(U) + D(P_U \| Q_U), \qquad (2.64)$$

where $D(\cdot \| \cdot)$ is the relative entropy [4] and Q_U is the probability distribution such that $Q_U(i) = \psi_i/\theta$.

Proof It can be easily verified that (2.7)–(2.9) and (2.63) hold. Furthermore, (2.64) follows from

$$H(X \mid U,R) = H(A \mid U,R) = \sum_{i=1}^{\ell} P_U(i) \log \psi_i \qquad (2.65)$$

together with (2.63).

THEOREM 2.7 (Achieving the minimal key consumption) *Suppose the probability mass $P_U(i)$ is rational for all $i = 1, \ldots, \ell$. Let θ be an integer such that $\theta \cdot P_U(i)$ is also an integer for all i, and let $\Psi = (\psi_1, \psi_2, \ldots, \psi_\ell)$ with $\psi_i = \theta \cdot P_U(i)$. Then the EPS system (R,U,X) induced by the partition code $\mathcal{C}(\Psi)$ achieves the lower bound in (2.59), namely $I(R;U,X) = H(U)$.*

In the following theorem, we prove that if $P_U(i)$ is irrational for any i, then no EPS can achieve zero key excess with finite X or R. Its proof is deferred to Section 2.5.

THEOREM 2.8 *Suppose (R,U,X) satisfies (2.7)–(2.9) and (2.62). If $P_U(u)$ is irrational for some $u \in \mathcal{U}$, then the support of X and R cannot be finite.*

Although it is difficult to construct codes satisfying (2.7)–(2.9) and (2.62) for P_U having irrational probability masses, Theorem 2.6 still gives a tight bound on $I(R;U,X)$, as shown in the following theorem.

THEOREM 2.9 *Suppose the support of P_U is a finite set of integers $\{1, \ldots, \ell\}$. Let $\Psi = (\psi_1, \ldots, \psi_{\ell+1})$, with*

$$\psi_i = \begin{cases} \lfloor P_U(i)\theta \rfloor, & 1 \leq i \leq \ell, \\ \theta - \sum_{i=1}^{\ell} \lfloor P_U(i)\theta \rfloor, & i = \ell+1. \end{cases}$$

Assume that θ is large enough such that $\lfloor P_U(i)\theta \rfloor \geq 1$ for all $1 \leq i \leq \ell$. For the partition code $\mathcal{C}(\Psi)$, $I(R;U,X) \to H(U)$ as $\theta \to \infty$.

Proof Consider a probability distribution Q_U with $Q_U(i) = \psi_i/\theta$ for $1 \leq i \leq \ell+1$. As $\theta \to \infty$, Q_U converges pointwise to P_U and hence $D(P_U \| Q_U) \to 0$ for finite ℓ. The theorem thus follows from (2.64).

In addition to minimizing the key consumption $I(R;U,X)$, we may also want to simultaneously minimize $H(X)$, which is the number of channel uses required to convey the ciphertext X. The following theorem and corollary illustrate that the zero key excess condition can be very harsh, requiring the EPS system to have a very large $H(R)$ and $H(X)$, even for very simple sources.

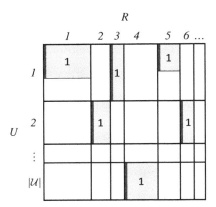

Figure 2.11 The relationship between (U, R, X) satisfying (2.7)–(2.9) and (2.62).

THEOREM 2.10 (EPS systems with minimal $I(R; U, X)$) *Let $\mathcal{X}$, $\mathcal{R}$, and $\mathcal{U}$ be the respective supports of random variables X, R, and U satisfying (2.7)–(2.9) and (2.62). Then*

$$\max_{x \in \mathcal{X}} P_X(x) \leq \min_{u \in \mathcal{U}} P_U(u), \qquad (2.66)$$

$$\max_{r \in \mathcal{R}} P_R(r) \leq \min_{u \in \mathcal{U}} P_U(u). \qquad (2.67)$$

Proof Consider any $u \in \mathcal{U}$ and $x \in \mathcal{X}$. By definition, $P_U(u) > 0$ and $P_X(x) > 0$. From (2.7), we have $P_{UX}(u,x) = P_U(u)P_X(x) > 0$. Consequently, there exists $r \in \mathcal{R}$ such that $P_{UXR}(u,x,r) > 0$. Notice that

$$P_{UXR}(u,x,r) = P_{XR}(x,r) \qquad (2.68)$$

$$= P_X(x)P_R(r), \qquad (2.69)$$

where (2.68) is due to (2.8) and (2.69) is due to (2.62). On the other hand,

$$P_{UXR}(u,x,r) \leq P_{UR}(u,r) \qquad (2.70)$$

$$= P_U(u)P_R(r), \qquad (2.71)$$

where (2.71) is due to (2.9). Finally, as $P_R(r) > 0$, we have $P_X(x) \leq P_U(u)$ and (2.66) follows. Due to the symmetric roles of X and R, the theorem is proved.

The proof of Theorem 2.10 can be readily seen from Fig. 2.11. Instead of partitioning each cell into columns proportional to $P_{X|U=u,R=r}$ as shown in Fig. 2.3(b), we partition each cell into rows proportional to $P_{X|U=u,R=r}$ to obtain Fig. 2.11. Since the same x cannot appear more than once in the same column but in different rows due to (2.8), the partition with label x in column r has length equal to $\frac{P_{RX}(r,x)}{P_R(r)} = P_X(x)$ due to (2.62). Thus, the thick line next to the number "1" in Fig. 2.11 in each column has the same length. Since $I(U;X) = 0$, the label x appears in each row and hence the length of a thick line must be less than $\min_u P_U(u)$. Therefore, $P_X(x) \leq \min_u P_U(u)$ for all x.

Theorem 2.10 is used to obtain bounds on $H(X)$ and $H(R)$ in the following corollary. We define the binary entropy function $h(\gamma) = -\gamma \log \gamma - (1-\gamma)\log(1-\gamma)$ for $0 < \gamma < 1$ and $h(0) = h(1) = 0$.

COROLLARY 2.3 *Let $\mathcal{X}$, $\mathcal{R}$, and $\mathcal{U}$ be the respective supports of random variables X, R, and U satisfying (2.7)–(2.9) and (2.62). Then*

$$\min\{H(X), H(R)\} \geq h(\pi \lfloor \pi^{-1} \rfloor) + \pi \lfloor \pi^{-1} \rfloor \log \lfloor \pi^{-1} \rfloor \qquad (2.72)$$

$$\geq \log \frac{1}{\pi}, \qquad (2.73)$$

where $\pi = \min_{u \in \mathcal{U}} P_U(u)$ and the right sides of (2.72) and (2.73) are equal if and only if π^{-1} is an integer.

Proof From (2.66), $\max_{x \in \mathcal{X}} P_X(x) \leq \min_{u \in \mathcal{U}} P_U(u)$. Together with [21, Theorem 10], this establishes (2.72). To prove (2.73), we first consider the case when π^{-1} is an integer. Then

$$h(\pi \lfloor \pi^{-1} \rfloor) + \pi \lfloor \pi^{-1} \rfloor \log \lfloor \pi^{-1} \rfloor = h(1) + \pi \pi^{-1} \log \frac{1}{\pi} = \log \frac{1}{\pi}.$$

If π^{-1} is not an integer, then $1 - \pi \lfloor \pi^{-1} \rfloor < \pi$. Hence,

$$h(\pi \lfloor \pi^{-1} \rfloor) + \pi \lfloor \pi^{-1} \rfloor \log \lfloor \pi^{-1} \rfloor$$

$$= \pi \lfloor \pi^{-1} \rfloor \log \frac{1}{\pi \lfloor \pi^{-1} \rfloor} + (1 - \pi \lfloor \pi^{-1} \rfloor) \log \frac{1}{1 - \pi \lfloor \pi^{-1} \rfloor} + \pi \lfloor \pi^{-1} \rfloor \log \lfloor \pi^{-1} \rfloor$$

$$> \pi \lfloor \pi^{-1} \rfloor \log \frac{1}{\pi} + (1 - \pi \lfloor \pi^{-1} \rfloor) \log \frac{1}{\pi}$$

$$= \log \frac{1}{\pi}.$$

Furthermore, the right-hand sides of (2.72) and (2.73) are equal only if π^{-1} is an integer. This proves the lower bounds on $H(X)$. Due to the symmetric roles of X and R in (2.7)–(2.9) and (2.62), the theorem is proved.

Suppose P_U is non-uniform so that $\min_{u \in \mathcal{U}} P_U(u) < |\mathcal{U}|^{-1}$. Corollary 2.1 shows that $\min\{H(X), H(R)\} \geq \log|\mathcal{U}|$. However, (2.73) further shows that

$$\min\{H(X), H(R)\} \geq \log\left(\min_{u \in \mathcal{U}} P_U(u)\right) > \log|\mathcal{U}|. \qquad (2.74)$$

Therefore, a longer initial key and a larger number of channel uses are required for systems that achieve the minimal expected key consumption. The following theorem shows that the lower bounds in (2.73) can be achieved for certain P_U, including the uniform distribution and D-adic distributions, $P_U(u) = D^{-i}$ for certain integers D and i.

THEOREM 2.11 *Let $\mathcal{U} = \{1, \ldots, \ell\}$ and let $P_U(\ell) \leq P_U(i)$ for $1 \leq i \leq \ell$. If there exists a set of positive integers $\Psi = \{\psi_i\}$ such that $P_U(i) = \psi_i P_U(\ell)$ for $1 \leq i \leq \ell$, then the partition code $\mathcal{C}(\Psi)$ simultaneously achieves the minimum $H(X)$ and $H(R)$ among all EPS systems achieving minimal key consumption.*

Proof Suppose (R, U, X) satisfies (2.7)–(2.9) and (2.62), so that $H(X) \geq \log \frac{1}{P_U(\ell)}$ from (2.73). Note that $P_U(\ell) = \left(\sum_{i=1}^{\ell} \psi_i \right)^{-1}$ from the definition of Ψ. Therefore,

$$H(X) \geq \log \left(\sum_{i=1}^{\ell} \psi_i \right). \tag{2.75}$$

The partition code $\mathcal{C}(\Psi)$ has $\theta = \sum_{i=1}^{\ell} \psi_i$ so that it can achieve equality in (2.75) from (2.63). Similarly, we can argue that the partition code $\mathcal{C}(\Psi)$ achieves the minimum $H(R)$.

For some other source distributions P_U, the partition code may not achieve the minimal number of channel uses $H(X)$, as illustrated in the following example.

EXAMPLE 2.5 *Consider an EPS system (R, U, X) satisfying (2.62) as follows. Since R and X are independent and U is deterministic from R and X, it is simpler to specify P_{URX} by specifying each term on the right-hand side of $P_{URX} = P_{U|RX} P_R P_X$.*

1. *X and R take values from the set $\{0, 1, 2, 3\}$.*
2. *$P_X(0) = P_R(0) = 2/5$, $P_X(i) = P_R(i) = 1/5$ for $i = 1, 2, 3$.*
3. *U is a function of (X, R) such that $U = 0$ if and only if (i) $X = 0$ and $R \neq 0$, or (ii) $R = 0$ and $X \neq 0$, or (iii) $X = R \neq 0$. Consequently, $P_{U|XR}(u \mid x, r)$ is well defined.*
4. *U can be verified as a binary random variable where $P_U(0) = 3/5$.*

It is straightforward to check that $\{U, X, R\}$ satisfies (2.7)–(2.9) and (2.62), and $H(X) = H(R) < \log 5$. However, $\theta = 5$ is the smallest integer such that $\theta \cdot P_U(u)$ is an integer. In this example, $H(X)$ is smaller than the value given in (2.63).

A couple of remarks can be made here.

1. Among all EPS systems achieving minimal key consumption, Theorem 2.11 shows that a partition code can simultaneously minimize $H(X)$ and $H(R)$ for some source distributions, but an exceptional case is shown in Example 2.5.
2. Usually, a cipher system uses a secret key that has a uniform distribution. It is interesting to note that a non-uniform P_R is used in Example 2.5.

Now, we consider the famous homophonic coding algorithm in [19] and use it to convert U in Example 2.5 into $(V_1, \ldots V_L)$. Without lost of generality, we consider binary V_i. We can check that $P_L(i) = \frac{1}{2^{1+i}}$ and $U = 0$ if and only if $L = 1 + 4i$ or $4 + 4i$ for $i \geq 0$. Although X can be constructed from applying a bit-to-bit one-time pad to $(V_1, \ldots V_L)$, the length L still discloses U. In this case, U can be fully recovered from L. The same problem appears if we use the interval algorithm in [20] instead. In fact, variable-length homophonic coding is useful for encoding a stream input. However, if U has finite length and $I(U; X) = 0$ is required, variable-length homophonic coding cannot be directly applied.

2.4.2 Minimal Number of Channel Uses

In the previous subsection, we proposed partition codes $\mathcal{C}(\Psi)$ which minimize the expected key consumption for error-free perfect secrecy systems. However, we also demonstrated that these codes do not guarantee the minimal number of channel uses $H(X)$, among all other EPS systems which also minimize the expected key consumption. Finding an EPS system that minimizes the number of channel uses for a given expected key consumption is a very challenging open problem. In this subsection, we aim to minimize $I(R;U,X)$ in the regime where $H(X)$ meets the lower bound in Corollary 2.1 (i.e., $H(X) = \log|\mathcal{U}|$). Unlike in Section 2.4.1, we can completely characterize this regime.

Using Theorem 2.2, we can show that, by using a one-time pad,

$$H(U) \leq \log|\mathcal{U}| = H(X) = H(R) = I(R;U,X).$$

Therefore, in this instance, the expected key consumption $I(R;U,X)$ is not minimal when the source U is not uniform. However, the following theorem shows that among all EPS systems that minimize the number of channel uses, the one-time pad minimizes the expected key consumption.

THEOREM 2.12   *Consider any EPS system (R,U,X) (e.g., one-time pad) with $H(X) = \log|\mathcal{U}|$. Then $I(R;U,X) = \log|\mathcal{U}|$ and $H(X|R,U) = 0$.*

Proof   If $H(X) = \log|\mathcal{U}|$, $P_X(x) = 1/|\mathcal{U}|$ for $x \in \mathcal{X}$ and

$$|\mathcal{X}| = |\mathcal{U}| \tag{2.76}$$

from Corollary 2.1. Let

$$\mathcal{X}_{ru} = \{x \in \mathcal{X} : P_{RUX}(r,u,x) > 0\}$$

be the set of possible values of X when $R=r$ and $U=u$. Due to (2.10), $\mathcal{X}_{ri} \cap \mathcal{X}_{rj} = \emptyset$ if $i \neq j$. Together with (2.76),

$$|\mathcal{U}| = |\mathcal{X}| \geq \left|\bigcup_u \mathcal{X}_{ru}\right| = \sum_u |\mathcal{X}_{ru}| \geq |\mathcal{U}| \min_u |\mathcal{X}_{ru}|. \tag{2.77}$$

On the other hand, for any $r \in \mathcal{R}$ and $u \in \mathcal{U}$,

$$\sum_{x \in \mathcal{X}_{ru}} P_{RUX}(r,u,x) = P_{UR}(u,r) = P_U(u)P_R(r) > 0$$

from (2.9), and hence $|\mathcal{X}_{ru}| \geq 1$. Substituting this result into (2.77) shows that $|\mathcal{X}_{ru}| = 1$. Therefore, X is a function of R and U, which verifies

$$H(X \mid UR) = 0. \tag{2.78}$$

Together with (2.7)–(2.9), we have $I(R;U,X) = H(X) = \log|\mathcal{U}|$.

2.4.3 The Fundamental Tradeoff

An important open problem is to find coding schemes that can achieve points on the tradeoff curve between Points 1 and 2 in Fig. 2.10. For a given source distribution P_U and number of channel uses $H(X) = \log|\mathcal{U}| + \gamma$, with $\gamma \geq 0$ we need to solve the following optimization problem:

$$f(\gamma) = \inf_{P_{RX|U} \in \mathcal{P}_\gamma} I(R;U,X), \qquad (2.79)$$

where

$$\mathcal{P}_\gamma = \{P_{RX|U} : I(R;U) = I(X;U) = H(U|X,R) = 0, H(X) = \log|\mathcal{U}| + \gamma\} \qquad (2.80)$$

is the set of feasible conditional distributions yielding an EPS system with the specified number of channel uses.

Recently, we have shown an algorithm [22] that can determine Ψ in a partition code $\mathcal{C}(\Psi)$ such that the minimal $I(R;U,X)$ is achieved for a given $H(X)$. However, solving (2.79) remains open in general. We end this section by showing two important structural properties of $f(\gamma)$ in the following proposition.

PROPOSITION 2.1 *Let P_U and $\gamma \geq 0$ be given. Then $\mathcal{P}_\gamma$ defined in (2.80) is non-empty for $\gamma \geq 0$, and $f(\gamma)$ defined in (2.79) is non-increasing in γ.*

Proof A non-vacuous feasible set is demonstrated as follows. Let (R,U,X) be a given EPS system. Define a second EPS system (R',U',X') as follows. Let $(R',U') = (R,U)$ and $X' = (X,A)$, where A is a random variable independent of (R,U,X) such that $H(A) = \delta$ for any given $\delta \geq 0$. In other words, (R',U',X') is constructed by adding some spurious randomness into the ciphertext of the EPS system (R,U,X). Setting $\delta = \gamma$ and supposing that (R,U,X) is a cipher system using a one-time pad yields $P_{R'X'|U'} \in \mathcal{P}_\gamma$.

By the same trick, we can show that $f(\gamma)$ is non-increasing. For any $\gamma > 0$ and $\epsilon > 0$, let (R,U,X) be an EPS system such that $P_{RX|U} \in \mathcal{P}_\gamma$ and

$$I(R;U,X) < f(\gamma) + \epsilon. \qquad (2.81)$$

It is easy to check that $P_{R'X'|U'} \in \mathcal{P}_{\gamma+\delta}$ and $H(X \mid U,R) = H(X' \mid U',R') - \delta$.
Then

$$f(\gamma+\delta) = \inf_{P_{\tilde{R}\tilde{X}|\tilde{U}} \in \mathcal{P}_{\gamma+\delta}} I(\tilde{X};\tilde{U},\tilde{R}) \qquad (2.82)$$

$$= \inf_{P_{\tilde{R}\tilde{X}|\tilde{U}} \in \mathcal{P}_{\gamma+\delta}} \left(H(\tilde{X}) - H(\tilde{X} \mid \tilde{U},\tilde{R})\right) \qquad (2.83)$$

$$= \log|\mathcal{U}| + \gamma + \delta - \sup_{P_{\tilde{R}\tilde{X}|\tilde{U}} \in \mathcal{P}_{\gamma+\delta}} H(\tilde{X} \mid \tilde{U},\tilde{R}) \qquad (2.84)$$

$$\leq \log|U| + \gamma + \delta - H(X' \mid U',R') \qquad (2.85)$$

$$= H(X) - H(X \mid U,R) \qquad (2.86)$$

$$< f(\gamma) + \epsilon, \qquad (2.87)$$

where (2.87) follows from (2.81). Since $\epsilon > 0$ is arbitrary, the second claim of the proposition is proved.

2.5 Proof of Theorem 2.8

Suppose there exists $u \in \mathcal{U}$ such that $P_U(u)$ is irrational. Define a new random variable U^* such that U^* is equal to 0 if $U = u$, and equal to 1 otherwise. Then $P_{U^*}(0)$ and $P_{U^*}(1)$ are irrational. As U^* is a function of U, by (2.7)–(2.9) and (2.62),

$$I(U^*;R) = I(U^*;X) = I(X;R) = H(U^* \mid X,R) = 0. \tag{2.88}$$

Therefore, it suffices to consider binary U.

Let $\mathcal{X}$ and $\mathcal{R}$ be the respective supports of X and R. Suppose to the contrary first that $|\mathcal{X}|$ and $|\mathcal{R}|$ are both finite. We can assume without loss of generality that $\mathcal{X} = \{1,\ldots,n\}$ and $\mathcal{R} = \{1,\ldots,m\}$. Let $x_i = P_X(i)$ for $i = 1,\ldots,n$, $r_j = P_R(j)$ for $j = 1,\ldots,m$, and $\mathbf{x}$ be the n-row vector with entries x_i. Similarly, we define the column vector $\mathbf{r}$.

As X and R are independent and $H(U \mid X,R) = 0$, there exists a function g such that $U = g(X,R)$. Hence, from X and R we induce an $n \times m$ *decoding matrix* G with entries $G_{i,j} = f(i,j)$ for $i = 1,\ldots,n$ and $j = 1,\ldots,m$. Then

$$\sum_{j=1}^{m} G_{i,j} r_j = P_U(1), \quad i = 1,\ldots,n \tag{2.89}$$

$$\sum_{i=1}^{n} x_i = \sum_{j=1}^{m} r_j = 1 \tag{2.90}$$

$$x_i \geq 0, r_j \geq 0, \quad i = 1,\ldots,n, j = 1,\ldots,m, \tag{2.91}$$

$$\sum_{i=1}^{m} x_i G_{i,j} = P_U(1), \quad j = 1,\ldots,m. \tag{2.92}$$

Here, (2.89) is due to the fact that $I(U;X) = 0$, (2.90), and (2.91) are required since P_X and P_R are probability distributions, and (2.92) follows from $I(U;R) = 0$.

In fact, for any $\mathbf{x}$, $\mathbf{r}$, and binary matrix G satisfying the above four conditions, one can construct random variables $\{U,R,X\}$ such that

$$I(U;R) = I(U;X) = I(X;R) = H(U \mid X,R) = 0, \tag{2.93}$$

where $U = f(X,R)$ and the probability distributions of X and R are specified by the vectors $\mathbf{x}$ and $\mathbf{r}$ respectively.

Next, we will prove that if the rows of G are not independent, then we can construct another random variable X^* with support $\mathcal{X}^*$, $|\mathcal{X}^*| < |\mathcal{X}|$, such that

$$I(U;R) = I(U;X^*) = I(X^*;R) = H(U \mid X^*,R) = 0. \tag{2.94}$$

To prove this claim, suppose that there exists disjoint subsets $\mathcal{A}$ and $\mathcal{B}$ of $\{1,\ldots,n\}$ and positive numbers $\alpha_i, i \in \mathcal{A} \cup \mathcal{B}$ such that

$$\sum_{i \in \mathcal{A}} \alpha_i G_i = \sum_{k \in \mathcal{B}} \alpha_k G_k, \tag{2.95}$$

where G_i is row i of G. Then we will claim that $\sum_{i \in \mathcal{A}} \alpha_i = \sum_{k \in \mathcal{B}} \alpha_k$. Multiplying both sides of (2.95) by $\mathbf{r}$,

$$\sum_{i \in \mathcal{A}} \alpha_i G_i \mathbf{r} = \sum_{k \in \mathcal{B}} \alpha_k G_k \mathbf{r} \tag{2.96}$$

$$\sum_{i \in \mathcal{A}} \alpha_i P_U(1) = \sum_{k \in \mathcal{B}} \alpha_k P_U(1) \tag{2.97}$$

$$\sum_{i \in \mathcal{A}} \alpha_i = \sum_{k \in \mathcal{B}} \alpha_k. \tag{2.98}$$

Let $\epsilon \triangleq \min_{i \in \mathcal{A} \cup \mathcal{B}} x_i / \alpha_i$. Assume without loss of generality that $n \in \mathcal{A}$ and that $\epsilon = x_n / \alpha_n$. Define

$$x_i^* = \begin{cases} x_i - \epsilon \alpha_i, & i \in \mathcal{A} \\ x_i + \epsilon \alpha_i, & i \in \mathcal{B} \\ x_i, & \text{otherwise.} \end{cases}$$

Note that $x_n^* = 0$. Suppose that the probability distribution of X is changed such that $P_X(i) = x_i^*$. Then it can be checked easily that (R, U, X) still satisfy (2.7)–(2.9) and (2.62). Furthermore, the size of the support of X is $|\mathcal{X}| \leq n - 1$.

Repeating this procedure, we can prove that for any random variable U, if there exist auxiliary random variables X, R satisfying (2.7)–(2.9) and (2.62), then there exist auxiliary random variables X^*, R^* such that (2.88) is satisfied and the rows and columns of the decoding matrix induced by X^* and R^* are all linearly independent. Hence, the decoding matrix G induced by X^* and R^* must be square (and thus $m = n$). Consequently, $\sum_{i=1}^n x_i G_{i,j} = P_U(1)$ for all $j = 1, \ldots, n$. There exists a unique solution $(z_1, \ldots, z_n)$ such that $\sum_{i=1}^n z_i G_{i,j} = 1$ for $j = 1, \ldots, n$. Clearly, $z_i = x_i / P_U(1)$. As all the entries in G are either 0 or 1, all the z_i are rational numbers. Therefore,

$$1 = \sum_{i=1}^n x_i = P_U(1) \sum_{i=1}^n z_i. \tag{2.99}$$

Hence, $P_U(1)$ must be rational and a contradiction occurs. Thus, $\mathcal{X}$ and $\mathcal{R}$ cannot both be finite. The case when only $\mathcal{X}$ or $\mathcal{R}$ is finite can be similarly proved.

2.6 Conclusion

This chapter studied EPS systems subject to the assumption that the message U and the secret key R are independent (i.e., $I(U;R) = 0$). Under this setup, we showed $P_R(r) \leq$

$\frac{1}{|\mathcal{U}|}$ on the key requirement, which implies $\log|\mathcal{U}| \leq H(R)$ and subsumes Shannon's fundamental bound $H(U) \leq H(R)$ for perfect secrecy.

For the ciphertext X, $P_X(x) \leq \frac{1}{|\mathcal{U}|}$ has been shown, and the lower bound on the minimum number of channel uses has been shown to be $\log|\mathcal{U}| \leq H(X)$. If the source distribution is defined on a countably infinite support, no security system can simultaneously achieve perfect secrecy and zero decoding error.

Our main contribution is to define and justify new concepts including *residual secret randomness*, *expected key consumption*, and *excess key consumption*. We have demonstrated the feasibility of extracting residual secret randomness in a collection of error-free perfect secrecy systems. We quantified the residual secret randomness as $H(R|U,X)$. We further distinguished between the size $H(R)$ of the secret key required prior to the commencement of transmission, and the expected key consumption $I(R;U,X)$ in a multi-round setting. In contrast to $H(R) \geq \log|\mathcal{U}|$, we showed that $I(R;U,X)$ is bounded below by $H(U)$, giving a more precise understanding of the role of source entropy in error-free perfect secrecy systems. The excess key consumption is quantified as $I(R;X)$, and is equal to 0 if and only if the minimal expected key consumption is achieved.

Another main objective of this chapter is to reveal the fundamental tradeoff between expected key consumption and the number of channel uses. For the regime where the minimal $I(R;U,X)$ is assumed, $H(X)$ and $H(R)$ are inevitably larger and corresponding lower bounds for $H(X)$ and $H(R)$ have been obtained. If the source distribution P_U has irrational numbers, the additional requirements on the alphabet sizes of X and R to achieve minimal $I(R;U,X)$ have been shown. We have proposed the *partition code*, which generalizes the one-time pad and can achieve minimal key consumption when all the probability masses in P_U are rational. In some cases, the partition code can simultaneously attain the minimal $H(X)$ and $H(R)$ in this regime.

At the other extreme, the regime where the minimal number of channel uses is assumed, the one-time pad has been shown to be optimal. For the intermediate regime, we have formulated an optimization problem for the fundamental tradeoff between $I(R;U,X)$ and $H(X)$.

This chapter has highlighted a few open problems. First, the complete characterization of the tradeoff between $I(R;U,X)$ and $H(X)$ remains open. Second, the partition code is only one class of codes designed to minimize expected key consumption. Codes achieving other points on the tradeoff curve are yet to be discovered. In particular, a code achieving minimal $H(X)$ and $H(R)$ in the regime of minimal expected key consumption is important for the design of efficient and secure systems.

Acknowledgments

The authors would like to thank an anonymous reviewer for his/her valuable comments.

Part of this work is supported by the Discovery Project Australian Research Council under Grant DP150103658.

References

[1] C. E. Shannon, "Communication theory of secrecy systems," *Bell Syst. Tech. J.*, vol. 28, no. 4, pp. 656–715, Oct. 1949.

[2] J. L. Massey, "An introduction to contemporary cryptology," *Proc. IEEE*, vol. 76, no. 5, pp. 533–549, May 1988.

[3] S.-W. Ho, "On the interplay between Shannon's information measures and reliability criteria," in *Proc. IEEE Int. Symp. Inf. Theory*, Seoul, Korea, Jun. 2009, pp. 154–158.

[4] T. M. Cover and J. A. Thomas, *Elements of Information Theory*, 2nd edn. Chichester: Wiley & Sons, 2006.

[5] B. Chor and E. Kushilevitz, "Secret sharing over infinite domains," *J. of Cryptology*, vol. 6, no. 2, pp. 87–96, Jun. 1993.

[6] L. Csirmaz, "Probabilistic infinite secret sharing," 2012. [Online]. Available: https://eprint.iacr.org/2012/412.pdf

[7] S.-W. Ho, T. Chan, and A. Grant, "Non-entropic inequalities from information constraints," in *Proc. IEEE Int. Symp. Inf. Theory*, Cambridge, MA, USA, Jul. 2012, pp. 1256–1260.

[8] C. Blundo, A. D. Santis, and U. Vaccaro, "On secret sharing schemes," *Inf. Process. Letters*, vol. 65, no. 1, pp. 25–32, Jan. 1998.

[9] C. Blundo, A. D. Santis, and A. G. Gaggia, "Probability of shares in secret sharing schemes," *Inf. Process. Letters*, vol. 72, no. 5–6, pp. 169–175, Dec. 1999.

[10] G. Vernam, "Cipher printing telegraph systems for secret wire and radio telegraphic communications," *J. American Inst. Elec. Eng.*, vol. 45, no. 2, pp. 295–301, Feb. 1926.

[11] G. V. Assche, *Quantum Cryptography and Secret-Key Distillation*. Cambridge: Cambridge University Press, 2006.

[12] C. Ye and P. Narayan, "Secret key and private key constructions for simple multiterminal source models," *IEEE Trans. Inf. Theory*, vol. 58, no. 2, pp. 639–651, Feb. 2012.

[13] I. Csiszár and P. Narayan, "Secrecy capacities for multiple terminals," *IEEE Trans. Inf. Theory*, vol. 50, no. 12, pp. 3047–3061, Dec. 2004.

[14] S.-W. Ho, J. Duan, and C. S. Chen, "Location-based information transmission systems using visible light communications," *Trans. Emerging Tel. Tech.*, vol. 28, no. 1, 2017.

[15] T. H. Chan and S.-W. Ho, "2-dimensional interval algorithm," in *Proc. IEEE Inf. Theory Workshop*, Paraty, Brazil, Oct. 2011, pp. 633–637.

[16] J. L. Massey, "On probabilistic encipherment," in *Proc. IEEE Inf. Theory Workshop*, Bellagio, Italy, 1987.

[17] H. N. Jendal, Y. J. B. Kuhn, and J. L. Massey, "An information-theoretic treatment of homophonic substitution," *Lecture Notes in Computer Science*, vol. 434, pp. 382–394, 1990.

[18] B. Ryabko and A. Fionov, "Efficient homophonic coding," *IEEE Trans. Inf. Theory*, vol. 45, no. 6, pp. 2083–2091, Sep. 1999.

[19] C. G. Günther and A. B. Boveri, "A universal algorithm for homophonic coding," *Lecture Notes in Computer Science*, vol. 330, pp. 405–414, 1988.

[20] M. Hoshi and T. S. Han, "Interval algorithm for homophonic coding," *IEEE Trans. Inf. Theory*, vol. 47, no. 3, pp. 1021–1031, Mar. 2001.

[21] S.-W. Ho and S. Verdú, "On the interplay between conditional entropy and error probability," *IEEE Trans. Inf. Theory*, vol. 56, no. 12, pp. 5930–5942, Dec. 2010.

[22] C. Uduwerelle, S.-W. Ho, and T. Chan, "Design of error-free perfect secrecy system by prefix codes and partition codes," in *Proc. IEEE Int. Symp. Inf. Theory*, Cambridge, MA, USA, Jul. 2012, pp. 1593–1597.

3 Secure Source Coding

Paul Cuff and Curt Schieler

This chapter assumes that a limited amount of secret key and reliable communication are available for use in encoding and transmitting a stochastic signal. The chapter starts at the beginning, with a more general proof of Shannon's "key must be as large as the message" result that holds even for stochastic encoders.

Three directions are explored. First, for lossless compression and perfect secrecy, variable length codes or key regeneration allow for a tradeoff between efficiency of compression and efficiency of secret key usage. Second, the relaxation to imperfect secrecy is studied. This is accomplished by measuring the level of secrecy either by applying a distortion metric to the eavesdropper's best possible reconstruction or by considering the eavesdropper's ability to guess the source realization. Finally, an additional relaxation is made to allow the compression of the source to be lossy.

The chapter concludes by showing how the oft-used equivocation metric for information theoretic secrecy is a particular special case of the rate–distortion theory contained herein.

3.1 Introduction

Source coding is the process of encoding information signals for transmission through digital channels. Since efficiency is a primary concern in this process, the phrase "source coding" is often used interchangeably with "data compression." The relationship between source coding and channel coding is that channel coding produces digital resources from natural resources (e.g., a physical medium), and source coding consumes digital resources to accomplish a task involving information, often simply moving it from one point to another.

The channel coding side of information theoretic security is referred to as physical layer security. This usually involves designing a communication system for a physical wiretap channel, introduced by Wyner in [1], which produces a provably secure digital communication link. Another important challenge in physical layer security is the production of a secret key based on common observations, such as channel fading parameters or prepared quantum states. Although a key agreement protocol does not necessarily involve a channel, it is consistent with the spirit of channel coding in that the objective is the production of digital resources. The digital resources generated by physical layer security or key agreement can be any of the following:

- common random bits
- common random bits unknown to an eavesdropper (secret key)
- a reliable digital communication link
- a reliable and secure digital communication link.

Secure source coding involves using these digital resources in an efficient way to produce a desired effect with regards to an information signal. The general setting of interest in this chapter is the Shannon cipher system [2] shown in Fig. 3.1, where $\mathbf{X}$ represents the information signal, K is a secret key, M is the encoded signal, which is transmitted over a reliable but non-secure digital channel, $\widehat{\mathbf{X}}$ is the reconstructed information, and A and B are the encoder and decoder that we wish to design in order to keep an eavesdropper, Eve, from learning about $\mathbf{X}$ and $\widehat{\mathbf{X}}$. A variation of this setting could incorporate reliable *and secure* communication channels, but the setting of Fig. 3.1 provides the necessary elements for fundamental understanding. For simplicity, we will mostly focus on asymptotic results in the case where $\mathbf{X}$ is a sequence of i.i.d. random variables with known distribution. This information signal is referred to herein as the "source."

Notice that secure source coding operates at a layer above the physical layer. The method with which the resources were produced is not the concern of source coding. If, for example, one has confidence in a key that was produced and secured by some cryptographic technique, a feasible implementation would be to encode the source using that key rather than an information theoretic secret key. However, to derive tight results, we assume that secure channels and secret keys are information theoretically secure.

One interesting feature of secure communication is the role of stochasticity in the encoder or decoder. The intentional inclusion of randomness in the encoding or decoding process can lead to superior designs in spite of the reliability issues it introduces. This effect is discussed throughout the chapter.

This chapter begins with the setting that demands the most in terms of digital resources: the source must be recovered without loss and the secrecy must be perfect. This setting is addressed in Section 3.2. Then, just as classic source coding must be generalized to answer questions such as how to encode continuous-valued sources, which cannot be encoded without loss, so a comprehensive view of secure source coding must

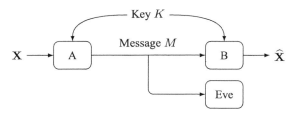

Figure 3.1 Shannon cipher system. An eavesdropper Eve intercepts the communication between the sender and receiver. The sender and receiver share a secret key, which can be used to encrypt the communication.

> **Box 3.1** Stochastic Encoder and Stochastic Decoder
>
> A stochastic encoder or stochastic decoder uses internal (i.e. local) randomness while performing its function. The encoder may not always produce the same message even when encountering the same source value with the same key. Likewise, the decoder may not produce the same reconstruction even if the message and the key are the same.

address how best to secure a source even when the resources needed for perfect secrecy are either not available or too expensive. Each of these generalizations provides rich new theory, highlighted throughout the chapter. In Section 3.3, several notions of imperfect secrecy are introduced, including causal disclosure, repeated and single guessing, and henchmen. Section 3.4 considers lossy compression and imperfect secrecy.

3.2 Lossless Compression, Perfect Secrecy

When encoded messages are transmitted over a non-secure digital channel, secrecy can be achieved by use of a secret key (represented as K in Fig. 3.1), known to the transmitter and receiver, but unknown to an eavesdropper.

The most ideal secrecy, "perfect secrecy," requires the transmitted message M to be independent of the source $\mathbf{X}$. That is, the conditional distribution of the message given the source, $P_{M|\mathbf{X}}$, does not actually depend on the source $\mathbf{X}$. This simple definition is more general than other notions of secrecy presented in this chapter because it is well defined even if the source is not modeled as random.

Even before the advent of information theory, it was already known that perfect secrecy can be achieved by using the Vernam cipher with a one-time-pad.

> **Box 3.2** One-Time Pad
>
> Message bits and secret key bits are aligned, and the transmission is computed by bit-wise exclusive-or operations. The key should be uniformly distributed and used only once.

The one-time pad requires that the space of secret key values $\mathcal{K}$, over which the key is distributed uniformly at random, be at least as large as the space of possible source values $\mathcal{X}$. Shannon showed this to be necessary in general.

THEOREM 3.1 ([2, Theorem 6]) *Perfect secrecy requires $|\mathcal{K}| \geq |\mathcal{X}|$.*

Proof Achievability is accomplished with the one-time pad.

Shannon's proof of the converse in [2] assumed that the encoder is deterministic. However, we will see that in general it can be beneficial to allow the encoder to be stochastic. For example, this allows for more efficient key usage in variable-length coding and key regeneration, as will be discussed in this section.[1]

A feature of perfect secrecy is that it is not affected by any assumption about the distribution of the source. For simplicity, let $\mathbf{X}$ be random with a support that covers the whole space $\mathcal{X}$. Let $\mathcal{M}_{x,k}$ be the conditional support of the message given the source and key values, and similarly for $\mathcal{M}_k$ and $\mathcal{M}_x$.

$$|\mathcal{M}| \geq \max_k |\mathcal{M}_k| \tag{3.1}$$

$$= \max_k \left| \bigcup_x \mathcal{M}_{x,k} \right| \tag{3.2}$$

$$\stackrel{(a)}{=} \max_k \sum_x |\mathcal{M}_{x,k}| \tag{3.3}$$

$$\geq \frac{1}{|\mathcal{K}|} \sum_{x,k} |\mathcal{M}_{x,k}| \tag{3.4}$$

$$= \frac{|\mathcal{X}|}{|\mathcal{K}|} \frac{1}{|\mathcal{X}|} \sum_{x,k} |\mathcal{M}_{x,k}| \tag{3.5}$$

$$\geq \frac{|\mathcal{X}|}{|\mathcal{K}|} \min_x \sum_k |\mathcal{M}_{x,k}| \tag{3.6}$$

$$\geq \frac{|\mathcal{X}|}{|\mathcal{K}|} \min_x \left| \bigcup_k \mathcal{M}_{x,k} \right| \tag{3.7}$$

$$= \frac{|\mathcal{X}|}{|\mathcal{K}|} \min_x |\mathcal{M}_x| \tag{3.8}$$

$$\stackrel{(b)}{=} \frac{|\mathcal{X}|}{|\mathcal{K}|} |\mathcal{M}|, \tag{3.9}$$

where (a) is a consequence of decodability, which requires that $\{\mathcal{M}_{x,k}\}_x$ are mutually exclusive for any k, and (b) is due to perfect secrecy, which requires that the conditional distribution of the message M be constant for all source values.

REMARK 3.1 *If the encoder is deterministic, then* $|\mathcal{M}_{x,k}| = 1$ *for all m and k. Therefore, (3.4) shows that in fact* $|\mathcal{K}| \geq |\mathcal{M}| \geq |\mathcal{X}|$ *is necessary in that case.*

REMARK 3.2 *The bit-wise exclusive-or implementation of the one-time pad described in Box 3.2 is common, but in general, for a key space of size* $|\mathcal{K}|$, *a single modulo* $|\mathcal{K}|$ *addition of the message and the key, after mapping them both to integer values, is an equally effective implementation of perfect secrecy that works even if the size of the key space is not a power of two.*

---

[1] In Wyner's wiretap channel [1], which is outside the scope of this chapter, a stochastic encoder is necessary to achieve the secrecy capacity.

3.2.1 Discrete Memoryless Source

When the source is modeled as a stochastic process, there is opportunity for compression. Assume the source is an i.i.d. sequence distributed according to P_X. That is,

$$\mathbf{X} = X^n \triangleq (X_1, \ldots, X_n). \tag{3.10}$$

In the block encoding framework, the definition of lossless encoding can be relaxed to near lossless by requiring the probability that the reconstruction does not equal the source sequence to be arbitrarily small. In this setting, the message size and secret key size are parameterized as rates. That is,

$$|\mathcal{M}| = 2^{nR}, \tag{3.11}$$

$$|\mathcal{K}| = 2^{nR_K}. \tag{3.12}$$

Under this relaxed notion of lossless compression, the minimum encoding rate R needed is the entropy of the source distribution, which can be achieved by enumerating the set of typical sequences.

Compression and secrecy can be combined by first compressing the source and then applying a one-time pad to the compressed message.[2] Indeed, this approach is optimal in the lossless setting, and there is no conflict between the two resources, communication rate and key rate.

THEOREM 3.2 *The closure of the set of rate pairs (R, R_K) for which near-lossless compression and perfect secrecy can be simultaneously achieved is all pairs satisfying*

$$R \geq H(X), \tag{3.13}$$

$$R_K \geq H(X). \tag{3.14}$$

Proof Achievability is accomplished by first compressing and then applying the one-time pad.

The converse for (3.13), well known from lossless compression, is not affected by a secret key.

The converse for (3.14) is as follows:

$$nR_K \geq H(K) \tag{3.15}$$

$$\geq I(K; X^n | M) \tag{3.16}$$

$$= I(K, M; X^n) - I(M; X^n) \tag{3.17}$$

$$\geq I(\widehat{X}^n; X^n) - I(M; X^n) \tag{3.18}$$

$$= H(X^n) - H(X^n | \widehat{X}^n) - I(M; X^n) \tag{3.19}$$

$$= nH(X) - H(X^n | \widehat{X}^n) - I(M; X^n). \tag{3.20}$$

The proof is completed by dividing both sides by n and noting that Fano's inequality makes $\frac{1}{n} H(X^n | \widehat{X}^n)$ vanish as the error probability vanishes. The other term in (3.20), $I(M; X^n)$, is zero due to the perfect secrecy requirement.

---

[2] This is also mentioned in [2], and the proof is omitted.

REMARK 3.3 *Just as lossless compression is relaxed to near-lossless compression, perfect secrecy can also be relaxed to near-perfect secrecy. However, this change does not affect the fundamental limits of Theorem 3.2. Notice that the last term in (3.20) vanishes as long as $I(M;X^n) \in o(n)$, a condition referred to as "weak secrecy" in the literature.*

REMARK 3.4 *The proof of Theorem 3.2 actually gives a stronger statement than Theorem 3.1. If we assume exact lossless compression and exact perfect secrecy, then (3.15) and (3.20) yield $H(K) \geq nH(X)$. Furthermore, exact lossless compression and exact perfect secrecy are not affected by the source distribution, so the bound is true for any distribution on the source space. By choosing the uniform distribution, we obtain $H(K) \geq n \log |\mathcal{X}| = \log |\mathcal{X}^n|$.*

Variable Length Coding

We now return to exact lossless compression but take advantage of the source distribution in a different way. Consider an encoder that varies the length of communication or secret key usage depending on the source value. This differs from the fixed-length block encoding model described by (3.11) and (3.12). In a variable-length coding setting, the resource usage is captured by the expected value of the length.

Consider the encoding of individual source symbols $X \sim P_X$. Reliable encoding requires two parts. The decoder should be able to uniquely decode without punctuation between the encoded symbols, and there should be a causal stopping rule by which the decoder is able to identify the number of key bits needed for decoding. In secrecy settings, pessimism is often prudent, so let us demand perfect secrecy even if an eavesdropper is given punctuation between the symbols. Such would be the case if there were detectable delays between symbol transmissions or if just a single symbol were transmitted.

It is immediately apparent that in order to achieve perfect secrecy, not even the codeword length can reveal information about the source. Thus, the codeword length, if it varies, must be a random variable independent of the source. Unsurprisingly, then, only codes with fixed-length codewords ultimately need to be considered. However, the variable length of the secret key can still be beneficial.

Prefix codes, of which the Huffman code has the shortest expected length, play an important role in lossless compression by providing unique decodability. It is shown in [3] and [4] that prefix codes also play an important role in this secrecy context. However, there is a tension between communication rate and key rate, which is expressed in the following achievability theorem.

THEOREM 3.3 ([3, Theorem 1], [4]) *A set of achievable rate pairs (R, R_K) for which lossless compression and perfect secrecy can be simultaneously achieved using variable length codes is the convex hull of the following set:*

$$\bigcup_{\mathfrak{e}:\, prefix\ code} \left\{ (R, R_K) : \begin{array}{l} R \geq l_{\mathfrak{e},\max}, \\ R_K \geq \mathbb{E} l_{\mathfrak{e}}(X). \end{array} \right\}. \tag{3.21}$$

Proof To achieve these rates, the encoder uses a prefix code and applies a one-time pad using the secret key. Then the encoder appends uniformly distributed random bits to make the transmission equal in length to the longest codeword. The decoder recovers one bit at a time using the secret key and the received transmission. Because of the prefix condition, the decoder can identify when the codeword is complete and declare a stop to the secret key usage. From the point of view of an eavesdropper, every transmission of length $l_{\max}$ is equally likely no matter which source symbol is encoded.

The convex hull is achieved by time sharing between codes.

REMARK 3.5 *There is tension between the two rates, with the key rate always the smaller of the two. The best code for the key rate is the Huffman code, which achieves a key rate within one bit of the entropy of the source, but which may unfortunately have a long maximum codeword length. At the other extreme of the tradeoff, the fixed-length code gives $R = R_K = \lceil \log |\mathcal{X}| \rceil$, which corresponds to Theorem 3.1.*

REMARK 3.6 *The savings in R_K stated in Theorem 3.3 are achieved by using a stochastic encoder. If the encoder is required to be deterministic, then instead one might use the secret key to pad each codeword to make them the same length. This is inefficient, and in fact the optimal rates reduce back to those implied by Theorem 3.1.*

Key Regeneration

Even in the block encoding model with fixed-length communication and a fixed-length key, there is some sense in which less key rate is needed than Theorem 3.1 implies. At the expense of a higher communication rate, and by using a stochastic encoder, it may be that for some realizations of the source not all of the secrecy of the key is exhausted, in the sense that a fresh key can be extracted by the encoder and decoder for future use without compromising security. This gives a way to take advantage of the known source distribution without compromising exact lossless compression or exact perfect secrecy, even in the fixed-length setting.

The easiest way to see this phenomenon is to apply the variable-length encoding scheme used for Theorem 3.3, one symbol at a time, to the fixed-length block encoding setting. This is done with a slight modification. The communication length is already constant in that scheme, but the key usage is variable and must be converted to a fixed-length scheme. To do this, let the key length be as long as the communication (i.e., the maximum codeword length of the prefix code), and perform a one-time pad on the entire transmission, which includes both the codeword from the prefix code and the random padding that follows it. Now this is a fixed-length scheme with perfect secrecy and lossless compression – both the communication and the secret key are as long as the longest codeword. However, notice that the padding itself can be considered a new random secret key. Both parties can compute it, and it is independent of all previous source symbols and transmissions.

In [4], a so-called partition code is used to achieve the same effect, though the extraction of the new secret key is not as straightforward. The overall conclusion is analogous to Theorem 3.3, though. If the communication rate is to be minimized, then the key rate and the communication rate should be the same and equal to $\log |\mathcal{X}|$,

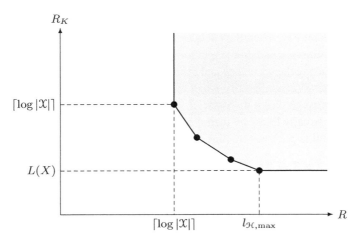

Figure 3.2 Region of achievable rate pairs for variable-length coding given by Theorem 3.3. Here, $L(X)$ denotes the average codeword length of a Huffman code (i.e., $\min_{\mathcal{C}} \mathbb{E} l_{\mathcal{C}}(X)$), and $l_{\mathcal{H},\max}$ is the maximum codeword length of the Huffman code.

corresponding to Theorem 3.1. There is no need for a stochastic encoder in that case. On the other hand, with a stochastic encoder and a higher transmission rate it is possible to regenerate a new secret key so that the usage of secret key rate is reduced to the entropy of the source. This results in essentially the same picture as Fig. 3.2 but without the round-off inefficiencies that arise from symbol-by-symbol encoding. That is, the extreme rates are $\log |\mathcal{X}|$ and $H(X)$ rather than $\lceil \log |\mathcal{X}| \rceil$ and the expected Huffman codeword length.

3.3 Lossless Compression, Imperfect Secrecy

It is natural to investigate the quality of source coding even when it falls between the extremes of perfect recovery and no information. In a secure source coding system, a high quality signal should be received by the intended recipient, and only a low quality signal should be obtainable by an eavesdropper.

3.3.1 Rate–Distortion Theory

It has become standard practice in information theory, beginning with Shannon's original paper [5], to measure the quality of an encoding by the average distortion that a reconstruction incurs. That is, for any distortion function $d(x, \hat{x})$, where x represents the value of the source and $\hat{x}$ represents the value of the reconstruction, consider the average distortion over a block of length n to be

$$d(x^n, \hat{x}^n) = \frac{1}{n} \sum_{t=1}^{n} d(x_t, \hat{x}_t). \tag{3.22}$$

The study of the tradeoff between this average distortion and the rate of compression is the celebrated *rate–distortion theory*.

In a secrecy system, we can explore the same tradeoff but with respect to the eavesdropper. To begin with, let us consider the case where the encoding is lossless for the intended receiver (more accurately, it is near lossless). Lossy compression will be explored in Section 3.4. Unlike the previous section, we will no longer require perfect secrecy. Instead, secrecy will be measured by the average distortion that an eavesdropper incurs if he tries to reconstruct the source.

One key feature in the theory of secure source coding is the role that past information about the values of the source can play in helping an eavesdropper reconstruct a signal. We refer to this as *causal disclosure*.

Box 3.3 Causal Disclosure

The eavesdropper obtains the past values of the source, or noisy observations of them, before making a reconstruction of the source at the present time. These observations are modeled as outputs W of a memoryless channel $P_{W|X}$.

Box 3.4 Full Causal Disclosure

Full causal disclosure is the worst-case assumption that $W = X$. Secrecy with respect to full causal disclosure is robust – that is, the analysis will be valid no matter what causal side information is obtained by an eavesdropper.

Causal disclosure means that at time index t, as the eavesdropper attempts to reconstruct the source symbol X_t, he has access to the eavesdropped message M and noisy observations $W^{t-1} = (W_1, \ldots, W_{t-1})$ of all of the past source symbols $X^{t-1} = (X_1, \ldots, X_{t-1})$, obtained as side information. The distribution of the noise $P_{W|X}$ must be modeled or assumed. In the full causal disclosure setting, there is no noise, so the eavesdropper may use M and the past source symbols X^{t-1} themselves while forming his estimate $\widehat{X}_t$.

Causal disclosure does not play a role in rate–distortion theory outside of the context of secrecy – if causal disclosure is available to the legitimate receiver, it does not change the rate–distortion theorem. But it fundamentally changes the nature of secrecy. If one thinks of the sequence of source symbols as a time sequence, then full causal disclosure, where $W = X$, is akin to the known-plaintext setting in cryptography. Without causal disclosure (e.g., if the channel is $P_{W|X} = P_W$), the encoder can essentially reuse the secret key indefinitely to force a high level of distortion upon the eavesdropper with a negligible key rate. This gives the impression of perfect secrecy for free, but it is really just an artifact of a presumptuous model that assumes an eavesdropper does not obtain additional side information.

DEFINITION 3.1 Given a memoryless source $\{X_t \sim P_X\}$, a causal disclosure channel $P_{W|X}$, and a distortion function $d(x,z)$, a rate–distortion triple (R, R_K, D) is achievable if for all $\epsilon > 0$ there exist a block length n and an encoder and decoder at compression rate R and key rate R_K as described in (3.11) and (3.12) such that

$$\mathbb{P}(X^n \neq \widehat{X}^n) < \epsilon, \tag{3.23}$$

$$\min_{\{Z_t = z_t(M, W^{t-1})\}_t} \mathbb{E} d(X^n, Z^n) \geq D. \tag{3.24}$$

THEOREM 3.4 ([6] Corollary 1) *The closure of achievable rate–distortion triples (R, R_K, D) for lossless compression with security robust to causal disclosure is the set of triples satisfying*

$$\bigcup_{U-X-W} \left\{ (R, R_K, D) : \begin{array}{l} R \geq H(X) \\ R_K \geq I(W; X|U) \\ D \leq \min_{z(u)} \mathbb{E} d(X, z(U)) \end{array} \right\}. \tag{3.25}$$

Achievability proof sketch To achieve the above rates, two messages are transmitted. For the first message, a codebook is generated from the P_U distribution to form a covering of the source space. The rate of this codebook is $I(X; U)$. Note that the key is not used in the encoding of the first message. The second message, at rate $H(X|U)$, is simply a uniformly distributed random mapping from each source sequence and key pair (x^n, k). These two messages, together with the key, allow for decoding of the source.

The case of full causal disclosure admits a simple secrecy argument. In that case, the bound on R_K in the theorem simplifies to $R_K \geq H(X|U)$, which is the rate of the second message. Therefore, a one-time pad can be used on the second message, which is effectively what the previously described encoding will yield anyway. Consequently, the eavesdropper receives exactly one piece of information, which is the codeword given by the first message.

The analysis is greatly assisted by the use of a likelihood encoder [12]. Let us refer to this codeword associated with the first message as $\bar{u}^n$, to which the eavesdropper has access. Due to the likelihood encoder, the posterior distribution of the source sequence given the first message is approximately

$$P_{X^n|M_1}(x^n|m_1) \approx \prod_{t=1}^{n} P_{X|U}(x_t|\bar{u}_t). \tag{3.26}$$

At least in the case of full causal disclosure, it is easy to see that this will lead to the distortion stated in the theorem, since the only meaningful information that the eavesdropper obtains is $\bar{u}^n$. The second message is fully protected by the key, and information leaked through causal disclosure has been sterilized by this encoding, as indicated in (3.26), rendering it useless to the eavesdropper.

REMARK 3.7 *The variable U can be interpreted as a part of the encoding that is not secured due to lack of secret key resource. This leaked information is designed to be* as useless as possible under the constraints of Theorem 3.4. In the case of full causal disclosure, the rest of the encoding is fully secured.

This decomposition of the encoding into a secured part and a part that is not secured is fortuitous. The encoder has two digital outputs. For the one that must be secured, this can be accomplished by using a one-time pad or even by using a secure digital channel. This opens the door to implement this source coding with cryptographic resources.

REMARK 3.8 *If there is no causal disclosure, then the theorem puts no lower bound on the key rate, and the optimal variable choice is $U = \emptyset$. As discussed previously, optimal distortion is forced on the eavesdropper without needing a positive secret key rate.*

REMARK 3.9 *Theorem 3.4 does not change even if the disclosure is assumed to be non-causal, meaning that the disclosure is available at all times except the one that is to be reconstructed. This is relevant to source coding settings where the sequence is not a time sequence and causality is unimportant.*

REMARK 3.10 *Theorem 3.4 does not change if (3.24) is changed to*

$$\max_{\{Z_t = z_t(M, W^{t-1})\}_t} \mathbb{P}\left(d(X^n, Z^n) < D\right) < \epsilon. \tag{3.27}$$

REMARK 3.11 *The optimization involved in finding the boundary points in Theorem 3.4 simplifies to a linear program.*

Example 3.1 *Hamming distortion.* Let distortion at the eavesdropper be measured by Hamming distortion (i.e., probability of symbol error), and consider full causal disclosure (i.e., set $W = X$). For any source distribution P_X, apply Theorem 3.4 to find the maximum distortion that can be forced on the eavesdropper as a function of the secret key rate.

Solution
Define the function $\phi(\cdot)$ as the linear interpolation of the points $(\log k, \frac{k-1}{k})$, $k \in \mathbb{N}$. This function $\phi(\cdot)$, which is the upper boundary of the convex hull of the set of all possible entropy–error-probability pairs for arbitrary distributions,[3] is found in [6] to be the main quantity of interest for this setting. Also, define

$$d_{\max} = 1 - \max_x P_X(x). \tag{3.28}$$

The maximum distortion as a function of the key rate, shown in Fig. 3.3, is

$$D(R_K) = \min\{\phi(R_K), d_{\max}\}. \tag{3.29}$$

---

[3] Fano's inequality relates to the lower boundary.

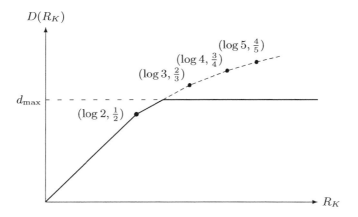

Figure 3.3 Illustration of $D(R_K)$.

The corner points of $\phi(\cdot)$ are achieved by constructing $P_{X|U}$ to be a uniform distribution over k values of X for each value of U. This can be shown to be possible as long as $\frac{k-1}{k} \leq d_{\max}$.

The two quantities $\phi(R_K)$ and $d_{\max}$ are complementary upper bounds on the distortion. The source distribution P_X only plays a role through $d_{\max}$, which is the best distortion one could hope for even with perfect secrecy, and the key rate is represented in $\phi(R_K)$.

3.3.2 Henchman

Another approach to characterizing partial secrecy is to consider how much extra information would be needed for an eavesdropper to form a good reconstruction of the source. In the henchman setting [7], this is formalized by supposing that a henchman is assisting the villainous eavesdropper. The henchman is aware of everything, including the source sequence realization, the ciphertext, and the key value. After secure communication is complete, the henchman uses rate-limited digital communication to help the eavesdropper achieve low distortion. This is illustrated in Fig. 3.4.

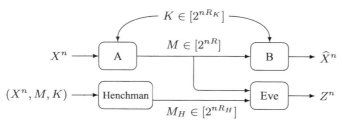

Figure 3.4 The henchman problem. A rate-limited henchman has access to the source sequence and the public message. The eavesdropper produces a reconstruction sequence Z^n based on the public message and the side information from the henchman.

Secure source coding should inflict high distortion on an eavesdropper in spite of side information. In the previous subsection, causal disclosure modeled the side information. In the henchman setting, the side information is the henchman's message, which is rate limited but not causal. It may be an appropriate worst-case analysis when decoding is not needed in real time, or when the source sequence is not a time sequence.

A distortion level D is considered achieved if the eavesdropper's expected distortion is above D no matter which rate R_H communication scheme is devised with the henchman, as defined below.

DEFINITION 3.2   Given a memoryless source $\{X_t \sim P_X\}$ and a distortion function $d(x,z)$, a rate–distortion quadruple (R, R_K, R_H, D) is achievable if for all $\epsilon > 0$ there exists a block length n and an encoder and decoder at compression rate R and key rate R_K as described in (3.11) and (3.12) such that

$$\mathbb{P}(X^n \neq \widehat{X}^n) < \epsilon, \tag{3.30}$$

$$\min_{\substack{Z^n = z^n(M, M_H), \\ M_H = m_H(X^n, M, K), \\ |\mathcal{M}_H| = 2^{nR_H}}} \mathbb{E}\, d(X^n, Z^n) \geq D. \tag{3.31}$$

THEOREM 3.5 ([7, Theorem 1])   *The closure of achievable rate–distortion tuples (R, R_K, R_H, D) for lossless compression with security robust to worst-case side information is the set of tuples satisfying*

$$R \geq H(X), \tag{3.32}$$

$$D \leq D(R_H) \cdot \mathbb{1}\{R_K > R_H\}, \tag{3.33}$$

where $\mathbb{1}$ is the indicator function and $D(r)$ is the point-to-point distortion-rate function:

$$D(r) \triangleq \min_{P_{Z|X}: r \geq I(X;Z)} \mathbb{E}\, d(X, Z). \tag{3.34}$$

Proof sketch   The converse is trivial. Consider that the henchman and the eavesdropper always have the option of compressing the source using a rate–distortion code or revealing the key if R_H is sufficiently high.

Achievability is proven using a random codebook construction. The important step in the secrecy analysis is to show that for large n a random sequence that is uniformly distributed on a random set (populated by sampling from the source distribution) is with high probability not compressible by the henchman beyond the rate–distortion function for memoryless sources, as long as the compression rate R_H is less than the exponential rate of the set size, which is R_K in this context.

REMARK 3.12   *If the henchman's rate is below the secret key rate, then the system effectively attains perfect secrecy. The henchman can only describe the source sequence as efficiently as rate–distortion theory allows. Conversely, if the henchman has enough rate to reveal the key, then the system provides no secrecy. Thus, in this model, secrecy is entirely dependent on whether or not the eavesdropper can obtain as much side information as the rate of the key.*

REMARK 3.13 *Another interpretation of the henchman setting is that of list decoding. The eavesdropper produces a list of reconstructions of size 2^{nR_H}, and performance is measured with respect to the best reconstruction in the list.*

REMARK 3.14 *Theorem 3.5 does not change if (3.31) is changed to*

$$\max_{\substack{Z^n = z^n(M,M_H), \\ M_H = m_H(X^n,M,K), \\ |\mathcal{M}_H| = 2^{nR_H}}} \mathbb{P}\left(d(X^n, Z^n) < D\right) < \epsilon. \tag{3.35}$$

3.3.3 Repeated Guessing

Another model for partial secrecy has the eavesdropper repeatedly guessing until he correctly identifies the source sequence [8]. Secrecy can be quantified by the expected number of guesses required (or any moment thereof). This setting addresses a situation where a password is being guessed and either confirmed or rejected upon each guess.

The dominant terms in the guessing moments do not come from source sequences that are statistically typical. Instead, a low probability set contributes almost all of the moment.

DEFINITION 3.3 Let $G(X^n|M)$ be the number of guesses made by the eavesdropper in order to guess X^n correctly upon observing message M, according to the optimal guessing strategy which orders the guesses from most probable to least probable.

Given a memoryless source $\{X_t \sim P_X\}$, a ρth-moment guessing exponent E_ρ, for $\rho > 0$, is achievable if for arbitrarily large n there exist an encoder and decoder that achieve exact lossless compression and use key rate R_K as described in (3.12) such that

$$\mathbb{E}\left[G(X^n|M)^\rho\right] \geq 2^{nE_\rho}. \tag{3.36}$$

THEOREM 3.6 ([8, Theorem 1]) *The supremum of achievable ρth-moment guessing exponents is*

$$E_\rho^* = \max_Q \{\min\{R_K, H(Q)\} \cdot \rho - D(Q\|P_X)\}. \tag{3.37}$$

Proof The method of types is appropriate for this analysis. Since each type contributes exponentially to the moment, and there are only a polynomial number of types, it can be argued that the guessing exponent remains unchanged even if the eavesdropper is told the type of X^n. Then, to calculate the moment, for each type we multiply the probability of the type by the conditional moment of guesses to find X^n within the type.

The relative entropy term $D(Q\|P_X)$ accounts for the probability of the type Q.

A priori, the distribution of source sequences within a type Q is uniform, and the size is $2^{nH(Q)}$ to first order in the exponent. With proper encoding, the posterior distribution of the source sequence given the type and the message is uniformly distributed over a set equal to the size of the key space, unless the key space is larger than the size of the type. One way to accomplish this is to enumerate the sequences and the key values and have the encoding simply add them modulo the size of the type.

Ultimately, the eavesdropper must either guess the uniformly distributed key value or guess the sequence uniformly from the type, whichever is easier. The ρth moment for doing this is asymptotically equivalent to the exponential size of the set to the ρth power.

REMARK 3.15 *Theorem 3.6 assumes that the encoding must be exact lossless compression. If near-lossless encoding is permitted, then perfect secrecy is achieved under this metric even without a key.*[4] *Non-typical sequences can all be encoded to one message value, and this low probability event will give the same guessing moment as perfect secrecy. The expression for the guessing exponent becomes*

$$E_\rho^* = \max_Q \{H(Q) \cdot \rho - D(Q\|P_X)\} = \rho H_{\frac{1}{\rho+1}}(X), \quad (3.38)$$

where $H_\alpha(\cdot)$ is the Rényi entropy of order α.[5]

REMARK 3.16 *If $R_K \leq H(X)$, then $E_\rho^* = \rho R_K$ according to Theorem 3.6, and the eavesdropper is essentially guessing the key. The effectiveness of the key begins to diminish at rates higher than this, and the formula in the theorem begins to be interesting.*

Recall that when $R_K > H(X)$ exact perfect secrecy is achievable for near-lossless compression, according to Theorem 3.2. On the other hand, $R_K \geq \log|\mathfrak{X}|$ is needed for perfect secrecy in the exact lossless compression setting according to Theorem 3.1. This repeated guessing setting is focused on the in-between region. Here, exact lossless compression is required, but the notion of perfect secrecy is relaxed.

REMARK 3.17 *By this guessing moment metric, the secret key is no longer effective once the key rate surpasses $R_K > \bar{H}\left(P_X^{\frac{1}{\rho+1}}\right)$, where the bar over H indicates that the argument is first normalized to be a probability distribution.*

REMARK 3.18 *A uniformly random code construction will achieve this secrecy performance with high probability.*

3.3.4 Best Guess

If the eavesdropper makes a single guess of the source sequence, we would hope that the probability of correctly guessing is exponentially small. This probability is proposed in [10] as a measure of partial secrecy. In fact, it turns out that under optimal encoding, the eavesdropper will be forced to either guess the value of the key or guess the a priori most likely source sequence, whichever gives the higher probability of success.

---

[4] Similarly, if a variable key rate is used, then a negligible expected key rate is needed to achieve perfect secrecy under this metric, even while achieving exact lossless encoding.

[5] This matches the guessing exponent derived in [9, Proposition 5] for guessing a source sequence in the absence of a message.

DEFINITION 3.4 Given a memoryless source $\{X_t \sim P_X\}$, a *best-guess exponent* E_G is achievable if for arbitrarily large n there exist an encoder and decoder that achieve lossless compression and use key rate R_K as described in (3.12) such that

$$\max_{x^n, m} P_{X^n|M}(x^n|m) \leq 2^{-nE_G}. \tag{3.39}$$

THEOREM 3.7 ([10]) *The supremum of achievable best-guess exponents is*

$$E_G^* = \min\{R_K, H_\infty(X)\}, \tag{3.40}$$

where

$$H_\infty(X) = \log \frac{1}{\max_x P_X(x)} \tag{3.41}$$

is the min-entropy or Rényi entropy of order ∞.

Proof The converse is trivial. The eavesdropper may always choose the better of two options: guess the key value without taking into account the source distribution, or guess the most likely source sequence while ignoring the message. These are the two upper bounds that the theorem achieves.

For achievability, the method of types is again convenient. First use a negligible amount of key to keep the type secret, then encode within each type exactly as was done for the proof of Theorem 3.6.

For a sequence of type Q, first consider what happens if the key space is smaller than the size of the type. Then the posterior probability of any one of these sequences is equivalent to the probability of correctly guessing both the type and the key value. This gives an upper bound of the probability of 2^{-nR_K}. On the other hand, if the key space is larger than the size of the type, then perfect secrecy is achieved for each of the sequences in that type, and

$$P_{X^n|M}(x^n, m) = P_{X^n}(x^n) \leq 2^{-nH_\infty(X)}. \tag{3.42}$$

REMARK 3.19 *The statement of Theorem 3.7 does not uncover any complexities or surprises in either the solution formula or the encoding needed to achieve it. Indeed, a uniform random code construction will achieve this secrecy performance with high probability. Perhaps the most interesting observation is the similarity of the mathematics between Theorems 3.6 and 3.7.*

Notice that E_ρ^* evaluated at $\rho = -1$ is equal to the negative of E_G^*. That is,

$$-E_{-1}^* = -\max_Q \{-\min\{R_K, H(Q)\} - D(Q\|P_X)\} \tag{3.43}$$

$$= \min_Q \{\min\{R_K, H(Q)\} + D(Q\|P_X)\} \tag{3.44}$$

$$= \min\{\min_Q \{R_K + D(Q\|P_X)\}, \min_Q \{H(Q) + D(Q\|P_X)\}\} \tag{3.45}$$

$$= \min\left\{R_K, \min_Q \mathbb{E}_Q \log \frac{1}{P_X(X)}\right\} \tag{3.46}$$

$$= \min\{R_K, H_\infty(X)\}. \tag{3.47}$$

This is not a coincidence. Even though the repeated guessing problem definition constrains $\rho > 0$, it can be extended to include $\rho < 0$ with a change in the direction of inequality in the definition of achievability:

$$\mathbb{E}[G(X^n|M)^\rho] \leq 2^{nE_\rho}. \tag{3.48}$$

This change is made because $(\cdot)^\rho$ is now monotonically decreasing on the positive domain when $\rho < 0$.

Under this change to Definition 3.3, when we consider $\rho = -1$, we get Definition 3.4. This is verified by the following inspection. Consider any random variable W on the positive integers with a monotonically non-increasing probability mass function [W will take the role of $G(X^n|M)$], and define $p_{\max} = P_W(1)$.

$$\mathbb{E}[W^{-1}] \geq p_{\max}. \tag{3.49}$$

$$\mathbb{E}\left[W^{-1}\right] \leq p_{\max} \sum_{k=1}^{\left\lceil \frac{1}{p_{\max}} \right\rceil} \frac{1}{k} \tag{3.50}$$

$$\leq p_{\max} \ln\left(\left\lceil \frac{1}{p_{\max}} \right\rceil + 1\right). \tag{3.51}$$

Therefore, $\mathbb{E}[W^{-1}]$ is exponentially equivalent to $p_{\max}$. By application, the same equivalence holds between $\mathbb{E}[G(X^n|m)^\rho]$ and $\max_{x^n} P_{X^n|M}(x^n|m)$ for any m.

REMARK 3.20 Unlike the repeated guessing setting, Theorem 3.7 holds true even if near-lossless compression is permitted or if a variable rate key is used.

REMARK 3.21 The first step of the encoding for achieving Theorem 3.7 uses negligible secret key to obscure the type of the sequence. This step was not needed in the proof of Theorem 3.6. In fact, the effect of this encoding step is to allow the result to be claimed for all messages m rather than averaged over the distribution of M.

3.4 Lossy Compression, Imperfect Secrecy

In full generality, secure source coding involves compression that may be lossy and secrecy that may be imperfect. In fact, compression of a source from a continuous distribution or a discrete distribution with infinite entropy must necessarily be lossy, and the theorems of the previous section are not relevant. In this section we outline how the theory is generalized.

New intricacies and tradeoffs are revealed when the theory allows for lossy reconstruction by the decoder, measured by average distortion. Under constrained resources, the objectives of source coding can be at odds with each other, forcing the system design to prioritize the quality of the reproduction against the level of security.

3.4.1 Rate–Distortion Theory

In the simplest case, we now have two measures of distortion to consider – the average distortion of the intended reconstruction at the legitimate receiver and the average distortion obtained by an eavesdropper's best reconstruction. The objective is to minimize the former and maximize the latter. However, a useful generalization of these pairwise distortion functions is to consider three-way distortion functions of the form $d(x, \hat{x}, z)$, which depend on the values of the source, legitimate reconstruction, and eavesdropper reconstruction together. We will make use of this generalization in Section 3.5.

As in the lossless compression case, the theory of secure source coding is enriched, and the security made robust, by taking into account causal disclosure. Since the reconstruction is now allowed to be lossy and hence different from the source sequence, we must now consider that both the source and the reconstruction may be obtained causally by an eavesdropper.

> **Box 3.5 Causal Disclosure**
>
> The eavesdropper obtains the past values of the source X and of the reconstruction $\widehat{X}$, or noisy observations of them, before making its reconstruction of the source at the present time. These observations are modeled as outputs W_x and $W_{\hat{x}}$ of two independent memoryless channels $P_{W_x|X}$ and $P_{W_{\hat{x}}|\widehat{X}}$. Full causal disclosure has noise-free channels.

DEFINITION 3.5 Given a memoryless source $\{X_t \sim P_X\}$, causal disclosure channels $P_{W|X}$ and $P_{W_{\hat{x}}|\widehat{X}}$, and a set of three-way distortion functions $\{d_j(x, \hat{x}, z)\}_{j=1}^J$, a rate–distortion tuple $(R, R_K, \{D_j\}_{j=1}^J)$ is achievable if there exist a block length n and an encoder and decoder at rates R and R_K as described in (3.11) and (3.12) such that

$$\min_{\{Z_t = z_t(M, W_x^{t-1}, W_{\hat{x}}^{t-1})\}_t} \mathbb{E} d_j(X^n, \widehat{X}^n, Z^n) \geq D_j \quad \forall j = 1, \ldots, J. \tag{3.52}$$

THEOREM 3.8 ([6, Theorem 1]) *The closure of achievable rate–distortion tuples $(R, R_K, \{D_j\}_{j=1}^J)$ for source coding with security robust to causal disclosure is the set of tuples satisfying*

$$\bigcup_{W_x - X - (U,V) - \widehat{X} - W_{\hat{x}}} \left\{ (R, R_K, \{D_j\}_{j=1}^J) : \begin{array}{l} R \geq I(X; U, V) \\ R_K \geq I(W_x, W_{\hat{x}}; V | U) \\ D_j \leq \min_{z(u)} \mathbb{E} d_j(X, \widehat{X}, z(U)) \quad \forall j \end{array} \right\}. \tag{3.53}$$

Achievability proof sketch As in the case of lossless compression, two messages are transmitted, and for the first message, a codebook is generated at rate $I(X; U)$ from the P_U distribution to form a covering of the source space. However, the second message uses a superposition code. For each u^n codeword (indexed by the first message) and key value, a codebook is generated at rate $I(X;V|U)$ from the $P_{V|U}$ distribution to form a conditional covering of the source. The decoder uses both messages and the key to identify the u^n and v^n codewords and produces $\widehat{X}^n$ memorylessly and stochastically according to $P_{\widehat{X}|U,V}$.

Again, the analysis is assisted by the use of likelihood encoders for both messages. Let us refer to the codeword associated with the first message as $\bar{u}^n$. The second message specifies a mapping of key values to codewords that we will call $\bar{v}^n(k)$. Due to the likelihood encoder, the key value is approximately independent of the two messages, and the posterior distribution of the source and reconstruction sequences given the key is approximately

$$P_{X^n,\widehat{X}^n|M,K}(x^n,\hat{x}^n|m,k) \approx \prod_{t=1}^{n} P_{X,\widehat{X}|U,V}(x_t,\hat{x}_t|\bar{u}_t,\bar{v}_t(k)). \tag{3.54}$$

The inequality constraint on R_K in (3.53), when made strict and combined with observation (3.54), assures that the uniformly distributed key cannot be decoded even with access to the causal disclosure. Consequently, further analysis shows that this random key causes the marginal distribution at each point in time to be

$$P_{X_t,\widehat{X}_t|M,W_x^{t-1},W_{\hat{x}}^{t-1}}(x,\hat{x}) \approx P_{X,\widehat{X}|U}(x,\hat{x}|\bar{u}_t). \tag{3.55}$$

This produces the claimed distortion levels.

REMARK 3.22 *The most basic application of Theorem 3.8 uses two distortion functions, where $-d_1(x,\hat{x})$ is the distortion of the decoder reconstruction and $d_2(x,z)$ is the distortion of the eavesdropper reconstruction.*

REMARK 3.23 *A stochastic decoder is crucial for achieving Theorem 3.8. Without it, causal disclosure of the reconstruction can reveal too much about the secret key and allow an eavesdropper to make inferences about future values of the source and reconstruction. This is dual to the wiretap channel setting, where the encoder must be stochastic.*

REMARK 3.24 *The variable U can be interpreted as a part of the encoding that is not secured due to lack of resources. The variable V represents the remainder of the encoding that is kept fully secure. However, to achieve secrecy of the V component, it is not sufficient to simply apply a one-time pad to the encoding (or simply transmit the encoding of V through a secure digital channel). The causal disclosure may still reveal the V codeword. Further randomization of the encoding by use of the key is needed.*

REMARK 3.25 *The optimal average score in a repeated game is solved by Theorem 3.8. Apply the theorem with full causal disclosure and with only one distortion function $d_1(x,\hat{x},z)$, which is the payoff function of the game. The interpretation of Definition 3.5 is that X^n is an i.i.d. state sequence, and a rate-limited communication system with rate-limited secret key is deployed to help one player of the game (with actions $\widehat{X}$) achieve the best possible performance against another player (with actions Z) by revealing the state X. The causal disclosure accounts for the fact that the players see each other's actions and the state after each play of the game.*

REMARK 3.26 *The source coding system that achieves Theorem 3.8 actually synthesizes a memoryless broadcast channel according to $P_{\widehat{X},U|X}$ in a very strong sense: An eavesdropper can decode the U^n sequence, and the conditional distribution of X^n and $\widehat{X}^n$ given U^n is precisely that induced by a memoryless broadcast channel, up to a negligible total variation distance.*

REMARK 3.27 *Theorem 3.8 does not change if (3.52) is changed to*

$$\max_{\{Z_t = z_t(M, W_x^{t-1}, W_{\hat{x}}^{t-1})\}_t} \mathbb{P}\left(d_j(X^n, \widehat{X}^n, Z^n) < D_j\right) < \epsilon \quad \forall j = 1, \ldots, J. \quad (3.56)$$

Example 3.2 *Binary jamming.* Consider an equiprobable binary source X, a single distortion function $d(x, \hat{x}, z) = \mathbb{1}\{x = \hat{x} \neq z\}$, and full causal disclosure (i.e., $W_x = X$ and $W_{\hat{x}} = \widehat{X}$). Apply Theorem 3.8 to find the maximum frequency with which $\widehat{X}$ can equal X without being jammed by Z, as a function of the compression rate and secret key rate.

Solution
Numerical optimization gives the tradeoff depicted in Fig. 3.5.

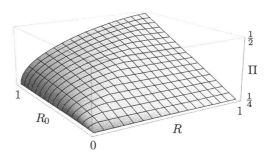

Figure 3.5 Binary jamming example.

Notice that increases in the key rate R_K continue to benefit the achievable performance even when $R_K > R$. This highlights that a one-time pad is not a sufficiently general encoder construction for secure source coding.

3.4.2 Henchman

We now return to the henchman setting of Fig. 3.4 and allow loss in the encoding, measured by average distortion. There are two performance guarantees to certify in this setting. The average distortion of the decoder reconstruction must be less than D, and the average distortion of the eavesdropper reconstruction must be greater than D_E, even with the assistance of the henchman.

The optimal encoding in the henchman setting is rather simple. For each secret key value, an independent source codebook is used (covering). However, tension between fidelity and security arises in the choice of distribution for which the codebook is designed. It is desirable to encode the source sequence in a way that is not easily refinable by the henchman.

DEFINITION 3.6   Given a memoryless source $\{X_t \sim P_X\}$ and a pair of distortion functions $d(x,\hat{x})$ and $d_E(x,z)$, a rate–distortion tuple (R, R_K, R_H, D, D_E) is achievable if for all $\epsilon > 0$ there exist a block length n and an encoder and decoder at compression rate R and key rate R_K as described in (3.11) and (3.12) such that

$$\mathbb{E}\, d(X^n, \widehat{X}^n) \leq D, \tag{3.57}$$

$$\min_{\substack{Z^n = z^n(M, M_H),\\ M_H = m_H(X^n, M, K),\\ |\mathcal{M}_H| = 2^{nR_H}}} \mathbb{E}\, d_E(X^n, Z^n) \geq D_E. \tag{3.58}$$

THEOREM 3.9   ([7]) *The closure of achievable rate–distortion tuples (R, R_K, R_H, D, D_E) for source coding with security robust to worst-case side information is the set of tuples satisfying*

$$R \geq I(X; \widehat{X}), \tag{3.59}$$

$$D \geq \mathbb{E}\, d(X, \widehat{X}), \tag{3.60}$$

$$D_E \leq \begin{cases} D_E(R_H) & \text{if } R_H < R_K, \\ \min\{D_E(R_H), D_E(R_H - R_K, P_{X,\widehat{X}})\} & \text{if } R_H \geq R_K, \end{cases} \tag{3.61}$$

for some $P_{X,\widehat{X}} = P_X P_{\widehat{X}|X}$, where $D_E(r)$ is the point-to-point distortion-rate function and $D_E(r, P_{X,\widehat{X}})$ is the point-to-point distortion-rate function with side information channel $P_{\widehat{X}|X}$ to the encoder and decoder:

$$D_E(r) \triangleq \min_{P_{Z|X}: r \geq I(X;Z)} \mathbb{E}\, d_E(X, Z), \tag{3.62}$$

$$D_E(r, P_{X,\widehat{X}}) \triangleq \min_{P_{Z|X,\widehat{X}}: r \geq I(X; Z|\widehat{X})} \mathbb{E}\, d_E(X, Z). \tag{3.63}$$

Proof sketch   The converse is again the easier proof. The henchman and the eavesdropper always have the option of compressing the source using a rate–distortion code or revealing the key as part of M_H, if R_H is sufficiently high. The non-trivial step in the converse, for $R_H > R_K$, involves conditioning on the joint empirical distribution of the source X^n and the reconstruction $\widehat{X}^n$ to claim the lower bound on the rate R and the upper bound $D_E \leq D_E(R_H - R_K, P_{X,\widehat{X}})$. However, the encoding need not produce a consistent empirical distribution for each instantiation of the source sequence. Careful use of empirical distribution bounds imposed by source coding are needed to account for this.

Achievability is proven using an independent codebook construction at rate R for each key value, generated according to the distribution $P_{\widehat{X}}$. A likelihood encoder is used to simplify the secrecy analysis. Analogous to the lossless case, the important step in the analysis is to show that for large n a random sequence X^n that is produced in two steps – first, select a sequence $\widehat{X}^n$ uniformly distributed on a random set (generated according to $P_{\widehat{X}}$), and second, produce X^n memorylessly according to $P_{X|\widehat{X}}$ – is with high probability not compressible by the henchman beyond the trivial schemes associated with explicitly describing the sequence $\widehat{X}^n$ from the codebook or treating X^n as a memoryless sequence.

REMARK 3.28 *If the henchman's rate is below the secret key rate, then the system effectively attains perfect secrecy. The henchman can only describe the source sequence as efficiently as rate–distortion theory allows. On the other hand, if the henchman has enough rate to reveal the secret key then he has a choice to make. He can transmit the key and use any additional rate to refine the source description further, but this will not always be more efficient then describing the source in the absence of the securely encoded message.*

REMARK 3.29 *If $R_K \geq R$, then Theorem 3.9 simplifies to*

$$D \geq D(R), \tag{3.64}$$

$$D_E \leq D_E(R_H). \tag{3.65}$$

This can be achieved by a one-time pad on the message, although the encoding outlined in the proof achieves this effect in a different way.

REMARK 3.30 *Theorem 3.9 does not change if (3.58) is changed to*

$$\max_{\substack{Z^n = z^n(M, M_H), \\ M_H = m_H(X^n, M, K), \\ |\mathcal{M}_H| = 2^{nR_H}}} \mathbb{P}\left(d_E(X^n, Z^n) < D_E\right) < \epsilon. \tag{3.66}$$

3.5 Equivocation

Equivocation, defined in Box 3.6, is the predominant metric for partial secrecy in the information theory literature. Shannon planted the seed by mentioning equivocation in his secrecy analysis in [2], and it took root when Wyner characterized the optimal equivocation rate for communication through the wiretap channel in [1].

It turns out that the use of equivocation as a secrecy metric is a particular special case of the secure source coding results already discussed in this chapter.

In hindsight, we can arrive at the equivocation rate metric of secrecy by applying Theorem 3.8 (the rate–distortion theorem) with the *log-loss* distortion function defined in Box 3.7. Applications of the log-loss distortion function to rate–distortion theory were first demonstrated in works such as [11].

DEFINITION 3.7 Given a memoryless source $\{X_t \sim P_X\}$ and a distortion function $d(x, \hat{x})$, a rate–distortion–equivocation quadruple (R, R_K, D, E) is achievable for *equivocation with respect to the source* if there exist a block length n and an encoder and

Box 3.6 Equivocation Rate

Equivocation is the conditional entropy of the protected information (usually the source) given everything known to the eavesdropper. Partial secrecy is often characterized by the equivocation rate, which is the equivocation divided by the length of the encoded source sequence. An equivalent quantity is the information leakage rate, which is the mutual information between what the eavesdropper knows and the protected information, again normalized by the block length. A high equivocation rate is a low information leakage rate.

Box 3.7 Logarithmic-Loss Distortion Function

Let $w \in \mathcal{W}$ be discrete valued and P be a distribution on $\mathcal{W}$. The log-loss function $d(w,P)$ is the logarithm of the inverse of the probability of w according to P,

$$d(w,P) \triangleq \log \frac{1}{P(w)}. \tag{3.67}$$

The log-loss function is zero if and only if P puts all probability on the outcome w. The optimal choice P^* to minimize the expected log-loss with respect to a random variable W is $P^* = P_W$, and the expected log-loss is $H(W)$. Similarly, if a correlated observation $V = v$ is available, then $P^* = P_{W|V=v}$, and the expected log-loss is $H(W|V)$.

decoder at compression rate R and key rate R_K as described in (3.11) and (3.12) such that

$$\mathbb{E} d(X^n, \hat{X}^n) \leq D, \tag{3.68}$$

$$\frac{1}{n} H(X^n | M) \geq E. \tag{3.69}$$

Likewise, (R, R_K, D, E) is achievable for *equivocation with respect to the reconstruction* if (3.69) is replaced by

$$\frac{1}{n} H(\hat{X}^n | M) \geq E, \tag{3.70}$$

and (R, R_K, D, E) is achievable for *equivocation with respect to both the source and reconstruction* if (3.69) is replaced by

$$\frac{1}{n} H(X^n, \hat{X}^n | M) \geq E. \tag{3.71}$$

COROLLARY 3.1 ([6] Corollary 5) *The closure of achievable rate–distortion–equivocation quadruples* (R, R_K, D, E) *for each of the three variants of equivocation are as follows:*

1. *For equivocation with respect to the source,*

$$\bigcup_{P_{\hat{X}|X}} \left\{ (R, R_K, D, E) : \begin{array}{l} R \geq I(X;\hat{X}) \\ D \geq \mathbb{E}d(X,\hat{X}) \\ E \leq H(X) - [I(X;\hat{X}) - R_K]_+ \end{array} \right\}, \qquad (3.72)$$

where $[x]_+ = \max\{0, x\}$.

2. *For equivocation with respect to the reconstruction,*

$$\bigcup_{X-U-\hat{X}} \left\{ (R, R_K, D, E) : \begin{array}{l} R \geq I(X;U) \\ D \geq \mathbb{E}d(X,\hat{X}) \\ E \leq H(\hat{X}) - [I(\hat{X};U) - R_K]_+ \end{array} \right\}. \qquad (3.73)$$

3. *For equivocation with respect to both the source and reconstruction,*

$$\bigcup_{X-U-\hat{X}} \left\{ (R, R_K, D, E) : \begin{array}{l} R \geq I(X;U) \\ D \geq \mathbb{E}d(X,\hat{X}) \\ E \leq H(X,\hat{X}) - [I(X,\hat{X};U) - R_K]_+ \end{array} \right\}. \qquad (3.74)$$

Proof sketch It must first be argued that Theorem 3.8 applies to this equivocation problem. Consider Definition 3.5 with two distortion constraints. For the first distortion constraint set $d_1(x,\hat{x},z) = -d(x,\hat{x})$ and $D_1 = -D$, where $d(x,\hat{x})$ and D are from Definition 3.7, the equivocation problem statement. The reason for the negative signs is that we intend to minimize this distortion instead of maximizing it.

The choice of the second distortion constraint and the causal disclosure depend on which of the three equivocation claims we wish to prove. In each case, set the second distortion constraint $D_2 = E$. For equivocation with respect to the source, set $d_2(x,\hat{x},z)$ to be the log-loss function with respect to the source X_t (i.e., set $W = X_t$ in Box 3.7), and let the causal disclosure be full causal disclosure of the source only (i.e., $W_x = X$ and $W_{\hat{x}} = \emptyset$).

Notice what happens to (3.52) with these specifications of log-loss and full causal disclosure of the source:

$$\min_{\{Z_t = z_t(M, W_x^{t-1}, W_{\hat{x}}^{t-1})\}_t} \mathbb{E} d_2(X^n, \widehat{X}^n, Z^n) = \min_{\{Z_t = z_t(M, X^{t-1})\}_t} \mathbb{E} \frac{1}{n} \sum_{t=1}^{n} \log \frac{1}{Z_t(X_t)} \qquad (3.75)$$

$$= \frac{1}{n} \sum_{t=1}^{n} \min_{Z_t = z_t(M, X^{t-1})} \mathbb{E} \log \frac{1}{Z_t(X_t)} \qquad (3.76)$$

$$= \frac{1}{n} \sum_{t=1}^{n} H(X_t | M, X^{t-1}) \qquad (3.77)$$

$$= \frac{1}{n} H(X^n | M). \qquad (3.78)$$

Thus, the second distortion constraint imposed by Definition 3.5 becomes precisely the equivocation constraint of Definition 3.7.

The choices of distortion constraint and causal disclosure for the other equivocation claims are similar. For equivocation with respect to the reconstruction, set $d_2(x,\hat{x},z)$ to be the log-loss function with respect to the reconstruction, and let the causal disclosure be full causal disclosure of the reconstruction only. For equivocation with respect to both the source and reconstruction, set $d_2(x,\hat{x},z)$ to be the log-loss function with respect to both the source and reconstruction, and let the causal disclosure be full causal disclosure of both the source and reconstruction. The equivalence between Definition 3.5 and Definition 3.7 follows as above.

What remains is to plug these choices of distortion functions and causal disclosure into Theorem 3.8 and show that the regions simplify to those stated in Corollary 3.1. This is done through simple manipulations that can be found in [6].

REMARK 3.31 *The causal disclosure present in the problem definition for Theorem 3.8 plays an essential role in the connection to equivocation rate. Without causal disclosure, the chain rule used in (3.78) would be invalid.*

REMARK 3.32 *Although Statement (1) of Corollary 3.1 is well understood and intuitive, the other two claims were newly introduced to the literature in [6]. Equivocation with respect to something other than the source is not commonly considered. The second two claims are qualitatively different from the first claim in that they use an auxiliary variable, and the accompanying encoding scheme is more complex. They retain some of the complexity of Theorem 3.8, which takes advantage of a stochastic decoder.*

REMARK 3.33 *While it appears quite difficult to expand the general rate–distortion theory, along the lines of Theorem 3.8, to settings with multiple correlated information sources (e.g., side information at the decoder or separated encoders), the special case of log-loss distortion is much more manageable. Notice how Corollary 3.1 has fewer auxiliary variables than Theorem 3.8. The coding scheme that achieves (3.72) for optimal equivocation with respect to the source is also considerably simpler. In some sense, log-loss distortion (i.e., the study of equivocation rate) allows one to treat partial secrecy as a quantity rather than a quality. The reduced intricacy has yielded tractable solutions in more complex secrecy settings.*

Acknowledgments

This work was supported by the National Science Foundation under grant CCF-1350595 and by the Air Force Office of Scientific Research under grant FA9550-15-1-0180.

References

[1] A. D. Wyner, "The wire-tap channel," *Bell Syst. Tech. J.*, vol. 54, pp. 1355–1387, Oct. 1975.

[2] C. E. Shannon, "Communication theory of secrecy systems," *Bell Syst. Tech. J.*, vol. 28, no. 4, pp. 656–715, Oct. 1949.

[3] Y. Kaspi and N. Merhav, "Zero-delay and causal secure source coding," *IEEE Trans. Inf. Theory*, vol. 61, no. 11, pp. 6238–6250, Nov. 2015.

[4] C. Uduwerelle, S.-W. Ho, and T. Chan, "Design of error-free perfect secrecy system by prefix codes and partition codes," in *Proc. IEEE Int. Symp. Inf. Theory*, Cambridge, MA, USA, Jul. 2012, pp. 1593–1597.

[5] C. E. Shannon, "A mathematical theory of communication," *Bell Syst. Tech. J.*, vol. 27, pp. 379–423, 623–656, Jul., Oct. 1948.

[6] C. Schieler and P. Cuff, "Rate-distortion theory for secrecy systems," *IEEE Trans. Inf. Theory*, vol. 60, no. 12, pp. 7584–7605, Dec. 2014.

[7] C. Schieler and P. Cuff, "The henchman problem: Measuring secrecy by the minimum distortion in a list," *IEEE Trans. Inf. Theory*, vol. 62, no. 6, pp. 3436–3450, Jun. 2016.

[8] N. Merhav and E. Arikan, "The Shannon cipher system with a guessing wiretapper," *IEEE Trans. Inf. Theory*, vol. 45, no. 6, pp. 1860–1866, Sep. 1999.

[9] E. Arıkan, "An inequality on guessing and its application to sequential decoding," *IEEE Trans. Inf. Theory*, vol. 42, no. 1, pp. 99–105, Jan. 1996.

[10] N. Merhav, "A large-deviations notion of perfect secrecy," *IEEE Trans. Inf. Theory*, vol. 49, no. 2, pp. 506–508, Feb. 2003.

[11] T. A. Courtade and T. Weissman, "Multiterminal source coding under logarithmic loss," *IEEE Trans. Inf. Theory*, vol. 60, no. 1, pp. 740–761, Jan. 2014.

[12] E. C. Song, P. Cuff and H. V. Poor, "The Likelihood Encoder for Lossy Compression," in *IEEE Trans. Inf. Theory*, vol. 62, no. 4, pp. 1836–1849, April 2016.

4 Networked Secure Source Coding

Kittipong Kittichokechai, Tobias J. Oechtering, and Mikael Skoglund

In this chapter we consider secure source coding problems in a network scenario consisting of multiple terminals. An information theoretic formulation using a mutual information-based secrecy/privacy measure is considered. We discuss how the networking aspects such as cooperation and interaction among nodes impact the fundamental tradeoff of the secure source coding system, i.e., the tradeoff among rate, distortion, and information leakage rate. Several problem settings are presented including the setting with a helper, the setting with an intermediate node, and the setting under a reconstruction privacy constraint. Results are presented together with some comprehensive discussions.

4.1 Introduction

The concept of information security and privacy is highly relevant today, and is expected to be even more so in future networks where almost everyone and everything will be connected (cf. Internet of Things (IoT), machine to machine (M2M), etc.). Users in the network may share information with one another for possible cooperation, but at the same time they wish to reveal as little information as possible to a certain network entity. Information security and privacy have emerged as necessary features and become standard requirements in modern communication systems. From the networking perspective, they may be addressed at different layers of the system. For example, encryption is used to protect information in the link layer. Nevertheless, a major security concern lies at the physical layer where the communication channel is vulnerable to eavesdropping. Addressing these issues at the physical layer is therefore essential and can effectively complement and improve the level of security in general.

Recently, due to potential applications in areas such as privacy in sensor networks and databases (see, e.g., [1]) and privacy of distributed storage (see, e.g., [2, 3] for privacy of genomic data), the concept of information security and privacy from a data compression perspective has been studied, under a common theme called *secure source coding*. In this chapter, we present some interesting instances of *networked secure source coding* problems that take into account information theoretic security and privacy requirements in the problem formulations. Different aspects of the network involving multiple terminals will be considered, e.g., coordination with a helper, cooperation through an intermediate node, and computation based on received information. In the problem setups, *side information* can be thought of as an abstraction for information received from other parts of the network or observed in previous transmissions. A *public helper* is a helping terminal, e.g., an intermediate terminal between source and destination, which operates according to the commonly agreed protocol but is also curious about private information extracted from the information it forwards. *Private links* can be viewed as links realized by encryption or a reliable network, while *public links* can be those where the encryption is weak and likely to be attacked.

4.1.1 Information Theoretic Measure for Privacy

Throughout the chapter we assume that there exists an external eavesdropper who has access to certain information that can be used to make inference on some private information. We assume that the eavesdropper is a passive adversary, i.e., it does not tamper with the transmission. There exist several information theoretic measures for privacy and security such as those based on equivocation, Kullback–Liebler (KL) divergence, Renyi's entropy and its variants, etc. In this chapter, we use *information leakage rate* or *equivocation rate* as a measure of information privacy (cf. [4, 5], etc.). The information leakage rate is defined as normalized mutual information between the targeted signal and all signals available at the eavesdropper. Similarly, the equivocation rate is defined as a normalized conditional entropy of the targeted signal conditioned on information available at the eavesdropper. Conceptually, they serve as asymptotic measures of average information leakage or remaining uncertainty of the targeted information at the eavesdropper. Unlike computational/complexity-based security measures, with the information theoretic measure we do not make any assumption on the computational capability of the eavesdropper, and even allow it to be informed about the joint distribution of relevant data. In this sense, it can be viewed as a worst-case assumption reasonably suited for the privacy and security assessment. Essentially, information theoretic privacy and security will rely on the condition that the eavesdropper does not have sufficient information to infer on the targeted information even when it has unlimited computational capabilities.

4.1.2 Fundamental Tradeoff

The networked secure source coding setup involves interaction among multiple terminals. Extension from a point-to-point setup to a multi-terminal setup provides an opportunity to improve information privacy/security since information can now flow in a decentralized manner. In order to attack the system with the same success rate, the attacker would need to rely on more complex strategies. Meanwhile, the distributed nature of the system allows the possibility of attacks on different parts of the network in which participating terminals and links may have different reliability and trust levels. It is therefore important for the network designer to understand how a system with multiple terminals can operate securely in different network topologies.

The main objective of designing networked secure source coding systems is to achieve the optimal source coding performance while ensuring a perfect security guarantee against any eavesdropper's inference. This objective is usually hard to realize in general as there is a fundamental tradeoff between achievable performance and privacy/security levels. For example, to reconstruct a source sequence with low average distortion at the receiver, a high-rate source description needs to be conveyed which could potentially lead to a high amount of information leakage if eavesdropped.

To understand such relations, we focus on characterizing the fundamental tradeoff in networked secure source coding systems, i.e., the tradeoff between coding performance metrics such as compression rate needed to describe the source and incurred distortion at the receiver, and the resulting information leakage rate at the eavesdropper. In particular, we are interested in how different networking aspects play a role in such a tradeoff. For instance, in a network with a helper node, the helper can provide extra side information to the receiver through a public communication channel. The helper's side information may be useful for the reconstruction at the receiver; however, it also poses some risks of leaking information over the public channel. Characterization of the fundamental tradeoff allows us to understand the optimal balance for operating points and corresponding performance limits that the transmitter, receiver, and helper can aim to achieve given the privacy risk they can tolerate. In this sense, it can serve as a privacy assessment framework for networked secure source coding. Moreover, development of the fundamental tradeoff involves investigation of several novel coding schemes that provide useful insights into designing practical coding schemes to achieve the optimal performance limit.

4.1.3 Models

We begin with a basic model of lossy source coding with side information at the receiver, the so-called Wyner–Ziv problem [6], and assume that there exists an eavesdropper observing the source description as well as side information correlated with the source, as shown in Fig. 4.1. A privacy constraint is imposed such that the normalized information leakage at the eavesdropper should be bounded below a certain value.

After becoming familiar with the basic model, we extend the problem to other multi-user settings including models of networks with one helper, networks with an intermediate node, networks with an untrusted node, and other interesting variations, to be discussed below. Since real-world large networks are usually complex and difficult to analyze rigorously, in this chapter we consider simpler small networks. Optimality

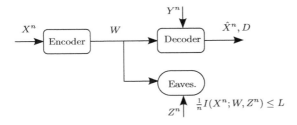

Figure 4.1 Secure source coding with side information.

results in such simplified models are useful and relevant for the larger network since they can be seen as necessary (person-by-person) optimality conditions. To learn and understand networked secure source coding, we look at different models capturing the essence of general networks.

- Networks with helpers exist in practical systems, with a prime example being the Internet. In the network with one transmitter, one helper, and one receiver, information from the helper can be one-sided (to a receiver) or two-sided (to both transmitter and receiver). In Section 4.3, we review recent problem settings where the helper's information is assumed to be public and can therefore be eavesdropped [7, 8] (see, e.g., Fig. 4.3). Also, we discuss problems of secure lossless source coding with a helper, studied, e.g., in [9–11], where the source description is eavesdropped (see, e.g., Fig. 4.5). Extensions to the lossy case were studied in [12] and [13] with complete characterizations of the optimal rate–distortion–leakage rate tradeoffs for some special cases.
- Transmitting data over the Internet commonly involves multi-hop transmission through a number of intermediate cooperative nodes. Cooperation can be of great benefit; however, it also raises privacy concerns when the intermediate nodes cannot be trusted. We present in Section 4.4 secure source coding problems under a privacy constraint at the intermediate helper, studied in [7, 8] in triangular (and cascade) settings. The tradeoff between cooperation and information privacy is captured by the fundamental tradeoff between compression rates, incurred distortion, and resulting information leakage at the intermediate helper. Related settings where the transmitter and receiver can use common side information (if available) to generate a secret key for *encrypting* the source description are also discussed.
- In a highly connected network, nodes usually receive information from several sources. In some scenarios, the receiver of information may wish to protect the privacy of its reconstruction sequence (computation, action, or decision made based on the received information) against any inference of the senders (see, e.g., Fig. 4.7). We present in Section 4.5 a new notion of privacy, called end-user privacy, studied in [14], where the end-user privacy constraint is required at different nodes in the system.

In the next sections, we provide an overview of the networked secure source problems described above. In each scenario, we formulate a problem from an information theoretic perspective and provide relevant results in terms of the fundamental tradeoffs for discrete memoryless sources. Our discussions mainly include pedagogical comments on how the results or proof techniques are derived and related to each other, and also on their novelty and interesting aspects. For more details/rigorous treatment of the problems, we refer the readers to the corresponding literature mentioned above and in each section.

Notation

We denote discrete random variables, their corresponding realizations or deterministic values, and their alphabets by upper case, lower case, and calligraphic letters, respectively. X_m^n denotes the sequence $\{X_m, \ldots, X_n\}$ when $m \leq n$, and the empty set otherwise. Also, we use the shorthand notation X^n for X_1^n. When a random variable X is constant we write $X = \emptyset$. Cardinality of the set $\mathcal{X}$ is denoted by $|\mathcal{X}|$. We use $[1:N]$ to denote the index set $\{1, 2, \ldots, N\}$. Finally, we use X–Y–Z to indicate that (X, Y, Z) forms a Markov chain, that is, their joint probability mass function (PMF) factorizes as $P_{X,Y,Z}(x,y,z) = P_{X,Y}(x,y)P_{Z|Y}(z|y)$ or $P_{X,Y,Z}(x,y,z) = P_{X|Y}(x|y)P_{Y,Z}(y,z)$.

4.2 Lossy Source Coding in the Presence of an Eavesdropper

In a lossy source coding problem, an encoder wishes to compress the source sequence into a source description and send it over a rate-limited link to a decoder. We assume that there is an eavesdropper who can observe the source description and some correlated side information, and thus may try to learn about the source as much as possible, as shown in Fig. 4.1. In this problem, the main goal is to efficiently communicate the source to the decoder, satisfying the average distortion criterion, and at the same time revealing only limited information about the source to the eavesdropper. Clearly, there exists a tradeoff between the reconstruction quality and the level of information privacy. For example, the encoder may send a high-rate source description to the decoder to achieve low average distortion. However, this can result in a large amount of information leaked to the eavesdropper. We are interested in characterizing the fundamental tradeoff between the minimum rate needed to describe the source, incurred distortion at the decoder, and the information leakage rate at the eavesdropper.

4.2.1 Problem Setup

The source and side information alphabets $\mathcal{X}, \mathcal{Y}, \mathcal{Z}$ are assumed to be finite. Let X^n be an n-length sequence which is i.i.d. according to $\prod_{i=1}^n P_X(x_i)$. Side information at the legitimate receiver and eavesdropper are given as outputs of the discrete memoryless channel with input X^n, i.e., $(Y^n, Z^n) \sim \prod_{i=1}^n P_{Y,Z|X}(y_i, z_i|x_i)$.

Given a source sequence X^n, an encoder generates a source description $W \in \mathcal{W}^{(n)}$ and sends it over the noise-free, rate-limited link to a decoder. Upon receiving W and side information Y^n, the decoder reconstructs the source as $\hat{X}^n$ subject to a distortion constraint. The eavesdropper is assumed to have access to W and side information Z^n correlated with the source. Information leakage rate at the eavesdropper is measured by the normalized mutual information $\frac{1}{n}I(X^n; W, Z^n)$.

DEFINITION 4.1 (Code) A $(|\mathcal{W}^{(n)}|, n)$ code for lossy secure source coding with side information consists of

- a stochastic encoder $F^{(n)}$ which takes X^n as input and generates $W \in \mathcal{W}^{(n)}$ according to a conditional PMF $p(w|x^n)$, and
- a decoder $g^{(n)} : \mathcal{W}^{(n)} \times \mathcal{Y}^{(n)} \to \hat{\mathcal{X}}^n$,

where $\mathcal{W}^{(n)}$ is a finite set.

Let $d: \mathcal{X} \times \hat{\mathcal{X}} \to [0, \infty)$ be the single-letter distortion measure. The distortion between the source sequence and its reconstruction at the decoder is defined as

$$d^{(n)}(X^n, \hat{X}^n) \triangleq \frac{1}{n} \sum_{i=1}^{n} d(X_i, \hat{X}_i),$$

where $d^{(n)}(\cdot)$ is the distortion function.

DEFINITION 4.2 (Achievability) A rate–distortion–leakage tuple $(R, D, L) \in \mathbb{R}_+^3$ is said to be *achievable* if for any $\delta > 0$ and all sufficiently large n there exists a $(|\mathcal{W}^{(n)}|, n)$ code such that

$$\frac{1}{n} \log |\mathcal{W}^{(n)}| \leq R + \delta,$$

$$E[d^{(n)}(X^n, g^{(n)}(W, Y^n))] \leq D + \delta,$$

and $\quad \dfrac{1}{n} I(X^n; W, Z^n) \leq L + \delta.$

The *rate–distortion–leakage region* $\mathcal{R}$ is the set of all achievable tuples.

4.2.2 Rate–Distortion–Leakage Tradeoff Result

The problem described above can be seen as an extension of the Wyner–Ziv rate–distortion problem [6] to include the leakage rate constraint (cf. Definition 4.2). This additional constraint adds another dimension to the problem, resulting in a new tradeoff among achievable rate R, distortion D, and leakage rate L. For example, when R is large, the leakage rate L can limit how small the distortion D can be. The following theorem provides the quantitative result which completely captures all possible tradeoffs.

THEOREM 4.1 ([12]) *The rate–distortion–leakage region $\mathcal{R}$ is the set of all tuples $(R, D, L) \in \mathbb{R}_+^3$ that satisfy*

$$R \geq I(X; V|Y), \tag{4.1a}$$

$$D \geq E\big[d\big(X, \tilde{g}(V, Y)\big)\big], \tag{4.1b}$$

$$L \geq I(X; V, Y) - I(X; Y|U) + I(X; Z|U) \tag{4.1c}$$

for some joint distributions of the form $P_X(x) P_{V|X}(v|x) P_{U|V}(u|v) P_{Y,Z|X}(y,z|x)$ *with* $|\mathcal{U}| \leq |\mathcal{X}| + 2$, $|\mathcal{V}| \leq |\mathcal{U}|(|\mathcal{X}| + 1)$, *and a function* $\tilde{g}: \mathcal{V} \times \mathcal{Y} \to \hat{\mathcal{X}}$.

The achievability proof is based on a random coding argument. It involves layered (superposition) coding with two codeword layers $\{U^n\}$ and $\{V^n\}$ and random binning on both layers to exploit side information available at the receiver. For encoding, the encoder looks for jointly typical codewords and sends the corresponding bin indices to the decoder. Analyses of achievable rate and distortion follow similarly as in the Wyner–Ziv problem [6], while the analysis of information leakage rate is shown below.

Let ϵ and δ_ϵ be positive real numbers where $\delta_\epsilon \to 0$ as $\epsilon \to 0$. Let $U^n(J)$ and $V^n(J, K)$ be the codewords chosen in the encoding, and M_1 and M_2 be the corresponding bin indices of the respective codewords sent over the public link, where $J \in [1 : 2^{n(I(X;U)+\delta_\epsilon)}]$, $K \in [1 : 2^{n(I(X;V|U)+\delta_\epsilon)}]$, $M_1 \in [1 : 2^{n(I(X;U|Y)+2\delta_\epsilon)}]$, and $M_2 \in [1 : 2^{n(I(X;V|U,Y)+2\delta_\epsilon)}]$.

Then, the information leakage averaged over all randomly chosen codebooks can be bounded as follows:

$$\begin{aligned}
I(X^n; M_1, M_2, Z^n) &= H(X^n) - H(X^n | M_1, M_2, Z^n) \\
&= H(X^n) - H(X^n | M_1, Z^n) + I(X^n; M_2 | M_1, Z^n) \\
&\overset{(a)}{\leq} H(X^n) - H(X^n | J, Z^n) + H(M_2) \\
&\leq H(X^n) - H(X^n, Z^n) + H(J) + H(Z^n | J) + H(M_2) \\
&\overset{(b)}{\leq} -nH(Z|X) + n(I(X;U) + \delta_\epsilon) + n(H(Z|U) + \delta_\epsilon) + n(I(X;V|U,Y) + 2\delta_\epsilon) \\
&\overset{(c)}{=} n(-H(Z|X,U) + I(X;U) + H(Z|U) + H(X|U,Y) - H(X|V,Y) + \delta'_\epsilon) \\
&= n(I(X;V,Y) - I(X;Y|U) + I(X;Z|U) + \delta'_\epsilon),
\end{aligned}$$

where (a) follows from the facts that given the codebook, M_1 is a function of J, that entropy is non-negative, and that conditioning reduces entropy; (b) follows from the memoryless property of the source and side information, from the codebook generation with $J \in [1 : 2^{n(I(X;U)+\delta_\epsilon)}]$, $M_2 \in [1 : 2^{n(I(X;V|U,Y)+2\delta_\epsilon)}]$, and from the bound $H(Z^n | J) \leq n(H(Z|U) + \delta_\epsilon)$ (see, e.g., [8, Lemma 4]), which follows from the fact that side information Z^n and the chosen codeword $U^n(J)$ are jointly typicality with high probability; and (c) follows from the Markov chain U–V–X–(Y,Z), which implies U–X–Z and U–(V,Y)–X, and $\delta'_\epsilon = 4\delta_\epsilon$. The information leakage constraint can be satisfied if

$$L \geq I(X;V,Y) - I(X;Y|U) + I(X;Z|U)$$

holds, where δ'_ϵ can be made arbitrary small with sufficiently large n. For an alternative and detailed treatment of the proof, we refer the readers to [12].

The above achievability scheme for Theorem 4.1 involves layered coding (superposition type coding), i.e., two layers of codewords U^n and V^n carry the description of X^n. This layered coding turns out to be essential for secure source coding problems as it provides some freedom for optimizing the codebook with respect to the presence of an eavesdropper in the general setting. However, when side information at the legitimate receiver and eavesdropper follow certain structures, e.g., one is degraded with respect to another, it can be shown that the original Wyner–Ziv coding scheme is optimal. That is, no extra codeword layer is needed to achieve the rate–distortion–leakage region.

We may interpret the right-hand side of the leakage rate bound (4.1c) as follows. It consists of a contribution from the remaining source uncertainty at the decoder after decoding codeword V^n, i.e., $H(X|V,Y)$, which can be utilized to reduce the leakage from the maximum level $H(X)$ to $I(X;V,Y)$. The term $I(X;Y|U) - I(X;Z|U)$ represents the potential gain in terms of leakage reduction due to different side information available

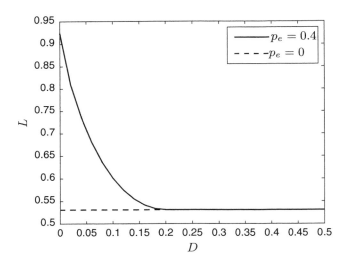

Figure 4.2 Leakage–distortion tradeoff with $p = 0.1$ and $p_e = 0.4$. The dashed line corresponds to $p_e = 0$, which yields the minimum leakage rate.

at the legitimate decoder and the eavesdropper, of which the amount can be tuned by the choice of codeword U.

From Theorem 4.1, it can be shown that the optimal tradeoff is always attained with the choice of U such that $I(U;Y) \leq I(U;Z)$. This implies that at the optimal design, codeword U^n can also be decoded by the eavesdropper. However, it should convey little information about X^n to the eavesdropper.

4.2.3 Example

To illustrate the tradeoff presented in Theorem 4.1, we consider the following binary example. The binary source X is assumed to be distributed according to Bernoulli(1/2). Side information at the legitimate decoder Y is an erased version of X with an erasure probability $p_e \in [0, 1]$, while side information at the eavesdropper Z is simply given as an output of a binary symmetric channel (BSC) with input X and crossover probability p. We assume the Hamming distortion measure, i.e., $d(x,\hat{x}) = 1$ if $x \neq \hat{x}$ and zero otherwise.

It can be shown that the rate–distortion–leakage region in Theorem 4.1 reduces to the set of all tuples (R, D, L) satisfying

$$R \geq p_e \cdot \big(1 - h(\alpha)\big),$$
$$D \geq p_e \cdot \alpha,$$
$$L \geq 1 - \big[p_e \cdot \big(h(\alpha) - h(p \star \alpha \star \beta) + h(p)\big)$$
$$+ (1 - p_e) \cdot \big(h(\alpha \star \beta) - h(p \star \alpha \star \beta) + h(p)\big)\big],$$

for some $\alpha, \beta \in [0, 1/2]$, where $h(\cdot)$ is the binary entropy function and $a \star b \triangleq a(1-b) + (1-a)b$.

For instance, if we are interested in the tradeoff between minimum leakage rate and incurred distortion, Fig. 4.2 shows minimum leakage rate L as a function of distortion D where the rate R is set to be sufficiently high such that the rate constraint is not active. It shows that the minimum leakage rate at the eavesdropper decreases with a small increase in distortion, especially in the low distortion region. There also exists a distortion level D^* such that for any distortion $D > D^*$ the minimum leakage rate cannot be improved further. This occurs when the distortion constraint is satisfied even without any information sent over the public link. The remaining leakage in this case is the minimum leakage due to the availability of side information at the eavesdropper.

4.3 Network with One Helper

Let us now extend the setup in the previous section to include a helper node, which might be seen as an abstraction for another part of the network. We consider a scenario where a coordinating helper provides side information to support the transmission from a transmitter to a receiver. It is possible that communication between the transmitter and receiver or that between the helper and receiver can be eavesdropped. If the link from the transmitter to receiver is not secure, the helper may offer to provide useful information over a secure link to avoid potential information leakage on the main channel. On the other hand, if the communication from the helper is over a public channel, it is interesting to see how one should utilize the helper given that there is a risk of information leakage. We are interested in the role of helper in providing side information and its consequences on the optimal tradeoff between compression rate, incurred distortion, and information leakage at the eavesdropper. In the following settings where the complete characterizations are known, we will see that it is optimal for the helper to first provide coarse-layered information using a standard rate–distortion code.

4.3.1 Eavesdropper Observes Helper Description Link

Let us consider lossy secure source coding problems with a helper, as depicted in Fig. 4.3. The setting is motivated by a scenario where the helper can only provide side information through a rate-limited communication link which is not secure due to its *public* nature, i.e., it can be eavesdropped by an external eavesdropper. In the "one-sided helper" setting the helper communicates through a public link only to the decoder, while in the "two-sided helper" case the helper *broadcasts* the same coded side information to both encoder and decoder. We provide an inner bound to the rate–distortion–leakage region for the one-sided helper case and show that it is tight under the logarithmic loss distortion measure [15] and for the Gaussian case with quadratic distortion and the Markov relation Y–X–Z. A Gaussian example illustrating the distortion–leakage tradeoff under different rate constraints for the one-sided helper is also given. For the two-sided helper case, the rate–distortion–leakage tradeoff is solved under general distortion. We note that the one-sided/two-sided helper settings considered in Fig. 4.3

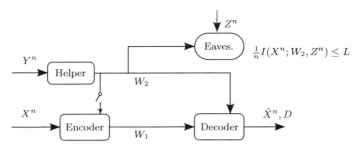

Figure 4.3 Secure source coding with one-sided/two-sided helper.

are essentially extensions of the one-helper problem [16–18] to include the presence of an eavesdropper.

One-Sided Helper

Let us consider the setting in Fig. 4.3 when the switch is open. The source, side information, and reconstruction alphabets $\mathcal{X}, \mathcal{Y}, \mathcal{Z}, \hat{\mathcal{X}}$ are assumed to be finite. Let (X^n, Y^n, Z^n) be the n-length sequences which are i.i.d. according to $P_{X,Y,Z}$. Given a source sequence X^n, an encoder generates a source description $W_1 \in \mathcal{W}_1^{(n)}$ and sends it over the noise-free, rate-limited link to a decoder. Meanwhile, a helper who observes the side information Y^n generates coded side information $W_2 \in \mathcal{W}_2^{(n)}$ and sends it to the decoder over another noise-free, rate-limited link. Given the source description and the coded side information, the decoder reconstructs the source as $\hat{X}^n$ subject to a distortion constraint. The eavesdropper in this case has access to the coded side information W_2 and extra side information Z^n. The information leakage at the eavesdropper is measured by the normalized mutual information $\frac{1}{n}I(X^n; W_2, Z^n)$.

Definitions for a code and achievability are similar to those of the secure lossy source coding problem discussed in Section 4.2 (cf. Definitions 4.1 and 4.2). The only differences are the inclusion of the helper who provides coded side information to the decoder and the new leakage rate constraint $\frac{1}{n}I(X^n; W_2, Z^n) \leq L + \delta$.

THEOREM 4.2 (Inner bound) *A tuple $(R_1, R_2, D, L) \in \mathbb{R}_+^4$ is achievable for secure source coding with one-sided helper if*

$$R_2 \geq I(Y; U), \tag{4.2a}$$

$$R_1 \geq I(X; V|U), \tag{4.2b}$$

$$D \geq E[d(X, \tilde{g}(U, V))], \tag{4.2c}$$

$$L \geq I(X; U, Z) \tag{4.2d}$$

for some joint distributions of the form $P_{X,Y,Z}(x,y,z)P_{U|Y}(u|y)P_{V|X}(v|x)$ with $|\mathcal{U}| \leq |\mathcal{Y}| + 4$, $|\mathcal{V}| \leq |\mathcal{X}| + 1$, and a function $\tilde{g} : \mathcal{U} \times \mathcal{V} \to \hat{\mathcal{X}}$.

An inner bound to the rate–distortion–leakage region is given by the convex hull of the set of all such tuples.

The proof idea behind Theorem 4.2 is based on the achievability scheme used in the original one-helper problem [16], i.e., first the helper communicates codeword U^n to the decoder using a rate–distortion code, then the encoder communicates V^n treating U^n as side information at the decoder. However, the resulting rate–distortion tradeoff can be different since the set of optimizing distributions may change due to an additional leakage constraint. For a complete proof, we refer readers to [8].

We note that the problem of characterizing the complete rate–distortion–leakage region under general distortion remains open. This is to be expected in view of the fact that the one-helper problem (without the leakage rate constraint) is still open. Nevertheless, the inner bound provided in Theorem 4.2 is shown to be tight for some special cases, namely, the setting under logarithmic loss distortion [15], and the Gaussian setting under quadratic distortion with Y–X–Z. Before stating the result of the rate–distortion–leakage region under logarithmic loss distortion, we restate the definition and highlight some important properties of the logarithmic loss distortion, for which more details can be found in [15].

DEFINITION 4.3 (Logarithmic loss) For the logarithmic loss distortion measure, we let the reconstruction alphabet $\hat{\mathcal{X}}$ be the set of probability distributions over the source alphabet $\mathcal{X}$, i.e., $\hat{\mathcal{X}} = \{p | p \text{ is a PMF on } \mathcal{X}\}$. For a sequence $\hat{X}^n \in \hat{\mathcal{X}}^n$, we denote $\hat{X}_i$, $i = 1, \ldots, n$, the ith element of $\hat{X}^n$. Then $\hat{X}_i, i = 1, \ldots, n$ is a probability distribution on $\mathcal{X}$, i.e., $\hat{X}_i : \mathcal{X} \to [0, 1]$, and $\hat{X}_i(x)$ is a probability distribution on $\mathcal{X}$ evaluated for the outcome $x \in \mathcal{X}$. The logarithmic loss distortion measure is defined as

$$d(x, \hat{x}) \triangleq \log\left(\frac{1}{\hat{x}(x)}\right) = D_{KL}(\mathbf{1}_{\{x\}} \| \hat{x}),$$

where $\mathbf{1}_{\{x\}} : \mathcal{X} \to \{0, 1\}$ is an indicator function such that, for $a \in \mathcal{X}$, $\mathbf{1}_{\{x\}}(a) = 1$ if $a = x$, and $\mathbf{1}_{\{x\}}(a) = 0$ otherwise. By using this definition for the symbol-wise distortion, the distortion between sequences is then defined as $d^{(n)}(x^n, \hat{x}^n) \triangleq \frac{1}{n} \sum_{i=1}^{n} d(x_i, \hat{x}_i)$.

In other words, under logarithmic loss distortion, the decoder generates "soft" estimates of the source sequence, and the distortion $d(x, \hat{x})$ is the Kullback–Leibler divergence between the empirical distribution of the event $X = x$ and the estimate $\hat{x}$.

Logarithmic loss has interesting properties, especially regarding the connection between the average distortion and conditional entropy. In the following, we present a couple of lemmas which appear in [15] and are essential in proving the result under the logarithmic loss distortion in this section and the next. Lemma 4.1 is used in the achievability proof, while Lemma 4.2 is used for bounding the conditional entropy in the converse proof.

LEMMA 4.1 Let U be the argument of the reconstruction function $g(\cdot)$, then under the log-loss distortion measure, we get $\min_{g(\cdot)} E[d(X, g(U))] = H(X|U)$.

LEMMA 4.2 Let Z be the argument of the reconstruction function $g^{(n)}(\cdot)$, then under the log-loss distortion measure, we get $E[d^{(n)}(X^n, g^{(n)}(Z))] \geq \frac{1}{n} H(X^n|Z)$.

THEOREM 4.3 (Logarithmic loss) *The rate–distortion–leakage region for the one-sided helper problem under logarithmic loss distortion is the set of all tuples $(R_1, R_2, D, L) \in \mathbb{R}_+^4$ that satisfy*

$$R_2 \geq I(Y; U), \qquad (4.3a)$$

$$R_1 \geq [H(X|U) - D]^+, \qquad (4.3b)$$

$$L \geq I(X; U, Z) \qquad (4.3c)$$

for some joint distributions of the form $P_{X,Y,Z}(x,y,z) P_{U|Y}(u|y)$ with $|\mathcal{U}| \leq |\mathcal{Y}| + 2$.

The proof of Theorem 4.3 is based on that of Theorem 4.2, where the interesting steps are the uses of Lemmas 4.1 and 4.2 which relate conditional entropy to the average distortion, allowing us establish the complete rate–distortion–leakage region. For example, by Lemma 4.1, choosing $\tilde{g}(\cdot)$ to be a conditional PMF $p(x|u,v)$ leads to $E[d(X, \tilde{g}(U, V))] = H(X|U, V)$. If $H(X|U) < D$, the encoder does not need to send anything, i.e., setting V constant. If $H(X|U) > D$, we define $V = X$ with probability $1 - \frac{D}{H(X|U)}$ and constant otherwise. Then we get $H(X|U, V) = D$ and thus $I(X; V|U) = H(X|U) - D$. For a complete proof and discussions regarding the use of Lemmas 4.1 and 4.2, we refer readers to [8].

Next, we present the distortion–rate–leakage function for the Gaussian setting under quadratic distortion. Let the sequences (X^n, Y^n, Z^n) be i.i.d. according to $P_{X,Y,Z}$. We assume that $Y \sim \mathcal{N}(0, \sigma_Y^2)$, $X = Y + N_1, N_1 \sim \mathcal{N}(0, \sigma_{N_1}^2)$ independent of Y, and $Z = X + N_2, N_2 \sim \mathcal{N}(0, \sigma_{N_2}^2)$ independent of (X, Y, N_1), where $\sigma_Y^2, \sigma_{N_1}^2, \sigma_{N_2}^2 > 0$. This satisfies the Markov chain assumption Y–X–Z.

THEOREM 4.4 (Gaussian case) *The minimum achievable distortion for given rates and leakage rate R_1, R_2, L under the Markov assumption Y–X–Z is given by*

$$D_{\min}(R_1, R_2, L) = \max\{2^{-2R_1}(2^{-2R_2}\sigma_Y^2 + \sigma_{N_1}^2), 2^{-2R_1}(\alpha^* \sigma_Y^2 + \sigma_{N_1}^2)\}, \qquad (4.4)$$

where $0 \leq \alpha^* = \frac{2^{-2L}(\sigma_Y^2+\sigma_{N_1}^2)(\sigma_{N_1}^2+\sigma_{N_2}^2) - \sigma_{N_1}^2\sigma_{N_2}^2}{\sigma_Y^2\sigma_{N_2}^2 - 2^{-2L}\sigma_Y^2(\sigma_Y^2+\sigma_{N_1}^2)} = \frac{\sigma_{N_2}^2}{\left(\frac{2^{2L}\sigma_{N_2}^2}{(\sigma_Y^2+\sigma_{N_1}^2)}-1\right)\sigma_Y^2} - \frac{\sigma_{N_1}^2}{\sigma_Y^2} \leq 1$, and

$$\tfrac{1}{2}\log\left(1 + \tfrac{\sigma_Y^2+\sigma_{N_1}^2}{\sigma_{N_2}^2}\right) \leq L \leq \tfrac{1}{2}\log\left(\tfrac{(\sigma_Y^2+\sigma_{N_1}^2)(\sigma_{N_1}^2+\sigma_{N_2}^2)}{\sigma_{N_1}^2\sigma_{N_2}^2}\right).$$

The proof of this theorem is based on the inner bound result in Theorem 4.2 specializing to the Gaussian case. While the inner bound result was proven for discrete memoryless sources, the extension to the quadratic Gaussian case is standard and it follows, for example, [19] and [20]. The converse proof involves the use of standard techniques such as entropy power inequality (EPI) [21]. For more details, see [8].

Next, we evaluate the minimum achievable distortion presented in Theorem 4.4 for given rates and leakage rate. For fixed $\sigma_Y^2 = 0.5$, $\sigma_{N_1}^2 = \sigma_{N_2}^2 = 0.25$, we plot $D_{\min}$ as a function of L for given R_1 and R_2 in Fig. 4.4.

We can see that, in general, for given R_1 and R_2, $D_{\min}$ is decreasing when L becomes larger. This is because the helper is able to transmit more information to the decoder without violating the leakage constraint. However, there exists an L^* such that for any

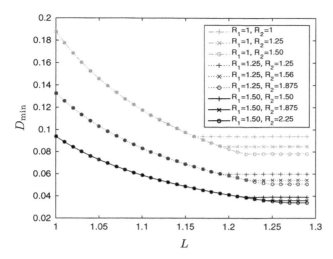

Figure 4.4 Gaussian example: Minimum achievable distortion as a function of leakage rate for given rates R_1, R_2. A tradeoff curve with a different parameter set can be found in [8].

$L > L^*$ we cannot improve $D_{\min}$ further since it is limited by the rate R_2. We note that L^* depends only on R_2; it can be seen from Fig. 4.4 that when $R_2 = 1.25$, we get the same L^* for different R_1, e.g., $R_1 = 1$ or 1.25. However, $D_{\min}$ still depends on R_1, i.e., it is saturated at a lower level for larger R_1. To this end, we conclude that for the high L region, R_2 is a limiting factor of $D_{\min}$.

On the other hand, when L is "small," the decreasing region is active, i.e., $D_{\min} = 2^{-2R_1}(\alpha^* \sigma_Y^2 + \sigma_{N_1}^2)$, and $D_{\min}$ depends only on R_1 and L (not on R_2). That is, in the "small" L region, $D_{\min}$ is limited by L so that we cannot improve $D_{\min}$ by increasing R_2. This can be seen from the plots: for a given R_1, three distortion–leakage curves with different R_2 coincide in the small L region.

Two-Sided Helper

Let us consider the setting in Fig. 4.3 when the switch is closed. Since the problem setting is similar to that of the one-sided helper case, for brevity we only point out the main differences. In the two-sided helper setting, the coded side information $W_2 \in \mathcal{W}_2^{(n)}$ is given to both encoder and decoder, and based on X^n and W_2, the encoder generates the source description $W_1 \in \mathcal{W}_1^{(n)}$. That is, the encoding function becomes $F_1^{(n)}$ taking (X^n, W_2) as input and generating W_1 according to $p(w_1|x^n, w_2)$. In this two-sided helper case, we can obtain the rate–distortion–leakage region under general distortion function.

THEOREM 4.5 *The rate–distortion–leakage region $\mathcal{R}_{\text{two-sided}}$ is the set of all tuples $(R_1, R_2, D, L) \in \mathbb{R}_+^4$ that satisfy*

$$R_2 \geq I(Y;U), \quad (4.5a)$$

$$R_1 \geq I(X;\hat{X}|U), \quad (4.5b)$$

$$D \geq E[d(X,\hat{X})], \quad (4.5c)$$

$$L \geq I(X;U,Z) \quad (4.5d)$$

for some joint distributions of the form $P_{X,Y,Z}(x,y,z)P_{U|Y}(u|y)P_{\hat{X}|X,U}(\hat{x}|x,u)$ *with* $|\mathcal{U}| \leq |\mathcal{Y}|+3$.

The achievability idea is similar to that of the one-sided helper case where we now treat U^n as side information available at both encoder and decoder, and the codeword $\hat{X}^n$ replaces codeword V^n. The interesting part of the proof is that we in fact proved an outer bound region with a larger set of joint distributions than the one presented in Theorem 4.5. However, it can be shown that the mutual information terms and average distortion remain unchanged when evaluated over these two sets of joint distributions. For more details, we refer readers to [8].

REMARK 4.1 *For the cases of logarithmic loss distortion and the Gaussian source with quadratic distortion specified before, it can be shown that the rate–distortion–leakage regions of the two-sided helper cases remain the same as those of the corresponding one-sided helper problems. This is reminiscent of the well-known result in the Wyner–Ziv source coding problem with Gaussian source and quadratic distortion that side information* Y^n *at the encoder does not improve the rate–distortion function, i.e.,* $R_{X|Y}(D) = R^{WZ}(D) = \frac{1}{2}\log(\frac{\text{var}(X|Y)}{D})$ *[19].*

4.3.2 Eavesdropper Observes Source Description Link

Another variant of the networked secure source coding with a helper is where the eavesdropper can listen in on the link between the transmitter and receiver instead. This line of work was studied in [11] and [12] for lossless and lossy settings, respectively. For the lossless setting, the rate–leakage region is fully characterized, while only inner and outer bounds are provided for the lossy case. Here we present the lossless settings with one-sided and two-sided helper.

One-Sided Helper

Let us consider the setting in Fig. 4.5 when the switch is open. The problem setup is similar to that in Section 4.3.1, except that the information leakage constraint becomes $\frac{1}{n}I(X^n;W_1) \leq L + \delta$.

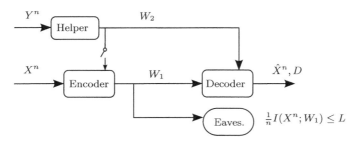

Figure 4.5 Secure source coding with one-sided/two-sided helper.

THEOREM 4.6 *The rate–leakage region $\mathcal{R}_{\text{one-sided}}$ is given as a set of all $(R_1, R_2, L) \in \mathbb{R}_+^3$ that satisfy*

$$R_2 \geq I(Y; U), \quad (4.6\text{a})$$

$$R_1 \geq H(X|U), \quad (4.6\text{b})$$

$$L \geq H(X|U) \quad (4.6\text{c})$$

for some joint distributions of the form $P_{X,Y}(x,y)P_{U|Y}(u|y)$ with $|\mathcal{U}| \leq |\mathcal{Y}| + 2$.

The achievability scheme used to prove Theorem 4.6 can be summarized as follows. First, the helper communicates codeword U^n to the decoder using a rate–distortion code. Then the encoder performs Slepian–Wolf binning [22] of the source X^n treating U^n as coded side information at the decoder. Note that this scheme is identical to that of [23] and [24], which shows that the original scheme is also optimal in the presence of an eavesdropper. The right-hand side of the leakage bound (4.6c) simply corresponds to the rate of the description W_1. This follows from the fact that the eavesdropper only observes the bin index of the source sequence, of which the remaining uncertainty at the eavesdropper depends on the number of source sequences in the bin. In this case, the bin size is approximately $2^{nI(X;U)}$ and thus the resulting leakage rate is equal to $H(X) - I(X; U)$. For a complete proof, we refer readers to [11].

REMARK 4.2 *The lossless setting was extended to the lossy reconstruction case in [12], where the reconstruction of the source at the decoder is subject to an average distortion constraint and additionally an eavesdropper is assumed to have side information correlated with the source. This problem was, however, not completely solved for a general distortion as it is essentially an extension of a variant of multi-terminal source coding problem which still remains open in general. The authors provided inner and outer bounds to the rate–distortion–leakage region, where the inner bound is proved based on the Berger–Tung coding scheme [25].*

Two-Sided Helper

In the following, we present the rate–leakage region for the problem of secure source coding with two-sided helper as shown in Fig. 4.5 when the switch is closed.

THEOREM 4.7 *The rate–leakage region $\mathcal{R}_{\text{two-sided}}$ is given as a set of all tuples $(R_1, R_2, L) \in \mathbb{R}_+^3$ that satisfy*

$$R_2 \geq I(Y; U), \quad (4.7\text{a})$$

$$R_1 \geq H(X|U), \quad (4.7\text{b})$$

$$L \geq H(X) - R_2 \quad (4.7\text{c})$$

for some joint distributions of the form $P_{X,Y}(x,y)P_{U|Y}(u|y)$ with $|\mathcal{U}| \leq |\mathcal{Y}| + 2$.

As before, the helper first communicates the codeword U^n using a rate–distortion code. However, unlike in previous cases where a binning type scheme is used, the encoder utilizes the common coded side information for encoding to reduce the leakage rate. It considers the set of conditionally typical sequences X^n given the coded side information from the helper, $T_\epsilon^{(n)}(X|u^n)$, and sends the index corresponding to the observed source sequence. An eavesdropper who does not know U^n has remaining uncertainty about the source sequence X^n that it could be in one of the 2^{nR_2} sets $T_\epsilon^{(n)}(X|u^n)$. Therefore, the resulting leakage rate is $H(X) - R_2$. It is interesting to note that the coded side information U^n plays a dual role here of reducing the rate from the encoder as well as providing security. A complete proof can be found in [11].

REMARK 4.3   *The lossless setting was extended to the lossy reconstruction case in [13], where the reconstruction of the source at the decoder is subject to an average distortion constraint and additionally an eavesdropper is assumed to have correlated side information. This problem is not completely solved for a general distortion as it is unclear how to optimally utilize code side information for both reducing the rate and protecting the source description. In some special cases where the dual roles are decoupled, i.e., side information at the encoder does not help to improve the rate–distortion tradeoff, the complete rate–distortion–leakage region can be obtained.*

4.4   Network with an Intermediate Node

Next, we review problems of secure triangular source coding with a public intermediate helper, as shown in Fig. 4.6. In contrast to the previous settings where the focus is on information leakage at an external eavesdropper, here we address the problem of information leakage at a legitimate user. The setting is motivated by a scenario where the helper is a public terminal that forwards the information as the protocol requests from the encoder to the decoder. However, the helper might be curious and not ignore the data which may not be intended for him. Clearly, there exists a tradeoff between the amount of information leaked to the helper and the helper's ability to support the source transmission (cooperation vs. privacy). Characterizing the optimal rate–distortion–leakage tradeoff for this problem is challenging due to the ambiguity of the helper's strategy and the role of side information at the encoder. In this section, we characterize the rate–distortion–leakage regions for various special cases based on different *side information patterns* available to the encoder, helper, and decoder.

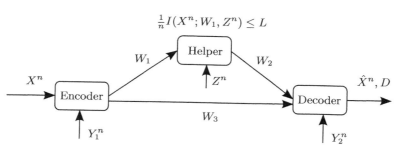

Figure 4.6   Secure triangular/cascade source coding with an intermediate public helper.

4.4.1 Models

- **Setting A:** As an initial study, we assume that Y_1 is constant, $Y_2 = Y$, and that X–Y–Z forms a Markov chain. The problem is solved under logarithmic loss distortion and for Gaussian sources with quadratic distortion where the forwarding scheme at the helper (setting $W_2 = W_1$) is optimal. Note that the Markov assumption X–Y–Z in this setting can be relevant in scenarios where the decoder is a fusion center collecting all correlated side information.
- **Setting B:** As an interesting and practically relevant variation of Setting A, we now assume that there exists common side information at the encoder and decoder, i.e., $Y_1 = Y_2 = Y$, and that X–Y–Z forms a Markov chain. The problem is solved under logarithmic loss distortion and for a Gaussian source with quadratic distortion.

  It is shown that side information at the encoder does not help to improve the rate–distortion tradeoff, and that the forwarding scheme at the helper is optimal. Interestingly, the availability of the side information at the encoder and decoder can be used for secret key generation. In our coding scheme, the secret key is used to scramble part of the message sent to the helper, and thus decrease the information leakage.
- **Setting C:** For completeness of the study in terms of availability of side information, we consider a different side information pattern. We assume that side information at the helper is also available at the encoder, i.e., $Y_1 = Z$, and let $Y_2 = Y$. In this case, we assume that X–Z–Y forms a Markov chain and solve the problem under general distortion. Due to X–Z–Y, the decode-and-reencode type scheme at the helper is shown to be optimal. That is, it is meaningful to take into account Z^n at the helper in relaying information to the decoder.

The settings considered in this section are different from the conventional triangular source coding problem in that the decoding constraint at the helper is replaced by the privacy leakage constraint. Also, all the triangular settings can reduce to the cascade ones when the private link from the encoder to the decoder is removed (i.e., setting W_3 constant). The results for the cascade settings can therefore be obtained straightforwardly from the triangular settings.

4.4.2 Problem Setup

Let us consider the setting in Fig. 4.6. The source and side information alphabets $\mathcal{X}, \mathcal{Y}_1, \mathcal{Y}_2, \mathcal{Z}$ are assumed to be finite. Let (X^n, Y_1^n, Y_2^n, Z^n) be the n-length sequences which are i.i.d. according to $P_{X,Y_1,Y_2,Z}$. Given the sequences (X^n, Y_1^n), an encoder generates a description $W_1 \in \mathcal{W}_1^{(n)}$ and sends it to the helper over a noise-free, rate-limited link. The encoder also generates a description $W_3 \in \mathcal{W}_3^{(n)}$ based on (X^n, Y_1^n) and sends it to the decoder over another noise-free, rate-limited link. Based upon the description W_1 and the side information Z^n, the helper generates a new description $W_2 \in \mathcal{W}_2^{(n)}$ and sends it to the decoder. Given W_2, W_3, and side information Y_2^n, the decoder reconstructs the source sequence as $\hat{X}^n$.

DEFINITION 4.4 (Code) A $(|\mathcal{W}_1^{(n)}|, |\mathcal{W}_2^{(n)}|, |\mathcal{W}_3^{(n)}|, n)$ code for secure triangular source coding with a public helper consists of

- a stochastic encoder $F_1^{(n)}$ that takes (X^n, Y_1^n) as input and generates $(W_1, W_3) \in \mathcal{W}_1^{(n)} \times \mathcal{W}_3^{(n)}$ according to a conditional PMF $p(w_1, w_3 | x^n, y_1^n)$,
- a stochastic helper $F_2^{(n)}$ that takes (W_1, Z_1^n) as input and generates $W_2 \in \mathcal{W}_2^{(n)}$ according to $p(w_2 | w_1, z^n)$,
- a decoder $g^{(n)} : \mathcal{W}_2^{(n)} \times \mathcal{W}_3^{(n)} \times \mathcal{Y}_2^{(n)} \to \hat{\mathcal{X}}^n$,

where $\mathcal{W}_1^{(n)}, \mathcal{W}_2^{(n)}$, and $\mathcal{W}_3^{(n)}$ are finite sets.

The information leakage at the helper who has access to W_1 and Z^n is measured by $\frac{1}{n} I(X^n; W_1, Z^n)$.

DEFINITION 4.5 (Achievability) A tuple $(R_1, R_2, R_3, D, L) \in \mathbb{R}_+^5$ is said to be *achievable* if for any $\delta > 0$ and all sufficiently large n there exists a $(|\mathcal{W}_1^{(n)}|, |\mathcal{W}_2^{(n)}|, |\mathcal{W}_3^{(n)}|, n)$ code such that

$$\frac{1}{n} \log |\mathcal{W}_i^{(n)}| \leq R_i + \delta, \; i = 1, 2, 3,$$

$$E[d^{(n)}(X^n, g^{(n)}(W_2, W_3, Y_2^n))] \leq D + \delta,$$

and $\quad \frac{1}{n} I(X^n; W_1, Z^n) \leq L + \delta.$

The *rate–distortion–leakage* region is defined as the set of all achievable tuples.

4.4.3 Rate–Distortion–Leakage Tradeoff Results

THEOREM 4.8 (Setting A, logarithmic loss) *The rate–distortion–leakage region* $\mathcal{R}_{\text{tri(A)},X-Y-Z,\text{log-loss}}$ *under logarithmic loss distortion and the X–Y–Z assumption is the set of all tuples* $(R_1, R_2, R_3, D, L) \in \mathbb{R}_+^5$ *that satisfy*

$$R_1 \geq [H(X|Y) - D - R_3]^+, \tag{4.8a}$$

$$R_2 \geq [H(X|Y) - D - R_3]^+, \tag{4.8b}$$

$$L \geq I(X;Z) + [H(X|Y) - D - R_3]^+. \tag{4.8c}$$

The proof idea is based on the achievability scheme utilizing the Wyner–Ziv binning, rate splitting for messages W_1 and W_3, and the forwarding scheme at the helper. The intuition behind the scheme is as follows. Since we assume that $X-Y-Z$ forms a Markov chain, it is optimal to perform the Wyner–Ziv coding with respect to side information Y^n, and ignore side information Z^n by simply forwarding the index received at the helper. This results in the same rate constraint for R_1 and R_2. Moreover, with this forwarding scheme at hand, rate splitting of the index sent over the cascade and private links turns out to be optimal. The terms on the right-hand side of the leakage rate constraint are simply the leakage rate due to the correlated side information Z^n, and the index received at the helper.

THEOREM 4.9 (Setting B, logarithmic loss) *The rate–distortion–leakage region* $\mathcal{R}_{\text{tri}(B),X-Y-Z,\text{log-loss}}$ *under logarithmic loss distortion and X–Y–Z is the set of all tuples* $(R_1,R_2,R_3,D,L) \in \mathbb{R}_+^5$ *that satisfy*

$$R_1 \geq [H(X|Y) - D - R_3]^+, \quad (4.9a)$$

$$R_2 \geq [H(X|Y) - D - R_3]^+, \quad (4.9b)$$

$$L \geq I(X;Z) + [H(X|Y) - D - R_3 - H(Y|X,Z)]^+. \quad (4.9c)$$

The proof idea for achievability is similar to the previous case, i.e., we use the achievability scheme which neglects side information for encoding at the encoder. The new and interesting step of the proof is the utilization of side information Y^n as a common randomness at the encoder and decoder to generate a secret key for encrypting the source description sent to the public helper.

REMARK 4.4 *We note that the availability of side information Y^n at the encoder does not improve the rate–distortion tradeoff under logarithmic loss distortion with respect to the Wyner–Ziv setting (as in the Gaussian case [19]). However, the common side information helps to reduce the leakage rate at the helper by allowing the encoder and decoder to generate a common secret key. We can see this from the leakage constraint (4.9c) above, where the leakage rate consists of contributions from the eavesdropper's side information $I(X;Z)$ and from the source description, which is now partially protected by the secret key of rate* $\min\{H(Y|X,Z), H(X|Y) - D - R_3\}$ *(cf. (4.8c) in Theorem 4.8 where there is no leakage reduction from the secret key).*

THEOREM 4.10 (Setting C, general distortion) *The rate–distortion–leakage region* $\mathcal{R}_{\text{tri}(C),X-Z-Y}$ *is the set of all tuples* $(R_1,R_2,R_3,D,L) \in \mathbb{R}_+^5$ *that satisfy*

$$R_1 \geq I(X;U|Z), \quad (4.10a)$$

$$R_2 \geq I(X,Z;U|Y), \quad (4.10b)$$

$$R_3 \geq I(X,Z;V|U,Y), \quad (4.10c)$$

$$D \geq E[d(X,\tilde{g}(U,V,Y))], \quad (4.10d)$$

$$L \geq I(X;U,Z) \quad (4.10e)$$

for some joint distributions of the form

$$P_{X,Z}(x,z)P_{Y|Z}(y|z)P_{U|X,Z}(u|x,z)P_{V|X,Z,U}(v|x,z,u),$$

with $|\mathcal{U}| \leq |\mathcal{X}||\mathcal{Z}|+3$, $|\mathcal{V}| \leq (|\mathcal{X}||\mathcal{Z}|+3)(|\mathcal{X}||\mathcal{Z}|+1)$, *and a function* $\tilde{g}: \mathcal{U} \times \mathcal{V} \times \mathcal{Y} \to \hat{\mathcal{X}}$.

The new step in the achievability proof of this case lies in the operation at the helper. Since we assume a new order of side information degradedness X–Z–Y, it is optimal for the helper to perform decoding and reencoding. In other words, side information Z^n at the helper is useful in providing extra information to the decoder. We note that if the leakage constraint at the helper is replaced by the decoding constraint under some distortion, the problem turns into the triangular source coding problem studied in [26].

For the complete proofs of Theorems 4.8–4.10, we refer readers to [8].

REMARK 4.5 *We discuss here the optimal operation at the helper in all the considered settings. Since the private link in the triangular setting can only provide additional information subject to its rate constraint, the processing ambiguity lies only in the cascade transmission, i.e., what is the best relaying strategy at the helper? For ease of discussion, we will for now neglect the private link, and argue that when the side information at the helper is degraded with respect to that at the decoder, the forwarding scheme at the helper is optimal; otherwise, it is optimal to employ the decode-and-reencode type scheme.*

- Let us consider Setting A, in which we assume that X–Y–Z forms a Markov chain (the discussion for Setting B follows similarly). On the cascade link, in order to attain low distortion at the decoder, we wish to compress the source so that the decoder, upon receiving (W_2, Y^n), can extract as much information about the source as possible, i.e., maximizing $I(X^n; W_2, Y^n)$. The joint PMF of this setting after summing out the reconstruction sequences is given by $P_{X^n,Y^n} P_{Z^n|Y^n} P_{W_1|X^n} P_{W_2|W_1,Z^n}$. The data processing inequality implies that $I(X^n; W_2, Y^n) \leq I(X^n; W_1, Y^n)$. This suggests that the forwarding scheme at the helper (setting $W_2 = W_1$) is a good strategy for this setting, and it is in fact optimal in this case.
- For the other case, Setting C, where we assume that X–Z–Y forms a Markov chain, the joint PMF after summing out the reconstruction sequences is given by $P_{X^n,Z^n} P_{Y^n|Z^n} P_{W_1|X^n,Z^n} P_{W_2|W_1,Z^n}$. To see if the forwarding scheme is still optimal, we derive from the joint PMF the inequality $I(X^n; W_2, Y^n) \leq I(X^n; W_1, Z^n)$. It suggests that, based on the information available, the helper can extract more information about X^n than the decoder does, regardless of the helper's scheme. Since W_2 can be generated based on (W_1, Z^n), it is reasonable that the helper takes into account Z^n in relaying the information, rather than just forwarding W_1. The decode-and-reencode type scheme is optimal in this case.

4.5 Network with Reconstruction Privacy

With the growing predominance of the Internet and the advance of cloud computing, a significant amount of data will be exchanged among users and service providers, which inevitably leads to a privacy concern. A user in the network may receive different information from different sources. Apart from being able to process the information efficiently, the user may also wish to protect the privacy of his/her computation or action taken based on the received information. In this section, we address the privacy concern of the final action taken at the end user. More specifically, we consider the problem of lossy source coding under the privacy constraint of the end user (decoder) whose goal is to reconstruct a sequence subject to a distortion criterion (see, e.g., Fig. 4.7). The privacy concern of the end user may arise due to the presence of an external eavesdropper

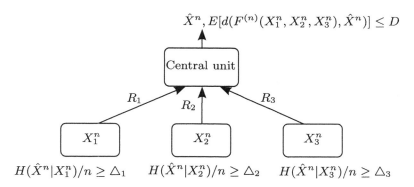

Figure 4.7 Multi-terminal source coding with end-user privacy.

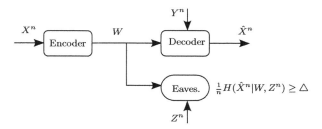

Figure 4.8 Source coding with end-user privacy at eavesdropper.

or a legitimate terminal such as a sender or a helper who is curious about the final reconstruction. We term the privacy criterion *end-user privacy*, and use the normalized equivocation of the reconstruction sequence at a particular terminal $H(\hat{X}^n|\cdot)/n$ as a privacy measure.

In this section, we focus on an implication of the end-user privacy constraint on the rate–distortion tradeoff where the constraint is imposed at the eavesdropper, as depicted in Fig. 4.8. This can be seen as a single-user case of Fig. 4.7. It corresponds to a scenario where there is an eavesdropper observing the source description and its side information, and we wish to prevent it from inferring the final reconstruction. We present inner and outer bounds to the rate–distortion–equivocation region for the cases where side information is available non-causally and causally at the decoder. In a special case where the decoder has no *memory*, i.e., each reconstruction symbol depends only on the source description and current side information symbol, the complete characterization of the rate–distortion–equivocation region is given.

We note also that the case where the end-user privacy is imposed at the encoder is included Fig. 4.8 when $Z^n = X^n$, since the encoder is a deterministic encoder. For more detailed discussions and other scenarios, we refer readers to [14].

4.5.1 End-User Privacy at Eavesdropper

We consider a setting in the presence of an external eavesdropper, as shown in Fig. 4.8. The source, side information, and reconstruction alphabets $\mathcal{X}, \mathcal{Y}, \mathcal{Z}, \hat{\mathcal{X}}$ are assumed to be finite. Let (X^n, Y^n, Z^n) be n-length sequences which are i.i.d. according to $P_{X,Y,Z}$. A function $F^{(n)}(X^n, Y^n)$ is assumed to be a component-wise function, where the ith component $F_i^{(n)}(X^n, Y^n) = F(X_i, Y_i)$ with $F : \mathcal{X} \times \mathcal{Y} \to \mathcal{F}$, for $i = 1, \ldots, n$ (cf., e.g., [27]). Given a source sequence X^n, an encoder generates a source description $W \in \mathcal{W}^{(n)}$ and sends it over the noise-free, rate-limited link to a decoder. Given the source description and the side information Y^n, the decoder generates $\hat{X}^n$ as an estimate of the value of the function $F^{(n)}(X^n, Y^n)$ such that it satisfies a distortion criterion. The eavesdropper has access to the source description and its own side information Z^n. The end-user privacy at the eavesdropper is then measured by the normalized conditional $H(\hat{X}^n | W, Z^n)/n$. We are interested in characterizing the optimal tradeoff between rate, distortion, and equivocation of the reconstruction sequence in terms of the rate–distortion–equivocation region.

The model in Fig. 4.8 is similar to the lossy secure source coding with side information in Section 4.2, except that the end-user privacy replaces the source privacy. The setting is also closely related to the model of side information privacy studied in [28], where the authors are interested in the privacy of side information at the second decoder who is also required to decode the source subject to a distortion constraint. As for the end-user privacy, a similar constraint was considered in [29] in the context of coding for watermarking and encryption.

DEFINITION 4.6 (Code) A $(|\mathcal{W}^{(n)}|, n)$ code for lossy source coding with end-user privacy consists of

- an encoder $f^{(n)} : \mathcal{X}^n \to \mathcal{W}^{(n)}$,
- a stochastic decoder[1] $G^{(n)}$ which maps $w \in \mathcal{W}^{(n)}$ and $y^n \in \mathcal{Y}^n$ to $\hat{x}^n \in \hat{\mathcal{X}}^n$ according to $p(\hat{x}^n | w, y^n)$,

where $\mathcal{W}^{(n)}$ is a finite set.

Let $d : \mathcal{F} \times \hat{\mathcal{X}} \to [0, \infty)$ be the single-letter distortion measure.[2] The distortion between the value of the function of the source sequence and side information and its estimate at the decoder is defined as

$$d^{(n)}(F^{(n)}(X^n, Y^n), \hat{X}^n) \triangleq \frac{1}{n} \sum_{i=1}^{n} d(F(X_i, Y_i), \hat{X}_i),$$

where $d^{(n)}(\cdot)$ is the distortion function.

---

[1] Since the goal of end-user privacy is to protect the reconstruction sequence generated at the decoder against any unwanted inferences, we allow the decoder to be a stochastic mapping, and it was shown by example, i.e., one in Section 4.5.3, that there exist cases where a stochastic decoder strictly enlarges the rate–distortion–equivocation region as compared to the one derived for deterministic decoders.

[2] Note that here $\hat{\mathcal{X}}$ does not denote an alphabet of the reconstruction of X, but of the outcome of the function $F(X, Y)$.

DEFINITION 4.7 (Achievability) A tuple $(R,D,\triangle) \in \mathbb{R}_+^3$ is said to be *achievable* if for any $\delta > 0$ and all sufficiently large n there exists a $(|\mathcal{W}^{(n)}|, n)$ code such that

$$\frac{1}{n}\log|\mathcal{W}^{(n)}| \leq R + \delta,$$

$$E[d^{(n)}(F^{(n)}(X^n, Y^n), \hat{X}^n)] \leq D + \delta,$$

and $\quad \frac{1}{n} H(\hat{X}^n | W, Z^n) \geq \triangle - \delta.$

The *rate–distortion–equivocation region* $\mathcal{R}_{\text{eve}}$ is the set of all achievable tuples.

DEFINITION 4.8 Let $\mathcal{R}_{\text{in}}^{(\text{eve})}$ be the set of all tuples $(R, D, \triangle) \in \mathbb{R}_+^3$ such that

$$R \geq I(X; U|Y), \tag{4.11}$$

$$D \geq E[d(F(X,Y), \hat{X})], \tag{4.12}$$

$$\triangle \leq H(\hat{X}|U, Y) + I(\hat{X}; Y|T) - I(\hat{X}; Z|T) - I(U; Z|T, Y, \hat{X}) \tag{4.13}$$

for some joint distributions of the form $P_{X,Y,Z}(x,y,z) P_{U|X}(u|x) P_{T|U}(t|u) P_{\hat{X}|U,Y}(\hat{x}|u,y)$ with $|\mathcal{T}| \leq |\mathcal{X}| + 5$, $|\mathcal{U}| \leq (|\mathcal{X}| + 5)(|\mathcal{X}| + 4)$.

In addition, let $\mathcal{R}_{\text{out}}^{(\text{eve})}$ be the same set as $\mathcal{R}_{\text{in}}^{(\text{eve})}$ except that the equivocation bound is replaced by

$$\triangle \leq H(\hat{X}|U, Y) + I(V, \hat{X}; Y|T) - I(V, \hat{X}; Z|T) \tag{4.14}$$

for some joint distributions $P_{X,Y,Z}(x,y,z) P_{U|X}(u|x) P_{T|U}(t|u) P_{V,\hat{X}|U,Y}(v,\hat{x}|u,y)$, where $H(T|V) = H(T|U) = 0$.

PROPOSITION 4.1 (Inner and outer bounds) *The rate–distortion–equivocation region* $\mathcal{R}_{\text{eve}}$ *for the problem in Fig. 4.8 satisfies* $\mathcal{R}_{\text{in}}^{(\text{eve})} \subseteq \mathcal{R}_{\text{eve}} \subseteq \mathcal{R}_{\text{out}}^{(\text{eve})}$.

The achievability scheme used to prove the inner bound is based on layered random binning with two codeword layers $\{T^n\}$ and $\{U^n\}$ together with the use of stochastic decoding mapping. For more details, please refer to [14].

In the equivocation bound of $\mathcal{R}_{\text{in}}^{(\text{eve})}$, the first term corresponds to uncertainty of $\hat{X}^n$ due to the use of a stochastic decoder. The difference $I(\hat{X}; Y|T) - I(\hat{X}; Z|T)$ can be considered as an additional uncertainty due to the fact that the eavesdropper observes Z^n but not Y^n, which is used for generating $\hat{X}^n$. The last mutual information term is related to the leakage of the second layer codeword U^n. However, the fact that its interpretation is not completely clear might be an indication that the bound is not tight.

REMARK 4.6 *We can relate this result to those of other settings where the function* $F^{(n)}(X^n, Y^n) = X^n$. *For example, the inner bound* $\mathcal{R}_{\text{in}}^{(\text{eve})}$ *can resemble the optimal result of the secure lossless source coding problem considered in [12]. To obtain the rate–equivocation region, we set* $\hat{X} = U = X$ *in* $\mathcal{R}_{\text{in}}^{(\text{eve})}$.

4.5.2 Causal Side Information

Next, we consider the variant of the problem depicted in Fig. 4.8 where side information Y^n is available only causally at the decoder. This could be relevant in delay-constrained applications as mentioned in [30] and references therein. We consider the following types of reconstructions.

- Causal reconstruction: $\hat{X}_i \sim p(\hat{x}_i|w, y^i, \hat{x}^{i-1})$ for $i = 1, \ldots, n$.
- Memoryless reconstruction: $\hat{X}_i \sim p(\hat{x}_i|w, y_i)$ for $i = 1, \ldots, n$.

DEFINITION 4.9   Let $\mathcal{R}_{\text{in}}^{(\text{eve,causal})}$ be the set of all tuples $(R, D, \Delta) \in \mathbb{R}_+^3$ such that

$$R \geq I(X; U), \tag{4.15}$$

$$D \geq E[d(F(X,Y), \hat{X})], \tag{4.16}$$

$$\Delta \leq H(\hat{X}|U, Z) \tag{4.17}$$

for some joint distributions of the form $P_{X,Y,Z}(x,y,z) P_{U|X}(u|x) P_{\hat{X}|U,Y}(\hat{x}|u,y)$ with $|\mathcal{U}| \leq |\mathcal{X}| + 3$.

In addition, let $\mathcal{R}_{\text{out}}^{(\text{eve,causal})}$ be the same set as $\mathcal{R}_{\text{in}}^{(\text{eve,causal})}$ except that the equivocation bound is replaced by

$$\Delta \leq H(\hat{X}|Z) \tag{4.18}$$

for some joint distributions $P_{X,Y,Z}(x,y,z) P_{U|X}(u|x) P_{\hat{X}|U,Y}(\hat{x}|u,y)$.

Causal Reconstruction

PROPOSITION 4.2 (Inner and outer bounds)   *The rate–distortion–equivocation region* $\mathcal{R}_{\text{eve}}$ *for the problem in Fig. 4.8 with* causal reconstruction *satisfies the relation* $\mathcal{R}_{\text{in}}^{(\text{eve,causal})} \subseteq \mathcal{R}_{\text{eve}} \subseteq \mathcal{R}_{\text{out}}^{(\text{eve,causal})}$.

Since the side information is only available causally at the decoder, it cannot be used for binning to reduce the rate. The achievable scheme follows that of source coding with causal side information [30], with the additional use of a stochastic decoder. The entropy term in the equivocation bound of $\mathcal{R}_{\text{in}}^{(\text{eve,causal})}$ corresponds to uncertainty of the reconstruction sequence given that the eavesdropper can decode the codeword U^n and has access to the side information Z^n.

Memoryless Reconstruction

PROPOSITION 4.3 (Rate–distortion–equivocation region)   *The rate–distortion–equivocation region* $\mathcal{R}_{\text{eve}}$ *for the problem in Fig. 4.8 with* memoryless reconstruction *is given by* $\mathcal{R}_{\text{in}}^{(\text{eve,causal})}$, *i.e.,* $\mathcal{R}_{\text{eve}} = \mathcal{R}_{\text{in}}^{(\text{eve,causal})}$.

When restricting the reconstruction symbol to depend only on the source description and the current side information symbol, the complete rate–distortion–equivocation region can be obtained. This restriction essentially allows us to overcome the difficulty in the converse proof of the general reconstruction case where, conditioned on the source description, the reconstruction process is not necessarily memoryless. For the complete proof, we refer readers to [14].

REMARK 4.7 *For the special case where $Y = \emptyset$, the rate–distortion–equivocation region is given by $\mathcal{R}_{\text{in}}^{(\text{eve,causal})}$ with the corresponding set of joint distributions such that $Y = \emptyset$. We can see that if the decoder is a deterministic mapping, the achievable equivocation rate is zero since the eavesdropper observes everything the decoder does. However, for some positive D, by using the stochastic decoder, we can achieve the equivocation rate of $H(\hat{X}|U,Z)$, which can be strictly positive. This shows that there exist cases where a stochastic decoder strictly enlarges the rate–distortion–equivocation region.*

REMARK 4.8 *Proposition 4.3 resembles the result of the special case in [31, Corollary 5] where there is no shared secret key.*

4.5.3 Example

In this section, we consider an example illustrating the potential gain from allowing the use of a stochastic decoder. Specifically, we consider the setting in Fig. 4.8 under memoryless reconstruction and assumptions that $Z = \emptyset$ and $F(X, Y) = X$. Then we evaluate the corresponding result in Proposition 4.3.

Let $\mathcal{X} = \hat{\mathcal{X}} = \{0, 1\}$ be binary source and reconstruction alphabets. We assume that the source symbol X is distributed according to Bernoulli(1/2), and side information $Y \in \{0, 1, e\}$ is an erased version of the source with an erasure probability p_e. The Hamming distortion measure is assumed, i.e., $d(x, \hat{x}) = 1$ if $x \neq \hat{x}$, and zero otherwise. Inspired by the optimal choice of U in the Wyner–Ziv result [6], we let U be the output of a BSC(p_u), $p_u \in [0, 1/2]$ with input X. The reconstruction symbol generated from a stochastic decoder is chosen such that $\hat{X} = Y$ if $Y \neq e$, otherwise $\hat{X} \sim P_{\hat{X}|U}$, where $P_{\hat{X}|U}$ is modeled as a BSC(p_2), $p_2 \in [0, 1/2]$. With these assumptions at hand, the inner bound to the rate–distortion–equivocation region in Proposition 4.3 can be expressed as

$$\mathcal{R}_{\text{in,random}} = \{(R, D, \Delta) | R \geq 1 - h(p_u)$$

$$D \geq p_e(p_u \star p_2)$$

$$\Delta \leq h(p_u(1 - p_e) + p_2 p_e)$$

$$\text{for some } p_u, p_2 \in [0, 1/2]\},$$

where $h(\cdot)$ is a binary entropy function and $a \star b \triangleq a(1 - b) + (1 - a)b$.

For comparison, we also evaluate the inner bound for the case of the Wyner–Ziv optimal deterministic decoder by setting $p_2 = 0$. We plot achievable minimum distortion as a function of equivocation rate for a fixed $R = 0.7136$, where $p_e = 0.5$. Figure 4.9 shows the tradeoff between achievable minimum distortion and equivocation rate for a fixed rate R. We can see that in general the minimum distortion is sacrificed for a higher equivocation. For the same particular structure of $P_{U|X}$ and the given deterministic decoder in this setting, it shows that, for a given rate R and distortion D, a higher equivocation rate Δ can be achieved by using a stochastic decoder.[3] As for

---

[3] Here we only evaluate and compare inner bounds on the rate–distortion–equivocation regions to illustrate a potential gain of allowing the use of a stochastic decoder.

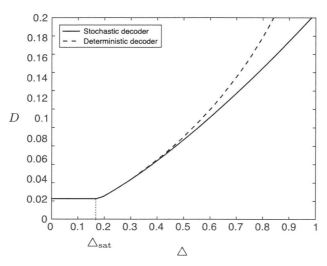

Figure 4.9 Achievable minimum distortion with respect to equivocation for a fixed rate $R = 0.7136$, and $p_e = 0.45$. A tradeoff curve with a different parameter set can be found in [14].

the low equivocation region, we observe a saturation of distortion because the minimum distortion is limited by the rate. The value Δ_{sat} at which the minimum distortion cannot be lowered by decreasing Δ can be specified as $\Delta_{\text{sat}} = h((1 - p_e)h^{-1}(1 - R))$, and the corresponding $D_{\min}(R, \Delta_{\text{sat}}) = p_e h^{-1}(1 - R)$ is the minimum distortion according to the Wyner–Ziv rate–distortion function.

Special case

In the special case where $Y = \emptyset$, the gain can be shown as follows (cf. Remark 4.7). If the decoder is a deterministic mapping, the achievable equivocation rate is always zero since the eavesdropper is as strong as the decoder. The corresponding distortion–rate function for this example is given by $D \geq h^{-1}(1 - R)$ [20, Chapter 3]. However, by using the stochastic decoder as above, we can achieve $D \geq h^{-1}(1 - R) \star h^{-1}(\Delta)$ (by letting $p_e = 1$ in $\mathcal{R}_{\text{in,random}}$). For $D = h^{-1}(1 - R) \star c$, where $c \in (0, 1/2]$, we can achieve a strictly positive equivocation rate $h(c)$.

4.6 Open Problems and Concluding Remark

In this chapter, we review works on networked secure source coding problems focusing on the fundamental tradeoff between coding performance and privacy guarantee. Apart from those included in our discussions, there exist several other works considering privacy in source coding networks with an eavesdropper. For instance, [13] considered privacy of the source in a setting with side information at the encoder, [32] considered settings where side information at the receiver and eavesdropper can be influenced by some cost-constrained action sequence, [33] considered source privacy in the central estimation officer (CEO) problem [34], and [35] considered source privacy in the successive refinement with degraded side information problem [36, 37]. In some cases, it is also important to impose an information privacy constraint at the legitimate user. For example, [38] considered multi-terminal source coding with a requirement of amplifying one source and masking another at the decoder.

Characterizing the fundamental tradeoff of networked secure source coding is an important task to understand the performance limit of secure systems. The results also provide useful insight into the optimal design as well as serving as a privacy assessment tool for designing network secure source coding systems. As seen in the previous discussions, several problems of networked secure source coding remain open in general. One reason is that many of them are in fact extensions of underlying open problems in network information theory. For example, lossy source coding with one helper under general distortion (a special case of multi-terminal source coding) is an open problem in network information theory. Networked secure source coding with a helper, which is essentially an extension to include a privacy constraint, therefore remains challenging in many cases. Problems of networked secure source coding with an intermediate helper also have cascade and triangular source coding problem as underlying problems, which largely remain open in general. Nevertheless, complete results can be obtained for several important special cases, e.g., cases under a degradedness assumption of the side information and cases under a specific distortion measure such as logarithmic loss distortion, or cases with Gaussian sources and quadratic distortion. Otherwise, general inner and outer bounds can be derived and they still provide useful insight for designing practical systems.

In some cases though, the challenge of characterizing completely the optimal tradeoff of networked secure source coding lies in the additional security/privacy constraint itself. For example, in the problem of networks with a two-sided helper under a privacy constraint, side information at the transmitter has a dual role. It can be used for encoding to reduce the rate, and also for security by generating a secret key. These two roles appear to be coupled and result in ambiguity of the optimal coding strategy. When the problem is specialized into cases where these two roles are decoupled, e.g., when side information at the encoder does not improve the rate–distortion tradeoff, the optimal tradeoff can be obtained. As for the setting with the end-user privacy constraint, the problem remains generally difficult since, conditioned on the source description, the reconstruction process is not necessarily memoryless. As seen in a special case of end-user privacy at the eavesdropper, when we restrict the reconstruction symbol to depend only on the source description and the current side information symbol, the complete rate–distortion–equivocation region can be given.

We see that these extra implications of the privacy constraint highlight interesting aspects of the coding scheme which is fundamentally different from those in the original problems in that it needs to fulfill both the (lossy) compression requirement and privacy criteria simultaneously. It raises challenges that may require novel approaches to solve these problems in a complete form.

References

[1] L. Sankar, S. R. Rajagopalan, and H. V. Poor, "Utility–privacy tradeoffs in databases: An information-theoretic approach," *IEEE Trans. Inf. Forensics Security*, vol. 8, no. 6, pp. 838–852, Jun. 2013.

[2] M. Gymrek, A. L. McGuire, D. Golan, E. Halperin, and Y. Erlich, "Identifying personal genomes by surname inference," *Science*, vol. 339, no. 6117, pp. 321–324, 2013.

[3] L. Kamm, D. Bogdanov, S. Laur, and J. Vilo, "A new way to protect privacy in large-scale genome-wide association studies," *Bioinformatics*, vol. 29, no. 7, pp. 886–893, Apr. 2013.

[4] A. D. Wyner, "The wire-tap channel," *Bell Syst. Tech. J.*, vol. 54, pp. 1355–1387, Oct. 1975.

[5] R. Ahlswede and I. Csiszár, "Common randomness in information theory and cryptography—Part I: Secret sharing," *IEEE Trans. Inf. Theory*, vol. 39, no. 4, pp. 1121–1132, Jul. 1993.

[6] A. D. Wyner and J. Ziv, "The rate–distortion function for source coding with side information at the decoder," *IEEE Trans. Inf. Theory*, vol. 22, no. 1, pp. 1–10, Jan. 1976.

[7] K. Kittichokechai, Y.-K. Chia, T. J. Oechtering, M. Skoglund, and T. Weissman, "Secure source coding with a public helper," in *Proc. IEEE Int. Symp. Inf. Theory*, Istanbul, Turkey, Jul. 2013, pp. 2209–2213.

[8] K. Kittichokechai, Y.-K. Chia, T. J. Oechtering, M. Skoglund, and T. Weissman, "Secure source coding with a public helper," *IEEE Trans. Inf. Theory*, vol. 62, no. 7, pp. 1–20, Jul. 2016.

[9] V. Prabhakaran and K. Ramchandran, "On secure distributed source coding," in *Proc. IEEE Inf. Theory Workshop*, Tahoe City, CA, USA, Sep. 2007, pp. 442–447.

[10] D. Gündüz, E. Erkip, and H. V. Poor, "Lossless compression with security constraints," in *Proc. IEEE Int. Symp. Inf. Theory*, Toronto, ON, Canada, Jul. 2008, pp. 111–115.

[11] R. Tandon, S. Ulukus, and K. Ramchandran, "Secure source coding with a helper," *IEEE Trans. Inf. Theory*, vol. 59, no. 4, pp. 2178–2187, Apr. 2013.

[12] J. Villard and P. Piantanida, "Secure multiterminal source coding with side information at the eavesdropper," *IEEE Trans. Inf. Theory*, vol. 59, no. 6, pp. 3668–3692, Jun. 2013.

[13] Y.-K. Chia and K. Kittichokechai, "On secure source coding with side information at the encoder," in *Proc. IEEE Int. Symp. Inf. Theory*, Istanbul, Turkey, Jul. 2013, pp. 2204–2208.

[14] K. Kittichokechai, T. J. Oechtering, and M. Skoglund, "Lossy source coding with reconstruction privacy," in *Proc. IEEE Int. Symp. Inf. Theory*, Honolulu, HI, USA, Jul. 2014, pp. 386–390.

[15] T. A. Courtade and T. Weissman, "Multiterminal source coding under logarithmic loss," *IEEE Trans. Inf. Theory*, vol. 60, no. 1, pp. 740–761, Jan. 2014.

[16] T. Berger, K. Housewright, J. Omura, S. Yung, and J. Wolfowitz, "An upper bound on the rate distortion function for source coding with partial side information at the decoder," *IEEE Trans. Inf. Theory*, vol. 25, no. 6, pp. 664–666, Nov. 1979.

[17] S. Jana and R. Blahut, "Partial side information problem: Equivalence of two inner bounds," in *Proc. 42nd Annual Conf. Inf. Sciences Systems*, Princeton, NJ, USA, Mar. 2008, pp. 1005–1009.

[18] H. Permuter, Y. Steinberg, and T. Weissman, "Two-way source coding with a helper," *IEEE Trans. Inf. Theory*, vol. 56, no. 6, pp. 2905–2919, Jun. 2010.

[19] A. D. Wyner, "The rate–distortion function for source coding with side information at the decoder—part II: General sources," *Inf. Control*, no. 38, pp. 60–80, Jan. 1978.

[20] A. El Gamal and Y.-H. Kim, *Network Information Theory*. Cambridge: Cambridge University Press, 2011.

[21] T. M. Cover and J. A. Thomas, *Elements of Information Theory*, 2nd edn. Chichester: Wiley & Sons, 2006.

[22] D. Slepian and J. Wolf, "Noiseless coding of correlated information sources," *IEEE Trans. Inf. Theory*, vol. 19, no. 4, pp. 471–480, Jul. 1973.

[23] R. Ahlswede and J. Körner, "Source coding with side information and a converse for degraded broadcast channels," *IEEE Trans. Inf. Theory*, vol. 21, no. 6, pp. 629–637, Nov. 1975.

[24] A. D. Wyner, "On source coding with side information at the decoder," *IEEE Trans. Inf. Theory*, vol. 21, no. 3, pp. 294–300, May 1975.

[25] T. Berger, *The Information Theory Approach to Communications*. Berlin, Heidelberg: Springer-Verlag, 1977, pp. 170–231.

[26] Y.-K. Chia, H. H. Permuter, and T. Weissman, "Cascade, triangular, and two-way source coding with degraded side information at the second user," *IEEE Trans. Inf. Theory*, vol. 58, no. 1, pp. 189–206, Jan. 2012.

[27] H. Yamamoto, "Wyner–Ziv theory for a general function of the correlated sources," *IEEE Trans. Inf. Theory*, vol. 28, no. 5, pp. 803–807, Sep. 1982.

[28] R. Tandon, L. Sankar, and H. V. Poor, "Discriminatory lossy source coding: Side information privacy," *IEEE Trans. Inf. Theory*, vol. 59, no. 9, pp. 5665–5677, Sep. 2013.

[29] N. Merhav, "On joint coding for watermarking and encription," *IEEE Trans. Inf. Theory*, vol. 52, no. 1, pp. 190–205, Jan. 2006.

[30] T. Weissman and A. El Gamal, "Source coding with limited-look-ahead side information at the decoder," *IEEE Trans. Inf. Theory*, vol. 52, no. 12, pp. 5218–5239, Dec. 2006.

[31] C. Schieler and P. Cuff, "Rate–distortion theory for secrecy systems," *IEEE Trans. Inf. Theory*, vol. 60, no. 12, pp. 7584–7605, Dec. 2014.

[32] K. Kittichokechai, T. J. Oechtering, M. Skoglund, and Y.-K. Chia, "Secure source coding with action-dependent side information," *IEEE Trans. Inf. Theory*, vol. 61, no. 12, pp. 6444–6464, Dec. 2015.

[33] F. Naghibi, S. Salimi, and M. Skoglund, "The CEO problem with secrecy constraints," *IEEE Trans. Inf. Forensics Security*, vol. 10, no. 6, pp. 1234–1249, Jun. 2015.

[34] T. Berger, Z. Zhang, and H. Viswanathan, "The CEO problem [multiterminal source coding]," *IEEE Trans. Inf. Theory*, vol. 42, no. 3, pp. 887–902, May 1996.

[35] D. Xu, K. Kittichokechai, T. J. Oechtering, and M. Skoglund, "Secure successive refinement with degraded side information," in *Proc. IEEE Int. Symp. Inf. Theory*, Honolulu, HI, USA, Jul. 2014, pp. 2674–2678.

[36] Y. Steinberg and N. Merhav, "On successive refinement for the Wyner–Ziv problem," *IEEE Trans. Inf. Theory*, vol. 50, no. 8, pp. 1636–1654, Aug. 2004.

[37] C. Tian and S. N. Diggavi, "On multistage successive refinement for Wyner–Ziv source coding with degraded side informations," *IEEE Trans. Inf. Theory*, vol. 53, no. 8, pp. 2946–2960, Aug. 2007.

[38] T. A. Courtade, "Information masking and amplification: The source coding setting," in *Proc. IEEE Int. Symp. Inf. Theory*, Cambridge, MA, USA, Jul. 2012, pp. 189–193.

Part II

Secure Communication

5 Secrecy Rate Maximization in Gaussian MIMO Wiretap Channels

Sergey Loyka and Charalambos D. Charalambous

Secrecy rate maximization in Gaussian MIMO wiretap channels is considered. While the optimality of Gaussian signaling and a general expression for the secrecy capacity have been well established, closed-form solutions for the optimal transmit covariance matrix are known for some special cases only, while the general case remains an open problem. This chapter reviews known closed-form solutions and presents a numerical algorithm for the general case with guaranteed convergence to the global optimum. The known solutions include full-rank and rank-1 cases (which, when combined, provide a complete solution for the case of two transmit antennas), the case of identical right singular vectors for the eavesdropper and legitimate channels, and the cases of weak, isotropic, and omnidirectional eavesdroppers, which also provide lower and upper bounds to the general case. Necessary optimality conditions and a tight upper bound for the rank of the optimal covariance matrix in the general case are discussed. Sufficient and necessary conditions for the optimality of three popular signaling strategies over MIMO channels, namely, isotropic and zero-forcing signaling as well as water-filling over the legitimate channel eigenmodes, are presented. The chapter closes with a detailed description of a numerical globally convergent algorithm to solve the general case, and gives some illustrative examples.

5.1 Introduction

Due to their high spectral efficiency, wireless MIMO (multiple input, multiple output) systems are widely adopted by academia and industry. The broadcast nature of wireless channels stimulated significant interest in their security aspects and the Gaussian MIMO wiretap channel (WTC) has emerged as a popular model to study information theoretic secrecy aspects of wireless systems [1]. A number of results have been obtained for this model, including the proof of optimality of Gaussian signaling [1–4], which is far from trivial and significantly more involved than that of the regular (no wiretap) MIMO channel. Once the functional form of the optimal input is established, the only unknown is its covariance matrix since the mean is always zero. This latter part has not been solved yet in the general case; only a number of special cases have been settled.

In this chapter, we review the well-known as well as recent results on an optimal transmit covariance matrix for the MIMO WTC. Several new results will be reported as well.

The optimal transmit covariance matrix under the total power constraint has been obtained for some special cases (low/high signal to noise ratio (SNR), multiple input, single output (MISO) channels, full-rank or rank-1 solutions) [2–12], but the general case is still open. The main difficulty lies in the fact that, unlike the regular MIMO channel, the underlying optimization problem for the MIMO WTC is not convex in general, in addition to the fact that the respective Karush–Kuhn–Tucker (KKT) optimality conditions are a system of non-linear matrix equalities and inequalities. It was conjectured in [4] and proved in [3] using an indirect approach (via the degraded channel) that the optimal signaling is on the positive directions of the difference channel. A direct proof (based on the necessary KKT conditions) has been obtained in [10], while the optimality of signaling on non-negative directions has been established in [7] via an indirect approach. Closed-form solutions for MISO and rank-1 MIMO channels have been obtained [2, 7, 10]. The low-SNR regime has been studied in detail in [9]. An exact full-rank solution for the optimal covariance matrix has been obtained in [10] and its properties have been characterized. In particular, unlike the regular channel (no eavesdropper), the optimal power allocation does not converge to a uniform one at high SNR and the latter remains sub-optimal at any finite SNR.

Finally, while no analytical solution is known in the general case, a globally convergent numerical algorithm was proposed in [13] to find an optimal covariance for any Gaussian MIMO wiretap channel (degraded or not), and its convergence to a global optimum, which takes only a moderate or small number of steps in practice, was proved.

The rest of this chapter is organized as follows. Section 5.2 introduces the MIMO WTC model. Rank-1 and full-rank solutions are discussed in Sections 5.3 and 5.4. The weak eavesdropper case is considered in Section 5.5, which is motivated by a scenario where the Tx–Rx distance is much smaller than the Tx–Ev one. Section 5.6 discusses an isotropic eavesdropper model, whereby the Tx does not know the directional properties of the Ev and hence assumes it is isotropic. Section 5.7 studies an omnidirectional eavesdropper, which may have a smaller number of antennas (and hence rank-deficient channel) and which has the same gain in any direction of a given sub-space. The case of identical right singular vectors of the Rx and Ev channels is investigated in Section 5.8. In Sections 5.9–5.11, we consider three popular signaling techniques: zero-forcing (ZF), standard water-filling (WF) over the eigenmodes of the legitimate channel, and isotropic signaling (whereby the covariance matrix is a scaled identity), and discuss sufficient and necessary conditions under which they are optimal for the MIMO WTC. These techniques are appealing for a number of reasons, including their lower complexity and existing solutions. Finally, Section 5.12 presents an algorithm for numerical evaluation of the Tx covariance matrix with guaranteed convergence to a global optimum in the general case.

Notation

$\lambda_i(\mathbf{W})$ denotes eigenvalues of a matrix $\mathbf{W}$; $(x)_+ = \max\{x, 0\}$ for a real scalar x; $\mathcal{N}(\mathbf{W})$ and $\mathcal{R}(\mathbf{W})$ are the null space and the range of a matrix $\mathbf{W}$; $(\mathbf{W})_+$ denotes positive eigenmodes of a Hermitian matrix $\mathbf{W}$:

$$(\mathbf{W})_+ = \sum_{i:\lambda_i(\mathbf{W})>0} \lambda_i \mathbf{u}_i \mathbf{u}_i^+, \qquad (5.1)$$

where λ_i is the ith largest eigenvalue of $\mathbf{W}$ and $\mathbf{u}_i$ is its corresponding eigenvector. $\mathbf{A} > \mathbf{B}$ means that $\mathbf{A} - \mathbf{B}$ is positive definite; $|\mathbf{A}|$ is the determinant of $\mathbf{A}$, while $\mathbf{A}'$ and $\mathbf{A}^+$ are its transposition and Hermitian conjugation.

5.2 MIMO Wiretap Channel

Let us consider the standard wiretap Gaussian MIMO channel model,

$$\mathbf{y}_1 = \mathbf{H}_1 \mathbf{x} + \boldsymbol{\xi}_1, \quad \mathbf{y}_2 = \mathbf{H}_2 \mathbf{x} + \boldsymbol{\xi}_2, \qquad (5.2)$$

where $\mathbf{x} = [x_1, x_2, \ldots x_m]' \in C^{m,1}$ is the transmitted complex-valued signal vector of dimension $m \times 1$, $\mathbf{y}_{1(2)} \in C^{n_{1(2)},1}$ are the received vectors at the receiver (eavesdropper), $\boldsymbol{\xi}_{1(2)}$ is the circularly symmetric additive white Gaussian noise at the receiver (eavesdropper; normalized to unit variance in each dimension), $\mathbf{H}_{1(2)} \in C^{n_{1(2)},m}$ is the $n_{1(2)} \times m$ matrix of the complex channel gains between each Tx and each receive (eavesdropper) antenna, and $n_{1(2)}$ and m are the numbers of Rx (eavesdropper) and Tx antennas respectively; see Fig. 5.1. The channels $\mathbf{H}_{1(2)}$ are assumed to be quasi-static (i.e., constant for a sufficiently long period of time so that the infinite horizon information theory assumption holds) and frequency-flat, with full channel state information (CSI) at the Rx and Tx ends. With slight modifications, this model can also include the case of spatially correlated noise.

The main performance indicator for a wiretap channel is its secrecy capacity, defined as follows. A secrecy rate is achievable if it satisfies the *secrecy* criterion (the information leakage to the eavesdropper approaches zero as the block length increases) in addition to the traditional *reliability* criterion (the error probability of the legitimate receiver approaches zero); see [1] for more details. The secrecy capacity is the supremum of all achievable secrecy rates, subject to the total power constraint.

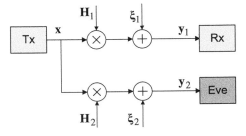

Figure 5.1 A block diagram of the Gaussian MIMO wiretap channel. Full channel state information is available at the transmitter. $\mathbf{H}_{1(2)}$ is the channel matrix to the legitimate receiver (eavesdropper); $\mathbf{x}$ is the transmitted signal and $\mathbf{y}_{1(2)}$ is the received (eavesdropper) signal; $\boldsymbol{\xi}_{1(2)}$ is the additive white Gaussian noise at the receiver (eavesdropper). The information leakage to the eavesdropper is required to approach zero asymptotically.

Gaussian signaling is known to be optimal for the Gaussian MIMO WTC (the proof of this is significantly more complicated than for the regular (no eavesdropper) Gaussian MIMO channel) [2–4], so that the only unknown part is its covariance (since the mean is always zero). For a given transmit covariance matrix $\mathbf{R} = E\{\mathbf{xx}^+\}$, where $E\{\cdot\}$ is statistical expectation, the maximum achievable secrecy rate between the Tx and Rx (so that the rate between the Tx and eavesdropper is zero) is [3, 4]

$$C(\mathbf{R}) = \ln \frac{|\mathbf{I} + \mathbf{W}_1 \mathbf{R}|}{|\mathbf{I} + \mathbf{W}_2 \mathbf{R}|} = C_1(\mathbf{R}) - C_2(\mathbf{R}), \quad (5.3)$$

where negative $C(\mathbf{R})$ is interpreted as zero rate, $\mathbf{W}_i = \mathbf{H}_i^+ \mathbf{H}_i$. The secrecy capacity subject to the total Tx power constraint is

$$C_s = \max_{\mathbf{R} \geq 0} C(\mathbf{R}) \text{ s.t. tr} \mathbf{R} \leq P_T, \quad (5.4)$$

where P_T is the total transmit power (also the SNR since the noise is normalized). It is well known that the problem in (5.4) is not convex in general and an explicit solution for the optimal Tx covariance is not known for the general case, but only for some special cases (e.g., low/high SNR, MISO channels, full-rank or rank-1 cases [2–13]). In fact, the problem in (5.4) is still open even for the degraded (and hence convex) but otherwise general case of $\mathbf{W}_1 \geq \mathbf{W}_2$.

The optimization problem in (5.4) is the main subject of this chapter. The following theorem gives the necessary optimality conditions in the general case, which are instrumental for further development and allow one to established closed-form solutions for the optimal covariance in many cases.

THEOREM 5.1 *Let $\mathbf{R}^*$ be an optimal covariance in (5.4),*

$$\mathbf{R}^* = \arg\max_{\mathbf{R} \geq 0} C(\mathbf{R}) \text{ s.t. tr} \mathbf{R} \leq P_T,$$

and let $\mathbf{u}_{i+}$ be its active eigenvector (i.e., corresponding to a positive eigenvalue). Then,

$$\mathbf{U}_{r+}^+ (\mathbf{W}_1 - \mathbf{W}_2) \mathbf{U}_{r+} > \mathbf{0}, \quad (5.5)$$

where the columns of semi-unitary matrix $\mathbf{U}_{r+}$ are the active eigenvectors $\{\mathbf{u}_{i+}\}$, so that $\mathbf{x}^+(\mathbf{W}_1 - \mathbf{W}_2)\mathbf{x} > 0 \ \forall \mathbf{x} \in \text{span}\{\mathbf{u}_{i+}\}$, i.e., a necessary condition for an optimal signaling strategy in (5.4) is to transmit into the positive directions of $\mathbf{W}_1 - \mathbf{W}_2$ (where the legitimate channel is stronger than the eavesdropper).

Proof Based on the necessary KKT optimality conditions ; see [10] for details.

It was demonstrated in [4] that rank($\mathbf{R}^*$) < m unless $\mathbf{W}_1 > \mathbf{W}_2$, i.e., an optimal transmission is of low rank over a non-degraded channel. The corollary below gives a more precise characterization based on the necessary optimality condition above.

COROLLARY 5.1 *Let $\mathbf{W}_1 - \mathbf{W}_2 = \mathbf{W}_+ + \mathbf{W}_-$, where $\mathbf{W}_{+(-)}$ collects positive (negative and zero) eigenmodes of $\mathbf{W}_1 - \mathbf{W}_2$ (found from its eigenvalue decomposition). Then,*

$$\text{rank}(\mathbf{R}^*) \leq \text{rank}(\mathbf{W}_+) \leq m, \quad (5.6)$$

i.e., the rank of an optimal covariance $\mathbf{R}^*$ *does not exceed the number of strictly positive eigenvalues of* $\mathbf{W}_1 - \mathbf{W}_2$ *(the rank of* $\mathbf{W}_+$*).*

5.3 Rank-1 Solution

Using Corollary 5.1, one immediately obtains an optimal covariance when $\text{rank}(\mathbf{W}_+) = 1$, e.g., for MISO or MIMO rank-1 channels.

COROLLARY 5.2 *Let* $\text{rank}(\mathbf{W}_+) = 1$. *The secrecy capacity and optimal covariance are*

$$C_s = \ln \lambda_1, \ \mathbf{R}^* = P_T \mathbf{u}_1 \mathbf{u}_1^+, \tag{5.7}$$

where λ_1, $\mathbf{u}_1$ *are the largest eigenvalue and corresponding eigenvector of* $(\mathbf{I} + P_T \mathbf{W}_2)^{-1}(\mathbf{I} + P_T \mathbf{W}_1)$ *or, equivalently, the largest generalized eigenvalue and corresponding eigenvector of* $(\mathbf{I} + P_T \mathbf{W}_1, \mathbf{I} + P_T \mathbf{W}_2)$, *so that transmit beamforming on* $\mathbf{u}_1$ *is the optimal strategy.*

Proof Corollary 5.1 ensures that $\text{rank}(\mathbf{R}^*) = 1$; the optimal covariance $\mathbf{R}^*$ in (5.7) follows in the same way as in [2].

Note that MISO channels (single-antenna channel at the receiver or eavesdropper) considered in [2, 6, 8] are special cases of this corollary with, e.g., $\mathbf{W}_1 = \mathbf{h}_1 \mathbf{h}_1^+$. The corollary allows not only MIMO channels with $\text{rank}(\mathbf{W}_1) = 1$ but also any higher-rank $\mathbf{W}_1$ and $\mathbf{W}_2$ provided that $\text{rank}(\mathbf{W}_+) = 1$.

Furthermore, the signaling in (5.7) is also optimal for any $\text{rank}(\mathbf{W}_+) \geq 1$ at sufficiently small SNR, where λ_1, $\mathbf{u}_1$ become the largest eigenvalue and corresponding eigenvector of the difference channel $\mathbf{W}_1 - \mathbf{W}_2$. The appeal of this signaling is due to its low complexity.

It should be emphasized that the solution in (5.7) is not zero-forcing (i.e., $\mathbf{W}_2 \mathbf{u}_1 \neq 0$) in general, i.e., the Tx does not form null in the Ev direction. Intuitively, doing so results in loss of power at the Rx and hence is not optimal in general. Such a solution may be optimal in some special cases; see Section 5.9.

5.4 Full-Rank Solution

The full-rank solution of the optimization problem in (5.4) is given by the following theorem.

THEOREM 5.2 *Let* $\mathbf{W}_1 > \mathbf{W}_2$ *and* $P_T > P_{T0}$, *where* P_{T0} *is a threshold power given by* (5.12). *Then,* $\mathbf{R}^*$ *is of full rank and is given by*

$$\mathbf{R}^* = \mathbf{U} \Lambda_1 \mathbf{U}^+ - \mathbf{W}_1^{-1}, \tag{5.8}$$

where the columns of the unitary matrix $\mathbf{U}$ are the eigenvectors of $\mathbf{Z} = \mathbf{W}_2 + \mathbf{W}_2(\mathbf{W}_1 - \mathbf{W}_2)^{-1}\mathbf{W}_2$, $\Lambda_1 = \mathrm{diag}\{\lambda_{1i}\} > \mathbf{0}$ is a diagonal positive-definite matrix,

$$\lambda_{1i} = \frac{2}{\lambda}\left(\sqrt{1 + \frac{4\mu_i}{\lambda}} + 1\right)^{-1}, \qquad (5.9)$$

and $\mu_i \geq 0$ are the eigenvalues of $\mathbf{Z}$; $\lambda > 0$ is found from the total power constraint $\mathrm{tr}\,\mathbf{R}^* = P_T$ as a unique solution of the equation

$$\frac{2}{\lambda}\sum_i \left(\sqrt{1 + \frac{4\mu_i}{\lambda}} + 1\right)^{-1} = P_T + \mathrm{tr}\,\mathbf{W}_1^{-1}. \qquad (5.10)$$

The corresponding secrecy capacity is

$$C_s = \ln \frac{|\mathbf{W}_1||\Lambda_1|}{|\mathbf{I} - \mathbf{W}_2(\mathbf{W}_1^{-1} - \mathbf{U}\Lambda_1\mathbf{U}^+)|} = \ln \frac{|\mathbf{W}_1|}{|\mathbf{W}_2|} + \ln \frac{|\Lambda_1|}{|\Lambda_2|}, \qquad (5.11)$$

where $\Lambda_2 = \Lambda_1 + \mathrm{diag}\{\mu_i^{-1}\}$ and the second equality holds when $\mathbf{W}_2$ is positive definite, $\mathbf{W}_2 > \mathbf{0}$. P_{T0} can be expressed as follows:

$$P_{T0} = \frac{2(\mu_1 + \lambda_{\min})}{\lambda_{\min}^2}\sum_i \left(\sqrt{1 + \frac{4\mu_i(\mu_1 + \lambda_{\min})}{\lambda_{\min}^2}} + 1\right)^{-1} - \mathrm{tr}\,\mathbf{W}_1^{-1}, \qquad (5.12)$$

where $\lambda_{\min}$ is the minimum eigenvalue of $\mathbf{W}_1$ and μ_1 is the maximum eigenvalue of $\mathbf{Z}$.

Proof Based on the KKT conditions, which are sufficient for optimality in this case (since the channel is degraded and hence the problem is convex); see [10] for details.

It should be pointed out that Theorem 5.2 gives an exact (not approximate) optimal covariance at finite SNR ($P_T \to \infty$ is not required) since P_{T0} is a finite constant that depends only on $\mathbf{W}_1$ and $\mathbf{W}_2$ and this constant is small in some cases: it follows from (5.12) that $P_{T0} \to 0$ if $\lambda_{\min} \to \infty$, i.e., P_{T0} is small if $\lambda_{\min}$ is large. In particular, P_{T0} can be bounded above as

$$P_{T0} \leq \frac{m\mu_1}{\lambda_{\min}^2} + \frac{m-1}{\lambda_{\min}}, \qquad (5.13)$$

and if $\lambda_{\min} \gg \mu_1$, then

$$P_{T0} \approx \frac{m}{\lambda_{\min}} - \mathrm{tr}\,\mathbf{W}_1^{-1} \leq \frac{m-1}{\lambda_{\min}} \leq 1, \qquad (5.14)$$

where the last inequality holds if $\lambda_{\min} \geq m - 1$. Figure 5.2 illustrates this case. On the other hand, when $\mathbf{W}_1 - \mathbf{W}_2$ approaches a singular matrix, it follows that $P_{T0} \to \infty$, so that P_{T0} is large iff $\mathbf{W}_1 - \mathbf{W}_2$ is close to singular.

Theorem 5.2, in combination with the rank-1 solution, provides the complete solution for the optimal covariance in the $m = 2$ case: if the channel is not strictly degraded or if the SNR is not above the threshold, the rank-1 solution in (5.7) applies; otherwise, Theorem 5.2 applies. Figure 5.2 illustrates this for the following channel:

$$\mathbf{W}_1 = \begin{bmatrix} 1.5 & 0.5 \\ 0.5 & 1.5 \end{bmatrix}, \quad \mathbf{W}_2 = \begin{bmatrix} 0.35 & 0.15 \\ 0.15 & 0.35 \end{bmatrix}. \tag{5.15}$$

Note that the transition to full-rank covariance takes place at low SNR of about $-6\,\mathrm{dB}$, i.e., P_{T0} is not high at all in this case.

We further observe that the first term in (5.11), $C_\infty = \ln\frac{|\mathbf{W}_1|}{|\mathbf{W}_2|}$, is independent of SNR, and the second one, $\Delta C = \ln\frac{|\Lambda_1|}{|\Lambda_2|} < 0$, monotonically increases with the SNR. Furthermore, $C_s \to C_\infty$, $\Delta C \to 0$ as $P_T \to \infty$, in agreement with Theorem 2 in [3]. This is also clear from Fig. 5.2.

Note also that the second term in (5.8) de-emphasizes weak eigenmodes of $\mathbf{W}_1$. Since λ is monotonically decreasing as P_T increases [this follows from (5.10)], λ_{1i} monotonically increases with P_T, and approaches $\lambda_{1i} \approx 1/\sqrt{\mu_i\lambda}$, $i = 1,\ldots,m$, at sufficiently high SNR, which is in contrast with the conventional WF solution, where the uniform power allocation is optimal at high SNR. Furthermore, it follows from (5.9) that λ_{1i} decreases with μ_i, i.e., stronger eigenmodes of $\mathbf{W}_2^{-1} - \mathbf{W}_1^{-1} = \mathbf{Z}^{-1}$ (which correspond to larger eigenmodes of $\mathbf{W}_1$ and weaker ones of $\mathbf{W}_2$) get allocated more power, which follows the same tendency as the conventional WF. It further follows from (5.8) that when $\mathbf{W}_1$ and $\mathbf{W}_2$ have the same eigenvectors, $\mathbf{R}^*$ also has the same eigenvectors, i.e., the optimal signaling is on the eigenvectors of $\mathbf{W}_{1(2)}$.

The case of singular $\mathbf{W}_1$ can also be included by observing that, under certain conditions, $\mathbf{R}^*$ puts no power on the null space of $\mathbf{W}_1$ so that all matrices can be projected, without loss of generality, on the positive eigenspace of $\mathbf{W}_1$ and Theorem 5.2 will apply to the projected channel.

It is instructive to consider the case when the required channel is much stronger than the eavesdropper one, $\mathbf{W}_1 \gg \mathbf{W}_2$, meaning that all eigenvalues of $\mathbf{W}_1$ are much larger than those of $\mathbf{W}_2$.

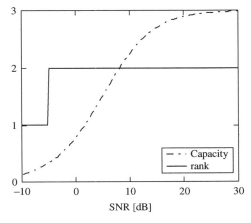

Figure 5.2 Secrecy capacity and the rank of $\mathbf{R}^*$ vs. SNR [dB] for the channel in (5.15). The transition to full-rank covariance takes place at about $-6\,\mathrm{dB}$.

COROLLARY 5.3 *Consider the MIMO WTC in (5.2) under the conditions of Theorem 5.2 and when the eavesdropper channel is much weaker than the required one,*

$$\lambda_1(\mathbf{W}_2) \ll m(P_T + \mathrm{tr}\,\mathbf{W}_1^{-1})^{-1}/4 \leq m/(4P_T), \tag{5.16}$$

where $\lambda_i(\mathbf{W}_2)$ is ith largest eigenvalue of $\mathbf{W}_2$, e.g., when $\mathbf{W}_2 \to 0$ with fixed $\mathbf{W}_1$. Then the optimal covariance in (5.8) becomes

$$\mathbf{R}^* \approx (\lambda^{-1}\mathbf{I} - \mathbf{W}_1^{-1}) - \lambda^{-2}\mathbf{W}_2. \tag{5.17}$$

Proof See [10].

The approximation in (5.17) should be understood in Frobenius or any other norm (since all norms are equivalent). An interpretation of (5.17) is immediate: the first term

$$\mathbf{R}_{\mathrm{WF}} = \lambda^{-1}\mathbf{I} - \mathbf{W}_1^{-1} \tag{5.18}$$

is the standard water-filling on the eigenmodes of $\mathbf{W}_1$ (which is the capacity-achieving strategy for the regular MIMO channel) and the second term is a correction due to the secrecy requirement: those modes that spill over into the eavesdropper channel get less power to accommodate the secrecy constraint.

Let us now consider the high-SNR regime.

COROLLARY 5.4 *When $\mathbf{W}_2 > 0$, the optimal covariance $\mathbf{R}^*$ in (5.8) in the high-SNR regime*

$$P_T \gg \mu_m^{-1/2} \sum_i \mu_i^{-1/2} \tag{5.19}$$

(e.g., when $P_T \to \infty$), where $\mu_m = \min_i \mu_i$, simplifies to

$$\mathbf{R}^* \approx \mathbf{U}\mathrm{diag}\{d_i\}\mathbf{U}^+, \quad d_i = \frac{P_T \mu_i^{-1/2}}{\sum_i \mu_i^{-1/2}}. \tag{5.20}$$

The corresponding secrecy capacity is

$$C_s \approx \ln\frac{|\mathbf{W}_1|}{|\mathbf{W}_2|} - \frac{1}{P_T}\left(\sum_i \frac{1}{\sqrt{\mu_i}}\right)^2, \tag{5.21}$$

where we have neglected the second- and higher-order effects in $1/P_T$.

Proof Follows from Theorem 5.2 along the same lines as that of Corollary 5.3.

Note that the optimal signaling is on the eigenmodes of $\mathbf{W}_2^{-1} - \mathbf{W}_1^{-1}$ with the optimal power allocation given by $\{d_i\}$. This somewhat resembles the conventional water-filling, but also has a remarkable difference: unlike the conventional WF, the secure WF in (5.20) does not converge to the uniform allocation in the high-SNR regime.[1] However, strong eigenmodes of $\mathbf{W}_2^{-1} - \mathbf{W}_1^{-1}$ (which correspond to weak modes of $\mathbf{W}_2$ and strong ones of $\mathbf{W}_1$) do get more power, albeit in a form different from that of the conventional WF.

---

[1] The sub-optimality of the isotropic signaling suggested in Theorem 2 of [3] is hidden in the $o(1)$ term. The second term of Eq. (5.21) above refines that $o(1)$ term.

5.5 Weak Eavesdropper

Motivated by a scenario where the legitimate receiver is closer to the transmitter than the eavesdropper so that its path loss is large, see, e.g., Fig. 5.4, the case of a weak eavesdropper is considered in this section. There are no additional assumptions here (e.g., for the channel to be degraded, etc.). The weak Ev case provides a lower bound to the secrecy capacity in the general case, which is tight when the eavesdropper path loss is large and hence serves as an approximation to the true capacity. It also captures the capacity saturation effect at high SNR observed in [3, 10].

THEOREM 5.3 *Consider the problem in* (5.4) *when the eavesdropper is weak,* $\lambda_i(\mathbf{W}_2\mathbf{R}) \ll 1$, *e.g., when* $\lambda_1(\mathbf{W}_2) \ll 1/P_T$. *The optimal covariance is given by*

$$\mathbf{R}^* \approx \mathbf{W}_\lambda^{-1/2}(\lambda^{-1}\mathbf{I} - \widehat{\mathbf{W}}_1^{-1})_+ \mathbf{W}_\lambda^{-1/2}, \tag{5.22}$$

where $\mathbf{W}_\lambda = \mathbf{I} + \lambda^{-1}\mathbf{W}_2$, $\widehat{\mathbf{W}}_1 = \mathbf{W}_\lambda^{-1/2}\mathbf{W}_1\mathbf{W}_\lambda^{-1/2}$, $\lambda \geq 0$ *is found from the total power constraint,*[2]

$$\operatorname{tr}\mathbf{R}^* = P_T \ \text{if}\ P_T < P_T^*, \tag{5.23}$$

and $\lambda = 0$ *otherwise; the threshold power*

$$P_T^* = \operatorname{tr}\mathbf{W}_2^{-1}(\mathbf{I} - \mathbf{W}_2^{1/2}\mathbf{W}_1^{-1}\mathbf{W}_2^{1/2})_+ \tag{5.24}$$

if $\mathbf{W}_2$ *is non-singular;* $P_T^* = \infty$ *if* $\mathbf{W}_2$ *is singular and* $\mathcal{N}(\mathbf{W}_2) \not\subseteq \mathcal{N}(\mathbf{W}_1)$. *The corresponding secrecy capacity is*

$$C_s \approx \sum_{i:\widehat{\lambda}_{1i}>\lambda} \ln(\widehat{\lambda}_{1i}/\lambda) - \operatorname{tr}\widehat{\mathbf{W}}_2(\mathbf{I} - \widehat{\mathbf{W}}_1^{-1})_+ \tag{5.25}$$

where $\widehat{\lambda}_{1i} = \lambda_i(\widehat{\mathbf{W}}_1)$, $\widehat{\mathbf{W}}_2 = \mathbf{W}_\lambda^{-1/2}\mathbf{W}_2\mathbf{W}_\lambda^{-1/2}$.

Proof The proof is based on the weak eavesdropper approximation

$$C(\mathbf{R}) \approx \ln|\mathbf{I} + \mathbf{W}_1\mathbf{R}| - \operatorname{tr}(\mathbf{W}_2\mathbf{R}), \tag{5.26}$$

which holds if $\lambda_i(\mathbf{W}_2\mathbf{R}) \ll 1$, and on the respective KKT optimality conditions. See [12] for details.

REMARK 5.1 *It may appear that* (5.22) *requires* $\widehat{\mathbf{W}}_1$ *and thus* $\mathbf{W}_1$ *to be positive definite, i.e., the singular case is not allowed. This is not so: the* $(\cdot)_+$ *operator makes sure that zero eigenmodes of* $\widehat{\mathbf{W}}_1$ *are eliminated so that singular* $\mathbf{W}_1$ *is allowed. The same observation also applies to* (5.24) *and* (5.25).

REMARK 5.2 *One way to ensure that the Ev is weak, i.e.,* $\lambda_i(\mathbf{W}_2\mathbf{R}) \ll 1$, *is to require*

$$\lambda_1(\mathbf{W}_2) \ll 1/P_T \tag{5.27}$$

---

[2] Here we implicitly assume that $\mathbf{W}_\lambda$ is non-singular, i.e., either $\mathbf{W}_2$ is non-singular or $\lambda > 0$ if it is singular. If this is not the case, a pseudo-inverse should be used instead.

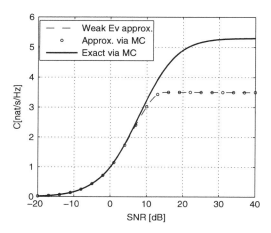

Figure 5.3 Weak eavesdropper approximation in (5.25) and exact secrecy capacity (via MC) versus SNR. $\mathbf{W}_{1,2}$ are as in (5.28), $\alpha = 0.1$. The approximation is accurate if SNR < 10 dB. Note the capacity saturation effect at high SNR in both cases.

[since $\lambda_i(\mathbf{W}_2\mathbf{R}) \leq \lambda_i(\mathbf{W}_2)\lambda_1(\mathbf{R}) \leq P_T\lambda_1(\mathbf{W}_2)$], from which it follows that this holds as long as the power (or SNR) is not too large, i.e., $P_T \ll 1/\lambda_1(\mathbf{W}_2)$; see also Fig. 5.3. It should be noted, however, that this approximation extends well beyond the low-SNR regime provided that the eavesdropper path loss is sufficiently large (i.e., $\lambda_1(\mathbf{W}_2)$ is small). For the scenario in Fig. 5.3, it works well up to about 10 dB and can extend to larger SNR for smaller α.

REMARK 5.3 *When the optimal covariance in (5.22) is full rank, it takes on the same form as in (5.17), thus revealing similarity with the standard water-filling over the channel eigenmodes in (5.18).*

To illustrate Theorem 5.3, and also to see how accurate the approximation is, Fig. 5.3 shows the secrecy capacity obtained from the theorem for

$$\mathbf{W}_1 = \begin{pmatrix} 2 & 0 \\ 0 & 1 \end{pmatrix}, \quad \mathbf{W}_2 = \alpha \begin{pmatrix} 2 & 1 \\ 1 & 1 \end{pmatrix}. \tag{5.28}$$

In addition, its exact values (without the weak eavesdropper approximation) obtained by brute force Monte Carlo (MC) based approach (where a large number of covariance matrices are randomly generated, subject to the total power constraint, and the best one is selected) are shown for comparison. To validate the analytical solution in Theorem 5.3, the approximate problem has also been solved by the MC-based approach. It is clear that the approximation is accurate in this case provided that SNR < 10 dB. Also note the capacity saturation effect, for both the approximate and exact values. This saturation effect has already been observed in [3, 10], and, in the case of $\mathbf{W}_1 > \mathbf{W}_2 > \mathbf{0}$, the saturation capacity is

$$C_s^* = \ln|\mathbf{W}_1| - \ln|\mathbf{W}_2|, \tag{5.29}$$

which follows directly from (5.3) by neglecting $\mathbf{I}$. In the weak eavesdropper approximation, the saturation effect is due to the fact that the second term in (5.26) is linear in P_T while the first is only logarithmic. So using the full available power is not optimal when that power is sufficiently high. Roughly speaking, the approximation is accurate before it reaches the saturation point, i.e., for $P_T < P_T^*$. The respective saturation capacity is obtained from (5.25) by setting $\lambda = 0$. In the case of $\mathbf{W}_1 > \mathbf{W}_2 > \mathbf{0}$, it is given by

$$C^* = \ln|\mathbf{W}_1| - \ln|\mathbf{W}_2| - \mathrm{tr}(\mathbf{I} - \mathbf{W}_2\mathbf{W}_1^{-1}). \tag{5.30}$$

By comparing (5.29) and (5.30), one concludes that the thresholds are close to each other when $\mathrm{tr}\,\mathbf{W}_2\mathbf{W}_1^{-1} \approx m$.

REMARK 5.4 *In the general case, the approximated capacity and corresponding optimal covariance in Theorem 5.3 provide a lower bound to the true secrecy capacity in (5.4) at any SNR/power and for any eavesdropper channel (weak or not):*

$$C_s \geq C(\mathbf{R}^*), \tag{5.31}$$

where $C(\mathbf{R})$ is as in (5.26), which follows from $\ln(1+x) \leq x\ \forall x \geq 0$. The sub-optimality gap can be bounded above as follows:

$$0 \leq C_s - C(\mathbf{R}^*) \leq \frac{P_T^2}{2}\lambda_1^2(\mathbf{W}_2), \tag{5.32}$$

so that the bound is tight for a weak eavesdropper or/and low SNR.

To obtain further insight in the weak eavesdropper regime, let us consider the case when $\mathbf{W}_{1,2}$ have the same eigenvectors. This is a broader case than it may first appear as it requires $\mathbf{H}_{1,2}$ to have the same right singular vectors while leaving the left ones unconstrained (see Section 5.8 for more details on this scenario). In this case, the results of Theorem 5.3 simplify as follows.

COROLLARY 5.5 *Let $\mathbf{W}_1$ and $\mathbf{W}_2$ have the same eigenvectors. Then, under the conditions of Theorem 5.3, the optimal covariance is*

$$\mathbf{R}^* = \mathbf{U}\boldsymbol{\Lambda}^*\mathbf{U}^+, \tag{5.33}$$

where $\mathbf{U}$ is found from the eigenvalue decompositions $\mathbf{W}_i = \mathbf{U}\boldsymbol{\Lambda}_i\mathbf{U}^+$ so that the eigenvectors of $\mathbf{R}^*$ are the same as those of $\mathbf{W}_{1,2}$. The diagonal matrix $\boldsymbol{\Lambda}^*$ collects the eigenvalues of $\mathbf{R}^*$:

$$\lambda_i(\mathbf{R}^*) = \left(\frac{1}{\lambda + \lambda_{2i}} - \frac{1}{\lambda_{1i}}\right)_+, \tag{5.34}$$

where λ_{ki} is ith eigenvalue of $\mathbf{W}_k$.

Note that the power allocation in (5.34) resembles the standard water-filling solution, except for the λ_{2i} term. In particular, only sufficiently strong eigenmodes are active:

$$\lambda_i(\mathbf{R}^*) > 0 \text{ iff } \lambda_{1i} > \lambda + \lambda_{2i}. \tag{5.35}$$

As P_T increases, λ decreases so that more eigenmodes become active; legitimate channel eigenmodes are active provided that they are stronger than those of the eavesdropper: $\lambda_{1i} > \lambda_{2i}$. Only the strongest eigenmode (for which the difference $\lambda_{1i} - \lambda_{2i}$ is largest) is active at low SNR.

5.6 Isotropic Eavesdropper and Capacity Bounds

The model above requires full eavesdropper CSI at the transmitter. This becomes questionable if the eavesdropper does not cooperate (e.g., when it is hidden in order not to compromise its eavesdropping ability). One approach to address this issue is via a compound channel model [14–16]. In this section, an alternative approach is considered in which the eavesdropper is characterized by a channel again assumed to be identical in all directions. We term this an "isotropic eavesdropper." This minimizes the amount of CSI to be available at the transmitter (one scalar parameter and no directional properties). Based on this, lower and upper (tight) capacity bounds are given for the general case, which are achievable by an isotropic eavesdropper.

A further physical justification for this model comes from an assumption that the eavesdropper cannot approach the transmitter too closely due to, e.g., some minimum protection distance, see Fig. 5.4. This ensures that the gain of the eavesdropper channel does not exceed a certain threshold in any transmit direction due to the minimum propagation path loss (induced by the minimum distance constraint). Since the channel power gain in the transmit direction $\mathbf{x}$ is $\mathbf{x}^+\mathbf{W}_2\mathbf{x} = |\mathbf{H}_2\mathbf{x}|^2$ (assuming $|\mathbf{x}| = 1$), and since $\max_{|\mathbf{x}|=1} \mathbf{x}^+\mathbf{W}_2\mathbf{x} = \epsilon_1$ (from the variational characterization of eigenvalues [17]), where ϵ_1 is the largest eigenvalue of $\mathbf{W}_2$, $\mathbf{W}_2 \leq \epsilon_1 \mathbf{I}$ ensures that the eavesdropper channel power gain does not exceed ϵ_1 in any direction.

In combination with the matrix monotonicity of the log-det function, the latter inequality ensures that $\epsilon_1 \mathbf{I}$ is the worst possible $\mathbf{W}_2$ that attains the capacity lower bound in (5.39), i.e., the isotropic eavesdropper with the maximum channel gain is the worst possible one among all eavesdroppers with a bounded spectral norm. Referring to Fig. 5.4, the eavesdropper channel matrix $\mathbf{H}_2$ can be presented in the form

$$\mathbf{H}_2 = \sqrt{\frac{\alpha}{R_2^\nu}} \widetilde{\mathbf{H}}_2, \tag{5.36}$$

where α/R_2^ν represents the average propagation path loss, R_2 is the eavesdropper–transmitter distance, ν is the path loss exponent (which depends on the propagation environment), α is a constant independent of distance (but dependent on frequency, antenna height, etc.) [18], and $\widetilde{\mathbf{H}}_2$ is a properly normalized channel matrix (includes local scattering/multipath effects but excludes the average path loss)

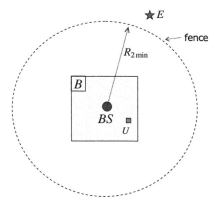

Figure 5.4 Physical scenario for a secret communication system: base station BS (the transmitter) is located on the rooftop of a secure building B, legitimate user U (the receiver) is inside the building B, and eavesdropper E is beyond the fence so that $R_2 \geq R_{2\min}$.

so that $\operatorname{tr} \widetilde{\mathbf{H}}_2^+ \widetilde{\mathbf{H}}_2 \leq n_2 m$ [19]. With this in mind, one obtains

$$\mathbf{W}_2 = \mathbf{H}_2^+ \mathbf{H}_2 = \frac{\alpha}{R_2^\nu} \widetilde{\mathbf{H}}_2^+ \widetilde{\mathbf{H}}_2 \leq \frac{\alpha}{R_{2\min}^\nu} \widetilde{\mathbf{H}}_2^+ \widetilde{\mathbf{H}}_2 \leq \frac{\alpha n_2 m}{R_{2\min}^\nu} \mathbf{I}, \qquad (5.37)$$

so that one can choose $\epsilon_1 = \alpha n_2 m R_{2\min}^{-\nu}$ in this scenario, where $R_{2\min}$ is the minimum transmitter–eavesdropper distance. Note that the model captures the impact of the number of transmit and eavesdropper antennas, in addition to the minimum distance and propagation environment. In our view, the isotropic eavesdropper model is more practically relevant than the full Tx CSI model.

The isotropic eavesdropper model is closely related to the parallel channel setting in [20,21]: even though the original channel is not parallel, it can be transformed (via an information-preserving transformation) into a parallel channel, for which independent signaling is known to be optimal [20,21]. This shows that signaling on the eigenvectors of $\mathbf{W}_1$ is optimal in this case, while an optimal power allocation is different from the standard water-filling [21]. These properties in combination with the bounds in (5.38) are exploited below.

While it is a challenging analytical task to evaluate the secrecy capacity in the general case, one can use the isotropic eavesdropper model given above to construct lower and upper capacity bounds for the general case using the standard matrix inequalities

$$\epsilon_m \mathbf{I} \leq \mathbf{W}_2 \leq \epsilon_1 \mathbf{I}, \qquad (5.38)$$

where $\epsilon_i = \lambda_i(\mathbf{W}_2)$ denotes the ith largest eigenvalue of $\mathbf{W}_2$, and the equalities are achieved when $\epsilon_1 = \epsilon_m$, i.e., by the isotropic eavesdropper. This is formalized below.

PROPOSITION 5.1 *The MIMO WTC secrecy capacity in (5.4) is bounded as follows:*

$$C^*(\epsilon_1) \leq C_s \leq C^*(\epsilon_m), \qquad (5.39)$$

where $C^*(\epsilon)$ is the secrecy capacity C_s when $\mathbf{W}_2 = \epsilon \mathbf{I}$, i.e., for the isotropic eavesdropper,

$$C^*(\epsilon) = \max_{\mathbf{R} \geq 0,\, \mathrm{tr}\mathbf{R} \leq P_T} \ln \frac{|\mathbf{I} + \mathbf{W}_1 \mathbf{R}|}{|\mathbf{I} + \epsilon \mathbf{R}|} = \sum_i \ln \frac{1 + g_i \lambda_i^*}{1 + \epsilon \lambda_i^*}, \quad (5.40)$$

$g_i = \lambda_i(\mathbf{W}_1)$, and $\lambda_i^* = \lambda_i(\mathbf{R}^*)$ are the eigenvalues of the optimal transmit covariance $\mathbf{R}^* = \mathbf{U}_1 \Lambda^* \mathbf{U}_1^\dagger$,

$$\lambda_i^* = \frac{\epsilon + g_i}{2\epsilon g_i} \left(\sqrt{1 + \frac{4\epsilon g_i}{(\epsilon + g_i)^2}\left(\frac{g_i - \epsilon}{\lambda} - 1\right)} - 1 \right)_+, \quad (5.41)$$

and $\lambda > 0$ is found from the total power constraint $\sum_i \lambda_i^* = P_T$.

The bounds gap in (5.39) can be bounded above as

$$\Delta C = C^*(\epsilon_m) - C^*(\epsilon_1) \leq m_+ \ln \frac{1 + \epsilon_1 P_T / m_+}{1 + \epsilon_m P_T / m_+} \leq m_+ \ln \frac{\epsilon_1}{\epsilon_m}, \quad (5.42)$$

where m_+ is the number of eigenmodes such that $g_i > \epsilon_m$. Both bounds are tight (achieved with equality) at high SNR if $g_{m+} > \epsilon_1$.

Proof Use the matrix monotonicity of the log-det function and a unitary transformation to put this model into the parallel channel setting; see [11] for details.

Thus, the optimal signaling is on the eigenvectors of $\mathbf{W}_1$ (or right singular vectors of $\mathbf{H}_1$), identically to the regular MIMO channel, with the optimal power allocation somewhat similar (but not identical) to conventional water-filling. The latter is further elaborated for the high and low SNR regimes below. Unlike the general case (of non-isotropic eavesdropper), the secrecy capacity of the isotropic eavesdropper case does not depend on the eigenvectors of $\mathbf{W}_1$ (but the optimal signaling does) but only on its eigenvalues, so that the optimal signaling problem here separates into two independent parts: (1) optimal signaling directions are selected as the eigenvectors of $\mathbf{W}_1$, and (2) optimal power allocation is done based on the eigenvalues of $\mathbf{W}_1$ and the eavesdropper channel gain ϵ. It is the lack of this separation that makes the optimal signaling problem so difficult in the general case.

The bounds in (5.39) coincide when $\epsilon_1 = \epsilon_m$, thus giving the secrecy capacity of the isotropic eavesdropper. Furthermore, as follows from (5.42), they are close to each other when the condition number ϵ_1/ϵ_m of $\mathbf{W}_2$ is not too large, thus providing a reasonable estimate of the secrecy capacity, see Fig. 5.5. Referring to Fig. 5.4, one can also set $\epsilon_1 = \alpha n_2 m R_{2\min}^{-\nu}$ and proceed with a conservative system design to achieve secrecy rate $C^*(\epsilon_1)$. Note that this design requires only the knowledge of n_2 and $R_{2\min}$ at the transmitter instead of the full CSI ($\mathbf{W}_2$), and hence is more realistic. This signaling strategy does not incur significant penalty (compared to the full CSI case) provided that the condition number ϵ_1/ϵ_m is not too large, as follows from (5.42). It can be further shown that $C^*(\epsilon_1)$ is the compound secrecy capacity for the class of eavesdroppers with bounded spectral norm (maximum channel gain), $\mathbf{W}_2 \leq \epsilon_1 \mathbf{I}$, and that signaling on the worst-case channel ($\mathbf{W}_2 = \epsilon_1 \mathbf{I}$) achieves the capacity [16].

We note that the power allocation in (5.41) has properties similar to those of the conventional water-filling, which are as follows.

PROPOSITION 5.2 *Properties of the optimum power allocation:*

1. λ_i^* *is an increasing function of* g_i *(strictly increasing unless* $\lambda_i^* = 0$ *or* $P_T = 0$*), i.e., stronger eigenmodes get allocated more power (as in the standard WF).*
2. λ_i^* *is an increasing function of* P_T *(strictly increasing unless* $\lambda_i^* = 0$*).* $\lambda_i^* = 0$ *for* $i > 1$ *and* $\lambda_1^* = P_T$ *as* $P_T \to 0$ *if* $g_1 > g_2$*, i.e., only the strongest eigenmode is active at low SNR, and* $\lambda_i^* > 0$ *if* $g_i > \epsilon$ *as* $P_T \to \infty$*, i.e., all sufficiently strong eigenmodes are active at high SNR.*
3. $\lambda_i^* > 0$ *only if* $g_i > \epsilon$*, i.e., only the eigenmodes stronger than the eavesdropper ones can be active.*
4. λ *is a strictly decreasing function of* P_T *and* $0 < \lambda < g_1 - \epsilon$*;* $\lambda \to 0$ *as* $P_T \to \infty$ *and* $\lambda \to g_1 - \epsilon$ *as* $P_T \to 0$.
5. *There are* m_+ *active eigenmodes if the following inequalities hold:*

$$P_{m_+} < P_T \le P_{m_++1}, \quad (5.43)$$

where P_{m_+} is a threshold power (to have at least m_+ active eigenmodes):

$$P_{m_+} = \sum_{i=1}^{m_+ - 1} \frac{\epsilon + g_i}{2\epsilon g_i} \left(\sqrt{1 + \frac{4\epsilon g_i}{(\epsilon + g_i)^2} \frac{g_i - g_{m_+}}{(g_{m_+} - \epsilon)_+}} - 1 \right), m_+ = 2, \ldots, m, \quad (5.44)$$

and $P_1 = 0$, so that m_+ is an increasing function of P_T.

It follows from Proposition 5.2 that there is only one active eigenmode, i.e., beamforming is optimal, if $g_2 > \epsilon$ and

$$P_T \le \frac{\epsilon + g_1}{2\epsilon g_1} \left(\sqrt{1 + \frac{4\epsilon g_1}{(\epsilon + g_1)^2} \frac{g_1 - g_2}{g_2 - \epsilon}} - 1 \right). \quad (5.45)$$

For example, this holds in the low SNR regime (note, however, that the single-mode regime extends well beyond low SNR if $\epsilon \to g_2$ and $g_1 > g_2$), or at any SNR if $g_1 > \epsilon$ and $g_2 \le \epsilon$.

While it is difficult to evaluate λ analytically from the power constraint, Property 4 ensures that any suitable numerical algorithm (e.g., the Newton–Raphson method) will do so efficiently.

As a side benefit of Proposition 5.2, one can use (5.43) as a condition for having m_+ active eigenmodes under the regular eigenmode transmission (no eavesdropper) with standard water-filling by taking $\epsilon \to 0$ in (5.44),

$$P_{m_+} = \sum_{i=1}^{m_+ - 1} \left(\frac{1}{g_{m_+}} - \frac{1}{g_i} \right), \quad (5.46)$$

and (5.46) approximates (5.44) when the eavesdropper is weak, $\epsilon \ll g_{m_+}$. To the best of our knowledge, the expression (5.46) for the threshold powers of the standard water-filling has not previously appeared in the literature.

5.6.1 High SNR Regime

Let us now consider the isotropic eavesdropper model when the SNR grows large, so that $g_i \lambda_i^* \gg 1$, $\epsilon \lambda_i^* \gg 1$. In this case, (5.40) simplifies to

$$C_\infty^* = \sum_{i_+} \ln \frac{g_i}{\epsilon}, \tag{5.47}$$

where the summation is over all active eigenmodes, $i_+ = \{i : g_i > \epsilon\}$, so that the secrecy capacity is independent of the SNR (saturation effect) and the impact of the eavesdropper is the multiplicative SNR loss, which is never negligible. To obtain a threshold value of P_T at which the saturation takes place, observe that $\lambda \to 0$ as $P_T \to \infty$, so that (5.41) becomes

$$\lambda_i^* = \frac{P_T \sqrt{\epsilon^{-1} - g_i^{-1}}}{\sum_{i_+} \sqrt{\epsilon^{-1} - g_i^{-1}}} (1 + o(1)), \tag{5.48}$$

where

$$\sqrt{\lambda} = \frac{1}{P_T} \sum_{i_+} \sqrt{\epsilon^{-1} - g_i^{-1}} (1 + o(1)) \tag{5.49}$$

from the total power constraint. Using (5.48), the secrecy capacity becomes

$$C^*(\epsilon) = \sum_{i_+} \ln \frac{g_i}{\epsilon} - \frac{1}{P_T} \left(\sum_{i_+} \sqrt{\frac{1}{\epsilon} - \frac{1}{g_i}} \right)^2 + o\left(\frac{1}{P_T}\right), \tag{5.50}$$

which is a refinement of (5.47). The saturation takes place when the second term is much smaller than the first one, so that

$$P_T \gg \frac{\sum_{i_+} \sqrt{\epsilon^{-1} - g_i^{-1}}}{\sum_{i_+} \ln \frac{g_i}{\epsilon}} \tag{5.51}$$

and $C^*(\epsilon) \approx C_\infty^*$ under this condition. This effect is illustrated in Fig. 5.5.

Note that, from (5.48), the optimal power allocation behaves almost like water-filling in this case, due to the $\sqrt{\epsilon^{-1} - g_i^{-1}}$ term.

Using (5.47), the gap ΔC_∞^* between the lower and upper bounds in (5.39) becomes

$$\Delta C_\infty^* = C_\infty^*(\epsilon_m) - C_\infty^*(\epsilon_1) = m_1 \ln \frac{\epsilon_1}{\epsilon_m} + \sum_{i=m_1+1}^{m_2} \ln \frac{g_i}{\epsilon_m}, \tag{5.52}$$

where $m_{1(2)}$ is the number of active eigenmodes when $\epsilon = \epsilon_{1(m)}$. Note that this gap is SNR independent, and if $m_1 = m_2 = m_+$, which is the case if $g_{m+} > \epsilon_1$, then

$$\Delta C_\infty^* = m_+ \ln \frac{\epsilon_1}{\epsilon_m}, \tag{5.53}$$

i.e., also independent of the eigenmode gains of the legitimate user, and is determined solely by the condition number of the eavesdropper channel and the number of active eigenmodes. Note that, in this case, the upper bounds in (5.42) are tight.

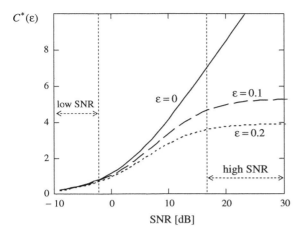

Figure 5.5 Secrecy capacity for the isotropic eavesdropper and the capacity of the regular MIMO channel (no eavesdropper, $\epsilon = 0$) vs. the SNR ($= P_T$ since the noise variance is unity); $g_1 = 2$, $g_2 = 1$. Note the saturation effect at high SNR, where the capacity strongly depends on ϵ but not on the SNR, and the negligible impact of the eavesdropper at low SNR.

5.6.2 When Is the Eavesdropper Negligible?

It is clear from (5.40) that under fixed $\{g_i\}$, P_T, the secrecy capacity converges to the conventional one $C^*(0)$ as $\epsilon \to 0$. However, no fixed ϵ (does not matter how small) can ensure by itself that the eavesdropper is negligible since one can always select sufficiently high P_T to make the saturation effect important (see Fig. 5.5). To answer the question in the section's title, we use (5.40) to obtain:

$$C^*(\epsilon) = \max_{\{\lambda_i\}} \sum_i \ln\left(1 + \frac{1 + (g_i - \epsilon)\lambda_i}{1 + \epsilon\lambda_i}\right) \text{ s.t. } \lambda_i \geq 0, \sum_i \lambda_i = P_T$$

$$\stackrel{(a)}{\approx} \max_{\{\lambda_i\}} \sum_i \ln(1 + (g_i - \epsilon)\lambda_i) \tag{5.54}$$

$$\stackrel{(b)}{\approx} \max_{\{\lambda_i\}} \sum_i \ln(1 + g_i\lambda_i) = C^*(0),$$

where (a) holds if

$$P_T \ll 1/\epsilon \tag{5.55}$$

(since $\lambda_i \leq P_T$), i.e., if the SNR is not too large, and (b) holds if

$$\epsilon \ll g_{i+}, \tag{5.56}$$

where i_+ is the set of active eigenmodes, i.e., if the eavesdropper eigenmodes are much weaker than the active eigenmodes of the legitimate channel. It is the combination of (5.55) and (5.56) that ensures that the eavesdropper is negligible. Neither condition alone is able to do so. Figure 5.5 illustrates this point. Equation (5.54) also indicates that the impact of the eavesdropper is the per-eigenmode gain loss of ϵ. Unlike the high

SNR regime in (5.47), where the loss is multiplicative (i.e., very significant and never negligible), here it is additive (mild or negligible in many cases).

5.6.3 Low SNR Regime

Let us now consider the low SNR regime, which is characteristic for CDMA-type systems [22]. Traditionally, this regime is defined via $P_T \to 0$. We, however, use a more relaxed definition requiring that $m_+ = 1$, which holds under (5.45). In this regime, assuming $g_1 > \epsilon$,

$$C^*(\epsilon) = \ln \frac{1 + g_1 P_T}{1 + \epsilon P_T} = \ln\left(1 + \frac{(g_1 - \epsilon)P_T}{1 + \epsilon P_T}\right) \overset{(a)}{\approx} \ln(1 + (g_1 - \epsilon)P_T), \quad (5.57)$$

where (a) holds when $P_T \ll 1/\epsilon$. It is clear from the last expression that the impact of the eavesdropper is an additive SNR loss of ϵP_T, which is negligible when $\epsilon \ll g_1$. Note a significant difference to the high SNR regime in (5.47), where this impact is never negligible. Figure 5.5 illustrates this difference.

Note further from (5.57) that the difference between the lower and upper bounds in (5.39) is the SNR gap of $(\epsilon_1 - \epsilon_m)P_T$, which is negligible if $g_1 \gg \epsilon_1 - \epsilon_m$. This may be the case even if the condition number ϵ_1/ϵ_m is large. Therefore, we conclude that the impact of the eavesdropper is more pronounced in the high SNR regime and is negligible in the low SNR one if its channel is weaker than the strongest eigenmode of the legitimate user.

When $g_1 - \epsilon \ll P_T$, (a) in (5.57) gives $C^*(\epsilon) \approx (g_1 - \epsilon)P_T$, which is linear in P_T. A similar capacity scaling at low SNR has been obtained in [23] for an i.i.d. block-fading single-input single-output (SISO) WTC, without, however, explicitly identifying the capacity but via establishing upper/lower bounds. Also note that the first two equalities in (5.57) do not require $P_T \to 0$ but only to satisfy (5.45).

5.7 Omnidirectional Eavesdropper

In this section, we consider a scenario where the eavesdropper has equal gain in all directions of a certain sub-space. This model accounts for two points: (1) when the transmitter has no particular knowledge about the directional properties of the eavesdropper, which is most likely from the practical perspective, it is reasonable to assume that its gain is the same in all directions; (2) on the other hand, when the eavesdropper has a small number of antennas (less than the number of transmit antennas), its channel rank, which does not exceed the number of transmit or receive antennas, is limited by this number so that the isotropic model of the previous section does not apply.[3]

---

[3] This was pointed out by A. Khisti.

For an omnidirectional eavesdropper, the channel gain is the same in all directions of its active sub-space, i.e.,

$$|\mathbf{H}_2\mathbf{x}|^2 = \mathbf{x}^+\mathbf{W}_2\mathbf{x} = \text{const.} \ \forall \mathbf{x} \in \mathcal{N}(\mathbf{W}_2)^\perp, \qquad (5.58)$$

where $\mathcal{N}(\mathbf{W}_2)^\perp$ is the sub-space orthogonal to the nullspace $\mathcal{N}(\mathbf{W}_2)$ of $\mathbf{W}_2$, i.e., its active sub-space, whose dimensionality is $r_2 = \text{rank}(\mathbf{W}_2)$. In particular, when the eavesdropper is isotropic, $\mathcal{N}(\mathbf{W}_2)$ is an empty set so that $\mathcal{N}(\mathbf{W}_2)^\perp$ is the entire space and $r_2 = m$. The condition in (5.58) implies that

$$\mathbf{W}_2 = \varepsilon \mathbf{U}_{2+}\mathbf{U}_{2+}^+, \qquad (5.59)$$

where $\mathbf{U}_{2+}$ is a semi-unitary matrix that collects active eigenvectors of $\mathbf{W}_2$, and $\mathcal{N}(\mathbf{W}_2)^\perp = \text{span}\{\mathbf{U}_{2+}\}$. Note that the model in (5.59) allows $\mathbf{W}_2$ to be rank-deficient: $r_2 < m$ is allowed. ε can be evaluated from, e.g., (5.37): $\varepsilon = \alpha n_2 m R_{2\min}^{-\nu}$.

THEOREM 5.4 *Consider the omnidirectional eavesdropper in (5.58), (5.59) and let $\mathcal{R}(\mathbf{W}_1) \subseteq \mathcal{R}(\mathbf{W}_2)$. Then the MIMO WTC secrecy capacity can be expressed as*

$$C_s = \max_{\text{tr}\mathbf{R} \leq P_T} \ln \frac{|\mathbf{I} + \mathbf{W}_1\mathbf{R}|}{|\mathbf{I} + \mathbf{W}_2\mathbf{R}|} = \max_{\text{tr}\mathbf{R} \leq P_T} \ln \frac{|\mathbf{I} + \mathbf{W}_1\mathbf{R}|}{|\mathbf{I} + \epsilon\mathbf{R}|} = C^*(\epsilon). \qquad (5.60)$$

Proof See Appendix.

Note that the secrecy capacity as well as the optimal signaling for an omnidirectional eavesdropper in Theorem 5.4 is the same as for the isotropic one in Proposition 5.1, i.e., the fact that the rank of the eavesdropper channel is low has no impact provided that $\mathcal{R}(\mathbf{W}_1) \subseteq \mathcal{R}(\mathbf{W}_2)$ holds (which is not the case in general, as can be shown via examples).

Since $\mathcal{R}(\mathbf{W})$ collects directions where the channel gain is not zero,

$$|\mathbf{H}\mathbf{x}|^2 = \mathbf{x}^+\mathbf{W}\mathbf{x} \neq 0 \ \forall \mathbf{x} \in \mathcal{R}(\mathbf{W}); \qquad (5.61)$$

the condition $\mathcal{R}(\mathbf{W}_1) \in \mathcal{R}(\mathbf{W}_2)$ means that $|\mathbf{H}_2\mathbf{x}| = 0$ implies $|\mathbf{H}_1\mathbf{x}| = 0$ (but the converse is not true in general) and hence $|\mathbf{H}_1\mathbf{x}| \neq 0$ implies $|\mathbf{H}_2\mathbf{x}| \neq 0$, i.e., the eavesdropper can "see" in any direction where the receiver can "see" (but there is no requirement here for the eavesdropper to be degraded with respect to the receiver, so that the channel is not necessarily degraded).

Further note that the condition in (5.58) does not require $\mathbf{U}_2 = \mathbf{U}_1$, i.e., the eigenvectors of the legitimate channel and of the eavesdropper can be different.

5.8 Identical Right Singular Vectors

In this section, we consider the case when $\mathbf{H}_{1,2}$ have the same right singular vectors (SV), so that their singular value decomposition takes the form

$$\mathbf{H}_k = \mathbf{U}_k \mathbf{\Sigma}_k \mathbf{V}^+, \qquad (5.62)$$

where the unitary matrices $\mathbf{U}_k, \mathbf{V}$ collect left and right singular vectors, respectively, and diagonal matrix $\mathbf{\Sigma}_k$ collects singular values of $\mathbf{H}_k$. In this model, the left singular vectors can be arbitrary. This is motivated by the fact that right singular vectors are determined by scattering around the Tx, while left ones are determined by scattering around the Rx and Ev, respectively. Therefore, when the Rx and Ev are spatially separated, their scattering environments may differ significantly (and hence the different left SVs) while the same scattering environment around the Tx induces the same right SVs. This is similar to the popular Kronecker MIMO channel correlation model [24], where the overall channel correlation is a product of the independent Tx and Rx parts, which are induced by respective sets of scatterers. In this section, we make no weak eavesdropper or any other assumptions.

After unitary (and thus information-preserving) transformations, this scenario can be put into the parallel channel setting of [20, 21]. In this case, the secrecy capacity and optimal covariance can be explicitly characterized.

PROPOSITION 5.3 *Consider the wiretap MIMO channel in (5.2), (5.62). The optimal Tx covariance for this channel takes the form*

$$\mathbf{R}^* = \mathbf{V}\mathbf{\Lambda}^*\mathbf{V}^+, \quad (5.63)$$

where the diagonal matrix $\mathbf{\Lambda}^*$ *collects its eigenvalues* λ_i^*:

$$\lambda_i^* = \frac{\lambda_{2i} + \lambda_{1i}}{2\lambda_{2i}\lambda_{1i}} \left(\sqrt{1 + \frac{4\lambda_{2i}\lambda_{1i}}{(\lambda_{2i} + \lambda_{1i})^2} \left(\frac{\lambda_{1i} - \lambda_{2i}}{\lambda} - 1 \right)_+ } - 1 \right) \quad (5.64)$$

where $\lambda_{ki} = \sigma_{ki}^2$ *and* σ_{ki} *denotes a singular value of* $\mathbf{H}_k$; $\lambda > 0$ *is found from the total power constraint* $\sum_i \lambda_i^* = P_T$.

Proof After a unitary transformation, the problem can be put into the parallel channel setting; see [12] for details.

In fact, Eq. (5.63) says that optimal signaling is on the right SVs of $\mathbf{H}_{1,2}$, and (5.64) implies that only those eigenmodes are active for which

$$\sigma_{1i}^2 > \sigma_{2i}^2 + \lambda. \quad (5.65)$$

If $\lambda_{2i} = 0$, then (5.64) reduces to

$$\lambda_i^* = \left(\frac{1}{\lambda} - \frac{1}{\lambda_{1i}} \right)_+, \quad (5.66)$$

i.e., as for standard WF. This implies that when $\lambda_{2i} = 0$ for all active eigenmodes, then the standard WF power allocation is optimal.

It should be stressed that the original channels in (5.62) are not parallel (diagonal). They become equivalent to a set of parallel independent channels after performing information-preserving transformations. Also, there is no assumption of degradedness here and no requirement for the optimal covariance to be of full rank or rank 1.

5.9 When Is ZF Signaling Optimal?

In this section, we consider the case when popular ZF signaling is optimal for the MIMO WTC, i.e., when active eigenmodes of optimal covariance $\mathbf{R}^*$ are orthogonal to those of $\mathbf{W}_2$: $\mathbf{W}_2\mathbf{R}^* = \mathbf{0}$.[4] It is clear that this does not hold in general. However, the importance of this scenario comes from the fact that such signaling does not require wiretap codes: since the eavesdropper gets no signal, regular coding on the required channel suffices. Hence, the system design follows the well-established standard framework and the secrecy requirement imposes no extra complexity penalty but is rather ensured by well-established ZF signaling.

PROPOSITION 5.4 *Consider the wiretap MIMO channel in (5.2) and let $\mathbf{W}_1$ and $\mathbf{W}_2$ have the same eigenvectors [so that $\mathbf{H}_1$ and $\mathbf{H}_2$ have the same right singular vectors as in (5.62)] and*

$$\lambda_{1i} \leq \lambda_{2i} + \lambda \text{ if } \lambda_{2i} > 0, \tag{5.67}$$

where λ is found from the total power constraint $\sum_i \lambda_i^* = P_T$,

$$\lambda_i^* = \lambda_i(\mathbf{R}^*) = \left(\frac{1}{\lambda} - \frac{1}{\lambda_{1i}}\right)_+ \text{ if } \lambda_{2i} = 0, \tag{5.68}$$

and 0 otherwise. Then, the Gaussian ZF signaling is optimal, i.e., $\mathbf{W}_2\mathbf{R}^* = \mathbf{0}$ so that active eigenmodes of $\mathbf{R}^*$ are orthogonal to those of $\mathbf{W}_2$ and the optimal covariance is as in (5.63), so that its eigenvectors are those of $\mathbf{W}_{1,2}$.

The necessary condition of ZF optimality is that active eigenvectors of $\mathbf{R}^*$ are also the active eigenvectors of $\mathbf{W}_1$ and inactive eigenvectors of $\mathbf{W}_2$, and that the power allocation is given by (5.68).

Proof Based on the necessary KKT conditions, which, under the ZF condition $\mathbf{W}_2\mathbf{R} = \mathbf{0}$, have a unique solution; see [12] for details.

REMARK 5.5 *The optimal power allocation in (5.68) is the same as standard WF. However, a subtle difference here is the condition for an eigenmode to be active, $\lambda_i^* > 0$: while standard WF requires $\lambda_{1i} > \lambda$, the solution above additionally requires $\lambda_{2i} = 0$, so that the set of active eigenmodes is generally smaller; the smaller the set of active eigenmodes, the larger the set of eavesdropper positive eigenmodes.*

It is gratifying to see that the standard WF over the eigenmodes of the required channel is optimal if ZF is optimal. In a sense, the optimal transmission strategy in this case is separated into two independent parts: part 1 ensures that the Ev gets no signal (via the ZF), and part 2 is the standard signaling and WF on the active eigenmodes of the legitimate channel as if the Ev were not there. No new wiretap codes need to be designed as regular coding on the required channel suffices, so that the secrecy requirement does

---

[4] This simply means that the Tx antenna array puts null in the direction of the eavesdropper, which is known as null forming in antenna array literature [25]. This can also be considered as a special case of interference alignment, so that Proposition 5.4 establishes its optimality.

not impose an extra complexity penalty (beyond the standard ZF). This is reminiscent of the classical source–channel coding separation [26].

5.10 When Is Standard Water-Filling Optimal?

Motivated by the fact that the transmitter may be unaware of the presence of an eavesdropper and hence uses the standard transmission on the eigenmodes of $\mathbf{W}_1$ with power allocated via the WF algorithm, we ask the question: is it possible for this strategy to be optimal for the MIMO WTC? The affirmative answer and conditions for this to happen are given below.

To this end, let $\mathbf{R}_{\mathrm{WF}}$ be the optimal Tx covariance matrix for transmission on $\mathbf{W}_1$ only, which is given by standard WF over the eigenmodes of $\mathbf{W}_1$:

$$\mathbf{R}_{\mathrm{WF}} = \mathbf{U}_1 \mathbf{\Lambda}^* \mathbf{U}_1^+, \; \lambda_i^* = \left\{ \frac{1}{\lambda} - \frac{1}{\lambda_{1i}} \right\}_+, \tag{5.69}$$

where $\mathbf{\Lambda}^* = \mathrm{diag}\{\lambda_i^*\}$ is a diagonal matrix of the eigenvalues of $\mathbf{R}_{\mathrm{WF}}$, and λ is found from the total power constraint $\sum_i \lambda_i^* = P_T$. Alternatively, one can consider P_T as parameterized by λ, where $P_T(\lambda)$ is monotonically decreasing in λ, with $P_T \to 0$ as $\lambda \to \lambda_{11}$ and $P_T \to \infty$ as $\lambda \to 0$.

THEOREM 5.5 *The standard WF transmit covariance matrix in* (5.69) *is also optimal for the Gaussian MIMO WTC if:*

1. *the eigenvectors of $\mathbf{W}_1$ and $\mathbf{W}_2$ are the same:* $\mathbf{U}_1 = \mathbf{U}_2$;
2. *for active eigenmodes $\lambda_i^* > 0$, their eigenvalues λ_{1i} and λ_{2i} are related as*

$$\lambda_{2i} = \frac{\lambda_{1i}}{1 + \alpha \lambda_{1i}} < \lambda_{1i}, \; \textit{for some } \alpha > 0, \tag{5.70}$$

or, equivalently, $\lambda_{2i}^{-1} = \lambda_{1i}^{-1} + \alpha$;
3. *for inactive eigenmodes $\lambda_i^* = 0$, the eigenvalues λ_{1i} and λ_{2i} are related either as in* (5.70) *or $\lambda_{1i} \leq \lambda_{2i}$.*

Proof See Appendix.

Note that the conditions of Theorem 5.5 do not require $\mathbf{W}_1 = a\mathbf{W}_2$ for some scalar $a > 1$; they also allow for the WTC to be non-degraded. However, the condition in (5.70) implies that larger λ_{1i} corresponds to larger λ_{2i}, so that, over the active signaling subspace, the channel is degraded.

The first condition in Theorem 5.5 implies that $\mathbf{H}_1$ and $\mathbf{H}_2$ have the same right singular vectors but imposes no constraints on their left singular vectors. This may represent a scenario where the transmitter is a base station and the legitimate channel as well as the eavesdropper experience the same scattering with their own individual scatterers around their own receivers (which determine the left singular vectors), as in Section 5.8.

5.11 When Is Isotropic Signaling Optimal?

In the regular MIMO channel ($\mathbf{W}_2 = \mathbf{0}$), isotropic signaling is optimal ($\mathbf{R}^* = a\mathbf{I}$) iff $\mathbf{W}_1 = b\mathbf{I}$, i.e., $\mathbf{W}_1$ has identical eigenvalues. Since this transmission strategy is appealing due to its low complexity (all antennas send independent, identically distributed codewords, so that no precoding, no Tx CSI, and thus no feedback is required), we consider isotropic signaling over the wiretap MIMO channel and characterize the set of channels on which it is optimal. It turns out to be much richer than that of the regular MIMO channel.

PROPOSITION 5.5 *Consider the MIMO wiretap channel in (5.2). The isotropic signaling is optimal, i.e., $\mathbf{R}^* = a\mathbf{I}$ in (5.4), for the set of channels $\{\mathbf{W}_1, \mathbf{W}_2\}$ that can be characterized as follows:*

1. *$\mathbf{W}_1$ and $\mathbf{W}_2$ have the same (otherwise arbitrary) eigenvectors, $\mathbf{U}_1 = \mathbf{U}_2$.*
2. *$\mathbf{W}_1 > \mathbf{W}_2$ so that $\lambda_i(\mathbf{W}_1) = a_i^{-1} > \lambda_i(\mathbf{W}_2) = b_i^{-1}$, where $\lambda_i(\mathbf{W})$ are the ordered eigenvalues of $\mathbf{W}$.*
3. *Take any $b_1 > 0$ and $a_1 < b_1$, and set*

$$\lambda = (a_1 + a)^{-1} - (b_1 + a)^{-1} > 0. \tag{5.71}$$

4. *For $i = 2, \ldots, m$, take any b_i such that*

$$b_i > \lambda a^2 (1 - \lambda a)^{-1} > 0, \tag{5.72}$$

and set

$$a_i = -a + (\lambda + (b_i + a)^{-1})^{-1} > 0. \tag{5.73}$$

This gives the complete characterization of the set of channels for which isotropic signaling is optimal.

Proof See [11].

Note that a special case of this proposition is when $\mathbf{W}_1$ and $\mathbf{W}_2$ have identical eigenvalues, as in the case of the regular MIMO channel. Unlike the regular channel, however, there is also a large set of channels with distinct eigenvalues which dictate the isotropic signaling as well. It is the interplay between the legitimate user and the eavesdropper that is responsible for this phenomenon, i.e., a non-isotropic nature of the first channel is compensated for by a carefully adjusted non-isotropy of the second one.

5.12 An Algorithm for Global Maximization of Secrecy Rates

Although a number of analytical solutions are available for an optimal Tx covariance matrix in the MIMO WTC (as discussed above), the general case remains an open problem. In this section, we introduce an algorithm for the global maximization of secrecy rates with guaranteed convergence [13].

Due to the non-convex nature of the optimization problem in (5.4) in the general case, constructing a numerical algorithm faces an immediate difficulty as global convergence cannot be guaranteed (see, e.g., [27, 28] for such algorithms).

Instead, we adopt an equivalent minimax reformulation of (5.4) [3]:

$$C_s = \max_{\mathbf{R}} \min_{\mathbf{K}} f(\mathbf{R}, \mathbf{K}) = \min_{\mathbf{K}} \max_{\mathbf{R}} f(\mathbf{R}, \mathbf{K}), \tag{5.74}$$

where[5]

$$f(\mathbf{R}, \mathbf{K}) = \frac{1}{2} \ln \frac{|\mathbf{I} + \mathbf{K}^{-1}\mathbf{H}\mathbf{R}\mathbf{H}'|}{|\mathbf{I} + \mathbf{W}_2\mathbf{R}|} \geq C(\mathbf{R}), \tag{5.75}$$

$$\mathbf{K} = \begin{pmatrix} \mathbf{I} & \mathbf{K}'_{21} \\ \mathbf{K}_{21} & \mathbf{I} \end{pmatrix} \geq \mathbf{0}, \ \mathbf{H} = \begin{pmatrix} \mathbf{H}_1 \\ \mathbf{H}_2 \end{pmatrix}, \tag{5.76}$$

and the optimization is over the set $\mathcal{S}$ of all feasible $\mathbf{R}, \mathbf{K}$:

$$\mathcal{S} = \{(\mathbf{R}, \mathbf{K}) : \mathrm{tr}\,\mathbf{R} \leq P, \ \mathbf{R}, \mathbf{K} \geq \mathbf{0}, \ \mathbf{K} \text{ as in } (5.76)\}. \tag{5.77}$$

Even if this reformulation may appear to be more difficult to solve (since it involves both min and max), this is not the case: the problem in (5.74) is convex (since $f(\mathbf{R}, \mathbf{K})$ is convex–concave in the right way) so that the KKT conditions are sufficient for global optimality and the global convergence of a numerical algorithm for this reformulated problem is within reach. The KKT conditions (of which there are two sets: one for min and one for max; see [13] for details) are solved below via a numerical algorithm, which is based on the barrier method and the primal/dual version of the Newton method.

To account for the positive semi-definite constraints $\mathbf{R}, \mathbf{K} \geq \mathbf{0}$, we use the barrier function $\psi_t(\mathbf{R}) = t^{-1} \ln |\mathbf{R}|$ so that the modified objective f_t is

$$f_t(\mathbf{R}, \mathbf{K}) = f(\mathbf{R}, \mathbf{K}) + \psi_t(\mathbf{R}) - \psi_t(\mathbf{K}), \tag{5.78}$$

where $t > 0$ is the barrier parameter (see [13, 29]). Let us introduce the following variables and gradients/Hessians [13]:

$$\mathbf{z} = \begin{bmatrix} \mathbf{x} \\ \mathbf{y} \end{bmatrix}, \ \nabla f_t = \begin{bmatrix} \nabla_x f_t \\ \nabla_y f_t \end{bmatrix}, \ \nabla^2 f_t = \begin{bmatrix} \nabla^2_{xx} f_t & \nabla^2_{xy} f_t \\ \nabla^2_{yx} f_t & \nabla^2_{yy} f_t \end{bmatrix}, \tag{5.79}$$

where $\mathbf{x} = \mathrm{vech}(\mathbf{R})$ and vech stacks column-wise all lower-triangular entries into a single column vector, and we use only $\mathbf{K}_{21}$ as independent variables: $\mathbf{y} = \mathrm{vec}(\mathbf{K}_{21})$. The expressions for gradients $\nabla_{x(y)} f_t$ and Hessians $\nabla^2_{xx(yy)} f_t$ can be found in [13] (alternatively, they can be evaluated numerically). The total power constraint $\mathrm{tr}\,\mathbf{R} = P_T$ can be expressed in the following form: $\mathbf{a}'\mathbf{z} = P_T$, where $\mathbf{a} = [\mathrm{vech}(\mathbf{I}_m)', \mathbf{0}']'$, $\mathbf{0}$ is the $n_1 n_2 \times 1$ all-zero vector, and $\mathbf{I}_m$ is the $m \times m$ identity matrix.

With this choice of variables and new objective in (5.78), the barrier method transforms the inequality-constrained problem in (5.74)–(5.77) into the following

---

[5] In this section, we use the real-valued channel model, so that all entries of $\mathbf{H}, \mathbf{R}, \mathbf{K}$ are real. Equivalently, one can consider real and imaginary parts as independent variables.

problem without inequality constraints:

$$\max_{\mathbf{x}} \min_{\mathbf{y}} f_t(\mathbf{x}, \mathbf{y}), \text{ s.t. } \mathbf{a}'\mathbf{z} = P_T. \qquad (5.80)$$

To solve this problem via the primal/dual Newton method, let $\mathbf{w} = [\mathbf{z}', \lambda]'$ be the vector of aggregated (primal/dual) variables, where $\lambda \geq 0$ is the dual variable (Lagrange multiplier responsible for the power constraint), and $\Delta \mathbf{w} = (\Delta \mathbf{z}', \Delta \lambda)'$ is its update, which is found as the solution of the following system of linear equations at each step:

$$\mathbf{T} \Delta \mathbf{w} = -\mathbf{r}(\mathbf{w}), \qquad (5.81)$$

where $\mathbf{T}$ is the KKT matrix,

$$\mathbf{T} = \begin{bmatrix} \nabla^2 f_t & \mathbf{a} \\ \mathbf{a}' & 0 \end{bmatrix}, \qquad (5.82)$$

and $\mathbf{r}(\mathbf{w})$ is the residual

$$\mathbf{r}(\mathbf{w}) = [(\nabla f_t + \mathbf{a}\lambda)', (\mathbf{a}'\mathbf{z} - P_T)]'. \qquad (5.83)$$

The solution of the KKT conditions corresponds to $\mathbf{r}(\mathbf{w}) = \mathbf{0}$. At each step, the variables are updated as $\mathbf{w} \to \mathbf{w} + \Delta \mathbf{w}$.

Since the algorithm requires an initial point to begin with, we use the following point:

$$\mathbf{R}_0 = m^{-1} P \mathbf{I} \to \mathbf{x}_0 = \text{vech}(\mathbf{R}_0), \ \mathbf{K}_0 = \mathbf{I} \to \mathbf{y}_0 = \mathbf{0}, \ \lambda_0 = 0, \qquad (5.84)$$

which is clearly feasible ($\mathbf{R}_0$ corresponds to isotropic signaling).

With this choice of variables and initial points, Algorithm 5.3, in combination with Algorithms 5.1 and 5.2, can now be used to solve numerically the minimax problem in (5.80); α, β in Algorithms 5.1–5.3 are backtracking line search parameters, s is the step size, ϵ is the desired accuracy, and μ is the barrier increase parameter [13, 29]. Global convergence of this algorithm has been proved in [13].

Algorithm 5.1 Backtracking line search

Require: $\mathbf{w}$, $0 < \alpha < 1/2$, $0 < \beta < 1$, $s = 1$.
   while $|\mathbf{r}(\mathbf{w} + s\Delta\mathbf{w})| > (1 - \alpha s)|\mathbf{r}(\mathbf{w})|$ **do** $s := \beta s$
   end while

Algorithm 5.2 Newton method for minimax optimization

Require: $\mathbf{z}_0, \lambda_0, \alpha, \beta, \epsilon$
   repeat
      1. Find $\Delta\mathbf{z}, \Delta\lambda$ using Newton step in (5.81).
      2. Find s using the backtracking line search (Algorithm 5.1).
      3. Update variables: $\mathbf{z}_{k+1} = \mathbf{z}_k + s\Delta\mathbf{z}, \lambda_{k+1} = \lambda_k + s\Delta\lambda$.
   until $|\mathbf{r}(\mathbf{z}_{k+1}, \lambda_{k+1})| \leq \epsilon$.

Algorithm 5.3 Barrier method

Require: $z, \lambda, \epsilon > 0, t > 0, \mu > 1$
repeat
   1. Solve the problem in (5.80) using the Newton method (Algorithm 5.2) starting at z, λ.
   2. Update variables: $z := z^*(t), \lambda := \lambda^*(t), t := \mu t$.
until $1/t < \epsilon$.

To demonstrate the algorithm's performance, we consider the following example:

$$\mathbf{H}_1 = \begin{bmatrix} 0.77 & -0.30 \\ -0.32 & -0.64 \end{bmatrix}, \mathbf{H}_2 = \begin{bmatrix} 0.54 & -0.11 \\ -0.93 & -1.71 \end{bmatrix}. \quad (5.85)$$

Convergence of the Newton method for different values of the barrier parameter t is demonstrated in Fig. 5.6, which shows the Euclidian norm of the residual $\mathbf{r}$ versus Newton steps. Even though this channel is not degraded, since the eigenvalues of $\mathbf{W}_1 - \mathbf{W}_2$ are $\{0.395, -3.293\}$, the algorithm does find the global optimum (this particular channel was selected because it is "difficult" for optimization; note also that the channel is not degraded so that the problem in (5.4) is not convex). Note the presence of two convergence phases: linear and quadratic, which is typical for Newton methods in general. After the quadratic phase is reached, the convergence is very fast (waterfall region). It takes 10–20 Newton steps to reach a very low residual (at the level of machine precision). This is in agreement with the observations in [29] (although obtained for different problems).

Figure 5.7 shows the corresponding secrecy rate evaluated via the upper bound in (5.75) and the actual achievable rate via $C(\mathbf{R}(t))$ in (5.3), where $\mathbf{R}(t)$ is an optimal

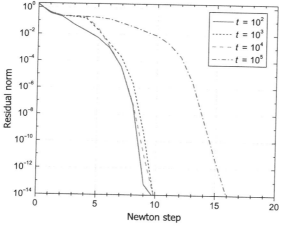

Figure 5.6 Convergence of the Newton method for different values of t; $m = 2$, $P = 10$, $\alpha = 0.3$, $\beta = 0.5$, $\mathbf{H}_1, \mathbf{H}_2$ as in (5.85). Note the presence of two convergence phases: linear and quadratic. It takes only 10 to 20 Newton steps to reach the machine precision level.

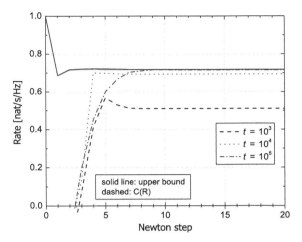

Figure 5.7 Secrecy rates for the same settings as in Fig. 5.6. Solid line: via the upper bound in (5.75) (the lines coincide for different t); dashed line: via $C(\mathbf{R})$ in (5.3). While $t = 10^3$ is sufficient for accurate computation of the upper bound, $t = 10^4$–10^5 is needed for the accurate computation of $C(\mathbf{R})$.

covariance at a particular step of the Newton method and for a given t. As the algorithm converges, they become almost equal if t is sufficiently large (in this case, about 10^4–10^5). While t has negligible impact on the upper bound, it significantly affects the corresponding $C(\mathbf{R}(t))$ so that the choice of t is not critical if the secrecy capacity is the only quantity of interest (since the upper bound is quite tight even for moderate t). However, if a transmitter is implemented with the optimal covariance $\mathbf{R}(t)$ returned by the algorithm, it is $C(\mathbf{R}(t))$ that determines the achievable rate and this choice is important. We attribute this fact to higher sensitivity of $C(\mathbf{R})$ to $\mathbf{R}$ compared to that of $f(\mathbf{R}, \mathbf{K})$. Similar observations apply to the number of Newton steps required to achieve a certain performance: if C_s is the quantity of interest, the upper bound converges to it in 3–5 steps. However, when using $\mathbf{R}$ in a system design, $C(\mathbf{R})$ should be used as a performance metric and, in addition to proper choice of t, it takes 5–10 steps to achieve the convergence for $C(\mathbf{R})$. Note that in both cases the number of steps is not large and the execution time is small (a few seconds). In general, larger t and m, n_1, n_2 require more steps to achieve the same accuracy. As expected, the behavior of the upper bound is not monotonic, while the residual norm decreases monotonically in each step.

5.13 Appendix

5.13.1 Proof of Theorem 5.4

First note that, for the omnidirectional eavesdropper, $\mathbf{W}_2 \leq \varepsilon \mathbf{I}$ so that $|\mathbf{I} + \mathbf{W}_2 \mathbf{R}| \leq |\mathbf{I} + \varepsilon \mathbf{R}|$ and hence

$$C_s = \max_{\text{tr}\mathbf{R} \leq P_T} \ln \frac{|\mathbf{I} + \mathbf{W}_1 \mathbf{R}|}{|\mathbf{I} + \mathbf{W}_2 \mathbf{R}|} \geq \max_{\text{tr}\mathbf{R} \leq P_T} \ln \frac{|\mathbf{I} + \mathbf{W}_1 \mathbf{R}|}{|\mathbf{I} + \varepsilon \mathbf{R}|} = C^*(\epsilon). \quad (5.86)$$

To prove the reverse inequality, let $\mathbf{P}_2$ be a projection matrix on $\mathcal{R}(\mathbf{W}_2)$, i.e., $\mathbf{P}_2 = \mathbf{U}_{2+}\mathbf{U}_{2+}^+$. Then, $\mathbf{P}_2 \mathbf{W}_k \mathbf{P}_2 = \mathbf{W}_k, k = 1, 2$, so that

$$C(\mathbf{R}) = \ln \frac{|\mathbf{I} + \mathbf{P}_2 \mathbf{W}_1 \mathbf{P}_2 \mathbf{R}|}{|\mathbf{I} + \mathbf{P}_2 \mathbf{W}_2 \mathbf{P}_2 \mathbf{R}|} = \ln \frac{|\mathbf{I} + \tilde{\mathbf{W}}_1 \tilde{\mathbf{R}}|}{|\mathbf{I} + \varepsilon \tilde{\mathbf{R}}|} = \tilde{C}(\tilde{\mathbf{R}}), \quad (5.87)$$

where $\tilde{\mathbf{R}} = \mathbf{U}_{2+}^+ \mathbf{R} \mathbf{U}_{2+}$ and likewise for $\tilde{\mathbf{W}}_k$, so that $\tilde{\mathbf{W}}_2 = \varepsilon \mathbf{I}$, where we used $|\mathbf{I} + \mathbf{AB}| = |\mathbf{I} + \mathbf{BA}|$. Further note that

$$\operatorname{tr} \tilde{\mathbf{R}} = \operatorname{tr} \mathbf{U}_{2+}^+ \mathbf{R} \mathbf{U}_{2+} = \sum_i \lambda_i(\mathbf{R}) |\mathbf{u}_{2+i}^+ \mathbf{u}_{Ri}|^2 \leq \sum_i \lambda_i(\mathbf{R}) = \operatorname{tr} \mathbf{R} \leq P_T, \quad (5.88)$$

where $\mathbf{u}_{2+i}$ and $\mathbf{u}_{Ri}$ are ith eigenvectors of $\mathbf{W}_2$ and $\mathbf{R}$, and we have used $\mathbf{R} = \sum_i \lambda_i(\mathbf{R}) \mathbf{u}_{Ri} \mathbf{u}_{Ri}^+$ and $|\mathbf{u}_{2+i}^+ \mathbf{u}_{Ri}|^2 \leq |\mathbf{u}_{2+i}|^2 |\mathbf{u}_{Ri}|^2 = 1$. Hence, $\tilde{\mathbf{R}}$ satisfies the power constraint if $\mathbf{R}$ does, and thus

$$C_s = \max_{\operatorname{tr} \mathbf{R} \leq P_T} C(\mathbf{R}) \leq \max_{\operatorname{tr} \tilde{\mathbf{R}} \leq P_T} \tilde{C}(\tilde{\mathbf{R}}) = \max_{\lambda_i \geq 0, \sum_i \lambda_i \leq P_T} \sum_i \ln \frac{1 + \tilde{g}_i \lambda_i}{1 + \epsilon \lambda_i} = \tilde{C}^*(\varepsilon), \quad (5.89)$$

where $\tilde{g}_i = \lambda_i(\tilde{\mathbf{W}}_1)$, and $\tilde{C}^*(\varepsilon)$ is the secrecy capacity under $\tilde{\mathbf{W}}_1$ and isotropic eavesdropper $\tilde{\mathbf{W}}_2 = \varepsilon \mathbf{I}$. Note that

$$\lambda_i(\tilde{\mathbf{W}}_1) = \lambda_i(\mathbf{U}_{2+}^+ \mathbf{W}_1 \mathbf{U}_{2+}) = \lambda_i([\mathbf{U}_2^+ \mathbf{W}_1 \mathbf{U}_2]_{r_2 \times r_2}) \leq \lambda_i(\mathbf{U}_2^+ \mathbf{W}_1 \mathbf{U}_2) = \lambda_i(\mathbf{W}_1), \quad (5.90)$$

where $[\mathbf{A}]_{k \times k}$ denotes the $k \times k$ principal sub-matrix of $\mathbf{A}$, $r_2 = \operatorname{rank}(\mathbf{W}_2)$. The inequality is due to the Cauchy eigenvalue interlacing theorem [17], and the last equality is due to the fact that $\mathbf{U}_2 \mathbf{W}_1 \mathbf{U}_2^+$ and $\mathbf{W}_1$ have the same eigenvalues. Based on this, one obtains

$$C_s \leq \tilde{C}^*(\varepsilon) \leq \max_{\lambda_i \geq 0, \sum_i \lambda_i \leq P_T} \sum_i \ln \frac{1 + g_i \lambda_i}{1 + \epsilon \lambda_i} = C^*(\varepsilon), \quad (5.91)$$

thus establishing $C_s = C^*(\varepsilon)$ under an omnidirectional eavesdropper with $\mathcal{R}(\mathbf{W}_1) \in \mathcal{R}(\mathbf{W}_2)$.

5.13.2 Proof of Theorem 5.5

We assume that $\mathbf{W}_1$ and $\mathbf{W}_2$ are non-singular; the singular case will follow from the standard continuity argument. The KKT conditions for the optimal covariance $\mathbf{R} = \mathbf{R}_{WF}$, which are necessary for optimality in (5.4), can be expressed as

$$(\mathbf{W}_1^{-1} + \mathbf{R})^{-1} - (\mathbf{W}_2^{-1} + \mathbf{R})^{-1} = \lambda' \mathbf{I} - \mathbf{M} \quad (5.92)$$

$$\lambda'(\operatorname{tr} \mathbf{R} - P_T) = 0, \quad \mathbf{MR} = 0 \quad (5.93)$$

$$\lambda' \geq 0, \quad \mathbf{M}, \mathbf{R} \geq 0, \quad \operatorname{tr} \mathbf{R} \leq P_T, \quad (5.94)$$

where $\mathbf{M} \geq 0$ is the Lagrange multiplier matrix responsible for the constraint $\mathbf{R} \geq 0$, while $\lambda' \geq 0$ is the Lagrange multiplier responsible for the total power constraint $\operatorname{tr} \mathbf{R} \leq P_T$. Multiplying both sides of (5.92) by $\mathbf{U}_1^+$ on the left and by $\mathbf{U}_1$ on the right, one obtains

$$(\mathbf{\Lambda}_1^{-1} + \mathbf{\Lambda}^*)^{-1} - (\mathbf{\Lambda}_2^{-1} + \mathbf{\Lambda}^*)^{-1} = \lambda' \mathbf{I} - \mathbf{U}_1^+ \mathbf{M} \mathbf{U}_1 = \lambda' \mathbf{I} - \mathbf{\Lambda}_M, \quad (5.95)$$

where $\Lambda_1, \Lambda_2, \Lambda_M$ are diagonal matrices of eigenvalues of $\mathbf{W}_1, \mathbf{W}_2, \mathbf{M}$. The last equality follows from the fact that all terms but $\mathbf{U}_1^+\mathbf{M}\mathbf{U}_1$ are diagonal so that the last term has to be diagonal too: $\mathbf{U}_1^+\mathbf{M}\mathbf{U}_1 = \Lambda_M$, i.e., $\mathbf{M}$ has the same eigenvectors as $\mathbf{W}_1, \mathbf{W}_2, \mathbf{R}$. The complementary slackness in (5.93) implies that $\lambda_i^* \lambda_{Mi} = 0$, where λ_{Mi} is the ith eigenvalue of $\mathbf{M}$, i.e., if $\lambda_i^* > 0$ (active eigenmode) then $\lambda_{Mi} = 0$ so that, after some manipulations, (5.95) can be expressed as

$$\lambda_i^* = \frac{1}{(\lambda_{2i}^{-1} + \lambda_i^*)^{-1} + \lambda'} - \frac{1}{\lambda_{1i}}$$
$$= \lambda^{-1} - \lambda_{1i}^{-1} \qquad (5.96)$$

for each $\lambda_i^* > 0$, where the second equality follows from (5.69). Therefore,

$$\lambda = (\lambda_{2i}^{-1} + \lambda_i^*)^{-1} + \lambda' \qquad (5.97)$$

and hence

$$\lambda_i^* = (\lambda - \lambda')^{-1} - \lambda_{2i}^{-1} = \lambda^{-1} - \lambda_{1i}^{-1}, \qquad (5.98)$$

so that $\lambda_{2i}^{-1} = \lambda_{1i}^{-1} + \alpha$ with $\alpha = (\lambda - \lambda')^{-1} - \lambda^{-1} > 0$ satisfies both equalities in (5.96). For inactive eigenmodes $\lambda_i^* = 0$, it follows from (5.95) that

$$\lambda_{1i} - \lambda_{2i} = \lambda' - \lambda_{Mi} \leq \lambda'. \qquad (5.99)$$

Observe that this inequality is satisfied when $\lambda_{1i} \leq \lambda_{2i}$ (since $\lambda' > 0$). To see that it also holds under (5.70), observe that

$$\lambda_{1i} - \lambda_{2i} = \frac{\alpha \lambda_{1i}^2}{1 + \alpha \lambda_{1i}} \leq \frac{\alpha \lambda^2}{1 + \alpha \lambda} = \lambda', \qquad (5.100)$$

where the inequality is due to $\lambda_{1i} \leq \lambda$ (which holds for inactive eigenmodes) and the fact that $\frac{\alpha \lambda_{1i}^2}{1+\alpha \lambda_{1i}}$ is increasing in λ_{1i}. Thus, one can always select $\lambda_{Mi} \geq 0$ to satisfy (5.99) and hence the KKT conditions in (5.92)–(5.94) have a unique solution which also satisfies (5.69). This proves the optimality of $\mathbf{R}_{\mathrm{WF}}$.

References

[1] M. Bloch and J. Barros, *Physical-Layer Security: From Information Theory to Security Engineering*. Cambridge: Cambridge University Press, 2011.
[2] A. Khisti and G. W. Wornell, "Secure transmission with multiple antennas i: The MISOME wiretap channel," *IEEE Trans. Inf. Theory*, vol. 56, no. 7, pp. 3088–3104, Jul. 2010.
[3] A. Khisti and G. W. Wornell, "Secure transmission with multiple antennas–Part II: The MIMOME wiretap channel," *IEEE Trans. Inf. Theory*, vol. 56, no. 11, pp. 5515–5532, Nov. 2010.
[4] F. Oggier and B. Hassibi, "The secrecy capacity of the MIMO wiretap channel," *IEEE Trans. Inf. Theory*, vol. 57, no. 8, pp. 4961–4972, Aug. 2011.
[5] T. Liu and S. Shamai (Shitz), "A note on the secrecy capacity of the multiple-antenna wiretap channel," *IEEE Trans. Inf. Theory*, vol. 55, no. 6, pp. 2547–2553, Jun. 2009.

[6] Z. Li, W. Trappe, and R. Yates, "Secret communication via multi-antenna transmission," in *Proc. Conf. Inf. Sciences Systems*, Baltimore, MD, USA, Mar. 2007, pp. 905–910.

[7] J. Li and A. Petropulu, "Transmitter optimization for achieving secrecy capacity in Gaussian MIMO wiretap channels," Sep. 2009. [Online]. Available: http://arxiv.org/abs/0909.2622

[8] S. Shafiee, N. Liu, and S. Ulukus, "Towards the secrecy capacity of the Gaussian MIMO wire-tap channel: The 2-2-1 channel," *IEEE Trans. Inf. Theory*, vol. 55, no. 9, pp. 4033–4039, Sep. 2009.

[9] M. C. Gursoy, "Secure communication in the low-SNR regime," *IEEE Commun.*, vol. 60, no. 4, pp. 1114–1123, Apr. 2012.

[10] S. Loyka and C. D. Charalambous, "On optimal signaling over secure MIMO channels," in *Proc. IEEE Int. Symp. Inf. Theory*, Cambridge, MA, USA, Jul. 2012, pp. 443–447.

[11] S. Loyka and C. D. Charalambous, "Further results on optimal signaling over secure MIMO channels," in *Proc. IEEE Int. Symp. Inf. Theory*, Istanbul, Turkey, Jul. 2013, pp. 2019–2023.

[12] S. Loyka and C. D. Charalambous, "Rank-deficient solutions for optimal signaling over secure MIMO channels," in *Proc. IEEE Int. Symp. Inf. Theory*, Honolulu, HI, USA, Jun. 2014, pp. 201–205.

[13] S. Loyka and C. D. Charalambous, "An algorithm for global maximization of secrecy rates in Gaussian MIMO wiretap channels," *IEEE Trans. Commun.*, vol. 63, no. 6, pp. 2288–2299, Jun. 2015.

[14] A. Khisti, "Interference alignment for the multiantenna compound wiretap channel," *IEEE Trans. Inf. Theory*, vol. 57, no. 5, pp. 2976–2993, May 2011.

[15] I. Bjelaković, H. Boche, and J. Sommerfeld, "Secrecy results for compound wiretap channels," *Probl. Inf. Transmission*, vol. 49, no. 1, pp. 73–98, Mar. 2013.

[16] R. F. Schaefer and S. Loyka, "The secrecy capacity of compound MIMO Gaussian channels," *IEEE Trans. Inf. Theory*, vol. 61, no. 10, pp. 5535–5552, Dec. 2015.

[17] R. A. Horn and C. R. Johnson, *Matrix Analysis*. Cambridge: Cambridge University Press, 1999.

[18] T. S. Rappaport, *Wireless Communications*, 2nd edn. Upper Saddle River, NJ: Prentice Hall, 2002.

[19] S. Loyka and G. Levin, "On physically-based normalization of MIMO channel matrices," *IEEE Trans. Wireless Commun.*, vol. 8, no. 3, pp. 1107–1112, Mar. 2009.

[20] A. Khisti, A. Tchamkerten, and G. W. Wornell, "Secure broadcasting over fading channels," *IEEE Trans. Inf. Theory*, vol. 54, no. 6, pp. 2453–2469, Jun. 2008.

[21] Z. Li, R. Yates, and W. Trappe, *Securing Wireless Communications at the Physical Layer*. Boston, MA: Springer US, 2010, pp. 1–18.

[22] D. N. C. Tse and P. Viswanath, *Fundamentals of Wireless Communication*. Cambridge: Cambridge University Press, 2005.

[23] Z. Rezki, A. Khisti, and M.-S. Alouini, "Ergodic secret message capacity of the wiretap channel with finite-rate feedback," *IEEE Trans. Wireless Commun.*, vol. 13, no. 6, pp. 3364–3379, Jun. 2014.

[24] J. P. Kermoal, L. Schumacher, K. I. Pedersen, P. E. Mogensen, and F. Frederiksen, "A stochastic MIMO radio channel model with experimental validation," *IEEE J. Sel. Areas Commun.*, vol. 20, no. 6, pp. 1211–1226, Jun. 2002.

[25] H. L. van Trees, *Optimum Array Processing*. Chichester: Wiley, 2002.

[26] T. M. Cover and J. A. Thomas, *Elements of Information Theory*, 2nd edn. Chichester: Wiley & Sons, 2006.

[27] Q. Li, M. Hong, H.-T. Wai, Y.-F. Liu, W.-K. Ma, and Z.-Q. Luo, "Transmit solutions for

MIMO wiretap channels using alternating optimization," *IEEE J. Sel. Areas Commun.*, vol. 31, no. 9, pp. 1714–1727, Sep. 2013.

[28] J. Steinwandt, S. A. Vorobyov, and M. Haardt, "Secrecy rate maximization for MIMO Gaussian wiretap channels with multiple eavesdroppers via alternating matrix POTDC," in *Proc. IEEE Int. Conf. Acoustics, Speech, Signal Process.*, Florence, Italy, May 2014, pp. 5686–5690.

[29] S. P. Boyd and L. Vandenberghe, *Convex Optimization*. Cambridge: Cambridge University Press, 2004.

6 MIMO Wiretap Channels

Mohamed Nafea and Aylin Yener

This chapter considers securing wireless communications at the physical layer using multiple antennas. In particular, the multiple antenna (MIMO) wiretap channel is presented and its secrecy capacity is provided. Further investigation of the high-SNR characterization of the secrecy capacity, i.e., secure degrees of freedom, reveals that the secrecy capacity of the MIMO wiretap channel does not scale with the transmit power when the eavesdropper has an advantage over the legitimate transmitter in the number of antennas. An external multi-antenna cooperative jammer terminal is introduced to the channel model in order to improve its secrecy capacity scaling with power. The secure degrees of freedom for this new multi-terminal multi-antenna channel is characterized. While for these models the eavesdropper channel state information is known at the legitimate terminals, the chapter next removes this assumption and presents results that utilize multiple antennas in order to provide secure communications irrespective of the eavesdropper channel state. An achievable strong secrecy rate and the secure degrees of freedom characterization for the MIMO wiretap channel when the eavesdropper channel is arbitrary, varying, and unknown at the legitimate terminals are presented. Finally, the extension of this model to the two-user multiple access MIMO wiretap channel is considered and its secure degrees of freedom region is characterized.

6.1 Introduction

The wiretap channel and the notion of secrecy capacity, introduced in [1] and generalized by [2], have provided the framework for physical layer design for wireless channels with information theoretic guarantees against eavesdropping. A natural model in wireless communications is one which utilizes multiple antennas for transmission and reception at each node. The improvement in secrecy rate that can be obtained using multiple antennas in a point-to-point channel with an eavesdropper has been studied extensively in several references, including [3, 4]. Specifically, the secrecy capacity of the Gaussian model consisting of a multiple antenna legitimate transmitter, a multiple antenna legitimate receiver, and a multiple antenna eavesdropper, termed the multiple-input multiple-output (MIMO) wiretap channel, has been identified in full generality in [5, 6]. Reference [5] has also provided the characterization of the secrecy capacity in the high signal-to-noise ratio (SNR) regime, revealing the fact that the secrecy capacity pre-log factor, i.e., the secure degrees of freedom (s.d.o.f.), of the

MIMO wiretap channel is equal to zero when the eavesdropper has more antennas than the legitimate transmitter. More recently, [7, 8] have considered the MIMO wiretap channel with a multi-antenna cooperative jammer (helper) and characterized its s.d.o.f. These references have assumed the channel state information of all channels, including that of the eavesdropper, to be available at the legitimate transmitter.

From a practical point of view, the eavesdropper channel state information (E-CSI) at the legitimate terminals is a strong assumption since the eavesdropper is, by nature, a passive entity. A step toward addressing this problem has been the proposal of a semi-blind transmission strategy for the MIMO wiretap channel in [3, 5, 9], where synthetic "artificial" noise is radiated in all directions that are orthogonal to the receiver. More recently, [10] has provided a concrete solution to the no E-CSI setting, by constructing a universal coding scheme that achieves a positive strongly secure rate for *any* eavesdropper channel for the MIMO wiretap channel with an eavesdropper whose channel is arbitrary, time varying, and completely unknown to the legitimate terminals, where an isotropic synthetic noise is radiated in all possible directions. References [11, 12] have extended this model to MIMO multiple access and broadcast wiretap channels and fully characterized their s.d.o.f. regions. Interestingly, the s.d.o.f. region for the two-user MIMO multiple access channel, with a complete knowledge of the eavesdropper channel at the legitimate terminals, is still an open problem. Noticeably, a special case of this model is the MIMO wiretap channel with a multi-antenna cooperative jammer [7, 8], where the second transmitter is dedicated as a helper terminal.

In this chapter, we provide a comprehensive overview of all these advances in the MIMO wiretap channel, focusing on the secrecy capacities of the models summarized above. More specifically, we focus on the recent advances that characterize the impact of cooperative jamming, i.e., intentional interference injection, as an enabler and an enhancer of security provided by multiple antennas, a recurrent theme in these models. Section 6.2 briefly reviews the secrecy capacity for the MIMO wiretap channel, and Section 6.3 describes its high-SNR characterization, to provide the background for what follows. Section 6.4 discusses in detail the MIMO wiretap channel when a multi-antenna external cooperative jammer node is available to the system. Section 6.5 considers the MIMO wiretap channel when the eavesdropper channel is unknown to the transmitter and receiver, and Section 6.6 describes the extension of this model to the multiple access setting, with these two models benefiting from cooperative jamming without having to designate an external entity as such. Section 6.7 concludes the chapter.

Notation

Logarithms are taken to be base 2. $\mathbf{0}_{m \times n}$ denotes an $(m \times n)$ matrix of zeros, and $\mathbf{I}_n$ denotes an $(n \times n)$ identity matrix. For matrix $\mathbf{A}$, null($\mathbf{A}$) denotes its null space, det($\mathbf{A}$) denotes its determinant, rank($\mathbf{A}$) denotes its rank, $||\mathbf{A}||^2$ denotes the sum of the Euclidean norm squared of all of its row vectors, and $||\mathbf{A}||_{\text{in}}$ denotes its induced norm. For vector $\mathbf{V}$, $||\mathbf{V}||$ denotes its Euclidean norm, and $\mathbf{V}_i^j$ denotes the ith to jth components in $\mathbf{V}$. For vector (matrix) $\mathbf{A}(i)$, $i = 1, 2, \ldots$, we use $\mathbf{A}^n$ to denote the n-letter extension of $\mathbf{A}(i)$, i.e., $\mathbf{A}^n = [\mathbf{A}(1) \cdots \mathbf{A}(n)]$. $\mathbf{A}^n \mathbf{B}^n$ denotes the row concatenation of the matrices $\{\mathbf{A}(i)\mathbf{B}(i), 1 \leq i \leq n\}$. The operators ^{T}, ^{H}, and † denote the transpose, Hermitian, and pseudo inverse operations. We use $p_{\mathbf{X}}(\mathbf{x})$ ($f_{\mathbf{X}}(\mathbf{x})$) to denote the probability mass (density) function of a discrete (continuous) random vector $\mathbf{X}$. $\mathbb{1}_{\{A\}}$ denotes the Kronecker delta function, which is equal to 1 when A is satisfied and zero otherwise. We use $[x]^+ = \max\{x,0\}$. A circularly symmetric complex Gaussian random vector with mean vector $\mathbf{m}$ and covariance matrix $\mathbf{K}$ is denoted by $\mathrm{CN}(\mathbf{m}, \mathbf{K})$.

6.2 Secrecy Capacity of the MIMO Wiretap Channel

In this section, we provide a brief overview of the secrecy capacity of the MIMO wiretap channel depicted in Fig. 6.1. The transmitter, receiver, and eavesdropper are equipped with N_{t}, N_{r}, and N_{e} antennas, respectively.

The signals received at the receiver and eavesdropper, at the ith channel use, are

$$\mathbf{Y}_{\mathrm{r}}(i) = \mathbf{H}\mathbf{X}(i) + \mathbf{Z}_{\mathrm{r}}(i) \tag{6.1}$$

$$\mathbf{Y}_{\mathrm{e}}(i) = \mathbf{G}\mathbf{X}(i) + \mathbf{Z}_{\mathrm{e}}(i), \tag{6.2}$$

where $\mathbf{X}(i)$ is the transmitted signal at the ith channel use, and $\mathbf{H}, \mathbf{G}$ are the channel gain matrices to the legitimate receiver and to the eavesdropper, respectively. $\mathbf{Z}_{\mathrm{r}}(i), \mathbf{Z}_{\mathrm{e}}(i)$ are the complex Gaussian noise vectors with zero mean and covariance matrices $\mathbf{I}_{N_r}$ and $\mathbf{I}_{N_e}$, i.e., independent and identically distributed (i.i.d.) components. Further, $\mathbf{Z}_{\mathrm{r}}(i)$ is independent from $\mathbf{Z}_{\mathrm{e}}(i)$, and both are independent across time. The channel gains $\mathbf{H}, \mathbf{G}$ are assumed to be static, i.e., fixed during the transmission period of n channel uses, and known to all terminals. The power constraint of the transmitter is given by

$$\frac{1}{n} \sum_{i=1}^{n} \mathrm{E}\{\mathbf{X}^{\mathrm{H}}(i)\mathbf{X}(i)\} \leq P. \tag{6.3}$$

The objective of communication is to transmit a message W reliably to the receiver, while keeping it secret from the eavesdropper. The transmitter uses a stochastic encoder

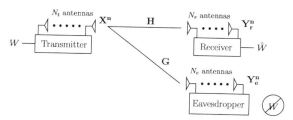

Figure 6.1 $(N_{\mathrm{t}} \times N_{\mathrm{r}} \times N_{\mathrm{e}})$ MIMO Gaussian wiretap channel.

$f_n : W \mapsto \mathbf{X}^n$. The decoder at the receiver maps its observation $\mathbf{Y}_r^n$ to an estimate $\hat{W}$ of the transmitted message. The probability of decoding error, $P_e^{(n)}$, at the legitimate receiver is thus defined as

$$P_e^{(n)} = \Pr\{\hat{W} \neq W\}. \tag{6.4}$$

Secrecy rate R_s is achievable if, for every $\epsilon > 0$, there exists a channel code $(2^{nR_s}, n)$ such that[1]

$$\lim_{n \to \infty} P_e^{(n)} = 0 \tag{6.5}$$

$$\lim_{n \to \infty} \frac{1}{n} I(W; \mathbf{Y}_e^n) = 0. \tag{6.6}$$

Secrecy capacity is defined as the supremum of all achievable secrecy rates.

The secrecy capacity for the general discrete memoryless channel described by the conditional distribution $p_{Y_r, Y_e | X}$ is expressed as [2]

$$C_s = \max_{p_{U,X}} I(U; Y_r) - I(U; Y_e), \tag{6.7}$$

where U is an auxiliary random variable with a bounded cardinality, which satisfies the Markov chain U–X–(Y_r, Y_e). This auxiliary random variable can be represented as a preprocessing on the channel input X which results in an effective channel to the eavesdropper that is degraded with respect to the legitimate channel. Although the secrecy capacity in (6.7) extends to continuous input alphabets with average power constraints [2], and hence applies for the generally non-degraded MIMO wiretap channel in question, the optimization problem in (6.7), over all feasible joint distributions $p_{U,X}$ satisfying U–X–(Y_r, Y_e), is difficult to solve. This fact has led researchers to search for an alternative way to characterize a computable expression for the secrecy capacity of the MIMO wiretap channel.

The secrecy capacities for channels with two-antenna transmitter, two-antenna receiver, and single-antenna eavesdropper, and N_t-antenna transmitter, single-antenna receiver, and N_e-antenna eavesdropper have been characterized in [4] and [3], respectively. The secrecy capacity for the general MIMO wiretap channel has been established in [5, 6]. Instead of directly solving the optimization problem in (6.7), the approach utilized in these papers is based on deriving a tight upper bound on the secrecy capacity, which gives an indication of the optimal choices for the channel input and the auxiliary random variable.

THEOREM 6.1 ([5, 6]) *The secrecy capacity for the MIMO wiretap channel is*

$$C_s = \max_{\mathbf{Q}_x \succeq 0, \mathrm{tr}(\mathbf{Q}_x) \leq P} \log \frac{\det(\mathbf{I}_{N_r} + \mathbf{H}\mathbf{Q}_x\mathbf{H}^H)}{\det(\mathbf{I}_{N_e} + \mathbf{G}\mathbf{Q}_x\mathbf{G}^H)}, \tag{6.8}$$

where $\mathbf{Q}_x$ is the covariance matrix of the channel input $\mathbf{X}$.

---

[1] We consider a weak secrecy constraint in this section, where the rate of mutual information vanishes as the block length $n \to \infty$; see Eq. (6.6). Strong secrecy will be considered in Section 6.5; see Eq. (6.87).

Proof sketch [5, 6] A genie-aided upper bound which considers an enhanced channel, with the eavesdropper's observation provided to the receiver, is derived. In particular, the secrecy capacity of the original channel is upper bounded as

$$C_s = \max_{p_U f_{\mathbf{X}|U}} I(U; \mathbf{Y}_r) - I(U; \mathbf{Y}_e) \quad (6.9)$$

$$\leq \max_{p_U f_{\mathbf{X}|U}} I(U; \mathbf{Y}_r \mathbf{Y}_e) - I(U; \mathbf{Y}_e) \quad (6.10)$$

$$= \max_{p_U f_{\mathbf{X}|U}} I(U; \mathbf{Y}_r | \mathbf{Y}_e) \quad (6.11)$$

$$\leq \max_{p_U f_{\mathbf{X}|U}} h(\mathbf{Y}_r | \mathbf{Y}_e) - h(\mathbf{Y}_r | \mathbf{Y}_e U \mathbf{X}) \quad (6.12)$$

$$= \max_{f_{\mathbf{X}}} I(\mathbf{X}; \mathbf{Y}_r | \mathbf{Y}_e), \quad (6.13)$$

where (6.13) follows from the Markov chain U–$\mathbf{X}\mathbf{Y}_e$–$\mathbf{Y}_r$.

Since the secrecy capacity of the original channel depends only on the marginal conditional distributions $f_{\mathbf{Y}_r|\mathbf{X}}$ and $f_{\mathbf{Y}_e|\mathbf{X}}$, introducing correlation between the Gaussian noise at the receiver and that at the eavesdropper, $\mathbf{Z}_r$ and $\mathbf{Z}_e$, does not change the secrecy capacity. This correlation can be used to tighten the upper bound in (6.13), as the mutual information term in (6.13) depends on the distribution $f_{\mathbf{Y}_r,\mathbf{Y}_e|\mathbf{X}}$. This results in an upper bound which is similar to Sato's upper bound for the sum capacity of the broadcast channel [13]. In addition, under the power constraint in (6.3), the mutual information $I(\mathbf{X}; \mathbf{Y}_r|\mathbf{Y}_e)$ is maximized by a Gaussian input distribution, $\mathbf{X} \sim \text{CN}(\mathbf{0}, \mathbf{Q}_x)$, since

$$I(\mathbf{X}; \mathbf{Y}_r|\mathbf{Y}_e) = h(\mathbf{Y}_r|\mathbf{Y}_e) - h(\mathbf{Z}_r|\mathbf{Z}_e) \quad (6.14)$$

$$\leq \log 2\pi e \det(\Lambda) - h(\mathbf{Z}_r|\mathbf{Z}_e), \quad (6.15)$$

where Λ is the covariance matrix of the linear minimum mean square estimation error (LMMSEE) of estimating $\mathbf{Y}_r$ from $\mathbf{Y}_e$, and hence the inequality in (6.15) is achieved with equality when $\mathbf{X}$ is Gaussian, which results in $\mathbf{Y}_r$ and $\mathbf{Y}_e$ being jointly Gaussian. The second term on the right-hand side of (6.15) does not depend on $f_{\mathbf{X}}$.

Let $\mathbf{Q}_z$ denote the covariance matrix of $\mathbf{Z} = \begin{bmatrix} \mathbf{Z}_r \\ \mathbf{Z}_e \end{bmatrix}$; we can express (6.13) as

$$C_s \leq \min_{\mathbf{Q}_z \in \mathcal{Q}_z} \max_{\mathbf{Q}_x \in \mathcal{Q}_x} I(\mathbf{X}; \mathbf{Y}_r|\mathbf{Y}_e), \quad (6.16)$$

where $\mathbf{X} \sim \text{CN}(\mathbf{0}, \mathbf{Q}_x)$, $\mathcal{Q}_x = \{\mathbf{Q}_x : \mathbf{Q}_x \succeq \mathbf{0}, \text{tr}(\mathbf{Q}_x) \leq P\}$, and

$$\mathcal{Q}_z = \left\{ \mathbf{Q}_z : \mathbf{Q}_z \succeq \mathbf{0}, \mathbf{Q}_z = \begin{bmatrix} \mathbf{I}_{N_r} & \Phi \\ \Phi^H & \mathbf{I}_{N_e} \end{bmatrix} \right\}; \quad (6.17)$$

Φ is the introduced correlation between $\mathbf{Z}_r$ and $\mathbf{Z}_e$.

Next, by showing that $I(\mathbf{X}; \mathbf{Y}_r|\mathbf{Y}_e)$ is convex in $\mathbf{Q}_z$ for each $\mathbf{Q}_x \in \mathcal{Q}_x$, and concave in $\mathbf{Q}_x$ for each $\mathbf{Q}_z \in \mathcal{Q}_z$, the existence of a saddle point solution, $(\mathbf{Q}_x^*, \mathbf{Q}_z^*)$, for the optimization problem in (6.16) follows. Reference [5] shows that this saddle point solution satisfies

$$\Phi^* \mathbf{H} \mathbf{S} = \mathbf{G} \mathbf{S} \text{ for all full rank } \mathbf{S} \text{ such that } \mathbf{S}\mathbf{S}^H = \mathbf{Q}_x^*, \quad (6.18)$$

when the secrecy capacity of the channel is non-zero. By exploiting this property, it is shown in [5] that the upper bound in (6.16) reduces to (6.8).

The achievability of (6.8) follows from the general expression (6.9) by setting $U = \mathbf{X} \sim CN(\mathbf{0}, \mathbf{Q}_x)$, which completes the proof for Theorem 6.1.

Now, we state a few remarks about the secrecy capacity of the MIMO wiretap channel.

REMARK 6.1 *The channel input, $\mathbf{X} \sim CN(\mathbf{0}, \mathbf{Q}_x^*)$, which achieves the secrecy capacity of the MIMO wiretap channel involves transmission of information only over the directions where the eavesdropper receives a degraded version of the signal received by the receiver. This results in an effective degraded channel to the eavesdropper with respect to the receiver. This can be readily seen from the property of the saddle point (optimal) solution in (6.18), which implies that the receiver can simulate the received signal at the eavesdropper by adding some synthetic noise. Conversely, the secrecy capacity is zero whenever the channel to the receiver is degraded with respect to the eavesdropper [5].*

REMARK 6.2 *The upper bound in (6.16) which corresponds to the secrecy capacity of an enhanced channel with the receiver observing its received signal and the signal received by the eavesdropper, minimized over all possible correlations between the two signals, is in fact achievable for the MIMO wiretap channel. This is due to the effective degraded channel resulting from the optimal transmission.*

REMARK 6.3 *An alternative approach for characterizing the secrecy capacity of the MIMO wiretap channel, using channel enhancement [14], was constructed and shown to have the same secrecy capacity as the original MIMO wiretap channel in [15] under a more general covariance matrix constraint on the channel input: $\frac{1}{n}\sum_{i=1}^{n} E(\mathbf{X}(i)\mathbf{X}^H(i)) \preceq \mathbf{Q}$, where $\mathbf{Q} \succeq \mathbf{0}$.*

REMARK 6.4 *For the special case of a single receive antenna $N_r = 1$ studied in [3], the solution of the optimization problem in (6.9) is such that $U = \mathbf{X} \sim CN(\mathbf{0}, \mathbf{Q}_x^*)$, where the optimal input covariance matrix $\mathbf{Q}_x^*$ is unit rank, which indicates that beamforming achieves secrecy capacity. In particular, the optimal transmission strategy is to beamform along the direction of the eigenvector which corresponds to the maximum eigenvalue of the matrix $(\mathbf{I}_{N_t} + P\mathbf{G}^H\mathbf{G})^{-1}(\mathbf{I}_{N_t} + P\mathbf{h}^H\mathbf{h})$ while the message is encoded by a scalar Gaussian wiretap code; $\mathbf{h}$ is the channel gain vector to the receiver. Note that the maximum eigenvalue of $(\mathbf{I}_{N_t} + P\mathbf{G}^H\mathbf{G})^{-1}(\mathbf{I}_{N_t} + P\mathbf{h}^H\mathbf{h})$ corresponds to the maximum ratio of the channel gain to the intended receiver to the channel gain to the eavesdropper, i.e., the optimal strategy for this case is beamforming along the direction which maximizes this ratio. Similarly, the solution for the special case $\{N_t = N_r = 2, N_e = 1\}$ developed in [4] employs a Gaussian input with a unit-rank covariance matrix. The optimal transmission strategy here is to beamform the signal along a direction which is orthogonal to the eavesdropper and as close as possible to the receiver's signal space.*

6.3 High-SNR Secrecy Capacity for the MIMO Wiretap Channel

In this section, we review the secrecy capacity results for the MIMO wiretap channel in the high-SNR regime developed in [3, 5]. These results provide some useful insights that we utilize in this chapter. The following theorem is derived in [5].

THEOREM 6.2 ([5]) *The secrecy capacity for the MIMO wiretap channel in the high-power limit, i.e., $P \to \infty$, takes the form*

$$C_s(P) = C_0(P) + \sum_{j:\sigma_j \geq 1} \log \sigma_j^2 - o(1), \qquad (6.19)$$

where

$$C_0(P) = \begin{cases} \log \det \left(\mathbf{I}_{N_r} + \frac{P}{p} \mathbf{H} \mathbf{G}^\sharp \mathbf{H}^H \right), & \text{if } N_t > N_e \\ 0, & \text{if } N_t \leq N_e, \end{cases} \qquad (6.20)$$

$\sigma_1 \leq \sigma_2 \leq \cdots \leq \sigma_s$ *are the generalized singular values of* $(\mathbf{H}, \mathbf{G})$, $\mathbf{G}^\sharp$ *is the projection matrix onto* null$(\mathbf{G})$, $s = \dim\{\text{null}(\mathbf{H})^\perp \cap \text{null}(\mathbf{G})^\perp\}$, and[2] $p = \dim\{\text{null}(\mathbf{H})^\perp \cap \text{null}(\mathbf{G})\}$.

Noticeably, (6.19) relies on the generalized singular value decomposition [16] of $(\mathbf{H}, \mathbf{G})$. Simultaneous diagonalization of $\mathbf{H}$ and $\mathbf{G}$ results in a set of N_t parallel channels, p of which have gains only to the receiver, and s of which have gains to both the receiver and eavesdropper, while the remaining $N_t - p - s$ parallel channels have either gains only to the eavesdropper or no gain to both receivers. Along with utilizing wiretap coding and Gaussian signals, the achievability at high SNR relies on signaling only over these p parallel channels which have gains only to the receiver, and those of the s parallel channels which have larger gains to the receiver than to the eavesdropper.

The pre-log factor of the secrecy capacity (secrecy rate $R_s(P)$), i.e., the secure degrees of freedom, for a Gaussian channel with complex-valued channel gains (the achievable s.d.o.f.), is defined as

$$D_s = \lim_{P \to \infty} \frac{R_s(P)}{\log P}. \qquad (6.21)$$

The s.d.o.f. for the MIMO wiretap channel can be deduced readily from Theorem 6.2 as follows [8, Section IV-A]. $\mathbf{H} \mathbf{G}^\sharp \mathbf{H}^H$ in (6.20) can be decomposed as [5, Eq. (107)]

$$\mathbf{H} \mathbf{G}^\sharp \mathbf{H}^H = \Psi_r \begin{bmatrix} \mathbf{0}_{(N_r - p) \times (N_r - p)} & \mathbf{0}_{(N_r - p) \times p} \\ \mathbf{0}_{p \times (N_r - p)} & \Omega \end{bmatrix} \Psi_r^H, \qquad (6.22)$$

---

[2] $\mathbf{A}^\perp$ denotes the space orthogonal to $\mathbf{A}$.

where Ψ_r is unitary and Ω is non-singular. Let $\Psi_r = [\Psi_{r,1} \ \Psi_{r,2}]$, where $\Psi_{r,1} \in \mathbb{C}^{N_r \times (N_r - p)}$ and $\Psi_{r,2} \in \mathbb{C}^{N_r \times p}$. For $N_t > N_e$, $C_0(P)$ in (6.20) can be expressed as[3]

$$C_0(P) = \log \det \left(\mathbf{I}_p + \frac{P}{p} \Omega \Psi_{r,2}^H \Psi_{r,2} \right) \qquad (6.23)$$

$$= \log P^p \det \left(\frac{1}{P} \mathbf{I}_p + \frac{1}{p} \Omega \right) = p \log P + O(1), \qquad (6.24)$$

and hence $C_s(P)$ in (6.19) is given by

$$C_s(P) = \begin{cases} p \log P + O(1), & N_t > N_e \\ O(1), & N_t \leq N_e. \end{cases} \qquad (6.25)$$

Using (6.21), the s.d.o.f. of the MIMO wiretap channel is

$$D_s = p = \dim\{\text{null}(\mathbf{H})^\perp \cap \text{null}(\mathbf{G})\}. \qquad (6.26)$$

Equation (6.26) was also derived in [17, Lemma 1].

When the channel gains are randomly generated according to a continuous distribution, then, for $N_t > N_e$, $\mathbf{x} \in \text{null}(\mathbf{G})$ implies that $\mathbf{x} \in \text{null}(\mathbf{H})^\perp$ almost surely (a.s.) [7]. Thus, $p = \dim(\text{null}(\mathbf{G})) = [N_t - N_e]^+$, and the s.d.o.f. is given by

$$D_s = [N_t - N_e]^+. \qquad (6.27)$$

In all of the aforementioned results, the eavesdropper channel is assumed to be known at the legitimate parties. A transmission scheme which involves transmitting information signals along the directions in the range of the receiver, and simultaneously sending artificial noise in all other directions, i.e., in the subspace orthogonal to the receiver, was proposed in [9], and its high-SNR performance was further studied in [3, 5]. Reference [3] proved that this scheme, when $N_r = 1$ and N_t, N_e are arbitrary, is near optimal at high SNR in the sense that the gap between its achievable rate, R_{AN}, and the secrecy capacity, $C_s(P)$, does not exceed $\log N_t$. Specifically, when $N_r = 1$, it is shown that [3]

$$\lim_{P \to \infty} C_s(P) - R_{AN}(P) = \begin{cases} \log N_t, & \text{if } \mathbf{h} \in \text{null}(\mathbf{G})^\perp \\ 0, & \text{if } \mathbf{h} \in \text{null}(\mathbf{G}). \end{cases} \qquad (6.28)$$

In contrast, [5] showed that the scheme can be arbitrarily far from optimal in the high-SNR regime for the general MIMO wiretap channel, where [5]

$$\lim_{P \to \infty} C_s(P) - R_{AN}(P) = \sum_{j: \sigma_j < 1} \log \frac{1}{\sigma_j^2}, \qquad (6.29)$$

which implies that the gap is large for small generalized singular values of $(\mathbf{H}, \mathbf{G})$.

The artificial noise scheme described above is referred to as a "semi-blind" scheme since selecting its signaling directions does not require the knowledge of the

---

[3] A function $f(P) = O(1)$ if and only if there exists a positive real number M and P_0 such that $f(P) \leq M$ for all $P \geq P_0$.

eavesdropper channel, but the transmission rate selection, and hence the Gaussian wiretap code construction, does. However, the rate selection step in this semi-blind scheme can be set with only a partial knowledge of the eavesdropper channel. This partial knowledge can be the knowledge of a finite set of possible channels, i.e., a compound setting, or the knowledge of bounds on the range of the possible channels, where the minimum rate which corresponds to the worst case eavesdropper channel can be chosen for all channels. When no such knowledge exists on the legitimate parties, a universal coding scheme irrespective of the channel state values of the eavesdropper is needed. In Section 6.5, we present an achievability scheme which radiates artificial noise isotropically in all directions, and achieves a positive strong secrecy rate for the MIMO wiretap channel when the eavesdropper channel states are arbitrary, time varying, and completely unknown to the legitimate terminals, as long as the number of the antennas at the eavesdropper is less than the number of antennas at any of the legitimate parties.

Artificial noise injection is a strategy that falls into a class of strategies termed *cooperative jamming* [18, 19], where intentional interference is introduced to the medium that is not necessarily unstructured noise [20, 21], is not necessarily orthogonal to the legitimate receiver, or carried out by the same transmitter who sends the message. Initially introduced for the single-antenna multiple access wiretap channel [18], where a transmitter who has a better channel to the eavesdropper than to the receiver, and hence is not capable of achieving a positive secrecy rate for itself, rather than staying silent, sends a jamming signal which reduces the reception capability of the eavesdropper more than it does for the legitimate receiver, cooperative jamming evolved to a general class of channel pre-fixing strategies for achieving improved secrecy rates for various multi-terminal models. The cooperative jammer can be either a legitimate transmitter in the system, or an external terminal that is employed in order to boost the secrecy rate, or even the s.d.o.f., of the system, as we will see in the next section. In particular, we notice from (6.27) that the s.d.o.f. for the $(N_t \times N_r \times N_e)$ MIMO wiretap channel is equal to zero when the number of eavesdropper antennas is greater than or equal to the number of transmitter antennas. In the next section, a multi-antenna cooperative jammer is introduced to the channel in order to increase its s.d.o.f.

6.4 MIMO Wiretap Channel with a Cooperative Jammer

In this section, we consider an $(N_t \times N_r \times N_e)$ MIMO wiretap channel with an external cooperative jammer that is equipped with N_c antennas, as depicted in Fig. 6.2. The s.d.o.f. of the channel for all possible values of the number of antennas at the cooperative jammer, N_c, has been characterized in [7], as we will detail next.

The received signals at the receiver and eavesdropper, at the ith channel use, are

$$\mathbf{Y}_r(i) = \mathbf{H}_t \mathbf{X}_t(i) + \mathbf{H}_c \mathbf{X}_c(i) + \mathbf{Z}_r(i) \tag{6.30}$$

$$\mathbf{Y}_e(i) = \mathbf{G}_t \mathbf{X}_t(i) + \mathbf{G}_c \mathbf{X}_c(i) + \mathbf{Z}_e(i), \tag{6.31}$$

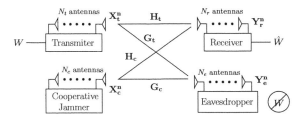

Figure 6.2 $(N_t \times N_r \times N_e)$ Gaussian MIMO wiretap channel with an N_c-antenna cooperative jammer.

where $\mathbf{X}_t, \mathbf{X}_c$ are the transmitter and cooperative jammer signals. The channel gains $\mathbf{H}_t$, $\mathbf{H}_c$, $\mathbf{G}_t$, and $\mathbf{G}_c$ are static and independently drawn from a continuous distribution. As in Sections 6.2 and 6.3, the channel gains are assumed to be known at all terminals. $\mathbf{Z}_r(i) \sim \mathrm{CN}(\mathbf{0}, \mathbf{I}_{N_r})$, $\mathbf{Z}_e(i) \sim \mathrm{CN}(\mathbf{0}, \mathbf{I}_{N_e})$ are the Gaussian noise at the receiver and eavesdropper at the ith channel use. $\mathbf{Z}_r(i)$ is independent from $\mathbf{Z}_e(i)$, and both are independent across time. The power constraints at the transmitter and cooperative jammer are $\mathrm{E}\{\mathbf{X}_t^H \mathbf{X}_t\} \leq P, \mathrm{E}\{\mathbf{X}_c^H \mathbf{X}_c\} \leq P$.

As in Section 6.2, the transmitter aims to send a message W reliably to the receiver, and to keep W secret from the eavesdropper. The receiver outputs an estimate $\hat{W}$ upon receiving its observation $\mathbf{Y}_r^n$. An achievable secrecy rate[4] R_s is defined as in (6.5) and (6.6), and its achievable s.d.o.f. is defined as in (6.21). The design criterion for the cooperative jamming and information signals should be such that the cooperative jamming causes the most harm (interference) possible at the eavesdropper, while causing the least possible interference at the receiver. The jamming signal does not carry any information, and there is no shared secret between the transmitter and cooperative jammer.

When the transmitter and the legitimate receiver have the same number of antennas, i.e., $N_t = N_r = N$, the s.d.o.f. for this channel is provided in the following theorem.

THEOREM 6.3 *[8, Theorem 1] The s.d.o.f. of the $(N \times N \times N_e)$ MIMO wiretap channel with an N_c-antenna cooperative jammer is given by*

$$D_s = \begin{cases} [N + N_c - N_e]^+, & 0 \leq N_c \leq N_e - N_{\min} \\ N - N_{\min}, & N_e - N_{\min} < N_c \leq N_{\max} \\ \frac{N + N_c - N_e}{2}, & N_{\max} < N_c \leq N + N_e \\ N, & N_c \geq N + N_e, \end{cases} \quad (6.32)$$

where $N_{\min} = \frac{\min\{N, N_e\}}{2}$ *and* $N_{\max} = \max\{N, N_e\}$.

---

[4] We again consider the weak secrecy constraint in this section.

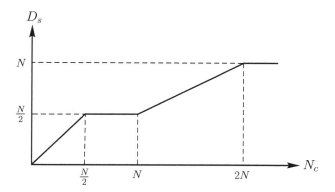

Figure 6.3 The s.d.o.f. for the $(N \times N \times N)$ MIMO wiretap channel with an N_c-antenna cooperative jammer.

The s.d.o.f. for the $(N \times N \times N)$ MIMO Gaussian wiretap channel with an N_c-antenna cooperative jammer is shown in Fig. 6.3, when N_c varies from 0 to $2N$.

REMARK 6.5 *Theorem 6.3 provides a complete characterization for the s.d.o.f. of the channel. The s.d.o.f. at $N_c = N + N_e$ is equal to N, which is the degrees of freedom (d.o.f.) of the $(N \times N)$ MIMO Gaussian channel with no secrecy constraint. Thus, increasing N_c over $N + N_e$ cannot increase the s.d.o.f. over N.*

Converse and achievability proofs of Theorem 6.3 are provided next in Sections 6.4.1 and 6.4.2, respectively.

6.4.1 Converse for Theorem 6.3

The converse for Theorem 6.3 is established by combining two upper bounds for the s.d.o.f. The first bound holds for $0 \leq N_c \leq N_e$ and relies on allowing for full cooperation between the transmitter and cooperative jammer, and then utilizing the high-SNR characterization in Theorem 6.2 for the resulting MIMO wiretap channel. The second bound extends the converse for the single-antenna case in [21] to the channel with multiple antenna terminals, and holds for $\max\{N, N_e\} \leq N_c \leq N + N_e$.

Upper Bound 1: $0 \leq N_c \leq N_e$

Allowing full cooperation between the transmitter and cooperative jammer, with the channel gains drawn from a continuous distribution, yields an $(N + N_c \times N \times N_e)$ MIMO Gaussian wiretap channel. Its s.d.o.f. provides an upper bound as cooperation between these two entities cannot decrease the secrecy rate. Using (6.27), we have

$$D_s \leq [N + N_c - N_e]^+, \qquad 0 \leq N_c \leq N_e. \tag{6.33}$$

Upper Bound 2: $\max\{N, N_e\} \leq N_c \leq N + N_e$

The converse proof for the single-antenna Gaussian wiretap channel with a single-antenna cooperative jammer in [21] is established by (1) deriving an upper bound for the achievable secrecy rate which represents the loss in the rate due to the secrecy constraint and provides a necessary condition for the jamming signal to achieve secrecy at high SNR, and (2) combining this bound with another upper bound on the secrecy rate which represents the maximum possible reliable rate when such a jamming signal is utilized. For the multi-antenna channel, the loss due to the secrecy constraint, the necessary condition for the jamming signal to achieve secrecy, and the reliable rate with the existence of such a jamming signal, vary according to the relative number of antennas at different terminals, and hence different cases need to be handled separately. In particular, when $N_t = N_r = N$, the cases $N \geq N_e$ and $N < N_e$ are considered. Additionally, when $N < N_e$, in order to obtain a tight upper bound, the reliable rate of the channel with a jamming signal that achieves the necessary condition for secrecy is further upper bounded by the reliable rate of an enhanced channel with fewer jamming antennas at the cooperative jammer [7, Section IV-B].

Let ϕ_i, $i = 1, 2, \ldots, 7$, denote constants that do not depend on P. Using the secrecy and reliability constraints, the secrecy rate R_s is upper bounded as follows [8].

Let $\tilde{\mathbf{X}}_t = \mathbf{X}_t + \tilde{\mathbf{Z}}_t$ and $\tilde{\mathbf{X}}_c = \mathbf{X}_c + \tilde{\mathbf{Z}}_c$ be noisy versions of the transmitted signals, where $\tilde{\mathbf{Z}}_t \sim \text{CN}(\mathbf{0}, \mathbf{K}_t)$ and $\tilde{\mathbf{Z}}_c \sim \text{CN}(\mathbf{0}, \mathbf{K}_c)$ are independent from $\{\mathbf{X}_t, \mathbf{X}_c, \mathbf{Z}_r, \mathbf{Z}_e\}$ and of each other, and i.i.d. across time. Define $\tilde{\mathbf{Z}}_1 = -\mathbf{H}_t \tilde{\mathbf{Z}}_t - \mathbf{H}_c \tilde{\mathbf{Z}}_c + \mathbf{Z}_r$ and $\tilde{\mathbf{Z}}_2 = -\mathbf{G}_t \tilde{\mathbf{Z}}_t - \mathbf{G}_c \tilde{\mathbf{Z}}_c + \mathbf{Z}_e$, where $\tilde{\mathbf{Z}}_1 \sim \text{CN}(\mathbf{0}, \Sigma_{\tilde{\mathbf{Z}}_1})$, $\tilde{\mathbf{Z}}_2 \sim \text{CN}(\mathbf{0}, \Sigma_{\tilde{\mathbf{Z}}_2})$, $\Sigma_{\tilde{\mathbf{Z}}_1} = \mathbf{H}_t \mathbf{K}_t \mathbf{H}_t^H + \mathbf{H}_c \mathbf{K}_c \mathbf{H}_c^H + \mathbf{I}_N$, and $\Sigma_{\tilde{\mathbf{Z}}_2} = \mathbf{G}_t \mathbf{K}_t \mathbf{G}_t^H + \mathbf{G}_c \mathbf{K}_c \mathbf{G}_c^H + \mathbf{I}_{N_e}$. $\tilde{\mathbf{Z}}_1^n, \tilde{\mathbf{Z}}_2^n$ are i.i.d. sequences of $\tilde{\mathbf{Z}}_1, \tilde{\mathbf{Z}}_2$. We have

$$nR_s = H(W) - H(W|\mathbf{Y}_e^n) + H(W|\mathbf{Y}_e^n) - H(W|\mathbf{Y}_r^n) + H(W|\mathbf{Y}_r^n) \tag{6.34}$$

$$\leq H(W|\mathbf{Y}_e^n) - H(W|\mathbf{Y}_r^n, \mathbf{Y}_e^n) + n\phi_1 \tag{6.35}$$

$$= I(W; \mathbf{Y}_r^n|\mathbf{Y}_e^n) + n\phi_1 \tag{6.36}$$

$$\leq h(\mathbf{Y}_r^n|\mathbf{Y}_e^n) - h(\mathbf{Y}_r^n|W, \mathbf{Y}_e^n, \mathbf{X}_t^n, \mathbf{X}_c^n) + n\phi_1 \tag{6.37}$$

$$= h(\mathbf{Y}_r^n, \mathbf{Y}_e^n) - h(\mathbf{Y}_e^n) + n\phi_2 \tag{6.38}$$

$$= h(\mathbf{Y}_r^n, \mathbf{Y}_e^n, \tilde{\mathbf{X}}_t^n, \tilde{\mathbf{X}}_c^n) - h(\tilde{\mathbf{X}}_t^n, \tilde{\mathbf{X}}_c^n|\mathbf{Y}_r^n, \mathbf{Y}_e^n) - h(\mathbf{Y}_e^n) + n\phi_2 \tag{6.39}$$

$$\leq h(\tilde{\mathbf{X}}_t^n, \tilde{\mathbf{X}}_c^n) + h(\mathbf{Y}_r^n, \mathbf{Y}_e^n|\tilde{\mathbf{X}}_t^n, \tilde{\mathbf{X}}_c^n) - h(\tilde{\mathbf{Z}}_t^n, \tilde{\mathbf{Z}}_c^n) - h(\mathbf{Y}_e^n) + n\phi_2 \tag{6.40}$$

$$= h(\tilde{\mathbf{X}}_t^n) + h(\tilde{\mathbf{X}}_c^n) + h(\tilde{\mathbf{Z}}_1^n|\tilde{\mathbf{X}}_t^n, \tilde{\mathbf{X}}_c^n) + h(\tilde{\mathbf{Z}}_2^n|\tilde{\mathbf{X}}_t^n, \tilde{\mathbf{X}}_c^n) - h(\mathbf{Y}_e^n) + n\phi_3 \tag{6.41}$$

$$\leq h(\tilde{\mathbf{X}}_t^n) + h(\tilde{\mathbf{X}}_c^n) + h(\tilde{\mathbf{Z}}_1^n) + h(\tilde{\mathbf{Z}}_2^n) - h(\mathbf{Y}_e^n) + n\phi_3 \tag{6.42}$$

$$= h(\tilde{\mathbf{X}}_t^n) + h(\tilde{\mathbf{X}}_c^n) - h(\mathbf{Y}_e^n) + n\phi_4, \tag{6.43}$$

where (6.35) follows from the secrecy and reliability constraints and Fano's inequality, (6.38) follows since $h(\mathbf{Y}_r^n|W, \mathbf{Y}_e^n, \mathbf{X}_t^n, \mathbf{X}_c^n) = h(\mathbf{Z}_r^n)$, where $\mathbf{Z}_r^n$ is independent from $\{W, \mathbf{Y}_e^n, \mathbf{X}_t^n, \mathbf{X}_c^n\}$, and (6.40) follows because

$$h(\tilde{\mathbf{X}}_t^n, \tilde{\mathbf{X}}_c^n|\mathbf{Y}_r^n, \mathbf{Y}_e^n) \geq h(\tilde{\mathbf{X}}_t^n, \tilde{\mathbf{X}}_c^n|\mathbf{Y}_r^n, \mathbf{Y}_e^n, \mathbf{X}_t^n, \mathbf{X}_c^n) = h(\tilde{\mathbf{Z}}_t^n, \tilde{\mathbf{Z}}_c^n). \tag{6.44}$$

The matrices $\mathbf{K}_t$ and $\mathbf{K}_c$ are chosen as $\mathbf{K}_t = \rho^2 \mathbf{I}_N$, $\mathbf{K}_c = \rho^2 \mathbf{I}_{N_c}$, with $0 < \rho \leq 1/\max\{\|\mathbf{H}_c^H\|_{\text{in}}, \sqrt{\|\mathbf{G}_t^H\|_{\text{in}}^2 + \|\mathbf{G}_c^H\|_{\text{in}}^2}\}$, to guarantee the finiteness of $\{\phi_i\}$, and to define stochastically equivalent versions of $\mathbf{Y}_r^n$ and $\mathbf{Y}_e^n$, which are needed in the analysis.

We now consider the following two cases.

Case 1: $N_e \leq N$

A stochastically equivalent form of $\mathbf{Z}_e$ is $\mathbf{Z}'_e = \mathbf{G}_t \tilde{\mathbf{Z}}_t + \tilde{\mathbf{Z}}_e$, where $\tilde{\mathbf{Z}}_e \sim \mathcal{CN}(\mathbf{0}, \mathbf{I}_{N_e} - \mathbf{G}_t \mathbf{K}_t \mathbf{G}_t^H)$ is independent from $\{\tilde{\mathbf{Z}}_t, \tilde{\mathbf{Z}}_c, \mathbf{X}_t, \mathbf{X}_c, \mathbf{Z}_r\}$. Thus,

$$\mathbf{Y}'^n_e = \mathbf{G}_t \tilde{\mathbf{X}}^n_t + \mathbf{G}_c \mathbf{X}^n_c + \tilde{\mathbf{Z}}^n_e \tag{6.45}$$

is a stochastically equivalent form of $\mathbf{Y}^n_e$. Let $\tilde{\mathbf{X}}_t = [\tilde{\mathbf{X}}^T_{t_1} \ \tilde{\mathbf{X}}^T_{t_2}]^T$, $\tilde{\mathbf{X}}_{t_1} \in \mathbb{C}^{N_e}$, $\tilde{\mathbf{X}}_{t_2} \in \mathbb{C}^{N-N_e}$. Let $\mathbf{G}_t = [\mathbf{G}_{t_1} \ \mathbf{G}_{t_2}]$, $\mathbf{G}_{t_1} \in \mathbb{C}^{N_e \times N_e}$, $\mathbf{G}_{t_2} \in \mathbb{C}^{N_e \times (N-N_e)}$. Hence,

$$h(\mathbf{Y}^n_e) = h(\mathbf{Y}'^n_e) \geq h(\mathbf{G}_t \tilde{\mathbf{X}}^n_t) \tag{6.46}$$

$$\geq h(\mathbf{G}_{t_1} \tilde{\mathbf{X}}^n_{t_1} | \tilde{\mathbf{X}}^n_{t_2}) \tag{6.47}$$

$$= h(\tilde{\mathbf{X}}^n_{t_1} | \tilde{\mathbf{X}}^n_{t_2}) + n \log |\det(\mathbf{G}_{t_1})|, \tag{6.48}$$

where (6.46) follows due to independence of $\mathbf{G}_t \tilde{\mathbf{X}}^n_t$ and $\mathbf{G}_c \mathbf{X}^n_c + \tilde{\mathbf{Z}}^n_e$. Substituting (6.48) in (6.43) yields

$$nR_s \leq h(\tilde{\mathbf{X}}^n_{t_2}) + h(\tilde{\mathbf{X}}^n_c) + n\phi_5. \tag{6.49}$$

Let $\tilde{\mathbf{X}}_c = [\tilde{\mathbf{X}}^T_{c_1} \ \tilde{\mathbf{X}}^T_{c_2}]^T$, $\tilde{\mathbf{X}}_{c_1} \in \mathbb{C}^N$, $\tilde{\mathbf{X}}_{c_2} \in \mathbb{C}^{N_c-N}$. Let $\mathbf{H}_c = [\mathbf{H}_{c_1} \ \mathbf{H}_{c_2}]$, $\mathbf{H}_{c_1} \in \mathbb{C}^{N \times N}$, $\mathbf{H}_{c_2} \in \mathbb{C}^{N \times (N_c-N)}$. Using the reliability constraint (6.5), we obtain a bound on R_s as [8]

$$nR_s \leq h(\mathbf{Y}^n_r) - h(\tilde{\mathbf{X}}^n_{c_1} | \tilde{\mathbf{X}}^n_{c_2}) - n \log |\det(\mathbf{H}_{c_1})|. \tag{6.50}$$

Summing (6.49) and (6.50) yields [7]

$$R_s \leq \frac{1}{2n} \left\{ h(\mathbf{Y}^n_r) + h(\tilde{\mathbf{X}}^n_{t_2}) + h(\tilde{\mathbf{X}}^n_{c_2}) \right\} + \phi_6 \tag{6.51}$$

$$\leq \frac{N + N_c - N_e}{2} \log P + \phi_7. \tag{6.52}$$

Thus, $D_s \leq (N + N_c - N_e)/2$.

Case 2: $N_e > N$

Let us write $\mathbf{X}_c$ and $\mathbf{H}_c$ as $\mathbf{X}_c = [\mathbf{X}'^T_{c_1} \ \mathbf{X}'^T_{c_2}]^T$, $\mathbf{H}_c = [\mathbf{H}'_{c_1} \ \mathbf{H}'_{c_2}]$, where $\mathbf{X}'_{c_1} \in \mathbb{C}^{N_e-N}$, $\mathbf{X}'_{c_2} = [\mathbf{X}'^T_{c_{21}} \ \mathbf{X}'^T_{c_{22}}]^T$, $\mathbf{X}'_{c_{21}} \in \mathbb{C}^N$, $\mathbf{X}'_{c_{22}} \in \mathbb{C}^{N_c-N_e}$, and $\mathbf{H}'_{c_1} \in \mathbb{C}^{N \times (N_e-N)}$, $\mathbf{H}'_{c_2} = [\mathbf{H}'_{c_{21}} \ \mathbf{H}'_{c_{22}}]$, $\mathbf{H}'_{c_{21}} \in \mathbb{C}^{N \times N}$, $\mathbf{H}'_{c_{22}} \in \mathbb{C}^{N \times (N_c-N_e)}$. Consider a modified channel where the cooperative jammer uses only the last $N_c + N - N_e$ of its N_c antennas. The transmitted signals in the modified channel are $\mathbf{X}^n_t$ and $\mathbf{X}'^n_{c_2}$, and the receiver gets

$$\bar{\mathbf{Y}}^n_r = \mathbf{H}_t \mathbf{X}^n_t + \mathbf{H}'_{c_2} \mathbf{X}'^n_{c_2} + \mathbf{Z}^n_r. \tag{6.53}$$

Since the cooperative jamming signal is a form of additive noise, the reliable rate of the modified channel, $\bar{R}$, is an upper bound for that of the original channel, R. Thus,

$$nR_s \leq nR \leq n\bar{R} \leq I(\mathbf{X}^n_t; \bar{\mathbf{Y}}^n_r) \tag{6.54}$$

$$\leq h(\bar{\mathbf{Y}}^n_r) - h(\tilde{\mathbf{X}}'^n_{c_{21}} | \tilde{\mathbf{X}}'^n_{c_{22}}) - n \log |\det(\mathbf{H}'_{c_{21}})|. \tag{6.55}$$

Let $\mathbf{G}_c = [\mathbf{G}_{c_1}\ \mathbf{G}_{c_2}]$, $\mathbf{G}_{c_1} \in \mathbb{C}^{N_e \times (N_e - N)}$, $\mathbf{G}_{c_2} \in \mathbb{C}^{N_e \times (N_c + N - N_e)}$. Another stochastically equivalent form of $\mathbf{Z}_e$ is $\mathbf{Z}_e'' = \mathbf{G}_t \tilde{\mathbf{Z}}_t + \mathbf{G}_c \tilde{\mathbf{Z}}_c + \tilde{\mathbf{Z}}_e'$, where $\tilde{\mathbf{Z}}_e' \sim \mathrm{CN}(0, \mathbf{I}_{N_e} - \mathbf{G}_t \mathbf{K}_t \mathbf{G}_t^H - \mathbf{G}_c \mathbf{K}_c \mathbf{G}_c^H)$. Thus,

$$h(\mathbf{Y}_e^n) = h(\mathbf{G}_t \tilde{\mathbf{X}}_t^n + \mathbf{G}_c \tilde{\mathbf{X}}_c^n + \tilde{\mathbf{Z}}_e'^n) \tag{6.56}$$

$$\geq h(\tilde{\mathbf{X}}_t^n) + h(\tilde{\mathbf{X}}_{c_1}'^n | \tilde{\mathbf{X}}_{c_2}'^n) + n \log |\det[\mathbf{G}_t\ \mathbf{G}_{c_1}]|. \tag{6.57}$$

Using (6.43), (6.55), and (6.57) yields

$$R_s \leq \frac{1}{2n}\{h(\bar{\mathbf{Y}}_r^n) + h(\tilde{\mathbf{X}}_{c22}'^n)\} + \phi_5 \leq \frac{N + N_c - N_e}{2} \log P + \phi_6. \tag{6.58}$$

Thus, $D_s \leq (N + N_c - N_e)/2$.

Obtaining the Bound

For $N_e \leq N$, upper bound 1 is equal to $N + N_c - N_e$ for $0 \leq N_c \leq N_e$, while upper bound 2, evaluated at $N_c = N$, is equal to $N - \frac{N_e}{2}$. Since $N - \frac{N_e}{2} < N + N_c - N_e$ for all $\frac{N_e}{2} < N_c \leq N$, then $D_s \leq N - \frac{N_e}{2}$ for all $\frac{N_e}{2} < N_c \leq N$. Using a similar argument when $N_e > N$, we have $D_s \leq \frac{N}{2}$ for all $N_e - \frac{N}{2} < N_c \leq N_e$. Combining these statements with the two bounds derived above, we obtain the upper bound in (6.32), which completes the converse proof for Theorem 6.3.

6.4.2 Achievability for Theorem 6.3

The achievability for Theorem 6.3 is established by proposing distinct achievable schemes for ten different cases of the relative number of antennas at the different terminals [7]. The need for these distinct schemes comes from the need for utilizing different signaling, precoding, and decoding strategies for different cases of the number of antennas. In particular, it is shown that Gaussian signaling, cooperative jamming, and linear receiver processing are sufficient to achieve an integer-valued[5] s.d.o.f., while achieving a non-integer s.d.o.f. requires utilizing structured signaling and jamming along with a combination of linear processing and signal scale alignment at the receiver [7, 22]. Additionally, for all cases, the precoders at the transmitter and cooperative jammer are chosen to perfectly align the information and cooperative jamming signals at the eavesdropper; however, the precoder design that achieves this condition is different for the cases $N \geq N_e$ and $N < N_e$.

Signal scale alignment considered in this proof is an extension of real interference alignment to complex channels [7, 8, 22], where different results than those used for real channels in [23] are utilized. In particular, the extension to complex channels is based on a result in the field of classification of transcendental complex numbers, which provides a bound on the absolute value of a complex algebraic number with rational coefficients in terms of its height (the maximum coefficient) [24, 25]. For

---

[5] From (6.32), the s.d.o.f. is not an integer when (1) $\min\{N, N_e\}$ is odd for $N_e - N_{\min} < N_c \leq N_{\max}$, (2) $N + N_c - N_e$ is odd for $N_{\max} < N_c \leq N + N_e$, and is integer valued otherwise.

complex channels, this result plays the same role as the Khintchine–Groshev theorem in Diophantine approximation [26] for channels with real coefficients.

The n-letter signals $\mathbf{X}_t^n, \mathbf{X}_c^n$ are i.i.d. sequences. Since $\mathbf{X}_c^n, \mathbf{X}_t^n$ are independent, we have a memoryless wiretap channel, and by setting $U = \mathbf{X}_t$ in (6.7), the secrecy rate

$$R_s = [I(\mathbf{X}_t; \mathbf{Y}_r) - I(\mathbf{X}_t; \mathbf{Y}_e)]^+ \tag{6.59}$$

is achievable. The transmitted signals are

$$\mathbf{X}_t = \mathbf{P}_t \mathbf{U}_t, \quad \mathbf{X}_c = \mathbf{P}_c \mathbf{V}_c, \tag{6.60}$$

where $\mathbf{U}_t = [U_1 \cdots U_d]^T$, $\mathbf{V}_c = [V_1 \cdots V_l]^T$ are the information and cooperative jamming streams, and $\mathbf{P}_t, \mathbf{P}_c$ are the transmitter and cooperative jammer precoders.

Case 1: $N_e \leq N, 0 \leq N_c \leq \frac{N_e}{2}$

For this case, $D_s = N + N_c - N_e$ is integer valued, for which $\mathbf{U}_t \sim \mathrm{CN}(\mathbf{0}, \bar{P} \mathbf{I}_d)$, $\mathbf{V}_c \sim \mathrm{CN}(\mathbf{0}, \bar{P} \mathbf{I}_l)$, with $d = N + N_c - N_e$ and $l = N_c$, are utilized, where $\bar{P} = \frac{1}{\alpha} P$, and α is a constant that satisfies the power constraints. Since $N_e \leq N$, the transmitter exploits this advantage by sending $N - N_e$ of its d streams orthogonal to the eavesdropper. In particular, choosing $\mathbf{P}_c = \mathbf{I}_{N_c}$, and designing $\mathbf{P}_t$ as

$$\mathbf{P}_t = \begin{bmatrix} \mathbf{P}_{t,a} & \mathbf{P}_{t,n} \end{bmatrix}, \tag{6.61}$$

with $\mathbf{P}_{t,a} = \mathbf{G}_t^\dagger \mathbf{G}_c$ in order to align the cooperative jamming and information streams at the eavesdropper, and the $N - N_e$ columns of $\mathbf{P}_{t,n}$ span $\mathrm{null}(\mathbf{G}_t)$, are sufficient.

Since $N_c \leq \frac{N_e}{2}$, the number of received streams at the receiver, $2N_c + N - N_e$, does not exceed its available spatial dimensions, N. Thus, the receiver can decode all the information and jamming streams at high SNR. Reference [7] has showed that $D_s = N + N_c - N_e$ is achievable for this case; see [7, Section V-A] for details.

Case 2: $N_e \leq N$, $\frac{N_e}{2} < N_c \leq N$, **and** N_e **Even**

The s.d.o.f. is equal to $D_s = N - \frac{N_e}{2}$ for all N_c in this case. For N_e even and $\frac{N_e}{2} < N_c \leq N$, this s.d.o.f. is achievable by the scheme for Case 1, with $d = N - \frac{N_e}{2}$ and the cooperative jammer using only $\frac{N_e}{2}$ of its N_c antennas.

Case 3: $N_e \leq N$, $\frac{N_e}{2} < N_c \leq N$, **and** N_e **Odd**

For N_e odd, $D_s = N - \frac{N_e}{2}$ is not an integer. Gaussian signaling cannot achieve fractional s.d.o.f. for the channel since fractional d.o.f. implies sharing at least one spatial dimension at the receiver between information and cooperative jamming signals, which provides zero d.o.f. when shared between Gaussian information and jamming signals with the same power scaling. Instead, $d = N - \frac{N_e - 1}{2}$ and $l = \frac{N_e + 1}{2}$ structured information and jamming streams are utilized, and *joint* signal space and signal scale alignment [8, 22] is employed.

The cooperative jammer utilizes only $\frac{N_e + 1}{2}$ of its N_c antennas, i.e.,

$$\mathbf{P}_c = \begin{bmatrix} \mathbf{I}_l \\ \mathbf{0}_{(N_c - l) \times l} \end{bmatrix}, \tag{6.62}$$

and $\mathbf{P}_t$ is chosen as

$$\mathbf{P}_t = \begin{bmatrix} \mathbf{P}_{t,a} & \mathbf{P}_{t,n} \end{bmatrix}, \tag{6.63}$$

$\mathbf{P}_{t,a} = \mathbf{G}_t^\dagger \mathbf{G}_c \mathbf{P}_c$, and $\mathbf{P}_{t,n}$ is defined as in (6.61).

The structured streams in (6.60) are described as follows: $U_i = U_{i,\text{Re}} + jU_{i,\text{Im}}$, $V_k = V_{k,\text{Re}} + jV_{k,\text{Im}}$, $i = 2, \ldots, d$, and $k = 2, \ldots, l$. $U_1, V_1, \{U_{i,\text{Re}}\}, \{U_{i,\text{Im}}\}, \{V_{i,\text{Re}}\}$, and $\{V_{i,\text{Im}}\}$ are i.i.d. uniform over $a\{-Q, \ldots, Q\}$, where a and integer Q satisfy the power constraints

$$Q = P^{\frac{1-\epsilon}{2+\epsilon}} - \nu, \qquad a = \gamma P^{\frac{3\epsilon}{2(2+\epsilon)}}, \tag{6.64}$$

with $\epsilon > 0$ arbitrarily small, and constants ν, γ. With these $\mathbf{U}_t$ and $\mathbf{V}_c$, it is shown that $I(\mathbf{X}_t; \mathbf{Y}_e)$ is bounded as [7]

$$I(\mathbf{X}_t; \mathbf{Y}_e) \leq N_e. \tag{6.65}$$

The decoding scheme at the receiver is described as follows. The receiver (1) projects its signal over a direction that is orthogonal to all streams but U_1, V_1, (2) decodes U_1, V_1 from the projection using the complex-field extension of real interference alignment, (3) removes U_1, V_1 from its received signal to decode the remaining information streams.

In particular, let $\mathbf{A} = \mathbf{H}_t \mathbf{P}_t = [\mathbf{a}_1 \cdots \mathbf{a}_d]$, $\mathbf{H}'_c = \mathbf{H}_c \mathbf{P}_c = [\mathbf{h}_{c,1} \cdots \mathbf{h}_{c,l}]$. Thus,

$$\mathbf{Y}_r = \mathbf{A}\mathbf{U}_t + \mathbf{H}'_c \mathbf{V}_c + \mathbf{Z}_r. \tag{6.66}$$

The receiver chooses $\mathbf{b} \in \mathbb{C}^N$ such that $\mathbf{b} \perp \text{span}\{\mathbf{a}_2, \ldots, \mathbf{a}_d, \mathbf{h}_{c,2}, \ldots, \mathbf{h}_{c,l}\}$ and obtains $\widetilde{\mathbf{Y}}_r = \mathbf{D}\mathbf{Y}_r = [\widetilde{Y}_{r_1} \ (\widetilde{\mathbf{Y}}_{r_2}^N)^T]^T$, where

$$\mathbf{D} = \begin{bmatrix} \mathbf{b}^H \\ \mathbf{0}_{(N-1)\times 1} & \mathbf{I}_{N-1,} \end{bmatrix} \tag{6.67}$$

$$\widetilde{Y}_{r_1} = \mathbf{b}^H \mathbf{a}_1 U_1 + \mathbf{b}^H \mathbf{h}_{c,1} V_1 + \mathbf{b}^H \mathbf{Z}_r, \tag{6.68}$$

$$\widetilde{\mathbf{Y}}_{r_2}^N = \widetilde{\mathbf{A}}\mathbf{U}_t + \widetilde{\mathbf{H}}_c \mathbf{V}_c + \mathbf{Z}_{r_2}^N, \tag{6.69}$$

$\widetilde{\mathbf{A}} = [\tilde{\mathbf{a}}_1 \cdots \tilde{\mathbf{a}}_d]$, $\tilde{\mathbf{a}}_i = \mathbf{a}_{i_2}^N$, and $\widetilde{\mathbf{H}}_c = [\tilde{\mathbf{h}}_{c,1} \cdots \tilde{\mathbf{h}}_{c,l}]$, $\tilde{\mathbf{h}}_{c,i} = \mathbf{h}_{c,i_2}^N$. Note that $\mathbf{b}$ is not orthogonal to $\mathbf{a}_1, \mathbf{h}_{c,1}$ a.s. due to the randomly generated channel gains.

Before continuing with the decoding scheme, let us first state the following definition.

DEFINITION 6.1 A set of complex numbers $\{z_1, \ldots, z_L\}$ are rationally independent if they are linearly independent over the field of rational numbers $\mathbb{Q}$, i.e., there does not exist $\{q_1, \ldots, q_L\}$ but the all-zero set such that $\sum_{i=1}^{L} q_i z_i = 0$.

The receiver uses $\widetilde{Y}_{r_1}$ in (6.68) to decode U_1, V_1 as follows: $\mathbf{b}^H \mathbf{a}_1$ and $\mathbf{b}^H \mathbf{h}_{c,1}$ are rationally independent a.s., and hence the mapping $(U_1, V_1) \mapsto \mathbf{b}^H \mathbf{a}_1 U_1 + \mathbf{b}^H \mathbf{h}_{c,1} V_1$ is invertible [23]. The receiver uses a hard decision decoder which maps $\widetilde{Y}_{r_1}$ to the nearest point in the constellation $\mathcal{R}_1 = \mathbf{b}^H \mathbf{a}_1 \mathcal{U}_1 + \mathbf{b}^H \mathbf{h}_{c,1} \mathcal{V}_1$, where $\mathcal{U}_1, \mathcal{V}_1 = a\{-Q, \ldots, Q\}$. Then, the receiver passes the output of the hard decision decoder through the invertible map $\mathbf{b}^H\mathbf{a}_1 U_1 + \mathbf{b}^H\mathbf{h}_{c,1} V_1 \mapsto (U_1, V_1)$ to decode both U_1 and V_1. Next, the receiver removes the decoded U_1, V_1 from $\widetilde{\mathbf{Y}}_{r_2}^N$ in (6.69), and performs zero-forcing to obtain

$$\widehat{\mathbf{Y}}_r = \begin{bmatrix} \mathbf{U}_{t2}^d \\ \mathbf{V}_{c2}^l \end{bmatrix} + \mathbf{B}^{-1} \mathbf{Z}_{r2}^N, \tag{6.70}$$

from which it decodes $U_2, \ldots, U_d$, where

$$\mathbf{B} = \begin{bmatrix} \tilde{\mathbf{a}}_2 \cdots \tilde{\mathbf{a}}_d \, \tilde{\mathbf{h}}_{c,2} \cdots \tilde{\mathbf{h}}_{c,l} \end{bmatrix} \in \mathbb{C}^{(N-1)\times(N-1)} \tag{6.71}$$

is shown to be full rank a.s. using [7, Lemma 1].

It is shown in [7, 8] that $I(\mathbf{X}_t; \mathbf{Y}_r)$ is lower bounded as

$$I(\mathbf{X}_t; \mathbf{Y}_r) \geq \left[2N - N_e - \exp\left(\frac{-d_{\min}^2}{4\|\mathbf{b}\|^2} \right) \right] \log(2Q+1) + o(\log P), \tag{6.72}$$

where $d_{\min}$ is the minimum distance between the points in $\mathcal{R}_1$. Thus, it remains to lower bound $d_{\min}$. To do so, we utilize the following result from number theory:

LEMMA 6.1 *[24, 25] For almost all $\mathbf{z} \in \mathbb{C}^n$ and for all $\epsilon > 0$,*

$$|p + \mathbf{z}.\mathbf{q}| > (\max_i q_i)^{-\frac{(n-1+\epsilon)}{2}} \tag{6.73}$$

holds for all $\mathbf{q} \in \mathbb{Z}^n$ and $p \in \mathbb{Z}$ except for finitely many of them.

Since the number of integers that violates the inequality in (6.73) is finite, there exists a constant κ such that, for almost all $\mathbf{z} \in \mathbb{C}^n$ and all $\epsilon > 0$, the inequality

$$|p + \mathbf{z}.\mathbf{q}| > \kappa (\max_i q_i)^{-\frac{(n-1+\epsilon)}{2}} \tag{6.74}$$

holds for all $\mathbf{q} \in \mathbb{Z}^n$ and $p \in \mathbb{Z}$.

Using (6.74), we have, for almost all channel gains, that $d_{\min}$ is lower bounded as

$$d_{\min} = \inf_{U_1, V_1 \in \{-2Q, \ldots, 2Q\}} a|\mathbf{b}^H \mathbf{a}_1| \left| U_1 + \frac{\mathbf{b}^H \mathbf{h}_{c,1}}{\mathbf{b}^H \mathbf{a}_1} V_1 \right| \tag{6.75}$$

$$\geq \kappa \frac{a|\mathbf{b}^H \mathbf{a}_1|}{(2Q)^{\frac{\epsilon}{2}}} \geq \kappa \gamma |f_1| 2^{-\frac{\epsilon}{2}} P^{\frac{\epsilon}{2}}. \tag{6.76}$$

Substituting (6.76) in (6.72) and using the result with (6.64) and (6.65), we get

$$R_s \geq \frac{1-\epsilon}{2+\epsilon}(2N - N_e)\log P + o(\log P), \tag{6.77}$$

and hence

$$D_s \geq \frac{(1-\epsilon)(2N - N_e)}{2+\epsilon}. \tag{6.78}$$

Since $\epsilon > 0$ is arbitrarily small, $D_s = N - \frac{N_e}{2}$ is achievable.

Case 4: $N_e \leq N$, $N < N_c \leq N + N_e$, **and** $N + N_c - N_e$ **Even**

For this case, $D_s = \frac{N+N_c-N_e}{2}$ is integer valued, and hence $\mathbf{U}_t \sim \text{CN}(\mathbf{0}, \bar{P}\mathbf{I}_d)$, $\mathbf{V}_c \sim \text{CN}(\mathbf{0}, \bar{P}\mathbf{I}_l)$, with $d = \frac{N+N_c-N_e}{2}$ and $l = \frac{N_c+N_e-N}{2}$, are utilized. The cooperative jammer chooses $\mathbf{P}_c$ such that $N_c - N$ of its streams are sent orthogonal to the receiver, leaving only $g = \frac{N_e+N-N_c}{2}$ streams visible, in order to allow for more space for the information streams. In particular, $\mathbf{P}_c$ is given by

$$\mathbf{P}_c = [\mathbf{P}_{c,I} \; \mathbf{P}_{c,n}], \tag{6.79}$$

where

$$\mathbf{P}_{c,I} = \begin{bmatrix} \mathbf{I}_g \\ \mathbf{0}_{(N_c-g) \times g} \end{bmatrix}, \tag{6.80}$$

and the columns of $\mathbf{P}_{c,n}$ span null($\mathbf{H}_c$). $\mathbf{P}_t$ is chosen as in (6.63). At high SNR, the receiver can decode the d information and the g cooperative jamming streams. Independence of the received directions at the receiver is shown by [7, Lemma 1], and $D_s = \frac{N_c+N_e-N}{2}$ is shown to be achievable [7, Section V-D].

Case 5: $N_e \leq N$, $N < N_c \leq N + N_e$, **and** $N + N_c - N_e$ **Odd**

The achievable scheme utilizes the precoding in Case 4, and the structured streams and decoding of Case 3. $D_s = \frac{N_c+N_e-N}{2}$ is shown to be achievable [7, Section V-E].

Case 6: $N_e > N$ **and** $N_e - N < N_c \leq N_e - \frac{N}{2}$

Since $N_e > N$, the transmitter, unlike the previous five cases, (1) cannot send any information streams orthogonal to the eavesdropper, and (2) is not sufficient for achieving perfect alignment at the eavesdropper alone. Thus, both $\mathbf{P}_t$ and $\mathbf{P}_c$ have to take part in achieving the alignment. For this case, $D_s = N + N_c - N_e$ is integer valued, and hence we utilize $\mathbf{U}_t, \mathbf{V}_c \sim \text{CN}(\mathbf{0}, \bar{P}\mathbf{I}_d)$, where $d = l = N + N_c - N_e$. The precoders $\mathbf{P}_t, \mathbf{P}_c$ are chosen as follows: Let $\mathbf{G}_{\text{com}} = [\mathbf{G}_t \; -\mathbf{G}_c]$, and let $\mathbf{Q}_{\text{com}}$ be randomly chosen such that its columns span null($\mathbf{G}_{\text{com}}$). Write $\mathbf{Q}_{\text{com}}$ as $\mathbf{Q}_{\text{com}} = [\mathbf{Q}_{\text{com},1}^T \; \mathbf{Q}_{\text{com},2}^T]^T$, $\mathbf{Q}_{\text{com},1} \in \mathbb{C}^{N \times d}$ and $\mathbf{Q}_{\text{com},2} \in \mathbb{C}^{N_c \times d}$. Set $\mathbf{P}_t = \mathbf{Q}_{\text{com},1}$ and $\mathbf{P}_c = \mathbf{Q}_{\text{com},2}$. This choice for $\mathbf{P}_t, \mathbf{P}_c$ results in $\mathbf{G}_t\mathbf{P}_t = \mathbf{G}_c\mathbf{P}_c$, and hence the alignment condition is achieved. $D_s = N + N_c - N_e$ is shown to be achievable [7, Section V-F].

Case 7: $N_e > N$, $N_e - \frac{N}{2} < N_c \leq N_e$, **and** N **Even**

The scheme in Case 6, with $N_c = N_e - \frac{N}{2}$, can be used to achieve the s.d.o.f. for all $N_e - \frac{N}{2} < N_c \leq N_e$ with one difference: here, the $\frac{N}{2}$ columns of $\mathbf{Q}_{\text{com}}$ are randomly chosen as linearly independent vectors from null($\mathbf{G}_{\text{com}}$). $D_s \geq \frac{N}{2}$ is achievable.

Case 8: $N_e > N$, $N_e - \frac{N}{2} < N_c \leq N_e$, **and** N **Odd**

For this case, $D_s = \frac{N}{2}$ is not an integer. Using precoders as in Case 6 and the structured streams and decoding scheme as in Case 3 achieves the s.d.o.f.

Case 9: $N_e > N$, $N_e < N_c \leq N + N_e$, **and** $N + N_c - N_e$ **Even**

For this case, d Gaussian information and d Gaussian jamming streams are sent, where $d = \frac{N+N_c-N_e}{2}$. Since $N_c > N_e$, the cooperative jammer designs its precoder so that it sends $N_c - N_e$ jamming streams orthogonal to the receiver, and, jointly with the transmitter, achieves the alignment condition.

In particular, let $\mathbf{P}_t = [\mathbf{P}_{t,1}\ \mathbf{P}_{t,2}]$, $\mathbf{P}_c = [\mathbf{P}_{c,1}\ \mathbf{P}_{c,2}]$, where $\mathbf{P}_{t,1} \in \mathbb{C}^{N \times g}$, $\mathbf{P}_{t,2} \in \mathbb{C}^{N \times (N_c - N_e)}$, $\mathbf{P}_{c,1} \in \mathbb{C}^{N_c \times g}$, $\mathbf{P}_{c,2} \in \mathbb{C}^{N_c \times (N_c - N_e)}$, and $g = \frac{N_e + N - N_c}{2}$. $\mathbf{P}_t$ and $\mathbf{P}_c$ are chosen as follows. Let $\mathbf{G}_{\text{com}} = [\mathbf{G}_t\ -\mathbf{G}_c]$, and let

$$\mathbf{G}'_{\text{com}} = \begin{bmatrix} \mathbf{G}_t & -\mathbf{G}_c \\ \mathbf{0}_{N \times N} & \mathbf{H}_c \end{bmatrix}. \tag{6.81}$$

Let $\mathbf{Q}'_{\text{com}}$ be chosen randomly so that its columns span $\text{null}(\mathbf{G}'_{\text{com}})$. Let the columns of $\mathbf{Q}_{\text{com}}$ be randomly chosen as linearly independent vectors in $\text{null}(\mathbf{G}_{\text{com}})$ and not in $\text{null}(\mathbf{G}'_{\text{com}})$. Write $\mathbf{Q}_{\text{com}} = [\mathbf{Q}^T_{\text{com},1}\ \mathbf{Q}^T_{\text{com},2}]^T$, $\mathbf{Q}'_{\text{com}} = [\mathbf{Q}'^T_{\text{com},1}\ \mathbf{Q}'^T_{\text{com},2}]^T$, and set $\mathbf{P}_{t,1} = \mathbf{Q}_{\text{com},1}$, $\mathbf{P}_{t,2} = \mathbf{Q}'_{\text{com},1}$, $\mathbf{P}_{c,1} = \mathbf{Q}_{\text{com},2}$, $\mathbf{P}_{c,2} = \mathbf{Q}'_{\text{com},2}$. This choice for $\mathbf{P}_t, \mathbf{P}_c$ results in $\mathbf{G}_t \mathbf{P}_t = \mathbf{G}_c \mathbf{P}_c$ and $\mathbf{H}_c \mathbf{P}_{c,2} = \mathbf{0}_{N \times (N_c - N_e)}$. $D_s \geq \frac{N+N_c-N_e}{2}$ is shown to be achievable [7, Section V-I].

Case 10: $N_e > N$, $N_e < N_c \leq N + N_e$, **and** $N + N_c - N_e$ **Odd**

The s.d.o.f. for this case is achieved by using the precoders in Case 9, and the structured streams and decoding schemes in Case 3.

This completes the proof for Theorem 6.3.

6.4.3 Interpretation of Results

Equation (6.32) in Theorem 6.1 shows the behavior of the s.d.o.f. associated with increasing N_c from 0 to $N + N_e$. For the N_c range from 0 to $N_e - \lceil \frac{\min\{N,N_e\}}{2} \rceil$, every additional antenna at the cooperative jammer provides the channel with one additional secure degrees of freedom. Perhaps surprisingly, increasing N_c in the range from $N_e - \lfloor \frac{\min\{N,N_e\}}{2} \rfloor$ to $\max\{N,N_e\}$ does not increase the s.d.o.f. When N_c exceeds $\max\{N,N_e\}$, the s.d.o.f. again increases by increasing N_c, until it arrives at its maximum value, N, where each additional antenna at the cooperative jammer in this range provides the channel with an additional $\frac{1}{2}$ secure degrees of freedom.

The reason for this behavior can be explained as follows. Achieving the secrecy constraint in (6.6) at high SNR requires that the entropy of the cooperative jamming signal, $\mathbf{X}_c^n$, is greater than or equal to the entropy of the information signal visible to the eavesdropper, and that the jamming signal completely covers the information signal at the eavesdropper. For $0 \leq N_c \leq N_e - \frac{\min\{N,N_e\}}{2}$, at high SNR, the receiver signal space is sufficient to decode both the information and cooperative jamming signals, and hence every additional antenna at the cooperative jammer results in utilizing two more spatial dimensions at the receiver, one for an information signal and the other for a cooperative jamming signal. With information and cooperative jamming signals that achieve the aforementioned necessary conditions for satisfying the secrecy constraint, the receiver signal space gets full at $N_e - \lfloor \frac{\min\{N,N_e\}}{2} \rfloor$. Thus, increasing N_c in

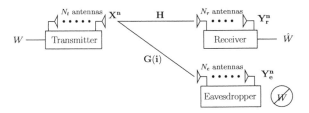

Figure 6.4 MIMO wiretap channel with arbitrary and varying eavesdropper channel.

the range from $N_e - \lfloor \frac{\min\{N,N_e\}}{2} \rfloor$ to $\max\{N, N_e\}$ does not increase the s.d.o.f. since the receiver signal space is already full, and no cooperative jamming signals can be sent orthogonal to the receiver for this case. When N_c exceeds $\max\{N, N_e\}$, the cooperative jammer starts sending signals orthogonal to the receiver, and every additional antenna at the cooperative jammer sets one dimension at the receiver free from interference, which is shared between extra information and jamming signals such that the secrecy constraint is satisfied at high SNR. When $N_e > N$, sending cooperative jamming signals orthogonal to the receiver, and yet satisfying the covering condition, requires that $N_c > N_e$.

6.5 MIMO Wiretap Channel with Unknown Eavesdropper Channel

So far, we have considered the models where the channel gain values of all terminals including those to the eavesdropper are known perfectly at the legitimate terminals. In the following, we shall remove the assumption on the knowledge of the channel state information to the eavesdropper (E-CSI). We will also henceforth consider the strong secrecy constraint. Both of these, namely removing possibly impractical CSI requirements and strengthening the secrecy guarantees, are steps toward bringing the vision of information theoretic security guarantees toward reality.

In this section, we consider the MIMO wiretap channel with the eavesdropper channel taking arbitrary values, varying from one channel use to the next, and known only to the eavesdropper, and under strong[6]. secrecy constraint; see Fig. 6.4. The results in this section are summarized from [10].

The received signals at the ith channel use are

$$\mathbf{Y}_r(i) = \mathbf{H}\mathbf{X}(i) + \mathbf{Z}_r(i), \tag{6.82}$$

$$\mathbf{Y}_e(i) = \mathbf{G}(i)\mathbf{X}(i), \tag{6.83}$$

where $\mathbf{X}(i)$ is the transmitted signal, $\mathbf{H}$ is the receiver channel, $\mathbf{G}(i)$ is the eavesdropper channel at the ith channel use, and $\mathbf{Z}_r(i)$ is the Gaussian noise at the receiver with

---

[6] The whole mutual information vanishes with increasing block length n; see Eq. (6.87)

i.i.d. zero mean and unit variance components. The channel $\mathbf{H}$ is static and known to all terminals, and the channel $\mathbf{G}(i)$ varies over the channel uses and is known only to the eavesdropper. The eavesdropper channel sequence $\mathbf{G}^n = [\mathbf{G}(1) \cdots \mathbf{G}(n)]$ is any arbitrary sequence; however, it is not adversarially chosen, i.e., the eavesdropper does not choose its channel states according to the signals it received from the transmitter over the previous channel uses. The eavesdropper channel in (6.83) is assumed to be noiseless as a worst-case assumption. Given an E-CSI realization, $\mathbf{G}^n = \mathbf{g}^n$, the wiretap channel is memoryless. The transmitter power constraint is expressed as in (6.3).

A message W, uniformly distributed over $\mathcal{W} = \{1, \ldots, 2^{nR_s}\}$, is to be sent reliably to the receiver, and kept secret from the eavesdropper for all possible realizations of the eavesdropper channel sequence. A $(2^{nR_s}, n)$ code for this model consists of:

1. Encoder: the encoder at the transmitter is given by

$$\mathbf{X}^n = f_n(W, \mathbf{H}), \tag{6.84}$$

where f_n is a stochastic encoder which does *not* depend on the eavesdropper channel.

2. Decoder: the decoder is a deterministic function ψ_n which outputs

$$\hat{W} = \psi_n(\mathbf{Y}_r^n, \mathbf{H}). \tag{6.85}$$

The probability of decoding error at the receiver $P_e^{(n)}$ is defined in (6.4). The secrecy rate R_s is achievable if there exists a sequence of channel codes $(2^{nR_s}, n)$ which satisfy

$$\lim_{n \to \infty} P_e^{(n)} = 0, \tag{6.86}$$

$$\lim_{n \to \infty} \sup_{\mathbf{g}^n} I(W; \mathbf{Y}_e^n | \mathbf{G}^n = \mathbf{g}^n) = 0. \tag{6.87}$$

The secrecy constraint in (6.87) is defined by the mutual information conditioned on $\mathbf{G}^n = \mathbf{g}^n$. While the modeling of $\mathbf{G}^n$ assumes it to be any arbitrary sequence of unknown – possibly not well defined – distribution, the joint distribution of $(W, \mathbf{Y}_e^n)$ conditioned on the realization of $\mathbf{G}^n$ is well defined. The supremum in (6.87) implies the independence of W and $\mathbf{Y}_e^n$ for any arbitrary $\mathbf{g}^n$. The following theorem provides an achievable *strong secrecy* rate for the channel described in (6.82) and (6.83).

THEOREM 6.4   [10, Proposition 1] *For the $(N_t \times N_r \times N_e)$ MIMO wiretap channel with legitimate channel $\mathbf{H}$, and eavesdropper channel $\mathbf{G}(i)$ that is time varying and known only to the eavesdropper, the following rate is achievable with strong secrecy:*

$$R_s \leq \begin{cases} \sup_{\sigma^2 > 0} \left[\sum_{i=1}^{N_m} \log\left(1 + \frac{s_i^2 P_t}{(1+s_i^2 \sigma^2) N_m}\right) - N_e \log\left(1 + \frac{P_t}{N_m \sigma^2}\right) \right]^+, & \text{if } N_m > N_e \\ 0, & \text{if } N_m \leq N_e, \end{cases} \tag{6.88}$$

where $N_m = \min\{N_t, N_r\}$, $s_i, i = 1, 2, \ldots, N_m$, are the singular values of $\mathbf{H}$, and $P_t = [P - N_m \sigma^2]^+$.

The proof is provided in Section 6.5.1.

REMARK 6.6 *The variable $N_m\sigma^2$ in (6.88) represents the power of artificial noise, i.e., the cooperative jamming signal[7] radiated isotropically in all directions by the transmitter. The transmission of this artificial noise is a crucial component of the achievable scheme for Theorem 6.4, which enables utilizing a signaling scheme that does not depend on the eavesdropper channel realization and yet achieves a positive strongly secure rate by generating sufficient noise at the eavesdropper in all directions. Note that P_t in (6.88) represents the remaining power used for transmitting the information signal.*

REMARK 6.7 *Unlike the semi-blind scheme in Section 6.3, where the artificial noise is transmitted orthogonal to the receiver, the achievable scheme for Theorem 6.4 involves transmitting artificial noise in all directions, including the direction to the receiver, since the directions of the eavesdropper channel can include a subset of the directions of the legitimate receiver's channel.*

REMARK 6.8 *The worst-case assumption of a noiseless eavesdropper channel is used to construct a universal coding scheme with a rate selection step that is, unlike for the semi-blind scheme in Section 6.3, independent from the eavesdropper channel realization. The secrecy rate in (6.88) achieved with this scheme is also achievable when additive noise is introduced to the eavesdropper channel.*

The achievable rate in (6.88) matches the converse in terms of the s.d.o.f.

THEOREM 6.5 *[10, Theorem 1] For a full-rank receiver channel $\mathbf{H}$, the s.d.o.f. for the MIMO wiretap channel defined in (6.82) and (6.83) is*

$$D_s = [N_m - N_e]^+, \quad (6.89)$$

where $N_m = \min\{N_t, N_r\}$.

Proof [10, Section IV-F] Theorem 6.5 follows by establishing a converse for the high-SNR characterization of the secrecy rate in (6.88). When $N_e \geq N_t$, since $\mathbf{G}(i)$ is arbitrary we can choose $\mathbf{G}(i) = [\mathbf{I}_{N_t}\ \mathbf{0}_{N_t \times (N_e - N_t)}]^T$ for all i, for which the eavesdropper receives $\mathbf{X}^n$ and the secrecy rate is zero. When $N_e < \min\{N_t, N_r\}$, since the channel is memoryless for a given eavesdropper channel, by using singular value decomposition of the full-rank channel $\mathbf{H}$, the secrecy rate can be upper bounded as

$$R_s \leq I(\mathbf{X}; \mathbf{Y}_r | \mathbf{Y}_e) \leq I(\mathbf{X}_1^{N_m}; \mathbf{Y}_{r_1}^{N_m} | \mathbf{X}_1^{N_e}) \quad (6.90)$$

$$\leq I(\mathbf{X}_{N_e+1}^{N_m}; \mathbf{Y}_{r N_e+1}^{N_m}). \quad (6.91)$$

Using (6.21), the s.d.o.f. of R_s in (6.88) and (6.91) are both equal to $[N_m - N_e]^+$.

6.5.1 Proof of Theorem 6.4

In this subsection, we provide the proof for Theorem 6.4 as presented in [10]. For a channel matrix $\mathbf{H}$, singular value decomposition of $\mathbf{H}$ results in $\mathbf{H} = \mathbf{U}_H \mathbf{\Sigma}_H \mathbf{V}_H$, where

---

[7] Here the transmitter provides cooperative jamming for its own secrecy rate improvement.

$\mathbf{U}_H \in \mathbb{C}^{N_r \times N_r}$, $\mathbf{V}_H \in \mathbb{C}^{N_t \times N_t}$ are unitary, and Σ_H is an $(N_r \times N_t)$ rectangular diagonal matrix. Using precoding and decoding matrices which cancel $\mathbf{U}_H, \mathbf{V}_H$, and discarding any channel input which does not reach the receiver or any channel output which contains noise only, the effective channel can always be represented by an $(N_m \times N_m)$ diagonal matrix, $N_m = \min\{N_t, N_r\}$. In other words, for $N_t \neq N_r$, extra antennas at the transmitter (receiver) can be discarded without decreasing the achievable secrecy rate given in (6.88). Hence, it is sufficient to consider the case $N_r = N_t$.

Additionally, only $N_e < N_m$ is considered, since the achievable rate is zero otherwise, cf. (6.88). Singular value decomposition of a full-rank $\mathbf{G}(i)$ yields

$$\mathbf{G}(i) = \mathbf{U}_{\mathbf{G}(i)} \Sigma_{\mathbf{G}(i)} \mathbf{V}_{\mathbf{G}(i)}, \tag{6.92}$$

where $\mathbf{U}_{\mathbf{G}(i)}, \mathbf{V}_{\mathbf{G}(i)}$ are unitary and $\Sigma_{\mathbf{G}(i)}$ can be written as $[\mathbf{D}_{\mathbf{G}(i)}\ \mathbf{0}_{N_e \times (N_t - N_e)}]$ with $\mathbf{D}_{\mathbf{G}(i)}$ being an $(N_e \times N_e)$ diagonal matrix. Multiplying $\mathbf{G}(i)$ in (6.92) from the left by $\mathbf{D}_{\mathbf{G}(i)}^{-1} \mathbf{U}_{\mathbf{G}(i)}^H$ yields

$$\tilde{\mathbf{G}}(i) = \begin{bmatrix} \mathbf{I}_{N_e} & \mathbf{0}_{N_e \times (N_t - N_e)} \end{bmatrix} \mathbf{V}_{\mathbf{G}(i)}. \tag{6.93}$$

Throughout the proof, we consider (6.93) for the eavesdropper channel at the ith channel use. Note that transforming the eavesdropper channel from (6.92) to (6.93) is not an actual channel transformation that is performed by the eavesdropper. Rather, (6.93) is an isomorphic[8] representation of the actual eavesdropper channel, which can be utilized in the proof since, as we will shortly see, neither the signaling directions nor the rate selection in the achievability scheme depend on the actual eavesdropper channel realizations, and the conditional mutual information term in (6.87), to be considered for analyzing the secrecy constraint, does not change by replacing $\mathbf{G}^n$ with $\tilde{\mathbf{G}}^n$ in the condition. Note that any eavesdropper channel which is not full rank represents a degraded version from the channel described in (6.93) for which the same secrecy rate can be achieved. Hence, it is sufficient to consider the form in (6.93).

We first describe the signaling scheme and the codebook generation. Next, we describe an achievability proof for the case when the eavesdropper channel is unknown and static, i.e., $\mathbf{G}(i) = \mathbf{G}$ for all i. Finally, we describe extending the proof to the case when the eavesdropper channel varies over the channel uses.

Signaling Scheme

In addition to the information, the transmitter sends artificial noise, i.e., a cooperative jamming signal, in all directions. The transmitted signal at the ith channel use is

$$\mathbf{X}(i) = \tilde{\mathbf{X}}(i) + \mathbf{N}(i), \tag{6.94}$$

where $\tilde{\mathbf{X}}(i)$ represents the information-carrying signal, and $\mathbf{N}(i) \sim \mathcal{CN}(\mathbf{0}, \sigma^2 \mathbf{I}_{N_t})$ is a circularly symmetric Gaussian artificial noise independent from $\tilde{\mathbf{X}}(i)$ and $\mathbf{Z}_r(i)$. The

---

[8] Two mathematical objects are called isomorphic if there exists an isomorphism, i.e., a structure-preserving map which admits an inverse, between them.

received signals at the ith channel use are

$$\mathbf{Y}_r(i) = \mathbf{H}\tilde{\mathbf{X}}(i) + \mathbf{Z}_{r,\text{eff}}(i), \tag{6.95}$$

$$\mathbf{Y}_e(i) = \tilde{\mathbf{G}}\tilde{\mathbf{X}}(i) + \mathbf{Z}_{e,\text{eff}}(i), \tag{6.96}$$

where

$$\mathbf{Z}_{r,\text{eff}}(i) = \mathbf{H}\mathbf{N}(i) + \mathbf{Z}_r(i), \tag{6.97}$$

$$\mathbf{Z}_{e,\text{eff}}(i) = \tilde{\mathbf{G}}(i)\mathbf{N}(i). \tag{6.98}$$

It is easy to show that $\mathbf{Z}_{r,\text{eff}}(i)$ and $\mathbf{Z}_{e,\text{eff}}(i)$ are both Gaussian vectors with zero mean and covariance matrices $\mathbf{I}_{N_r} + \sigma^2 \mathbf{H}\mathbf{H}^H$ and $\sigma^2 \mathbf{I}_{N_e}$, respectively.

Codebook Generation

Coding is performed over the information-carrying signal $\tilde{\mathbf{X}}^n$ in (6.94), and then the artificial noise sequence, $\mathbf{N}^n$, is added to $\tilde{\mathbf{X}}^n$ to generate the n-letter transmitted signal $\mathbf{X}^n$. A random coding argument is considered, where the existence of a codebook which achieves the reliability and secrecy constraints in (6.86) and (6.87) is proved by averaging over all codebooks in the ensemble. In particular, let $\mathcal{C}$ be a random variable which represents the generated codebook. The codewords in the codebook $\mathcal{C}$ are independently drawn according to an n-letter truncated Gaussian distribution, $f_{\tilde{\mathbf{X}}_T^n}$, which is described as follows [27, Section 7.3]: Let $\tilde{\mathbf{X}}_G \sim \text{CN}\left(\mathbf{0}, \frac{(1-\tilde{\epsilon})P_t}{N_t}\mathbf{I}_{N_t}\right)$, where P_t is defined as in (6.88) and $0 < \tilde{\epsilon} < 1$. For $\mathbf{x}^n = [\mathbf{x}(1)\cdots\mathbf{x}(n)]$, we have

$$f_{\tilde{\mathbf{X}}_T^n}(\mathbf{x}^n) = \frac{1}{\nu_{n,\tilde{\epsilon}}} \mathbb{1}_{\left\{\frac{\|\mathbf{x}^n\|^2}{n} \leq P_t\right\}} \prod_{i=1}^n f_{\tilde{\mathbf{X}}_G}(\mathbf{x}(i)), \tag{6.99}$$

where

$$\nu_{n,\tilde{\epsilon}} = \int \mathbb{1}_{\left\{\frac{\|\mathbf{x}^n\|^2}{n} \leq P_t\right\}} \prod_{i=1}^n f_{\tilde{\mathbf{X}}_G}(\mathbf{x}(i)) d\mathbf{x}^n \tag{6.100}$$

satisfies $0 < \nu_{n,\tilde{\epsilon}} < 1$. It has been shown in [28] that for any $0 < \tilde{\epsilon} < 1$, there exists a $\phi(\tilde{\epsilon}) > 0$ which satisfies

$$1 - \nu_{n,\tilde{\epsilon}} \leq \exp(-n\phi(\tilde{\epsilon})), \tag{6.101}$$

$$\lim_{\tilde{\epsilon} \to 0} \phi(\tilde{\epsilon}) = 0. \tag{6.102}$$

REMARK 6.9 *We sample the codewords from an n-letter truncated Gaussian distribution in order to provide a uniform upper bound on the average power of each codeword in the codebook. This uniform upper bound is needed to construct an approximation argument, as in [29], which enables quantizing the eavesdropper channel into a finite subset of values. The quantization step renders the analysis of the secrecy constraint tractable.*

Let $\tilde{\mathbf{Y}}_{r,G}$ and $\tilde{\mathbf{Y}}_{e,G}$ denote the corresponding received signals at the receiver and eavesdropper when the transmitted signal is $\mathbf{X} = \tilde{\mathbf{X}}_G + \mathbf{N}$.

Each codebook $\mathcal{C}$ consists of 2^{nR} codewords, drawn from the n-letter truncated Gaussian distribution $f_{\tilde{\mathbf{X}}_T^n}$ described in (6.99), and the code rate R is chosen as

$$R = I(\tilde{\mathbf{X}}_G; \tilde{\mathbf{Y}}_{r,G}) - \delta', \tag{6.103}$$

where $\delta' > 0$ is arbitrarily small. Note that

$$I(\tilde{\mathbf{X}}_G; \tilde{\mathbf{Y}}_{r,G}) \leq I(\mathbf{X}; \tilde{\mathbf{Y}}_{r,G}) \leq \max_{p_\mathbf{X}} I(\mathbf{X}; \mathbf{Y}_r) \tag{6.104}$$

due to the Markov chain $\tilde{\mathbf{X}}_G$–$\mathbf{X}$–$\tilde{\mathbf{Y}}_{r,G}$. Thus, the code rate R is less than the capacity of the main channel.

Encoder at the Transmitter

Recall that the information message W is uniformly distributed over the set $\mathcal{W} = \{1, 2, \ldots, 2^{nR}\}$. Consider a dummy message $\tilde{W} \in \tilde{\mathcal{W}} = \{1, 2, \ldots, 2^{n\tilde{R}_s}\}$. The stochastic encoder in (6.84) is described as follows: Each codebook $\mathcal{C}$ is partitioned into $|\mathcal{W}| = 2^{nR_s}$ disjoint sub-codebooks, $\{\mathcal{C}_w\}_{w=1}^{|\mathcal{W}|}$, with each sub-codebook $\mathcal{C}_w$ containing $|\tilde{\mathcal{W}}| = 2^{n\tilde{R}_s}$ codewords, where

$$R_s = I(\tilde{\mathbf{X}}_G; \tilde{\mathbf{Y}}_{r,G}) - I(\tilde{\mathbf{X}}_G; \tilde{\mathbf{Y}}_{e,G}) - \delta' - \delta, \tag{6.105}$$

$$\tilde{R}_s = I(\tilde{\mathbf{X}}_G; \tilde{\mathbf{Y}}_{e,G}) + \delta, \tag{6.106}$$

where $\delta > 0$ is arbitrarily small. In order to send a message $W = w$, the encoder (1) randomly selects a dummy message $\tilde{W} = \tilde{w}$ from the sub-codebook $\mathcal{C}_w$, (2) outputs the $\tilde{w}$th codeword in the sub-codebook $\mathcal{C}_w$, (3) adds this codeword to the artificial noise sequence $\mathbf{N}^n$, and (4) transmits the sum over the channel.

Due to the isomorphic representation (6.93) of the eavesdropper channel, and the worst-case assumption of a noiseless channel to the eavesdropper, we have that $I(\tilde{\mathbf{X}}_G; \tilde{\mathbf{Y}}_{e,G})$ does not depend on the value of $\tilde{\mathbf{G}}^n$, since the multiplication by the unitary matrix $\mathbf{V}_\mathbf{G}$ does not change the mutual information. Thus, the transmission and randomization rates in (6.105) and (6.106) do not depend on the eavesdropper channel realization.

Decoder at the Legitimate Receiver

The probability of decoding error for a given codebook $\mathcal{C} = C$ is given by

$$P_e^{(n)}(C) = \frac{1}{|\mathcal{W}||\tilde{\mathcal{W}}|} \sum_{w,\tilde{w}} \Pr\left\{\psi_C(\mathbf{y}^n) \neq (w, \tilde{w}) \Big| \tilde{\mathbf{X}}^n = \mathbf{x}_{w,\tilde{w}}^n\right\} \tag{6.107}$$

where $\psi_C(\mathbf{y}^n)$ is the decoder output based on received signal $\mathbf{y}^n$, and $\mathbf{x}_{w,\tilde{w}}^n$ is the $\tilde{w}$th codeword in the sub-codebook $\mathcal{C}_w$. We utilize the maximum likelihood decoder at the legitimate receiver due to the truncated Gaussian codewords, as in [27, Section 7.3].

Achievability for Unknown Static Eavesdropper Channel: $\tilde{\mathbf{G}}(i) = \tilde{\mathbf{G}}$

Let $\tilde{\mathbf{Y}}^n_{e,G}$, $\tilde{\mathbf{Y}}^n_{e,T}$, $\tilde{\mathbf{Y}}^n_{e,\mathcal{C}}$ denote the received signals at the eavesdropper when the transmitted signals are $\tilde{\mathbf{X}}^n_G + \mathbf{N}^n$, $\tilde{\mathbf{X}}^n_T + \mathbf{N}^n$, and $\tilde{\mathbf{X}}^n_{\mathcal{C}} + \mathbf{N}^n$, respectively; $\tilde{\mathbf{X}}^n_T$ is distributed according to the n-letter truncated Gaussian distribution $f_{\tilde{\mathbf{X}}^n_T}$, and $\tilde{\mathbf{X}}^n_{\mathcal{C}}$ is uniformly distributed over the codewords of the codebook $\mathcal{C}$. We have a constant eavesdropper channel, i.e., $\tilde{\mathbf{G}}^n = \tilde{\mathbf{g}}^n$, where $\tilde{\mathbf{g}}^n = [\tilde{\mathbf{g}} \cdots \tilde{\mathbf{g}}]$ consists of n copies of $\tilde{\mathbf{g}}$. Let $f_{\tilde{\mathbf{g}}^n, \mathbf{A}^n}$ denote the distribution of $\mathbf{A}^n$ when the eavesdropper channel is $\tilde{\mathbf{g}}^n$.

Before we provide the proof, we state the necessary definitions and results. We begin with the definition of *variational distance* between two probability distributions.

DEFINITION 6.2 [30, Section 4.1] The variational distance between two distributions q and q', defined on the same probability space, is defined as

$$d(q, q') = \int_x |q(x) - q'(x)| dx \tag{6.108}$$

$$= 2 \sup_A \int_{x \in A} (q(x) - q'(x)) dx, \tag{6.109}$$

where the two definitions (6.108) and (6.109) are equivalent.

LEMMA 6.2 *Data processing inequality for variational distance [30, Lemma 2] [10, Lemma 11]: Let Q, Q' be two probability distributions over the domain Ω, and let F be any randomized function on Ω. The distribution $F(Q)$ is defined as follows: A draw from $F(Q)$ is obtained by drawing independently x from Q and f from F and then outputting $f(x)$ (likewise for $F(Q')$). Then, we have*

$$d\left(F(Q), F(Q')\right) \leq d\left(Q, Q'\right). \tag{6.110}$$

Now, we define the *information density*.

DEFINITION 6.3 [31, Definition 2] For the random vectors $\tilde{\mathbf{X}}^n_G$ and $\tilde{\mathbf{Y}}^n_{e,G}$ with the joint distribution $f_{\tilde{\mathbf{X}}^n, \mathbf{Y}^n_e}(\mathbf{x}^n, \mathbf{y}^n)$ and marginal distributions $f_{\tilde{\mathbf{X}}^n_G}(\mathbf{x}^n)$ and $f_{\tilde{\mathbf{Y}}^n_{e,G}}(\mathbf{y}^n)$, the information density is the random variable defined as

$$i_{\tilde{\mathbf{g}}^n, \tilde{\mathbf{X}}^n_G, \tilde{\mathbf{Y}}^n_{e,G}}\left(\tilde{\mathbf{X}}^n, \mathbf{Y}^n_e\right) = \log \frac{\prod_{i=1}^n f_{\tilde{\mathbf{g}}^n, \mathbf{Y}_e | \tilde{\mathbf{X}}}\left(\mathbf{Y}_e(i) | \tilde{\mathbf{X}}(i)\right)}{f_{\tilde{\mathbf{g}}^n, \tilde{\mathbf{Y}}^n_{e,G}}\left(\mathbf{Y}^n_e\right)}. \tag{6.111}$$

Next, we define *channel resolvability*, which is an important concept in the proof.

DEFINITION 6.4 [31, Definition 7] A rate R is an achievable resolution rate for the channel $p_{\mathbf{Y}|\mathbf{X}}$ if for every input process $\mathbf{X}$, and every $\epsilon, \gamma > 0$, there exists an $\tilde{\mathbf{X}}$ with a resolution rate[9]

$$\frac{1}{n} R(\tilde{\mathbf{X}}^n) \leq R + \gamma, \tag{6.112}$$

---

[9] The resolution rate of a distribution P on Ω, $R(P)$, is defined as the minimum $\log K$ such that P is K-type, i.e., $P(\omega) \in \{\frac{1}{K}, \frac{2}{K}, \ldots, 1\}$ for all $\omega \in \Omega$, where K is a positive integer. Note that if P is equiprobable, then the resolution rate of P, $R(P)$, is equal to the entropy rate $H(P)$.

such that

$$d(\mathbf{Y}; \tilde{\mathbf{Y}}) \leq \epsilon, \tag{6.113}$$

where $\mathbf{Y}$ ($\tilde{\mathbf{Y}}$) is the output of the channel when the input is $\mathbf{X}$ ($\tilde{\mathbf{X}}$). The resolvability of the channel $p_{\mathbf{Y}|\mathbf{X}}$ is the minimum achievable resolution rate R.

Now, we are ready to describe the proof, which consists of the following steps.

Step 1: For any eavesdropper channel $\tilde{\mathbf{g}}^n$, it is shown that the average over the random code $\mathcal{C}$ of the variational distance, $d_{\tilde{\mathbf{g}}^n,\mathcal{C}}$, between the joint distribution of the message W and the eavesdropper observation, and the product of their marginals, decreases uniformly and exponentially with increasing block length n.

Step 2: The eavesdropper channel is quantized to a finite subset of the eavesdropper channel values. It is shown that, for this finite subset of quantized eavesdropper channels, there must exist a good codebook which achieves small $\mathrm{E}_{\mathcal{C}}\{d_{\tilde{\mathbf{g}}^n,\mathcal{C}}\}$.

Step 3: Following the approximation argument in [29], it is shown that when the eavesdropper channel, $\tilde{\mathbf{g}}^n$, is not in the finite subset, then $d_{\tilde{\mathbf{g}}^n,\mathcal{C}}$ can be approximated by the variational distance of an eavesdropper channel that is in the finite subset, and hence is also small.

Step 4: Using [32, Lemma 1], which relates the variational distance to mutual information, it is shown that the secrecy constraint (6.87) is satisfied.

We begin with Step 1. In the following, for a given eavesdropper channel $\tilde{\mathbf{g}}^n$, and when a codeword $\tilde{\mathbf{X}}^n_{\mathcal{C}}$ in codebook $\mathcal{C}$ is transmitted with added artificial noise, we provide an upper bound for $\mathrm{E}_{\mathcal{C}}\{d_{\tilde{\mathbf{g}}^n,\mathcal{C}}\}$, where $d_{\tilde{\mathbf{g}}^n,\mathcal{C}}$ is the variational distance between the distributions $p_W f_{\tilde{\mathbf{g}}^n, \tilde{\mathbf{Y}}^n_{e,\mathcal{C}}}$ and $p_W f_{\tilde{\mathbf{g}}^n, \tilde{\mathbf{Y}}^n_{e,\mathcal{C}}|W}$. The upper bound on $\mathrm{E}_{\mathcal{C}}\{d_{\tilde{\mathbf{g}}^n,\mathcal{C}}\}$ consists of two components. The first component represents the distance between the eavesdropper output distributions when the inputs are drawn from $f_{\tilde{\mathbf{X}}^n_T}$ and $f_{\tilde{\mathbf{X}}^n_G}$, and is shown to vanish exponentially with n by the construction of $f_{\tilde{\mathbf{X}}^n_T}$. The other component represents the distance between the joint distribution of W and $\tilde{\mathbf{Y}}^n_{e,\mathcal{C}}$, and the product of the marginals when the input is drawn from $f_{\tilde{\mathbf{X}}^n_G}$, and its average over $\mathcal{C}$ is shown to vanish exponentially with n using a channel resolvability approach as in [31, 33, 34], which determines a lower bound on the rate of randomization in (6.106) that is required to approximate the joint distribution of W and $\tilde{\mathbf{Y}}^n_{e,\mathcal{C}}$ with the product of the marginals. Hence, independence of W and the eavesdropper observation is guaranteed.

The variational distance $d_{\tilde{\mathbf{g}}^n,\mathcal{C}}$ is upper bounded as follows:

$$d_{\tilde{\mathbf{g}}^n,\mathcal{C}} = d\left(p_W f_{\tilde{\mathbf{g}}^n, \tilde{\mathbf{Y}}^n_{e,\mathcal{C}}}, p_W f_{\tilde{\mathbf{g}}^n, \tilde{\mathbf{Y}}^n_{e,\mathcal{C}}|W}\right) \tag{6.114}$$

$$= \sum_w p_W(w) \int \left| f_{\tilde{\mathbf{g}}^n, \tilde{\mathbf{Y}}^n_{e,\mathcal{C}}}(\mathbf{y}^n) - f_{\tilde{\mathbf{g}}^n, \tilde{\mathbf{Y}}^n_{e,\mathcal{C}}|W}(\mathbf{y}^n|w) \right| d\mathbf{y}^n \tag{6.115}$$

$$\leq 2\sum_{w} p_W(w) \int \left| f_{\tilde{\mathbf{g}}^n, \tilde{\mathbf{Y}}^n_{e,T}} (\mathbf{y}^n) - f_{\tilde{\mathbf{g}}^n, \tilde{\mathbf{Y}}^n_{e,e}|W} (\mathbf{y}^n|w) \right| d\mathbf{y}^n \tag{6.116}$$

$$\leq 2\sum_{w} p_W(w) \Bigg[\int \left| f_{\tilde{\mathbf{g}}^n, \tilde{\mathbf{Y}}^n_{e,G}} (\mathbf{y}^n) - f_{\tilde{\mathbf{g}}^n, \tilde{\mathbf{Y}}^n_{e,T}} (\mathbf{y}^n) \right| d\mathbf{y}^n$$

$$+ \int \left| f_{\tilde{\mathbf{g}}^n, \tilde{\mathbf{Y}}^n_{e,G}} (\mathbf{y}^n) - f_{\tilde{\mathbf{g}}^n, \tilde{\mathbf{Y}}^n_{e,e}|W} (\mathbf{y}^n|w) \right| d\mathbf{y}^n \Bigg], \tag{6.117}$$

where (6.116) follows from [33, Appendix II-D] and (6.117) follows from applying the triangle inequality to the integrand in (6.116).

In order to upper bound the first integral in (6.117), we utilize the data processing inequality in Lemma 6.2. Using (6.110), the first integral in (6.117) is upper bounded as

$$\int \left| f_{\tilde{\mathbf{g}}^n, \tilde{\mathbf{Y}}^n_{e,G}} (\mathbf{y}^n) - f_{\tilde{\mathbf{g}}^n, \tilde{\mathbf{Y}}^n_{e,T}} (\mathbf{y}^n) \right| d\mathbf{y}^n \leq \int \left| f_{\tilde{\mathbf{X}}^n_G} (\mathbf{x}^n) - f_{\tilde{\mathbf{X}}^n_T} (\mathbf{x}^n) \right| d\mathbf{x}^n \tag{6.118}$$

$$= \int_{\frac{1}{n}\|\mathbf{x}^n\|^2 > P_t} \left| f_{\tilde{\mathbf{X}}^n_G} (\mathbf{x}^n) - f_{\tilde{\mathbf{X}}^n_T} (\mathbf{x}^n) \right| d\mathbf{x}^n + \int_{\frac{1}{n}\|\mathbf{x}^n\|^2 \leq P_t} \left| f_{\tilde{\mathbf{X}}^n_G} (\mathbf{x}^n) - f_{\tilde{\mathbf{X}}^n_T} (\mathbf{x}^n) \right| d\mathbf{x}^n \tag{6.119}$$

$$\leq \int_{\frac{1}{n}\|\mathbf{x}^n\|^2 > P_t} f_{\tilde{\mathbf{X}}^n_G} (\mathbf{x}^n) d\mathbf{x}^n + \int_{\frac{1}{n}\|\mathbf{x}^n\|^2 > P_t} f_{\tilde{\mathbf{X}}^n_T} (\mathbf{x}^n) d\mathbf{x}^n$$

$$+ \int_{\frac{1}{n}\|\mathbf{x}^n\|^2 \leq P_t} \left| f_{\tilde{\mathbf{X}}^n_G} (\mathbf{x}^n) - f_{\tilde{\mathbf{X}}^n_T} (\mathbf{x}^n) \right| d\mathbf{x}^n \tag{6.120}$$

$$\leq (1 - v_{n,\tilde{\varepsilon}}) + \int_{\frac{1}{n}\|\mathbf{x}^n\|^2 \leq P_t} \left| f_{\tilde{\mathbf{X}}^n_G} (\mathbf{x}^n) - v_{n,\tilde{\varepsilon}}^{-1} f_{\tilde{\mathbf{X}}^n_G} (\mathbf{x}^n) \right| d\mathbf{x}^n \tag{6.121}$$

$$\leq (1 - v_{n,\tilde{\varepsilon}}) + v_{n,\tilde{\varepsilon}}^{-1} - 1 \tag{6.122}$$

$$= v_{n,\tilde{\varepsilon}}^{-1} - v_{n,\tilde{\varepsilon}} \leq 4 e^{-n\phi(\tilde{\varepsilon})}, \tag{6.123}$$

where (6.118) follows from Lemma 6.2, (6.120) follows from the triangle inequality, (6.121) follows from the fact that $f_{\tilde{\mathbf{X}}^n_T} = 0$ when $\frac{1}{n}\|\mathbf{x}^n\|^2 > P_t$ and the definitions of $v_{n,\tilde{\varepsilon}}$ and $f_{\tilde{\mathbf{X}}^n_T}$ in (6.100) and (6.99), and the inequality in (6.123) follows from a simple manipulation; see [10, Appendix A].

Next, we get an upper bound for the average over $\mathcal{C}$ of the second integral in (6.117) using the following series of lemmas.

LEMMA 6.3 *For any $\varepsilon > 0$, there exists $\phi'(\varepsilon) > 0$ such that*

$$\Pr\left[\frac{1}{n} i_{\tilde{\mathbf{g}}^n, \tilde{\mathbf{X}}^n_G \tilde{\mathbf{Y}}^n_{e,G}} \left(\tilde{\mathbf{X}}^n_G, \tilde{\mathbf{Y}}^n_{e,G} \right) > I\left(\tilde{\mathbf{X}}_G; \tilde{\mathbf{Y}}_{e,G} \right) + \varepsilon \right] \leq e^{-n\phi'(\varepsilon)}. \tag{6.124}$$

Proof Lemma 6.3 is proved for a general time-varying eavesdropper channel, i.e., $\tilde{\mathbf{G}}(i) = \tilde{\mathbf{g}}(i)$, and hence it can also be applied to the static eavesdropper channel case. The proof utilizes the fact that the probability that the norm of a length-n Gaussian distributed random vector is larger than its variance is negligible [28]. For a given eavesdropper channel $\tilde{\mathbf{g}}^n$, we have from (6.96) that

$$\prod_{i=1}^n f_{\tilde{\mathbf{g}}^n,\mathbf{Y}_{\mathrm{e}}|\tilde{\mathbf{X}}}(\mathbf{Y}_{\mathrm{e}}(i)|\tilde{\mathbf{X}}(i)) = \frac{1}{(\pi\sigma^2)^{nN_e}} \exp\left(-\frac{\|\mathbf{Y}_{\mathrm{e}}^n - \tilde{\mathbf{g}}^n\tilde{\mathbf{X}}^n\|^2}{\sigma^2}\right), \qquad (6.125)$$

since $\mathbf{Z}_{\mathrm{e,eff}}(i) \sim \mathcal{CN}(\mathbf{0}, \sigma^2 \mathbf{I}_{N_e})$ for $i = 1, 2, \ldots, n$ are independent. Recall that $\tilde{\mathbf{g}}^n \tilde{\mathbf{X}}^n$ denotes $[\tilde{\mathbf{g}}(1)\tilde{\mathbf{X}}(1) \cdots \tilde{\mathbf{g}}(n)\tilde{\mathbf{X}}(n)]$. In addition, for $\tilde{\mathbf{X}}^n \triangleq \tilde{\mathbf{X}}_G^n$ and $\tilde{\mathbf{G}}^n = \tilde{\mathbf{g}}^n$, $\tilde{\mathbf{Y}}_{\mathrm{e},G}^n$ is distributed as

$$f_{\tilde{\mathbf{g}}^n, \tilde{\mathbf{Y}}_{\mathrm{e},G}^n}(\mathbf{Y}_{\mathrm{e}}^n) = \frac{1}{(\pi P_t'')^{nN_e}} \exp\left(-\frac{\|\mathbf{Y}_{\mathrm{e}}^n\|^2}{P_t''}\right), \qquad (6.126)$$

where P_t'' is defined as

$$P_t'' = P_t' + \sigma^2, \qquad P_t' = \frac{P_t(1-\tilde{\epsilon})}{N_t}, \qquad (6.127)$$

and P_t is as defined in (6.88). Thus, the information density in (6.111) is given by

$$\frac{1}{n} i_{\tilde{\mathbf{g}}^n, \tilde{\mathbf{X}}_G^n \tilde{\mathbf{Y}}_{\mathrm{e},G}^n}(\tilde{\mathbf{X}}_G^n, \tilde{\mathbf{Y}}_{\mathrm{e},G}^n) =$$
$$= N_e \log \frac{P_t''}{\sigma^2} + \left\{\frac{1}{n}\left(\frac{\|\tilde{\mathbf{Y}}_{\mathrm{e},G}^n\|^2}{P_t''}\right) - \frac{1}{n}\left(\frac{\|\tilde{\mathbf{Y}}_{\mathrm{e},G}^n - \tilde{\mathbf{g}}^n \tilde{\mathbf{X}}_G^n\|^2}{\sigma^2}\right)\right\} \log e. \qquad (6.128)$$

Let $\tilde{\mathbf{N}}^n = \tilde{\mathbf{Y}}_{\mathrm{e},G}^n - \tilde{\mathbf{g}}^n \tilde{\mathbf{X}}_G^n$, $\mathbf{N}' = \frac{\tilde{\mathbf{N}}}{\sigma}$, $\varepsilon' = \frac{\varepsilon}{\log e}$, and $\varepsilon'' = \frac{P_t''}{P_t'}\varepsilon'$. Using (6.128), we upper bound the probability in (6.124) as

$$\Pr\left(\frac{1}{n} i_{\tilde{\mathbf{g}}^n, \tilde{\mathbf{X}}_G^n \tilde{\mathbf{Y}}_{\mathrm{e},G}^n}(\tilde{\mathbf{X}}_G^n, \tilde{\mathbf{Y}}_{\mathrm{e},G}^n) > I(\tilde{\mathbf{X}}_G; \tilde{\mathbf{Y}}_{\mathrm{e},G}) + \varepsilon\right)$$

$$= \Pr\left(\frac{1}{n}\frac{\|\tilde{\mathbf{Y}}_{\mathrm{e},G}^n\|^2}{P_t''} > \frac{1}{n}\|\mathbf{N}'^n\|^2 + \varepsilon'\right) \qquad (6.129)$$

$$\leq \Pr\left(\frac{1}{n}\frac{\|\tilde{\mathbf{g}}^n \tilde{\mathbf{X}}_G^n\|^2}{P_t'} > \frac{1}{n}\|\mathbf{N}'^n\|^2 + \varepsilon'\right) \qquad (6.130)$$

$$= \Pr\left(\frac{1}{n}\frac{\|\tilde{\mathbf{g}}^n \tilde{\mathbf{X}}_G^n\|^2 + \|\tilde{\mathbf{N}}^n\|^2}{P_t''} > \frac{1}{n}\|\mathbf{N}'^n\|^2 + \frac{P_t''}{P_t'}\varepsilon'\right) \qquad (6.131)$$

$$\leq \Pr\left(\frac{1}{n}\|\mathbf{N}'^n\|^2 < N_e(1-\varepsilon_1)\right) + \Pr\left(\frac{1}{n}\|\mathbf{N}'^n\|^2 \geq N_e(1-\varepsilon_1)\right)$$

$$\times \Pr\left(\frac{1}{n}\frac{\|\tilde{\mathbf{g}}^n \tilde{\mathbf{X}}_G^n\|^2}{P_t'} > \frac{1}{n}\|\mathbf{N}'^n\|^2 + \varepsilon'' \left| \frac{1}{n}\|\mathbf{N}'^n\|^2 \geq N_e(1-\varepsilon_1)\right.\right) \qquad (6.132)$$

$$\leq e^{-\phi(\varepsilon_1)} + \Pr\left(\frac{1}{n}\frac{\|\tilde{\mathbf{g}}^n\tilde{\mathbf{X}}_G^n\|^2}{P_t'} > N_e(1-\varepsilon_1) + \varepsilon''\left|\frac{1}{n}\|\mathbf{N}'^n\|^2 \geq N_e(1-\varepsilon_1)\right.\right) \quad (6.133)$$

$$= e^{-\phi(\varepsilon_1)} + \Pr\left(\frac{1}{nN_e}\left\|\frac{\tilde{\mathbf{g}}^n\tilde{\mathbf{X}}_G^n}{\sqrt{P_t'}}\right\| > (1-\varepsilon_1) + \frac{\varepsilon''}{N_e}\right) \quad (6.134)$$

$$\leq 2e^{-n\phi(\varepsilon_1)}, \quad (6.135)$$

where (6.129) follows since $I(\tilde{\mathbf{X}}_G; \tilde{\mathbf{Y}}_{e,G}) = N_e \log \frac{P_t''}{\sigma^2}$, (6.130) follows from the triangle inequality, (6.133) follows since $\mathbf{N}'(i) \sim \text{CN}(\mathbf{0}, \mathbf{I}_{N_e})$, $i = 1, 2, \ldots, n$, and hence, as shown in [28], there exists $\phi(\varepsilon_1) > 0$ such that the first term in (6.132) is upper bounded by $\exp(-n\phi(\varepsilon_1))$, and finally (6.135) follows since, for any $\tilde{\mathbf{g}}^n$, $\frac{\tilde{\mathbf{g}}(i)\tilde{\mathbf{X}}_G(i)}{\sqrt{P_t'}} \sim \text{CN}(\mathbf{0}, \mathbf{I}_{N_e})$, $i = 1, 2, \ldots, n$, due to the special form of $\tilde{\mathbf{g}}(i)$ in (6.93), and hence the second term in (6.134) can also be upper bounded by $\exp(-n\phi(\varepsilon_1))$ [28].

LEMMA 6.4 *[10, Appendix E] [33, Appendix II-D]: By the symmetry of the random code construction, we have*

$$\mathbb{E}_{\mathcal{C}}\left[\sum_w p_W(w) \int \left|f_{\tilde{\mathbf{g}}^n, \tilde{\mathbf{Y}}_{e,G}^n}(\mathbf{y}^n) - f_{\tilde{\mathbf{g}}^n, \tilde{\mathbf{Y}}_{e,\mathcal{C}}^n|W}(\mathbf{y}^n|w)\right| d\mathbf{y}^n\right]$$

$$= \mathbb{E}_{\mathcal{C}}\left[\int \left|f_{\tilde{\mathbf{g}}^n\tilde{\mathbf{Y}}_{e,G}^n}(\mathbf{y}^n) - f_{\tilde{\mathbf{g}}^n, \tilde{\mathbf{Y}}_{e,\mathcal{C}}^n|W}(\mathbf{y}^n|1)\right| d\mathbf{y}^n\right]. \quad (6.136)$$

Due to Lemma 6.4, we proceed with upper bounding the integral on the right-hand side of (6.136) instead. This integral is upper bounded in the following lemma.

LEMMA 6.5 *[31, Lemma 5]: For any sequence $\{\mu_n > 0\}$ and a codebook $\mathcal{C}$, we have*

$$\int \left|f_{\tilde{\mathbf{g}}^n, \tilde{\mathbf{Y}}_{e,G}^n}(\mathbf{y}^n) - f_{\tilde{\mathbf{g}}^n, \tilde{\mathbf{Y}}_{e,\mathcal{C}}^n|W}(\mathbf{y}^n|1)\right| d\mathbf{y}^n$$

$$\leq \frac{2}{\log e}\mu_n + 2\Pr\left[\log \frac{f_{\tilde{\mathbf{g}}^n, \tilde{\mathbf{Y}}_{e,\mathcal{C}}^n|W}(\tilde{\mathbf{Y}}_{e,\mathcal{C},W=1}^n|1)}{f_{\tilde{\mathbf{g}}^n, \tilde{\mathbf{Y}}_{e,G}^n}(\tilde{\mathbf{Y}}_{e,\mathcal{C},W=1}^n)} > \mu_n\right], \quad (6.137)$$

where $\tilde{\mathbf{Y}}_{e,\mathcal{C},W=1}^n$ is the random vector with the distribution $f_{\tilde{\mathbf{g}}^n, \tilde{\mathbf{Y}}_{e,\mathcal{C}}^n|W}(\mathbf{y}^n|1)$.

The second term on the right-hand side of (6.137) is upper bounded in the following lemma.

LEMMA 6.6 *For any $\{\mu_n > 0\}$ and τ_n defined as*

$$\tau_n = \frac{e^{\mu_n \ln 2} - \mu_{n,\tilde{\varepsilon}}^{-1}}{2}, \quad (6.138)$$

we have, for $\tau_n > 0$, that

$$\mathbb{E}_{\mathcal{C}}\left[\Pr\left[\log \frac{f_{\tilde{\mathbf{g}}^n, \tilde{\mathbf{Y}}_{e,\mathcal{C}}^n|W}(\tilde{\mathbf{Y}}_{e,\mathcal{C},W=1}^n|1)}{f_{\tilde{\mathbf{g}}^n, \tilde{\mathbf{Y}}_{e,G}^n}(\tilde{\mathbf{Y}}_{e,\mathcal{C},W=1}^n)} > \mu_n\right]\right] \leq \frac{A+B+C}{\nu_{n,\tilde{\varepsilon}}}, \quad (6.139)$$

where

$$A \triangleq \Pr\left[\frac{1}{n}i_{\tilde{g}^n,\tilde{X}_G^n\tilde{Y}_{e,G}^n}\left(\tilde{X}_G^n,\tilde{Y}_{e,G}^n\right) > I\left(\tilde{X}_G;\tilde{Y}_{e,G}\right) + \delta + \frac{1}{n}\log\tau_n\right], \quad (6.140)$$

$$B \triangleq \Pr\left[\frac{1}{n}i_{\tilde{g}^n,\tilde{X}_G^n\tilde{Y}_{e,G}^n}\left(\tilde{X}_G^n,\tilde{Y}_{e,G}^n\right) > I\left(\tilde{X}_G;\tilde{Y}_{e,G}\right) + \delta\right], \quad (6.141)$$

$$C \triangleq \frac{1}{\tau_n^2}\left(\Pr\left[\frac{1}{n}i_{\tilde{g}^n,\tilde{X}_G^n\tilde{Y}_{e,G}^n}\left(\tilde{X}_G^n,\tilde{Y}_{e,G}^n\right) > I\left(\tilde{X}_G;\tilde{Y}_{e,G}\right) + \frac{\delta}{2}\right] + 2^{-\frac{n\delta}{2}}\right), \quad (6.142)$$

and δ is the arbitrarily small parameter used in choosing the transmission and randomization rates in (6.105) and (6.106).

Proof [10, Appendix F] The proof for Lemma 6.6 follows similar steps to the achievability proof for the channel resolvability theorem in [31]. The difference here is that the codewords of the random code $\mathcal{C}$, over which the expectation in (6.139) is taken, are drawn from the truncated Gaussian distribution $f_{\tilde{X}_T^n}$, and not from a Gaussian distribution as in [31]. This results in the term $\nu_{n,\tilde{\epsilon}}$ in the denominator of the right-hand side of (6.139). The idea of the proof is to construct an input process (by selecting an n-letter distribution and a certain rate for the input) with the minimum possible rate to generate an output at the eavesdropper which approximates an output that is independent from the message W; the approximation here is in the sense of a variational distance which vanishes with the block length n. Note that constructing the input process here corresponds to constructing the codewords of one sub-codebook $\mathcal{C}_w$ of the random code $\mathcal{C}$, with a rate that corresponds to the minimum rate of randomization we need to induce an output distribution at the eavesdropper which is independent from W. By defining an input $\tilde{X}^n \sim f_{\tilde{X}_T^n}$, an output Y_e^n which is generated by $\tilde{X}^n$ according to the eavesdropper channel, and $2^{n\tilde{R}_s}$ i.i.d. codewords each distributed according to $f_{\tilde{X}_T^n}$, it is shown in [10, Appendix F] that for $\tilde{R}_s > I(\tilde{X}_G;\tilde{Y}_{e,G})$, the condition (6.139) is satisfied.

Now, we upper bound the second term in (6.117), averaged over $\mathcal{C}$, in the following lemma.

LEMMA 6.7 *For a positive δ in (6.105) and (6.106) and a sufficiently large n, there exists a positive constant $\tilde{c}$ such that*

$$\mathbb{E}_{\mathcal{C}}\left[\int \left|f_{\tilde{g}^n,\tilde{Y}_{e,\mathcal{C}}^n}(\mathbf{y}^n) - f_{\tilde{g}^n,\tilde{Y}_{e,\mathcal{C}}^n|W}(\mathbf{y}^n|w)\right|d\mathbf{y}^n\right] \leq \exp(-\tilde{c}n). \quad (6.143)$$

Proof Using proper choices for $\delta > 0$ and $\mu_n > 0$, and the result in Lemma 6.3, it can be shown that each term on the right-hand side of (6.139) is upper bounded by a term which exponentially decreases with n; see [10, Appendix C]. By using the results in Lemmas 6.4 and 6.5, we obtain the desired bound in (6.143).

Finally, by combining the bounds in (6.123) and (6.143) via (6.117), the variational distance $d_{\tilde{g}^n,\mathcal{C}}$, averaged over the random code $\mathcal{C}$, is upper bounded as

$$\mathbb{E}_{\mathcal{C}}\{d_{\tilde{g}^n,\mathcal{C}}\} \leq \exp(-c'n), \quad (6.144)$$

where $c' > 0$ is a constant. This completes the proof for Step 1.

Now, we proceed with Step 2 of the proof. We construct the finite set $\mathcal{S}_M$ of the quantized eavesdropper channels. From (6.93), we have

$$\tilde{\mathbf{G}}\tilde{\mathbf{G}}^{\mathrm{H}} = \mathbf{I}_{N_e}. \tag{6.145}$$

It follows from (6.145) that the absolute value of the real and imaginary parts of the elements of $\tilde{\mathbf{G}}$ does not exceed 1. Note that this property holds only for the isomorphic representation of the eavesdropper channel, (6.93), utilized in the proof, and does not generally hold for any arbitrary realization of the actual eavesdropper channel.

For an integer M, let us define $\bar{\mathbf{G}}$ as a matrix whose elements have real and imaginary parts which take values from the set $\{-1, \frac{-M+1}{M}, \ldots, \frac{M-1}{M}\}$. For each $\bar{\mathbf{G}}$, let us define the hypercube, $\text{cube}_{\bar{\mathbf{G}}}$, of the $(N_e \times N_t)$ matrices adjacent to $\bar{\mathbf{G}}$ as

$$\text{cube}_{\bar{\mathbf{G}}} = \left\{ \begin{array}{l} \mathbf{G}: \quad 0 \leq \text{Re}(M\mathbf{G}_{i,j} - M\bar{\mathbf{G}}_{i,j}) \leq 1 \\ \phantom{\mathbf{G}:\quad} 0 \leq \text{Im}(M\mathbf{G}_{i,j} - M\bar{\mathbf{G}}_{i,j}) \leq 1 \end{array} \right\}. \tag{6.146}$$

Thus, we have $\bigcup_{\bar{\mathbf{G}}} \text{cube}_{\bar{\mathbf{G}}}$ represents all matrices whose elements have real and imaginary parts within the interval $[-1, 1]$.

We construct the set $\mathcal{S}_M$ as follows. For each $\bar{\mathbf{G}}$, we select from $\text{cube}_{\bar{\mathbf{G}}}$ only one matrix, if it exists, which satisfies (6.145), and include it in the set $\mathcal{S}_M$. Thus, the number of elements in the set $\mathcal{S}_M$ is upper bounded[10] by the number of hypercubes $\text{cube}_{\bar{\mathbf{G}}}$ (and hence the number of the matrices $\bar{\mathbf{G}}$), which is equal to $\{2M+1\}^{2N_t \times N_e}$. Note that M represents the accuracy of the quantization.

Since we have generated the random code $\mathcal{C}$ and selected the decoder as in [27], then the probability of error averaged over the random code $\mathcal{C}$ is upper bounded as [27]

$$\mathbb{E}_{\mathcal{C}}\left\{P_e^{(n)}(\mathcal{C})\right\} \leq 5 \exp(-nE(R(\delta'))), \tag{6.147}$$

where the error exponent $E(R(\delta'))$ is a positive constant. Thus, using (6.144), (6.147), the union bound, and the Markov inequality, there must exist a good codebook C^* which satisfies that (1) $P_e^{(n)}(C^*) \to 0$ as $n \to \infty$, and (2) for each $\tilde{\mathbf{g}} \in \mathcal{S}_M$, we have

$$d_{\tilde{\mathbf{g}}^n, C^*} \leq 2 \sum_w p_W(w) \int_{\mathbf{y}^n} \left| f_{\tilde{\mathbf{g}}^n, \tilde{\mathbf{Y}}_{e,G}^n}(\mathbf{y}^n) - f_{\tilde{\mathbf{g}}^n, \tilde{\mathbf{Y}}_{e,C^*}|W}(\mathbf{y}^n|w) \right| d\mathbf{y}^n \tag{6.148}$$

$$\leq 3 \times 2(2M+1)^{2N_t N_e} e^{-c'n}. \tag{6.149}$$

This completes Step 2 of the achievability proof for the static eavesdropper channel case.

We now begin with Step 3, where we show that $d_{\tilde{\mathbf{g}}^n, \mathcal{C}}$ for $\tilde{\mathbf{g}} \notin \mathcal{S}_M$ is also small. Let us fix a good codebook $\mathcal{C} = C^*$ which satisfies (6.149) and that $P_e^{(n)}(C^*) \to 0$ as $n \to \infty$, and let $\tilde{\mathbf{g}} \notin \mathcal{S}_M$. Let $\mathbf{A}_i$ denote the ith row of the matrix $\mathbf{A}$. By construction of the set $\mathcal{S}_M$, there must exist a $\mathbf{g}' \in \mathcal{S}_M$ such that, for $\mathbf{g}_\Delta(k) = \tilde{\mathbf{g}}(k) - \mathbf{g}'(k)$, $k = 1, \ldots, n$, we have

$$\|\mathbf{g}_{\Delta,i}(k)\|^2 \leq \frac{2N_t}{M^2}, \qquad \text{for all } i = 1, 2 \ldots, N_e,\ k = 1, 2, \ldots, n. \tag{6.150}$$

---

[10] It possible that a hypercube $\text{cube}_{\bar{\mathbf{G}}}$ does not contain any matrix which satisfies (6.145).

Let $\mathbf{x}_\Delta^n = \mathbf{g}_\Delta^n \mathbf{x}^n$, where $\mathbf{x}^n \in C^*$. Using simple algebra, it can be shown that [10, Section IV-D]

$$\frac{1}{n}\|\mathbf{x}_\Delta^n\|^2 \leq \frac{2N_tN_e}{M^2}P_t. \tag{6.151}$$

Note that for large M, i.e., for finer quantization of the channel, the norm of the difference $\mathbf{x}_\Delta^n$ is small.

Let $\mathcal{S}_M^n = \mathcal{S}_M \times \cdots \times \mathcal{S}_M$ be the n-fold Cartesian product of $\mathcal{S}_M$; for $\varepsilon > 0$, define

$$(r')^2 = \frac{2N_tN_e}{M^2}\frac{P_t}{\sigma^2}, \tag{6.152}$$

$$r = r' + \sqrt{N_e(1+\varepsilon)}, \tag{6.153}$$

$$g(r,r') = r'(2r+r'). \tag{6.154}$$

Note that, since $\mathbf{Y}_e(i) - \tilde{\mathbf{g}}\tilde{\mathbf{X}}(i) \sim \mathrm{CN}(\mathbf{0}, \sigma^2 \mathbf{I}_{N_e})$ for $i = 1, 2, \ldots, n$, and $r^2 > N_e(1+\varepsilon) > N_e$, there exists $\phi(\varepsilon) > 0$ such that, for all values of $\tilde{\mathbf{g}}^n$, we have [28]

$$\Pr\left(\frac{1}{n\sigma^2}\|\mathbf{Y}_e^n - \tilde{\mathbf{g}}^n\tilde{\mathbf{X}}^n\|^2 \geq r^2 \,\Big|\, \tilde{\mathbf{X}}^n = \mathbf{x}^n\right) \leq e^{-n\phi(\varepsilon)}. \tag{6.155}$$

By choosing M, with respect to n, large enough such that

$$ng(r,r') < 1, \tag{6.156}$$

then for any $\tilde{\mathbf{g}}^n \notin \mathcal{S}_M^n$, there must exist a $\mathbf{g}'^n \in \mathcal{S}_M^n$ which satisfies [10, Appendix G]

$$d_{\tilde{\mathbf{g}}^n, C^*} \leq 2\sum_w p_W(w) \int \left| f_{\tilde{\mathbf{g}}^n, \tilde{\mathbf{Y}}_{e,G}^n}(\mathbf{y}^n) - f_{\tilde{\mathbf{g}}^n, \tilde{\mathbf{Y}}_{e,C^*}^n|W}(\mathbf{y}^n|w) \right| d\mathbf{y}^n + 8e^{-n\phi(\tilde{\varepsilon})} \tag{6.157}$$

$$\leq 2\sum_w p_W(w) \int \left| f_{\mathbf{g}'^n, \tilde{\mathbf{Y}}_{e,G}^n}(\mathbf{y}^n) - f_{\mathbf{g}'^n, \tilde{\mathbf{Y}}_{e,C^*}^n|W}(\mathbf{y}^n|w) \right| d\mathbf{y}^n + 8e^{-n\phi(\tilde{\varepsilon})}$$
$$+ 4e^{-n\phi(\varepsilon)} + 4ng(r,r') \tag{6.158}$$

$$\leq 12(2M+1)^{2N_tN_e}e^{-c'n} + 8e^{-n\phi(\tilde{\varepsilon})} + 4e^{-n\phi(\varepsilon)} + 4ng(r,r'), \tag{6.159}$$

where $\phi(\varepsilon)$ is defined as in (6.155), (6.157) follows from (6.117) and (6.123), and (6.158) follows from upper bounding $|\log f_{\tilde{\mathbf{g}}^n, \mathbf{Y}_e^n|\tilde{\mathbf{X}}^n}(\mathbf{y}^n|\mathbf{x}^n) - \log f_{\mathbf{g}'^n, \mathbf{Y}_e^n|\tilde{\mathbf{X}}^n}(\mathbf{y}^n|\mathbf{x}^n)|$ by $ng(r,r')$, and using this bound to upper bound the integral in (6.157) for the two cases $\frac{1}{n\sigma^2}\|\mathbf{y}^n - \tilde{\mathbf{g}}^n\mathbf{x}^n\|^2 \geq r^2$ and $\frac{1}{n\sigma^2}\|\mathbf{y}^n - \tilde{\mathbf{g}}^n\mathbf{x}^n\|^2 < r^2$ separately. The reason we generate the codewords of the random code $\mathcal{C}$ according to the truncated Gaussian distribution in (6.99) is that upper bounding the integral in (6.157) for the second case requires that the average power of each codeword does not exceed P_t. Equation (6.159) follows since the good code we consider here, C^*, satisfies (6.149). With a proper choice for M, the upper bound in (6.159) can be shown to decay exponentially with n, and hence $d_{\tilde{\mathbf{g}}^n, C^*}$ vanishes with n for any $\tilde{\mathbf{g}}^n \notin \mathcal{S}_M^n$. This completes the proof for Step 3.

In particular, since $g(r,r')$ in (6.154) decays as $\frac{1}{M}$, there must exist a constant $c_M > 0$ such that, for $M = \exp(nc_M)$, we have both $(2M+1)^{2N_tN_e}e^{-c'n}$ and $ng(r,r')$ decay exponentially with n. Applying this to (6.149) and (6.159), we have, for the good codebook C^* and a constant $c_0 > 0$, that

$$d_{\tilde{\mathbf{g}}^n, C^*} \leq \exp(-c_0 n), \qquad \text{for all } \tilde{\mathbf{g}}^n. \tag{6.160}$$

So far, we have shown the existence of a good codebook, C^*, which satisfies $P_e^{(n)}(C^*) \to 0$ as $n \to \infty$, and $d_{\tilde{\mathbf{g}}^n, C^*} \leq \exp(-c_0 n)$ for all $\tilde{\mathbf{g}}^n$, and a constant $c_0 > 0$.

We now begin with Step 4. In order to show that the scheme described above achieves the secrecy constraint in (6.87), we utilize the following lemma.

LEMMA 6.8 *[32, Lemma 1] For the joint distribution* $p_W f_{\tilde{\mathbf{g}}^n, \mathbf{Y}_e^n | W}$, *and the good codebook* C^*, *we have*

$$I(W; \mathbf{Y}_e^n | \tilde{\mathbf{G}}^n = \tilde{\mathbf{g}}^n) \leq d_{\tilde{\mathbf{g}}^n, C^*} \log \frac{|\mathcal{W}|}{d_{\tilde{\mathbf{g}}^n, C^*}}. \tag{6.161}$$

Since $\log |\mathcal{W}| = nR_s$ is linear in n, it follows from (6.161) that $I(W; \mathbf{Y}_e^n | \tilde{\mathbf{G}}^n = \tilde{\mathbf{g}}^n)$ decreases exponentially fast with n for any eavesdropper channel $\tilde{\mathbf{g}}^n$, and hence the secrecy constraint in (6.87) is satisfied, and R_s in (6.105) is indeed an achievable strong secrecy rate. By evaluating R_s in (6.105), we obtain the achievable secrecy rate in (6.88), which completes the proof of Theorem 6.4 for the static eavesdropper channel case.

Achievability for Unknown and Varying Eavesdropper Channel $\tilde{\mathbf{G}}(i)$

Next, we consider the more general case when the eavesdropper channel varies from one channel use to the next, and is completely unknown to the legitimate parties. For this case, using the same analysis as in the static case, [10] shows that, for any eavesdropper channel sequence $\tilde{\mathbf{g}}^n = [\tilde{\mathbf{g}}(1) \cdots \tilde{\mathbf{g}}(n)]$, Eq. (6.144) holds, i.e., the variational distance averaged over the random code decays exponentially fast with n. However, the size of the set $\mathcal{S}_M^n$ is equal to $|\mathcal{S}_M^n| = (2M+1)^{2N_t N_e n}$. Thus, choosing $M = \exp(nc_M)$, as in the static case, so that $ng(r, r')$ in (6.159) decays exponentially fast with n, results in $|\mathcal{S}_M^n|$ increasing *doubly exponentially* with n. Hence, unlike for the static case, cf. (6.149), the union bound and Markov inequality are not sufficient to show the existence of a good codebook that satisfies the reliability and secrecy constraints. In order to solve this problem, [10] utilizes the elimination of correlation argument from [35], where another level of randomization with negligible rate is introduced by considering a small set of codebooks over which the averaged variational distance is shown to be small. This is proved by showing that the probability that the variational distance averaged over these codebooks is larger than any given constant is doubly exponentially small with respect to n. This small set of codebooks is then used to construct a two-stage scheme, where the index of the codebook, $k = 1, \ldots, K$, is transmitted via a reliable, but not necessarily secure, code to the receiver in the second stage.

In particular, consider K independent random codebooks, $\{\mathcal{C}_k\}_{k=1}^K$, each generated as in (6.99)–(6.103), and define $d'_{\tilde{\mathbf{g}}^n, \mathcal{C}_k}$ as

$$d'_{\tilde{\mathbf{g}}^n, \mathcal{C}_k} = \frac{1}{2} \sum_w p_W(w) \int \left| f_{\tilde{\mathbf{g}}^n, \tilde{\mathbf{Y}}_{e,G}^n}(\mathbf{y}^n) - f_{\tilde{\mathbf{g}}^n, \tilde{\mathbf{Y}}_{e, \mathcal{C}_k}^n | W}(\mathbf{y}^n | w) \right| d\mathbf{y}^n. \tag{6.162}$$

Note that, from (6.117) and (6.123), we have

$$d_{\tilde{g}^n, \mathcal{C}_k} \leq 4 d'_{\tilde{g}^n, \mathcal{C}_k} + 8 e^{-n\phi(\tilde{\epsilon})}. \tag{6.163}$$

For a positive sequence $\{\epsilon_n\}$ and a sufficiently large n such that $1 + e^2 e^{-c'n} \leq e^{\epsilon_n}$, where c' is a positive constant defined as in (6.144), [35, Eqs. (4.1)–(4.5)] show that

$$\Pr\left(\frac{1}{K}\sum_{k=1}^{K} d'_{\tilde{g}^n, \mathcal{C}_k} \geq \epsilon_n\right) \leq e^{-\epsilon_n K}. \tag{6.164}$$

Thus, using the union bound, we have

$$\Pr\left(\frac{1}{K}\sum_{k=1}^{K} d'_{\tilde{g}^n, \mathcal{C}_k} \geq \epsilon_n, \text{ for all } \tilde{g}^n \in \mathcal{S}_M^n\right) \geq 1 - |\mathcal{S}_M^n| \Pr\left(\frac{1}{K}\sum_{k=1}^{K} d'_{\tilde{g}^n, \mathcal{C}_k} \geq \epsilon_n\right) \tag{6.165}$$

$$\geq 1 - |\mathcal{S}_M^n| e^{-\epsilon_n K}. \tag{6.166}$$

As in Step 3 in the previous case, when $\tilde{g}^n \notin \mathcal{S}_M^n$ and if M is sufficiently large such that $ng(r, r') < 1$, we have, from (6.157) and (6.158), that there must exist a $g'^n \in \mathcal{S}_M^n$ which satisfies

$$\frac{1}{K}\sum_{k=1}^{K} d'_{\tilde{g}^n, \mathcal{C}_k} \leq \frac{1}{K}\sum_{k=1}^{K} d'_{g'^n, \mathcal{C}_k} + e^{-n\phi(\varepsilon)} + ng(r, r'). \tag{6.167}$$

In addition, using the union bound, Markov inequality, and (6.147),

$$\Pr\left(\mathbb{E}_{\mathcal{C}_k}\left\{P_e^{(n)}(\mathcal{C}_k)\right\} > 5nKe^{-nE(R(\delta'))}, \text{ for some } k\right)$$

$$\leq \sum_{k=1}^{K} \Pr\left(\mathbb{E}_{\mathcal{C}_k}\left\{P_e^{(n)}(\mathcal{C}_k)\right\} > 5nKe^{-nE(R(\delta'))}\right) \leq \frac{K}{nK} = \frac{1}{n}, \tag{6.168}$$

and hence

$$\Pr\left(\mathbb{E}_{\mathcal{C}_k}\left\{P_e^{(n)}(\mathcal{C}_k)\right\} \leq 5nKe^{-nE(R(\delta'))}, k = 1, 2, \ldots, k\right) \geq 1 - \frac{1}{n}. \tag{6.169}$$

By choosing ϵ_n, K, and M such that, for $0 < \varepsilon' < \min\{\phi(\varepsilon), \phi(\tilde{\epsilon}), \frac{E(R(\delta'))}{2}\}$, we have

$$\epsilon_n = e^{-\varepsilon'n}, \quad K = e^{2\varepsilon'n}, \quad \text{and } M = e^{2\varepsilon'n}; \tag{6.170}$$

we can easily show that the conditions

$$\lim_{n \to \infty} |\mathcal{S}_M^n| e^{-\epsilon_n K} = 0, \quad ng(r, r') < \epsilon_n, \quad e^{-n\phi(\varepsilon)} < \epsilon_n \tag{6.171}$$

are satisfied, and hence, from (6.163), (6.166)–(6.171), and for any eavesdropper channel sequence $\tilde{g}^n$, there must exist K good codebooks, $\{C_k^*\}_{k=1}^{K}$, which satisfy

$$\lim_{n \to \infty} \mathbb{E}_{C_k^*}\left\{P_e^{(n)}(C_k^*)\right\} = 0 \quad \text{for all } k, \tag{6.172}$$

$$\frac{1}{K}\sum_{k=1}^{K} d_{\tilde{g}^n, C_k^*} < 24\epsilon_n. \tag{6.173}$$

Finally, these K good codebooks are used to construct a two-stage transmission scheme as follows: In the first stage, the transmitter uniformly chooses $K' \in \{1,\ldots,K\}$, and in order to send a message W, a randomly selected codeword from the Wth sub-codebook of $\mathcal{C}_K^*$ is transmitted, after being added to the artificial noise. In the second stage, the index K' is transmitted to the receiver using a good channel code for the legitimate channel. The decoder at the receiver first decodes K', and then decodes W using $\psi_{n,\mathcal{C}_{K'}^*}$. Using this scheme, the same secrecy rate in (6.88) is achievable [10, Section IV-E], which completes the proof for Theorem 6.4. Next, we will extend the model considered in this section to the two-user multiple access channel (MAC).

6.6 MIMO MAC Gaussian Wiretap Channel with Unknown Eavesdropper Channel

Consider a two-user ($N_{t,1} \times N_{t,2} \times N_r \times N_e$) MIMO MAC Gaussian wiretap channel, as in Fig. 6.5, with the same assumptions on the legitimate and eavesdropper channels as in Section 6.5. This section summarizes the characterization of the s.d.o.f. region of this channel under a strong secrecy constraint which has been established in [11].

The received signals are given by

$$\mathbf{Y}_r(i) = \mathbf{H}_1\mathbf{X}_1(i) + \mathbf{H}_2\mathbf{X}_2(i) + \mathbf{Z}_r(i), \qquad (6.174)$$

$$\mathbf{Y}_e(i) = \mathbf{G}_1(i)\mathbf{X}_1(i) + \mathbf{G}_2(i)\mathbf{X}_2(i), \qquad (6.175)$$

where $i = 1,\ldots,n$, $\mathbf{X}_k, k = 1,2$, is the signal transmitted by the kth transmitter, and $\mathbf{Z}_r(i) \sim \mathrm{CN}(\mathbf{0},\mathbf{I}_{N_r})$ is the Gaussian noise at the receiver. As in Section 6.5, the legitimate channels $\mathbf{H}_1,\mathbf{H}_2$ are static and known to all terminals, while the eavesdropper channels $\{\mathbf{G}_k(i), k = 1,2\}$ are arbitrary, time varying, and revealed only to the eavesdropper. It is assumed that the channels $\{\mathbf{G}_k^n, k = 1,2\}$ are independent from the signals $\{\mathbf{X}_k, k = 1,2\}$, and that the number of eavesdropper antennas, N_e, is known to the legitimate parties. The power constraint at the kth transmitter is expressed as in (6.3), with $\mathbf{X}_k, P_k$ instead of $\mathbf{X}, P$.

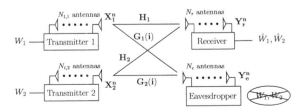

Figure 6.5 Two-user MIMO MAC Gaussian wiretap channel.

Transmitter k sends a message $W_k \in \{1,\ldots,2^{nR_{s,k}}\}$, $k=1,2$, using a stochastic encoder $f_{n,k}: W_k \mapsto \mathbf{X}_k^n$, to the receiver, which must be kept confidential from the eavesdropper. The receiver utilizes its observation to output estimates $\hat{W}_1, \hat{W}_2$ of both messages. The probability of decoding error at the receiver is given by

$$P_e^{(n)} = \Pr\{(\hat{W}_1, \hat{W}_2) \neq (W_1, W_2)\}. \tag{6.176}$$

A secrecy rate pair $(R_{s,1}, R_{s,2})$ is achievable if for every $\epsilon > 0$, there exists a channel code $(2^{nR_{s,1}}, 2^{nR_{s,2}}, n)$ such that

$$\lim_{n \to \infty} P_e^{(n)} = 0, \tag{6.177}$$

$$\lim_{n \to \infty} \sup_{\tilde{\mathbf{g}}_k^n, k=1,2} I(W_1, W_2; \mathbf{Y}_e^n | \tilde{\mathbf{G}}_k^n = \tilde{\mathbf{g}}_k^n, k=1,2) = 0, \tag{6.178}$$

where $\tilde{\mathbf{G}}_k(i), k=1,2$, and $i=1,\ldots,n$ is defined as in (6.93). The s.d.o.f., $D_{s,k}$, associated with the rate $R_{s,k}$ is defined as in (6.21).

The number of transmit dimensions at the kth transmitter is defined as $r_k = \text{rank}(\mathbf{H}_k)$, $k = 1, 2$, while the number of receive dimensions at the receiver is given by $r_0 = \text{rank}([\mathbf{H}_1 \ \mathbf{H}_2])$. The characterization of the s.d.o.f. region for the channel model in (6.174) and (6.175) comes in terms of these transmit and receive dimensions, as described by the following theorem.

THEOREM 6.6 *[11, Theorem 1] The s.d.o.f. region for the MIMO MAC Gaussian wiretap channel described in (6.174) and (6.175) is expressed as the convex hull of the following five points:*

$$p_0 = (0,0), \tag{6.179}$$

$$p_1 = ([r_1 - N_e]^+, 0), \tag{6.180}$$

$$p_2 = (0, [r_2 - N_e]^+), \tag{6.181}$$

$$p_3 = ([r_1 - N_e]^+, [r_0 - r_1 - N_e]^+), \tag{6.182}$$

$$p_4 = ([r_0 - r_2 - N_e]^+, [r_2 - N_e]^+). \tag{6.183}$$

The s.d.o.f. regions for three different cases of the number of eavesdropping antennas, namely (a) $0 \leq N_e \leq \min\{r_0 - r_1, r_0 - r_2\}$, (b) $\min\{r_0 - r_1, r_0 - r_2\} \leq N_e \leq \max\{r_0 - r_1, r_0 - r_2\}$, and (c) $N_e \geq \max\{r_0 - r_1, r_0 - r_2\}$, are shown in Fig. 6.6.

Proof sketch [11] The proof for Theorem 6.6 is established by first considering the case of parallel receiver channels, which is described as follows: Let $X_{1,i}, X_{2,i}, Z_{r,i}$, and $Y_{r,i}$ denote the ith element in $\mathbf{X}_1, \mathbf{X}_2, \mathbf{Z}_r$, and $\mathbf{Y}_r$, respectively. In the parallel channel model, the ith antenna at the receiver, for $i = 1, \ldots, N_r$, observes

$$Y_{r,i} = \begin{cases} X_{1,i} + Z_{r,i}, & i \in \mathcal{A} \\ X_{1,i} + X_{2,i} + Z_{r,i}, & i \in \mathcal{B} \\ X_{2,i} + Z_{r,i}, & i \in \mathcal{C}, \end{cases} \tag{6.184}$$

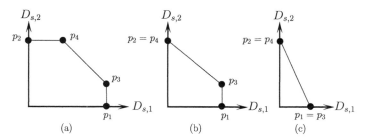

Figure 6.6 The s.d.o.f. region for the MIMO MAC wiretap channel for different cases of the number of eavesdropper antennas, N_e.

where $Z_{r,i} \in \text{CN}(0,1)$, $i = 1, \ldots, N_r$, are independent, and $\mathcal{A}, \mathcal{B}, \mathcal{C}$ are a disjoint partition of the set of indices $\{1, 2, \ldots, N_r\}$.

The achievability for Theorem 6.6 for this parallel channel model follows by orthogonalizing the signals from the two transmitters at the receiver, and then applying the result in Theorem 6.5 to each of these non-interfering users. In particular, the achievability of the point p_3 in Fig. 6.6(a) follows from restricting transmitter 2 to send signals over the channels in $\mathcal{C}$ and restricting transmitter 1 to send signals over the channels in $\{\mathcal{A} \cup \mathcal{B}\}$. Achievability of point p_4 is achieved by exchanging the roles of the two transmitters. Since showing the achievability of p_3 and p_4 in Fig. 6.6(a) is sufficient to show the achievability of the whole region, then the region in Theorem 6.6 is achievable for the case of parallel channels.

The converse for the parallel channel model relies on the fact that the secrecy capacity of the compound channel, i.e., a finite number of possible eavesdroppers, serves as an upper bound for the secrecy capacity of the eavesdropper channel model considered in this section. In particular, for each of the cases (1) $N_e \leq \min\{|\mathcal{A}|, |\mathcal{C}|\}$, (2) $N_e \geq \max\{|\mathcal{A}|, |\mathcal{C}|\}$, and (3) $\min\{|\mathcal{A}|, |\mathcal{C}|\} \leq N_e \leq \max\{|\mathcal{A}|, |\mathcal{C}|\}$, a set of eavesdroppers, each monitoring a subset of the parallel channels, is selected, and the resulting secrecy constraints are combined to established a tight upper bound. For the second case, the selection of the set of eavesdroppers is carried out through an induction process [11, Section V].

For a general, not parallel, channel model, a channel transformation which is performed via generalized singular value decomposition of $\mathbf{H}_1$ and $\mathbf{H}_2$, and a channel enhancement argument, are utilized [11, Section VI] to reduce the channel to a parallel channel model, establishing the achievability and converse for the general case.

REMARK 6.10 *The result in Theorem 6.6 can be interpreted as follows. Since the eavesdropper aims to maximize its mutual information with the transmitted signals, when she selects which parallel channels to observe, she gives precedence to the parallel channels over which only one user can transmit, which are the major contributor to her mutual information with the transmitted signals. In fact, the s.d.o.f. region in Theorem 6.6 can be expressed as the convex hull of a set of rectangles, each parametrized by the dimensions of the sub-space occupied by the transmission from the two users, since the single parallel channels are monitored by the eavesdropper.*

6.7 Conclusions

In this chapter, we have presented a comprehensive review of the secrecy capacity results for multiple antenna (MIMO) Gaussian wiretap channels and their high-SNR characterization. We have introduced an external multi-antenna cooperative jammer to the MIMO wiretap channel model, where the cooperative jammer serves as a helper terminal to improve the high-SNR performance of the channel, and presented the characterization of the secure degrees of freedom (s.d.o.f.) for this model. We have also presented results for the MIMO wiretap and MIMO MAC wiretap models that provide secrecy with no knowledge of the eavesdropper channel. In particular, we have reviewed the universal coding scheme which achieves a strongly secure rate for the MIMO wiretap channel with an eavesdropper channel that is arbitrary, time varying, and only known to the eavesdropper. In all the models considered, the common thread in the signaling schemes is the use of judicious injection of signals by the legitimate terminal(s) introduced solely to improve secrecy rather than carrying information, i.e., cooperative jamming. As a concluding note, we remark that multiple antennas have also been utilized to improve the secrecy rates and s.d.o.f. for several multi-terminal models not covered in this chapter, such as broadcast and interference channels, for which we refer the reader to [12, 36–44].

Acknowledgments

This work was supported by NSF Grants CCF 13-19338 and CNS 13-14719.

References

[1] A. D. Wyner, "The wire-tap channel," *Bell Syst. Tech. J.*, vol. 54, pp. 1355–1387, Oct. 1975.
[2] I. Csiszár and J. Körner, "Broadcast channels with confidential messages," *IEEE Trans. Inf. Theory*, vol. 24, no. 3, pp. 339–348, May 1978.
[3] A. Khisti and G. W. Wornell, "Secure transmission with multiple antennas i: The MISOME wiretap channel," *IEEE Trans. Inf. Theory*, vol. 56, no. 7, pp. 3088–3104, Jul. 2010.
[4] S. Shafiee, N. Liu, and S. Ulukus, "Towards the secrecy capacity of the Gaussian MIMO wire-tap channel: The 2-2-1 channel," *IEEE Trans. Inf. Theory*, vol. 55, no. 9, pp. 4033–4039, Sep. 2009.
[5] A. Khisti and G. W. Wornell, "Secure transmission with multiple antennas–Part II: The MIMOME wiretap channel," *IEEE Trans. Inf. Theory*, vol. 56, no. 11, pp. 5515–5532, Nov. 2010.
[6] F. Oggier and B. Hassibi, "The secrecy capacity of the MIMO wiretap channel," *IEEE Trans. Inf. Theory*, vol. 57, no. 8, pp. 4961–4972, Aug. 2011.

[7] M. Nafea and A. Yener, "Secure degrees of freedom of $N \times N \times M$ wiretap channel with a K-antenna cooperative jammer," in *Proc. IEEE Int. Conf. Commun.*, London, UK, Jun. 2015, pp. 4169–4174.

[8] M. Nafea and A. Yener, "Secure degrees of freedom for the MIMO wiretap channel with a multiantenna cooperative jammer," in *Proc. IEEE Inf. Theory Workshop*, Hobart, TAS, Australia, Nov. 2014, pp. 626–630.

[9] S. Goel and R. Negi, "Guaranteeing secrecy using artificial noise," *IEEE Trans. Wireless Commun.*, vol. 7, no. 6, pp. 2180–2189, Jun. 2008.

[10] X. He and A. Yener, "MIMO wiretap channels with unknown and varying eavesdropper channel states," *IEEE Trans. Inf. Theory*, vol. 60, no. 11, pp. 6844–6869, Nov. 2014.

[11] X. He, A. Khisti, and A. Yener, "MIMO multiple access channel with an arbitrarily varying eavesdropper: Secrecy degrees of freedom," *IEEE Trans. Inf. Theory*, vol. 59, no. 8, pp. 4733–4745, Aug. 2013.

[12] X. He, A. Khisti, and A. Yener, "MIMO broadcast channel with an unknown eavesdropper: Secrecy degrees of freedom," *IEEE Trans. Commun.*, vol. 62, no. 1, pp. 246–255, Jan. 2014.

[13] H. Sato, "An outer bound to the capacity region of broadcast channels," *IEEE Trans. Inf. Theory*, vol. 24, no. 3, pp. 374–377, May 1978.

[14] H. Weingarten, Y. Steinberg, and S. Shamai (Shitz), "The capacity region of the Gaussian multiple-input multiple-output broadcast channel," *IEEE Trans. Inf. Theory*, vol. 52, no. 9, pp. 3936–3964, Sep. 2006.

[15] T. Liu and S. Shamai (Shitz), "A note on the secrecy capacity of the multiple-antenna wiretap channel," *IEEE Trans. Inf. Theory*, vol. 55, no. 6, pp. 2547–2553, Jun. 2009.

[16] C. C. Paige and M. A. Saunders, "Towards a generalized singular value decomposition," *SIAM Journal on Numerical Analysis*, vol. 18, no. 3, pp. 398–405, 1981.

[17] M. Yuksel and E. Erkip, "Diversity–multiplexing tradeoff for the multiple-antenna wire-tap channel," *IEEE Trans. Wireless Commun.*, vol. 10, no. 3, pp. 762–771, Mar. 2011.

[18] E. Tekin and A. Yener, "Achievable rates for the general Gaussian multiple access wire-tap channel with collective secrecy," in *Proc. 44th Annual Allerton Conf. Commun., Control, Computing*, Monticello, IL, Sep. 2006, pp. 809–816.

[19] E. Tekin and A. Yener, "The general Gaussian multiple-access and two-way wiretap channels: Achievable rates and cooperative jamming," *IEEE Trans. Inf. Theory*, vol. 54, no. 6, pp. 2735–2751, Jun. 2008.

[20] X. He and A. Yener, "Providing secrecy with structured codes: Two-user Gaussian channels," *IEEE Trans. Inf. Theory*, vol. 60, no. 4, pp. 2121–2138, Apr. 2014.

[21] J. Xie and S. Ulukus, "Secure degrees of freedom of one-hop wireless networks," *IEEE Trans. Inf. Theory*, vol. 60, no. 6, pp. 3359–3378, Jun. 2014.

[22] M. A. Maddah-Ali, "On the degrees of freedom of the compound MISO broadcast channels with finite states," in *Proc. IEEE Int. Symp. Inf. Theory*, Austin, TX, USA, Jun. 2010, pp. 2273–2277.

[23] A. S. Motahari, S. O. Gharan, M. A. Maddah-Ali, and A. K. Khandani, "Real interference alignment: Exploiting the potential of single antenna systems," *IEEE Trans. Inf. Theory*, vol. 60, no. 8, pp. 4799–4810, Aug. 2014.

[24] V. Sprindzuk, "On Mahler's conjecture," *Doklady Akademii Nauk SSSR*, vol. 154, pp. 783–786, 1964, (in Russian); English translation in *Soviet Math. Dokl.* vol. 5, (1964), pp. 183–186.

[25] D. Kleinbock, "Baker–Sprindzhuk conjectures for complex analytic manifolds," Oct. 2002. [Online]. Available: http://arxiv.org/abs/math/0210369

[26] W. M. Schmidt, *Diophantine approximation*. Berlin, Heidelberg: Springer-Verlag, 1980.
[27] R. G. Gallager, *Information Theory and Reliable Communication*. Chichester: Wiley & Sons, 1968.
[28] G. Poltyrev, "On coding without restrictions for the AWGN channel," *IEEE Trans. Inf. Theory*, vol. 40, no. 2, pp. 409–417, Mar. 1994.
[29] D. Blackwell, L. Breiman, and A. J. Thomasian, "The capacity of a class of channels," *Ann. Math. Stat.*, vol. 30, no. 4, pp. 1229–1241, Dec. 1959.
[30] D. A. Levin, Y. Peres, and E. L. Wilmer, *Markov Chains and Mixing Times*. Providence, RI: American Mathematical Society, 2009.
[31] T. S. Han and S. Verdú, "Approximation theory of output statistics," *IEEE Trans. Inf. Theory*, vol. 39, no. 3, pp. 752–772, May 1993.
[32] I. Csiszár, "Almost independence and secrecy capacity," *Probl. Pered. Inform.*, vol. 32, no. 1, pp. 48–57, 1996.
[33] M. R. Bloch and J. N. Laneman, "On the secrecy capacity of arbitrary wiretap channels," in *Proc. 46th Annual Allerton Conf. Commun., Control, Computing*, Monticello, IL, USA, Sep. 2008, pp. 818–825.
[34] M. Bloch and J. N. Laneman, "Information-spectrum methods for information-theoretic security," *Proc. IEEE Inf. Theory Workshop*, pp. 23–28, Jun. 2009.
[35] R. Ahlswede, "Elimination of correlation in random codes for arbitrarily varying channels," *Z. Wahrscheinlichkeitstheorie verw. Gebiete*, vol. 44, pp. 159–175, 1978.
[36] H. D. Ly, T. Liu, and Y. Liang, "Multiple-input multiple-output Gaussian broadcast channels with common and confidential messages," *IEEE Trans. Inf. Theory*, vol. 56, no. 11, pp. 5477–5487, Nov. 2010.
[37] E. Ekrem and S. Ulukus, "The secrecy capacity region of the Gaussian MIMO multi-receiver wiretap channel," *IEEE Trans. Inf. Theory*, vol. 57, no. 4, pp. 2083–2114, Apr. 2011.
[38] E. Ekrem and S. Ulukus, "Capacity region of Gaussian MIMO broadcast channels with common and confidential messages," *IEEE Trans. Inf. Theory*, vol. 58, no. 9, pp. 5669–5680, Sep. 2012.
[39] R. Liu, T. Liu, H. V. Poor, and S. Shamai (Shitz), "Multiple-input multiple-output Gaussian broadcast channels with confidential messages," *IEEE Trans. Inf. Theory*, vol. 56, no. 9, pp. 4215–4227, Sep. 2010.
[40] R. Liu, T. Liu, H. V. Poor, and S. Shamai (Shitz), "New results on multiple-input multiple-output broadcast channels with confidential messages," *IEEE Trans. Inf. Theory*, vol. 59, no. 3, pp. 1346–1359, Mar. 2013.
[41] X. He and A. Yener, "The Gaussian interference wiretap channel when the eavesdropper channel is arbitrarily varying," in *Proc. IEEE Int. Symp. Inf. Theory*, Cambridge, MA, USA, Jul. 2012, pp. 2316–2320.
[42] X. He and A. Yener, "The interference wiretap channel with an arbitrarily varying eavesdropper: Aligning interference with artificial noise," Monticello, IL, USA, Sep. 2012, pp. 204–211.
[43] A. Khisti, "Interference alignment for the multiantenna compound wiretap channel," *IEEE Trans. Inf. Theory*, vol. 57, no. 5, pp. 2976–2993, May 2011.
[44] E. Ekrem and S. Ulukus, "Degraded compound multi-receiver wiretap channels," *IEEE Trans. Inf. Theory*, vol. 58, no. 9, pp. 5681–5698, Sep. 2012.

7 MISO Wiretap Channel with Strictly Causal CSI: A Topological Viewpoint

Zohaib Hassan Awan and Aydin Sezgin

Physical-layer security offers an alternative to cryptology, where instead of using the secrecy keys, randomness of the wireless channel is utilized to conceal information. In this chapter, we give an overview of different state-of-the-art multi-terminal information theoretic models to secure information with varying quality of channel state information at the transmitter. We first introduce the essential elements of information theoretic security, and later on discuss its application from a degrees of freedom viewpoint. In particular, we focus our attention on the models in which the topology of the network is known at the transmitter.

7.1 Introduction

Wireless communication has completely revolutionized the lives of common users. Modern communication systems provide high data rates, which allow users to perform different day-to-day tasks wirelessly. For example, users can access their bank account details and monitor their home security remotely. However, due to the broadcast nature of the medium, the information exchange between these communication nodes can be overheard by unintended users in the network. The leakage of information to unintended users may have serious consequences. In this chapter, we intend to answer this fundamental question of how to secure wireless communication.

7.2 State of the Art and Preliminaries

In conventional systems, cryptographic encryption is used to conceal information from adversaries. Figure 7.1 shows a simple example which sheds light on the problem of secret key sharing studied in cryptology. The source message at Alice is converted to ciphertext with the help of a secret key which is shared between Alice and the legitimate receiver, Bob. The ciphertext is then communicated over the wireless channel and is received by both Bob and Eve. Upon getting the ciphertext and having access to the secret key, Bob can easily recover Alice's message [1]. It is assumed that Eve, the eavesdropper, is of limited computational complexity or has limited resources. Thus, it is unable to generate all combinations of secret key, and consequently cannot recover Alice's message. In this framework, secrecy relies on the difficulty level of some

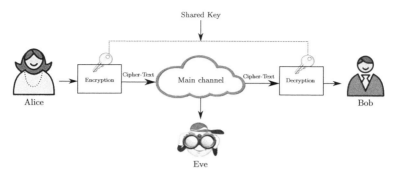

Figure 7.1 A typical secret key sharing setting.

hard mathematical problem which the eavesdropper is unable to solve due to limited computational resources. Thus, it is not provably secure, mathematically. In ad hoc wireless networks, key sharing becomes more expensive due to the requirement of key storage at each node. Conventional communication architectures are built on a layered approach, where separate layers are used to satisfy two important constraints, reliability and secrecy.

Reliability means that the information sent by Alice can be decoded at the intended node Bob with an arbitrarily small probability of error. Mathematically, it means that

$$\lim_{n \to \infty} \Pr\{\widehat{W}_i \neq W_i\} = 0. \tag{7.1}$$

Reliability-related issues are usually dealt at the physical layer by means of error correcting codes.

Secrecy means that the information destined for Bob from Alice cannot be decoded by any other node in the network. Secrecy issues are handled in the upper layers of the protocol stack, by means of authentication and encryption.

As stated before, using current technologies, security is considered as part of the upper layers of the protocol stack and completely ignores the physical layer attributes of the wireless channel. This leads to an important question. Can we use the physical layer of the wireless channel to secure information? It turns out that by exploiting the random fluctuations of the wireless channel, for instance differences in the fading gains between the legitimate channel, i.e., the channel from source to legitimate receiver, and the eavesdropper channel, i.e., the channel from source to adversary, it is possible to transmit information securely to the legitimate receiver. In what follows, we revisit the essentials of physical layer security and give an overview of the state of the art.

7.3 Secrecy Capacity Characterization

Wyner in his seminal work introduced a basic information theoretic model to study security by taking the uncertainty of the physical channel into account, where the

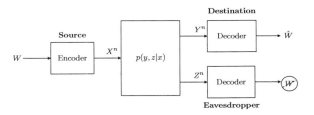

Figure 7.2 The wiretap channel.

channel input–output relationship is modeled by a fixed probability distribution [2], as shown in Fig. 7.2. Wyner's wiretap model consists of three nodes: a source, a legitimate receiver, and an eavesdropper. The source wishes to send confidential messages to the legitimate receiver and wishes to conceal them from the eavesdropper. Wyner showed that secrecy capacity is established when the eavesdropper channel (the channel from source to eavesdropper) is a degraded version of the main channel (the channel from source to legitimate receiver).[1] In information theoretic security, also dubbed "physical layer security," there are no assumptions placed on the computational power of the eavesdropper. The eavesdropper has unlimited resources and tries to decode the codewords being transmitted. Furthermore, the secrecy level is *quantifiable* and is proven mathematically. *Equivocation* captures the secrecy level of the communication system and is defined as the conditional entropy of the confidential message W given the channel output at the eavesdropper,

$$R_e = \frac{H(W|Z^n)}{n}, \qquad (7.2)$$

where R_e denotes the equivocation rate, the operator $H(\cdot)$ denotes the entropy, W denotes the confidential message intended for the legitimate receiver, and Z^n denotes the channel output at the eavesdropper for n channel uses [2]. (The reader can refer to Appendix 7.7.1 for more details about definitions.) Perfect secrecy for the channel is obtained when the eavesdropper gets no information about the confidential message W from Z^n (asymptotically), i.e.,

$$\lim_{n \to \infty} \frac{I(W;Z^n)}{n} = \frac{H(W)}{n} - \frac{H(W|Z^n)}{n} = 0. \qquad (7.3)$$

A rate–equivocation pair (R, R_e) indicates that the communication rate R is achievable with a secrecy level (secure transmission rate) of R_e. The secrecy capacity region, which is the set of all achievable rate–equivocation pairs (R, R_e) for the wiretap channel, is shown in Fig. 7.3, from which it can be easily seen that the secrecy rate (R_e) increases linearly with the communication rate (R) till C_s, where C_s denotes

---

[1] The term "secrecy capacity" is a counterpart of "channel capacity," which incorporates additional secrecy constraints.

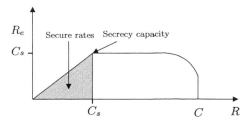

Figure 7.3 Secrecy capacity region of the wiretap channel, showing the secure transmission rate (R_e) as a function of communication rate (R) [3].

the secrecy capacity of the wiretap channel. At the point (C_s, C_s), all information is transmitted securely and perfect secrecy capacity is established. If we further increase R beyond C_s, it is not possible to secure all information and only C_s bits are transmitted securely. Roughly speaking, this means that for any communication model there exists a fundamental performance limit up to which we can transmit information securely. In the last decade, the wiretap channel has attracted significant interest in the research community and has been extended to study a variety of multi-user channels, e.g., the broadcast channel [3, 4], the multi-access channel [5–9], the relay channel [10–12], the interference channel [13, 14], and the multi-antenna channel [15–18]. For a review of other related contributions the reader may refer to [19] (and references therein).

7.4 Secure Degrees of Freedom

Characterizing the complete secrecy capacity region is a challenging task in general. Recently, a number of contributions have focused on characterizing the *approximate* capacity of these networks. The approximate capacity is measured by the notion of secure degrees of freedom (SDoF). Similar to degrees of freedom (DoF) [20], SDoF captures the way secrecy capacity (rate) scales asymptotically with the logarithm of the signal-to-noise ratio (SNR). A precise description of DoF and SDoF metrics are elucidated in Appendix 7.7.2.

7.4.1 Impact of CSIT

In conventional communication systems it is generally assumed that channel state information (CSI) is perfectly known at all nodes. The CSI at the receiver side is obtained through well-known channel estimation techniques and is then conveyed to the transmitter, where it is utilized to allocate the system resources, for instance input power, optimally. From a secrecy viewpoint, if the CSI (or statistics) from both the legitimate receiver and the eavesdropper is not perfectly conveyed to the transmitter, the transmitter is then unable to assign the system resources appropriately. This leads to information leakage to the eavesdropper, and consequently reduces the secrecy rate. The problem of conveying CSI to the transmitter has perturbed communication designers for a long time. This is in particular true for mobile networks.

In a mobile network, multiple nodes communicate with each other over a shared medium. This in turn leads to a fundamental problem of interference in networks. Interference in the network can be eradicated by different means, for example by interference alignment schemes (interference cancelation) [21], which mostly require perfect knowledge of channel state information at the transmitter (CSIT). As the network grows, providing perfect CSI to the transmitter requires extra overhead in terms of power or channel uses, i.e., more bits are required to convey the channel coefficients. Motivated by these scenarios, the focus has recently shifted to studying models with reduced quality of CSI at the transmitter side. The main inspiration behind studying these models stems from the fact that, since the receiver can estimate the channel with a certain accuracy, it can then decide how many bits are sufficient to feed back based on certain cost constraints. In [22], Maddah-Ali and Tse introduced a multi-input single-output (MISO) broadcast channel with strictly causal (delayed) CSI available at the transmitter and showed that delayed CSI is useful in the sense that it enlarges the DoF relative to the same model with no CSI. For the K-user MISO broadcast channel with delayed CSIT, an achievable sum DoF is given by [22, Theorem 1]

$$\text{DoF}(M,K) \geq \frac{K}{1 + \frac{1}{2} + \cdots + \frac{1}{K}}, \qquad (7.4)$$

where M denotes the number of antennas at the transmitter, for $M \geq K$. The coding scheme carefully utilizes the feedback information to learn the side information which later on is used to help the receivers to decode their respective symbols. The main idea of the coding scheme is elucidated with the help of the following example.

Example 7.1 Consider the two-user MISO broadcast channel with two transmit antennas and a single antenna at each receiver. The transmitter wants to send two symbols (v_1, v_2) to receiver 1 and two symbols (w_1, w_2) to receiver 2. The coding scheme consists of three time slots and is shown in Fig. 7.4.

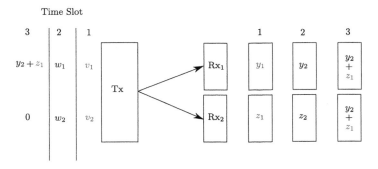

Figure 7.4 Maddah-Ali and Tse scheme with delayed CSIT.

In the first and second time slots, the transmitter sends fresh information to both receivers, respectively. At the end of the second time slot, each receiver gets a linear combination of two desired symbols and requires one extra equation to decode their respective symbols. More specifically, receiver 1 requires one extra equation to decode (v_1, v_2). This equation is available as side information at receiver 2 (z_1). Similarly, receiver 2 requires one extra equation to decode (w_1, w_2), available as side information at receiver 1 (y_2). Due to the availability of past CSIT, the transmitter can learn these pieces of side information and in the third time slot multicasts them to both receivers. At the end of the third time slot, since receiver 1 knows the CSI and y_2, it can subtract out the contribution of y_2 from $y_2 + z_1$ to get z_1 and subsequently decodes (v_1, v_2) through channel inversion. Receiver 2 can also perform similar operations to decode (w_1, w_2). Thus, 2 symbols are sent to each receiver over a total of 3 time slots, which yields a DoF of $\frac{2}{3}$ at each receiver.

The model studied by Maddah-Ali and Tse in [22] is extended to a variety of CSIT models, for instance the two- and three-user multi-input multi-output (MIMO) broadcast channel in [23,24] and mixed CSIT (perfect delayed CSI along with imperfect instantaneous CSI) in [25], all from a DoF perspective. Based on the insights obtained from the communication models studied in the existing literature, some effort has been made to study the corresponding SDoF region. In [15], Khisti and Wornell studied a Gaussian MIMO wiretap channel in which perfect CSI of the legitimate receiver and eavesdropper is available at the transmitter, and establish the SDoF. For the two-user MISO broadcast channel with two transmit antennas and a single antenna at each of the two receivers, the optimal sum SDoF is 2, and is obtained by zero-forcing the confidential messages at the unintended receivers. Yang et al. in [26] study the MIMO broadcast channel and show that strictly causal CSI is still useful from an SDoF perspective since it enlarges the secrecy region, in comparison to models with no CSIT. The SDoF of the MIMO wiretap channel with delayed CSIT is given by [26, Theorem 1]

$$d_s(N_1, N_2, M) = \begin{cases} M - N_2 & \text{if } N_2 \leq M \leq N_1 \\ \frac{N_1 M (M - N_2)}{N_1 N_2 + M(M - N_2)} & \text{if } \min\{N_1, N_2\} \leq M \leq N_1 + N_2 \\ \frac{N_1 (N_1 + N_2)}{N_1 + 2N_2} & \text{if } M \geq N_1 + N_2, \end{cases} \quad (7.5)$$

where M denotes the number of transmit antennas and N_1 and N_2 denote the number of antennas at the legitimate receiver and eavesdropper, respectively. The coding scheme in [26] follows by an appropriate extension of the Maddah Ali–Tse scheme [22] with additional noise injection to account for security constraints. We illustrate the coding scheme with the following example.

Example 7.2 Consider the $(2,1,1)$ MISO wiretap channel with two antennas at the transmitter and a single antenna at each of the legitimate receiver and eavesdropper. The transmitter wants to send two symbols to the legitimate receiver and wishes to conceal

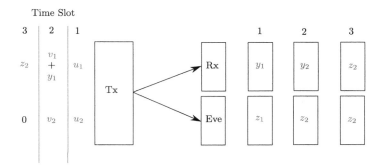

Figure 7.5 Yang *et al.* scheme with delayed CSIT and security constraints.

them from the eavesdropper. The coding scheme consists of three time slots. In the first time slot the transmitter injects artificial noise u_1 and u_2 from both antennas, as shown in Fig. 7.5. At the end of the first time slot, each receiver gets a linear combination of artificial noise. The channel output at the legitimate receiver can be interpreted as a secret key which later on is used to secure messages in the next time slot. Due to the availability of past CSIT, the transmitter is able to learn the channel output at the legitimate receiver and in the second time slot sends v_1 and v_2 along with the past channel output at the legitimate receiver in time slot 1 (y_1). At the end of time slot 2, since the legitimate receiver knows y_1 and also the CSI, it subtracts out the contribution of y_1 from y_2 to get one equation with two unknowns. Thus, the legitimate receiver requires one extra equation to be available as interference (side information) at the eavesdropper to successfully decode the intended symbols. In time slot 3, due to the availability of delayed CSIT, the transmitter can learn z_2 and sends it to the legitimate receiver, which helps to decode both symbols securely. Thus, 2 symbols are sent to the legitimate receiver over a total of 3 time slots, which yields an SDoF of $\frac{2}{3}$.

In another related work, Zaidi *et al.* in [27] studied the two-user MIMO X-channel with asymmetric feedback and delayed CSIT, and characterized the complete sum SDoF region. The sum SDoF region of the two-user (M,M,N,N) MIMO X-channel with asymmetric output feedback and delayed CSIT is given by the set of all non-negative pairs $(d_{11}+d_{21}, d_{12}+d_{22})$ satisfying [27, Theorem 1]

$$\frac{d_{11}+d_{21}}{d_s(N,N,2M)} + \frac{d_{12}+d_{22}}{\min(2M,2N)} \leq 1,$$

$$\frac{d_{11}+d_{21}}{\min(2M,2N)} + \frac{d_{12}+d_{22}}{d_s(N,N,2M)} \leq 1. \quad (7.6)$$

We now illustrate the main ingredients of this scheme with the following example. The coding scheme sheds light on how to utilize the leverage provided by the output feedback to secure information.

Example 7.3 Consider a two-user X-channel which consists of two transmitters and two receivers, where each node has a single antenna as shown in Fig. 7.6. Transmitter 1 wants to transmit symbols v_{11} and w_{12} to receiver 1 and receiver 2, respectively. Similarly, transmitter 2 wants to transmit symbols v_{21} and w_{22} to receiver 1 and receiver 2, respectively. Each receiver is assumed to have perfect instantaneous knowledge of its channel coefficients, and receiver i is allowed to feed back its channel output to transmitter i, for $i = \{1,2\}$. In addition to this, the symbols that are destined for each receiver are meant to be kept secret from the other receiver. The communication takes place in four channel uses. For this model, we show that a $\frac{1}{2}$ SDoF is achievable at each receiver.

In the first time slot, both transmitters inject artificial noise u_1 and u_2. At the end of time slot 1, receiver i feeds back the channel output to transmitter i, for $i \in \{1,2\}$. In the second time slot, both transmitters send fresh information to receiver 1. More specifically, transmitter 1 sends v_{11} along with the past channel output at receiver 1 in time slot 1 (y_1), and transmitter 2 transmits v_{21}. At the end of time slot 2, since receiver 1 knows y_1 and also the CSI, it subtracts out the contribution of y_1 from y_2 to get one equation with two unknowns. Thus, receiver 1 requires one extra equation to be available as interference (side information) at receiver 2 to successfully decode the intended symbols. At the end of time slot 2, receiver i feeds back the channel output to transmitter i, for $i \in \{1,2\}$. The operations in time slot 3 are similar to those in time slot 2, where the roles of transmitter 1 and 2 as well as receiver 1 and 2 are reversed. In the third time slot, both transmitters send fresh information to receiver 2, where transmitter 2 sends w_{22} along with the past channel output at receiver 2 in time slot 1 (z_1) and transmitter 1 transmits w_{12}. At the end of time slot 3, since receiver 2 knows z_1 and also the CSI, it subtracts out the contribution of z_1 from z_3 to get one equation with two unknowns. Thus, it requires one extra equation to be available as interference (side information) at receiver 1 to decode the symbols correctly. At the end of time

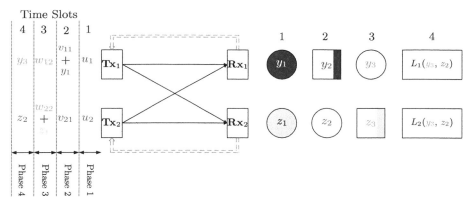

Figure 7.6 Illustration of the coding scheme for the $(1,1,1,1)$ MIMO X-channel with output feedback and delayed CSIT.

slot 3, receiver i feeds back the channel output to transmitter i, for $i \in \{1,2\}$. At the end of time slot 3, due to the availability of feedback, transmitter 1 can learn y_3 and transmitter 2 can learn z_2, which are the necessary equations required by receiver 1 and receiver 2 to decode the intended symbols securely; in time slot 4, they are multicast to both receivers. At the end of time slot 4, each receiver gets a linear combination of y_3 and z_2. Since receiver 1 knows the CSI and y_3, it subtracts out the contribution of y_3 from the channel output $L_1(y_3, z_2)$ to obtain z_2, and then by using (y_2, z_2) decodes (v_{11}, v_{21}) through channel inversion. Receiver 2 can also perform similar operations to decode (w_{12}, w_{22}).

7.4.2 Topological Diversity

Interactions between mobile users in a cellular network are subjected to different external effects, such as inter-cell interference, wave propagation path loss, or jamming. These external factors influence links in an asymmetric manner, which leads to some links being stronger than others statistically. A fundamental issue with the DoF (SDoF) metric is that it ignores the diversity of link strengths and assumes that all non-zero channels are equally strong in the sense that each link is capable of carrying 1 DoF, irrespective of the actual magnitude of the channel coefficient. This issue was first solved by [28] in the context of an interference channel by introducing a new metric called generalized degrees of freedom (GDoF) (see Appendix 7.7.3 for more details). This metric provides a richer approximation of channel capacity and takes into account the relative strengths of interfering signals. For broadcast channels this metric was recently investigated in [29], where the authors characterized the sum GDoF. As mentioned before, similar to DoF, the SDoF metric does not capture the relative strengths of the communication links. A natural direction is to explore the corresponding generalized secure degree of freedom (GSDoF). The GSDoF takes the following parameters into account:

- relative strength of communication links
- reliability issues (7.1)
- secrecy constraints (7.3).

In what follows, we provide an overview of the MISO wiretap channel model where the system performance is measured by the GSDoF metric.

7.5 GSDoF of MISO Wiretap Channel with Delayed CSIT

In this section, we consider a MISO wiretap channel with topology information conveyed at the transmitter. The Gaussian $(2,1,1)$ MISO wiretap channel consists of three nodes: a transmitter, a legitimate receiver, and an eavesdropper, as shown in Fig. 7.7. The transmitter is equipped with two antennas and each of the legitimate receiver and eavesdropper is equipped with a single antenna. The transmitter wants to reliably transmit message $W \in \mathcal{W} = \{1, \ldots, 2^{nR(A_1, \rho)}\}$ to the legitimate receiver and wishes to conceal it from the eavesdropper. Each receiver is allowed to convey the past CSI to the transmitter. Although the legitimate receiver can convey the past CSIT, in practice the eavesdropper – whose primary role is to decode the intended communication – is not willing to convey its own CSI to the transmitter. In this chapter, we assume that the eavesdropper is a part of the communication system and in its desire to learn some information it agrees to convey the past CSIT. Let $A_1 \in \{1, \alpha\}$ denote the link power exponent from the transmitter to the legitimate receiver and $A_2 \in \{1, \alpha\}$ denote the link power exponent from the transmitter to the eavesdropper, respectively, for $0 \leq \alpha \leq 1$, where we denote the stronger link by $A_i := 1$ and the weaker link by $A_i := \alpha$, $i = 1, 2$. Then, based on the topology of the network, this model can alternate between the following possible states:

1. $(A_1, A_2) = (1, 1)$
2. $(A_1, A_2) = (1, \alpha)$
3. $(A_1, A_2) = (\alpha, 1)$
4. $(A_1, A_2) = (\alpha, \alpha)$.

Let us now focus on (2), where the legitimate receiver is stronger than the eavesdropper. For this case, we now examine the interplay between CSIT and topology. As alluded to before, although numerous forms of CSIT are available we restrict ourselves to three fundamental types, namely

- perfect CSIT
- no CSIT
- delayed CSIT.

First, consider the two extreme cases where receivers can convey either perfect CSI or no CSI to the transmitter. For the case in which perfect CSI is conveyed by both receivers, i.e., there are abundant resources available in the system, a GSDoF of 1 is achievable by zero-forcing [15]. Next, we consider the other extreme case of no CSIT. Notice that, since the link to the legitimate receiver is stronger than the one to the eavesdropper $(A_1 > A_2)$, by simple uncoded transmission the legitimate receiver gets 1 DoF and the eavesdropper gets α DoF. Thus, only $1 - \alpha$ GSDoF is achievable. Now, the only case left is when both receivers convey only the past or delayed CSIT. For this interesting case, the authors in [34] showed that in comparison to no CSIT, a strictly better GSDoF is achievable when delayed CSI is conveyed by both receivers. We now formally revisit this system model as shown in Fig. 7.7 and state the main results.

Let $\lambda_{A_1 A_2}$ be the fraction of time topology state $A_1 A_2$ occurs, such that

$$\sum_{(A_1, A_2) \in \{1, \alpha\}^2} \lambda_{A_1 A_2} = 1. \tag{7.7}$$

The channel input–output relationship at time instant t is

$$y_t = \sqrt{\rho^{A_{1,t}}} \mathbf{h}_t \mathbf{x}_t + n_{1t},$$

$$z_t = \sqrt{\rho^{A_{2,t}}} \mathbf{g}_t \mathbf{x}_t + n_{2t}, \ t = 1, \ldots, n, \tag{7.8}$$

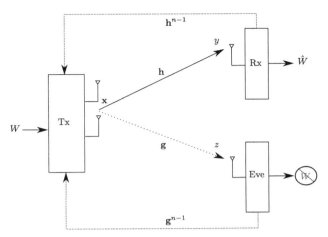

Figure 7.7 $(2,1,1)$ MISO wiretap channel, where the link to the eavesdropper is weaker than the link to the legitimate receiver.

where $\mathbf{x} \in \mathbb{C}^{2 \times 1}$ is the channel input vector, $\mathbf{h} \in \mathcal{H} \subseteq \mathbb{C}^{1 \times 2}$ is the channel vector connecting the intended receiver to the transmitter, and $\mathbf{g} \in \mathcal{G} \subseteq \mathbb{C}^{1 \times 2}$ is the channel vector connecting the eavesdropper to the transmitter, respectively. The parameter ρ is subject to the input power constraint, and n_i is assumed to be independent and identically distributed (i.i.d.) white Gaussian noise, with $n_i \sim \mathcal{CN}(0,1)$ for $i = 1, 2$. A code for the Gaussian $(2, 1, 1)$ MISO wiretap channel with delayed CSIT and alternating topology consists of a sequence of stochastic encoders at the transmitter,

$$\{\phi_t : \mathcal{W} \times \mathcal{H}^{t-1} \times \mathcal{G}^{t-1} \longrightarrow \mathcal{X}_1 \times \mathcal{X}_2\}_{t=1}^n, \quad (7.9)$$

where the message W is drawn uniformly over $\mathcal{W}$; and a decoding function at the intended receiver,

$$\psi : \mathcal{Y}^n \times \mathcal{H}^n \times \mathcal{G}^{t-1} \longrightarrow \widehat{\mathcal{W}}. \quad (7.10)$$

A GSDoF $d(A_1)$ is said to be achievable if there exists a sequence of codes satisfying

- reliability condition (7.1),
- perfect secrecy condition (7.3), and
- communication rate condition

$$\lim_{\rho \to \infty} \liminf_{n \to \infty} \frac{\log |\mathcal{W}(n, \rho, A_1)|}{n \log \rho} \geq d(A_1). \quad (7.11)$$

We now provide the upper and lower bounds on the GSDoF of the MISO wiretap channel with delayed CSIT and alternating topology.

7.5.1 Upper Bound

The following theorem provides an upper bound on the GSDoF of the MISO wiretap channel with delayed CSIT and alternating topology.

THEOREM 7.1 ([34, Theorem 1]) *For the (2,1,1) MISO wiretap channel with delayed CSIT and alternating topology ($\lambda_{A_1 A_2}$), an upper bound on the GSDoF is given by*

$$d(\lambda_{A_1 A_2}) \leq \frac{(3-\alpha)\lambda_{1\alpha} + 2(\lambda_{11} + \alpha\lambda_{\alpha\alpha}) + (1+\alpha)\lambda_{\alpha 1}}{3}. \tag{7.12}$$

The upper bound is based on the property of entropy symmetry of channel outputs.

Next, we provide some optimal encoding schemes for fixed topological states. For simplicity of analysis and in accordance with the DoF framework, we ignore the additive Gaussian noise and only mention the asymptotic behavior of the inputs by ignoring the exact power allocations.

7.5.2 Fixed Topology ($\lambda_{11} = 1$)

By removing the topology consideration, i.e., by setting $(A_{1t}, A_{2t}) := (1, 1)\ \forall t$, the model reduces to the MISO wiretap channel with delayed CSIT studied in [26], for which the optimal SDoF is given by 2/3 SDoF. (See Example 1.2 for a sketch of the coding scheme.)

7.5.3 Fixed Topology ($\lambda_{1\alpha} = 1$)

For the case in which the legitimate receiver is stronger than the eavesdropper ($\lambda_{1\alpha} = 1$), the GSDoF is given by

$$d = \frac{3-\alpha}{3}. \tag{7.13}$$

The converse follows immediately from the upper bound established in Theorem 7.1 by setting $\lambda_{1\alpha} := 1$ in (7.12). We now provide an overview of the encoding scheme that is used to establish (7.13). In this scheme, the transmitter wants to send three symbols (v_1, v_2, v_3) to the legitimate receiver and wishes to conceal them from the eavesdropper. The communication scheme consists of three phases, each comprising of only one time slot.

In the first time slot, the transmitter injects structured artificial noise (via lattice coding), $\mathbf{u} := [u_1, u_2]^T$, along with the confidential symbol (v_1) intended for the legitimate receiver. In this phase, the leverage provided by the topology of the network is utilized as follows. Similar to the coding scheme that is used to achieve (7.5), the transmitter injects artificial noise from both antennas, where due to the topology of the network the output at the eavesdropper is obtained at a lower power level ($\mathcal{O}(\rho^\alpha)$) compared to the legitimate receiver ($\mathcal{O}(\rho)$). One can improve this scheme by reducing the transmission power of the artificial noise to the order of the eavesdropper ($\mathcal{O}(\rho^\alpha)$), so that the transmitter can then use the remaining power ($\mathcal{O}(\rho^{1-\alpha})$) to send a fresh symbol to the legitimate receiver. The eavesdropper will receive the fresh symbol embedded in artificial noise but at a reduced power level, and hence cannot decode it. In this phase, the transmitter sends

$$\mathbf{x}_1 = \begin{bmatrix} u_1 \\ u_2 \end{bmatrix} + \begin{bmatrix} v_1 \rho^{-\alpha/2} \\ \phi \end{bmatrix}, \tag{7.14}$$

where the channel input–output relationship is given by

$$y_1 = \underbrace{\sqrt{\rho}\mathbf{h}_1\mathbf{u}}_{\mathcal{O}(\rho)} + \underbrace{\sqrt{\rho^{(1-\alpha)}}h_{11}v_1}_{\mathcal{O}(\rho^{1-\alpha})}, \tag{7.15a}$$

$$z_1 = \underbrace{\sqrt{\rho^\alpha}\mathbf{g}_1\mathbf{u}}_{\mathcal{O}(\rho^\alpha)} + \underbrace{\sqrt{\rho^0}g_{11}v_1}_{\mathcal{O}(\rho^0)}. \tag{7.15b}$$

At the end of phase 1, each receiver conveys the past CSI to the transmitter. At the end of phase 1, the eavesdropper gets a linear combination of artificial noise along with the confidential symbol (v_1) at noise level, and is unable to decode it. The legitimate receiver gets the confidential symbol embedded in a linear combination of artificial noise ($\mathbf{h}_1\mathbf{u}$). It first computes $\mathbf{h}_1\mathbf{u}$ from the channel output (y_1) by treating v_1 as noise, within bounded noise distortion. Afterwards, it subtracts out the contribution of $\mathbf{h}_1\mathbf{u}$ from y_1 and decodes v_1 through channel inversion. The decoding of v_1 is possible at the legitimate receiver since it is received at a higher power level compared to the eavesdropper. Thus, it can be readily shown that $(1-\alpha)\log(\rho)$ bits are sent via the v_1 symbol to the legitimate receiver securely.

In the second phase, the transmitter sends new symbols ($\mathbf{v} := [v_2, v_3]^T$) to the legitimate receiver along with a linear combination of the channel output ($\mathbf{h}_1\mathbf{u}$) at the legitimate receiver during the first phase. The transmitter can easily learn ($\mathbf{h}_1\mathbf{u}$), since it knows $\mathbf{u}$ and the past CSI of the legitimate receiver ($\mathbf{h}_1$), and sends

$$\mathbf{x}_2 = \begin{bmatrix} v_2 \\ v_3 \end{bmatrix} + \begin{bmatrix} \mathbf{h}_1\mathbf{u} \\ \phi \end{bmatrix}, \tag{7.16}$$

where the channel input–output relationship is given by

$$y_2 = \sqrt{\rho}\mathbf{h}_2\mathbf{v} + \sqrt{\rho}h_{21}\mathbf{h}_1\mathbf{u}, \tag{7.17a}$$

$$z_2 = \sqrt{\rho^\alpha}\mathbf{g}_2\mathbf{v} + \sqrt{\rho^\alpha}g_{21}\mathbf{h}_1\mathbf{u}. \tag{7.17b}$$

At the end of phase 2, each receiver conveys the past CSI to the transmitter. Since the legitimate receiver knows the CSI ($\mathbf{h}_2$) and also the channel output y_1 from phase 1, it subtracts out the contribution of $\mathbf{h}_1\mathbf{u}$ from the channel output y_2, to obtain one equation with two unknowns ($\mathbf{v} := [v_2, v_3]^T$). Thus, the legitimate receiver requires one extra equation to successfully decode the intended variables, being available as interference (side information) at the eavesdropper but with a *reduced* power level.

In the third phase, due to the availability of past CSIT, the transmitter can construct the side information (z_2) at the eavesdropper and sends

$$\mathbf{x}_3 = \begin{bmatrix} \mathbf{g}_2\mathbf{v} + g_{21}\mathbf{h}_1\mathbf{u} \\ \phi \end{bmatrix}, \tag{7.18}$$

where the channel input–output relationship is given by

$$y_3 = \sqrt{\rho}h_{31}\mathbf{g}_2\mathbf{v} + \sqrt{\rho}h_{31}g_{21}\mathbf{h}_1\mathbf{u}, \tag{7.19a}$$

$$z_3 = \sqrt{\rho^\alpha}g_{31}\mathbf{g}_2\mathbf{v} + \sqrt{\rho^\alpha}g_{31}g_{21}\mathbf{h}_1\mathbf{u}. \tag{7.19b}$$

At the end of phase 3, by using y_1, the legitimate receiver subtracts out the contribution of $\mathbf{h}_1\mathbf{u}$ from (y_2, y_3) and decodes $\mathbf{v}$ through channel inversion. It can be readily seen that $2 + (1 - \alpha)$ bits are sent to the legitimate receiver over a total of three time slots, which yields the desired GSDoF in (7.13).

The result in (7.13) provides some interesting insights. If we set $\alpha := 0$, the expression in (7.13) gives a GSDoF of 1. For this setting, the link between the transmitter and the eavesdropper is absent, and thus the model reduces to the point-to-point channel between the transmitter and the legitimate receiver, where 1 DoF is achievable. Conversely, if $\alpha := 1$, the expression in (7.13) gives a GSDoF of 2/3. In this case both links are equally strong and (7.13) recovers the result established in the context of the MISO wiretap channel without topology in (7.5).

7.6 Discussion and Directions for Future Work

We now discuss the possible extensions of this class of models and give some directions for future research.

7.6.1 MISO Broadcast Channel with Topology

Throughout this chapter, we have considered the simplest case of MISO wiretap model with perfect delayed CSIT and alternating topology, where each receiver acts as either a legitimate receiver or an eavesdropper. An interesting case arises when each receiver performs both of these two roles, simultaneously. This class of problems are commonly referred to as broadcast channel. For this setup, the transmitter wants to send message W_1 to receiver 1 and W_2 to receiver 2. The message intended for receiver 1 needs to be kept secret from receiver 2, and the message W_2 intended for receiver 2 needs to be kept secret from receiver 1. Thus, each receiver plays two different roles; for example, receiver 1 not only acts as a legitimate receiver for the messages W_1 intended for it, but it also acts as an eavesdropper for the message W_2 intended for the other receiver. For this model, the additional topological information can be utilized to construct efficient schemes which can improve the GSDoF of these channels. In related models without secrecy constraints, for instance [29], instead of transmitting the analog side information (constructed due to the availability of delayed CSIT), digitized side information is multi-cast, which provides some extra resources to send information. It will be interesting to see whether one can harness the same benefits for models with security constraints.

7.6.2 Synergistic Benefits of CSIT with Topology

In this chapter, we assume that only delayed CSIT is conveyed by both receivers. Due to random fluctuations of the wireless medium, it becomes difficult to convey the same quality of CSIT over time. In [30], the authors studied a model in which asymmetric CSIT is available from both receivers. For this model, the authors showed that joint coding across different states provides larger rates compared to coding independently over different states. From s security viewpoint, this model was recently studied in [31, 32], where the authors showed the synergistic benefits of alternating CSIT in terms of SDoF. An interesting extension of the MISO wiretap channel with topology considered in this chapter is to generalize it to the asymmetric CSIT setting.

7.6.3 Extension to K-User Case

A practical communication network generally contains more than two receivers. A natural extension to the problem described in this chapter is to generalize it to the K-user setting. Here, the transmitter wants to send a confidential message to each receiver and wishes to conceal it from the remaining set of users. Thus, the message W_k intended for the kth user needs to be kept secret from the remaining $K-1$ users. The extension to the K-user case is quite a challenging task since the existing techniques for obtaining upper bounds do not provide a tight upper bound. Furthermore, establishing a lower bound is also far from trivial. The key difficulty in establishing the coding scheme is how to securely multicast the common side information overheard at different receivers. It would be interesting to see how topological information can be utilized to close the performance gap for this setting.

7.7 Appendix

In this section, we provide an overview of some basic definitions which complement the contents of this chapter. The reader can refer to [21, 33] for details.

7.7.1 Entropy

Let X denote a discrete random variable with the alphabet set $\mathcal{X} = \{x_1, x_2, \ldots, x_n\}$ and probability mass function $p_X(x_i) = p(X = x_i)$, for $i = 1, \ldots, n$. For convenience, we denote the probability mass function $p_X(x_i)$ as $p(x_i)$.

DEFINITION 7.1   [33] The entropy of a discrete random variable X is defined as

$$H(X) = -\sum_{i=1}^{n} p(x_i) \log p(x_i).$$

DEFINITION 7.2   [33] The conditional entropy of the discrete random variable Y given X is defined as

$$H(Y|X) = -\sum_{x \in \mathcal{X}} p(x) \sum_{y \in \mathcal{Y}} p(y|x) \log p(y|x)$$
$$= -\sum_{x \in \mathcal{X}} \sum_{y \in \mathcal{Y}} p(x,y) \log p(y|x),$$

where $p(x,y)$ denotes the joint probability distribution and $p(y|x)$ denotes the conditional probability distribution of random variable Y given X.

7.7.2 Degrees of Freedom

Consider a communication channel which consists of a single point-to-point channel. The transmitter wants to send message $W \in \mathcal{W} = \{1,\ldots,2^{nR}\}$ to the receiver. The encoder encodes the messages W into n codewords and sends them over the channel. The communication rate R is said to be achievable if, for arbitrarily large n, the receiver is able to decode the message W with a small error probability. The maximum achievable rate is defined as the channel capacity. The channel input–output relationship for the Gaussian point-to-point channel is given by

$$y = \sqrt{\rho}hx + z, \tag{7.20}$$

where y is the channel output, h is the channel connecting the receiver to the transmitter, z is the Gaussian noise with $\mathcal{CN}(0,\sigma^2)$, x is the channel input symbol, $||x||^2 \leq 1$, and the parameter ρ is subject to the input power constraint. The capacity of (7.20) is given by

$$C = \log\left(1 + \frac{\rho ||h||^2}{\sigma^2}\right) \quad \text{bits/channel use.} \tag{7.21}$$

The DoF metric measures the behavior of the capacity when the transmit power approaches infinity with fixed channel gain and noise power. Mathematically, the capacity can be expressed in terms of DoF as

$$C = d\log(\rho) + o(\log(\rho)), \tag{7.22}$$

where d denotes the DoF [$d = 1$ for (7.20)], and $o(\cdot)$ denotes the Landau notation.[2] Notice that in the DoF approximation the channel strength $||h||^2$ and noise power σ^2 are irrelevant, since they do not scale with ρ. The SDoF metric is obtained along similar lines by incorporating the additional security constraint 7.3.

7.7.3 Generalized Degrees of Freedom

A fundamental issue with the DoF metric is that it assumes that all channels are equally strong. This implies that in high power regimes the ratio of channel power of two different links approaches one. Thus, it does not provide much insight into how to assign resources optimally for channels with different link strengths. The GDoF solves this issue and takes the relative strength of the channel into account. Consider the channel

$$y = \sqrt{\rho^\alpha}hx + z, \tag{7.23}$$

where the parameter α indicates the relative strength of the communication link. The capacity of (7.23) is

$$C = \log\left(1 + \frac{\rho^\alpha ||h||^2}{\sigma^2}\right) \quad \text{bits/channel use,} \tag{7.24}$$

---

[2] Let $f(x)$ and $g(x)$ be some functions defined over real x, then $f(x) = o(g(x))$ implies that $\frac{f(x)}{g(x)} \to 0$ as $x \to \infty$.

and the corresponding GDoF is given by

$$C = \alpha \log(\rho) + o(\log(\rho)), \tag{7.25}$$

where α denotes the GDoF of (7.23).

Acknowledgments

This work is supported by the German Research Foundation, Deutsche Forschungsgemeinschaft (DFG), Germany, under grant SE 1697/11. The results in this chapter are largely based on "On MISO Wiretap Channels with Delayed CSIT and Topology," 10th International ITG Conference on Systems, Communications and Coding, Hamburg, Germany, Feb. 2015 [34].

References

[1] W. Stallings, *Cryptography and Network Security: Principles and Practice*, 2nd edn. Upper Saddle River, NJ: Prentice-Hall, Inc., 1999.
[2] A. D. Wyner, "The wire-tap channel," *Bell Syst. Tech. J.*, vol. 54, pp. 1355–1387, Oct. 1975.
[3] I. Csiszár and J. Körner, "Broadcast channels with confidential messages," *IEEE Trans. Inf. Theory*, vol. 24, no. 3, pp. 339–348, May 1978.
[4] R. Liu, T. Liu, H. V. Poor, and S. Shamai (Shitz), "New results on multiple-input multiple-output broadcast channels with confidential messages," *IEEE Trans. Inf. Theory*, vol. 59, no. 3, pp. 1346–1359, Mar. 2013.
[5] E. Tekin and A. Yener, "The Gaussian multiple access wire-tap channel," *IEEE Trans. Inf. Theory*, vol. 54, no. 12, pp. 5747–5755, Dec. 2008.
[6] Y. Liang and H. V. Poor, "Multiple access channels with confidential messages," *IEEE Trans. Inf. Theory*, vol. 54, no. 3, pp. 976–1002, Mar. 2008.
[7] E. Tekin and A. Yener, "The general Gaussian multiple-access and two-way wiretap channels: Achievable rates and cooperative jamming," *IEEE Trans. Inf. Theory*, vol. 54, no. 6, pp. 2735–2751, Jun. 2008.
[8] Z. H. Awan, A. Zaidi, and L. Vandendorpe, "On multiaccess channel with unidirectional cooperation and security constraints," in *Proc. 50th Annual Allerton Conf. Commun., Control, Computing*, Monticello, IL, USA, Sep. 2012, pp. 982–987.
[9] Z. H. Awan, A. Zaidi, and L. Vandendorpe, "Multiaccess channel with partially cooperating encoders and security constraints," *IEEE Trans. Inf. Forensics Security*, vol. 8, no. 7, pp. 1243–1254, Jul. 2013.
[10] L. Lai and H. El Gamal, "The relay eavesdropper channel: Cooperation for secrecy," *IEEE Trans. Inf. Theory*, vol. 54, no. 9, pp. 4005–4019, Sep. 2008.
[11] Z. H. Awan, A. Zaidi, and L. Vandendorpe, "On secure transmission over parallel relay eavesdropper channel," in *Proc. 48th Annual Allerton Conf. Commun., Control, Computing*, Monticello, IL, USA, Sep. 2010, pp. 859–866.
[12] Z. H. Awan, A. Zaidi, and L. Vandendorpe, "Secure communication over parallel relay channel," *IEEE Trans. Inf. Forensics Security*, vol. 7, no. 2, pp. 359–371, Apr. 2012.

[13] O. O. Koyluoglu and H. E. Gamal, "Cooperative encoding for secrecy in interference channels," *IEEE Trans. Inf. Theory*, vol. 57, no. 9, pp. 5682–5694, Sep. 2011.

[14] Z. Li, R. Yates, and W. Trappe, "Secrecy capacity of a class of one-sided interference channel," in *Proc. IEEE Int. Symp. Inf. Theory*, Toronto, ON, Canada, Jul. 2008, pp. 379–383.

[15] A. Khisti and G. W. Wornell, "Secure transmission with multiple antennas–Part II: The MIMOME wiretap channel," *IEEE Trans. Inf. Theory*, vol. 56, no. 11, pp. 5515–5532, Nov. 2010.

[16] F. Oggier and B. Hassibi, "The secrecy capacity of the MIMO wiretap channel," *IEEE Trans. Inf. Theory*, vol. 57, no. 8, pp. 4961–4972, Aug. 2011.

[17] T. Liu and S. Shamai (Shitz), "A note on the secrecy capacity of the multiple-antenna wiretap channel," *IEEE Trans. Inf. Theory*, vol. 55, no. 6, pp. 2547–2553, Jun. 2009.

[18] R. Bustin, R. Liu, H. V. Poor, and S. Shamai (Shitz), "An MMSE approach to the secrecy capacity of the MIMO Gaussian wiretap channel," *EURASIP J. Wireless Commun. Netw.*, p. 8, Nov. 2009.

[19] Y. Liang, H. V. Poor, and S. Shamai (Shitz), "Information theoretic security," *Foundations and Trends in Communications and Information Theory*, vol. 5, no. 4–5, pp. 355–580, 2009.

[20] V. R. Cadambe and S. A. Jafar, "Interference alignment and the degrees of freedom of wireless X networks," *IEEE Trans. Inf. Theory*, vol. 55, no. 9, pp. 3893–3908, Sep. 2009.

[21] S. A. Jafar, "Interference alignment – A new look at signal dimensions in a communication network," *Foundations and Trends in Communications and Information Theory*, vol. 7, no. 1, pp. 1–134, 2010.

[22] M. A. Maddah-Ali and D. Tse, "Completely stale transmitter channel state information is still very useful," *IEEE Trans. Inf. Theory*, vol. 58, no. 7, pp. 4418–4431, Jul. 2012.

[23] C. S. Vaze and M. K. Varanasi, "The degrees of freedom region of the two-user and certain three-user MIMO broadcast channel with delayed CSI," Jan. 2011. [Online]. Available: http://arxiv.org/abs/1101.0306

[24] M. J. Abdoli, A. Ghasemi, and A. K. Khandani, "On the degrees of freedom of three-user MIMO broadcast channel with delayed CSIT," in *IEEE Int. Symp. Inf. Theory*, St. Petersburg, Russia, Jul. 2011, pp. 209–213.

[25] G. Tiangao and S. A. Jafar, "Optimal use of current and outdated channel state information: Degrees of freedom of the MISO BC with mixed CSIT," *IEEE Commun. Letters*, vol. 16, no. 7, pp. 1084–1087, Jul. 2012.

[26] S. Yang, M. Kobayashi, P. Piantanida, and S. Shamai (Shitz), "Secrecy degrees of freedom of MIMO broadcast channels with delayed CSIT," *IEEE Trans. Inf. Theory*, vol. 59, no. 9, pp. 5244–5256, 2013.

[27] A. Zaidi, Z. H. Awan, S. Shamai (Shitz), and L. Vandendorpe, "Secure degrees of freedom of MIMO X-channels with output feedback and delayed CSIT," *IEEE Trans. Inf. Forensics Security*, vol. 8, no. 11, pp. 1760–1774, Nov. 2013.

[28] R. H. Etkin, D. N. C. Tse, and H. Wang, "Gaussian interference channel capacity to within one bit," *IEEE Trans. Inf. Theory*, vol. 54, no. 12, pp. 5534–5562, Dec. 2008.

[29] J. Chen, P. Elia, and S. A. Jafar, "On the vector broadcast channel with alternating CSIT: A topological perspective," Feb. 2014. [Online]. Available: http://arxiv.org/abs/1402.5912

[30] R. Tandon, S. A. Jafar, S. Shamai (Shitz), and H. V. Poor, "On the synergistic benefits of alternating CSIT for the MISO broadcast channel," *IEEE Trans. Inf. Theory*, vol. 59, no. 7, pp. 4106–4128, Jul. 2013.

[31] Z. H. Awan, A. Zaidi, and A. Sezgin, "Achievable secure degrees of freedom of MISO broadcast channel with alternating CSIT," in *IEEE Int. Symp. Inf. Theory*, Honolulu, HI, USA, Jun. 2014, pp. 31–35.

[32] P. Mukherjee, R. Tandon, and S. Ulukus, "MISO broadcast channels with confidential messages and alternating CSIT," in *IEEE Int. Symp. Inf. Theory*, Honolulu, HI, USA, Jun. 2014, pp. 216–220.

[33] T. M. Cover and J. A. Thomas, *Elements of Information Theory*, 2nd edn. Chichester: Wiley & Sons, 2006.

[34] Z. H. Awan and A. Sezgin, "On MISO wiretap channel with delayed CSIT and alternating topology," in *Proc. 10th Int. ITG Conf. Systems, Communications and Coding*, Hamburg, Germany, Feb. 2015, pp. 1–6.

8 Physical-Layer Security with Delayed, Hybrid, and Alternating Channel State Knowledge

Pritam Mukherjee, Ravi Tandon, and Sennur Ulukus

In this chapter, we will discuss how the quality and availability of channel state information (CSI) affects secrecy in wireless networks. In particular, we study how the delay in the availability of CSI affects secrecy in the context of the two-user broadcast channel with confidential messages. We adopt a secure degrees of freedom perspective and investigate various CSI scenarios, including cases when the availability of CSI at the transmitter varies across users and with time. We discuss how to leverage such variabilities in CSI for secrecy and highlight the differences between the optimal degrees of freedom with or without secrecy constraints.

8.1 Introduction

The availability of channel state information at the transmitters (CSIT) plays a crucial role in securing wireless communication at the physical layer. Various well-known physical-layer security techniques such as coding for the fading wiretap channel [1–4], coding for the multiple-antenna wiretap channel [5–8], artificial noise injection [9], cooperative jamming [10, 11], cooperation for secrecy [12–18], secure signal alignment [19–22], and other related techniques rely upon the assumption of timely availability of precise CSIT; see also a recent review article in [23]. In most practical scenarios, the channel gains are measured by the receivers and then fed back to the transmitters. The measurement and feedback process necessarily introduces imprecision and delay into the CSI. Motivated by this fact, in this chapter we explore the fundamental limits of physical layer security when the CSIT is imperfect. For concreteness, we will focus on a particular wireless network model: the multiple-input single-output (MISO) broadcast channel with confidential messages (BCCM). This effectively models practical systems such as a cellular downlink network where each user wants not only reliability but also confidentiality of the information intended for it.

The focus of this chapter is on the secure degrees of freedom (s.d.o.f.) region of the fading two-user MISO BCCM, in which the transmitter with two antennas has two confidential messages, one for each of the single antenna users (see Fig. 8.1). The secrecy capacity region of the MISO broadcast channel (BC) for the case of perfect and instantaneous CSI at all terminals (transmitter and the receivers) has been characterized in [24, 25]. Using these results, it follows that for the two-user MISO BCCM, the sum s.d.o.f. is 2 with perfect and instantaneous CSIT. Since the assumption of perfect and

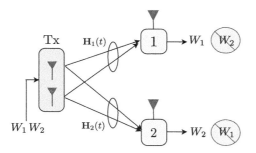

Figure 8.1 MISO broadcast channel with confidential messages (BCCM).

instantaneous CSIT may be too optimistic in practice, we consider settings where CSIT may be delayed, imprecise, or not available at all.

Let us now briefly summarize some of the results in the literature for delayed or imprecise CSIT with no security constraints. With perfect, i.e., precise and instantaneous CSIT (P), the sum d.o.f. for the two-user MISO broadcast channel is 2. With no CSIT (N), however, [26] showed that the sum d.o.f. collapses to 1. With delayed,[1] i.e., completely stale CSIT (D), it is shown in [27] that the sum d.o.f. for the two-user MISO BC increases to $\frac{4}{3}$. This result is generalized to the MIMO BC in [28]. Other channel models besides the BC have also been investigated. Reference [29] establishes the d.o.f. region of the two-user MIMO interference channel with delayed CSIT, while [30] extends the result by incorporating output feedback. Reference [31] determines the capacity region of the binary interference channel with delayed CSIT. For the X-channel, [32, 33] show that the optimal sum d.o.f. is $\frac{4}{3}$ with perfect channel knowledge. With delayed CSIT the optimal sum d.o.f. of the X-channel remains unknown in general. However, with a restriction of the transmission policies to linear schemes, [34] determines the sum d.o.f. of the X-channel to be $\frac{6}{5}$ (see also [35–37] and the references therein). With global feedback, where each transmitter receives output feedback from every receiver, [38] shows that the sum d.o.f. of the two-user X-channel with delayed CSIT is the same as that of the two-user MISO broadcast channel with two antennas at the transmitter and the optimal sum d.o.f. is $\frac{4}{3}$. Reference [39] provides achievable schemes for layered interference networks with delayed CSIT and provides an example where the d.o.f. scales with the number of users even with distributed transmitters and delayed CSIT.

In this chapter, we focus on a case with security constraints: the two-user MISO BC with confidential messages (MISO BCCM) in the presence of imperfect CSIT. For the two-user MISO BCCM with no CSIT, the sum s.d.o.f. is zero as the two users are statistically equivalent and hence no secrecy is possible. On the other hand, with completely outdated CSIT from both users, [40] shows that the sum s.d.o.f. increases to 1 (see also [41], where generalization to the K-user case is presented). In the first part

---

[1] By delayed CSIT, we refer to the standard assumption as in [27] in which the delay in acquiring CSIT is larger than the channel coherence time.

of the chapter, we will present the key ideas behind this surprising result which shows the usefulness of delayed CSIT for secrecy.

We next consider the setting of hybrid (or heterogeneous) CSIT, which models the variability in the quality/delay of channel knowledge supplied by different users. The complete characterization of the d.o.f. of all fixed heterogeneous CSIT configurations is only known for the two-user MISO BC with and without security constraints. References [42, 43] determine the optimal sum d.o.f. without security constraints for state PD to be 3/2, and [44] recently settled the states PN and DN through a novel converse proof showing that the optimal sum d.o.f. is 1. For the MISO BCCM, [45, 46] establishes the optimal sum s.d.o.f. for the states PD, PN, and DN to be 1, 1, and $\frac{1}{2}$, respectively. We will elaborate on these results in the second part of this chapter. Beyond these results, partial results are available for the three-user MISO BC with hybrid CSIT in [47, 48]. Recently, [49] determined the *linear* d.o.f. region for the three-user MISO BC, but by and large the problems of heterogeneous CSIT both with and without secrecy constraints remain open.

In the third part of this chapter, we discuss the setting of alternating CSIT, which captures the variability of channel knowledge across time and across users. Such variability can arise either naturally (due to the time variation in tolerable feedback overhead from a user) or it can be artificially induced (by deliberately altering the channel feedback mechanism over time). For example, instead of requiring perfect CSIT from one user and delayed CSIT from the other user throughout the duration of communication, one may require that for half of the time, the first user provide perfect CSIT while the second user provide delayed CSIT (state PD), and the roles of the users are reversed for the remaining half of the time (state DP), the total network feedback overhead being the same in both cases. The alternating CSIT framework, in which multiple CSIT states, for instance, PD and DP in the above example, arise over time, was introduced in [50], where the d.o.f. region was characterized for the two-user MISO BC. It was shown that synergistic gains in d.o.f. are possible by jointly coding across these states. In this chapter, we focus on alternating CSIT in the context of security, following [46, 51, 52], which characterizes the s.d.o.f. region of the MISO BCCM with alternating CSIT.

We start by presenting results on the MISO BCCM with homogeneous CSIT. Next, we focus on the setting of hybrid or heterogeneous CSIT, which captures the variability of CSIT quality across users. Finally, we present results on the alternating CSIT model, which captures the variability of CSIT quality across users and over time. These results highlight the interplay between the quality of CSIT and the s.d.o.f. We conclude the chapter with some future directions and a few open problems.

8.2 The MISO Broadcast Channel

We consider a two-user MISO BC, shown in Fig. 8.1, where the transmitter Tx, equipped with two antennas, wishes to send independent confidential messages to two single-antenna receivers, 1 and 2. The input–output relations at time t are given by

$$Y(t) = \mathbf{H}_1(t)\mathbf{X}(t) + N_1(t), \tag{8.1}$$

$$Z(t) = \mathbf{H}_2(t)\mathbf{X}(t) + N_2(t), \tag{8.2}$$

where $Y(t)$ and $Z(t)$ are the channel outputs of receivers 1 and 2, respectively. The 2×1 channel input $\mathbf{X}(t)$ is power constrained as $\mathbb{E}[||\mathbf{X}(t)||^2] \leq P$, and $N_1(t)$ and $N_2(t)$ are circularly symmetric complex white Gaussian noises with zero mean and unit variance. The 1×2 channel vectors $\mathbf{H}_1(t)$ and $\mathbf{H}_2(t)$ of receivers 1 and 2, respectively, are independent and identically distributed (i.i.d.) with continuous distributions, and are also i.i.d. over time. We denote $\mathbf{H}(t) = \{\mathbf{H}_1(t), \mathbf{H}_2(t)\}$ the collective channel vectors at time t and $\mathbf{H}^n = \{\mathbf{H}(1), \ldots, \mathbf{H}(n)\}$ the sequence of channel vectors up to time n.

8.2.1 Modeling the Quality of CSIT

In practice, the receivers estimate the channel coefficients and feed them back to the transmitter. In general, the receiver can choose to send not only the current measurements, but rather any function of all the channel measurements it has taken up to that time. The CSIT at time t can thus be any function of the measured channel coefficients up to time t. There are two key aspects to the CSIT: precision and delay. Precision captures the fact that the measurements made at the receivers and sent to the transmitter are imprecise (usually, quantized) and noisy. Delay is introduced since making measurements and feeding them back to the transmitter takes time. While the imprecise nature of CSIT has been considered in the literature [53–56], here, we focus on the delay aspect of CSIT, and assume that the CSIT, when available, has infinite precision.

In order to model the delay in CSIT, we assume that at each time t, there are three possible CSIT states for each user:

- *Perfect CSIT* (P): This denotes the availability of precise and instantaneous CSI of a user at the transmitter. In this state, the transmitter has precise channel knowledge before the start of the communication.
- *Delayed CSIT* (D): In this state, the transmitter does not have the CSI at the beginning of the communication. In slot t, the receiver may send any function of all the channel coefficients up to and including time t as CSI to the transmitter. However, the CSIT becomes available only after a delay such that the CSI is completely outdated, that is, independent of the current channel realization.
- *No CSIT* (N): In this state, there is no CSI of the user available at the transmitter.

Denote the CSIT of user 1 by I_1 and the CSIT of user 2 by I_2. Then,

$$I_1, I_2 \in \{\text{P}, \text{D}, \text{N}\}. \tag{8.3}$$

Thus, for the two-user MISO BC, we have nine CSIT states, namely PP, DD, NN, PD, DP, PN, NP, DN, and ND. Further, we assume that perfect and global CSI is available at both receivers, i.e., both receivers know $\mathbf{H}^n$.

The states PP, DD, and NN are *homogeneous* in the sense that the CSIT of both receivers are of the same quality. On the other hand, the quality of CSIT varies across the receivers in states PD, PN, and DN; hence, they can be termed *hybrid* or *heterogeneous* states.

8.2.2 Security Requirements

In an information theoretic context, the secrecy of communication can be measured by the conditional entropy of the confidential message W given the channel output Z^n of the unintended receiver, i.e., $H(W|Z^n)$. We adopt the so-called *weak secrecy* notion introduced in [57, 58], which requires

$$\frac{1}{n}H(W|Z^n) \to \frac{1}{n}H(W) \tag{8.4}$$

as the length of the code $n \to \infty$. Intuitively, the remaining uncertainty of the message given the observation at the unintended receiver approaches the original uncertainty of the message without any observations. Alternately, the weak secrecy condition can be recast as

$$\frac{1}{n}I(W;Z^n) \to 0 \tag{8.5}$$

as $n \to \infty$. If we ignore the normalization by n, we can intuitively interpret this condition as requiring the channel output at the unintended receiver to be asymptotically independent of the confidential message.

For the two-user MISO BCCM, we impose two secrecy conditions: first, the confidentiality of message W_1 from the second user, given as

$$\frac{1}{n}I(W_1;Z^n,\mathbf{H}^n) \to 0, \tag{8.6}$$

and second, the confidentiality of W_2 from the first user, given as

$$\frac{1}{n}I(W_2;Y^n,\mathbf{H}^n) \to 0 \tag{8.7}$$

as $n \to \infty$.

A secure rate pair (R_1,R_2) is achievable if there exists a sequence of codes which satisfy the reliability constraints at the receivers, namely, $\Pr\left[W_i \neq \widehat{W}_i\right] \to 0$ as $n \to \infty$, for $i = 1, 2$, and also the above confidentiality constraints. The secrecy capacity region is defined as the union of all achievable secure rate pairs (R_1,R_2).

8.2.3 A Degrees-of-Freedom Perspective

For general multi-user channel models, characterizing the exact secrecy capacity region is a hard problem. Despite the difficulty of exactly characterizing the capacity regions, we can still gain significant insights by studying the secure degrees of freedom region of the channel in the high signal-to-noise (SNR) regime. Recall that the capacity of the additive complex Gaussian channel with unit noise variance is $\log(1+P) \approx \log P$ for large P, where P is the average power constraint. The secure degrees of freedom measures the secure rate normalized by the Gaussian capacity at high SNR. For the two-user MISO BCCM, an s.d.o.f. pair (d_1,d_2) is achievable, if there exists a secure rate pair (R_1,R_2) such that

$$d_1 = \lim_{P \to \infty} \frac{R_1}{\log P}, \quad d_2 = \lim_{P \to \infty} \frac{R_2}{\log P}. \tag{8.8}$$

The s.d.o.f. region is the union of all achievable s.d.o.f. pairs (d_1, d_2). We will focus on the s.d.o.f. region of the MISO BCCM as we investigate the impact of imperfect CSIT on security.

8.3 The MISO BCCM in Homogeneous CSIT States

8.3.1 Perfect CSIT from Both Users (PP)

Let us first consider the MISO BCCM with perfect CSIT from both users, i.e., in state PP [24]. In state PP, the exact secrecy capacity region is known for the MIMO BCCM in general [25]. Perhaps surprisingly, even in the general MIMO case, both confidential messages W_1 and W_2 can be transmitted *simultaneously* at their maximal secrecy rates as if over two separate Gaussian wiretap channels [25]. If we focus on the s.d.o.f. region and specialize to the two-user MISO BCCM with two transmitter antennas and one receiver antenna, we find that the s.d.o.f. region is the square formed by the points $(0,0)$, $(0,1)$, $(1,0)$, and $(1,1)$, as shown in Fig. 8.2.

It is easy to see the optimality of this region. Since each receiver has only one antenna, the d.o.f. for each user is bounded above by 1, even without any confidentiality constraints. Let us then focus on the achievability of this s.d.o.f. region. Heuristically, the additive Gaussian noise has only negligible impact in the high-SNR regime; thus, we can effectively ignore it. Further, achieving an s.d.o.f. of d can be interpreted as sending d Gaussian symbols per time slot. Thus, to achieve an s.d.o.f. pair $(d_1, d_2) = (1, 1)$ on the MISO BCCM, we wish to send a total of two Gaussian symbols (u, v), u intended for the first receiver and v intended for the second receiver, in one time slot. This can be accomplished as follows:

The transmitter sends

$$\mathbf{X} = u\mathbf{H}_2^\perp + v\mathbf{H}_1^\perp, \tag{8.9}$$

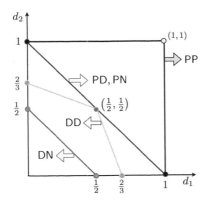

Figure 8.2 Secure degrees of freedom regions for various CSIT states.

where $\mathbf{H}_i^\perp$ is a 2×1 vector such that $\mathbf{H}_i\mathbf{H}_i^\perp = 0$. Ignoring the additive Gaussian noise at high SNR, the outputs at the receivers are given by

$$Y = u\mathbf{H}_1\mathbf{H}_2^\perp, \qquad (8.10)$$

$$Z = v\mathbf{H}_2\mathbf{H}_1^\perp. \qquad (8.11)$$

Since the receivers have perfect CSI, the first receiver can decode u from Y, while the second receiver can decode v from Z. Confidentiality of the messages u and v is also guaranteed since u is not observed at the second receiver, while v is not observed at the first receiver.

Note that this scheme also achieves the optimal sum d.o.f. of 2 when there are no confidentiality constraints. Thus, in state PP, adding security constraints incurs no additional penalty on the s.d.o.f. of the MISO BCCM.

8.3.2 No CSIT from Any User (NN)

This is the opposite extreme case when no CSIT is available from either receiver, i.e., state NN. As in state PP, the exact secrecy capacity region is known for the two-user MISO BCCM [59]. In fact, the region consists of only one point, $(0,0)$; thus, no positive secure rate is achievable. Intuitively, this is because in the absence of any CSIT, both receivers are *statistically equivalent*; thus, each receiver can decode the message intended for the other receiver as well. Since no positive secure rate is feasible, the s.d.o.f. region also consists of the single point $(0,0)$.

8.3.3 Delayed CSIT from Both Users (DD)

In this case, the MISO BCCM is in state DD, i.e., completely stale CSIT is available from both receivers. Perhaps surprisingly, such completely stale CSIT can also be very useful for *retrospective interference alignment* both with or without secrecy constraints. Reference [27] shows that for the MISO BC with no secrecy constraints, delayed CSIT increases the optimal sum d.o.f. from 1 to $\frac{4}{3}$. The benefit of delayed CSIT is even more pronounced when confidentiality constraints are imposed. While the optimal sum s.d.o.f. is 0 in state NN, it increases to 1 in state DD [40].

The following theorem characterizes the s.d.o.f. region of the MISO BCCM in state DD.

THEOREM 8.1 ([40])   *The s.d.o.f. region of the two-user MISO BCCM in state DD is the set of all non-negative pairs (d_1, d_2) satisfying*

$$3d_1 + d_2 \leq 2, \qquad (8.12)$$

$$d_1 + 3d_2 \leq 2. \qquad (8.13)$$

The region is shown in Fig. 8.2. The optimal sum s.d.o.f. is 1, achieved at the point $(\frac{1}{2},\frac{1}{2})$. On the other hand, the maximum single-user s.d.o.f. is $\frac{2}{3}$, which is achieved when the BCCM is treated as a wiretap channel with a message for one of the two receivers with the other receiver being the eavesdropper. Clearly, in order to achieve the full region, it suffices to achieve the points $(\frac{2}{3},0)$ and $(\frac{1}{2},\frac{1}{2})$. Let us now look at the achievable schemes proposed in [40] for these two points, and the tools that enable a tight converse.

Achievable Schemes
Scheme Achieving the Point $(\frac{2}{3},0)$
In this scheme, see Fig. 8.3, the transmitter securely sends two information symbols (u_1,u_2) to the first receiver in three time slots.

1. At $t=1$, the transmitter sends artificial noise symbols q_1 and q_2 using two antennas. The transmitted signal is

$$\mathbf{X}(1) = \begin{bmatrix} q_1 & q_2 \end{bmatrix}^T, \quad (8.14)$$

where q_1 and q_2 are i.i.d. as $\mathcal{CN}(0,P)$. The received signals are:

$$Y(1) = h_{11}(1)q_1 + h_{12}(1)q_2 \triangleq L_1(q_1,q_2) \triangleq K, \quad (8.15)$$

$$Z(1) = h_{21}(1)q_1 + h_{22}(1)q_2 \triangleq L_2(q_1,q_2). \quad (8.16)$$

The received signal at user 1 acts as a "key" K. Since there is delayed CSIT, the transmitter can reconstruct this key K and use it in the next slots.

2. At $t=2$, the transmitter sends the two information symbols (u_1,u_2) intended for the first receiver linearly combined with the first user's key. The transmitted signal is

$$\mathbf{X}(2) = \begin{bmatrix} u_1+K & u_2+K \end{bmatrix}^T. \quad (8.17)$$

The received signals are:

$$Y(2) = h_{11}(2)u_1 + h_{12}(2)u_2 + (h_{11}(2)+h_{12}(2))K \triangleq LC_1(u_1,u_2,K), \quad (8.18)$$

$$Z(2) = h_{21}(2)u_1 + h_{22}(2)u_2 + (h_{21}(2)+h_{22}(2))K \triangleq LC_2(u_1,u_2,K). \quad (8.19)$$

Thus, the first user can retrieve a linear combination of just its intended symbols (u_1,u_2). However, the second user gets a linear combination $LC_2(u_1,u_2,K)$. Due to delayed CSIT however, the transmitter can reconstruct LC_2 and use it in the

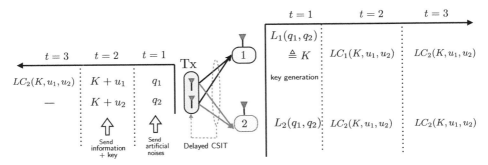

Figure 8.3 Achieving the s.d.o.f. pair $(\frac{2}{3},0)$ in state DD.

following slots. Note that $LC_2(u_1, u_2, K)$ is also useful for the first receiver, since it can eliminate K and get a second linear combination of its intended symbols (u_1, u_2), allowing it to decode (u_1, u_2). Further, if $LC_2(u_1, u_2, K)$ is transmitted, it leaks no additional information to the second receiver.

3. At $t = 3$, the transmitter sends $LC_2(u_1, u_2, K)$ as

$$\mathbf{X}(3) = [LC_2(u_1, u_2, K) \quad 0]^T. \tag{8.20}$$

The received signals are:

$$Y(3) = h_{11}(3) LC_2(u_1, u_2, K), \tag{8.21}$$

$$Z(3) = h_{21}(3) LC_2(u_1, u_2, K). \tag{8.22}$$

As mentioned already, $LC_2(u_1, u_2, K)$ enables the first receiver to decode its desired symbols (u_1, u_2), while leaking no further information to the second receiver. The symbols (u_1, u_2) are buried in the artificial noise at the second receiver, and thus the information leakage is $o(\log P)$.

Scheme Achieving the Point $(\frac{1}{2}, \frac{1}{2})$

Here we present the scheme presented for state DD in [40]. In this scheme, the transmitter sends four information symbols, (u_1, u_2) to the first receiver and (v_1, v_2) to the second receiver, in four time slots. Figure 8.4 shows the scheme. It is as follows:

1. At $t = 1$, the transmitter sends artificial noise symbols q_1 and q_2 using two antennas. The transmitted signal is

$$\mathbf{X}(1) = \begin{bmatrix} q_1 & q_2 \end{bmatrix}^T, \tag{8.23}$$

where q_1 and q_2 are i.i.d. as $\mathcal{CN}(0, P)$. The received signals are:

$$Y(1) = h_{11}(1) q_1 + h_{12}(1) q_2 \triangleq K_1, \tag{8.24}$$

$$Z(1) = h_{21}(1) q_1 + h_{22}(1) q_2 \triangleq K_2. \tag{8.25}$$

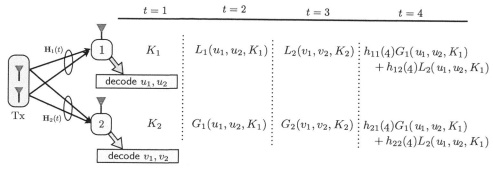

Figure 8.4 Achieving the s.d.o.f. pair $(\frac{1}{2}, \frac{1}{2})$ using state DD.

The received signals act as keys K_1 and K_2 for the respective users 1 and 2. Since there is delayed CSIT, the transmitter can reconstruct these keys and use them in the next slots.

2. At $t = 2$, the transmitter sends the two information symbols (u_1, u_2) intended for the first receiver linearly combined with the first user's key. The transmitted signal is

$$\mathbf{X}(2) = [u_1 + K_1 \quad u_2 + K_1]^T. \tag{8.26}$$

The received signals are:

$$Y(2) = h_{11}(2)u_1 + h_{12}(2)u_2 + (h_{11}(2) + h_{12}(2))K_1 \triangleq L_1(u_1, u_2, K_1), \tag{8.27}$$

$$Z(2) = h_{21}(2)u_1 + h_{22}(2)u_2 + (h_{21}(2) + h_{22}(2))K_1 \triangleq G_1(u_1, u_2, K_1). \tag{8.28}$$

Thus, the first user can retrieve a linear combination of just its intended symbols (u_1, u_2). However, the second user gets a linear combination $G_1(u_1, u_2, K_1)$. Due to delayed CSIT, however, the transmitter can reconstruct G_1 and use it in the following slots.

3. At $t = 3$, the roles of the receivers are reversed and the transmitter sends the second user's symbols (v_1, v_2) linearly combined with the second user's key K_2, i.e.,

$$\mathbf{X}(3) = [v_1 + K_2 \quad v_2 + K_2]^T. \tag{8.29}$$

The received signals are:

$$Y(3) = h_{11}(3)v_1 + h_{12}(3)v_2 + (h_{11}(3) + h_{12}(3))K_2 \triangleq L_2(v_1, v_2, K_2), \tag{8.30}$$

$$Z(3) = h_{21}(3)v_1 + h_{22}(3)v_2 + (h_{21}(3) + h_{22}(3))K_2 \triangleq G_2(v_1, v_2, K_2). \tag{8.31}$$

This allows the second user to retrieve a linear combination of just its information symbols, which remain secure at the first user, which receives $L_2(v_1, v_2, K_2)$.

4. At $t = 4$, the transmitter sends a linear combination of G_1 and L_2, i.e.,

$$\mathbf{X}(4) = [G_1(u_1, u_2, K_1) \quad L_2(v_1, v_2, K_2)]^T. \tag{8.32}$$

The received signals are:

$$Y(4) = h_{11}(4)G_1(u_1, u_2, K_1) + h_{12}(4)L_2(v_1, v_2, K_2), \tag{8.33}$$

$$Z(4) = h_{21}(4)G_1(u_1, u_2, K_1) + h_{22}(4)L_2(v_1, v_2, K_2). \tag{8.34}$$

Essentially this provides the first user with G_1, from which it can eliminate K_1 to get another independent linear combination of (u_1, u_2). A similar situation takes place at the second user. Finally, each user has two linearly independent combinations of two symbols and thus can decode the information symbols intended for it. The information leakage is only $o(\log P)$, as shown in [40].

Converse Tools

The main tool in establishing the converse is the *statistical equivalence property* (SEP), which can be stated for the MISO BCCM[2] as:

---

[2] For a general MIMO version, see [40, Property 2.1].

DEFINITION 8.1 (Statistical equivalence property) With delayed CSIT from both receivers, the differential entropy of the channel outputs are equal when conditioned on the same past outputs, i.e.,

$$h(Y(t)|Y^{t-1}, Z^{t-1}, \mathbf{H}^n) = h(Z(t)|Y^{t-1}, Z^{t-1}, \mathbf{H}^n). \tag{8.35}$$

This property has an intuitive explanation: since the CSIT is delayed, the channel input $\mathbf{X}(t)$ does not depend on the current channel gains which are themselves i.i.d. over time and across the users. Thus, the outputs have the same differential entropy conditioned on the same past outputs.

The SEP allows us to prove the following useful lemma:

LEMMA 8.1 *For the two-user MISO BCCM,*

$$2h(Y^n) \stackrel{.}{\geq} h(Y^n, Z^n), \tag{8.36}$$

$$2h(Z^n) \stackrel{.}{\geq} h(Y^n, Z^n), \tag{8.37}$$

where $a \stackrel{.}{\geq} b$ *is used to denote that* $\lim_{P \to \infty} \frac{a}{\log P} \geq \lim_{P \to \infty} \frac{b}{\log P}$.

For a brief sketch of the proof of these inequalities, consider (8.36); the other inequality will follow by interchanging the roles of the receivers. Consider a virtual output $\tilde{Y}$ at the first receiver, such that the channel gains and the additive Gaussian noise are i.i.d. with that of the actual output Y. With delayed CSIT, the conditional entropies of $Y(t)$ and $\tilde{Y}(t)$ are the same when conditioned on the past outputs Y^{t-1}. Thus, $2h(Y^n) \geq \sum_{t=1}^n h(Y(t)|Y^{t-1}) + h(\tilde{Y}(t)|Y^{t-1}) \geq \sum_{t=1}^n h(Y(t), \tilde{Y}(t)|Y^{t-1})$. Further, given the two observations $Y(t)$ and $\tilde{Y}(t)$, the channel input $\mathbf{X}(t)$ can be estimated to within noise variance by solving the equation

$$\begin{bmatrix} Y(t) \\ \tilde{Y}(t) \end{bmatrix} = \begin{bmatrix} \mathbf{H}_1(t) \\ \tilde{\mathbf{H}}_1(t) \end{bmatrix} X(t), \tag{8.38}$$

where $\tilde{\mathbf{H}}_1$ is the 1×2 channel matrix corresponding to the output $\tilde{Y}(t)$. Now, given $X(t)$, both $Y(t)$ and $Z(t)$ can be estimated to within noise variance as well. Therefore, $h(Y(t), \tilde{Y}(t)|Y^{t-1}) \geq h(Y(t), Z(t)|Y^{t-1}) \geq h(Y(t), Z(t)|Y^{t-1}, Z^{t-1})$. Equation (8.36) then follows immediately.

As a simple corollary to the lemma, we have the following inequalities:

$$2h(Y^n) \stackrel{.}{\geq} h(Z^n), \tag{8.39}$$

$$h(Y^n) \stackrel{.}{\geq} h(Z^n|Y^n). \tag{8.40}$$

These two inequalities are used along with the decodability requirements and the security constraints to prove the bounds in (8.12) and (8.13).

In this section, we have discussed the three homogeneous CSIT states that arise out of our modeling of CSIT quality. However, in practice, the CSIT states can vary across users giving rise to hybrid CSIT states. In the next section, we discuss the three heterogeneous CSIT states that arise in the two-user MISO BCCM.

8.4 The MISO BCCM in Hybrid CSIT States

Three hybrid CSIT states arise in the two-user MISO BCCM with our modeling of CSIT quality: PD, PN, and DN. We first state two theorems that characterize the s.d.o.f. regions of the MISO BCCM for these states.

THEOREM 8.2 ([45]) *The s.d.o.f. region of the two-user MISO BCCM in state* PN *or state* PD *is the set of all non-negative pairs* (d_1, d_2) *satisfying*

$$d_1 + d_2 \leq 1. \tag{8.41}$$

THEOREM 8.3 ([45]) *The s.d.o.f. region of the two-user MISO BCCM in state* DN *is the set of all non-negative pairs* (d_1, d_2) *satisfying*

$$d_1 + d_2 \leq \frac{1}{2}. \tag{8.42}$$

Figure 8.2 shows the s.d.o.f. regions established in the above theorems. Note that the s.d.o.f. region of the MISO BCCM in state PD is the same as that in state PN. This suggests, perhaps surprisingly, that when one of the users provides instantaneous CSI, delayed CSI from the other user does not increase the s.d.o.f. at all; it might as well not provide any CSI. Further, the optimal sum s.d.o.f. in state PD, 1, is the same as that of the optimal sum s.d.o.f. in state DD. Thus, from a sum s.d.o.f. perspective, if one of the users provides delayed CSIT, there is no benefit if the other user provides instantaneous CSIT instead of delayed CSIT. However, this is not true from the s.d.o.f. region perspective. The s.d.o.f. region for state DD is strictly contained in the s.d.o.f. region for state PD; for example, the maximum individual s.d.o.f. for a user in state DD is $\frac{2}{3}$, while it is 1 in state PD.

Theorem 8.3 also establishes the optimal s.d.o.f. of the wiretap channel with delayed CSIT from the legitimate receiver but no CSI from the eavesdropper to be $\frac{1}{2}$. An achievable scheme for this model is provided in [40]. It is further shown to be optimal when only linear achievable schemes are admissible, that is, in terms of linear s.d.o.f., in [60]. Theorem 8.3 strengthens that result by establishing the optimality of $\frac{1}{2}$ s.d.o.f. without any linearity assumptions. Further, it also establishes the optimal s.d.o.f. of the wiretap channel with no CSI from the legitimate receiver and delayed CSI from the eavesdropper.

Let us now focus on the proofs of Theorems 8.2 and 8.3. Note that an achievable scheme for state PN suffices as an achievable scheme for state PD. Thus, we present achievable schemes for states PN and DN. Further, note that a converse for state PD suffices as an upper bound for state PN. We will briefly sketch the converses for states PD and DN.

8.4.1 Achievable Schemes for States PN and DN

State PN
Note that it is sufficient to show the achievability of only two points: $(d_1, d_2) = (1, 0)$ and $(d_1, d_2) = (0, 1)$. The achievability of these corner points follows in straightforward

manner as follows: sending the message to user 1 by superimposing it with artificial noise in a direction orthogonal to user 1's channel to achieve the pair $(1,0)$, and sending the message to user 2 in a direction orthogonal to user 1's channel to achieve the pair $(0,1)$. This completes the proof of the achievability of the region in (8.41).

State DN

To prove the achievability of the s.d.o.f. region in (8.42), it suffices to achieve only the two points $(d_1,d_2) = \left(\frac{1}{2},0\right)$ and $(d_1,d_2) = \left(0,\frac{1}{2}\right)$. The entire region can be obtained by time sharing. A scheme to achieve the point $(d_1,d_2) = \left(\frac{1}{2},0\right)$ was presented in [40]. We present it briefly here for completeness.

Scheme Achieving $(d_1,d_2) = \left(\frac{1}{2},0\right)$

We wish to send one symbol u securely to the first user in two time slots. This can be done as follows:

At time $t = 1$, the transmitter does not have any channel knowledge. It sends

$$\mathbf{X}(1) = [q_1 \quad q_2]^T, \tag{8.43}$$

where q_1 and q_2 denote independent artificial noise symbols distributed as $\mathcal{CN}(0,P)$. Both receivers receive linear combinations of the two symbols q_1 and q_2. The outputs are:

$$Y(1) = h_{11}(1)q_1 + h_{12}(1)q_2 \stackrel{\Delta}{=} L(q_1,q_2), \tag{8.44}$$

$$Z(1) = h_{21}(1)q_1 + h_{22}(1)q_2 \stackrel{\Delta}{=} G(q_1,q_2), \tag{8.45}$$

where we drop the Gaussian noise at high SNR. Due to delayed CSIT from receiver 1, the transmitter can reconstruct $L(q_1,q_2)$ in the next time slot and use it for transmission.

At time $t = 2$, the transmitter sends

$$\mathbf{X}(2) = \begin{bmatrix} u & L(q_1,q_2) \end{bmatrix}^T. \tag{8.46}$$

The received signals are:

$$Y(2) = h_{11}(2)u + h_{12}(2)L(q_1,q_2), \tag{8.47}$$

$$Z(2) = h_{21}(2)u + h_{22}(2)L(q_1,q_2). \tag{8.48}$$

Since the receivers have full channel knowledge, receiver 1 can recover u by eliminating $L(q_1,q_2)$ from $Y(1)$ and $Y(2)$. Therefore, $I(u;Y(1),Y(2)) = \log P + o(\log P)$. The information leakage to the second user is bounded by $o(\log P)$, see [40]. Thus, $(d_1,d_2) = \left(\frac{1}{2},0\right)$ is achievable in this scheme.

Scheme Achieving $(d_1,d_2) = \left(0,\frac{1}{2}\right)$

In this scheme, we wish to send one symbol u securely to the second user in two time slots. This can be done as follows:

At time $t = 1$, the transmitter does not have any channel knowledge. It sends

$$\mathbf{X}(1) = [u \quad q]^T, \tag{8.49}$$

where q denotes an independent artificial noise symbol distributed as $\mathcal{CN}(0,P)$. Both receivers receive linear combinations of the two symbols u and q. The receivers' outputs are:

$$Y(1) = h_{11}(1)u + h_{12}(1)q \triangleq L(u,q), \tag{8.50}$$

$$Z(1) = h_{21}(1)u + h_{22}(1)q \triangleq G(u,q). \tag{8.51}$$

Due to delayed CSIT from receiver 1, the transmitter can reconstruct $L(u,q)$ and use it for transmission in the next slot.

At time $t=2$, the transmitter sends

$$\mathbf{X}(2) = \begin{bmatrix} L(u,q) & 0 \end{bmatrix}^T. \tag{8.52}$$

The received signals are:

$$Y(2) = h_{11}(2)L(u,q), \tag{8.53}$$

$$Z(2) = h_{21}(2)L(u,q). \tag{8.54}$$

Since the receivers have full channel knowledge, receiver 2 can recover u by eliminating q from $L(u,q)$ and $G(u,q)$. On the other hand, u is buried in the artificial noise q at the first receiver, and therefore the information leakage to the first receiver is $o(\log P)$. Thus, $(d_1, d_2) = (0, \frac{1}{2})$ is achievable in this scheme.

8.4.2 Converse Tools

A key ingredient in the converse proofs for both states PD and DN is a variant of the SEP that we have already introduced: a property we call *local statistical equivalence* (LSEP) [51].

DEFINITION 8.2 (Local statistical equivalence property) Let us focus on the channel output of a receiver that provides either delayed or no CSIT, say receiver 2, corresponding to the states PD, DD, or DN at time t:

$$Z(t) = \mathbf{H}_2(t)\mathbf{X}(t) + N_2(t). \tag{8.55}$$

Now consider $(\tilde{\mathbf{H}}_2(t), \tilde{N}_2(t))$, which are i.i.d. as $(\mathbf{H}_2(t), N_2(t))$, respectively. Using these random variables, we define artificial channel outputs as

$$\tilde{Z}(t) = \tilde{\mathbf{H}}_2(t)\mathbf{X}(t) + \tilde{N}_2(t). \tag{8.56}$$

Let $\Omega = (\mathbf{H}^n, \tilde{\mathbf{H}}^n)$. Now the local statistical equivalence property is the following:

$$h(Z(t)|Z^{t-1}, \Omega) = h(\tilde{Z}(t)|Z^{t-1}, \Omega). \tag{8.57}$$

This property shows that if we consider the outputs of a receiver for such states in which it supplies delayed CSIT, then the entropy of the channel outputs conditioned on the past outputs is the same as that of another artificial receiver whose channel is distributed identically as the original receiver. Note that we focus only on the receiver that provides delayed CSIT; hence we call it *local*. The original and artificial receivers have *statistically equivalent* channels in the sense that the conditional differential entropies of the outputs at the real and the artificial receivers given the past outputs are equal.

As with SEP, we have the following lemma due to LSEP:

LEMMA 8.2 *With the second receiver with channel output Z providing delayed or no CSIT, we have*

$$2h(Z^n|\Omega) \stackrel{.}{\geq} h(Y^n|\Omega), \quad (8.58)$$

$$h(Z^n|\Omega) \stackrel{.}{\geq} h(Y^n|Z^n,\Omega). \quad (8.59)$$

This lemma along with usual converse techniques are sufficient to prove the converse for state PD. To prove the converse for state DN, we need an additional ingredient. In a recent result [44], it was shown that for the MISO BCCM in state PN, the maximum of $h(Y^n|W_2) - h(Z^n|W_2)$ is less than $no(\log P)$ over all possible channel inputs, i.e., the entropy of the channel output of the receiver which provides no CSIT is at least as large as the entropy of the other user's channel output. This is akin to the *least alignment lemma* in [60] for *linear* transmission strategies. Intuitively, no alignment is possible at the receiver that does not provide CSIT, and hence the channel output at that receiver has the maximum entropy. This result, along with the LSEP and the decodability and security requirements suffice to prove the converse for state DN.

8.5 The MISO BCCM in Alternating CSIT States

In the previous section, we explored the impact of channel knowledge variations across the users. The quality of CSIT, however, remained fixed for the full duration of the communication. With long codewords and mobile environments, this assumption may not hold. Indeed, as we will show, such variation over time can be leveraged to achieve gains in s.d.o.f.; thus, such variation may be part of the network design itself. We can model this variation as follows [50]:

Let $\lambda_{I_1 I_2}$ be the fraction of the time the state $I_1 I_2$ occurs. Then,

$$\sum_{I_1,I_2} \lambda_{I_1 I_2} = 1. \quad (8.60)$$

We also assume symmetry: $\lambda_{I_1 I_2} = \lambda_{I_2 I_1}$ for every $I_1 I_2$. Specifically,

$$\lambda_{\text{PD}} = \lambda_{\text{DP}}, \quad \lambda_{\text{DN}} = \lambda_{\text{ND}}, \quad \lambda_{\text{PN}} = \lambda_{\text{NP}}. \quad (8.61)$$

Let us further define the following:

$$\lambda_{\text{P}} \triangleq \lambda_{\text{PP}} + \lambda_{\text{PD}} + \lambda_{\text{PN}} \quad (8.62)$$

$$\lambda_{\text{D}} \triangleq \lambda_{\text{PD}} + \lambda_{\text{DD}} + \lambda_{\text{DN}} \quad (8.63)$$

$$\lambda_{\text{N}} \triangleq \lambda_{\text{PN}} + \lambda_{\text{DN}} + \lambda_{\text{NN}}. \quad (8.64)$$

Using these definitions, it is easy to verify that

$$\lambda_P + \lambda_D + \lambda_N = 1. \tag{8.65}$$

Here, we can interpret these three quantities as follows:

- λ_P represents the total fraction of time the CSIT of a user is in the P state.
- λ_D represents the total fraction of time the CSIT of a user is delayed, that is, the state D.
- λ_N represents the total fraction of time a user supplies no CSIT.

Given the probability mass function (pmf) $\lambda_{I_1 I_2}$, our goal is to characterize the s.d.o.f. region of the two-user MISO BCCM. We have the following theorem:

THEOREM 8.4 *The s.d.o.f. region for the two-user MISO BCCM with alternating CSIT, $\mathcal{D}(\lambda_{I_1 I_2})$, is the set of all non-negative pairs (d_1, d_2) satisfying*

$$d_1 \leq \min\left(\frac{2 + 2\lambda_P - \lambda_{PP}}{3}, 1 - \lambda_{NN}\right), \tag{8.66}$$

$$d_2 \leq \min\left(\frac{2 + 2\lambda_P - \lambda_{PP}}{3}, 1 - \lambda_{NN}\right), \tag{8.67}$$

$$3d_1 + d_2 \leq 2 + 2\lambda_P, \tag{8.68}$$

$$d_1 + 3d_2 \leq 2 + 2\lambda_P, \tag{8.69}$$

$$d_1 + d_2 \leq 2(\lambda_P + \lambda_D). \tag{8.70}$$

We next make a few remarks highlighting the consequences and interesting aspects of this theorem.

Sum s.d.o.f. ($\max(d_1 + d_2)$)

From the region stated in (8.66)–(8.70), it is clear that the sum s.d.o.f. is given by

$$\text{sum s.d.o.f.} = \min\left(2\left(\frac{2 + 2\lambda_P - \lambda_{PP}}{3}\right), 2(1 - \lambda_{NN}), 2(\lambda_P + \lambda_D), 1 + \lambda_P\right). \tag{8.71}$$

The sum s.d.o.f. expression in (8.71) can be significantly simplified by noting that the first two terms in the minimum are inactive due to the inequalities $1 + \lambda_P \leq 2\left(\frac{2+2\lambda_P - \lambda_{PP}}{3}\right)$ and $2(\lambda_P + \lambda_D) = 2(1 - \lambda_N) \leq 2(1 - \lambda_{NN})$. These inequalities follow directly from (8.62)–(8.65). Using these inequalities, the sum s.d.o.f. expression above is equivalent to

$$\text{sum s.d.o.f.} = \min\left(2(\lambda_P + \lambda_D), 1 + \lambda_P\right) \tag{8.72}$$

$$= \min\left(2(\lambda_P + \lambda_D), 2\lambda_P + \lambda_D + \lambda_N\right) \tag{8.73}$$

$$= 2\lambda_P + \lambda_D + \min(\lambda_D, \lambda_N). \tag{8.74}$$

Figure 8.5 shows the sum s.d.o.f. as a function of λ_P and λ_D.

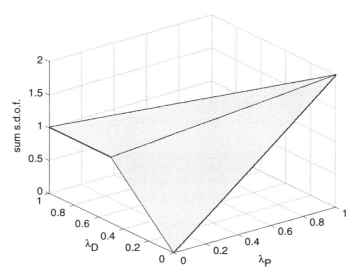

Figure 8.5 The sum s.d.o.f. as a function of λ_P and λ_D.

Same Marginals Property

From (8.74), we notice that the marginal probabilities λ_P, λ_D, and λ_N are sufficient to determine the sum s.d.o.f. Thus, for any given pmf $\lambda_{I_1 I_2}$ satisfying the symmetry conditions (8.61), there exists an *equivalent* alternating CSIT problem having only three states: PP, DD, and NN occurring for λ_P, λ_D, and λ_N fractions of the time, respectively, that has the same sum s.d.o.f. This observation is similar to the case when there is no secrecy [50]. However, unlike in [50], the s.d.o.f. region *does not* have the same property in general, as we can see the explicit dependence of the s.d.o.f. region in (8.66)–(8.70) on λ_{PP} and λ_{NN}.

Channel Knowledge Equivalence

We next highlight an interesting property which shows that from the sum s.d.o.f. perspective, no CSIT is equivalent to delayed CSIT when $\lambda_D \geq \lambda_N$, and delayed CSIT is equivalent to perfect CSIT when $\lambda_D < \lambda_N$.

Equivalence of Delayed and No CSIT When $\lambda_D \geq \lambda_N$

From a sum s.d.o.f. perspective, we see that when $\lambda_D \geq \lambda_N$, the sum s.d.o.f. depends only on λ_P. Hence, as long as $\lambda_D \geq \lambda_N$ holds, the N states behave as D states in the sense that, if the N states were enhanced to D states, the sum s.d.o.f. would not increase. Essentially, the N states can be combined with various D states and we obtain the same sum s.d.o.f. as if every N state were replaced by a D state. Consider an example, where the states PD, DP, and NN occur for $\frac{2}{5}$, $\frac{2}{5}$, and $\frac{1}{5}$ of the time, respectively. Note that $\lambda_D = \frac{2}{5} > \lambda_N = \frac{1}{5}$ in this case. The sum s.d.o.f., from (8.74), is $2\lambda_P + \lambda_D + \lambda_N = \frac{7}{5}$. Now, if we enhance the N states to D states, we get the states PD, DP, and DD occurring for $\frac{2}{5}$, $\frac{2}{5}$, and $\frac{1}{5}$ of the time, respectively. The sum s.d.o.f. of this enhanced system is still $\frac{7}{5}$.

Equivalence of Delayed and Perfect CSIT When $\lambda_D \leq \lambda_N$

From a sum s.d.o.f. perspective, we see that when $\lambda_D \leq \lambda_N$, the sum s.d.o.f. depends only on λ_N. Hence, in this case, if $\lambda_D \leq \lambda_N$, the delayed CSIT is as good as perfect CSIT, that is, every D state can be enhanced to a P state without any increase in the sum s.d.o.f. For example, consider a system where the states PD, DP, and NN occur for $\frac{1}{5}$, $\frac{1}{5}$, and $\frac{3}{5}$ of the time, respectively. Note that $\lambda_D = \frac{1}{5} < \lambda_N = \frac{3}{5}$ in this case. The sum s.d.o.f. for this system is $\frac{4}{5}$, from (8.74). By enhancing the D states to P states, we get a system where the states PP and NN occur for $\frac{2}{5}$ and $\frac{3}{5}$ of the time, respectively. The sum s.d.o.f. for this enhanced system is still $\frac{4}{5}$.

Minimum CSIT Required for a Sum s.d.o.f. Value

Figure 8.6 shows the tradeoff between λ_P and λ_D for a given value of sum s.d.o.f. The highlighted corner point in each curve shows the most *efficient* point in terms of CSIT requirement. Any other feasible point either involves redundant CSIT or unnecessary instantaneous CSIT where delayed CSIT would have sufficed. For example, the following are the minimum CSIT requirements for various sum s.d.o.f. values:

$$\text{sum s.d.o.f.} = 2 : (\lambda_P, \lambda_D)_{\min} = (1, 0), \tag{8.75}$$

$$\text{sum s.d.o.f.} = \frac{3}{2} : (\lambda_P, \lambda_D)_{\min} = \left(\frac{1}{2}, \frac{1}{4}\right), \tag{8.76}$$

$$\text{sum s.d.o.f.} = \frac{4}{3} : (\lambda_P, \lambda_D)_{\min} = \left(\frac{1}{3}, \frac{1}{3}\right), \tag{8.77}$$

$$\text{sum s.d.o.f.} = 1 : (\lambda_P, \lambda_D)_{\min} = \left(0, \frac{1}{2}\right). \tag{8.78}$$

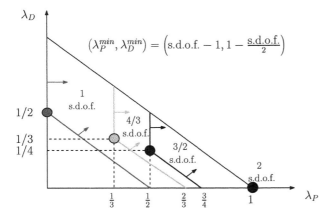

Figure 8.6 Tradeoff between delayed and perfect CSIT.

In general, for a given value of sum s.d.o.f. $= s$, the minimum CSIT requirements are given by:

$$(\lambda_P, \lambda_D)_{\min} = \begin{cases} (s-1, 1-\frac{s}{2}), & \text{if } 1 \leq s \leq 2 \\ (0, \frac{s}{2}), & \text{if } 0 \leq s \leq 1. \end{cases} \quad (8.79)$$

Cost of Security

We recall that in the case with no security [50], the sum d.o.f. is given by

$$\text{sum d.o.f.} = 2 - \frac{2\lambda_N}{3} - \frac{\max(\lambda_N, 2\lambda_D)}{3}. \quad (8.80)$$

Comparing with (8.74), we see that the loss in d.o.f. that must be incurred to incorporate secrecy constraints is given by

$$(\text{sum d.o.f.}) - (\text{sum s.d.o.f.}) \triangleq \text{loss} = \begin{cases} \lambda_N, & \text{if } \lambda_N \geq 2\lambda_D \\ \frac{2}{3}(2\lambda_N - \lambda_D), & \text{if } 2\lambda_D \geq \lambda_N \geq \lambda_D \\ \frac{1}{3}(\lambda_N + \lambda_D), & \text{if } \lambda_D \geq \lambda_N. \end{cases} \quad (8.81)$$

If we define $\alpha = \lambda_D/(\lambda_D + \lambda_N)$, we can rewrite (8.81) as

$$\text{loss} = (\lambda_D + \lambda_N) \times \begin{cases} (1-\alpha), & \text{if } \alpha \leq \frac{1}{3} \\ \left(\frac{4}{3} - 2\alpha\right), & \text{if } \frac{1}{2} \geq \alpha \geq \frac{1}{3} \\ \frac{1}{3}, & \text{if } \alpha \geq \frac{1}{2}. \end{cases} \quad (8.82)$$

We show this loss as a function of α in Fig. 8.7. Note that $\lambda_D + \lambda_N$ is the fraction of the time a user feeds back imperfect (delayed or none) CSIT. If this fraction is fixed, increasing the fraction of the delayed CSIT decreases the penalty due to the security constraints, but only to a certain extent. When $\lambda_N \geq \lambda_D$, increasing the fraction of delayed CSIT leads to a decrease in the penalty due to the security constraints. However, once the fraction of the delayed CSIT (state D) matches that of no CSIT (N), that is, $\lambda_D \geq \lambda_N$, increasing the fraction of delayed CSIT further does not reduce the penalty any more.

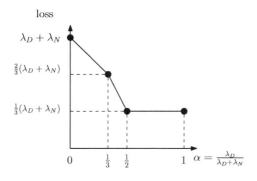

Figure 8.7   Cost of security.

Synergistic Benefits

It was shown in [50] that by coding across different states one can achieve higher sum d.o.f. than by optimal encoding for each state separately and time sharing. A similar result holds true in our case as well. We illustrate this with the help of a few examples.

Example 1

Consider a special case where only states PD and DP occur, each for half of the time. The optimal sum s.d.o.f. in this case is $\frac{3}{2}$ [46, 51]; see also (8.74). The optimal sum s.d.o.f. for the PD state alone is 1, as we have already seen. Thus, by encoding for each state separately and time sharing between the PD and DP states, we can achieve only 1 sum s.d.o.f., whereas joint encoding across the states achieves a sum s.d.o.f. of $\frac{3}{2}$. Thus, we have a synergistic benefit of 50% in this case.

Example 2

Consider another special case with three states, PD, DP, and NN, each occurring for one-third of the time. The optimal sum s.d.o.f. is $\frac{4}{3}$. If we encode for each state separately and time share between them, we can achieve a sum s.d.o.f. of $\frac{1}{3} \times 1 + \frac{1}{3} \times 1 + \frac{1}{3} \times 0 = \frac{2}{3}$, since the NN state does not provide any secrecy. If we encode across the PD and DP states optimally and then time share with the NN state, we can achieve $\frac{2}{3} \times \frac{3}{2} + \frac{1}{3} \times 0 = 1$ sum s.d.o.f. Thus, in this case too we get a synergistic benefit by coding across all the states together.

Example 3

Now, assume we have the following three states: PN, NP, and DD, each occurring for one-third of the time. The optimal sum s.d.o.f. for this case is $\frac{4}{3}$. On the other hand, the optimal sum s.d.o.f. of the PN state alone is 1 [44], and that of the DD state alone is also 1 [40]. Thus, by separately encoding for each state and time sharing, we can achieve $\frac{1}{3} \times 1 + \frac{1}{3} \times 1 + \frac{1}{3} \times 1 = 1$ sum s.d.o.f. Note that the optimal sum s.d.o.f. for PN and NP states, each occurring for half of the time, is also 1, using (8.74). Thus, optimal encoding for PN and NP together and time sharing with the DD state also yields sum s.d.o.f. of 1. Therefore, there is a synergistic benefit to be gained by coding across all the states together in this case too.

Example 4

Consider the case where the two states DD and NN occur for equal fractions of the time. The optimal sum s.d.o.f. of the DD state alone is 1 [40]. The NN state by itself does not provide any secrecy and its s.d.o.f. $= 0$. Thus, by encoding for the individual states and time sharing, at most $1 \times \frac{1}{2} + 0 \times \frac{1}{2} = \frac{1}{2}$ sum s.d.o.f. is achievable. However, by jointly encoding across both the DD and NN states, the optimal sum s.d.o.f. of 1 is achievable. Thus, we have a synergistic benefit of 100% in terms of sum s.d.o.f. in this case.

Example 5

Finally, consider the case where the two states DN and ND occur for equal fractions of time. We have seen that the optimal sum s.d.o.f. for the DN state is $\frac{1}{2}$. Thus, by separately encoding across the individual states, only $\frac{1}{2}$ sum s.d.o.f. is achievable. However, by jointly encoding across both the DN and ND states, the optimal sum s.d.o.f. of 1 is achievable. Thus, we have a synergistic benefit of 100% in terms of sum s.d.o.f. in this case.

Lack of Synergistic Benefits

There are some situations where joint encoding across alternating states does not yield any benefit in terms of the s.d.o.f. region. For example, consider a case with only two states, PN and NP, each occurring for half of the time. The optimal sum s.d.o.f. for the PN state alone is 1, which is achieved by zero-forcing. The optimal sum s.d.o.f. of both PN and NP states together is also 1; thus, encoding for each state separately is optimal in this case. Indeed, separable encoding for each individual state suffices to achieve the full s.d.o.f. region as well. This result is perhaps surprising, since in the case with no security, we do get synergistic benefits of joint encoding across the PN and NP states. The optimal sum s.d.o.f. with joint encoding is $\frac{3}{2}$, while that for each state alone is 1 [50].

We now sketch a proof of the achievability for Theorem 8.4. The converse uses the LSEP and we skip it here for brevity [46].

8.5.1 Achievability

The achievability for Theorem 8.4 involves first identifying a few key constituent schemes and then time sharing among them appropriately based on the given fractions $\lambda_{I_1 I_2}$.

Constituent Schemes

We summarize these key schemes in Table 8.1. Note that a particular sum s.d.o.f. value can be achieved in various ways through alternation between different possible sets of CSIT states. To this end, we use the following notation: if there are r schemes achieving a particular s.d.o.f. value, we denote these schemes as $S_1^{\text{sum s.d.o.f.}}, S_2^{\text{sum s.d.o.f.}}, \ldots, S_r^{\text{sum s.d.o.f.}}$. For example, in Table 8.1, to achieve the sum s.d.o.f. value of 1, we present $r = 3$ distinct schemes and these are denoted as S_1^1, S_2^1, and S_3^1.

We have already seen some of the schemes in Table 8.1: scheme S^2 for state PP, scheme S_1^1 for state DD, and scheme $S_1^{2/3}$ for state DD. Further, schemes $S_2^{2/3}$ and $S_3^{2/3}$ are reinterpretations of $S_1^{2/3}$ in the following way. Observe that $S_1^{2/3}$ does not require delayed CSIT from both users in every time slot; indeed, it requires delayed CSIT from the first and second users in the first and second time slots, respectively. No delayed CSIT is required for the third time slot. This CSIT requirement translates to having the DN, ND, and NN states each occur for one-third of the time. Thus, the scheme $S_1^{2/3}$ can be reinterpreted as the scheme $S_3^{2/3}$. It can also be interpreted as scheme $S_2^{2/3}$ by noting that $S_2^{2/3}$ requires better CSIT than $S_3^{2/3}$. Note that the scheme $S_2^{3/2}$ also suffices as $S_1^{3/2}$.

Table 8.1 Constituent schemes.

Sum s.d.o.f.	CS notation	CSIT states	Fractions of states	(d_1, d_2)
	\multicolumn{4}{c}{Summary of constituent schemes (CS)}	
2	S^2	PP	1	$(1,1)$
3/2	$S_1^{3/2}$	PD, DP	$\left(\frac{1}{2}, \frac{1}{2}\right)$	$\left(\frac{3}{4}, \frac{3}{4}\right)$
	$S_2^{3/2}$	PD, DP, PN, NP	$\left(\frac{1}{4}, \frac{1}{4}, \frac{1}{4}, \frac{1}{4}\right)$	$\left(\frac{3}{4}, \frac{3}{4}\right)$
4/3	$S_1^{4/3}$	PD, DP, NN	$\left(\frac{1}{3}, \frac{1}{3}, \frac{1}{3}\right)$	$\left(\frac{2}{3}, \frac{2}{3}\right)$
	$S_2^{4/3}$	PN, NP, DD	$\left(\frac{1}{3}, \frac{1}{3}, \frac{1}{3}\right)$	$\left(\frac{2}{3}, \frac{2}{3}\right)$
1	S_1^1	DD	1	$\left(\frac{1}{2}, \frac{1}{2}\right)$
	S_2^1	DD, NN	$\left(\frac{1}{2}, \frac{1}{2}\right)$	$\left(\frac{1}{2}, \frac{1}{2}\right)$
	S_3^1	DN, ND	$\left(\frac{1}{2}, \frac{1}{2}\right)$	$\left(\frac{1}{2}, \frac{1}{2}\right)$
2/3	$S_1^{2/3}$	DD	1	$\left(\frac{2}{3}, 0\right)$
	$S_2^{2/3}$	DD, NN	$\left(\frac{2}{3}, \frac{1}{3}\right)$	$\left(\frac{2}{3}, 0\right)$
	$S_3^{2/3}$	DN, ND, NN	$\left(\frac{1}{3}, \frac{1}{3}, \frac{1}{3}\right)$	$\left(\frac{2}{3}, 0\right)$

Thus, we need five new schemes: $S_2^{3/2}$, $S_1^{4/3}$, $S_2^{4/3}$, S_2^1, and S_3^1. Here, we present two representative schemes, $S_2^{3/2}$ and S_3^1, in detail and refer the interested reader to [46] for details of the remaining schemes.

Scheme $S_2^{3/2}$

This scheme is presented in [51] as $S_1^{3/2}$. The same scheme can be reinterpreted as $S_2^{3/2}$. We wish to send three confidential symbols from the transmitter to each of the receivers in four channel uses at high P (that is, negligible noise). Let us denote by (u_1, u_2, u_3) and (v_1, v_2, v_3) the confidential symbols intended for receivers 1 and 2, respectively. The scheme is as follows:

1. At time $t = 1$, $S(1) = \text{PD}$. As the transmitter knows $\mathbf{H}_1(1)$, it sends

$$\mathbf{X}(1) = [u_1 \quad 0]^T + q\mathbf{H}_1(1)^\perp, \tag{8.83}$$

where $\mathbf{H}_1(1)\mathbf{H}_1(1)^\perp = 0$, and q denotes an artificial noise distributed as $\mathcal{CN}(0, P)$. Here, $\mathbf{H}_1(1)^\perp$ is a 2×1 beamforming vector orthogonal to the 1×2 channel vector $\mathbf{H}_1(1)$ of receiver 1, which ensures that the artificial noise q does not create interference at receiver 1. The receivers' outputs are:

$$Y(1) = h_{11}(1)u_1, \tag{8.84}$$

$$Z(1) = h_{21}(1)u_1 + q\mathbf{H}_2(1)\mathbf{H}_1(1)^\perp \stackrel{\Delta}{=} K. \tag{8.85}$$

Thus, receiver 1 has observed u_1 while receiver 2 gets a linear combination of u_1 and q, which we denote as K. Due to delayed CSIT from receiver 2, the transmitter can reconstruct K in the next channel use and use it for transmission.

2. At time $t = 2$, $S(2) = $ DP. The transmitter knows $\mathbf{H}_2(2)$ and K. It sends

$$\mathbf{X}(2) = [v_1 + K \quad v_2 + K]^\mathrm{T} + u_2 \mathbf{H}_2(2)^\perp. \tag{8.86}$$

The received signals are:

$$Y(2) = h_{11}(2)v_1 + h_{12}(2)v_2 + (h_{11}(2) + h_{12}(2))K + u_2 \mathbf{H}_1(2)\mathbf{H}_2(2)^\perp, \tag{8.87}$$

$$= L_1(v_1, v_2, K) + u_2 \mathbf{H}_1(2)\mathbf{H}_2(2)^\perp, \tag{8.88}$$

$$Z(2) = h_{21}(2)v_1 + h_{22}(2)v_2 + (h_{21}(2) + h_{22}(2))K \tag{8.89}$$

$$\triangleq L_2(v_1, v_2, K), \tag{8.90}$$

where we have defined $L_1(v_1, v_2, K)$ and $L_2(v_1, v_2, K)$ as linear combinations of v_1, v_2, and K at receivers 1 and 2, respectively.

3. At time $t = 3$, $S(3) = $ NP. The transmitter knows $\mathbf{H}_2(3)$ and $L_1(v_1, v_2, K)$ (via delayed CSIT from $t = 2$). Using these, it transmits

$$\mathbf{X}(3) = [L_1(v_1, v_2, K) \quad 0]^\mathrm{T} + u_3 \mathbf{H}_2(3)^\perp, \tag{8.91}$$

and the channel outputs are:

$$Y(3) = h_{11}(3)L_1(v_1, v_2, K) + u_3 \mathbf{H}_1(3)\mathbf{H}_2(3)^\perp, \tag{8.92}$$

$$Z(3) = h_{21}(3)L_1(v_1, v_2, K). \tag{8.93}$$

At the end of this step, note that receiver 2 can decode v_1 and v_2 by first eliminating K using $Z(1)$ and $Z(3)$ to get a linear combination of v_1 and v_2, which it can then use with $Z(2)$ to solve for v_1 and v_2.

4. At time $t = 4$, $S(4) = $ PN. The transmitter knows $\mathbf{H}_1(4)$ and it sends

$$\mathbf{X}(4) = [L_1(v_1, v_2, K) \quad 0]^\mathrm{T} + v_3 \mathbf{H}_1(4)^\perp. \tag{8.94}$$

The channel outputs are:

$$Y(4) = h_{11}(4)L_1(v_1, v_2, K), \tag{8.95}$$

$$Z(4) = h_{21}(4)L_1(v_1, v_2, K) + v_3 \mathbf{H}_2(4)\mathbf{H}_1(4)^\perp. \tag{8.96}$$

Thus, at the end of these four steps the outputs at the two receivers can be summarized (see Fig. 8.8) as:

$$\mathbf{Y} = \begin{bmatrix} u_1 \\ \alpha_1 L_1(v_1, v_2, K) + u_2 \\ \alpha_2 L_1(v_1, v_2, K) + u_3 \\ L_1(v_1, v_2, K) \end{bmatrix}, \quad \mathbf{Z} = \begin{bmatrix} K \\ L_2(v_1, v_2, K) \\ L_1(v_1, v_2, K) \\ \beta L_1(v_1, v_2, K) + v_3 \end{bmatrix}.$$

Using $\mathbf{Y}$, receiver 1 can decode all three symbols (u_1, u_2, u_3), and using $\mathbf{Z}$, receiver 2 can decode (v_1, v_2, v_3). It can be shown that the information leakage is only $o(\log P)$ [46,51].

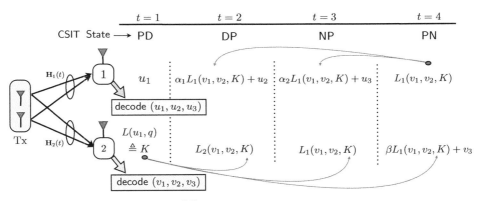

Figure 8.8 Achieving $\frac{3}{2}$ s.d.o.f. using scheme $S_2^{3/2}$.

Scheme S_3^1

The scheme S_3^1 uses the states (DN, ND) with fractions $(\frac{1}{2}, \frac{1}{2})$ to achieve $(d_1, d_2) = (\frac{1}{2}, \frac{1}{2})$. In particular, the scheme, first presented in [46], achieves the s.d.o.f. pair $(d_1, d_2) = \left(\frac{2n}{4n+1}, \frac{2n}{4n+1}\right)$ as a function of the block length n. Taking the limit $n \to \infty$ yields the s.d.o.f. pair $\left(\frac{1}{2}, \frac{1}{2}\right)$.

The scheme is shown in Fig. 8.9. Unlike the other schemes we have encountered where the optimal sum s.d.o.f. can be achieved within a finite number of time slots, this scheme cannot achieve a sum s.d.o.f. of 1 in a finite number of slots. We wish to send $2n$

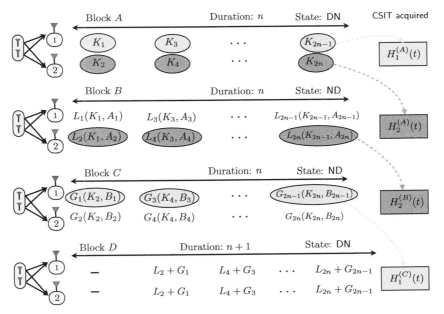

Figure 8.9 Achieving a sum s.d.o.f. of $4n/(4n+1)$ using scheme S_3^1.

symbols to each receiver in $4n+1$ time slots. The scheme involves transmission in four blocks where the first three blocks, say A, B, and C, each have n time slots, while the last block D has $n+1$ slots; thus, a total of $4n+1$ time slots are required in the scheme. Broadly, the scheme resembles the scheme achieving the s.d.o.f. pair $\left(\frac{1}{2},\frac{1}{2}\right)$ using the channel state DD presented in Section 8.3.3. The first block, with channel state DN, is used to generate shared secret *keys* by transmitting artificial noise symbols. However, unlike the case of channel state DD, the transmitter cannot reconstruct the keys created at the second receiver immediately. The transmitter acquires the required CSI of the second user corresponding to the first block in the second block when the channel state is ND. The second, third, and fourth blocks are also similar in spirit to the second, third, and fourth slots in the scheme for channel state DD, and are used to transmit the first user's message symbols, the second user's message symbols, and exchange the first and second users' channel outputs from the third and second blocks, respectively. We now describe the scheme in more detail as follows:

1. In block A, $S(t) = $ DN. In each time slot i in block A, the transmitter generates two artificial noise symbols and sends them using its two antennas. The receivers receive different linear combinations of the two artificial noise symbols K_{2i-1} and K_{2i}, as shown in Fig. 8.9. Due to delayed CSIT from the first user, the transmitter can reconstruct each of $K_{2i-1}, i = 1,\ldots,n$, by the end of block A. Thus, they can act as shared keys between the transmitter and the first receiver. However, since the second receiver does not feedback any CSIT (due to the fact that the state in the block is DN), the transmitter cannot reconstruct the observations of the second receiver at the end of block A.

2. In block B, $S(t) = $ ND. At the beginning of this slot, the transmitter has the keys $K_{2i-1}, i = 1,\ldots,n$, shared with the first user. It uses these keys to send information intended for the first user. It creates $2n$ linearly independent combinations of the $2n$ symbols intended for the first receiver: $a_1,\ldots,a_{2n}$. In slot i, it transmits

$$\mathbf{X}^B(i) = \begin{bmatrix} a_{2i-1} + K_{2i-1} & a_{2i} + K_{2i-1} \end{bmatrix}^T. \tag{8.97}$$

The first and second receivers receive linearly independent combinations $L_{2i-1}(A_{2i-1}, K_{2i-1})$ and $L_{2i}(A_{2i}, K_{2i-1})$ in slot i, where A_i denotes the ith linear combination of the first user's symbols, as shown in Fig. 8.9. Since the state is ND, the second user provides delayed CSIT to the transmitter. In the ith slot, the second user feeds back $\mathbf{H}_2^A(i)$, that is, the channel coefficients of the second user in slot i within block A. *Note that this is unlike any other achievable scheme we have encountered so far; in all other schemes, the receiver feeds back the channel coefficients of the current slot, which appears as delayed CSIT at the beginning of the next slot.* Thus, at the end of slot B, the transmitter has all the channel coefficients of the second user from block A, so it can reconstruct the outputs of the second receiver in block A, $K_{2i}, i = 1,\ldots,n$, which now act as shared keys between the transmitter and the second receiver.

3. In block C, $S(t) = $ ND. At the beginning of this slot, the transmitter has the keys $K_{2i}, i = 1,\ldots,n$ shared with the second user. It uses these keys to send information

securely to the second user. It creates $2n$ linearly independent combinations of the $2n$ symbols intended for the second receiver: $b_1, \ldots, b_{2n}$. In slot i, it transmits

$$\mathbf{X}^C(i) = \begin{bmatrix} b_{2i-1} + K_{2i-1} & b_{2i} + K_{2i-1} \end{bmatrix}^T. \quad (8.98)$$

The first and second receivers receive linearly independent combinations $G_{2i-1}(B_{2i-1}, K_{2i})$ and $G_{2i}(B_{2i}, K_{2i})$ in slot i, where B_i denotes the ith linear combination of the second user's symbols, as shown in Fig. 8.9. As CSIT, in the ith slot the second user feeds back the channel coefficients $\mathbf{H}_2^B(i)$, which allows the transmitter to reconstruct $L_{2i}(A_{2i}, K_{2i-1})$. Note that now if $L_{2i}(A_{2i}, K_{2i-1})$ and $G_{2i-1}(B_{2i-1}, K_{2i})$ could be exchanged, each of the receivers would receive $2n$ linear combinations of the $2n$ symbols intended for it, thus allowing both receivers to decode their own messages. However, $G_{2i-1}(B_{2i-1}, K_{2i})$ is not known to the transmitter yet, since the first user has not fed back its channel in block C. This CSIT will be obtained in the next block.

4. In block D, $S(t) = \text{ND}$. The transmitter wishes to send the symbols $L_{2i}(A_{2i}, K_{2i-1}) + G_{2i-1}(B_{2i-1}, K_{2i}), i = 1, \ldots, n$, in this block. To do so, the transmitter does not transmit anything in the first slot in this block. It only acquires the channel coefficients $\mathbf{H}_1^C(i)$ from the first user who is supplying delayed CSIT in this block. In the ith slot, $i = 1, \ldots, n$, the transmitter acquires the channel coefficients $\mathbf{H}_1^C(i)$ and transmits

$$\mathbf{X}^D(i) = \begin{bmatrix} L_{2i-2}(A_{2i-2}, K_{2i-3}) + G_{2i-3}(B_{2i-3}, K_{2i-2}) & 0 \end{bmatrix}^T, \ i = 2, \ldots, n+1. \quad (8.99)$$

The first user can now obtain $L_{2i-1}(A_{2i-1}, K_{2i-1})$ and $L_{2i}(A_{2i}, K_{2i-1})$ for every $i = 1, \ldots, n$, while the second user obtains $G_{2i-1}(B_{2i-1}, K_{2i})$ and $G_{2i}(B_{2i}, K_{2i})$ for $i = 1, \ldots, n$. Now, by eliminating the respective keys, each user can decode the $2n$ symbols intended for it from the $2n$ linearly independent combinations available to it. Also, the keys ensure confidentiality, and the information leakage is only $o(\log P)$, as is shown in [46].

8.6 Conclusions and Open Problems

In this chapter, we explored how the quality of CSIT affects the s.d.o.f. region of the two-user MISO BCCM. The delay aspect of CSIT was modeled as either being perfect (instantaneous), delayed (completely stale), or none (no CSIT at all). With this simple model, the full s.d.o.f. region is established for homogeneous, hybrid, and alternating CSIT settings.

We conclude by pointing out several open directions for future research.

Impact of Partially Delayed and Imprecise CSIT

In our modeling of delayed CSIT, we assume that the delay in CSIT is larger than the channel coherence. This is perhaps too pessimistic, and in practical scenarios the delay in CSIT can be smaller than the coherence interval, yet large enough that it cannot be modeled effectively as instantaneous CSIT. Reference [61] is a work in this direction where it is shown that for the two-user X-channel, the CSIT can be delayed by two-thirds of the coherence interval for a block fading model and yet the optimal s.d.o.f. of $\frac{4}{3}$ for perfect CSIT can be achieved. Also, we have ignored the precision aspect of delay in this chapter. This is clearly too optimistic. The recent paper [44] offers some insight on how the precision of CSIT affects the d.o.f. in the MISO BC. With security constraints, [62] determines the s.d.o.f. region of the MISO BCCM with precise delayed CSIT and imprecise instantaneous CSIT.

Extensions to Multi-Receiver (> 2) Broadcast Channels

Another line of research is extending the results for the two-user MISO BCCM to the K-user MISO BCCM. A recent paper [49] characterizes completely the linear d.o.f. region with hybrid CSIT states and presents new outer bounds for the K-user case. With security constraints, the case of homogeneous delayed CSIT has recently been settled by [41]; the cases of hybrid or alternating CSIT remain open to the best of our knowledge.

Generalizations for Other Multi-User Networks

Lastly, imperfect CSIT situations occur for multi-transmitter models as well. While some results are known in homogeneous CSIT settings without security constraints [29, 35, 36, 39], the settings with hybrid or alternating CSIT remain largely unexplored. To the best of our knowledge, in the hybrid CSIT setting and distributed transmitters, the d.o.f. region is known only for the two-user interference channel with hybrid CSIT [63]. When security constraints are introduced, [60] determined the optimal linear sum s.d.o.f. for the wiretap channel with a helper with delayed CSIT from the legitimate receiver and no CSIT from the eavesdropper to be $\frac{1}{3}$. Recently, [64] determined the optimal sum s.d.o.f. for the wiretap channel with M helpers, the K-user multiple access wiretap channel, and the K-user interference channel with an external eavesdropper, when the eavesdropper's CSIT is not available. However, the s.d.o.f. regions of most multi-transmitter models remain unknown in hybrid and alternating CSIT settings.

References

[1] Y. Liang, H. V. Poor, and S. Shamai (Shitz), "Secure communication over fading channels," *IEEE Trans. Inf. Theory*, vol. 54, no. 6, pp. 2470–2492, Jun. 2008.
[2] Z. Li, R. Yates, and W. Trappe, "Achieving secret communication for fast Rayleigh fading channels," *IEEE Trans. Wireless Commun.*, vol. 9, no. 9, pp. 2792–2799, Sep. 2010.
[3] Z. Li, R. Yates, and W. Trappe, *Securing Wireless Communications at the Physical Layer*. Boston, MA: Springer US, 2010, pp. 1–18.
[4] P. K. Gopala, L. Lai, and H. El Gamal, "On the secrecy capacity of fading channels," *IEEE Trans. Inf. Theory*, vol. 54, no. 10, pp. 4687–4698, Oct. 2008.

[5] S. Shafiee, N. Liu, and S. Ulukus, "Towards the secrecy capacity of the Gaussian MIMO wire-tap channel: The 2-2-1 channel," *IEEE Trans. Inf. Theory*, vol. 55, no. 9, pp. 4033–4039, Sep. 2009.
[6] F. Oggier and B. Hassibi, "The secrecy capacity of the MIMO wiretap channel," *IEEE Trans. Inf. Theory*, vol. 57, no. 8, pp. 4961–4972, Aug. 2011.
[7] A. Khisti and G. W. Wornell, "Secure transmission with multiple antennas–Part II: The MIMOME wiretap channel," *IEEE Trans. Inf. Theory*, vol. 56, no. 11, pp. 5515–5532, Nov. 2010.
[8] E. Ekrem and S. Ulukus, "The secrecy capacity region of the Gaussian MIMO multi-receiver wiretap channel," *IEEE Trans. Inf. Theory*, vol. 57, no. 4, pp. 2083–2114, Apr. 2011.
[9] S. Goel and R. Negi, "Guaranteeing secrecy using artificial noise," *IEEE Trans. Wireless Commun.*, vol. 7, no. 6, pp. 2180–2189, Jun. 2008.
[10] E. Tekin and A. Yener, "The general Gaussian multiple-access and two-way wiretap channels: Achievable rates and cooperative jamming," *IEEE Trans. Inf. Theory*, vol. 54, no. 6, pp. 2735–2751, Jun. 2008.
[11] E. Tekin and A. Yener, "The Gaussian multiple access wire-tap channel," *IEEE Trans. Inf. Theory*, vol. 54, no. 12, pp. 5747–5755, Dec. 2008.
[12] Y. Oohama, "Coding for relay channels with confidential messages," in *Proc. IEEE Inf. Theory Workshop*, Cairns, Australia, Sep. 2001, pp. 87–89.
[13] L. Lai and H. El Gamal, "The relay eavesdropper channel: Cooperation for secrecy," *IEEE Trans. Inf. Theory*, vol. 54, no. 9, pp. 4005–4019, Sep. 2008.
[14] X. He and A. Yener, "Cooperation with an untrusted relay: A secrecy perspective," *IEEE Trans. Inf. Theory*, vol. 56, no. 8, pp. 3807–3827, Aug. 2010.
[15] E. Ekrem and S. Ulukus, "Secrecy in cooperative relay broadcast channels," *IEEE Trans. Inf. Theory*, vol. 57, no. 1, pp. 137–155, Jan. 2011.
[16] L. Dong, Z. Han, A. P. Petropulu, and H. V. Poor, "Improving wireless physical layer security via cooperating relays," *IEEE Trans. Signal Process.*, vol. 58, no. 3, pp. 1875–1888, Mar. 2010.
[17] R. Bassily and S. Ulukus, "Deaf cooperation and relay selection strategies for secure communication in multiple relay networks," *IEEE Trans. Signal Process.*, vol. 61, no. 6, pp. 1544–1554, Mar. 2013.
[18] R. Bassily and S. Ulukus, "Secure communication in multiple relay networks through decode-and-forward strategies," *J. Commun. Networks*, vol. 14, no. 4, pp. 352–363, Aug. 2012.
[19] O. O. Koyluoglu, H. El Gamal, L. Lai, and H. V. Poor, "Interference alignment for secrecy," *IEEE Trans. Inf. Theory*, vol. 57, no. 6, pp. 3323–3332, Jun. 2011.
[20] R. Bassily and S. Ulukus, "Ergodic secret alignment," *IEEE Trans. Inf. Theory*, vol. 58, no. 3, pp. 1594–1611, Mar. 2012.
[21] J. Xie and S. Ulukus, "Secure degrees of freedom of one-hop wireless networks," *IEEE Trans. Inf. Theory*, vol. 60, no. 6, pp. 3359–3378, Jun. 2014.
[22] J. Xie and S. Ulukus, "Secure degrees of freedom of K-user Gaussian interference channels: A unified view," *IEEE Trans. Inf. Theory*, vol. 61, no. 5, pp. 2647–2661, May 2015.
[23] R. Bassily, E. Ekrem, X. He, E. Tekin, J. Xie, M. Bloch, S. Ulukus, and A. Yener, "Cooperative security at the physical layer: A summary of recent advances," *IEEE Signal Process. Mag.*, vol. 30, no. 5, pp. 16–28, Sep. 2013.
[24] R. Liu and H. V. Poor, "Secrecy capacity region of a multiple-antenna Gaussian broadcast channel with confidential messages," *IEEE Trans. Inf. Theory*, vol. 55, no. 3, pp. 1235–1249, Mar. 2009.

[25] R. Liu, T. Liu, H. V. Poor, and S. Shamai (Shitz), "Multiple-input multiple-output Gaussian broadcast channels with confidential messages," *IEEE Trans. Inf. Theory*, vol. 56, no. 9, pp. 4215–4227, Sep. 2010.

[26] C. S. Vaze and M. K. Varanasi, "The degrees of freedom regions of MIMO broadcast, interference, and cognitive radio channels with no CSIT," *IEEE Trans. Inf. Theory*, vol. 58, no. 8, pp. 5354–5374, Aug. 2012.

[27] M. A. Maddah-Ali and D. Tse, "Completely stale transmitter channel state information is still very useful," *IEEE Trans. Inf. Theory*, vol. 58, no. 7, pp. 4418–4431, Jul. 2012.

[28] C. S. Vaze and M. K. Varanasi, "The degrees of freedom region of the two-user and certain three-user MIMO broadcast channel with delayed CSI," Jan. 2011. [Online]. Available: http://arxiv.org/abs/1101.0306

[29] C. S. Vaze and M. K. Varanasi, "The degrees of freedom region and interference alignment for the MIMO interference channel with delayed CSIT," *IEEE Trans. Inf. Theory*, vol. 58, no. 7, pp. 4396–4417, Jul. 2012.

[30] R. Tandon, S. Mohajer, H. V. Poor, and S. Shamai (Shitz), "Degrees of freedom region of the MIMO interference channel with output feedback and delayed CSIT," *IEEE Trans. Inf. Theory*, vol. 59, no. 3, pp. 1444–1457, Mar. 2013.

[31] A. Vahid, M. A. Maddah-Ali, and A. S. Avestimehr, "Capacity results for binary fading interference channels with delayed CSIT," *IEEE Trans. Inf. Theory*, vol. 60, no. 10, pp. 6093–6130, Oct. 2014.

[32] S. A. Jafar and S. Shamai (Shitz), "Degrees of freedom region of the MIMO X channel," *IEEE Trans. Inf. Theory*, vol. 54, no. 1, pp. 151–170, Jan. 2008.

[33] M. A. Maddah-Ali, A. S. Motahari, and A. K. Khandani, "Communication over MIMO X-channels: Interference alignment, decomposition, and performance analysis," *IEEE Trans. Inf. Theory*, vol. 54, no. 8, pp. 3457–3470, Aug. 2008.

[34] S. Lashgari, A. S. Avestimehr, and C. Suh, "Linear degrees of freedom of the X-channel with delayed CSIT," *IEEE Trans. Inf. Theory*, vol. 60, no. 4, pp. 2180–2189, Apr. 2014.

[35] D. T. H. Kao and A. S. Avestimehr, "Linear degrees of freedom of the MIMO X-channel with delayed CSIT," in *Proc. IEEE Int. Symp. Inf. Theory*, Honolulu, HI, USA, Jun. 2014, pp. 366–370.

[36] A. Ghasemi, A. S. Motahari, and A. K. Khandani, "On the degrees of freedom of X-channel with delayed CSIT," in *Proc. IEEE Int. Symp. Inf. Theory*, St. Petersburg, Russia, Jul. 2011, pp. 767–770.

[37] M. J. Abdoli, A. Ghasemi, and A. K. Khandani, "On the degrees of freedom of K-user SISO interference and X channels with delayed CSIT," *IEEE Trans. Inf. Theory*, vol. 59, no. 10, pp. 6542–6561, Oct. 2013.

[38] R. Tandon, S. Mohajer, H. V. Poor, and S. Shamai (Shitz), "On X-channels with feedback and delayed CSI," in *Proc. IEEE Int. Symp. Inf. Theory*, Cambridge, MA, USA, Jul. 2012, pp. 1877–1881.

[39] M. J. Abdoli and A. S. Avestimehr, "Layered interference networks with delayed CSI: DoF scaling with distributed transmitters," *IEEE Trans. Inf. Theory*, vol. 60, no. 3, pp. 1822–1839, Mar. 2014.

[40] S. Yang, M. Kobayashi, P. Piantanida, and S. Shamai (Shitz), "Secrecy degrees of freedom of MIMO broadcast channels with delayed CSIT," *IEEE Trans. Inf. Theory*, vol. 59, no. 9, pp. 5244–5256, 2013.

[41] S. Yang and M. Kobayashi, "Secure communication in K-user multi-antenna broadcast channel with state feedback," in *Proc. IEEE Int. Symp. Inf. Theory*, Hong Kong, Jun. 2015, pp. 1976–1980.

[42] H. Maleki, S. A. Jafar, and S. Shamai (Shitz), "Retrospective interference alignment over interference networks," *IEEE J. Sel. Topics Signal Process.*, vol. 6, no. 3, pp. 228–240, Jun. 2012.

[43] R. Tandon, M. A. Maddah-Ali, A. Tulino, H. V. Poor, and S. Shamai (Shitz), "On fading broadcast channels with partial channel state information at the transmitter," in *Proc. 9th Int. Symp. Wireless Commun. Systems*, Paris, France, Aug. 2012, pp. 1004–1008.

[44] A. G. Davoodi and S. A. Jafar, "Aligned image sets under channel uncertainty: Settling a conjecture by Lapidoth, Shamai and Wigger on the collapse of degrees of freedom under finite precision CSIT," Mar. 2014. [Online]. Available: http://arxiv.org/abs/1403.1541

[45] P. Mukherjee, R. Tandon, and S. Ulukus, "Secrecy for MISO broadcast channels with heterogeneous CSIT," in *Proc. IEEE Int. Symp. Inf. Theory*, Hong Kong, Jun. 2015, pp. 1966–1970.

[46] P. Mukherjee, R. Tandon, and S. Ulukus, "Secure degrees of freedom region of the two-user MISO broadcast channel with alternating CSIT," Feb. 2015. [Online]. Available: http://arxiv.org/abs/1502.02647

[47] S. Amuru, R. Tandon, and S. Shamai (Shitz), "On the degrees-of-freedom of the 3-user MISO broadcast channel with hybrid CSIT," in *Proc. IEEE Int. Symp. Inf. Theory*, Honolulu, HI, USA, Jun. 2014, pp. 2137–2141.

[48] K. Mohanty and M. K. Varanasi, "On the dof region of the K-user MISO broadcast channel with hybrid CSIT," Nov. 2013. [Online]. Available: http://arxiv.org/abs/1311.6647

[49] S. Lashgari, R. Tandon, and A. S. Avestimehr, "MISO broadcast channel with hybrid CSIT: Beyond two users," Apr. 2015. [Online]. Available: http://arxiv.org/abs/1504.04615

[50] R. Tandon, S. A. Jafar, S. Shamai (Shitz), and H. V. Poor, "On the synergistic benefits of alternating CSIT for the MISO broadcast channel," *IEEE Trans. Inf. Theory*, vol. 59, no. 7, pp. 4106–4128, Jul. 2013.

[51] P. Mukherjee, R. Tandon, and S. Ulukus, "MISO broadcast channels with confidential messages and alternating CSIT," in *IEEE Int. Symp. Inf. Theory*, Honolulu, HI, USA, Jun. 2014, pp. 216–220.

[52] P. Mukherjee, R. Tandon, and S. Ulukus, "Secrecy for MISO broadcast channels via alternating CSIT," in *Proc. IEEE Int. Conf. Commun.*, London, UK, Jun. 2015, pp. 4157–4162.

[53] S.-C. Lin and P.-H. Lin, "On secrecy capacity of fast fading multiple-input wiretap channels with statistical CSIT," *IEEE Trans. Inf. Forensics Security*, vol. 8, no. 2, pp. 414–419, Feb. 2013.

[54] Z. Rezki, A. Khisti, and M. S. Alouini, "On the secrecy capacity of the wiretap channel with imperfect main channel estimation," *IEEE Trans. Commun.*, vol. 62, no. 10, pp. 3652–3664, Oct. 2014.

[55] X. Zhou and M. R. McKay, "Secure transmission with artificial noise over fading channels: Achievable rate and optimal power allocation," *IEEE Trans. Veh. Technol.*, vol. 59, no. 8, pp. 3831–3842, Oct. 2010.

[56] A. Mukherjee and A. L. Swindlehurst, "Robust beamforming for security in MIMO wiretap channels with imperfect CSI," *IEEE Trans. Signal Process.*, vol. 59, no. 1, pp. 351–361, Jan. 2011.

[57] A. D. Wyner, "The wire-tap channel," *Bell Syst. Tech. J.*, vol. 54, pp. 1355–1387, Oct. 1975.

[58] I. Csiszár and J. Körner, "Broadcast channels with confidential messages," *IEEE Trans. Inf. Theory*, vol. 24, no. 3, pp. 339–348, May 1978.

[59] P. Mukherjee and S. Ulukus, "Fading wiretap channel with no CSI anywhere," in *Proc. IEEE Int. Symp. Inf. Theory*, Istanbul, Turkey, Jul. 2013, pp. 1347–1351.

[60] S. Lashgari and A. S. Avestimehr, "Blind wiretap channel with delayed CSIT," in *Proc. IEEE Int. Symp. Inf. Theory*, Honolulu, HI, USA, Jun. 2014, pp. 36–40.

[61] N. Lee, R. Tandon, and R. W. Heath, "Distributed space–time interference alignment with moderately delayed CSIT," *IEEE Trans. Wireless Commun.*, vol. 14, no. 2, pp. 1048–1059, Feb. 2015.

[62] Z. Wang, M. Xiao, M. Skoglund, and H. V. Poor, "Secrecy degrees of freedom of the two-user MISO broadcast channel with mixed CSIT," in *Proc. IEEE Inf. Theory Workshop*, Jerusalem, Israel, Apr. 2015, pp. 1–5.

[63] K. Mohanty, C. S. Vaze, and M. K. Varanasi, "The degrees of freedom region of the MIMO interference channel with hybrid CSIT," *IEEE Trans. Wireless Commun.*, vol. 14, no. 4, pp. 1837–1848, Apr. 2015.

[64] P. Mukherjee, J. Xie, and S. Ulukus, "Secure degrees of freedom of one-hop wireless networks with no eavesdropper CSIT," Jun. 2015. [Online]. Available: http://arxiv.org/abs/1506.06114

9 Stochastic Orders, Alignments, and Ergodic Secrecy Capacity

Pin-Hsun Lin and Eduard A. Jorswieck

We investigate the relation between different stochastic orders and the degradedness of a fast fading wiretap channel with statistical channel state information at the transmitter (CSIT). In particular, we derive sufficient conditions to identify the ergodic secrecy capacities for both single and multiple antenna cases even though there is only statistical CSIT.

9.1 Introduction

To design a wiretap code with higher secrecy rate, channel state information at the transmitter of the legitimate and eavesdropper's channels[1] should be known to a certain degree. When there is perfect CSIT of both channels, the secrecy capacity of a Gaussian wiretap channel can be achieved by a Gaussian input. However, due to several practical issues such as (1) being a malicious user, Eve is by no means intending to feed back the correct CSI to Alice; (2) limited feedback bandwidth; (3) the delay caused by channel estimation; (4) the speed of channel variation, etc., it is more reasonable to consider cases where perfect CSIT is unavailable.

In this chapter we consider the cases when only statistical CSIT of both channels from Alice to Bob and Alice to Eve, respectively, are available. One possible way to get such information from Eve virtually is as follows. For some bounded space like indoor parking lots or malls, it is possible to collect the channel statistics offline of most of the positions within that space, where the channel variation may be due to the movement of people or cars, etc. When Alice wants to transmit, she can use the known statistics measured offline of the channel between her and the unknown user closest to her as Eve's channel information to design her wiretap code for the worst-case scenario.

Although the secrecy capacity formula for discrete memoryless non-degraded wiretap channels was proved in [1], the optimal selection of the auxiliary random variable and channel prefixing are still unknown for additive white Gaussian noise (AWGN) channels with partial CSIT in general. Only a few capacity results under the considered scenario are known, and are summarized as follows. When there are multiple antennas at the transmitter and both Bob's and Eve's channels are Rayleigh distributed (but

---

[1] In the following, to simplify the discussion we use Alice, Bob, and Eve to replace the transmitter, the legitimate receiver, and the eavesdropper, respectively.

with different statistics) where the entries of each channel vector are identically and independently distributed (i.i.d.), the ergodic secrecy capacity is proved [2]. This result is further extended to cases in which both Bob and Eve have multiple antennas with total or per-antenna power constraints [3]. For the extreme case in which the channel gains are unknown to all parties, with Bob's and Eve's channels being fast Rayleigh fading, the secrecy capacity can be achieved by a discrete channel input [4]. However, for more general settings, e.g., Bob's and Eve's channels do not belong to the same type of distribution, only some lower and upper bounds of the secrecy capacity are known [5–8].

In this chapter we try to partly unveil the capacity of this unknown region by investigating the relation between different stochastic orders and the ergodic secrecy capacity of the wiretap channel under fast fading with statistical CSIT. More specifically, even if we cannot directly compare the two channels by trichotomy since there is no perfect CSIT, we may still be able to *stochastically* compare the *channel orders*. Furthermore, for cases with channel distributions satisfying certain conditions, we show that Gaussian input without channel prefixing is optimal. Then we do not bother to design and transmit the artificial noise that is commonly used in multiple antenna wiretap channels to deteriorate Eve's signal-to-noise ratio (SNR) [5, 8].

The main results covered in this chapter are as follows.

1. For cases with a single antenna, we characterize the relation between different stochastic orders and the existence of the equivalent degraded wiretap channel with positive ergodic secrecy capacity. The considered stochastic orders include the usual stochastic order, the convex order, and the increasing convex order. Combining stochastic orders and the same marginal property, we are able to *align* the channel realizations of the fast fading channel without affecting the secrecy capacity such that we can compare the order between the two random channels to attain the degradedness. Several examples with practical channel distributions are also illustrated.
2. We prove the secrecy capacity of binary erasure fading wiretap channels.
3. We derive two secrecy capacity upper bounds for different scenarios.
4. We propose a layered signaling scheme and derive the achievable ergodic secrecy rate. When the wiretap channel is not degraded, the layered scheme enables transmissions only on those layers where Bob's channels are better than Eve's, which cannot be attained by the traditional Gaussian codebook.
5. We show by several numerical examples that under Nakagami-m fading channels, with only three binary layers, the proposed layered scheme can outperform the Gaussian codebook.
6. We extend the characterization between the usual stochastic order and the existence of the equivalent degraded wiretap channel with positive ergodic secrecy capacity from single-antenna cases to multiple-antenna cases. We derive sufficient conditions for several fading channels which are degraded. We also provide some cases in which the channel enhancement [9], which was originally designed for perfect CSIT cases, can be applied to get the optimal input distribution and therefore the explicit secrecy capacity, both under statistical CSIT.

The rest of the chapter is organized as follows. In Section 9.2 we review some important preliminary knowledge on wiretap channels and stochastic orders. In Section 9.3 we introduce the system model to be considered. In Section 9.4, we discuss our first main result, which partially characterizes the relation between the existence of the degraded wiretap channel with positive secrecy capacity and several stochastic orders among Bob's and Eve's channels. In Section 9.5.1 we prove that for a layered erasure wiretap channel, the binary expansion superposition coding scheme can achieve the secrecy capacity with statistical CSIT of both channels. Secrecy capacity characterization with examples is provided in Section 9.5.2. In the same section we also derive both lower and upper capacity bounds for the fast fading Gaussian wiretap channel. In Section 9.6, numerical simulation illustrates our results. In Section 9.7, we extend the above discussion to cases with multiple antennas and provide some practical wiretap channels as examples of when channel enhancement can be used. Finally, Section 9.8 concludes the chapter.

9.2 Preliminaries

In this section we introduce the underlying background knowledge for this work, which includes the properties of wiretap channels and stochastic orders. Logarithms used in this chapter are to base 2.

9.2.1 Properties of Wiretap Channels

We first introduce the following definitions and properties for wiretap channels, which are important in deriving the main results in this work. Define the received signals at Bob and Eve as Y_b and Y_e, respectively, with channel input X.

DEFINITION 9.1   [10, p. 373] The wiretap channel is *physically degraded* if the transition distribution function satisfies $p_{Y_b Y_e | X}(\cdot | \cdot) = p_{Y_b | X}(\cdot | \cdot) p_{Y_e | Y_b}(\cdot | \cdot)$, i.e., X, Y_b, and Y_e form a Markov chain $X \to Y_b \to Y_e$. The wiretap channel is *stochastically degraded* if its conditional marginal distribution is the same as that of a physically degraded wiretap channel, i.e., there exists a distribution $\tilde{p}_{Y_e | Y_b}(\cdot | \cdot)$ such that $p_{Y_e | X}(y_e | x) = \sum_{y_b} p_{Y_b | X}(y_b | x) \tilde{p}_{Y_e | Y_b}(y_e | y_b)$.

LEMMA 9.1   *[10, Lemma 2.1] Two wiretap channels have the same capacity equivocation region if they have the same marginal channel transition probability distributions $p_{Y|X}(\cdot|\cdot)$ and $p_{Z|X}(\cdot|\cdot)$.*

DEFINITION 9.2   [10, p. 372] Bob's channel is *more capable* than Eve's channel if $I(X;Y) \geq I(X;Z)$ for all input p_X. Bob's channel is *less noisy* than Eve's channel if $I(U;Y) \geq I(U;Z)$ for every U that satisfies the Markov chain $U \to X \to (Y,Z)$.

In the following, we call a stochastically degraded channel simply a degraded channel because capacities of wiretap channels only depend on marginal distributions.

9.2.2 Properties of Stochastic Orders

Now we introduce several stochastic orders followed by the technique of coupling.

DEFINITION 9.3 [11, (1.A.7), (3.A.1), (4.A.1), (5.A.1)] For given random variables X, Y, the *usual stochastic order* (st), the *convex order* (cx), the *concave order* (cv), the *increasing convex order* (icx), the *increasing concave order* (icv), and the *Laplace transform order* (Lt) are respectively defined as

$X \leq_{\text{st}} Y$ if $E[f(X)] \leq E[f(Y)]$ for all increasing f,

$X \leq_{\text{cx (cv)}} Y$ if $E[f(X)] \leq E[f(Y)]$ for all convex (concave) f,

$X \leq_{\text{icx (icv)}} Y$ if $E[f(X)] \leq E[f(Y)]$ for all increasing convex (concave) f,

$X \leq_{\text{Lt}} Y$ if $E[\exp(-sX)] \leq E[\exp(-sY)]$ for all $s > 0$.

Note that the stochastic orders in Definition 9.3 can be further derived as the following relations, which are easier for evaluation. Denote the complementary cumulative distribution function (CCDF) by $\bar{F}$

THEOREM 9.1 [11, (1.A.3), (3.A.7), (4.A.5)] For random variables X and Y, $X \leq_{\text{st}} Y$ if and only if $\bar{F}_X(x) \leq \bar{F}_Y(x)$, for all x; $X \leq_{\text{icx}} Y$ if and only if

$$\int_t^\infty \bar{F}_X(h)dh \leq \int_t^\infty \bar{F}_Y(h)dh \tag{9.1}$$

for all t; and $X \leq_{\text{cx}} Y$ if and only if (9.1) is valid for all t and $E[X] = E[Y]$. Similarly, $X \leq_{\text{icv}} Y$ if and only if

$$\int_{-\infty}^t \bar{F}_X(h)dh \leq \int_{-\infty}^t \bar{F}_Y(h)dh \tag{9.2}$$

for all t, and $X \leq_{\text{cv}} Y$, if and only if (9.2) is valid for all t and $E[X] = E[Y]$.

Note that when X and Y are non-negative, the condition $E[X] = E[Y]$ can be further expressed by $\int_0^\infty \bar{F}_X(h)dh = \int_0^\infty \bar{F}_Y(h)dh$ [12, (1.12)]. Compared with the original form in Def. 9.3, the integral form of the complementary cumulative distribution functions (CCDFs) which unifies the expression of the considered stochastic orders by the functions of CCDFs only, highly simplifies the following derivations.

LEMMA 9.2 [13, Lemma 9.2.1] Let F and G be continuous distribution functions. If X has distribution F then the random variable $G^{-1}(F(X))$ has distribution G.

DEFINITION 9.4 [14, Definition 1] A *coupling* of a collection of random variables $\{X_i, i \in I\}$, where I denotes some index set, is a family of jointly distributed random variables $(X_i' : i \in I)$ such that X_i and X_i' are equal in distribution.

PROPOSITION 9.1 [13, Proposition 9.2.2] If $X \geq_{\text{st}} Y$, then there exist random variables X' and Y' having the same distributions as X and Y, respectively, such that $P(X' \geq Y') = 1$.

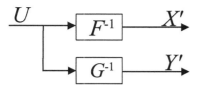

Figure 9.1 The explicit generation of $X' \sim F$, $Y' \sim G$, and $P(X' > Y') = 1$, where $U \sim \text{Unif}(0, 1)$.

The proof is restated as follows. Let $U \sim \text{Unif}(0,1)$ and F be a distribution function. From Lemma 9.2, we know that $F^{-1}(U)$ has the distribution as F which is denoted by $F^{-1(U)} \sim F$, where F^{-1} is the inverse mapping. Furthermore, if $X \geq_{\text{st}} Y$, $X \sim F$, and $Y \sim G$, we can construct X' and Y' which have the same distributions as X and Y, respectively, by $X' = F^{-1}(U)$ and $Y' = G^{-1}(U)$, as shown in Fig. 9.1. In addition to the assumption $X \geq_{\text{st}} Y$, i.e., $F(x) \leq G(x) \, \forall x$, it is clear that $P(X' \geq Y') = 1$.

The importance of coupling is that, if we know $X \geq_{\text{st}} Y$, even though the order of x and y may be different in each realization, it is still sufficient to justify the degradedness. In particular, by the same marginal property, as long as we can find the random variables X' and Y' which follow the same distributions of X and Y, respectively, and for each realization $x' > y'$, we attain the degradedness. Therefore, coupling together with the usual stochastic order form a sufficient condition to identify the degradedness of a wiretap channel. Proposition 9.1 can be directly extended to the vector case [11, Theorem 6.B.1] coined as the *usual multivariate stochastic order*, where, in the corresponding expressions $P(X' \geq Y') = 1$, the inequalities are element-wise.

9.3 System Model

We first consider the case with single antennas at all terminals and then extend to multiple antennas. The considered fast fading wiretap channel [1] is

$$Y_b = \sqrt{H_r} X + Z_r, \tag{9.3}$$

$$Y_e = \sqrt{H_e} X + Z_e, \tag{9.4}$$

where H_r and H_e are real-valued non-negative independent random variables denoting the square of Bob's and Eve's fading channels with CCDFs $\bar{F}_{H_r}$ and $\bar{F}_{H_e}$, respectively. The channel input is denoted by X. Without loss of generality, we normalize the channel input power constraint as $E[X^2] \leq 1$. Note that for cases in which Bob's and Eve's channel realizations are negative can be modeled by an additional phase rotation, which can be de-rotated at receivers because of the full CSI assumption at receivers. Since this phase rotation is independent of other variables, we can construct the equivalent channel model as (9.3) and (9.4) without changing the capacity. We assume that Alice knows only the distributions but not the instantaneous realizations of H_r and H_e. The noises Z_r

and Z_e at Bob and Eve, respectively, are independent AWGN's with zero mean and unit variance.

From [1], with the assumption that Alice does not know the instantaneous h_r and h_e, Bob knows only h_r, and Eve knows both h_r and h_e (where H_r is independent of H_e), we know that the secrecy capacity of this channel can be represented by

$$C_s \stackrel{(a)}{=} \max_{p(x|u), p(u)} I(U; Y_b, H_r) - I(U; Y_e, H_e, H_r)$$

$$\stackrel{(b)}{=} \max_{p(x|u), p(u)} I(U; Y_b|H_r) + I(U; H_r) - I(U; Y_e|H_e, H_r) - I(U; H_e, H_r)$$

$$\stackrel{(c)}{=} \max_{p(x|u), p(u)} I(U; Y_b|H_r) - I(U; Y_e|H_e), \quad (9.5)$$

where in (a) we start from the result for the discrete memoryless wiretap channel [1] and treat H_r and $\{H_e, H_r\}$ as channel outputs [15] at Bob and Eve, respectively; in (b) we use the chain rule of mutual information; in (c) we use the fact that H_r and H_e are independent of U since there is only statistical CSIT. Finally, after applying the quantization scheme used in the proof of Theorem 3.3 and Remark 3.8 in [16], the capacity for Gaussian wiretap channels can be derived.

9.4 The Relation between Degradedness and Stochastic Orders

In this section we will show how to *align* Bob's and Eve's channel realizations such that the considered wiretap channel is degraded. To attain this alignment, we investigate the relation between the existence of the equivalent degraded wiretap channel and different stochastic orders between Bob's and Eve's channels. This relation is helpful to distinguish the existence of positive ergodic secrecy capacity.

Based on the definitions in Sections 9.2.1 and 9.2.2, we give the following definitions. For the ease of the following discussion, we define three sets as follows.

DEFINITION 9.5 $\mathcal{S}_{st} = \{(H_r, H_e) : H_r \geq_{st} H_e\}$, $\mathcal{S}_{cx} = \{(H_r, H_e) : H_r \geq_{cx} H_e\}$, and $\mathcal{S}_{icx} = \{(H_r, H_e) : H_r \geq_{icx} H_e\}$.

DEFINITION 9.6 Define the set of pairs of fast fading channels (H_r, H_e) as $\mathcal{S}_{\mathcal{D}+} = \{(H_r, H_e) :$ the fast fading Gaussian wiretap channel is degraded and $C_s > 0\}$.

From Fig. 9.2 we can see the relation between the set $\mathcal{S}_{\mathcal{D}+}$ and other channel orders used in information theory [17]: degraded, less noisy, and more capable. From [1] we know that for all three sets, $U = X$ is optimal to achieve the secrecy capacity and there is no need to do rate splitting. In this chapter we mainly focus on investigating the relation between $\mathcal{S}_{\mathcal{D}+}$ and the sets of the aforementioned stochastic orders, i.e., $\mathcal{S}_{st}$, $\mathcal{S}_{cx}$, and $\mathcal{S}_{icx}$.

THEOREM 9.2 *With statistical CSI of both channels at the transmitter and the power constraint $E[X^2] \leq 1$, if $(H_r, H_e) \in \mathcal{S}_{\mathcal{D}+}$, the ergodic secrecy capacity is*

$$C_s = \frac{1}{2} \left\{ E_{H_r}[\log(1 + H_r)] - E_{H_e}[\log(1 + H_e)] \right\}. \quad (9.6)$$

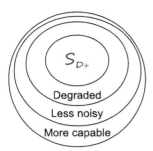

Figure 9.2 The relation between different orders and the set S_{D+}.

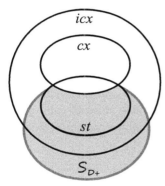

Figure 9.3 The relation between different stochastic orders and the set S_{D+}, which is shown by the shaded area.

The proof follows the standard ones for the case with full CSIT under proper modification to accommodate the statistical CSIT. Please refer to [18, Appendix I] for the details.

In the following, we classify the relation between S_{D+} and different stochastic orders. The results are illustrated in Fig. 9.3.

LEMMA 9.3 *The usual stochastic order $H_r \geq_{st} H_e$ is sufficient but not necessary to generate an equivalent degraded wiretap channel.*

Sketch of proof To show that the usual stochastic order is not necessary, we can construct an example in which H_1 has non-zero probability at zero magnitude while H_2 has no zero component. Let the support of the probability density function (PDF) of H_1 be $\in [h_0, h_1]$, where $h_1 > 0$ is the crossing point of $\bar{F}_{H_r}$ and $\bar{F}_{H_e}$. By definition, this example does not satisfy the usual stochastic order. However, we can prove that this example satisfies $P(r \leq \tilde{H}_r \leq r + \epsilon, e \leq \tilde{H}_e) = 0$, where $\epsilon > 0$ is arbitrarily small and $e > r + \epsilon$, by constructing an equivalent joint CCDF $\bar{F}_{\tilde{H}_r, \tilde{H}_e}(r, e) = \min\{\bar{F}_{\tilde{H}_r}(r), \bar{F}_{\tilde{H}_e}(e)\}$ [19]. Therefore, this example is degraded.

REMARK 9.1 *We provide two methods to prove the sufficient part of this result. Both methods have explicit construction of the equivalent channels which are degraded. The* first method is by applying Proposition 9.1 to the wiretap channel. That is, if $H_r \geq_{st} H_e$, there exist H'_r and H'_e which have the same distributions as H_r and H_e, respectively (from the same marginal property we know that the two wiretap channels formed by (H_r, H_e) and (H'_r, H'_e) have the same capacity), such that $P(H'_r \geq H'_e) = 1$. Therefore, the wiretap channel formed by (H'_r, H'_e) is degraded. The second method can be found in the proof of [18, Lemma 1]. More specifically, we prove that if $H_r \geq_{st} H_e$, then we have an equivalent degraded wiretap channel. We can slightly modify the proof of [19, Lemma 3] and adapt it to the wiretap channel. The advantage of the second one is the capability to derive a counterexample to show that $H_r \geq_{st} H_e$ is not necessary to generate an equivalent degraded wiretap channel, which is the second part of Lemma 9.3. A counterexample is shown in Fig. 9.4. It is clear that the example does not follow the usual stochastic order by definition.

REMARK 9.2 Since the order of the realizations of random channels H_r and H_e changes many times within a codeword length, at first glance we are not able to claim that it is a degraded wiretap channel. However, with the aid of the same marginal property, in addition to coupling, we know that we can equivalently align all the channel realizations within a codeword length such that the channel gain of each of Bob's channels is no worse than Eve's if $H_r \geq_{st} H_e$. Then we can claim that there exists an equivalent degraded wiretap channel.

REMARK 9.3 In [9], the wiretap channel with multiple input multiple output multiple antennas eavesdropper is enhanced in the space domain to equivalently form a degraded channel. Here we align/reorder the channel realizations in the time domain to form a sequence of degraded channels.

REMARK 9.4 When the usual stochastic order is strict for any non-zero interval, it implies $\mathcal{S}_{\mathcal{D}+}$. From Theorem 9.1 it can be easily proved that the intersection of $\mathcal{S}_{st}$ and

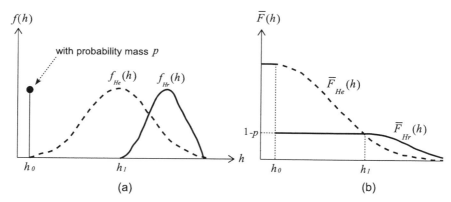

Figure 9.4 An example where $H_r \not\geq_{st} H_e$ that results in a degraded wiretap channel. (a) PDFs of the two channels. Please note that $f_{H_r}(h)$ is a mixed distribution, which has probability mass $p > 0$ at h_0. (b) CCDFs of the two channels. Note that A_e and A_r are the area between $\bar{F}_{H_e}$ and $\bar{F}_{H_r}$ before and after the crossing point h_1, respectively.

$\mathcal{S}_{cx}$ happens only when $\bar{F}_{H_r}(h) = \bar{F}_{H_e}(h)$ for all h, which results in $C_s = 0$. Note that the usual stochastic order is not a subset of $\mathcal{S}_{\mathcal{D}+}$ since the usual stochastic order can result in a zero secrecy capacity.

REMARK 9.5 *Since the degradedness is stricter than less noisy and more capable [16], once $H_r \geq_{st} H_e$ is valid, the wiretap channel is also less noisy and more capable.*

Note that Lemma 9.3 does not reveal the relation between the degradedness and other stochastic orders. In the following lemmas we characterize the relation first by an impossibility result in Lemma 9.4, and then by partial characterizations in Lemmas 9.5 and 9.6.

LEMMA 9.4 *The increasing convex order is not sufficient to guarantee $(H_r, H_e) \in \mathcal{S}_{\mathcal{D}+}$, i.e., $\mathcal{S}_{\mathcal{D}+} \cap \mathcal{S}_{icx} \neq \emptyset$ and $\mathcal{S}_{\mathcal{D}+} \not\supseteq \mathcal{S}_{icx}$. $(H_r, H_e) \in \mathcal{S}_{\mathcal{D}+}$ does not necessarily imply $H_r \geq_{icx} H_e$, i.e., $\mathcal{S}_{icx} \not\supseteq \mathcal{S}_{\mathcal{D}+}$.*

The proof is given in [18, Appendix II].

Sketch of proof To show the second statement of this lemma, we can exploit the example used in the proof of Lemma 9.3 with the additional condition $\int_0^{h_0} |\bar{F}_{H_1}(h) - \bar{F}_{H_2}(h)| dh \leq \int_{h_0}^{\infty} |\bar{F}_{H_1}(h) - \bar{F}_{H_2}(h)| dh$. Then we have an example which satisfies the increasing convex order. That means, there exists $(H_1, H_2) \in \mathcal{S}_{icx}$ which is degraded.

LEMMA 9.5 *If $(H_r, H_e) \in \mathcal{S}_{\mathcal{D}+}$, then $H_r \not\geq_{cx} H_e$. If $H_r \geq_{cx} H_e$, then $(H_r, H_e) \notin \mathcal{S}_{\mathcal{D}+}$. That is, $\mathcal{S}_{\mathcal{D}+} \cap \mathcal{S}_{cx} = \emptyset$.*

The proof is given in [18, Appendix III].

Sketch of proof To show the third statement, we can show the existence of cases with orders more general than the increasing convex order that can result in a degraded broadcast channel by constructing an example in which there are several discrete channel values with non-zero probabilities. By using the proof that usual stochastic order is not necessary for degradedness, we can complete the proof.

In the following we provide a scenario where $H_r \geq_{icx} H_e$ is a necessary condition for $(H_r, H_e) \in \mathcal{S}_{\mathcal{D}+}$.

LEMMA 9.6 *Assume that there is only one crossing point between $\bar{F}_{H_r}$ and $\bar{F}_{H_e}$ at $h = h_1$, and $\bar{F}_{H_r}(h) > \bar{F}_{H_e}(h)$ for all $h > h_1$, and $\bar{F}_{H_r}(h) \leq \bar{F}_{H_e}(h)$ for all $h \leq h_1$. If H_r and H_e form a degraded wiretap channel and $H_r \not\geq_{icx} H_e$, then the ergodic secrecy capacity is zero.*

REMARK 9.6 *If we exchange the labels of H_r and H_e in Fig. 9.4 with the assumption that $A_e > A_r$, then it satisfies $H_r \geq_{icv} H_e$, i.e., $\int_{-\infty}^h \bar{F}_{H_r}(u) du \geq \int_{-\infty}^h \bar{F}_{H_e}(u) du$ for all h. Note that in this case we may not be able to claim the existence of the degradedness. However, we can easily derive that if $H_r \geq_{icv} H_e$, then the ergodic secrecy rate is non-negative by substituting $U = X$ into (9.5). This property can be extended to the more general case $H_r \geq_{Lt} H_e$ [20]: if $H_r \geq_{Lt} H_e$, then $E[\log(1 + H_r)] \geq E[\log(1 + H_e)]$, which can be further interpreted as $R_s \geq 0$ by Gaussian signaling. Since $C_s \geq R_s$, we can claim that: if $H_r \geq_{Lt} H_e$, $C_s \geq 0$.*

REMARK 9.7 *The above discussion can be easily extended from real to complex cases. Assume that the noise at Bob and Eve are now circularly symmetric complex Gaussian noises with zero mean and unit variance. Since both Bob and Eve know their CSI, the phase rotation due to the complex channel can be compensated at Bob and Eve without changing the secrecy capacity. After this compensation we can form an equivalent complex wiretap channel as*

$$Y'_{r,I} + iY'_{r,Q} = \sqrt{H_r}X_I + Z'_{r,I} + i(\sqrt{H_r}X_Q + Z'_{r,Q}),$$
$$Y'_{e,I} + iY'_{e,Q} = \sqrt{H_e}X_I + Z'_{e,I} + i(\sqrt{H_e}X_Q + Z'_{e,Q}), \quad (9.7)$$

where $X = X_I + iX_Q$, $Z'_r = Z'_{r,I} + iZ'_{r,Q}$, and $Z'_e = Z'_{e,I} + iZ'_{e,Q}$ are the rotated versions of Z_r and Z_e. Since Z_r and Z_e are circularly symmetric complex Gaussian noise, Z'_r and Z'_e have the same distributions as Z_r and Z_e, respectively. Then it can be seen that the in-phase and quadrature channels form a pair of identical parallel real wiretap channels as shown in (9.3) and (9.4).

9.5 The Fast Fading Wiretap Channel with Statistical CSIT

Under statistical CSIT, for cases in which we are not able to find the equivalent degraded wiretap channels, Gaussian codebooks may not be optimal [21]. Thus, in this section we instead propose to generate sub-channels by layered signaling. The idea of the layered signaling is to mimic the operation of the case with perfect CSIT such that Alice can choose sub-channels for transmission according to the ergodic secrecy rate of each sub-channel even if Alice only has statistical CSI. In the beginning of this section we verify the performance of the fast fading wiretap channel with statistical CSIT where the channel gain is binary expanded as multiple layers. We prove that independent coding over each sub-channel can achieve the secrecy capacity under this scenario. We then apply the same coding scheme to fast fading Gaussian wiretap channels with statistical CSIT where $(H_r, H_e) \notin \mathcal{S}_{\mathcal{D}^+}$. We also derive a general upper bound for this case. Finally, we provide several examples under practical fast fading channels which have positive ergodic secrecy capacities.

9.5.1 The Layered Erasure Wiretap Channel

In this section we adapt the scheme from [19] to analyze the performance of the fast fading wiretap channel under the deterministic model as shown in Fig. 9.5. By the deterministic model [22], we assume that Alice transmits a vector of q bits which are i.i.d. Bernoulli random sequences. Each bit is a layer of the transmitted signal. The fading effect is modeled by the erasure of the less significant bits which cannot be transmitted successfully. The number of bits that can be received by Bob and Eve without being erased are $N_r[t]$ and $N_e[t]$, respectively, from the most significant bit (MSB) of the transmit signal. $N_r[t]$ and $N_e[t]$ are with PMFs $P_{N_r}(n)$ and $P_{N_e}(n)$, respectively, and $t \in \mathbb{N}$ is the time index. We also assume that both Bob and Eve perfectly

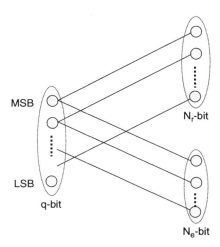

Figure 9.5 System model of the layered erasure wiretap channel.

know their channels. With the i.i.d. and memoryless assumptions of the channel states, the time index is dropped to simplify the expression.

THEOREM 9.3 *The secrecy capacity of the layered erasure wiretap channel with statistical CSIT and full CSI at the receivers is*

$$C_s = \sum_{n \in I_r} \bar{F}_{N_r}(n) - \bar{F}_{N_e}(n), \tag{9.8}$$

where $I_r = \{n | \bar{F}_{N_r}(n) > \bar{F}_{N_e}(n)\}$, $\bar{F}_{N_r}(n)$ is the CCDF of Bob's channel, and $\bar{F}_{N_e}(n)$ is the CCDF of Eve's channel.

The detailed proof is given in [18, Appendix IV]. The proof sketch is as follows. The achievable scheme is to transmit the nth bit X_n, which is i.i.d. Bernoulli with $p = 1/2$ in layer n. In particular, we select $U_n = X_n$ for each n. Then, the capacity from Alice to Bob on layer n, $n = 1, \ldots, q$, is

$$H\left(\frac{1}{2}\right) \cdot P(N_r \geq n) + 0 \cdot P(N_r < n) = P(N_r \geq n) = \bar{F}_{N_r}(n).$$

Similarly, the capacity from Alice to Eve on each layer n is $\bar{F}_{N_e}(n)$. Therefore, the achievable secrecy rate of the wiretap channel on layer n is $\bar{F}_{N_r}(n) - \bar{F}_{N_e}(n)$. With independent coding on each level, the expected number of bits transmitted without erasure can be expressed as the right-hand side (RHS) of (9.8), which is the summed rate of all layers belonging to the set I_r, i.e., only layers with positive rates.

To prove the upper bound, we first enhance Bob's channel as follows:

$$\bar{F}_{\tilde{N}_r}(n) = \max\{\bar{F}_{N_r}(n), \bar{F}_{N_e}(n)\} \text{ for all } n. \tag{9.9}$$

Clearly, this enhancement scheme results in $\tilde{N}_r \geq_{\text{st}} N_e$ and we have a degraded wiretap channel with secrecy capacity $\max_{p(x)} I(X; Y_r) - I(X; Y_e)$. After substituting $X = X^q$,

$Y_r = X^{\tilde{N}_r}$, and $Y_e = X^{N_e}$ into the capacity expression, with some manipulations we can prove that the upper bound is identical to the lower bound.

9.5.2 The Fast Fading Gaussian Wiretap Channel with Statistical CSIT

Based on the stochastic orders introduced in Section 9.3, in this subsection we first discuss several practical examples of the set $\mathcal{S}_{\mathcal{D}+}$. From the capacity result with the layered signaling scheme in Section 9.5.1, we then introduce two capacity upper bounds for the channel considered. According to the observations from the upper bound, we propose an achievable scheme and derive the achievable ergodic secrecy rate for the cases in which the wiretap channels do not belong to $\mathcal{S}_{\mathcal{D}+}$.

Examples of the Set $\mathcal{S}_{\mathcal{D}+}$

Based on Theorem 9.2, several examples with practical fading channels are provided in the following to explain how to determine the existence of the positive ergodic secrecy capacity given the distributions of the fading channels.

Example 9.1 Assume the magnitudes of Bob's and Eve's channels are two independent Rayleigh random variables with variances σ_r^2 and σ_e^2, respectively. Assume $\sigma_r^2 > \sigma_e^2$. Since the square of the Rayleigh distributed random variable X with variance σ_X^2 is exponentially distributed with CCDF $\exp\left(\frac{-x}{2\sigma_X^2}\right)$, we have $\bar{F}_{H_r}(x) > \bar{F}_{H_e}(x)$ for all x. Then, from Theorem 9.2 we know that the capacity is given in (9.6). Note that this result coincides with [2, Lemma 1] with a single antenna.

Example 9.2 Assume the magnitudes of Bob's and Eve's channels are two independent Nakagami-m random variables with shape parameters m_r and m_e, and spread parameters w_r and w_e, respectively. Then the squares of their amplitudes follow an Erlang distribution. From Theorem 9.2 we know that the capacity is given in (9.6) if

$$\gamma\left(m_r, \frac{m_r}{w_r}x\right)\Gamma(m_e) \geq \gamma\left(m_e, \frac{m_e}{w_e}x\right)\Gamma(m_r) \quad \text{for all } x > 0, \tag{9.10}$$

where $\gamma(.)$ is the incomplete gamma function, and $\Gamma(.)$ is the ordinary gamma function defined as $\Gamma(h) = \int_0^\infty t^{s-1}e^{-t}dt$. An example of the inequality in (9.10) being valid is $(m_r, w_r) = (3, 2)$ with $(m_e, w_e) = (1, 2)$.

Example 9.3 Assume the magnitudes of Bob's and Eve's channels are Nakagami-m and Rayleigh random variables, respectively. From Theorem 9.2, we know that the capacity is given in (9.6) if

$$\gamma\left(m_r, \frac{m_r}{w_r}x\right)/\Gamma(m_r) \geq \exp\left(\frac{-x}{2\sigma_e^2}\right) \quad \text{for all } x > 0.$$

For example, $(m_r, w_r) = (3, 2)$ and $\sigma_e^2 = 3$ satisfy the requirement and result in channels belonging to $\mathcal{S}_{\mathcal{D}+}$.

Example 9.4 For full CSIT cases under block fading, if Alice knows h_r and h_e then the CCDFs are given by $\bar{F}_{H_r} = 1 - \mathrm{u}(h - h_r)$ and $\bar{F}_{H_e} = 1 - \mathrm{u}(h - h_e)$, where u is the unit step function. From Lemma 9.3 we know that $h_r > h_e$ results in a positive secrecy capacity, which is consistent with the traditional way to check the positivity of single-input, single-output (SISO) secrecy capacity when there is perfect CSIT.

Example 9.5 If Alice knows Bob's channel perfectly and only the CCDF of Eve's channel, as long as $h_r > \mathrm{supp}(H_e)$, from Lemma 9.3 we know that the secrecy capacity of this wiretap channel is positive. On the other hand, when H_e is not bounded, we are not able to claim the degradedness.

Upper Bounds of the Ergodic Secrecy Capacity

In the following we provide an explicit upper bound of the fast fading Gaussian wiretap channel with statistical CSIT. If the channel distribution has non-zero probability at zero magnitude, we further derive a tighter upper bound.

THEOREM 9.4 *The capacity of a fast fading Gaussian wiretap channel with power constraint $E[X^2] \leq 1$ and statistical CSIT is bounded above by*

$$C_s \leq C_s^{\mathrm{UB}} = \frac{1}{2}\left\{\int_{h_r \in I_r} \log(1 + h_r) f_{H_r}(h_r) dh_r - \int_{h_e \in I_r} \log(1 + h_e) f_{H_e}(h_e) dh_e\right\}, \quad (9.11)$$

where $I_r = \{h | \bar{F}_{H_r}(h) > \bar{F}_{H_e}(h)\}$.

A tighter upper bound can be described in the following example with one crossing point if H_r has non-zero probability at $H_r = 0$. More specifically, under that condition, we can use a better enhancement scheme to attain an equivalent degraded wiretap channel.

LEMMA 9.7 *For $H_r \geq_{\mathrm{icx}} H_e$ with one crossing point between $\bar{F}_{H_r}$ and $\bar{F}_{H_e}$ and non-zero probability $p > 0$ at $H_r = 0$, a tighter upper bound of the ergodic secrecy capacity can be attained by the following enhancement:*

$$\bar{F}_{\tilde{H}_r}(h) = \begin{cases} 1 - p, & h < \bar{F}_{H_e}^{-1}(1-p), \\ \max\{\bar{F}_{H_r}(h), \bar{F}_{H_e}(h)\}, & h \geq \bar{F}_{H_e}^{-1}(1-p), \end{cases} \quad (9.12)$$

where $\bar{F}_{H_e}^{-1}$ is the inverse function of $\bar{F}_{H_e}$.

An illustrative example is shown in Fig. 9.6. From this example it can be easily seen that the upper bound from Lemma 9.7 is tighter than that from Theorem 9.4. Proofs are available in [18, Section IV-B].

Achievable Ergodic Secrecy Rate

From the analysis of the upper bound, we observe that if we do not transmit within the interval of h where Eve's channel has higher CCDF than Bob's, we will have a higher ergodic secrecy rate. Using a Gaussian codebook we are not able to attain this goal since there is no concept of channel partition. To overcome the caveat of using Gaussian

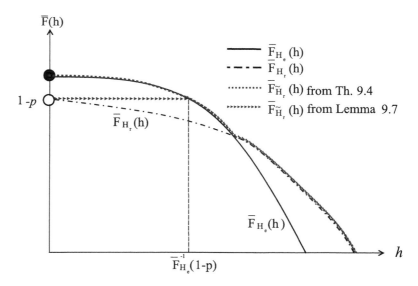

Figure 9.6 Comparison of the upper bounds from Theorem 9.4 and Lemma 9.7.

signaling, we consider the binary expansion scheme [22] of the uniform transmit signal as follows. To improve the performance, we construct sub-channels such that we can avoid the transmission of messages on those sub-channels with zero rates. The concept of *layer* in the binary expansion scheme is one possibility to attain the goal.

After applying binary expansion [22] to the received signal, we have

$$Y = \sqrt{H}X + Z = \sum_{n=1}^{\infty}\sqrt{H}\left(\sqrt{3}X_n 2^{-n}\right) + Z, \quad (9.13)$$

where H denotes the square of the channel magnitude, and Z is AWGN with zero mean and unit variance; $X_n = 2b_n - 1$ is the nth digit of X, with $b_n \sim \text{Bern}(1/2)$. We assume $b_1, b_2, \ldots$ are i.i.d. It can be seen that X is continuous and uniformly distributed in $\{-1, 1\}$. The factor $\sqrt{3}$ is to make the transmit signal satisfy the unit channel input power constraint introduced in Section 9.3, since the variance of $\sum_{n=1}^{\infty} X_n 2^{-n}$ is $1/3$. By the assumption of full CSI at the receiver, we can rewrite (9.13) without changing the capacity as

$$Y^{(h)} = \sum_{n=1}^{\infty} X_n 2^{-n} + \frac{Z}{\sqrt{3H}} = \underbrace{\sum_{j=1}^{n} X_j 2^{-j}}_{V_n} + \underbrace{\sum_{j=n+1}^{\infty} X_j 2^{-j}}_{U_n} + \frac{Z}{\sqrt{3H}}, \quad (9.14)$$

where $U_n \sim \text{Unif}(-2^{-n}, 2^{-n})$ is independent of V_n and Z. Bob's and Eve's channels therefore are modeled by $Y_b^{(h)} = X + Z_r/\sqrt{3H_r} \triangleq X + Z_r^{(h)}$ and $Y_e^{(h)} = X + Z_e/\sqrt{3H_e} \triangleq X + Z_e^{(h)}$, respectively. In the following we show our result of the ergodic secrecy rate based on the layered transmission scheme.

THEOREM 9.5 *The following ergodic secrecy rate is achievable:*

$$R_s = \sum_{n \in \{k | R_k > 0\}} R_n, \qquad (9.15)$$

where the ergodic secrecy rate of the nth layer is

$$R_n = \left(E_{H_r} \left[h\left(X + Z_r^{(h)}\right) - h\left(V_n'^{(h)}\right) \right] - E_{H_e} \left[h\left(X + Z_e^{(h)}\right) - h\left(V_n'^{(h)}\right) \right] \right)^+, \qquad (9.16)$$

and $V_n'^{(h)}$ follows the PDF

$$f_{V_n'^{(h)}}(v) = \sum_{k=0}^{2^{n-1}-1} \left\{ \bar{F}_G^{(h)}\left(v - 1 + 2^{-(n-2)}k + 2^{-n}\right) - \bar{F}_G^{(h)}\left(v - 1 + 2^{-(n-2)}k + 3 \cdot 2^{-n}\right) \right\}, \qquad (9.17)$$

with $\bar{F}_G^{(h)}$ being the CCDF of the Gaussian random variable with zero mean and variance $1/(3h)$.

REMARK 9.8 We can observe that the rate expression in (9.16) is similar to the fast fading ergodic secrecy capacity with full CSIT [23], i.e.,

$$C_s = \frac{1}{2} E_{H_r, H_e} \left[(\log(1 + P(h_r, h_e) H_r) - \log(1 + P(h_r, h_e) H_e))^+ \right], \qquad (9.18)$$

where $P(h_r, h_e)$ is the power allocation based on the knowledge of h_r and h_e at the transmitter, in the sense that in both cases the summation/expectation is over positive terms only. More specifically, in [23] each sub-channel corresponds to one specific channel gain after quantizing, which is similar to each layer in the layered signaling scheme. That is, for these two scenarios Alice can determine on which sub-channel to transmit according to the statistical CSIT and the perfect CSIT, respectively.

REMARK 9.9 Note that from (9.15) it seems that the layered signaling scheme dominates the normal Gaussian signaling since layer selection can be performed in the former scheme. However, this is not precise because in the layered signaling scheme the channel input distribution is assumed to be uniform for the ease of the binary decomposition, which may not be optimal. It can be shown in numerical results that sometimes the Gaussian signaling may outperform the proposed layered signaling scheme.

REMARK 9.10 Discrete signaling, e.g., the quadrature amplitude modulation (QAM) used in [21], is a special case of the proposed layered scheme in the complex field. More specifically, the transmission of M-pulse amplitude modulation (PAM) signaling with $M = 2^n$, $n \in \mathbb{N}$, is equivalent to the proposed scheme with only the first n layers being transmitted, while layers with indices larger than n are discarded. For example, if we set $n = 2$ in the expansion of X and discard U_2, we obtain $X = \{-3/4, -1/4, 1/4, 3/4\}$ with equal probability, which is the same as a 4-PAM. Moreover, QAM constellations can be straightforwardly attained by extending the domain of X from real to complex.

REMARK 9.11 *For additive exponentially distributed noise, which can happen for cases of non-coherent reception due to fast varying phase noise [24], we can additionally expand the noise into a binary expression, where each digit is an independent Bernoulli random variable with probability as a function of the level of that digit [25]. Then the polar coding for binary symmetric channels [26] can simply be plugged into each layer to implement the system practically.*

9.6 Numerical Results

In this section we consider several examples to show the performance of the proposed upper and lower bounds. As a comparison baseline, we use $X \sim N(0,1)$, with ergodic secrecy rate $R_s^G = \frac{1}{2}(E_{H_r}[\log(1+H_r)] - E_{H_e}[\log(1+H_e)])^+$. To avoid rate loss due to further bounding (9.16), we resort to numerically computing the differential entropies in (9.16) based on the known PDFs of X, V_{n-1}, $V_{r,n}^{(h)}$, $V_{e,n}^{(h)}$, $Z_r^{(h)}$, and $Z_e^{(h)}$. The calculated results are listed in Table 9.1. For simplicity, the number of layers considered is up to three. Therefore, the numerically computed achievable rate is a lower bound of the proposed ergodic secrecy rate. In the first example we assume that both Bob's and Eve's channels are Nakagami-m distributed with parameters $m_r = 10$, $w_r = 1$ and $m_e = 1$, $w_e = 1.2$, respectively. It is clear that this wiretap channel does not follow the usual stochastic order because there exists a crossing point between the CCDFs of the two channels. However, we can use Theorem 9.4 to get an upper bound of the ergodic secrecy capacity by only transmitting in the intervals that Bob's channel has a larger CCDF than Eve's channel, which is 0.095 bits/channel use (bpcu). Using numerical integration, the proposed scheme is bounded below by 0.0318 bpcu. In contrast, the Gaussian signaling results in $R_s^G = 0.0057$ bpcu, which is much lower than that of the proposed achievable scheme. We also consider the case with $m_r = 10$, $w_r = 1$, $m_e = 1$, and $w_e = 1.4$. Note that in this case $R_s^G = 0$ and our proposed upper and lower bounds are 0.0786 and 0.0131 bpcu, respectively. Since the regime with non-zero secrecy rate is enlarged, the connectivity of the network can be significantly improved by the proposed scheme. In another example, we set $m_r = 1$, $w_r = 1$, $m_e = 0.2$, and $w_e = 1$. It can be observed that the proposed upper and lower bounds are 0.1625 and 0.1292 bpcu, respectively, where $R_s^G = 0.1247$. In this case, even if the achievable ergodic secrecy rate of the proposed scheme is only slightly larger than Gaussian signaling, it is still closer to the upper bound.

To further show the advantage of using the layered signaling, we use the following example. From the channels with $m_r = 5$, $w_r = 1$ and $m_e = 1$, $w_e = 1.4$, we can observe that the second layer wiretap channel contributes negative values to the ergodic secrecy rate. Thus, by discarding the second layer we can obtain a higher achievable rate. Another example is $m_r = 10$, $w_r = 1$, $m_e = 1$, and $w_e = 1.6$, where only the third layer provides a positive ergodic secrecy rate. Therefore, using layered signaling, we can partially achieve the same effect as the case with full CSIT even if here we only have the statistical CSIT.

Table 9.1 Parameters and numerical results of layered signaling.

(m_r, w_r)	$E_{H_r}\left[h(X+Z_r^{(h)})\right]$	$E_{H_r}\left[h(V_1'^{(h)})\right]$	$E_{H_r}\left[h(V_2'^{(h)})\right]$	$E_{H_r}\left[h(V_3'^{(h)})\right]$
(10, 1)	1.764	1.4317	1.6905	1.7451
(5, 1)	1.7746	1.4485	1.7036	1.7575
(1, 1)	1.8953	1.6034	1.8307	1.8796
(m_e, w_e)	$E_{H_e}\left[h(X+Z_e^{(h)})\right]$	$E_{H_e}\left[h(V_1'^{(h)})\right]$	$E_{H_e}\left[h(V_2'^{(h)})\right]$	$E_{H_e}\left[h(V_3'^{(h)})\right]$
(1, 1.2)	1.8581	1.5496	1.79	1.8418
(1, 1.4)	1.8278	1.5047	1.7563	1.8108
(0.2, 1)	3.8633	3.6738	3.8201	3.853
(1, 1.6)	1.8023	1.4664	1.7278	1.7846

9.7 Multiple-Antenna Fading Wiretap Channel with Statistical CSIT

In this section we characterize the ergodic secrecy capacity of the fast fading wiretap channel with multiple antennas at Alice and single or multiple antennas at both Bob and Eve. We still assume that there is only statistical CSIT of both channels at Alice. First, we characterize the relation between the existence of the equivalent degraded wiretap channel and the usual multivariate stochastic order. Wiretap channels with several commonly considered practical channel distributions are investigated and the conditions to attain the degradedness are derived. Then we partially characterize cases in which the channel enhancement technique [9], which is originally proposed for the case with perfect CSIT, can be applied even if there is only statistical CSIT.

9.7.1 Multiple Antennas without Channel Enhancement

In this section we consider the case with a multiple-input single-output single-antenna eavesdropper (MISOSE). The received signals at Bob and Eve can be respectively represented as

$$Y_b = \boldsymbol{H_r}^H \boldsymbol{X} + Z_r, \tag{9.19}$$

$$Y_e = \boldsymbol{H_e}^H \boldsymbol{X} + Z_e, \tag{9.20}$$

where $\boldsymbol{H_r} \in \mathbb{C}^{n_T \times 1}$ and $\boldsymbol{H_e} \in \mathbb{C}^{n_T \times 1}$; Z_r and Z_e are independent circularly symmetric complex AWGN with zero mean and unit variance. In the following we show the condition that implies that the fast fading MISOSE channel is a degraded one. Note that contrary to the full CSIT case, where channel enhancement [9] can construct an equivalent degraded wiretap channel, the way to construct an equivalent degraded wiretap channel for statistical CSIT is unknown. Thus the following characterization of degradedness focuses on the intrinsic property of the channels without an additional transformation like the channel enhancement [9].

LEMMA 9.8 *Assume the density function of Eve's channel is f_{H_e}. Then only if the density of Bob's channel satisfies $f_{H_r}(h_r) = |a|^{-2n_T} f_{H_e}\left(\frac{h_r}{a}\right)$ with $a \in \mathbb{C}, |a| \geq 1$, the MISOSE wiretap channel is degraded.*

The detailed proof can be found in [27]. The proof sketch is as follows. From Definition 9.1, we need to find $f_{Y_e|Y_b'}(y_e|y_b)$ such that $X \to Y_b' \to Y_e$ forms a Markov chain. However, the input of the stochastic operation described by $f_{Y_e|Y_b'}(y_e|y_b)$ is a scalar, which hinders us from manipulating the individual entries of X such that given Y_b', X and Y_e are independent. Thus the only possibility is that after applying the technique of coupling there exists a realization H_r of the equivalent Bob's channel H_r which is aH_e with the scaling factor $|a| \geq 1$. Then the resulting density of H_r can be derived from the following fact: if Z is a complex random vector with density p_Z and if A is a complex non-singular matrix, then $W = AZ$ is a complex random vector with density $p_W(w) = |\det(A)|^{-2} p_Z(A^{-1}w)$.

REMARK 9.12 *For the full CSIT case, we can prove that only $H_r = aH_e$ with $|a| \geq 1$ results in degradedness as follows. Under full CSIT, $f_{Y_b|X}(y_b|x)$ and $f_{Y_e|X}(y_e|x)$ are Gaussian. From [28, Corollary 3.5.5], which states if X and Y are independent n-dimensional random vectors and $X + Y$ is Gaussian, then X and Y are Gaussian as well, we know that the operation $f_{Y_e|Y_b'}(y_e|y_b)$ just introduces a Gaussian noise such that given X the variance of Y_b is the same as that of Y_e plus the variance of the additional noise. Then, by the same argument used in the proof of Lemma 9.8, we know that $H_r = aH_e$ with $|a| \geq 1$ is the only degraded case.*

REMARK 9.13 *Given the distributions of H_r and H_e, assume Lemma 9.8 is fulfilled. For cases in which the optimal input covariance matrix Σ_x is unit rank, i.e., $\Sigma_x^* = P_T u^* (u^*)^H$, where $u^* \in \mathbb{C}^{n_T}$ and $\operatorname{tr}(u^*(u^*)^H) \leq 1$, is*

$$u^* = \arg \max_{u:\operatorname{tr}(u^*(u^*)^H) \leq 1} E[\log(1 + |H_r^H u|^2 P_T) - \log(1 + |H_e^H u|^2 P_T)]. \quad (9.21)$$

One such example is shown in [6], in which $H_r \sim CN(0, \Sigma_r)$, $H_e \sim CN(0, \Sigma_e)$. It is proved that at high SNR, unit rank Σ_x is optimal. With this fact, we can degenerate the MISOSE channel to a SISOSE one. More specifically, the received signals at Bob and Eve can be represented by

$$Y_b = H_r^H u^* V + Z_r \triangleq AV + Z_r, \quad (9.22)$$

$$Y_e = H_e^H u^* V + Z_e \triangleq BV + Z_e, \quad (9.23)$$

where $V \sim CN(0, P_T)$, and A and B are independent random variables. The equivalent SISOSE wiretap channel described by (9.22) and (9.23) is degraded because of the assumption that Lemma 9.8 is fulfilled. Then the secrecy capacity can be expressed as the RHS of (9.21). Since we assume full CSIR and both Bob and Eve know u, then A and B are known to Bob and Eve, respectively. Then we can compensate the phase rotation of the channels A and B at Bob and Eve, respectively, to form another equivalent wiretap channel which is composed of real Bob's and Eve's channels. And therefore $|A|^2 \geq_{\text{st}} |B|^2$ is a sufficient condition that (A,B) forms a degraded wiretap channel.

REMARK 9.14 *If $H_r \geq_{\text{st}} H_e$ but Lemma 9.8 is not satisfied, we can still use transmitter beamforming to form a degraded SISOSE wiretap channel. Note that the ergodic secrecy capacity of this new channel is smaller than that of the original one. More specifically, by [11, Section 6.B.5], $H_r \geq_{\text{st}} H_e$ implies $\Sigma_{i=1}^{n_T} u_i H_{r,i} \geq_{\text{st}} \Sigma_{i=1}^{n_T} u_i H_{e,i}$, whenever $u_i \geq 0$, $i = 1, 2, \ldots, n_T$. Note that when Lemma 9.8 is not fulfilled, this transformation may no longer be optimal in general. Note also that the optimal u should be solved in order to maximize the ergodic secrecy rate, which can be done by plugging in the constraints $u_i \geq 0$, $i = 1, 2, \ldots, n_T$ to (9.21).*

Based on the observation from MISOSE cases, we aim to extend the description of this relation to cases in which Bob and Eve have multiple antennas, i.e., the case with a multiple-input multiple-output multiple-antenna eavesdropper (MIMOME). We assume that all nodes are equipped with the same number of antennas n_T. The received signals at Bob and Eve can then be respectively expressed as

$$Y_b = H_r X + Z_r, \tag{9.24}$$

$$Y_e = H_e X + Z_e, \tag{9.25}$$

where $Z_r \sim \text{CN}(0, I_{n_T})$ and $Z_e \sim \text{CN}(0, I_{n_T})$, and H_r and $H_e \in \mathbb{C}^{n_T \times n_T}$ with entries varying for each code symbol. For the MIMOME case, we apply the vector version of Proposition 9.1 to the eigenvalues of $H_r^{-1} H_r^{-H}$ and $H_e^{-1} H_e^{-H}$, which are real. Therefore there is no such issue of using real expressions as in MISOSE.

Assume the random matrix is composed of i.i.d. entries with continuous distribution and finite variance. Then such a random matrix is invertible with probability one. In particular, it is valid in the special case when the entries are real Gaussian. On the other hand, we can also construct an alternative channel with full rank which does not change the capacity [9]. Thus we may assume the channel matrices are invertible when n_T is large enough but not infinity. In addition, with the assumption of full channel state information at receiver (CSIR), we can normalize (9.24) and (9.25) equivalently as

$$Y_b' = X + Z_r', \tag{9.26}$$

$$Y_e' = X + Z_e', \tag{9.27}$$

where $Z_r' \sim \text{CN}(0, A)$ and $Z_e' \sim \text{CN}(0, B)$, $A \triangleq H_r^{-1} H_r^{-H}$, $B \triangleq H_e^{-1} H_e^{-H}$. For the full CSIT and full CSIR cases, to make the Markov chain $X \to Y_b' \to Y_e'$ valid, i.e., it is a (stochastically) degraded wiretap channel, the constraint $B - A \succcurlyeq 0$ is sufficient.[2] In the considered scenario we have full CSIR but only statistical CSIT, so we aim to construct an equivalent degraded channel by showing $P(B - A \succcurlyeq 0) = 1$ according to the coupling. In the following we aim to find the relation of the degradedness and the

---

[2] The reason that it is not necessary is that we may be able to use the channel enhancement scheme to obtain a degraded channel with $B \not\succcurlyeq A$.

stochastic order among the eigenvalues of A and B. Note that in [11, Theorem 6.B.1] the usual stochastic order in the vector (but not matrix) version is considered, where in the expression of $\text{vec}(B) \leq_{\text{st}} \text{vec}(A)$, the inequality is element-wise, i.e., $b'_i \leq a'_i \,\forall i$, for $P(\text{vec}(B') \leq \text{vec}(A')) = 1$. However, we cannot directly apply the multivariate usual stochastic order to our scenario since it does not guarantee the positive definiteness of $B - A$. Instead, it is sufficient to check the stochastic order of the eigenvalues of $A - B$, i.e., $\Lambda_B \geq_{\text{st}} \Lambda_A$, to attain the existence of A' and B' such that $P(B' - A' \succcurlyeq 0) = 1$ after using coupling.

We first transform $B' - A' \succcurlyeq 0$ into a form such that we can simply connect it to the eigenvalues by the following lemmas.

LEMMA 9.9 *[29, 10.50(b)] Let both $Y \succ 0$ and $X \succcurlyeq 0$ be Hermitian. Then $Y - X \succcurlyeq 0$ if and only if the eigenvalues of XY^{-1} all satisfy $\lambda_i \leq 1$.*

We then use the following lemma to connect the eigenvalues of XY^{-1} to those of X and Y.

LEMMA 9.10 *[30, Theorem 9H.1] If X and Y are $n \times n$ positive semidefinite Hermitian matrices, then*

$$\lambda_{\max}(XY) \leq \lambda_{\max}(X)\lambda_{\max}(Y). \qquad (9.28)$$

Then, from Lemma 9.9 and Lemma 9.10 we can derive the following theorem, which stochastically compares the minimum and maximum eigenvalues of the covariance matrices of Bob's and Eve's channels.

THEOREM 9.6 *A sufficient condition to have a degraded MIMOME wiretap channel is*

$$\lambda_{\min}(H_r H_r^H) \geq_{\text{st}} \lambda_{\max}(H_e H_e^H). \qquad (9.29)$$

The detailed proof can be found in [27]. The proof is sketched as follows. From Lemma 9.10 we know that $\lambda_{\max}(AB^{-1}) \leq \lambda_{\max}(A)\lambda_{\max}(B^{-1})$. If we enforce the upper bound of $\lambda_{\max}(AB^{-1})$ to be less than 1, then from Lemma 9.9 we know that $B - A \succcurlyeq 0$ is valid. After some manipulations we can get the sufficient condition of the degraded MIMO wiretap channel.

REMARK 9.15 *To obtain a degraded wiretap channel, (9.29) is a sufficient condition. The reasons are: (1) $A \preccurlyeq B$ may not be necessary for the existence of a degraded wiretap channel. More specifically, for the full CSIT case, for arbitrary covariance matrices A and B, [9] proves that such a channel can be transformed into a degraded one by the channel enhancement technique; (2) the usual stochastic order is sufficient but not necessary, which can be seen from the SISOSE case in Lemma 9.3.*

Example 9.6 Consider a $2 \times 2 \times 2$ Gaussian wiretap channel. A Rician fading channel with factor K can be described by

$$H = \frac{\sqrt{K}}{\sqrt{1+K}}\bar{H} + \frac{1}{\sqrt{1+K}}H_w, \qquad (9.30)$$

with mean $\sqrt{K}/\sqrt{1+K}\bar{H}$ and covariance matrix $I_N/(1+K)$, where H_w is a 2×2 random matrix with i.i.d. $\text{CN}(0,1)$ entries. From [31, Theorems 1 and 2,] we know that the CDFs of the minimum and maximum eigenvalues of a non-central Wishart matrix HH^H are respectively

$$F_{\min}(x) = 1 - \frac{\det(\Psi(x))}{\det(\Psi(0))}, \ F_{\max}(x) = \frac{\det(\Xi(x))}{\det(\Psi(0))}, \quad (9.31)$$

where

$$\{\Psi(x)\}_{i,j} = \begin{cases} 2^{(2i-s-t)/2} Q_{s+t-2i+1,t-s}(\sqrt{2\lambda_j}, \sqrt{2x}), \ j=1,\ldots,L \\ \Gamma(t+s-i-j+1, x), \ j=L+1,\ldots,s, \end{cases} \quad (9.32)$$

$$\{\Xi(x)\}_{i,j} = \begin{cases} 2^{(2i-s-t)/2} \big[Q_{s+t-2i+1,t-s}(\sqrt{2\lambda_j}, 0) - \\ \qquad Q_{s+t-2i+1,t-s}(\sqrt{2\lambda_j}, \sqrt{2x}) \big], \ j=1,\ldots,L \\ \gamma(t+s-i-j+1, x), \ j=L+1,\ldots,s, \end{cases}$$

where L is the number of non-zero eigenvalues $\{\lambda_j\}$ of $K\bar{H}\bar{H}^H$, $s = \min(n_T, n_R)$, $t = \max(n_T, n_R)$, n_T and n_R are the numbers of transmit and receive antennas, respectively, $Q_{p,q}(a,b) = \int_b^\infty x^m \exp\left(-\frac{x^2+a^2}{x}\right) I_n(ax) dx$ is the Nuttall Q-function [32, Eq. (86)], where $0 < a, b < \infty$, and m, n are non-negative integers. Γ and γ are the upper and lower incomplete gamma functions, respectively. We consider two cases of $\{\lambda_j\}$ in the following. We first set $(\lambda_1, \lambda_2) = (2.1K, 1.9K)$, where K is the Rician K-factor. From Fig. 9.7 we can find that if we set the K-factor of Bob's channel to 15, then that value of Eve's channel should not be larger than 6.9 to satisfy (9.29). Note that the thick normal and dashed curves in Fig. 9.7 are the CDFs of the minimum and maximum eigenvalues of the Wishart matrix formed by Eve's channel matrix, respectively. We then show the ergodic secrecy capacity versus different K-factors in Fig. 9.8 for cases with Bob's K-factor greater than or equal to 15 such that (9.29) is satisfied under different transmit power P_T's. Since there is no analytical solution of the input covariance matrix for such a channel, we exhaustively search for the 2×2 optimal input covariance matrix. The number of random channel realizations is 10^4. From this figure we can find that the ergodic secrecy capacity increases with increasing Bob's K-factor, which means under fixed Eve's channel, the reduction of the uncertainty of Bob's CSI at Alice indeed improves the ergodic secrecy capacity. In addition, we can also observe that due to imperfect CSIT, the efficiency to increase the ergodic secrecy capacity by increasing the transmit power is low. Moreover, we can see that the ergodic secrecy capacity of $P_T = 500$ is quite close to that with infinite P_T. Note that when Bob's K-factor approaches infinity, Alice knows Bob's CSI perfectly. In Fig. 9.9 we show another result with the same setting as Fig. 9.8, except that here $(\lambda_1, \lambda_2) = (3K, K)$, to illustrate the effect resulting from one of the directions of the channel mean being much stronger than the other. From Fig. 9.8 we can see that under the trace constraint $\text{tr}(\bar{H}\bar{H}^H) = 4$, when the eigenvalue spread is larger, Bob's and Eve's channels will have stronger paths in the same direction, which results in lower secrecy capacity.

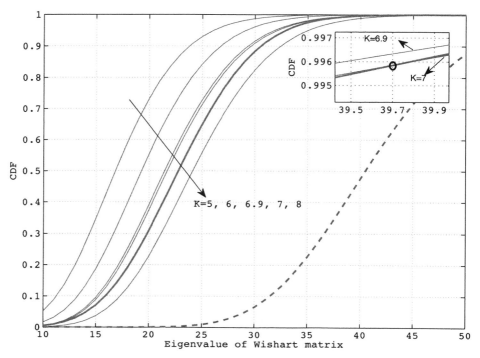

Figure 9.7 The relation of (9.29) and different K-factors of Eve's channel when that of Bob's channel is set to 15. The thick solid and dashed curves denote the CDFs of the minimum and maximum eigenvalues of the Wishart matrix formed by Eve's channel matrix, respectively.

REMARK 9.16 Note that when the fast fading wiretap channel with only statistical CSIT is verified as a degraded one, i.e., there exists A' and B' such that $A' \succcurlyeq B'$ for each channel realization, where A' and B' are the covariance matrices of the equivalent noise at Bob and Eve, by [38] we know that solving the optimal covariance matrices of the channel input and the worst case noise is a min-max problem. For full CSIT cases we can use convex optimization tools to solve it numerically or some partial analytical results can be seen in [33, Theorem 2], [34, 35], etc.

In the following we show another sufficient condition to have a degraded channel.

THEOREM 9.7 Let $U_r D_r V_r^H$ and $U_e D_e V_e^H$ be the singular value decompositions of H_r and H_e, respectively. Assume V_r is independent of D_r and U_r, and V_e is independent of D_e and U_e. Also assume that V_r^H and V_e^H have the same distribution. If $D_r \geq_{st} D_e$, then there exists an equivalent degraded wiretap channel.

REMARK 9.17 An example of V_r being independent of D_r and U_r as required in the proof of Theorem 9.7 is as follows: if the channel matrix has i.i.d. Gaussian entries, by LQ decomposition (LQD) we can get the right singular vector V_r which is the Q of the LQD and is independent of the L [36, Theorem 2.3.18], and the random matrix Q of the

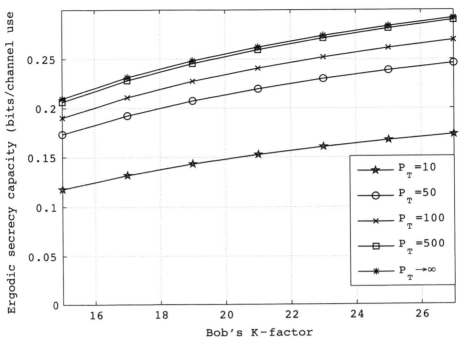

Figure 9.8 The ergodic secrecy capacities versus K-factors of Bob's channel. The K-factor of Eve's channel is set as 6.9 under different transmit power (linear) with $(\lambda_1, \lambda_2) = (2.1K, 1.9K)$.

LQD follows the isotropic distribution (i.d.) with PDF [37]

$$f(Q) = \frac{\Pi_{k=1}^{n_T}\Gamma(k)}{\pi^{\frac{n_T(n_T+1)}{2}}}\delta(Q^H Q - I_{n_T}), \quad (9.33)$$

where Γ is the gamma function and δ is the Dirac-delta function. Therefore, if H_r and H_e are random matrices with i.i.d. Gaussian entries, after LQD we can get V_r and V_e which are both i.d. having the same distribution. Then the requirements of Theorem 9.7 are automatically satisfied.

In the following, we consider another condition on the structure of the random matrix. We prove that if the channels can be decomposed into i.d. unitary matrices, the MIMOME wiretap channel is equivalent to a degraded one.

THEOREM 9.8 *Let $H_r = \Sigma_r^{1/2} H_1$, $H_e = \Sigma_e^{1/2} H_2$. If H_1 and H_2 are i.d. and $\Sigma_r \succeq \Sigma_e \succ 0$, then it is equivalent to a degraded wiretap channel.*

REMARK 9.18 *For Theorems 9.6, 9.7, and 9.8, we transform the original channels to an equivalent one, such that for all code symbols (channel realizations) the channels are degraded.*

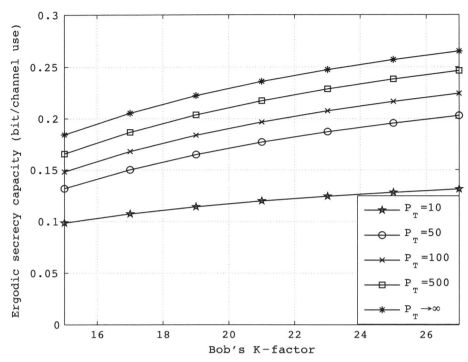

Figure 9.9 The ergodic secrecy capacities versus K-factors of Bob's channel when that of Eve's channel is set as 6.9 under different transmit power (linear) with $(\lambda_1, \lambda_2) = (3K, 1K)$.

REMARK 9.19 *The constraint $\Sigma_r \succeq \Sigma_e$ in Theorem 9.8 may be relaxed to $\Sigma_r \not\succeq \Sigma_e$ by deterministic channel enhancement. Examples will be shown in the next subsection.*

9.7.2 Multiple Antennas with Channel Enhancement

In this subsection we discuss how to apply the channel enhancement argument [9], which was originally designed for channels with full CSIT, to the model considered where the transmitter only has statistical CSIT and the channels are fast fading. For fading channels which are not isotropically distributed, we use the following examples to show that it is still possible to use channel enhancement to attain the secrecy capacity.

Example 9.7 Assume for a fast fading wiretap channel that $\boldsymbol{H}_r = \boldsymbol{\Sigma}_r^{1/2} \boldsymbol{H}$ and $\boldsymbol{H}_e = \boldsymbol{\Sigma}_e^{1/2} \boldsymbol{H}$, where $\boldsymbol{H}$ has two different realizations, i.e., $\boldsymbol{H} \in \{\boldsymbol{0}, \boldsymbol{H}_0\}$ and $P(\boldsymbol{H} = \boldsymbol{0}) = p$, $P(\boldsymbol{H} = \boldsymbol{H}_0) = 1 - p$. It is obvious that the considered $\boldsymbol{H}$ is not isotropically distributed. When $\boldsymbol{H} = \boldsymbol{0}$, it is clear that no matter how we select the channel input, the secrecy capacity is zero. Thus we can enhance the wiretap channel according to $\boldsymbol{H} = \boldsymbol{H}_0$. Then, following the standard procedure of the channel enhancement scheme, we can get the ergodic secrecy capacity as

$$C_s = \max_{\Sigma_X : \operatorname{tr}(\Sigma_X) \leq P_T} (1-p) \big(\log |I_{n_T} + H^H \Sigma_r H \Sigma_X| - \log |I_{n_T} + H^H \Sigma_e H \Sigma_X| \big)^+. \quad (9.34)$$

In the following two examples we illustrate cases in which the number of realizations of the random channels are not restricted to two.

Example 9.8 For the wiretap channel

$$Y_b = H\Sigma_r^{1/2} X + Z_r, \quad (9.35)$$

$$Y_e = H\Sigma_e^{1/2} X + Z_e, \quad (9.36)$$

assume that the fading channel H has realizations $\{H_0, \{AH_0\} : A \in U(n_r^2)\}$, where $U(n)$ is a unitary group with degree n. The distribution of H can be more general than that in Example 9.7. Then it can be easily seen that we can apply the channel enhancement to the pair of channel realizations $(H_0 \Sigma_r^{1/2}, H_0 \Sigma_e^{1/2})$ to achieve the secrecy capacity. A simple way to see it is that the receivers know A and multiplying Y_b and Y_e by unitary A does not change the capacity.

Example 9.9 For the wiretap channel

$$Y_b = \Sigma_r^{1/2} HX + Z_r, \quad (9.37)$$

$$Y_e = \Sigma_e^{1/2} HX + Z_e, \quad (9.38)$$

assume that the fading channel H has realizations $\{H_0, \{H_0 B\} : B \in U(n_r^2)\}$, where $U(n)$ is a unitary group with degree n. The distribution of H can be more general than that in Example 9.7. Then it can be seen that we can apply the channel enhancement to the pair of channel realizations $(\Sigma_r^{1/2} H_0, \Sigma_e^{1/2} H_0)$ to achieve the secrecy capacity. It can also be seen that B can be absorbed into X when the optimal distribution of X is Gaussian, which is in fact the case when the channel enhancement is considered.

9.8 Conclusion

In this chapter we have investigated some important issues about fast fading wiretap channels with only statistical CSIT. For example, when can we achieve the secrecy capacity, and when not? What is a better signaling scheme than the Gaussian one? We have partially answered the above questions by characterizing the relation between the stochastic orders among Bob's and Eve's channels and the degradedness and the positive ergodic secrecy capacity of a fast fading wiretap channel with only statistical CSI of both channels at Alice. For more general orders other than the usual stochastic order, we derived a capacity lower bound by layered signaling and upper bounds. By numerical results we illustrated that the proposed achievable scheme outperforms a Gaussian codebook under Nakagami-m channels. We extended the above discussion to cases in which the transmitter has multiple antennas and the legitimate receiver and the eavesdropper both have single or multiple antennas. More specifically, based on the technique of coupling we derived criteria to check the degradedness of several practical wiretap channels. We also partially characterized cases in which the channel enhancement technique can be applied even if there is only statistical CSIT. One example of a Rician MIMOME $2 \times 2 \times 2$ channel was illustrated where the secrecy capacity can be derived under different K-factors. With these results, we can provide confidential data transmission for certain relevant scenarios with partial CSIT.

References

[1] I. Csiszár and J. Körner, "Broadcast channels with confidential messages," *IEEE Trans. Inf. Theory*, vol. 24, no. 3, pp. 339–348, May 1978.
[2] S.-C. Lin and P.-H. Lin, "On ergodic secrecy capacity of multiple input wiretap channel with statistical CSIT," *IEEE Trans. Inf. Forensics Security*, vol. 8, no. 2, pp. 414–419, Feb. 2013.
[3] S.-C. Lin and C.-L. Lin, "On secrecy capacity of fast fading MIMOME wiretap channels with statistical CSIT," *IEEE Trans. Wireless Commun.*, vol. 13, no. 6, pp. 3293–3306, Jun. 2014.
[4] P. Mukherjee and S. Ulukus, "Fading wiretap channel with no CSI anywhere," in *Proc. IEEE Int. Symp. Inf. Theory*, Istanbul, Turkey, Jul. 2013, pp. 1347–1351.
[5] S. Goel and R. Negi, "Guaranteeing secrecy using artificial noise," *IEEE Trans. Wireless Commun.*, vol. 7, no. 6, pp. 2180–2189, Jun. 2008.
[6] J. Li and A. Petropulu, "On ergodic secrecy rate for Gaussian MISO wiretap channels," *IEEE Trans. Wireless Commun.*, vol. 10, no. 4, pp. 1176–1187, Apr. 2011.
[7] M. R. Bloch and J. N. Laneman, "Exploiting partial channel state information for secrecy over wireless channels," *IEEE J. Sel. Areas Commun.*, vol. 31, no. 9, pp. 1840–1849, Sep. 2013.
[8] P.-H. Lin, S.-H. Lai, S.-C. Lin, and H.-J. Su, "On Secrecy Rate of the Generalized Artificial-Noise Assisted Secure Beamforming for Wiretap Channels," *IEEE J. Sel. Areas Commun.*, vol. 31, no. 9, pp. 1728–1740, Sep. 2013.
[9] T. Liu and S. Shamai (Shitz), "A note on the secrecy capacity of the multiple-antenna wiretap channel," *IEEE Trans. Inf. Theory*, vol. 55, no. 6, pp. 2547–2553, Jun. 2009.
[10] Y. Liang, H. V. Poor, and S. Shamai (Shitz), "Information theoretic security," *Found. Trends Commun. Inf. Theory*, vol. 5, no. 4–5, pp. 355–580, 2009.
[11] M. Shaked and J. G. Shanthikumar, *Stochastic Orders*. New York: Springer, 2007.
[12] B. Hajek, *Notes for ECE 534: An Exploration of Random Processes for Engineers*. 2014. [Online]. Available: www.ifp.illinois.edu/ hajek/Papers/randomprocJan14.pdf
[13] S. M. Ross, *Stochastic Process*, 2nd edn. New York: John Wiley & Sons, 1996.
[14] H. Colonius, "An invitation to coupling and copulas: With applications to multisensory modeling," Nov. 2015. [Online]. Available: http://arxiv.org/abs/1511.05303
[15] G. Caire and S. Shamai (Shitz), "On the capacity of some channels with channel state information," *IEEE Trans. Inf. Theory*, vol. 45, no. 6, pp. 2007–2019, Sep. 1999.
[16] A. El Gamal and Y.-H. Kim, *Network Information Theory*. Cambridge: Cambridge University Press, 2011.
[17] J. Körner and K. Marton, "Comparison of two noisy channels," in *Colloquia Mathematica Societatis, János Bolyai, 16, Topics in Info. Th.*, North Holland, 1977, pp. 411–424.
[18] P.-H. Lin and E. A. Jorswieck, "On the fading Gaussian wiretap channel with statistical channel state information at transmitter," *IEEE Trans. Inf. Forensics Security*, vol. 11, no. 1, pp. 46–58, Jan. 2016.

[19] D. N. C. Tse and R. D. Yates, "Fading broadcast channels with state information at the receivers," *IEEE Trans. Inf. Theory*, vol. 58, no. 6, pp. 3453–3471, Jun. 2012.

[20] A. Rajan and C. Tepedelenlioglu, "Stochastic ordering of fading channels through the Shannon transform," *IEEE Trans. Inf. Theory*, vol. 61, no. 4, pp. 1619–1628, Apr. 2015.

[21] Z. Li, R. Yates, and W. Trappe, "Achieving secret communication for fast Rayleigh fading channels," *IEEE Trans. Wireless Commun.*, vol. 9, no. 9, pp. 2792–2799, Sep. 2010.

[22] A. S. Avestimehr, S. Diggavi, and D. Tse, "Wireless network information flow: A deterministic approach," *IEEE Trans. Inf. Theory*, vol. 57, no. 4, pp. 1872–1905, Apr. 2011.

[23] Y. Liang, H. V. Poor, and S. Shamai (Shitz), "Secure communication over fading channels," *IEEE Trans. Inf. Theory*, vol. 54, no. 6, pp. 2470–2492, Jun. 2008.

[24] A. Martinez, "Communication by energy modulation: The additive exponential noise channel," *IEEE Trans. Inf. Theory*, vol. 57, no. 6, pp. 3333–3351, Jun. 11.

[25] O. O. Koyluoglu, K. Appaiah, and S. Vishwanath, "Expansion coding: Achieving the capacity of an AEN channel," in *Proc. IEEE Int. Symp. Inf. Theory*, Cambridge, MA, USA, Jul. 2012, pp. 1932–1936.

[26] E. Arıkan, "Channel polarization: A method for constructing capacity-achieving codes for symmetric binary-input memoryless channels," *IEEE Trans. Inf. Theory*, vol. 55, no. 7, pp. 3051–3073, Jul. 2009.

[27] P.-H. Lin, E. A. Jorswieck, R. F. Schaefer, and M. Mittelbach, "On the degradedness of fast fading Gaussian multiple-antenna wiretap channels with statistical channel state information at the transmitter," in *Proc. IEEE Global Commun. Conf. Workshops*, San Diego, CA, USA, Dec. 2015, pp. 1–5.

[28] Z. Sasvari, *Multivariate Characteristic and Correlation Functions*. Berlin: De Gruyter Studies in Mathematics, 2013.

[29] G. A. F. Seber, *A Matrix Handbook for Statisticians*. Hoboken, NJ: Wiley Series in Probability and Statistics, 2008.

[30] A. W. Marshall and I. Olkin, *Inequalities: Theory of Majorization and its Application*, 2nd edn. London: Academic Press, 1979.

[31] S. Jin, M. R. McKay, X. Gao, and I. B. Collings, "MIMO multichannel beamforming: SER and outage using new eigenvalue distributions of complex noncentral Wishart matrices," *IEEE Trans. Commun.*, vol. 56, no. 3, pp. 424–434, Mar. 2008.

[32] A. H. Nuttall, "Some integrals involving the Q-function," Naval Underwater Systems Center, New London Lab., New London, CT, Tech. Rep. 4297, Apr. 1972.

[33] S. Loyka and C. D. Charalambous, "On optimal signaling over secure MIMO channels," in *Proc. IEEE Int. Symp. Inf. Theory*, Cambridge, MA, USA, Jul. 2012, pp. 443–447.

[34] J. Li and A. Petropulu, "Optimality of beamforming and closed form secrecy capacity of MIMO wiretap channels with two transmit antennas," in *Proc. IEEE 16th Int. Symp. Wireless Personal Multimedia Commun.*, Jun. 2013, pp. 1–5.

[35] S. A. A. Fakoorian and A. L. Swindlehurst, "Full rank solutions for the MIMO Gaussian wiretap channel with an average power constraint," *IEEE Trans. Signal Process.*, vol. 61, no. 10, pp. 2620–2631, May 2013.

[36] A. Gupta and D. Nagar, *Matrix Variate Distributions*. London: CRC Press, 2000.

[37] B. Hassibi and T. L. Marzetta, "Multiple antennas and isotropically random unitary inputs: The received signal density in closed form," *IEEE Trans. Inf. Theory*, vol. 48, no. 6, pp. 1473–1484, Jun. 2002.

[38] S. Loyka and C.D. Charalambos, "An Algorithm for Global Maximization of Secrecy Rates in Gaussian MIMO Wiretap Channels," *IEEE Trans. Commun.*, vol. 63, no.6, pp. 2288–12299, 2015.

10 The Discrete Memoryless Arbitrarily Varying Wiretap Channel

Janis Nötzel, Moritz Wiese, and Holger Boche

We describe recent results which bring together two central models of information theory: the arbitrarily varying channel (AVC) and the wiretap channel. This leads to the arbitrarily varying wiretap channel (AVWC): A sender (Alice) would like to send messages to a receiver (Bob) through a noisy channel. In addition to the usual noise, communication over this channel is subject to two difficulties. First, there is a second receiver (Eve), sometimes also called an eavesdropper. Eve obtains her own noisy version of the channel inputs and should not be able to decode any of the messages sent by the legal parties Alice and Bob. Second, the state of the channels both to the intended receiver as well as to the eavesdropper can vary arbitrarily over consecutive channel uses. Neither the sender nor the intended receiver know the true channel state, which may be thought of as being selected by a jammer (James) that tries to prevent Alice and Bob from communicating reliably. Thus, the communication system is subjected to two attacks at the same time: one active (James) and one passive (Eve). All inputs and outputs of this channel will be assumed to be taken from finite alphabets.

10.1 Introduction

10.1.1 System Model

The alphabet of the sender is $\mathcal{X}$, the receiver's alphabet is $\mathcal{Y}$. The eavesdropper receives letters taken from an alphabet $\mathcal{Z}$ and the jammer's inputs are assumed to be taken from the alphabet $\mathcal{S}$. The probabilistic law governing the transmission is described by a stochastic matrix R with entries $r(y,z|s,x)$, where the pair $(y,z) \in \mathcal{Y} \times \mathcal{Z}$ is the row index and the pair $(s,x) \in \mathcal{S} \times \mathcal{X}$ is the column index.

The channel is assumed to be memoryless. For n uses of the channel, this means that the probability of the intended receiver obtaining the output sequence $y^n = (y_1, \ldots, y_n)$ and the eavesdropper receiving $z^n = (z_1, \ldots, z_n)$ given that $x^n = (x_1, \ldots, x_n)$ and $s^n = (s_1, \ldots, s_n)$ were input to the channel is

$$\mathbb{P}(y^n, z^n | s^n, x^n) := \prod_{i=1}^{n} r(y_i, z_i | s_i, x_i). \qquad (10.1)$$

We assume that the legal receiver and the eavesdropper do not perform any joint decoding strategies, so that another notation provides more useful here: effectively, the output distributions that are seen by the two parties are given by $w(y|s,x) := \sum_{z\in\mathcal{Z}} r(y,z|s,x)$ and $v(z|s,x) := \sum_{y\in\mathcal{Y}} r(y,z|s,x)$, so that (10.1) simplifies to

$$\mathbb{P}(y^n, z^n|s^n, x^n) = \prod_{i=1}^{n} w(y_i|s_i, x_i) v(z_i|s_i, x_i). \tag{10.2}$$

Within the limited scope of this chapter we will put emphasis solely on those cases where the asymptotic behavior of the system is considered as $n \to \infty$, and moreover it is only required that the average decoding error vanishes asymptotically but without any preference concerning the speed at which this happens. At the same time it is in the legal parties' interest to make sure that the amount of information that is being conveyed to Eve approaches zero, again in the limit $n \to \infty$ and without any constraints regarding speed of convergence. Throughout, the amount of information that Eve can obtain regarding the messages is measured on the basis of non-normalized mutual information. We treat different cases regarding additional information that is shared between Alice and Bob: this information can be secret both from James and Eve, it can be secret from James but known to Eve, it can be that no additional information is shared by Alice and Bob, and, finally, it can be that no additional information is shared by Alice and Bob and, in addition, they only keep a part of the messages secret while another part is made public so that Eve can decode it as well.

Throughout, any form of backwards communication from Bob or Eve to Alice and/or James is forbidden. The system can then be schematically depicted as follows.

The AVWC treats its two predecessors, the arbitrarily varying channel and the wiretap channel, in a unified manner. Historically, the arbitrarily varying channel and the wiretap channel have evolved separately. This corresponds to decomposing Fig. 10.1 as depicted in Fig. 10.2.

The probabilistic law (10.1) implies that the action of the channel is completely described by the pair (W, V) of matrices $W = (w(y|s,x))_{y\in\mathcal{Y}, s\in\mathcal{S}, x\in\mathcal{X}}$, $V = (v(z|s,x))_{z\in\mathcal{Z}, s\in\mathcal{S}, x\in\mathcal{X}}$ of conditional probabilities. With respect to the historical development we will use a description via the pair $(\mathfrak{W}, \mathfrak{V}) = ((w(\cdot|\cdot,s))_{s\in\mathcal{S}}, (v(\cdot|\cdot,s))_{s\in\mathcal{S}})$.

In order to treat activation of capacities later on in the chapter we introduce a notion of parallel use of two channels. Let W_1 be a channel going from $\mathcal{X}_1$ to $\mathcal{Y}_1$ and W_2 be a channel going from $\mathcal{X}_2$ to $\mathcal{Y}_2$. The respective transition probability matrices are denoted

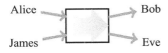

Figure 10.1 The arbitrarily varying wiretap channel.

Figure 10.2 Arbitrarily varying channel (left) and wiretap channel (right).

$(w_1(y_1|x_1))_{x_1 \in \mathcal{X}_1, y_1 \in \mathcal{Y}_1}$ and $(w_2(y_2|x_2))_{y_2 \in \mathcal{Y}_2, x_2 \in \mathcal{X}_2}$. The transition probability matrix of $W_1 \otimes W_2$ is defined by $w(y_1, y_2|x_1, x_2) := w_1(y_1|x_1) \cdot w_2(y_2|x_2)$ (for all $x_1 \in \mathcal{X}_1$, $x_2 \in \mathcal{X}_2$, $y_1 \in \mathcal{Y}_1$, and $y_2 \in \mathcal{Y}_2$). It is understood that $W_1 \otimes W_2$ describes all statistical effects arising from parallel use of the two channels W_1 and W_2. The notation then carries over to arbitrarily varying channels, where we set

$$\mathfrak{W} \otimes \mathfrak{W}' := (W_s \otimes W'_{s'})_{s \in \mathcal{S}, s' \in \mathcal{S}'}. \tag{10.3}$$

10.1.2 Historical Background

The Arbitrarily Varying Channel

This model was introduced by Blackwell, Breiman, and Thomasian [1] in 1960. The authors distinguished two cases: one where sender and receiver use only one specific coding strategy during message transmission, the other where they are allowed to pick from one of many such strategies. In both cases it is assumed that the jammer knows all the possible strategies, but is completely ignorant about the particular message being transmitted and about the particular choice of code that is utilized in the second case. We will call these the *uncorrelated*[1] case and the *shared randomness assisted* case, respectively.

The authors of [1] proved a coding theorem for the shared randomness assisted case. They also gave an explicit example of an arbitrarily varying channel that has zero capacity in the first case and non-zero capacity in the second. Thus, one outcome of their analysis was a channel model that clearly distinguished between randomized encoding and decoding as a method of proof and its physical significance for actual message transmission.

In 1978, Ahlswede [2] proved that the example provided in [1] is generic: the uncorrelated coding capacity of an arbitrarily varying channel is either zero, or it equals the shared randomness assisted capacity. In this work, Ahlswede was also concerned with the amount of shared randomness that has to be used to reach the full capacity of the system. His "Random Code Reduction Lemma" (to be found under that name e.g., in [3]) provided the insight that the distributed use of a small set of no more than n^2 randomly chosen permutations is already enough to achieve this goal. In 1997, he and Cai [4] were able to prove that another, even stronger, reduction is possible: they proved that the observation of (secret) correlated events by sender and receiver alone is already sufficient to operate the system at any rate close to capacity.

In 1985, Ericson [5] gave a necessary condition for the uncorrelated coding capacity of an arbitrarily varying channel to be zero. His analysis included some very practical aspects: Hard decision decoding of a real-valued signal under the influence of Gaussian noise and a power-limited jammer. The hard decision decoder transforms the analog

---

[1] Note that, due to the presence of an eavesdropper, it makes sense to allow the use of randomized encodings. Using, in such cases, the term "random code" is much too imprecise due to the potential presence of shared randomness between sender and receiver. Thus, we prefer to use the term "uncorrelated codes." The random choice of codewords within an uncorrelated code represents lack of knowledge both for Eve and James.

channel into a discrete binary symmetric channel where the probability of the input bit being flipped depends on the power of the (legal) sender, the power of the jammer's signal, and the variance of the assumed Gaussian noise at the receiver. The binary symmetric channel had also been considered before by Stiglitz [6] in his work on communication over non-statistically describable channels; the impact of jamming on the AWGN was later studied in [7] by Narayan and extended by Csiszár in [8].

One very specific concept that highlights the importance of shared secret randomness in a practical mobile communications deployment is described in the 1982 work by Pickholtz, Schilling, and Milstein [9]. They considered the spreading of signals over the frequency domain in a randomized fashion which is kept secret from a jammer. The approach has the very natural consequence that any jammer with only limited power available for his jamming signal has a high probability of missing out on one of the orthogonal components which can then be successfully detected at the receiver (who knows how the spreading is performed). Whether the use of shared randomness in such a system can make a difference as large as the jump from zero to (large) non-zero capacity as it is observed in the discrete arbitrarily varying channel model is not known to the authors yet.

We will now explain how the arbitrarily varying channel incorporates two very different design aspects. Namely, this model is suited very well to developing a clean understanding of the similarities and differences between stationary and non-stationary memoryless channels (as has been noted, e.g., in [6]).

On n channel uses, a discrete memoryless channel acts as

$$\mathbb{P}(y^n \text{ received}|x^n \text{ was sent}) = \prod_{i=1}^{n} w(y_i|x_i) \quad \forall\, x^n \in \mathcal{X}^n,\, y^n \in \mathcal{Y}^n, \qquad (10.4)$$

while the probabilistic law governing transmission over a non-stationary channel can take forms such as

$$\mathbb{P}(y^n \text{ received}|x^n \text{ was sent}) = \prod_{i=1}^{n} \delta(y_i, x_i \oplus s_i) \quad \forall x^n \in \mathcal{X}^n,\, y^n \in \mathcal{Y}^n, \qquad (10.5)$$

where $\mathcal{S} = \mathcal{X} = \mathcal{Y} = \{1,\ldots,L\}$ for some $L \in \mathbb{N}$, addition is modulo L, and $(s_i)_{i=1}^{\infty}$ is a sequence with elements taken from $\mathcal{S}$ that is unknown to both the sender and receiver. Of course, reliable communication over the latter channel is impossible. This fact reflects itself in the theory of arbitrarily varying channels: The shared randomness assisted capacity of the modulo L adder channel is equal to zero, let alone the capacity using uncorrelated codes.

In more elaborate models [10], some further restrictions may apply to the possible actions of the jammer. As one readily sees, (10.1) allows a smooth interpolation between the two extreme cases outlined above. It therefore allows for an adequate modeling of different types of influence on the communication line, and what remains is to deliver a quantification of the severity that is attributed to these types of noise. This is done as follows:

The communication between sender and receiver has to work, no matter what action the jammer takes. This part of the model is closer to combinatorics and models used in the theory of error correcting codes, where a limited number of errors is corrected no matter which part of the codeword is affected. The other part ($w(\cdot|s,\cdot)$ when s is fixed) is closer to channel models that are typically used in mobile communications and that may apply to noise in a receiving antenna. The quantification of error for the latter part of the system is less severe in the sense that it only delivers an estimate of the *probability of erroneous transmission*. One way of quantifying the two types of error is described next.

The Arbitrarily Varying Channel under the Average Error Criterion

Let the formula according to which the performance of the system is being evaluated when an uncorrelated code $\mathcal{K}$ is put to use be

$$\mathrm{err}(\mathcal{K}) := 1 - \min_{s^n \in \mathcal{S}^n} \frac{1}{M} \sum_{m=1}^{M} w^{\otimes n}(D_m | s^n, x_m^n), \tag{10.6}$$

where the union $\cup_{m=1}^{M} D_m \subset \mathcal{Y}^m$ is over pairwise disjoint sets, each $x_m^n \in \mathcal{X}^m$ is a codeword, and $w^{\otimes n}(D_m|s^n, x_m^n) := \sum_{y^n \in D_m} w^{\otimes n}(y^n|s^n, x_m^n)$ for all $D_m \subset \mathcal{Y}^n$ and codewords x_m^n.

Let the corresponding performance criterion when using shared randomness assisted codes be

$$\mathrm{err}(\mathcal{K}) := 1 - \min_{s^n \in \mathcal{S}^n} \frac{1}{M} \sum_{m=1}^{M} \sum_{(a,b) \in \mathbf{A} \times \mathbf{B}} \gamma(a,b) w^{\otimes n}(D_{m,a}|s^n, x_{m,b}^n), \tag{10.7}$$

where the symbols a,b denote the random choice between different encoding and decoding schemes according to a distribution γ on $\mathbf{A} \times \mathbf{B}$. When speaking about the shared randomness assisted capacity of an AVC, it is understood that the optimization is carried out not only over the encoding and decoding but also over the choice of γ. One uses the term "common randomness" when $\mathbf{A} = \mathbf{B} = \{1,\ldots,\Gamma_n\}$ and $\gamma(a,b) = 1$ if and only if $a = b$. Optimizing only over this particular subset of codes yields the common randomness assisted capacity.

The influence of shared or common randomness for the transmission over an AVC can now be sketched as follows. Let $\mathbf{A} = \mathbf{B} = S_n$, where S_n is the group of permutations on $\{1,\ldots,n\}$. Let $\gamma(\tau,\tau') = (n!)^{-1}$ if and only if $\tau = \tau'$. Asymptotically, this allows sender and receiver to transform the jammer's input into a probabilistic choice: his choice s^n gets transformed into the uniform distribution on the set $T_{N(\cdot|s^n)} := \{\hat{s}^n : N(s|\hat{s}^n) = N(s|s^n) \ \forall \ s \in \mathcal{S}\}$, where $N(s|s^n)$ is the number of times that the letter s occurs in the string s^n. After some calculation it turns out that for large numbers n this distribution is close enough to $q^{\otimes n}$, where $q(s) := \frac{1}{n} N(\cdot|s^n)$, that the *effective* channel that the sender and receiver use for transmission is the memoryless channel w_q defined by $w_q(y|x) := \sum_{s \in \mathcal{S}} q(s) w(y|s,x)$ for all $x \in \mathcal{X}$ and $y \in \mathcal{Y}$. The degrees of freedom of the jammer are reduced to a choice of probability distribution q on $\mathcal{S}$. This gives us the connection to the discrete memoryless case, but *only up to the choice of* q. It turns out that the capacity of the AVC under error criterion (10.7) is given by

$$C_r(w) = \max_{p \in \mathcal{P}(\mathcal{X})} \min_{q \in \mathcal{P}(\mathcal{S})} I(p; w_q). \tag{10.8}$$

The quantity $I(p; w_q)$ is the usual mutual information $I(X; Y)$ between the random variables (X, Y) defined by $\mathbb{P}(X = x, Y = y) := p(x)w_q(y|x)$. Equation 10.8 equals the capacity of certain compound channels $\{w_q\}_q$. We take a short look at this model below.

A variation of the arbitrarily varying channel that has been studied in [11] delivers an even closer connection to burst error correction then the model presented in this chapter. In [11], the authors considered the case where the abilities of the jammer get reduced to using only certain maximal amounts of jamming signals. Application of such a reduction to the case of the channel in Eq. (10.5) for, e.g., $L = 2$ restricts the jammer's abilities such that he can only change a fixed number k within n transmitted bits. This approach connects the AVC to (burst) error correction and coding schemes that are widely used in modern communication systems [12].

Connection to the Compound Channel

This model collects analysis of asymptotically reliable message transmission under criteria such as

$$\mathrm{err}_{\mathrm{comp}}(\mathcal{C}) := 1 - \min_{s \in \mathcal{S}} \frac{1}{M} \sum_{m=1}^{M} w_s^{\otimes n}(D_m | x_m^n), \tag{10.9}$$

where $\mathcal{S}$ may be an arbitrary set of channels. The capacity of compound channels was determined by Wolfowitz [13, 14] and independently by Blackwell, Breiman, and Thomasian [15].

The compound channel models situations where the specification of channel conditions via, e.g., pilot signals delivered an estimate of the channel parameters plus a range of possible deviations. An important distinction [13, 14] to be made is where either sender or receiver know the channel index $s \in \mathcal{S}$ exactly. The use of (noiseless) feedback from receiver to sender establishes the key to understanding the transition between the cases. In the asymptotic setting treated here, there are two possible values for the capacity: the capacity with an informed sender can be strictly larger than that where both sender and receiver are equally ignorant about the channel state.

With feedback, the sender may send pilot signals on a fraction of the channel uses. In the asymptotic setting, this leads to faithful transmission at any rate below $\min_{s \in \mathcal{S}} C(w_s)$, where $C(w_s)$ is the capacity of the memoryless channel w_s. The former capacity equals the capacity of the compound channel with state knowledge at the sender. A connection to the broadcast channel was established by Cover [16] in 1972; a more general model including memory was studied, for example, in [17].

Symmetrizability

After the shared randomness assisted capacity of the AVC was found to equal (10.8) in [1], where it was also observed that there exist examples where the shared randomness assisted capacity is larger than zero while the uncorrelated coding capacity equals zero, the question that remained open was how exactly the two quantities are related.

This question was only partly answered by Ahlswede in [2]. It was Ericson [5] who proved that symmetrizability is a sufficient condition for the uncorrelated coding capacity to equal zero. An AVC w is said to be *symmetrizable* [5] if there is a channel q with conditional probability distribution $(q(s|x))_{s \in \mathcal{S}, x \in \mathcal{X}}$ such that, for all $x, x' \in \mathcal{X}$,

$$\sum_{s \in \mathcal{S}} q(s|x) w(\cdot|s, x') = \sum_{s \in \mathcal{S}} q(s|x') w(\cdot|s, x). \tag{10.10}$$

Later, Csiszár and Narayan [11] provided the last missing link by showing that Ericson's condition was also necessary. It was also this work that offered a completely different approach to (random) code construction for the arbitrarily varying channel as compared to [2]. A more detailed comparison of conditions under which the capacity of an AVC is non-zero can also be found in [18].

A very interesting model that we must consider in order to relate major parts of our work to the existing literature is that of an AVC under the maximal error criterion and, going even further, to the discrete memoryless channel under the zero error criterion.

The Arbitrarily Varying Channel under the Maximal Error Criterion and its Connection to the Zero Error Capacity

The arbitrarily varying channel under the maximal error criterion

$$\text{err}_{\max}(\mathcal{C}) := 1 - \min_{s^n \in \mathcal{S}^n} \min_{1 \leq m \leq M} w^{\otimes n}(D_m | s^n, x_m^n) \tag{10.11}$$

was investigated by Kiefer and Wolfowitz in [19] and then by Ahlswede in [20] and [21].

While the latter work delivered a capacity formula for the case where the output alphabet is binary, [20] contains an interesting connection to the zero-error capacity of discrete memoryless channels: Let the set of all functions from $\mathcal{X}$ to $\mathcal{Y}$ be denoted $\mathcal{S}$. To every $s \in \mathcal{S}$, associate the $|\mathcal{Y}| \times |\mathcal{X}|$ matrix w_s defined by $w_s(y|x) := \delta(s(x), y)$. Then every channel v from $\mathcal{X}$ to $\mathcal{Y}$ can be written as $v = \sum_s q(s) w_s$. The zero-error capacity of the discrete memoryless channel v equals the (uncorrelated coding) capacity of the AVC defined by $w(y|x, s) := w_s(y|x)$ under the maximal error probability criterion. Note that, when the use of shared randomness is permitted, the maximal error probability criterion is always equivalent to the average error probability criterion [3]. The zero-error capacity C_0 of a discrete memoryless channel is a quantity which is strongly related to graph theory and combinatorics. Today, about six decades after its introduction by Shannon [22], it is still poorly understood – there is no closed formula available for it. An exception is the zero-error capacity with noiseless feedback, for which a closed form expression was established by Shannon in [22]. In that work, it was also shown that there exist channels for which the capacity with feedback is strictly larger than that without.

The relation between the AVC under the maximal error probability criterion and the complicated problem of finding a computable formula for the zero-error capacity highlights the complexity of the coding problem for the AVC under maximal error. Since the work of Ahlswede, there has not been much effort in that direction.

Activation

Since we will be concerned with the effect of super-activation of the secrecy capacity of arbitrarily varying wiretap channels, we have to quote some of the history of additivity questions in Shannon information theory. In [22] it was conjectured that the zero-error capacity C_0 is additive, meaning that for all channels u and v the equality $C_0(u \otimes v) = C_0(u) + C_0(v)$ is true, where $(u \otimes v)(y, y'|x, x') := u(y|x) \cdot v(y'|x')$.

The work of Haemers [23, 24] and later of Alon [25] made it clear that even strict non-additivity holds: For the product $u \otimes v$ of two channels it can happen that $C_0(u \otimes v) > C_0(u) + C_0(v)$, an effect that we will call "activation" in the following. This result disproved the conjecture from Shannon's original paper [22] that the equality $C_0(u \otimes v) = C_0(u) + C_0(v)$ would hold for all channels. Prior to that, it had been noted by Lovász in his seminal paper [26, Problem 2] that additivity of the zero-error capacity was a non-trivial question. Apart from the effect of activation, research on the zero-error capacity led to the development of the Lovász Θ function [26], which in some cases (for example the pentagon – see [26]) yields the capacity, while failing for others [27].

The effect of activation of the ability to perform certain communication tasks has also been observed in other scenarios: for secret key agreement in [28], for example, and of course the Ahlswede dichotomy itself falls (to some extent) into this category, as does conferencing between two senders that use an arbitrarily varying multiple-access channel [29]. The activation that is observed for the zero-error capacity uses only identical resources, the other three effects mix, e.g., backward and forward communication, shared randomness, and a channel, or a channel together with a conferencing protocol.

Observation of these activation effects leads to one of our core results: we are able to give equivalent conditions for the super-activation of the secrecy capacity of two arbitrarily varying channels. This effect was first observed in [30].

Prior to explaining our results we draw attention to the early developments regarding the wiretap channel itself.

The Wiretap Channel

This model has been studied widely in the literature. The initial analysis and modeling that built a bridge between cryptography and information theoretic quantities was due to Shannon in [31], where an intuitively appealing derivation of equivocation as a "theoretical secrecy index" was given. The analysis of what is nowadays called "the" wiretap channel started with the celebrated work [32] of Wyner in 1975. Instead of equivocation itself, Wyner made use of its normalized version, where normalization is with respect to the number of letters that are sent over the channel. Such normalized measures are now summarized under the term "weak" secrecy, compared to "strong" secrecy measures that are not normalized. A comparative analysis of these measures has, for example, been given in [33], as well as in [34] and later in [35]. An important follow-up of [32] was [36], by Csiszar and Körner in 1978. While Wyner considered only the degraded wiretap channel, [36] took a much more general perspective. The wiretap channel in the presence of common randomness that is kept secret from Eve (one could equally well speak of a secret key) has been studied only recently by Kang and Liu in [37]. A connection to rate distortion theory was investigated for example in [38]; the impact of side information received attention in [39].

Lately, substantial research efforts have been put into physical embodiments of the information theoretic concepts behind the wiretap channel.

From Information Theoretic Secrecy to Physical Layer Security

This topic is explained in more detail in books such as [34], [40], or [41], the review article [42], or publications such as [43, 44]. The impact on today's wireless industry is evident from, e.g., [45]. An obvious reason for this increasing importance is the increasing pervasion of our everyday lives by wireless communication. In order to understand why the focus is put now on the physical layer we quote [34]:

Given the importance of modern communication systems, secrecy should be embedded into them at all layers where this can be done in a cost-effective manner. It is common sense that secrecy typically relies on the availability of randomness, and there is no obvious reason why this randomness should not be harvested at the physical layer already, where it is available in abundance.

The collective presence of both mathematical capabilities and human needs led to a variety of publications that take into account various modeling aspects and can be located on different layers of the wireless communication medium. Starting roughly within this millennium, there has been a growing interest into such topics from the academic community. First, models were published that were not formulated based on finite alphabets and probability-preserving maps acting on them but on the discrete-time complex baseband representation of wireless communication channels [46]. For example, Goel and Negi [47, 48] studied the multiple input, multiple output (MIMO) wiretap channel, a scenario where a legal sender and receiver try to hide their information from an eavesdropper when all three parties use multi-antenna systems. This problem was independently approached by Li, Hwu, and Ratazzi [49], who considered the case of a wireless system with multiple transmit and receive antennas that is being eavesdropped on by an eavesdropper equipped with only a limited number of receive antennas. They provided a more in-depth analysis of their approach in [50]. The MIMO wiretap channel was then investigated in a lot of papers, e.g., from Khisti and Wornell [51], Shafiee, Liu, and Ulukus [52], Oggier and Hassibi [53], or Kim and Poor [54]. Among the most recent results on that topic are those found by He, Khisti, and Yener [55] and He and Yener [56]. Varying degrees of uncertainty concerning the system parameters were imposed, the main difference from the arbitrarily varying wiretap channel being the quantification of these parameters. While the model treated in this chapter only specifies the size of the eavesdropper's alphabet and any number of possible channels linking the sender to the eavesdropper, the MIMO wiretap models in the baseband are so far mostly concerned with the cases where the number of antennas at the eavesdropper is limited, an assumption that is roughly comparable to the limitation of the alphabet size in the information theoretic model, and the link to the eavesdropper can be any memoryless channel otherwise.

While it may seem at first sight that any meaningful capacity of such a system should generically be equal to zero, one important assumption of the above series of models make them a very fruitful object of study: once the illegitimate receiver has strictly fewer antennas than each of the two legitimate parties, it can only access a fraction of the signals. Upon suitable randomization at the sender, this opens the possibility of secret communication between the legal parties even when a fraction of the signals are received completely without error – similar results hold for spread spectrum communications [9].

The Arbitrarily Varying Wiretap Channel
We now come back to the main topic of this chapter. Starting from 2007 with the conference publication [57] of Liang, Kramer, Poor, and Shamai (Shitz) that was followed by their journal version [58], there has been a growing interest in models that combine state uncertainty and secrecy problems.

In the work by Boche, Bjelaković, and Sommerfeld [59], a coding theorem for the compound wiretap channel with channel state information at the transmitter was obtained. In addition to that, a multi-letter characterization of the capacity without state information at the encoder was provided. Their follow-up work [60] dealt with the arbitrarily varying wiretap channel. A single-letter lower bound on the random code secrecy capacity for the average error criterion and what we call the mean secrecy criterion here in the case of a best channel to the eavesdropper was given, together with both a single- and a multi-letter expression for an upper bound.

We additionally show that in the case of a non-symmetrizable channel to the legitimate receiver the uncorrelated coding secrecy capacity equals the random code secrecy capacity, a result similar to Ahlswede's dichotomy result for ordinary AVCs.

The lower bound on the secrecy capacity as well as other results had already been derived earlier in [61], but for a weaker secrecy criterion. Earlier treatment of the AVWC can also be found in [60, 62], which could not, however, give a complete characterization of the secrecy capacity achieved by correlated random coding. In their work [63], Wiese and Boche studied the multiple-access channel under the strong secrecy criterion.

10.1.3 New Approaches and New Results

In the following, we present and discuss results that were obtained in [64] and [65] in a unified context. In particular, the presentation of proofs is similar to that in [64, 65].

Common Randomness
To the channel model as depicted in Fig. 10.1, we add the option of Alice and Bob having access to perfect copies of the outcomes of a random experiment $\mathcal{G}$ (a source of common randomness). We first consider the case where Eve gets an exact copy of the outcomes received by Alice and Bob. Based on the results obtained for that case, we extend our study to the case where no common randomness is present and the case where Eve remains completely ignorant of the values of the common randomness.

The only party which has no access to $\mathcal{G}$ in all the scenarios we study is James. Three important capacities which we derive from the two scenarios are the "correlated random coding mean secrecy capacity" (if Eve has information about $\mathcal{G}$), the "secret common

Figure 10.3 Secure coding schemes for correlated random coding (left) and secret common randomness assisted coding (right).

randomness assisted secrecy capacity" (if Eve has no information about $\mathcal{G}$), and the "uncorrelated coding secrecy capacity" (when there is no common randomness). The situation is depicted in Fig. 10.3.

We also define a "capacity with public side information," which is the data transmission benchmark for systems where Eve gets to know some of the messages, and the "correlated random coding maximum secrecy capacity" that introduces a more severe secrecy criterion.

We use the label C_S for the uncorrelated coding secrecy capacity and $C_{S,ran}^{mean}$ for the correlated random coding mean secrecy capacity. The secret common randomness assisted secrecy capacity is labeled C_{key}, the capacity with public side information C_{pp}, and the correlated random coding maximum secrecy capacity $C_{S,ran}^{max}$.

Our treatment of the subject allows one to observe the behavior of the system under changes in the amount of and access to the common randomness: for common randomness set to zero, one observes instabilities of the system (in the sense that the capacity is not a continuous function of the channel parameters anymore) and the effect of super-activation. If common randomness is used between Alice and Bob but Eve gets to know it as well, we see that even a relatively small (logarithmic in the block length) number of bits of common randomness resolve the instabilities (in the sense that the correlated random coding capacity is continuous in the channel parameters). It remains unknown whether super-activation is possible when common randomness is present.

The full advantage from common randomness can only be gained if Eve is kept ignorant of it. In that case, the amount of common randomness that is shared between Alice and Bob needs to be quantified, as it adds to the capacity.

Relation between Capacities when Common Randomness Is Not Kept Secret

The earlier literature has come up with an elegant geometric characterization for those AVCs which can at best have a positive shared randomness assisted capacity: the *symmetrizability* condition (10.10).

We prove the following non-trivial result: If $\mathfrak{W}$ is non-symmetrizable, then $C_S(\mathfrak{W}, \mathfrak{V}) = C_{S,ran}^{mean}(\mathfrak{W}, \mathfrak{V})$ for all possible $\mathfrak{V}$. We do not attempt to give a necessary and sufficient condition for C_S to be positive, since a geometric characterization in the spirit of the symmetrizability condition (10.10) is not even known for the usual wiretap channel. Rather, when speaking about the wiretap channel one usually refers to the concept of "less noisy" channels that was developed in [66] and used in, e.g., [36].

This part of our work makes heavy use of the results that were obtained in [11] by extending their Lemma 3 to the situation where Eve gets some information via $\mathfrak{V}$.

When No Common Randomness Is Present: Super-Activation

A surprising result that was discovered only recently by Boche and Schaefer in [30] is that of super-activation of AVWCs. We will explain this example in more detail in Remark 10.5. This effect was until then only known for information transmission capacities in quantum information theory, where it was proven by Smith and Yard in [67] that there exist channels which have the property that each of them alone has zero capacity, but together, when used in parallel, have a positive capacity.

Boche and Schaefer [30] gave an explicit example of super-activation which we repeat in Remark 10.5. Based on our finer analysis, we are able to provide the following: First, we give a much clearer characterization of super-activation of C_S in Theorem 10.4. Second, our proof of Theorem 10.4 together with Example 10.1 yields additional insights regarding the coding strategies that may be employed in order to achieve rates arbitrarily close to C_S. These differ substantially from the ones that were used in [30].

Third, and more for the sake of a clean discussion of coding and secrecy concepts, we define the capacity C_{pp} which explicitly keeps some of the messages public (such that Eve is able to decode them). We do not attempt to give a further characterization of C_{pp} in this work, but we show that this capacity also shows super-activation by use of the code concepts that were developed in [30]. Details are given in Section 10.2.2, together with the exact definition of C_{pp}. In Section 10.3, we relate the super-activation of C_{pp} to that of C_S. The focus of this work will, however, be on the interplay between C_S, $C_{S,\mathrm{ran}}^{\mathrm{mean}}$, and C_{key}.

We will now clarify our notation regarding additivity, super-additivity, activation, and super-activation. The inequality

$$C_S(\mathfrak{W}_1 \otimes \mathfrak{W}_2, \mathfrak{V}_1 \otimes \mathfrak{V}_2) \geq C_S(\mathfrak{W}_1, \mathfrak{V}_1) + C_S(\mathfrak{W}_2, \mathfrak{V}_2) \qquad (10.12)$$

follows trivially from the definition of C_S. It is common to all notions of capacity known to the authors. When equality holds in (10.12) we speak of additivity. When the inequality in (10.12) is strict, we speak of super-additivity. If the inequality in (10.12) is strict and, additionally, it holds that

$$C_S(\mathfrak{W}_1, \mathfrak{V}_1) = C_S(\mathfrak{W}_2, \mathfrak{V}_2) = 0, \qquad (10.13)$$

we speak of *super-activation*. The term "activation" is reserved for cases where the inequality in (10.12) is strict and, additionally, $C_S(\mathfrak{W}_1, \mathfrak{V}_1) = 0$ and $C_S(\mathfrak{W}_2, \mathfrak{V}_2) > 0$. An example of the latter effect is the Ahlswede dichotomy: just set $\mathfrak{V}_1 = \mathfrak{V}_2 = \mathfrak{T}$, where $\mathfrak{T} = (T)$ and $t(z|s,x) = |\mathcal{Z}|^{-1}$ for all z, s, and x, and use $\mathfrak{W}_2$ for the transmission of (approximate) common randomness.

While it is clear from explicit examples that super-activation of C_S is possible, it turns out in our work via Theorem 10.7 that the effect is connected to the super-activation of $C_{S,\mathrm{ran}}^{\mathrm{mean}}$, if the latter occurs.

As a last introductory statement concerning super-additivity, let us mention the connection of super-activation to information transmission in networks. Consider two orthogonal channels $\mathfrak{W}_1, \mathfrak{W}_2$ in a mobile communication network. Both transmit data from Alice to Bob. In addition to that, let $\mathfrak{V}_1, \mathfrak{V}_2$ be channels from Alice to Eve. The surprising result then is that, while it may be completely impossible to send information securely if Alice and Bob use only $\mathfrak{W}_1$ or only $\mathfrak{W}_2$, there exist coding schemes that enable Alice to send her information securely if both she and Bob have access to both $\mathfrak{W}_1$ and $\mathfrak{W}_2$!

Continuity and Discontinuity Questions

We will now give a broad sketch of our results concerning C_S and $C_{S,\text{ran}}^{\text{mean}}$, before we start concentrating on C_{key}. It was proven in [64] that $C_{S,\text{ran}}^{\text{mean}}$ is a continuous quantity. The continuous dependence of the performance of a communication system on the relevant system parameters is of central importance. As an example, consider recent efforts to build what are called "smart grids" – such systems certainly do have high requirements concerning both reliability and stability of the communication in order to avoid potentially damaging consequences for their users.

Also, from a mathematical point of view, the question whether a capacity is continuous is interesting: there is no immediately obvious way to deduce this statement directly just from the definition of the capacity. In [64], $C_{S,\text{ran}}^{\text{mean}}$ was proven to be continuous based on proving first the validity of the following capacity formula:

$$C_{S,\text{ran}}^{\text{mean}}(\mathfrak{W}, \mathfrak{V}) = \lim_{n \to \infty} \frac{1}{n} \max_{p \in \mathcal{P}(\mathcal{U}_n)} \max_{U \in C(\mathcal{U}_n, \mathcal{X}^n)}$$
$$\left(\min_{q \in \mathcal{P}(S)} I(p; W_q^{\otimes n} \circ U) - \max_{s^n \in S^n} I(p; V_{s^n} \circ U) \right). \qquad (10.14)$$

Another result in [64] was that in the above formula one may set $\mathcal{U}_n = \mathcal{X}^n$. While one may argue that this is not an efficient description since one is forced to compute the limit of a series of convex optimization problems, it turns out to be incredibly useful. Not only can continuity be proven to hold based on (10.14), but together with another useful characterization from [64],

$$C_{S,\text{ran}}^{\text{mean}}(\mathfrak{W}, \mathfrak{V}) = \lim_{n \to \infty} \frac{1}{n} \max_{p \in \mathcal{P}(\mathcal{X}^n)} \max_{U \in C(\mathcal{U}_n, \mathcal{X}^n)}$$
$$\left(\min_{q \in \mathcal{P}(S^n)} I(p; W_q \circ U) - \max_{s^n \in S^n} I(p; V_{s^n} \circ U) \right), \qquad (10.15)$$

one gets into the position of using simple blocking arguments when proving achievability results for C_{key}. The nature of such blocking arguments can be picked up in Section 10.4.4, around Eqs. (10.148) and (10.149).

In contrast to $C_{S,\text{ran}}^{\text{mean}}$, the capacity C_S is not a continuous function of the channel. This highlights the importance of distributed resources in communication networks – in this case, the use of small amounts of common randomness. While one may now be tempted to think that the transmission of messages over AVWCs without the use of common randomness is a rather adventurous task, we are also able to prove that such a perception is wrong: Our analysis shows that C_S is continuous around its positivity points (this has been observed for classical quantum arbitrarily varying channels in [68]), and we give an exact characterization of the discontinuity points as well. An example of a discontinuity point of C_S has already been given in [69].

Moreover, our characterization of discontinuity relies purely on the computation of functions that are *continuous* themselves, so that a calculation of such points is at least within reach from a computational point of view.

The Use of Secret Common Randomness

We extend earlier research to the case where *lots* of common randomness can be used (exponentially many random bits, to be precise) during our investigation of C_{key}. If common randomness is used at a non-zero rate, this rate adds linearly to the capacity of the system. All the capacity formulas that can be proven to hold in the various non-trivial scenarios are given by multi-letter expressions. Only if the common randomness exceeds the maximal amount of information that can be leaked to Eve do we recover a single-letter description. At that point, the linear increase in capacity stops, as shown in Fig. 10.4.

We do not dive into the issues arising when sub-exponentially many random bits are available, although the repeated appearance of the activating effect of common randomness in arbitrarily varying systems seems to deserve closer study. Our method of proving the direct part again yields nothing more than the statement that any number of random bits which scales asymptotically as $\text{const.} + (1 + \epsilon) \log(n)$ (for some $\epsilon > 0$) is sufficient for evading all issues that may arise from symmetrizable $\mathfrak{W}$.

Our restriction to positive rates G of common randomness allows us to give an elegant formula for C_{key} as follows: For every $G > 0$,

$$C_{\text{key}}(\mathfrak{W}, \mathfrak{V}, G) = \min\{C_{S,\text{ran}}^{\text{mean}}(\mathfrak{W}, \mathfrak{V}) + G, C_{S,\text{ran}}^{\text{mean}}(\mathfrak{W}, \mathfrak{T})\}. \qquad (10.16)$$

Again, $\mathfrak{T}$ denotes the AVC consisting only of the memoryless "trash" channel T that maps every legal input x and jamming input s onto an arbitrary element of $\mathcal{Z}$ with equal probability ($t(z|s,x) = |\mathcal{Z}|^{-1}$). While a reader that is familiar with the topic would certainly have guessed the validity of the formula, it is worth noting that this formula is generally "hard to compute" in the sense that it requires one to calculate the limit in

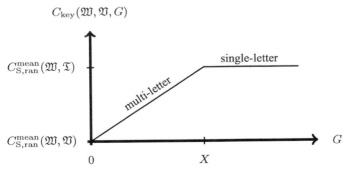

Figure 10.4 Scaling of secrecy capacity with the rate G of secret common randomness. $X = C_{S,\text{ran}}^{\text{mean}}(\mathfrak{W}, \mathfrak{T}) - C_{S,\text{ran}}^{\text{mean}}(\mathfrak{W}, \mathfrak{V})$.

the formula (10.14) – as long as $G < C_{\text{S,ran}}^{\text{mean}}(\mathfrak{W}, \mathfrak{T}) - C_{\text{S,ran}}^{\text{mean}}(\mathfrak{W}, \mathfrak{V})$. If this condition is not met, then $C_{\text{key}}(\mathfrak{W}, \mathfrak{V}) = C_{\text{S,ran}}^{\text{mean}}(\mathfrak{W}, \mathfrak{T})$. Since the latter is the usual capacity of the AVC $\mathfrak{W}$, we conclude that if enough common randomness is available, the capacity of the system can be described by a single-letter formula.

10.2 Notation and Definitions

10.2.1 Basic Notation

In the context presented in this work, every finite set is equivalently called an alphabet. Alphabets are denoted by script letters: $\mathcal{A}, \mathcal{B}, \mathcal{S}, \mathcal{X}, \mathcal{Y}, \mathcal{Z}$. The cardinality of a set $\mathcal{A}$ is denoted by $|\mathcal{A}|$. Every natural number $N \in \mathbb{N}$ defines a set $[N] := \{1, \ldots, N\}$. The set of all permutations on such $[N]$ is written S_N. The function $\exp : \mathbb{R} \to \mathbb{R}_+$ is defined with respect to base 2: $\exp(t) := 2^t$. The logarithm log is defined with respect to the same base. For any $c \in \mathbb{R}$ we define $|c|^+$ by setting $|c|^+ := c$ if $c > 0$ and $|c|^+ := 0$ otherwise. A function $f : \mathcal{A} \to \mathbb{R}$ is non-negative ($f \geq 0$) if $f(a) \geq 0$ holds for all $a \in \mathcal{A}$. With each finite set $\mathcal{A}$ we associate the corresponding set $\mathcal{P}(\mathcal{A}) := \{p : \mathcal{A} \to [0,1] : p \geq 0, \sum_{a \in \mathcal{A}} p(a) = 1\}$ of probability distributions on $\mathcal{A}$. Each random variable A with values in $\mathcal{A}$ is associated with the unique $p \in \mathcal{P}(\mathcal{A})$ satisfying $\mathbb{P}(A = a) = p(a)$ for all $a \in \mathcal{A}$. An important subset of $\mathcal{P}(\mathcal{A})$ is the set of its extreme points. Every such extreme point is a point measure $\delta_a(a') := \delta(a, a')$, where $\delta(\cdot, \cdot)$ is the usual Kronecker delta. The one-norm (or total variation) distance between two probability distributions $p, p' \in \mathcal{P}(\mathcal{A})$ is $\|p - p'\|_1 = \sum_{a \in \mathcal{A}} |p(a) - p'(a)|$.

The expectation of a function $f : \mathcal{A} \to \mathbb{R}$ with respect to a distribution $p \in \mathcal{P}(\mathcal{A})$ is written $\mathbb{E}_p f := \sum_{s \in \mathcal{A}} p(a) f(a)$ or, if p is clear from the context, simply $\mathbb{E} f$.

For each alphabet $\mathcal{A}$ and natural number $n \in \mathbb{N}$ we can build the n-fold Cartesian product $\mathcal{A}^n := \mathcal{A} \times \cdots \times \mathcal{A}$. Elements of $\mathcal{A}^n$ are denoted $a^n = (a_1, \ldots, a_n)$. Each such element gives rise to the corresponding empirical distribution or *type* $\bar{N}(\cdot|a^n) \in \mathcal{P}(\mathcal{A})$ defined via $N(a|a^n) := |\{i : a_i = a\}|$ and $\bar{N}(\cdot|a^n) := \frac{1}{n} N(\cdot|a^n)$. Given $\mathcal{A}$ and $n \in \mathbb{N}$, the set of all empirical distributions arising from an element $a^n \in \mathcal{A}^n$ is $\mathcal{P}_0^n(\mathcal{A}) := \{\bar{N}(\cdot|a^n) : a^n \in \mathcal{A}^n\}$. Each type $p \in \mathcal{P}_0^n(\mathcal{A})$ defines the *type class* $T_p := \{a^n : \bar{N}(\cdot|a^n) = p(\cdot)\}$.

Channels are given by affine maps $W : \mathcal{P}(\mathcal{A}) \to \mathcal{P}(\mathcal{B})$. The set of channels is denoted $C(\mathcal{A}, \mathcal{B})$. Every channel is uniquely represented (and can therefore be identified with) its set $\{w(b|a)\}_{a \in \mathcal{A}, b \in \mathcal{B}}$ of transition probabilities, which are defined via $w(b|a) := W(\delta_a)(b)$. It acts as

$$W(p) := \sum_{a \in \mathcal{A}} \sum_{b \in \mathcal{B}} w(b|a) p(a) \delta_b, \qquad (10.17)$$

where both $W(p) \in \mathcal{P}(\mathcal{B})$ and $\{\delta_b\}_{b \in \mathcal{B}} \mathcal{P}(\mathcal{B})$ [another way of writing the above formula would be to set $W(p)(\cdot) = \sum_{a \in \mathcal{A}} \sum_{b \in \mathcal{B}} w(b|a) p(a) \delta_b(\cdot)$]. As a shorthand, we will also write Wp to denote $W(p)$, in analogy to linear algebra (every channel is naturally associated with its representing stochastic matrix $(w(a|b))_{a,b}$, and can therefore be extended to a linear map on the appropriate vector spaces).

When operating on product alphabets such as $\mathcal{A} \times \mathcal{B}$ we define $p \otimes q \in \mathcal{P}(\mathcal{A} \times \mathcal{B})$ to be the distribution defined by $(p \otimes q)(a,b) := p(a)q(b)$. Correspondingly, $p^{\otimes n} \in \mathcal{P}(\mathcal{A}^n)$ is defined via $p^{\otimes n}(a^n) := \prod_{i=1}^n p(a_i)$. The same conventions hold for channels: if $V : \mathcal{P}(\mathcal{A}) \to \mathcal{P}(\mathcal{B})$ and $W : \mathcal{P}(\mathcal{A}') \to \mathcal{P}(\mathcal{B}')$, then $V \otimes W : \mathcal{P}(\mathcal{A} \times \mathcal{A}') \to \mathcal{P}(\mathcal{B} \times \mathcal{B}')$ is defined via its transition probabilities as $(v \otimes w)((b,b')|(a,a')) := v(b|a)w(b'|a')$, and the notation carries over to n-fold products $W^{\otimes n}$ of $W : \mathcal{P}(\mathcal{A}) \to \mathcal{P}(\mathcal{B})$ as before by setting $w^{\otimes n}(b^n|a^n) := \prod_{i=1}^n w(b_i|a_i)$.

For channels $W \in C(\mathcal{A} \times \mathcal{B}, \mathcal{C})$, it will become important to derive a short notation for cases where one input remains fixed while the other is arbitrary. Such induced channels will be denoted, in cases where this is unambiguously possible, by W_p, where

$$W_p(\delta_a) := W(\delta_a \otimes p). \tag{10.18}$$

At times, we will also use $w_p(b|a)$ or $w(b|a,p)$ for the corresponding transition probabilities.

The Shannon entropy of $p \in \mathcal{P}(\mathcal{A})$ is $H(p) := -\sum_{a \in \mathcal{A}} p(A) \log p(a)$; the relative entropy between two probability distributions $p, q \in \mathcal{P}(\mathcal{A})$ is

$$D(p\|q) := \begin{cases} \sum_{a \in \mathcal{A}} p(a) \log(p(a)/q(a)), & \text{if } q \gg p, \\ \infty, & \text{else,} \end{cases} \tag{10.19}$$

where a pair (p,q) is said to satisfy $q \gg p$ if $q(a) = 0$ implies $p(a) = 0$ for all $a \in \mathcal{A}$. Every $p \in \mathcal{P}(\mathcal{A})$ and channel $W : \mathcal{P}(\mathcal{A}) \to \mathcal{P}(\mathcal{B})$ define a pair of random variables (A,B) defined by $\mathbb{P}((A,B) = (a,b)) = p(a)w(b|a)$ (for all $a \in \mathcal{A}$, $b \in \mathcal{B}$). This enables us to use an equivalent formulation for the mutual information:

$$I(p;W) := I(A;B). \tag{10.20}$$

We define the mutual information on pairs of sequences $a^n \in \mathcal{A}^n$, $b^n \in \mathcal{B}^n$ by defining a pair of random variables (A,B) with values in $\mathcal{A} \times \mathcal{B}$ via $\mathbb{P}((A,B) = (a,b)) := \bar{N}(a,b|a^n,b^n)$ and then setting

$$I(a^n;b^n) := I(A;B). \tag{10.21}$$

In addition, a suitable measure of distance between AVWCs is required. Our object of choice is the Hausdorff distance, which we define as follows: For two channels $W, \tilde{W} \in C(\mathcal{A}, \mathcal{B})$, set

$$\|W - \tilde{W}\| := \max_{a \in \mathcal{A}} \|W(\delta_a) - \tilde{W}(\delta_a)\|. \tag{10.22}$$

Now we define, for a given $\mathfrak{W} = (W_s)_{s \in \mathcal{S}}$ and $\mathfrak{W}' = (W'_{s'})_{s' \in \mathcal{S}'}$,

$$g(\mathfrak{W}, \mathfrak{W}') := \max_{s \in \mathcal{S}} \min_{s' \in \mathcal{S}'} \|W_s - W'_{s'}\|. \tag{10.23}$$

Finally, we define, for $\mathfrak{V}, \mathfrak{V}'$ being AVCs with output alphabet $\mathcal{C}$, the quantity

$$d((\mathfrak{W}, \mathfrak{V}), (\mathfrak{W}', \mathfrak{V}')) := \max\{g(\mathfrak{W} \otimes \mathfrak{V}, \mathfrak{W}' \otimes \mathfrak{V}'), g(\mathfrak{W}' \otimes \mathfrak{V}', \mathfrak{W} \otimes \mathfrak{V})\}. \tag{10.24}$$

This is a metric on the set of finite-state AVWCs with the corresponding alphabets $\mathcal{A}$ for Alice, $\mathcal{B}$ for Bob, and $\mathcal{C}$ for Eve. The convex hull of an AVC $\mathfrak{W} = (W_s)_{s \in \mathcal{S}}$ is

$$\mathrm{conv}(\mathfrak{W}) := \left\{ W = \sum_{s \in \mathcal{S}} q(s) W_s : q \in \mathcal{P}(\mathcal{S}) \right\}. \quad (10.25)$$

Finally, for any given $W \in C(\mathcal{A}, \mathcal{B})$, $a \in \mathcal{A}$, and subset $\mathcal{B}' \subset \mathcal{B}$, we use the notation

$$w(\mathcal{B}'|a) := \sum_{b \in \mathcal{B}'} w(b|a). \quad (10.26)$$

10.2.2 Models and Operational Definitions

At first, we give a formal definition of an arbitrarily varying channel.

DEFINITION 10.1 (AVWC) Let $\mathcal{X}, \mathcal{Y}, \mathcal{Z}, \mathcal{S}$ be finite sets and for each $s \in \mathcal{S}$, let $W_s \in C(\mathcal{X}, \mathcal{Y})$ and $V_s \in C(\mathcal{X}, \mathcal{Z})$. Define $\mathfrak{W} := (W_s)_{s \in \mathcal{S}}$ and $\mathfrak{V} := (V_s)_{s \in \mathcal{S}}$. The corresponding arbitrarily varying wiretap channel is denoted $(\mathfrak{W}, \mathfrak{V})$. Its action is specified by the sequence $(\{W_{s^n}, V_{s^n}\}_{s^n \in \mathcal{S}^n})_{n \in \mathbb{N}}$, where $W_{s^n} := W_{s_1} \otimes \cdots \otimes W_{s_n}$ and $V_{s^n} := V_{s_1} \otimes \cdots \otimes V_{s_n}$.

REMARK 10.1 *For $n \in \mathbb{N}$ and $q \in \mathcal{P}(\mathcal{S}^n)$ we will use the abbreviations*

$$W_q^n := \sum_{s^n \in \mathcal{S}^n} q(s^n) W_{s^n}, \qquad V_q^n := \sum_{s^n \in \mathcal{S}^n} q(s^n) V_{s^n}. \quad (10.27)$$

The corresponding probabilities are defined for all $x^n \in \mathcal{X}^n$, $y^n \in \mathcal{Y}^n$, $z^n \in \mathcal{Z}^n$ as

$$w_q^n(y^n|x^n) := W_q^n(\delta_{x^n})(y^n), \qquad v_q^n(z^n|x^n) := V_q^n(\delta_{x^n})(z^n). \quad (10.28)$$

A central part of our work is to study AVWCs under joint use. Mathematically, this is described as follows. Let $(\mathfrak{W}_1, \mathfrak{V}_1)$ and $(\mathfrak{W}_2, \mathfrak{V}_2)$ be two AVWCs. Assume without loss of generality that they have a joint state set $\mathcal{S}$. Set

$$(\mathfrak{W}_1 \otimes \mathfrak{W}_2, \mathfrak{V}_1 \otimes \mathfrak{V}_2) := ((W_1(\cdot|\cdot,s) \otimes W_2(\cdot|\cdot,s'), V_1(\cdot|\cdot,s) \otimes V_2(\cdot|\cdot,s')))_{s,s' \in \mathcal{S}}.$$

We introduce three different classes of codes, which are related to each other as follows: The class of shared randomness assisted codes contains those which use common randomness and these again contain the uncorrelated codes. Formal definitions are as follows:

DEFINITION 10.2 (Shared randomness assisted code) A shared randomness assisted code $\mathcal{K}_n$ for the AVWC $(\mathfrak{W}, \mathfrak{V})$ consists of a set $[K]$ of messages, two finite alphabets $[\Gamma], [\Gamma']$, and a set of stochastic encoders $e^\gamma \in C([K], \mathcal{X}^n)$ (one for every value $\gamma \in [\Gamma]$) together with a collection $(\{D_k^{\gamma'}\}_{k=1}^K)_{\gamma'=1}^{\Gamma'}$ of sets satisfying $\bigcup_{k=1}^K D_k^{\gamma'} \subset \mathcal{Y}^n$ and $D_k^{\gamma'} \cap D_{k'}^{\gamma'} = \emptyset$ for all $k \neq k'$ and for all γ'. In addition to that, there is a probability distribution $\mu \in \mathcal{P}([\Gamma] \times [\Gamma'])$. Every such code defines the joint random variables $\mathfrak{S}_{s^n} := (\mathfrak{K}_n, \mathfrak{K}'_n, \Gamma_n, \Gamma'_n, \mathfrak{Z}_{s^n}, \mathfrak{X}_n, \mathfrak{Y}_{s^n})$, $s^n \in \mathcal{S}^n$, which are distributed according to

$$\mathbb{P}(\mathfrak{S}_{s^n} = (k, k', \gamma, \gamma', z^n, x^n, y^n)) = \frac{1}{K} \mathbb{1}_{D_{k'}^{\gamma'}}(y^n) w_{s^n}(y^n|x^n) v_{s^n}(z^n|x^n).$$

The average error of $\mathcal{K}_n$ is

$$\text{err}(\mathcal{K}_n) = 1 - \min_{s^n \in \mathcal{S}^n} \sum_{k,\gamma,\gamma'=1}^{K,\Gamma,\Gamma'} \frac{\mu(\gamma,\gamma')}{K} \sum_{x^n \in \mathcal{X}^n} e^\gamma(x^n|k) w_{s^n}(D_k^{\gamma'}|x^n).$$

A special and important class of shared randomness assisted codes are the *common randomness* assisted codes. These are defined as follows:

DEFINITION 10.3 (Common randomness assisted code) A common randomness assisted code $\mathcal{K}_n$ for the AVWC $(\mathfrak{W}, \mathfrak{V})$ consists of a set $[K]$ of messages, a set $[\Gamma]$ of values for the common randomness, and a set of stochastic encoders $e^\gamma \in C([K], \mathcal{X}^n)$ (one for each element $\gamma \in [\Gamma]$), together with a collection $(D_k^\gamma)_{k,\gamma=1}^{K,\Gamma}$ of subsets D_k^γ of $\mathcal{Y}^n$ satisfying $D_k^\gamma \cap D_{k'}^\gamma = \emptyset$ for all $\gamma \in [\Gamma]$, whenever $k \neq k'$. Every such code defines the joint random variables $\mathfrak{S}_{s^n} := (\mathfrak{K}_n, \mathfrak{K}'_n, \Gamma_n, \mathfrak{X}_n, \mathfrak{Y}_{s^n}, \mathfrak{Z}_{s^n})$, $s^n \in \mathcal{S}^n$, which are distributed according to

$$\mathbb{P}(\mathfrak{S}_{s^n} = (k,k',\gamma,x^n,y^n,z^n)) = \frac{1}{\Gamma \cdot K} e^\gamma(x^n|k) \mathbb{1}_{D_{k'}^\gamma}(y^n) w_{s^n}(y^n|x^n) v_{s^n}(z^n|x^n).$$

The average error of $\mathcal{K}_n$ is

$$\text{err}(\mathcal{K}_n) = 1 - \min_{s^n \in \mathcal{S}^n} \frac{1}{K \cdot \Gamma} \sum_{k,\gamma=1}^{K,\Gamma} \sum_{x^n \in \mathcal{X}^n} e^\gamma(x^n|k) w_{s^n}(D_k^\gamma|x^n).$$

For technical reasons, we also define, for all state sequences s^n, the corresponding average success probability of the code:

$$d_{s^n}(\mathcal{K}_n) = \frac{1}{K \cdot \Gamma} \sum_{k,\gamma=1}^{K,\Gamma} \sum_{x^n \in \mathcal{X}^n} e^\gamma(x^n|k) w_{s^n}(D_k^\gamma|x^n).$$

If a whole communication network is utilized for secret message transmission it may be possible to use one part of the network to establish common randomness between the legal parties (one could equally well speak of a secret key here), which is then used to send messages over another part of the system that may be symmetrizable. This idea was first established in [30]. Here, we give a more careful analysis of the underlying structure, an undertaking which motivates the following definition:

DEFINITION 10.4 (Private/public code) A private/public code $\mathcal{K}_n$ for the AVWC $(\mathfrak{W}, \mathfrak{V})$ consists of two sets $[K]$, $[L]$ of messages, an encoder $E \in C([K] \times [L], \mathcal{X}^n)$, and a collection $(D_{kl})_{k,l=1}^{K,L}$ of subsets of $\mathcal{Y}^n$ satisfying $D_{kl} \cap D_{k'l'} = \emptyset$ whenever $(k,l) \neq (k',l')$. Every such code defines the joint random variables $\mathfrak{S}_{s^n} := (\mathfrak{K}, \mathfrak{L}, \mathfrak{K}', \mathfrak{L}', \mathfrak{X}^n, \mathfrak{Y}_{s^n}, \mathfrak{Z}_{s^n})$, $s^n \in \mathcal{S}^n$, which are distributed according to

$$\mathbb{P}(\mathfrak{S}_{s^n} = (k,l,k',l',x^n,y^n,z^n)) = \frac{1}{K \cdot L} e(x^n|k,l) \mathbb{1}_{D_{k'l'}}(y^n) w_{s^n}(y^n|x^n) v_{s^n}(z^n|x^n).$$

The average error of $\mathcal{K}_n$ is

$$\text{err}(\mathcal{K}_n) = 1 - \min_{s^n \in \mathcal{S}^n} \sum_{k,l=1}^{K,L} \sum_{x^n \in \mathcal{X}^n} \frac{1}{K \cdot L} e(x^n|k,l) w_{s^n}(D_{k,l}|x^n).$$

With this definition we can formalize the idea of "wasting" a few bits in order to guarantee secret communication. We would like to compare this approach to the case of a compound channel, where a sender that knows the channel parameters can send pilot sequences to the receiver in order to let him estimate the channel. The pilot sequences do not carry information. Using such a scheme, a sender with state information can reliably transmit at strictly higher rates than a sender that has no state information. The higher capacity is reached by "wasting" some transmissions for the estimation. Since the number of channel uses that have to be used for estimation grows only sub-exponentially in the number of channel uses, there is no negative impact on the message transmission rate in asymptotic scenarios. Definition 10.4 naturally brings with it the following coding scheme.

DEFINITION 10.5 (Private/public coding scheme) A private/public coding scheme operating at rates $(R_{\text{pri}}, R_{\text{pub}})$ consists of a sequence $(\mathcal{K}_n)_{n \in \mathbb{N}}$ of private/public codes such that

$$\lim_{n \to \infty} \text{err}(\mathcal{K}_n) = 0, \qquad \limsup_{n \to \infty} \max_{s^n \in \mathcal{S}^n} I(\mathfrak{K}_n; \mathfrak{Z}_{s^n} | \mathfrak{L}_n) = 0,$$

$$\liminf_{n \to \infty} \frac{1}{n} \log(L_n) = R_{\text{pub}}, \qquad \liminf_{n \to \infty} \frac{1}{n} \log(K_n) = R_{\text{pri}}.$$

Another class of codes arises when all messages ought to be kept secret, and in addition one may use common randomness.

DEFINITION 10.6 (Common randomness assisted coding scheme satisfying mean secrecy criterion) A common randomness assisted coding scheme satisfying the mean secrecy criterion operating at rate R consists of a sequence $(\mathcal{K}_n)_{n \in \mathbb{N}}$ of common randomness assisted codes such that

$$\lim_{n \to \infty} \text{err}(\mathcal{K}_n) = 0,$$

$$\liminf_{n \to \infty} \frac{1}{n} \log(K_n) = R,$$

$$\limsup_{n \to \infty} \max_{s^n \in \mathcal{S}^n} I(\mathfrak{K}_n; \mathfrak{Z}_{s^n} | \Gamma_n) = 0.$$

DEFINITION 10.7 (Common randomness assisted coding scheme satisfying maximum secrecy criterion) A common randomness assisted coding scheme satisfying the maximum secrecy criterion operating at rate R consists of a sequence $(\mathcal{K}_n)_{n \in \mathbb{N}}$ of common randomness assisted codes such that

$$\lim_{n \to \infty} \text{err}(\mathcal{K}_n) = 0,$$

$$\liminf_{n \to \infty} \frac{1}{n} \log(K_n) = R,$$

$$\limsup_{n \to \infty} \max_{s^n \in \mathcal{S}^n} \max_{1 \le \gamma \le \Gamma} I(\mathfrak{K}_n; \mathfrak{Z}_{s^n} | \Gamma_n = \gamma) = 0.$$

DEFINITION 10.8 (Secure uncorrelated coding scheme) A secure uncorrelated coding scheme operating at rate R consists of a sequence $(\mathcal{K}_n)_{n\in\mathbb{N}}$ of common randomness assisted codes with $\Gamma_n = 1$ for all $n \in \mathbb{N}$ such that

$$\lim_{n\to\infty} \mathrm{err}(\mathcal{K}_n) = 0,$$

$$\liminf_{n\to\infty} \frac{1}{n}\log(K_n) = R,$$

$$\limsup_{n\to\infty} \max_{s^n \in \mathcal{S}^n} I(\mathfrak{K}_n; \mathfrak{Z}_{s^n}) = 0.$$

DEFINITION 10.9 (Secure secret common randomness assisted coding scheme) A secure secret coding common randomness assisted coding scheme $\mathfrak{K}$ operating at rate R and using an amount $G_{\mathfrak{K}} > 0$ of common randomness consists of a sequence $\mathfrak{K} := (\mathcal{K}_n)_{n\in\mathbb{N}}$ of common randomness assisted codes with $\lim_{n\to\infty} \frac{1}{n}\log\Gamma_n = G_{\mathfrak{K}}$ such that

$$\lim_{n\to\infty} \mathrm{err}(\mathcal{K}_n) = 0,$$

$$\liminf_{n\to\infty} \frac{1}{n}\log(J_n) = R,$$

$$\limsup_{n\to\infty} \max_{s^n \in \mathcal{S}^n} I(\mathfrak{K}_n; \mathfrak{Z}_{s^n}) = 0.$$

REMARK 10.2 *Common randomness is only quantified for secrecy schemes where it is kept secret. The reason for this follows from [70], where it is proven that any shared randomness needed in order to achieve the correlated random coding mean secrecy capacity can always be assumed to not be larger than polynomially many bits of common randomness. This argument was sharpened in [64].*

We now define the five different capacities that we relate in this work:

DEFINITION 10.10 (Secrecy capacities) Given $(\mathfrak{W}, \mathfrak{V})$, we define for every $G > 0$ the secret common randomness assisted secrecy capacity as

$$C_{\mathrm{key}}(\mathfrak{W}, \mathfrak{V}, G) := \sup \left\{ R : \begin{array}{l} \text{There exists a secure secret common randomness} \\ \text{assisted coding scheme } \mathfrak{K} \text{ at rate } R \text{ with } G_{\mathfrak{K}} = G \end{array} \right\}.$$

The other capacities are defined as:

$$C_{\mathrm{S}}(\mathfrak{W}, \mathfrak{V}) := \sup \left\{ R : \begin{array}{l} \text{There exists a secure uncorrelated} \\ \text{coding scheme operating at rate } R \end{array} \right\},$$

$$C_{\mathrm{S,ran}}^{\mathrm{mean}}(\mathfrak{W}, \mathfrak{V}) := \sup \left\{ R : \begin{array}{l} \text{There exists a common randomness assisted} \\ \text{coding scheme satisfying the mean secrecy} \\ \text{criterion operating at rate } R \end{array} \right\},$$

$$C_{\mathrm{S,ran}}^{\max}(\mathfrak{W}, \mathfrak{V}) := \sup \left\{ R : \begin{array}{l} \text{There exists a common randomness assisted} \\ \text{coding scheme satisfying the maximal secrecy} \\ \text{criterion operating at rate } R \end{array} \right\},$$

$$C_{\mathrm{pp}}(\mathfrak{W}, \mathfrak{V}) := \sup \left\{ R : \begin{array}{l} \text{There exists a private/public coding scheme at} \\ \text{rates } (R_{\mathrm{pub}}, R_{\mathrm{pri}}) \text{ such that } R = R_{\mathrm{pri}} \end{array} \right\}.$$

We now state the formal definition of super-activation:

DEFINITION 10.11 (Super-activation) Let $(\mathfrak{W}_1, \mathfrak{V}_1)$ and $(\mathfrak{W}_2, \mathfrak{V}_2)$ be AVWCs. Then $(\mathfrak{W}_1, \mathfrak{V}_1)$, $(\mathfrak{W}_2, \mathfrak{V}_2)$ are said to show super-activation if $C_S(\mathfrak{W}_1, \mathfrak{V}_1) = C_S(\mathfrak{W}_2, \mathfrak{V}_2) = 0$ but $C_S(\mathfrak{W}_1 \otimes \mathfrak{W}_2, \mathfrak{V}_1 \otimes \mathfrak{V}_2) > 0$.

At last, let us introduce a quantity which will be of importance during our proofs and when quantifying how close an AVC is to being symmetrizable: We let M_{fin} denote the set of all finite sets of elements of $C(\mathcal{X}, \mathcal{Y})$.

DEFINITION 10.12 The function $F : M_{\text{fin}} \to \mathbb{R}_+$ is defined by setting

$$F(\mathfrak{W}') := \min_{Q \in C(\mathcal{X}, \mathcal{S})} \max_{x \neq x'} \left\| \sum_{s \in \mathcal{S}} q(s|x) W'(\delta_{x'} \otimes \delta_s) - \sum_{s \in \mathcal{S}} q(s|x') W'(\delta_x \otimes \delta_s) \right\|_1,$$

for every $\mathfrak{W}' = (W'(\cdot|\cdot, s))_{s \in \mathcal{S}} \in M_{\text{fin}}$.

This function obviously has the property that for every AVWC $\mathfrak{W}'$, the statement $F(\mathfrak{W}') = 0$ is equivalent to $\mathfrak{W}'$ being symmetrizable.

10.3 Main Results and Insights

For the remainder of this section, let $(\mathfrak{W}, \mathfrak{V})$ be an AVWC. We start with the coding theorems for $C_{S,\text{ran}}^{\text{mean}}$ and $C_{S,\text{ran}}^{\text{max}}$, and note that they are both continuous. Afterwards, we show how C_{key} and C_S relate to $C_{S,\text{ran}}^{\text{mean}}$. This connection reveals that C_{key} is a continuous function as well. In contrast, C_S has discontinuity points. These points are quantified in Theorems 10.6 and 10.5, and brought into context with the symmetrizability condition (10.10) and Definition 10.12.

Our last result relates to [30], which showed a very surprising effect that has so far not been observed for classical information-carrying systems: super-activation. We give a precise characterization of the conditions which lead to super-activation of C_S in Theorem 10.7, and relate it to the potential super-activation of $C_{S,\text{ran}}^{\text{mean}}$.

We conclude with a discussion of the relation between C_{pp} and C_S, and state some open questions.

10.3.1 Assisted Capacities: Coding Theorems for $C_{S,\text{ran}}^{\text{mean}}$, $C_{S,\text{ran}}^{\text{max}}$, and C_{key}

THEOREM 10.1 *Define*

$$C^*(\mathfrak{W}, \mathfrak{V}) := \lim_{n \to \infty} \frac{1}{n} \max_{p \in \mathcal{P}(\mathcal{X}^n)} \max_{U \in C(\mathcal{X}^n, \mathcal{X}^n)}$$

$$\left(\min_{q \in \mathcal{P}(\mathcal{S})} I(p; W_q^{\otimes n} \circ U) - \max_{s^n \in \mathcal{S}^n} I(p; V_{s^n} \circ U) \right). \tag{10.29}$$

Both the correlated random coding mean secrecy and the correlated random coding maximum secrecy capacities are given by the formula

$$C_{S,\text{ran}}^{\text{mean}}(\mathfrak{W}, \mathfrak{V}) = C_{S,\text{ran}}^{\text{max}}(\mathfrak{W}, \mathfrak{V}) = C^*(\mathfrak{W}, \mathfrak{V}). \tag{10.30}$$

For the sake of a more streamlined discussion, let us define the following:

DEFINITION 10.13 For every $n \in \mathbb{N}$, $p \in \mathcal{P}(\mathcal{X}^n)$, and $U \in C(\mathcal{X}^n, \mathcal{X}^n)$, set

$$C_n^*(p, U, \mathfrak{W}, \mathfrak{V}) := \frac{1}{n}\left(\min_{q \in \mathcal{P}(\mathcal{S})} I(p; W_q^{\otimes n} \circ U) - \max_{s^n \in \mathcal{S}^n} I(p; V_{s^n} \circ U)\right). \quad (10.31)$$

Obviously,

$$C^*(\mathfrak{W}, \mathfrak{V}) = \lim_{n \to \infty} \max_{p \in \mathcal{P}(\mathcal{X}^n)} \max_{U \in C(\mathcal{X}^n, \mathcal{X}^n)} C_n^*(p, U, \mathfrak{W}, \mathfrak{V}). \quad (10.32)$$

An important non-trivial observation [64] is the following:

THEOREM 10.2 *The function $(\mathfrak{W}, \mathfrak{V}) \mapsto C^*(\mathfrak{W}, \mathfrak{V})$ is continuous with respect to the metric d. Thus, $(\mathfrak{W}, \mathfrak{V}) \mapsto C_{S,\mathrm{ran}}^{\mathrm{mean}}(\mathfrak{W}, \mathfrak{V})$ and $(\mathfrak{W}, \mathfrak{V}) \mapsto C_{S,\mathrm{ran}}^{\max}(\mathfrak{W}, \mathfrak{V})$ are continuous functions of the channel.*

The proof of this theorem only requires minor changes compared to that of [69, Theorem 2], where the continuity of the capacity of the corresponding compound wiretap channel is shown.

As was shown in [69] using an example with small alphabets and a state set of no more than two elements, the capacity C_S is discontinuous. Hence, the continuity of the correlated random coding secrecy capacity becomes even more remarkable, especially as it turns out in the proof of Theorem 10.1 that only very little correlated randomness is required to cause such a qualitative change of capacity functions. The exact characterization of the discontinuity points of the uncorrelated coding secrecy capacity $C_S(\mathfrak{W}, \mathfrak{V})$ is the content of Theorems 10.5 and 10.6.

The following theorem addresses the question of what happens if Eve does not get to know the values of the common randomness.

THEOREM 10.3 (Coding theorem for secret common randomness assisted secrecy capacity) *For every $G > 0$,*

$$C_{\mathrm{key}}(\mathfrak{W}, \mathfrak{V}, G) = \min\{C^*(\mathfrak{W}, \mathfrak{V}) + G, C^*(\mathfrak{W}, \mathfrak{T})\}, \quad (10.33)$$

with $\mathfrak{T} = (T)$ again denoting the AVC consisting only of the memoryless channel that assigns the uniform output distribution to every input symbol.

Of course, $C^*(\mathfrak{W}, \mathfrak{T})$ is the capacity of the AVC $\mathfrak{W}$ under average error. This capacity has a single-letter description. Since the first argument in the above minimum is not single letter, it may be speculated whether there is room for improvement in this characterization or, if not, for which values of G the description in terms of a single-letter quantity is possible and for which not. Apart from the complicated multi-letter form, an important corollary of the above formula is that the following is true:

COROLLARY 10.1 *For every $G > 0$, the function $(\mathfrak{W}, \mathfrak{V}) \mapsto C_{\mathrm{key}}(\mathfrak{W}, \mathfrak{V}, G)$ is continuous.*

REMARK 10.3 C_{key} shows the following behavior: If $G = 0$ in the sense that $\Gamma_n = 1$ for all $n \in \mathbb{N}$, then for all AVWCs $(\mathfrak{W}, \mathfrak{V})$ we have that $C_{\text{key}}(\mathfrak{W}, \mathfrak{V}, G) = C_S(\mathfrak{W}, \mathfrak{V})$. On the other hand, if one considers the limit of G going to zero, one sees that

$$C_{S,\text{ran}}^{\text{mean}}(\mathfrak{W}, \mathfrak{V}) = \lim_{G \to 0} C_{\text{key}}(\mathfrak{W}, \mathfrak{V}, G) = C^*(\mathfrak{W}, \mathfrak{V}) \tag{10.34}$$

holds. This justifies our choice of defining C_{key} only for $G > 0$.

10.3.2 The Non-Assisted Capacity

The Impact of Symmetrizability

For those cases where the secret or partially secret common randomness Γ is set to one for every number of channel uses, we have to deal with the symmetrizability properties of the legal link $\mathfrak{W}$ from Alice to Bob. Initial statements in that case are as follows:

THEOREM 10.4 (Symmetrizability properties of C_S)

1. If $\mathfrak{W}$ is symmetrizable, then $C_S(\mathfrak{W}, \mathfrak{V}) = 0$.
2. If $\mathfrak{W}$ is non-symmetrizable, then $C_S(\mathfrak{W}, \mathfrak{V}) = C_{S,\text{ran}}^{\text{mean}}(\mathfrak{W}, \mathfrak{V})$.

The proof of Theorem 10.4 is based on Lemma 10.4 (with $\Gamma = 1$), which assures the existence of secure coding schemes at certain rates – if $\mathfrak{W}$ is non-symmetrizable. A positive outcome of this approach is that there are generic communication systems (AVWCs with a symmetrizable legal link and $C_{S,\text{ran}}^{\text{mean}}(\mathfrak{W}, \mathfrak{V}) > 0$) for which it is much easier to design codes that convey little information to Eve than codes which ensure robust communication. However, a drawback of a direct application of Lemma 10.4 is that one is only able to prove achievability of rates up to $\max_{p \in \mathcal{P}(\mathcal{X})} C_1^*(p, Id, \mathfrak{W}, \mathfrak{V})$.

In order to derive from Lemma 10.3 the connection between symmetrizability and the capacity C_S, one needs to take into account the effects arising from precoding. More precisely, it is desirable to prove achievability of all rates up to $C_n^*(p, U, \mathfrak{W}, \mathfrak{V})$ for every $n \in \mathbb{N}$, $p \in \mathcal{P}(\mathcal{X}^n)$, and every $U \in C(\mathcal{X}^n, \mathcal{X}^n)$ (cf. Definition 10.13). Such a process of adding precoding may unfortunately cause the AVWC arising from the concatenation of precoding and the original AVWC to be symmetrizable. This highly interesting interplay of precoding and symmetrizability is quantified in Lemma 10.1 and Example 10.1.

LEMMA 10.1 Let $\mathfrak{W}$ be an arbitrarily varying channel with input alphabet $\mathcal{A}$, output alphabet $\mathcal{B}$, and state set $\mathcal{S}$. Let $T \in C(\mathcal{A}', \mathcal{A})$ be a channel. Let $\mathfrak{W}'$ be the arbitrarily varying channel with input alphabet $\mathcal{A}'$, output alphabet $\mathcal{B}$, and transition probabilities defined by $w'(b|a', s) := \sum_{a \in \mathcal{A}} w(b|a, s) t(a|a')$ (or, equivalently, via setting $W_s' := W_s \circ T$ for all $s \in \mathcal{S}$).

If $\mathfrak{W}$ is symmetrizable then $\mathfrak{W}'$ is symmetrizable as well.

The point that, even for channels T whose associated matrix $(t(a|a')_{a' \in \mathcal{A}', a \in \mathcal{A}}$ has full rank, the reverse implication "$\mathfrak{W}'$ is symmetrizable $\Rightarrow$ $\mathfrak{W}$ is symmetrizable" does not hold came as a surprise and is proven here by explicit example.

EXAMPLE 10.1 For a number $x \in [0,1]$ we use the convention $x' := 1-x$. Define an AVC $\mathfrak{W} \subset C(\{x_1,x_2\},\{1,2,3\})$ by setting

$$w(\cdot|s_1,x_1) := \delta_1, \qquad (10.35)$$

$$w(\cdot|s_1,x_2) := 0.4\delta_1 + 0.5\delta_2 + 0.1\delta_3, \qquad (10.36)$$

$$w(\cdot|s_2,x_1) := \delta_2, \qquad (10.37)$$

$$w(\cdot|s_2,x_2) := \delta_3, \qquad (10.38)$$

where $\delta_i(j) = 1$ if and only if $i = j$ holds for $i,j \in [3]$. Then W is non-symmetrizable: the equation

$$\lambda \cdot w(\cdot|s_1,x_1) + \lambda' \cdot w(\cdot|s_2,x_1) = \mu \cdot w(\cdot|s_1,x_2) + \mu' \cdot w(\cdot|s_2,x_2) \qquad (10.39)$$

cannot have a solution with $\lambda, \mu \in [0,1]$ because δ_3 appears only on the right-hand side and with strictly positive weights.

However, if we add precoding by a binary-symmetric channel N_p with parameter $p \in [0,1]$ denoting the probability of correct transmission of the bit, we obtain the new AVC $\mathfrak{W}'$ defined via $W'_s := W_s \circ N_p$. It has output distributions

$$w'(\cdot|s_1,x_1) = p\delta_1 + p'(0.4\delta_1 + 0.5\delta_2 + 0.1\delta_3), \qquad (10.40)$$

$$w'(\cdot|s_1,x_2) = p(0.4\delta_1 + 0.5\delta_2 + 0.1\delta_3) + p'\delta_1, \qquad (10.41)$$

$$w'(\cdot|s_2,x_1) = p\delta_2 + p'\delta_3, \qquad (10.42)$$

$$w'(\cdot|s_2,x_2) = p\delta_3 + p'\delta_2. \qquad (10.43)$$

Set $p = 0.4$. The equation

$$\lambda \cdot w'(\cdot|s_1,x_1) + \lambda' \cdot w'(\cdot|s_2,x_1) = \mu \cdot w'(\cdot|s_1,x_2) + \mu' \cdot w'(\cdot|s_2,x_2) \qquad (10.44)$$

is solved by setting

$$\lambda = 38/45, \qquad \mu = 32/45. \qquad (10.45)$$

This shows that $\mathfrak{W}'$ is symmetrizable. The situation is depicted in Fig. 10.5.

In order to derive the direct part of Theorem 10.4 from Lemma 10.4 we can therefore not use a simple blocking strategy. Rather, we will present two methods of proof. The first employs reasoning along the lines of Eqs. (10.49)–(10.55). This approach is based on the concept of using a few non-secret bits in order to guarantee secrecy for the actual data. While this is highly interesting from a practical point of view, it does not utilize the full strength of Lemma 10.4. This proof uses a set of public messages that can be read by Eve but not by James; secrecy is only obtained for the (exponentially larger) set of private messages.

Our second proof of Theorem 10.4 is based on lifting the optimal precodings for n channel uses to $n+1$ channel uses by using no precoding on the $(n+1)$th channel use. This type of precoding preserves non-symmetrizability. The second proof makes almost full use of the statements of Lemma 10.4, as we still set $\Gamma = 1$. No public messages are used in the construction of the code.

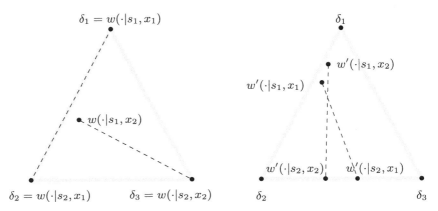

Figure 10.5 Black dots denote probability distributions. Light gray lines are the edges of the probability simplex $\mathcal{P}(\{1,2,3\})$. The sets $\mathrm{conv}(\{w(\cdot|s_1,x_i),w(\cdot|s_2,x_i)\})$ where $i=1,2$ are visualized as dashed lines. The intersection of the dashed lines in the right polytope shows that $\mathfrak{W}'$ is symmetrizable.

Although there is no advantage in terms of any of the capacities defined here, our second approach is conceptually different from the first. It offers the additional insight that one may design optimal codes for the AVWC without having to split the information transmission task into two parts, where one of them clearly leaks information to Eve.

It remains an interesting open question whether, for n channel uses, the optimal channel U arising from the nth term of the optimization problem (10.14) does in fact symmetrize $(W_{s^n})_{s^n \in S^n}$ or not. So far, it can only be said that "injection" of noise at the encoder has to be treated with care when the channel that is to be transmitted over is an AVWC (or an AVC).

The Unassisted Case: Continuity and Discontinuity

We now take a slightly different point of view, under which the AVWC becomes an object that has some parameters that may vary. When considering practical deployment aspects, such a point of view is necessary as all the information one may have gathered about the channel during some training phase might not be accurate enough to model the real-world behavior. Thus one needs to understand whether a slight error in the parameters may lead to catastrophic events, and this is the content of our next theorem.

THEOREM 10.5 (Stability of C_S) *If* $(\mathfrak{W},\mathfrak{V})$ *satisfies* $C_S(\mathfrak{W},\mathfrak{V}) > 0$ *then there is an* $\epsilon > 0$ *such that for all* $(\mathfrak{W}',\mathfrak{V}')$ *satisfying* $d((\mathfrak{W},\mathfrak{V}),(\mathfrak{W}',\mathfrak{V}')) \le \epsilon$ *we have* $C_S(\mathfrak{W}',\mathfrak{V}') > 0$.

However, despite the reassuring statement of Theorem 10.5, care has to be taken at the points that are characterized below.

THEOREM 10.6 (Discontinuity properties of C_S) *Let* $(\mathfrak{W},\mathfrak{V})$ *be an AVWC.*

1. *The function C_S is discontinuous at the point $(\mathfrak{W},\mathfrak{V})$ if and only if the following hold:*

a $C_{S,\text{ran}}^{\text{mean}}(\mathfrak{W},\mathfrak{V}) > 0$, and
b $F(\mathfrak{W}) = 0$ but for all $\epsilon > 0$ there is $\mathfrak{W}_\epsilon$ such that $d(\mathfrak{W},\mathfrak{W}_\epsilon) < \epsilon$ and $F(\mathfrak{W}_\epsilon) > 0$.

2. If C_S is discontinuous in the point $(\mathfrak{W},\mathfrak{V})$ then it is discontinuous for all $\hat{\mathfrak{V}}$ for which $C_{S,\text{ran}}^{\text{mean}}(\mathfrak{W},\hat{\mathfrak{V}}) > 0$.

Note that $F(\mathfrak{W}) = 0$ is equivalent to $\mathfrak{W}$ being symmetrizable – a property which is defined in Eq. (10.10). The function F itself is the content of Definition 10.12.

An immediate corollary of this result is the following:

COROLLARY 10.2 *For every $\mathfrak{W}$, the function $\mathfrak{V} \mapsto C_S(\mathfrak{W},\mathfrak{V})$ is continuous.*

Regarding future research, it may be of interest to quantify the degree of continuity of the capacity of arbitrarily varying channels in those regions where it is continuous. Note that discontinuity is caused both by the legal link $\mathfrak{W}$ (see statement 1) and the link $\hat{\mathfrak{V}}$ to Eve (statement 2), but depends on $\hat{\mathfrak{V}}$ only insofar as the capacity $C_{S,\text{ran}}^{\text{mean}}(\mathfrak{W},\hat{\mathfrak{V}})$ has to stay above zero in order for a discontinuity to occur.

REMARK 10.4 *It is necessary to request the existence of the $\mathfrak{W}_\epsilon$ in the first statement of Theorem 10.6, and an easy example why this is so is the following:*
Define $W_{i,\epsilon} \in C(\{1,2\},\{1,2,3\})$ for $i = 1,2$ and $\epsilon \in [0,1/2]$ by

$$W_{1,\epsilon} := \begin{pmatrix} 0 & 1-\epsilon \\ \epsilon & 0 \\ 1-\epsilon & \epsilon \end{pmatrix}, \quad W_{2,\epsilon} := \begin{pmatrix} 1-\epsilon & 0 \\ \epsilon & 1-\epsilon \\ 0 & \epsilon \end{pmatrix}. \tag{10.46}$$

For every $\epsilon \in [0,1/2]$, these AVCs are symmetrizable with $u(1|1) = \epsilon/(1-\epsilon)$ and $u(1|2) = (1-2\cdot\epsilon)/(1-\epsilon)$. The reason for this is that for every $\epsilon \in [0,1/2]$ the convex sets $\text{conv}(\{W_{1,\epsilon}(\delta_1), W_{2,\epsilon}(\delta_1)\})$ and $\text{conv}(\{W_{1,\epsilon}(\delta_2), W_{2,\epsilon}(\delta_2)\})$ have non-empty intersections. It is also geometrically clear that for any $\epsilon \in (0,1/2)$, there will be a small vicinity of AVCs which share this property. Thus, around such a $\mathfrak{W}_\epsilon$, all other AVCs are symmetrizable as well, and for every $\mathfrak{V}$ we therefore have both $C_S(\mathfrak{W}_\epsilon,\mathfrak{V}) = 0$ and $C_S(\mathfrak{W}',\mathfrak{V}) = 0$ whenever $d(\mathfrak{W}_\epsilon,\mathfrak{W}')$ is small enough. A look at Fig. 10.6 shows that C_S is continuous around all points $(\mathfrak{W}_\epsilon,\mathfrak{V})$ for which $\epsilon \in (0,1/2]$.

Of course, however nice a characterization of a set of interesting objects is, it is pretty useless if the set turns out to be empty. Fortunately, it has been proven in [69] that the function mapping an AVC $\mathfrak{W}$ to its capacity has discontinuity points by an explicit example.

Such an example is also given by $(\mathfrak{W}_0,\mathfrak{T})$ with $\mathfrak{W}_0$ taken from above.

We will now explain how super-activation occurs in a two-fold way: first in detail for C_{pp}, and then by providing the key argument for the super-activation of C_S.

Set $\mathfrak{W} := \mathfrak{W}_1 \otimes \mathfrak{W}_2$ and $\mathfrak{V} := \mathfrak{V}_1 \otimes \mathfrak{V}_2$. In order to simplify the discussion, one may additionally set $\mathfrak{V}_2 = \mathfrak{W}_2 = (Id)$, where $Id \in C([2],[2])$ is the identity channel, and assume that $\mathfrak{W}_1$ is symmetrizable but that $C_{S,\text{ran}}^{\text{mean}}(\mathfrak{W}_1,\mathfrak{V}_1) = \alpha > 0$. It follows that $C_{\text{pp}}(\mathfrak{W}_1,\mathfrak{V}_1) = C_{\text{pp}}(\mathfrak{W}_2,\mathfrak{V}_2) = 0$, because of symmetrizability and since decoding of the messages that are sent via $(\mathfrak{W}_2,\mathfrak{V}_2)$ is possible without any error both for Bob

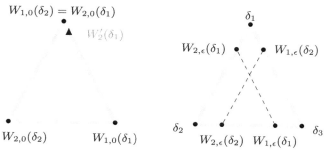

Figure 10.6 Continuity. Light gray lines depict edges of the probability simplex $\mathcal{P}(\{1,2,3\})$. The left-hand side shows the symmetrizable AVC $\mathfrak{W}_0$. The triangle indicates how an arbitrarily small change of $\mathfrak{W}_0$ can make the resulting AVC $\mathfrak{W}'$ non-symmetrizable. The right-hand side shows $\mathfrak{W}_\epsilon$ for $\epsilon = 1/4$. It is evident that small enough changes of $\mathfrak{W}_{\frac{1}{4}}$ cannot make the channel non-symmetrizable.

and for Eve. These messages may therefore be treated as common randomness that is known by Eve. We know that already with the choice $L_n = n^2$ we have enough common randomness to remove any effect arising from symmetrizability of $\mathfrak{W}_1$. Since the code arising from the combination of sending and decoding public messages via (Id, Id) and private messages via $(\mathfrak{W}_1, \mathfrak{V}_1)$ is a coding scheme that fits under Definition 10.5, we get $C_{\text{pp}}(\mathfrak{W}_1 \otimes \mathfrak{W}_2, \mathfrak{V}_1 \otimes \mathfrak{V}_2) \geq \alpha > 0$.

That such a scheme also works when C_S is considered instead of C_{pp} can be understood as follows:

Let $(\mathfrak{W}_1, \mathfrak{V}_1)$ and $(\mathfrak{W}_2, \mathfrak{V}_2)$ be two AVCs. Let $\mathfrak{W}_1$ be symmetrizable, but such that $C_{S,\text{ran}}^{\text{mean}}(\mathfrak{W}_1, \mathfrak{V}_1) = \alpha > 0$. Since $\mathfrak{W}_1$ is symmetrizable we have $C_S(\mathfrak{W}_1, \mathfrak{V}_1) = 0$. If no additional resources are available, the surplus α in the common randomness assisted secrecy capacity cannot be put to use. Now let $C_S(\mathfrak{W}_2, \mathfrak{V}_2) = 0$ but $C_S(\mathfrak{W}_2, \mathfrak{T}) = \beta > 0$ ($\mathfrak{T}$ denotes the trash channel, so this just means that it is possible to reliably transmit messages over $\mathfrak{W}_2$). Then

$$C_S(\mathfrak{W}_1 \otimes \mathfrak{W}_2, \mathfrak{V}_1 \otimes \mathfrak{V}_2) \geq \alpha > 0, \tag{10.47}$$

and the reason for this effect is that (as before when we considered C_{pp}) a small number of messages can be sent over $\mathfrak{W}_2$ and are then used as common randomness, therefore increasing the rate of messages that can be sent reliably over $\mathfrak{W}_1$ from zero to α. Of course, the messages sent over $\mathfrak{W}_2$ can be read by Eve. That this causes no problems with the security requirements can be seen by defining a toy model where only two parallel channels with respective adversarially controlled channel states are considered. This is done as follows:

Let us define random variables $\mathfrak{R}_{s,\hat{s}} = (\mathfrak{M}, \hat{\mathfrak{M}}, \mathfrak{Z}_{1,s}, \hat{\mathfrak{Z}}_{2,\hat{s}})$ where

$$\mathbb{P}(\mathfrak{R} = (m, \hat{m}, z, \hat{z})) = \frac{1}{M}\frac{1}{\hat{M}} w_{1,s}(z|m,\hat{m}) \hat{w}_{2,\hat{s}}(\hat{z}|\hat{m}) \tag{10.48}$$

and the channels $\{W_{1,s}\}_{s\in S}$ and $\{\hat{W}_{\hat{s}}\}_{\hat{s}\in\hat{S}}$ can be controlled by James separately. It is understood that m are the messages, whereas $\hat{m}$ are the values of the shared randomness that is distributed between Alice and Bob by using $\{\hat{W}_{2,\hat{s}}\}_{\hat{s}\in\hat{S}}$. We assume that for some small $\epsilon \geq 0$ we have

$$\max_{\hat{s}\in\hat{S}} I(\mathfrak{M}; \hat{\mathfrak{Z}}_{2,\hat{s}}|\hat{\mathfrak{M}}) \leq \epsilon. \tag{10.49}$$

Observe that $\hat{\mathfrak{Z}}_{2,\hat{s}}$ depends solely on $\hat{\mathfrak{M}}$ via the channel $\hat{W}_{2,\hat{s}}$ (here we use the fact that the two AVCs are used in parallel), so that the data processing inequality yields for every $s, \hat{s}$ that

$$I(\mathfrak{M}; \mathfrak{Z}_{1,s}, \hat{\mathfrak{Z}}_{2,\hat{s}}) \leq I(\mathfrak{M}; \mathfrak{Z}_{1,s}, \hat{\mathfrak{M}}). \tag{10.50}$$

Since $\mathfrak{M}$ and $\hat{\mathfrak{M}}$ are independent we have, for every s and $\hat{s}$,

$$I(\mathfrak{M}; \mathfrak{Z}_{1,s}, \hat{\mathfrak{Z}}_{2,\hat{s}}) \leq I(\mathfrak{M}; \mathfrak{Z}_{1,s}, \hat{\mathfrak{M}}) \tag{10.51}$$

$$= H(\mathfrak{M}, \hat{\mathfrak{M}}) + H(\mathfrak{Z}_{1,s}, \hat{\mathfrak{M}}) - H(\mathfrak{M}, \mathfrak{Z}_{1,s}, \hat{\mathfrak{M}}) - H(\hat{\mathfrak{M}}) \tag{10.52}$$

$$= I(\mathfrak{M}; \mathfrak{Z}_{1,s}|\hat{\mathfrak{M}}) \tag{10.53}$$

$$\leq \epsilon. \tag{10.54}$$

Thus it is clear that, in addition,

$$\max_{s\in S, \hat{s}\in\hat{S}} I(\mathfrak{M}; \mathfrak{Z}_{1,s}, \hat{\mathfrak{Z}}_{2,\hat{s}}) \leq \epsilon. \tag{10.55}$$

It is also evident that this argument ceases to hold true when the channels that are used for transmission of M and of $\hat{M}$ are not orthogonal anymore. Our sketch indicates why the protocol developed in [30] is able to meet the secrecy requirement in Definition 10.8.

The Unassisted Capacities: Super-Activation

Our next result is potentially the most interesting in this work, since it sheds additional light on a rather new phenomenon: the super-activation of "the" secrecy capacity of AVWCs.

THEOREM 10.7 (Characterization of super-activation of C_S via properties of $C_{S,\text{ran}}^{\text{mean}}$) Let $(\mathfrak{W}_i, \mathfrak{V}_i)_{i=1,2}$ be AVWCs.

1. Assume that $C_S(\mathfrak{W}_1, \mathfrak{V}_1) = C_S(\mathfrak{W}_2, \mathfrak{V}_2) = 0$. Then

$$C_S(\mathfrak{W}_1 \otimes \mathfrak{W}_2, \mathfrak{V}_1 \otimes \mathfrak{V}_2) > 0 \tag{10.56}$$

if and only if $\mathfrak{W}_1 \otimes \mathfrak{W}_2$ is not symmetrizable and $C_{S,\text{ran}}^{\text{mean}}(\mathfrak{W}_1 \otimes \mathfrak{W}_2, \mathfrak{V}_1 \otimes \mathfrak{V}_2) > 0$.
If $(\mathfrak{W}_i, \mathfrak{V}_i)_{i=1,2}$ can be super-activated,

$$C_S(\mathfrak{W}_1 \otimes \mathfrak{W}_2, \mathfrak{V}_1 \otimes \mathfrak{V}_2) = C_{S,\text{ran}}^{\text{mean}}(\mathfrak{W}_1 \otimes \mathfrak{W}_2, \mathfrak{V}_1 \otimes \mathfrak{V}_2). \tag{10.57}$$

2. There exist AVWCs which exhibit the above behavior.
3. If $C_{S,\text{ran}}^{\text{mean}}$ shows super-activation for $(\mathfrak{W}_1, \mathfrak{V}_1)$ and $(\mathfrak{W}_2, \mathfrak{V}_2)$, then C_S shows super-activation for $(\mathfrak{W}_1, \mathfrak{V}_1)$ and $(\mathfrak{W}_2, \mathfrak{V}_2)$ if and only if at least one of $\mathfrak{W}_1$ or $\mathfrak{W}_2$ is non-symmetrizable.

4. If $C_{S,ran}^{mean}$ shows no super-activation for $(\mathfrak{W}_1,\mathfrak{V}_1)$ and $(\mathfrak{W}_2,\mathfrak{V}_2)$ then super-activation of C_S can only happen if $\mathfrak{W}_1$ is non-symmetrizable and $\mathfrak{W}_2$ is symmetrizable and $C_{S,ran}^{mean}(\mathfrak{W}_1,\mathfrak{V}_1) = 0$ and $C_{S,ran}^{mean}(\mathfrak{W}_2,\mathfrak{V}_2) > 0$. The statement is independent of the specific labeling.

REMARK 10.5 *Of course, for $\mathfrak{W}_1 \otimes \mathfrak{W}_2$ to be non-symmetrizable, at least one out of $\mathfrak{W}_1, \mathfrak{W}_2$ is non-symmetrizable.*

While Theorem 10.7 offers a complete characterization, it does not give any explicit examples. Fortunately this has already been done in [30], where two AVWCs were used as follows: The first legal link is modeled by an AVC $\mathfrak{W}_1 = (W_{1,1}, W_{1,2})$ with $W_{1,1}, W_{1,2} \in C(\{1,2\},\{1,2,3\})$. The transition probabilities were given by

$$W_{1,1} = \begin{pmatrix} 1 & 0 & 0 \\ 0 & 0 & 1 \end{pmatrix}^\top, \quad W_{1,2} = \begin{pmatrix} 0 & 0 & 1 \\ 0 & 1 & 0 \end{pmatrix}^\top \tag{10.58}$$

(note that we assume that the columns of a channel matrix sum to one, not the rows!) and the first link to the eavesdropper by $\mathfrak{V}_1 = (V_1)$ (no influence from the jammer on that link). For the purpose of this example, it would even be sufficient to let $\mathfrak{V}_1 = \mathfrak{T}$. This channel has the property that $\mathfrak{W}_1$ is symmetrizable. The second link was chosen to consist of two binary symmetric channels W_2, V_2 where W_2 was a degraded version of V_2, but both had non-zero capacity. Thus, $C_S(\mathfrak{W}_2, \mathfrak{V}_2) = 0$, but nonetheless it is possible to transmit (non-secret) messages via $\mathfrak{W}_2$. This example fits into the third class of pairs of AVWCs described in Theorem 10.7.

While this explicit example is very interesting, our analysis provides a more systematic view.

All our arguments apply to the strong secrecy criterion. The weak secrecy criterion can be handled differently, and will be the scope of future work.

The Relation between C_{pp} and C_S

As a last point in this subsection, we would like to discuss relations between C_{pp} and C_S. At first, let us observe a similarity: the former shows super-activation if and only if the latter shows super-activation. To see this, we argue as follows: By definition, the class of codes that transmit public and privat messages (as defined in Definition 10.5) includes that class from Definition 10.8, where no public information is transmitted. Therefore it holds that $C_{pp}(\mathfrak{W}, \mathfrak{V}) \geq C_S(\mathfrak{W}, \mathfrak{V})$ for all AVWCs $(\mathfrak{W}, \mathfrak{V})$. Further, the definition of private/public codes according to Definition 10.5 is more narrow than the one of a common randomness assisted code according to Definition 10.3. Therefore, every private/public code is at the same time also a common randomness assisted code. In particular, the public messages may be treated as if they were common randomness if $L = \Gamma$.

Consequently, $C_{pp}(\mathfrak{W}, \mathfrak{V}) > 0$ implies that $C_{S,ran}^{mean}(\mathfrak{W}, \mathfrak{V}) > 0$ for all $(\mathfrak{W}, \mathfrak{V})$. We conclude from Theorem 10.4 that $C_{pp}(\mathfrak{W}, \mathfrak{V}) > 0$ implies $C_S(\mathfrak{W}, \mathfrak{V}) > 0$ for all $(\mathfrak{W}, \mathfrak{V})$. This yields

$$\forall (\mathfrak{W}, \mathfrak{V}): \quad C_{pp}(\mathfrak{W}, \mathfrak{V}) > 0 \quad \Leftrightarrow \quad C_S(\mathfrak{W}, \mathfrak{V}) > 0. \tag{10.59}$$

Now, let C_{pp} show super-activation on $((\mathfrak{W}_1,\mathfrak{V}_1),(\mathfrak{W}_2,\mathfrak{V}_2))$. Then (10.59) implies that both $C_S(\mathfrak{W}_1,\mathfrak{V}_1) = C_S(\mathfrak{W}_2,\mathfrak{V}_2) = 0$ and $C_S(\mathfrak{W},\mathfrak{V}) > 0$. Therefore, super-activation of C_{pp} implies super-activation of C_S.

In the reverse direction, let C_S show super-activation on the pair $((\mathfrak{W}_1,\mathfrak{V}_1),(\mathfrak{W}_2,\mathfrak{V}_2))$. By Eq. (10.10) we see that C_{pp} shows super-activation as well.

10.3.3 Open Questions

From the above results, the following questions arise naturally:

1. Are there AVWCs $\mathfrak{W},\mathfrak{V}$ such that $C_{pp}(\mathfrak{W},\mathfrak{V}) > C_{S,ran}^{mean}(\mathfrak{W},\mathfrak{V})$ holds?
   This question is of huge practical importance, as it allows the quantification of the interplay between private and public communication in interfering networks when i.i.d. assumptions are not met, which is often the case.
2. Is the effect of super-activation (or super-additivity) of a capacity connected to its discontinuity?
   From a study of the classical literature [23–25, 30, 65] the answer seems to be "yes," but a look at super-activation for quantum channels [67] reveals that this is not the case. One may of course restrict the question to classical systems. Then, a study of the question of whether $C_{S,ran}^{mean}$ shows super-activation seems to be a good way to obtain further insights.
3. Connected to above question is that of a single-letter formula for $C_{S,ran}^{mean}$, which would of course have to be given by a non-additive quantity.
4. We would like to formulate the following conjecture:

   CONJECTURE 10.1 *There exist pairs $(\mathfrak{W}_1,\mathfrak{V}_1)$ and $(\mathfrak{W}_2,\mathfrak{V}_2)$ of (finite) AVWCs such that $C_{S,ran}^{mean}(\mathfrak{W}_1,\mathfrak{V}_1) = C_{S,ran}^{mean}(\mathfrak{W}_1,\mathfrak{V}_1) = 0$, but*

   $$C_{S,ran}^{mean}(\mathfrak{W}_1 \otimes \mathfrak{W}_2, \mathfrak{V}_1 \otimes \mathfrak{V}_2) > 0. \tag{10.60}$$

5. From our discussion of the relation (10.59) between C_{pp} and C_S it turned out that both quantities are strongly related, but a remaining question is to give an exact formula for C_{pp}.
6. Our definition of C_{pp} can actually be regarded as an optimization problem inside the capacity region of an arbitrarily varying broadcast channel with public and private messages. Here, public messages are viewed as a parameter to be optimized over in order to increase the number of private messages. Of course, it would be interesting to study the capacity region of the arbitrarily varying broadcast channel with public and private messages in a broader sense.

10.4 Proofs and Intermediate Technical Results

This section contains the main steps of the proofs of our results, along with additional technical results and definitions.

10.4.1 Technical Definitions, Results, and Facts

An important part of our results builds on the mathematical structure that was developed in [11]. The structure of the codes developed there relies on randomly sampling codewords that are randomly and uniformly chosen from the set T_p, for a fixed $p \in \mathcal{P}_0^n(\mathcal{X})$. This probabilistic law was used in [11], such that seamless connectivity to [11] is guaranteed.

We also need a notion of typicality, and we shall use the following: for arbitrary finite sets $\mathcal{A}, \mathcal{B}, \mathcal{C}$, every $p \in \mathcal{P}(\mathcal{A})$, $\tilde{V} \in C(\mathcal{A} \times \mathcal{B}, \mathcal{C})$, $\delta > 0$, and a given $(a^n, b^n) \in \mathcal{A}^n \times \mathcal{B}^n$, we define $p_{ABC} \in \mathcal{P}(\mathcal{A} \times \mathcal{B} \times \mathcal{C})$ via $p_{ABC}(a,b,c) := \bar{N}(a^n, b^n)\tilde{v}(c|a,b)$, and

$$T_{p,\delta}^n := \{a^n \in \mathcal{A}^n : D(\bar{N}(\cdot|a^n) \| p) \leq \delta\}, \tag{10.61}$$

$$T_{\tilde{V},\delta}(a^n, b^n) := \{c^n : D(\bar{N}(\cdot|a^n, b^n, c^n) \| p_{ABC}) \leq \delta\}. \tag{10.62}$$

These definitions are only valid for $\delta > 0$. Each $T_{\tilde{V},\delta}(s^n, x^n)$ obeys the estimate

$$\tilde{v}^{\otimes n}(T_{\tilde{V},\delta}(a^n)|a^n) \geq 1 - 2^{-n\cdot\delta/2} \tag{10.63}$$

for all $n \in \mathbb{N}$ such that $|\mathcal{A} \times \mathcal{B}| \frac{1}{n} \log(2n) \leq \delta$. We set, for every $p \in \mathcal{P}(\mathcal{X})$,

$$E(p) := \max_{q \in \mathcal{P}(\mathcal{S})} I(p; V_q) \quad \text{and} \quad B(p) := \min_{q \in \mathcal{P}(\mathcal{S})} I(p; W_q). \tag{10.64}$$

For the technical part of our proofs, the most important tool will be the Chernoff–Hoeffding bound:

LEMMA 10.2 *Let b be a positive number. Let $Z_1, \ldots, Z_L$ be i.i.d. random variables with values in $[0,b]$ and expectation $\mathbb{E}Z_l = v$, and let $0 < \varepsilon < \frac{1}{2}$. Then*

$$\mathbb{P}\left(\left\{\frac{1}{L}\sum_{l=1}^{L} Z_l \notin [(1 \pm \varepsilon)v]\right\}\right) \leq 2\exp\left(-L \cdot \frac{\varepsilon^2 \cdot v}{3 \cdot b}\right), \tag{10.65}$$

where $[(1 \pm \varepsilon)v]$ denotes the interval $[(1-\varepsilon)v, (1+\varepsilon)v]$.

The proof can be found in [71, Theorem 1.1] and in [72].

The next lemma is essential to proving the direct part of Theorem 10.3. It quantifies how many messages L and how many different values Γ for the common randomness are needed in order to make the output distributions at Eve's site independent of the chosen message k.

LEMMA 10.3 *For every $\tau > 0$ there exists a value $v(\tau) > 0$ and an $N_0(\tau) \in \mathbb{N}$ such that for all $n \geq N_0(\tau)$ and $K, L, \Gamma \in \mathbb{N}$ and type $p \in \mathcal{P}_0^n(\mathcal{X})$ there exist codewords $(\mathbf{x}_{kl\gamma})_{k,l,\gamma=1}^{K,L,\Gamma}$ in $T_p \subset \mathcal{X}^n$ and decoding sets $D_{kl}^\gamma \subset \mathcal{Y}^n$ obeying $D_{kl}^\gamma \cap D_{k'l'}^\gamma = \emptyset$ if $(k,l) \neq (k',l')$, such that we have:*

If $\frac{1}{n}\log(K \cdot L) \leq \min_{q \in \mathcal{P}(\mathcal{S})} I(p; W_q) - v(\tau)$ and $\Gamma \geq 2^{n \cdot 5 \cdot v(\tau)}$ then

$$\min_{s^n} \sum_{\gamma=1}^{\Gamma} \frac{1}{\Gamma} \sum_{k,l=1}^{K,L} \frac{1}{K \cdot L} w_{s^n}(D_{kl}^\gamma|\mathbf{x}_{kl\gamma}) \geq 1 - 2^{-n \cdot v(\tau)}. \tag{10.66}$$

If $\frac{1}{n}\log(L \cdot \Gamma) \geq \max_q I(p; V_q) + \tau$, then

$$\max_{s^n,k} \left\| \frac{1}{L \cdot \Gamma} \sum_{l,\gamma=1}^{L,\Gamma} V_{s^n}(\cdot|\tau(\mathbf{x}_{kl\gamma})) - \mathbb{E}V_{s^n}(\cdot|X^n) \right\|_1 \leq 2^{-n \cdot \nu(\tau)}, \qquad (10.67)$$

where X^n is distributed according to $\mathbb{P}(X^n = x^n) := \frac{1}{|T_p|} \mathbb{1}_{T_p}(x^n)$, and the dependence of ν on τ is such that $\lim_{\tau \to 0} \nu(\tau) = 0$.

For our purpose, Lemma 10.3 delivers the correct asymptotic scaling of the numbers of secret messages K, the number of additional messages L that are just being sent in order to obfuscate Eve, and the number of values for the (secret) common randomness Γ. However, it is insufficient for dealing with the case $\Gamma = 1$ and can therefore not be used to prove results for C_S. For this case, we employ the following extension of Lemma 3 in [11]:

LEMMA 10.4 *For any $\tau > 0$ and $\beta > 0$, there exists a value $\nu(\tau) > 0$ and an $N_0(\tau)$ such that for all $n \geq N_0(\tau)$, natural numbers K, L satisfying $K \cdot L \geq 2^{n \cdot \tau}$, and type $p \in \mathcal{P}_0^n(\mathcal{X})$ satisfying $\min_{x:p(x)>0} p(x) \geq \beta$, there exist codewords $(\mathbf{x}_{kl\gamma})_{k,l,\gamma=1}^{K,L,\Gamma}$ in $T_p \subset \mathcal{X}^n$ and a $c' > 0$ such that if $\Gamma^{-1} > \exp(-2^{n \cdot c'})$ and upon setting $R = \frac{1}{n}\log(K \cdot L)$ we have for every $x^n, \hat{x}^n \in T_p$:*

$$\max_{s^n,\gamma} |\{(k,l) : (x^n, \mathbf{x}_{kl\gamma}, s^n) \in T_{\bar{N}(\cdot|x^n, \hat{x}^n, s^n)}\}| \leq 2^{n(|R - I(\hat{x}^n; x^n, s^n)|^+ + \tau)}, \qquad (10.68)$$

$$\max_{s^n,\gamma} |\{(k,l) : I(\mathbf{x}_{kl\gamma}; s^n) \geq \tau\}| \leq K \cdot L \cdot 2^{-n \cdot \tau}, \qquad (10.69)$$

$$\max_{s^n,\gamma} \left| \left\{ (k,l) : \begin{array}{l} \text{There is } (k',l') \neq (k,l) \\ \text{such that } I(x^n; \hat{x}^n, s^n) - \\ -|R - I(\hat{x}^n; s^n)|^+ > \tau \end{array} \right\} \right| \leq K \cdot L \cdot 2^{-n \cdot \tau/2}, \qquad (10.70)$$

$$\frac{\log L}{n} \geq \max_{q \in \mathcal{P}(\mathcal{S})} I(p; V_q) + \tau \quad \Rightarrow$$

$$\max_{s^n,k,\gamma} \left\| \frac{1}{L} \sum_{l=1}^{L} V_{s^n}(\cdot|\mathbf{x}_{kl\gamma}) - \mathbb{E}V_{s^n}(\cdot|X^n) \right\|_1 \leq 2^{-n \cdot \nu(\tau)}, \qquad (10.71)$$

where X^n is distributed according to $\mathbb{P}(X^n = x^n) := \frac{1}{|T_p|} \mathbb{1}_{T_p}(x^n)$ and the dependence of ν on τ is such that $\lim_{\tau \to 0} \nu(\tau) = 0$.

We will apply this lemma to AVWCs for which the link between Alice and Bob is non-symmetrizable.

REMARK 10.6 *The properties (10.68), (10.69), and (10.70) of the code are identical to those stated in [11, Lemma 3]. This lemma again is the main ingredient of the proof of [11] that non-symmetrizability (10.10) is sufficient for message transmission over AVCs if the average error criterion and non-randomized codes are used. Accordingly, our strategy is to use the properties (10.68), (10.69), and (10.70) in Lemma 10.4 to ensure successful message transmission over the legal link, if $\mathfrak{W}$ is non-symmetrizable.*

The main tool used by Csiszár and Narayan for proving properties (10.68), (10.69), *and* (10.70) *in [11] was large deviation theory. This is where we can make the connection to our work and prove the additional properties via application of the Chernoff–Hoeffding bound.*

Roughly speaking, this method of proof amounts to adding some additional requirements in a situation where any exponential number of additional requirements can be satisfied simultaneously.

10.4.2 Basic Quantities and Estimates

We now have to prove some core results from which all the other statements can be deduced. The idea is to make a random selection of the codewords $\mathbf{x}_{kl\gamma}$ where k are the messages, l are non-secret messages which are only being sent in order to obfuscate the received signal at Eve, and γ are the values of the common randomness. In order to increase readability, the superscript n is dropped for the elements $\mathbf{x}_{kl\gamma}$ of a codebook. When applying the results to AVWCs, the decoder is the one defined in [11] whenever we study C_S and is defined here according to our needs for the study of C_{key}.

We define events E_1, E_2, E_3 which describe certain desirable properties of our codewords, depending on $(\mathfrak{W}, \mathfrak{V})$ and the numbers K, L, Γ of available indices k, l, γ. We then use Chernoff bounds. This guarantees that the random selection of codewords has each single property we would like them to have with probability bounded below by $1 - \exp(-2^{nc})$ for some positive constant $c > 0$ and all large enough n under some conditions on Γ, L, and K. These conditions depend on the AVWC $(\mathfrak{W}, \mathfrak{V})$. Application of a union bound then reveals the existence of one particular choice of codewords that has all the desired properties simultaneously.

This method was used by Csiszár and Narayan [11, Lemma 3] to prove properties (10.68), (10.69), and (10.70). Thus, what remains to be shown is that the event (10.71) has high probability.

Let $p \in \mathcal{P}_0^n(\mathcal{A})$, $q \in \mathcal{P}_0^n(\mathcal{S})$. Throughout, tweak asymptotic quantities such that they are calculated with respect to the random variables (S, X, Z) defined via $\mathbb{P}((S, X, Z) = (s, x, z)) := p(x)q(s)v(z|x,s)$. The distribution of (S, X, Z) is labeled p_{SXZ}. The variable p remains fixed, and q will always denote a type corresponding to one of the choices of James.

The proof will require us to draw codewords at random. As stated already, we adapt this procedure to the one chosen in [11]. This is done as follows: We define the random variables $\mathbf{X}_{kl\gamma}$ ($1 \leq k \leq K$, $1 \leq l \leq L$, $1 \leq \gamma \leq \Gamma$) by $\mathbb{P}(\mathbf{X}_{kl\gamma} = x^n) := \frac{1}{|T_p|} \mathbb{1}_{T_p}(x^n)$ for all $k \in [K]$, $l \in [L]$, $\gamma \in [\Gamma]$, and $x^n \in \mathcal{X}^n$, where $K, L, \Gamma \in \mathbb{N}$. We write $\mathbf{x}_{kl\gamma}$ for the realizations of $\mathbf{X}_{kl\gamma}$, instead of $x_{kl\gamma}^n$. The random variable

$$\mathbf{X} := (\mathbf{X}_{kl\gamma})_{k,l,\gamma=1}^{K,L,\Gamma} \qquad (10.72)$$

is defined such that its components are mutually independent. The realizations of $\mathbf{X}$ are written $\mathbf{x}$. We use the projections $\pi_{kl\gamma}$ defined by $\pi_{kl\gamma}(\mathbf{x}) := \mathbf{x}_{kl\gamma}$. Further, we define projections as, e.g., $\mathbf{x}_\gamma := \pi_\gamma(\mathbf{x}) := (\mathbf{x}_{kl\gamma})_{k,l=1}^{K,L}$.

When calculating expectations of any of the $X_{kl\gamma}$ we need no reference to k, l, γ due to the independence of our random variables. We therefore add another random variable, X^n, distributed as $\mathbb{P}(X^n = x^n) = \frac{1}{|T_p|} \mathbb{1}_{T_p}(x^n)$.

A first and crucial step in the proofs of the technical lemmas 10.3 and 10.4 is to fix some $\delta > 0$ and $p \in \mathcal{P}(\mathcal{X})$ and define, for all $s^n \in \mathcal{S}^n$ and $z^n \in \mathcal{Z}^n$, the functions $\Theta_{s^n, z^n} : \mathcal{X}^n \to [0, b]$ (where $b := 2^{-n(H(Z|X,S) - f_2(n,\delta))}$ for some function f_2 to be subsequently stated) by

$$M(s^n, z^n) := \{x^n \in T_p : D(\bar{N}(\cdot|s^n, x^n, z^n) \| p_{SXZ}) \leq \delta\}, \tag{10.73}$$

$$\Theta_{s^n, z^n}(x^n) := \nu^{\otimes n}(z^n | s^n, x^n) \mathbb{1}_{M(s^n, z^n)}. \tag{10.74}$$

In order to further enhance readability, the dependence of both M and Θ on δ is suppressed here and in the following. All our proofs rely on a common strategy, which only deviates in one point: the codes that ensure reliable transmission. For non-symmetrizable AVWCs we rely on [11] and use the codes defined therein. This is sufficient to obtain all the results for the uncorrelated coding secrecy capacity.

The coding theorem for secret common randomness assisted secrecy capacity requires an additional definition of codes. This definition is as follows:

For every $n \in \mathbb{N}$, set $\Xi_n := \mathcal{P}_0^n(\mathcal{S})$. For every x^n, define (not necessarily disjoint) "decoding" sets by

$$\hat{D}_{x^n} := \bigcup_{\xi \in \Xi_n} T_{W_\xi, \delta}(x^n), \tag{10.75}$$

and, for a collection $\mathbf{x}_\gamma := (x_{kl\gamma})_{k,l=1}^{K,L}$ of codewords with fixed value of γ, set

$$D(\mathbf{x}_\gamma)_{kl} := \hat{D}_{x_{kl\gamma}} \cap \left(\bigcup_{k' \neq k} \bigcup_{l' \neq l} \hat{D}_{x_{k'l'\gamma}} \right)^c. \tag{10.76}$$

This defines the code $\mathcal{K}_n$. This definition allows the decoder to decode the randomization index l as well, an approach which works for AVWCs and compound (wiretap) channels with convex state sets via the minimax theorem. This code ensures reliable transmission only if Γ is sufficiently large.

In order to give a joint treatment of the subject we define the following events, where we implicitly assume a functional dependence $\delta = \delta(\tau)$ that we specify later during our proofs. Keep in mind that each $\mathbf{x}$ denotes a choice $(\mathbf{x}_{kl1})_{k,l=1}^{K,L}, \ldots, (\mathbf{x}_{kl\Gamma})_{k,l=1}^{K,L}$ of codebooks.

$$E_1 := \left\{ \mathbf{x} \,\middle|\, \forall s^n, z^n, k : \sum_{l,\gamma=1}^{L} \frac{\Theta_{s^n, z^n}(\mathbf{x}_{kl\gamma})}{L \cdot \Gamma} \in [(1 \pm 2^{-n \cdot \tau/4}) \mathbb{E} \Theta_{s^n, z^n}] \right\}, \tag{10.77}$$

$$E_2 := \left\{ \mathbf{x} \,\middle|\, \min_{s^n} \frac{1}{\Gamma} \sum_{\gamma=1}^{\Gamma} d_{s^n}(\mathcal{K}_\gamma) \geq 1 - 2 \cdot 2^{-n\delta/4} \right\}, \tag{10.78}$$

$$E_3 := \left\{ \mathbf{x} \,\middle|\, (10.68), (10.69), \text{ and } (10.70) \text{ hold for all } x^n, \hat{x}^n \in T_p \right\}. \tag{10.79}$$

The average success probability $d_{s^n}(\mathcal{K}_\gamma)$ (cf. Definition 10.3) is defined as

$$d_{s^n}(\mathcal{K}_\gamma) = \frac{1}{K \cdot L} \sum_{k,l=1}^{K,L} \sum_{x^n \in \mathcal{X}^n} e^\gamma(x^n|k) w_{s^n}(D_k^\gamma|x^n). \tag{10.80}$$

The event E_3 was proven to have high probability in [11], which leads to the following statement:

LEMMA 10.5 (Cf. [11])   *There is $c' > 0$ such that, if $\mathfrak{W}$ is non-symmetrizable, we have*

$$\mathbb{P}(E_3) \geq 1 - \Gamma \cdot \exp(-2^{n \cdot c'}). \tag{10.81}$$

Our main effort in the following will be to show that a similar bound is true for $\mathbb{P}(E_1)$ and $\mathbb{P}(E_2)$ under the right conditions on K, L, and Γ. With respect to these conditions, any of the intersections $E_i \cap \cdots \cap E_j$ will then have very high probability as well.

For the proofs of both Lemmas 10.3 and 10.4 it will be important to control the amount of information which leaks out to Eve. This will require us to prove that a careful random choice of codewords will be provably secure, and this is the main content of the following lemma [which CN the message transmission capabilities of the common randomness assisted codes defined in (10.75) and (10.76)].

LEMMA 10.6   *Let $K, L, \Gamma \in \mathbb{N}$. Let the random variable $\mathbf{X}$ be as defined in (10.72). Then for every $\tau > 0$ and $\beta > 0$ there is a $\delta > 0$ and $N \in \mathbb{N}$ such that for all $n \geq N$ and types $p \in \mathcal{P}_0^n(\mathcal{X})$, the following statements are true:*

1. *If $\frac{1}{n}\log(L \cdot \Gamma) \geq E(p) + \tau$ and $\min_{x: p(x) > 0} p(x) \geq \beta$, then $\mathbb{P}(E_1) \geq 1 - 2 \cdot |\mathcal{S} \times \mathcal{X} \times \mathcal{Z}|^n \cdot \exp(-2^{n \cdot \tau/6})$.*
2. *If $\frac{1}{n}\log(K \cdot L) \leq B(p) - \delta - 2 \cdot f_1(\sqrt{2 \cdot \delta})$, then $\mathbb{P}(E_2) \geq 1 - \exp(n \cdot \log(|\mathcal{S}|) - \Gamma \cdot 2^{-n\delta})$.*
3. *For every $\beta > 0$, $|\mathcal{X}|$, $|\mathcal{S}|$, and $|\mathcal{Z}|$, a functional dependence between δ and τ can be chosen such that $\lim_{\tau \to 0} \delta(\tau) = 0$.*

The number N depends on $|\mathcal{X}|$, $|\mathcal{S}|$, $|\mathcal{Z}|$ as well as on p (via the quantity $\beta := \min_{x \in \mathcal{X}: p(x) > 0} p(x)$) and on δ. The function $f_1 : [0, 1/2] \to \mathbb{R}_+$ is defined via $f_1(x) := -\sqrt{x/2} \cdot \log(x \cdot |\mathcal{Z}|^2)$.

The proof of Lemma 10.6 is based on the Chernoff bound Lemma 10.2 for the second statement. The proof of statement 1 makes additional use of Ahlswede's robustification technique, that we state here for the reader's convenience.

LEMMA 10.7 ([73, 74])   *Let $f : \mathcal{S}^n \to [0, 1]$. Then for all $\varepsilon \in [0, 1]$,*

$$\min_{q \in \mathcal{P}_0^n(\mathcal{S})} \sum_{s^n \in \mathcal{S}^n} f(s^n) q(s_1) \cdots q(s_n) \geq 1 - \varepsilon \Rightarrow \tag{10.82}$$

$$\frac{1}{n!} \sum_{\pi \in \Pi_n} f(\pi(s^n)) \geq 1 - 3 \cdot (n+1)^{|\mathcal{S}|} \cdot \varepsilon.$$

We set $f(s^n) := w_{s^n}(\hat{D}_{x^n}|x^n)$ for some fixed but arbitrary $x^n \in T_p$. This is enough to prove that the codes $\mathcal{K}_\gamma$ defined in (10.76) satisfy

$$\min_{s^n \in \mathcal{S}^n} \mathbb{E} d_{s^n}(\mathbf{X}_\gamma) \geq 1 - 2^{-n \cdot \delta/4} \tag{10.83}$$

for all large enough $n \in \mathbb{N}$ (the average success probability d_{s^n} is defined in Definition 10.3), whenever

$$K \cdot L \leq 2^{n(\min_q I(p;W_q) - \delta - 2 \cdot f_1(\sqrt{2 \cdot \delta}))} \leq 2^{n(\min_\xi I(p;W_\xi) - \delta - 2 \cdot f_1(\sqrt{2 \cdot \delta}))}. \tag{10.84}$$

The lower bound (10.83) is entirely independent of the choice of $s^n \in \mathcal{S}^n$. It thus follows from the Chernoff bound Lemma 10.2 that

$$\mathbb{P}\left(\forall s^n : \frac{1}{\Gamma} \sum_{\gamma=1}^{\Gamma} d_{s^n}(\mathcal{K}_\gamma) \leq (1-\epsilon)\mathbb{E} d_{s^n}(\mathcal{K})\right)$$
$$\leq \exp(n \cdot \log(|\mathcal{S}|) - \Gamma \cdot \epsilon^2 \cdot (1 - 2^{-n \cdot \delta/4})/3). \tag{10.85}$$

Choose $\epsilon = 2^{-n \cdot \delta/4}$ to obtain the statement. The functional dependence $\tau \mapsto \delta(\tau)$ may be set to $\delta(\tau) = \tau$.

10.4.3 Proofs of Lemmas

Proof of Lemma 10.3

We know from Lemma 10.6 that (if $\frac{1}{n}\log(K \cdot L) \leq B(p) - \delta - 2 \cdot f_1(\sqrt{2 \cdot \delta})$ for some $\delta > 0$ and n is large enough)

$$\mathbb{P}(E_2) \geq 1 - \exp\left(n \cdot \log(|\mathcal{S}|) - \Gamma \cdot \epsilon^2 \cdot \left(\frac{1}{3} - 2^{-n \cdot \delta/2}\right)\right). \tag{10.86}$$

Let $p \in \mathcal{P}_0^n(\mathcal{X})$, $\tau > 0$, and $\delta = \delta(\tau)$. Then Eq. (10.85) together with Lemma 10.6 and a union bound delivers

$$\frac{1}{n}\log(\Gamma_n \cdot L_n) \geq E(p) + \tau, \quad B(p) - \delta - 2 \cdot f_1(\sqrt{2 \cdot \delta}) \geq \frac{1}{n}\log(K_n \cdot L_n). \tag{10.87}$$

It then follows that, for all large enough n,

$$\mathbb{P}(E_1 \cap E_2) > 0. \tag{10.88}$$

Thus, there is a realization $\mathbf{x}$ of $\mathbf{X}$ such that for this particular realization we have

$$\forall s^n, z^n, k : \frac{1}{L \cdot \Gamma} \sum_{l,\gamma=1}^{L,\Gamma} \Theta_{s^n,z^n}(\mathbf{x}_{kl\gamma}) \in [(1 \pm 2^{-n\tau/4})\mathbb{E}\Theta_{s^n,z^n}], \tag{10.89}$$

$$\min_{s^n \in \mathcal{S}^n} \frac{1}{\Gamma} \sum_{\gamma=1}^{\Gamma} d_{s^n}(\mathcal{K}_\gamma) \geq 1 - 2 \cdot 2^{-n\delta/2}. \tag{10.90}$$

By application of the triangle inequality for $\|\cdot\|_1$ and using (10.89), we additionally get that

$$\frac{1}{L \cdot \Gamma} \sum_{l,\gamma=1}^{L,\Gamma} \|v_{s^n}(\cdot|\mathbf{x}_{kl\gamma}) - \Delta(s^n,\cdot,\mathbf{x}_{kl\gamma})\|_1 \leq 2^{-n \cdot \nu(\tau)} \qquad (10.91)$$

holds uniformly in $k \in [K]$ for all large enough n and setting $\nu(\tau) := \min\{\delta(\tau), \tau\}/5$ (note that $\nu(\tau) = \tau/5$ is a valid choice). This completes the proof of Lemma 10.3.

Proof of Lemma 10.4

The proof is similar to that of Lemma 10.3. As we know already that for some $c' > 0$ and all large enough $n \in \mathbb{N}$

$$\mathbb{P}(E_3) \geq 1 - \Gamma \cdot \exp(2^{-n \cdot c'}) \qquad (10.92)$$

holds from [11], there is not much left to prove. Only $\mathbb{P}(E_1)$ needs to be controlled in order to get (10.71). From Lemma 10.6 we have

$$\mathbb{P}(E_1^{\complement}) \leq 2 \cdot |\mathcal{X} \times \mathcal{S} \times \mathcal{Z}|^n \cdot \exp(-2^{n \cdot \tau/2}), \qquad (10.93)$$

if we choose $\delta = \delta(\tau)$. Combining the above we get

$$\mathbb{P}(E_1 \cap E_3) \geq 1 - (2 + \Gamma) \cdot \exp(-2^{n \cdot c''}), \qquad (10.94)$$

for some $c'' > 0$ and for all large enough n. If Γ scales at most exponentially there will thus exist $N_0 \in \mathbb{N}$ such that for all $n \geq N_0$ there exists a choice $\mathbf{x} = (\mathbf{x}_{kl\gamma})_{k,l,\gamma=1}^{K,L,\Gamma}$ satisfying all conditions in Lemma 10.4 and, in addition, the estimate

$$\forall s^n, z^n, k : \frac{1}{L \cdot \Gamma} \sum_{l,\gamma=1}^{L,\Gamma} \Theta_{s^n,z^n}(\mathbf{x}_{kl\gamma}) \notin [(1 \pm 2^{-n\tau/4})\mathbb{E}\Theta_{s^n,z^n}]. \qquad (10.95)$$

Repeated use of the triangle inequality for $\|\cdot\|_1$ proves validity of (10.71). Thus Lemma 10.4 is proven.

Proof of Lemma 10.1

Let $\mathfrak{U}$ be symmetrizable. Let $Q \in C(\mathcal{A}, \mathcal{T})$ be the symmetrizing channel, meaning that for all $a, a' \in \mathcal{A}$ the equality

$$(U \circ (Id \otimes Q))(a, a') = (U \circ (Id \otimes Q))(a', a) \qquad (10.96)$$

holds true. Then, again for all $a, a' \in \mathcal{A}$,

$$(U \circ (T \otimes QT))(a, a') = \sum_{a'', a''' \in \mathcal{A}} \sum_{r \in \mathcal{R}} u(\cdot|a'', r) t(a''|a) q(r|a''') t(a'''|a') \qquad (10.97)$$

$$= \sum_{a'', a''' \in \mathcal{A}} \sum_{r \in \mathcal{R}} u(\cdot|a''', r) t(a''|a) q(r|a'') t(a'''|a') \qquad (10.98)$$

$$= (U \circ (T \otimes QT))(a', a). \qquad (10.99)$$

Thus, $\mathfrak{U}'$ is symmetrizable and the symmetrizing map is QT.

10.4.4 Proofs of Theorems

Proof of Theorem 10.1

We first sketch the direct part of the proof. Take any finite alphabet $\mathcal{U}$, natural number $r \in \mathbb{N}$, and any channel $U \in C(\mathcal{X}^r, \mathcal{X}^r)$. We define an AVWC $(\hat{\mathfrak{W}}, \hat{\mathfrak{V}})$ via $\hat{W}_{s^r} := W_{s^r} \circ U$ and $\hat{V}_{s^r} := V_{s^r} \circ U$. Then for $K, L, m \in \mathbb{N}$ and any $p \in \mathcal{P}_0^m(\mathcal{U})$ with $\min_{u:p(u)>0} = \beta > 0$ we choose codewords $(\mathbf{u}_{kl})_{k,l=1}^{K,L}$ out of $\mathcal{U}^m$ independently at random according to $|T_p|^{-1} \mathbb{1}_{T_p}$ and use the decoder that was defined in (10.76), but based on the alphabet $\mathcal{U}$ instead of $\mathcal{X}$. Using the same techniques as those presented for the proof of Lemma 10.6 we show that for $\log(K \cdot L) \le \min_q I(p; W_q^{\otimes r}) - \delta'$ and $\log(K) \ge \max_{s^r \in \mathcal{S}^r} I(p; V_{s^r} \circ U) + \delta'$ (where $\delta' = \delta'(\delta)$ with $\lim_{\delta \to 0} \delta'(\delta) = 0$) there is a choice of $(\mathbf{x}_{kl})_{k,l=1}^{K,L}$ such that for $n = r \cdot m$ we have

$$\frac{1}{K \cdot L} \sum_{k,l=1}^{K,L} (w_q^{\otimes r} \circ u)^{\otimes m}(D_{kl}|\mathbf{x}_{kl}) \ge 1 - 2^{-n \cdot \delta} \tag{10.100}$$

and

$$\max_{k \in [K], s^n} \left\| \frac{1}{L} \sum_{l=1}^{L} V_{s^n}(\delta_{\mathbf{x}_{kl}}) - V_{s^n}(|T_p|^{-1} \mathbb{1}_{T_p}) \right\|_1 \le 2^{-n \cdot \delta}. \tag{10.101}$$

To every permutation $\rho \in S_n$ we associate the permuted decoder $\mathcal{K}_\rho$ arising from permuting both the codewords and the decoding sets using ρ. We draw a number $\Gamma \in \mathbb{N}$ of permutations independently at random, according to the uniform distribution. We define the function $\rho \mapsto D_{s^n}(\rho)$ via

$$D_{s^n}(\rho) := \frac{1}{K \cdot L} \sum_{k,l=1}^{K,L} (w_{s^r} \circ U)_{s^m}(\rho(D_m)|\rho(\mathbf{x}_{kl})), \tag{10.102}$$

and give a lower bound on its expected value via Ahlswede's robustification technique as

$$\mathbb{E} D_{s^n} = \frac{1}{n!} \sum_{\rho \in S_n} D_{s^n}(\rho) \ge 1 - 3 \cdot (n+1)|\mathcal{S}| \cdot 2^{-n \cdot \delta}. \tag{10.103}$$

Then, applying Lemma 10.2, we have, for every $s^n \in \mathcal{S}^n$,

$$\mathbb{P}\left(\frac{1}{\Gamma} \sum_{\gamma=1}^{\Gamma} \frac{1}{K \cdot L} \sum_{k,l=1}^{K,L} w_{s^n}(\rho_\gamma(D_m)|\rho_\gamma(\mathbf{x}_{k,l})) \notin [(1 \pm \epsilon) \mathbb{E} D_{s^n}(\mathcal{K}_\gamma)] \right)$$

$$\le \exp(-\Gamma \cdot \epsilon^2 \cdot \mathbb{E} d_{s^n}(\mathcal{K})/3) \tag{10.104}$$

$$\le \exp(-\Gamma \cdot \epsilon^2 \cdot 2/3). \tag{10.105}$$

Here, the probability is calculated with respect to our random choice of permutations. Setting $\Gamma = n^4$ and $\epsilon = n^{-1}$ gives

$$\mathbb{P}\left(\forall s^n \in \mathcal{S}^n : \frac{1}{\Gamma}\sum_{\gamma=1}^{\Gamma} \frac{1}{M}\sum_{m=1}^{M} w_{s^n}(\rho_\gamma(D_m)|\rho_\gamma(\mathbf{x}_m)) \geq 1 - \frac{2}{n}\right)$$

$$\leq \exp(n \cdot \log|\mathcal{S}| - n^2 \cdot 2/3). \tag{10.106}$$

This implies the existence of a reliable coding scheme that asymptotically needs only n^4 permutations and operates at rates

$$\frac{1}{r}\min_{q \in \mathcal{P}(\mathcal{S})} I(p; W_q^{\otimes r} \circ U) - \max_{s^r \in \mathcal{S}^r} I(p; V_{s^r} \circ U) - 2\delta' \tag{10.107}$$

for every $m, r \in \mathbb{N}$, $p \in \mathcal{P}_0^m(\mathcal{U})$, and $\delta' > 0$. Continuity of the mutual information in p then lets us conclude the even stronger statement that for every $p \in \mathcal{P}(\mathcal{U})$ and $\delta' > 0$, the number

$$\frac{1}{r}\min_{q \in \mathcal{P}(\mathcal{S})} I(p; W_q^{\otimes r} \circ U) - \max_{s^r \in \mathcal{S}^r} I(p; V_{s^r} \circ U) - 3\delta' \tag{10.108}$$

is an achievable rate under the maximum secrecy criterion. It is then straightforward to prove from here that

$$C_{\mathrm{S,ran}}^{\mathrm{mean}}(\mathfrak{W},\mathfrak{V}) \geq C_{\mathrm{S,ran}}^{\mathrm{max}}(\mathfrak{W},\mathfrak{V}) \geq C^*(\mathfrak{W},\mathfrak{V}) \tag{10.109}$$

holds. For the converse, it is sufficient to develop the corresponding upper bound on $C_{\mathrm{S,ran}}^{\mathrm{mean}}(\mathfrak{W},\mathfrak{V})$. This uses standard tools, with the exception of one conceptual difference: the fact that common randomness is used prohibits a straightforward application of the data processing inequality. It is thus necessary to limit the amount of shared randomness of an arbitrary shared randomness assisted code in order to overcome this difficulty. This is the content of the following lemma that we state here in a slightly simplified version as compared to its original version in [30].

LEMMA 10.8 ([30]) *Let* $R < C_{\mathrm{S,ran}}^{\mathrm{mean}}(\mathfrak{W},\mathfrak{V})$ *and* $\varepsilon > 0$. *There exists a positive integer* $L = L(R,\varepsilon)$ *such that for all sufficiently large n there exists a common randomness assisted code* $\mathcal{K}_n$ *satisfying*

$$\frac{1}{n}\log K \geq R - \varepsilon, \tag{10.110}$$

$$\mathrm{err}(\mathcal{K}_n) \leq \lambda, \tag{10.111}$$

$$\max_{s^n \in \mathcal{S}^n} I(\mathfrak{K}_n; \mathfrak{Z}_{s^n}^n | \Gamma_n) \leq \delta, \tag{10.112}$$

$$|\Gamma| \leq L. \tag{10.113}$$

Let $R < C_{\mathrm{S,ran}}^{\mathrm{mean}}(\mathfrak{W},\mathfrak{V})$ and $\mathcal{K}_n$ be the common randomness assisted code from Lemma 10.8. By [3, Lemma 12.3], the average error incurred by any uncorrelated code $\mathcal{K}_n$ used over the AVC $\mathfrak{W}$ equals the average error of $\mathcal{K}_n$ over the AVC determined by the convex hull of $\{W_s : s \in \mathcal{S}\}$, i.e., the AVC $\{W_q : q \in \mathcal{P}(\mathcal{S})\}$. This is a simple consequence of the fact that the average error is affine in the channel and carries over to correlated random codes. Hence, (10.111) implies

$$\max_{q \in \mathcal{P}(\mathcal{S})} \frac{1}{K} \sum_{k=1}^{K} \sum_{\gamma \in \Gamma_n} \sum_{x^n \in \mathcal{A}^n} E^\gamma(x^n|k) w_q^{\otimes n}\big((\mathcal{D}_j^\gamma)^c|x^n\big) P_{G_n}(\gamma) \leq \varepsilon. \quad (10.114)$$

Due to Fano's inequality [3, Lemma 3.8], (10.114) implies, for every $q \in \mathcal{P}(\mathcal{S})$,

$$H(\mathfrak{K}_n|\mathfrak{K}'_{n,q}\Gamma_n) = \sum_{\gamma=1}^{\Gamma} H(\mathfrak{K}_n|\mathfrak{K}'_{n,q}, \Gamma_n = \gamma) \frac{1}{\Gamma} \quad (10.115)$$

$$\leq 1 + \sum_{\gamma=1}^{\Gamma} \mathbb{P}(\mathfrak{K}_n \neq \mathfrak{K}'_n|G_n = \gamma) \frac{1}{\Gamma} \log K \quad (10.116)$$

$$= 1 + \varepsilon \log K. \quad (10.117)$$

The distribution of $(\mathfrak{K}_n, \mathfrak{K}'_{n,q}, \mathfrak{Y}_{n,q}, \Gamma_n)$ is given by

$$\mathbb{P}\big((\mathfrak{K}_n, \mathfrak{K}'_n, \mathfrak{Y}_{n,q}, \Gamma_n) = (k, k', y^n, \gamma)\big) = \sum_{x^n} \frac{e^\gamma(x^n|k)}{K \cdot \Gamma} \mathbb{1}_{D_{k'}^\gamma}(y^n) w_q^{\otimes n}(y^n|x^n). \quad (10.118)$$

Hence, the independence of $\mathfrak{K}_n$ and Γ_n yields

$$\log K = I(\mathfrak{K}_n; \mathfrak{K}'_{n,q}|\Gamma_n) + H(\mathfrak{K}_n|\mathfrak{K}'_{n,q}, \Gamma_n) \quad (10.119)$$

$$\leq I(\mathfrak{K}_n; \mathfrak{K}'_{n,q}|\Gamma_n) + 1 + \varepsilon \log K, \quad (10.120)$$

so by rearranging, using the data processing inequality, and taking (10.112) into account, we have, for every $q \in \mathcal{P}(\mathcal{S})$ and $s^n \in \mathcal{S}^n$,

$$(1 - \varepsilon) \log K \leq I(\mathfrak{K}_n; \mathfrak{K}'_{n,q}|\Gamma_n) - I(\mathfrak{K}_n; \mathfrak{Z}_{s^n}^n|\Gamma_n) + 1 + \varepsilon. \quad (10.121)$$

We have to get rid of Γ_n. The only reasonable way to achieve this seems to be through the use of the convexity of the mutual information in the channel argument. But while this is a valid choice for the "secrecy term," it is certainly invalid for the "legal" term. This is due to the fact that Γ_n is independent of $\mathfrak{K}_n$, but not of $\mathfrak{K}'_{n,q}$ or $\mathfrak{Y}_{n,q}$. An application of the data processing inequality is thus only possible conditioned on Γ_n. It is here where the importance of Lemma 10.8 becomes evident: The number Γ is bounded and independent of n for n sufficiently large, and hence we can write

$$I(\mathfrak{K}_n; \mathfrak{K}'_{n,q}|\Gamma_n) = H(\mathfrak{K}_n) - H(\mathfrak{K}_n|\mathfrak{Y}_{n,q}\Gamma_n) \quad (10.122)$$

$$\leq H(\mathfrak{K}_n) - H(\mathfrak{K}_n|\mathfrak{Y}_{n,q}) + H(\Gamma_n) \quad (10.123)$$

$$\leq I(\mathfrak{K}_n; \mathfrak{Y}_{n,q}) + \log L(R, \varepsilon), \quad (10.124)$$

where we employed the fact that $H(S) \leq H(S,T) = H(S|T) + H(T)$. Thus, if n is sufficiently large, we obtain

$$\frac{1}{n}\log K \leq \frac{1}{n(1-\varepsilon)}\left(\min_{q\in\mathcal{P}(\mathcal{S})} I(\mathfrak{K}_n;\mathfrak{K}_{n,q}|\Gamma_n) - \max_{s^n\in\mathcal{S}^n} I(\mathfrak{K}_n;\mathfrak{Z}^n_{s^n}|\Gamma_n) + 1 + \varepsilon\right)$$

$$\leq \frac{1}{n(1-\varepsilon)}\left(\min_{q\in\mathcal{P}(\mathcal{S})} I(\mathfrak{K}_n;\mathfrak{Y}_{q,n}) - \max_{s^n\in\mathcal{S}^n} I(\mathfrak{K}_n;\mathfrak{Z}_{s^n})\right) \quad (10.125)$$

$$+ \frac{\log L(R,\varepsilon) + 1 + \varepsilon}{n(1-\varepsilon)}.$$

For n sufficiently large, as $L(R,\varepsilon)$ is independent of n, the second term of (10.126) is bounded above by ε. In combination with (10.110), and as ε was arbitrary, this implies $R \leq C^*(\mathfrak{W},\mathfrak{V})$, hence $C^{\text{mean}}_{\text{S,ran}}(\mathfrak{W},\mathfrak{V}) \leq C^*(\mathfrak{W},\mathfrak{V})$, and therefore also $C^{\max}_{\text{S,ran}}(\mathfrak{W},\mathfrak{V}) \leq C^*(\mathfrak{W},\mathfrak{V})$. This completes the proof of Theorem 10.1.

REMARK 10.7 *As the average error is affine in the channel, one can even pass to a minimum over $\tilde{q}\in\mathcal{P}(\mathcal{S}^n)$ in (10.114), while replacing $W_q^{\otimes n}$ with $W_{\tilde{q}}^n$. Then, the rest of the proof can be performed as above, and as a result one obtains that even the inequality*

$$C^{\text{mean}}_{\text{S,ran}}(\mathfrak{W},\mathfrak{V}) \leq \lim_{r\to\infty}\frac{1}{r}\max_{p\in\mathcal{P}(\mathcal{X}^r)}\max_{U\in C(\mathcal{X}^r,\mathcal{X}^r)} \quad (10.126)$$

$$\left(\min_{\tilde{q}\in\mathcal{P}(\mathcal{S}^r)} I(p;W_{\tilde{q}}^n \circ U) - \max_{s^r\in\mathcal{S}^r} I(p;V_{s^r}\circ U)\right)$$

holds, thus proving that an equivalent way of writing C^* is via the equality

$$C^*(\mathfrak{W},\mathfrak{V}) = \lim_{r\to\infty}\frac{1}{r}\max_{p\in\mathcal{P}(\mathcal{X}^r)}\max_{U\in C(\mathcal{X}^r,\mathcal{X}^r)} \quad (10.127)$$

$$\left(\min_{q\in\mathcal{P}(\mathcal{S}^r)} I(p;W_q^r \circ U) - \max_{s^r\in\mathcal{S}^r} I(p;V_{s^r}\circ U)\right).$$

Proof of Theorem 10.3

We start with the converse part of the coding theorem. The main ingredients of this proof are Fano's inequality [3, Lemma 3.8], data processing, and the almost convexity of the entropy. Let a sequence $\mathcal{K} = (\mathcal{K}_n)_{n=1}^{\infty}$ of common randomness assisted codes be given such that for all $n\in\mathbb{N}$ we have

$$\min_{s^n\in\mathcal{S}^n}\frac{1}{\Gamma_n\cdot K_n}\sum_{\gamma,k=1}^{\Gamma_n,K_n} e^{\gamma}(x^n|k)w_{s^n}(D_k^{\gamma}|x^n) \geq 1-\epsilon_n, \quad (10.128)$$

$$\max_{s^n\in\mathcal{S}^n} I(\mathfrak{K}_n;\mathfrak{Z}_{s^n}) \leq \epsilon_n, \quad (10.129)$$

where $\limsup_{n\to\infty}\epsilon_n = 0$. Set $R := \liminf_{n\to\infty}\frac{1}{n}\log K_n$ and $G := \lim_{n\to\infty}\frac{1}{n}\log\Gamma_n$. In addition to the random variable defined in Definition 10.3, consider $(\mathfrak{K}_n,\mathfrak{K}'_{q,n},\Gamma_n)$ distributed as

$$\mathbb{P}((\mathfrak{K}_n,\mathfrak{Y}_q^n,\mathfrak{K}'_{q,n},\Gamma_n) = (k,k',\gamma)) = \sum_{s^n\in\mathcal{S}^n} q^{\otimes n}(s^n)\mathbb{P}(\mathfrak{K}_n,\mathfrak{Y}_{q,n},\mathfrak{K}'_n,\Gamma_n). \quad (10.130)$$

Then for all $n \in \mathbb{N}$, $q \in \mathcal{P}(\mathcal{S})$, and $s^n \in \mathcal{S}^n$, Fano's inequality implies

$$(1 - \epsilon_n) \log K_n \le I(\mathfrak{K}_n; \mathfrak{K}'_{q,n} | \Gamma_n) - I(\mathfrak{K}_n; \mathfrak{Z}_{s^n}) + 1 + \epsilon_n. \quad (10.131)$$

Applying the data processing inequality, we get

$$(1 - \epsilon_n) \log K_n \le I(\mathfrak{K}_n; \mathfrak{Y}_q^n | \Gamma_n) - I(\mathfrak{K}_n; \mathfrak{Z}_{s^n}) + 1 + \epsilon_n. \quad (10.132)$$

From, e.g., [3, Lemma 3.4] and the independence of the random variables $\mathfrak{K}_n$ and $\mathfrak{G}_n$, it follows that the asymptotic scaling of the rate $\liminf_{n \to \infty} \frac{1}{n} \log K_n$ is bounded above as

$$(1 - \epsilon_n) \log K_n \le I(\mathfrak{K}_n; \mathfrak{Y}_q^n) - I(\mathfrak{K}_n; \mathfrak{Z}_{s^n} | \Gamma_n) + H(\Gamma_n) + 1 + \epsilon_n. \quad (10.133)$$

Since this bound holds for all $q \in \mathcal{P}(\mathcal{S})$ and $s^n \in \mathcal{S}^n$, we further have

$$\log K_n \le \frac{1}{1 - \epsilon_n} \left(\min_{q \in \mathcal{P}(\mathcal{S})} I(\mathfrak{K}_n; \mathfrak{Y}_q^n) - \max_{s^n \in \mathcal{S}^n} I(\mathfrak{K}_n; \mathfrak{Z}_{s^n}) \right)$$

$$+ \frac{1 + \epsilon_n}{1 - \epsilon_n} + \frac{\log \Gamma_n}{1 - \epsilon_n}. \quad (10.134)$$

Define the distribution $p \in \mathcal{P}([K_n])$ and the channel $U \in C([K_n], \mathcal{X}^n)$ by

$$p(k) = \frac{1}{K_n}, \quad U(x^n | k) := \sum_{\gamma=1}^{\Gamma_n} \frac{1}{\Gamma} e^{\gamma}(x^n | k) \quad (k \in [K_n], \, x^n \in \mathcal{X}^n). \quad (10.135)$$

Then we arrive at

$$\log K_n \le \frac{1}{1 - \epsilon_n} \left(\min_{q \in \mathcal{P}(\mathcal{S})} I(p; W_q^{\otimes n} \circ U) - \max_{s^n \in \mathcal{S}^n} I(p; V_{s^n} \circ U) \right)$$

$$+ \frac{1 + \epsilon_n}{1 - \epsilon_n} + \frac{\log \Gamma_n}{1 - \epsilon_n}. \quad (10.136)$$

A more relaxed upper bound can be derived by optimizing over all $p \in \mathcal{P}([K_n])$ and $U \in C([K_n], \mathcal{X}^n)$. We then obtain, by further increasing the size of the input alphabet from K_n to $\mathcal{X}^n$, that

$$R \le \lim_{n \to \infty} \frac{1}{n} \max_{p \in \mathcal{X}^n} \max_{U \in C(\mathcal{X}^n, \mathcal{X}^n)}$$

$$\left(\min_{q \in \mathcal{P}(\mathcal{S})} I(p; W_q^{\otimes n} \circ U) - \max_{s^n \in \mathcal{S}^n} I(p; V_{s^n} \circ U) \right) + G. \quad (10.137)$$

This bound is valid since $K_n \le |\mathcal{X}^n|$ for every reliably working code and, therefore, $\mathcal{P}([K_n]) \subset \mathcal{P}(\mathcal{X}^n)$ under a suitable embedding $[K_n] \subset \mathcal{X}^n$.

From Theorem 10.1 we know that the capacity $C_{S,\text{ran}}^{\text{mean}}$ equals the left term in the above sum. Therefore we have proven the desired result.

Another obvious bound on the capacity arises by ignoring all security issues: since $\mathcal{K}$ ensures an asymptotically perfect transmission, we have

$$\lim_{n \to \infty} \frac{1}{n} \log K_n \le \max_{p \in \mathcal{P}(\mathcal{X})} \min_{q \in \mathcal{P}(\mathcal{S})} I(p; W_q). \quad (10.138)$$

This establishes the converse part of the coding theorem.

We now focus on the direct part of the proof of the coding theorem. Fix $G > 0$ and define $p := \arg\max_{p \in \mathcal{P}(\mathcal{X})}(B(p) - E(p))$, where $E(p)$ and $B(p)$ are defined in (10.64). Set $G' := \max\{E(p), G\}$. Intuitively, this is the amount of common randomness which is put to use in the obfuscation of Eve. Choose a $\tau > 0$ such that $\nu(\tau)$ from Lemma 10.3 satisfies $\nu(\tau) < G'$. Let $n \in \mathbb{N}$ be such that for all $n \geq N$ there is $p_n \in \mathcal{P}_0^n(\mathcal{X})$ such that $|B(p_n) - B(p)| \leq \max\{\tau, \nu(\tau)\}$ and $|E(p_n) - E(p)| \leq \max\{\tau, \nu(\tau)\}$. This can be achieved by approximating p through types p_n as in [15, Lemma 4] and using the fact that both B and E are continuous functions. Take three sequences $(K_n)_{n=1}^\infty$, $(L_n)_{n=1}^\infty$, $(\Gamma_n)_{n=1}^\infty$ of natural numbers. Without loss of generality, we can ensure that $(\Gamma_n)_{n \in \mathbb{N}}$ satisfies both $\Gamma_n \leq 2^{n \cdot G'}$ for all $n \in \mathbb{N}$ and $\lim_{n \to \infty} \frac{1}{n} \log \Gamma_n = G'$. Now let $n \in \mathbb{N}$ satisfying $n \geq N$ be fixed but also large enough such that in addition

$$E(p) - G' + 4\tau \geq \frac{1}{n}\log(L_n) \geq E(p) - G' + 2\tau, \tag{10.139}$$

$$B(p) - E(p) + G' - 4(\tau + \nu(\tau)) \geq \frac{1}{n}\log(K_n) \tag{10.140}$$

$$\geq B(p) - E(p) + G' - 2(\tau + \nu(\tau)) \tag{10.141}$$

hold. This implies both

$$\frac{1}{n}\log(K_n \cdot L_n) \leq B(p_n) - \nu(\tau) \tag{10.142}$$

and

$$\frac{1}{n}\log(L_n \cdot \Gamma_n) \geq E(p) + 2\tau \geq E(p_n) + \tau. \tag{10.143}$$

Asymptotically, we therefore get the inequality

$$\liminf_{n \to \infty} \frac{1}{n}\log(K_n) \geq B(p) - E(p) + G - 4 \cdot (\tau + \nu(\tau)). \tag{10.144}$$

At the same time, the prerequisites of Lemma 10.3 are met such that a reliable sequence of codes that is also secure with respect to $\|\cdot\|_1$ exists. For all large enough $n \in \mathbb{N}$ we have

$$\min_{s^n} \sum_{\gamma=1}^{\Gamma} \frac{1}{\Gamma} \sum_{k,l=1}^{K,L} \frac{1}{K \cdot L} w_{s^n}(D_{kl}^\gamma | \mathbf{x}_{kl\gamma}) \geq 1 - 2^{-n \cdot \nu(\tau)}, \tag{10.145}$$

$$\max_{s^n, k} \left\| \frac{1}{L \cdot \Gamma} \sum_{l,\gamma=1}^{L,\Gamma} v_{s^n}(\cdot | \mathbf{x}_{kl\gamma}) - \mathbb{E} v_{s^n}(\cdot | X^n) \right\|_1 \leq 2^{-n \cdot \nu(\tau)}. \tag{10.146}$$

It can already be seen that this yields reliable communication at any rate strictly below $B(p) - E(p) + G$. We have thus proved the achievability of rates close to $B(p) - E(p) + G$. It is clear that time sharing between a trivial strategy where only one codeword is being transmitted (which is then automatically perfectly secure) and the strategy that was proven to work in the above implies the achievability of all rates $R \in [0, |B(p) - E(p) + G|^+]$. From [3, Lemma 2.7], we know that our exponential bound (10.67) asymptotically leads to fulfillment of the strong secrecy criterion.

We have thus proven that, for each $\tau' > 0$, the number

$$\max_{p \in \mathcal{P}(\mathcal{X})} \left(\min_{q \in \mathcal{P}(\mathcal{S})} I(p; W_q) - \max_{q \in \mathcal{P}(\mathcal{S})} I(p; V_q) \right) + G - \tau' \quad (10.147)$$

is an achievable rate. We proceed by adding channels U at the sender and using blocks of the original channels together. For every $r \in \mathbb{N}$, $G > 0$, $\delta > 0$, and $p \in \mathcal{P}(\mathcal{X}^r)$, let $U \in C(\mathcal{X}^r, \mathcal{X}^r)$ be any channel. From the previous it is clear that there exist sequences $\mathcal{K} = (\mathcal{K}_m)_{m=1}^{\infty}$ such that for every $s^{r \cdot m} \in (\mathcal{S}^r)^m = \mathcal{S}^{r \cdot m}$ we have

$$\frac{1}{K_m} \frac{1}{\Gamma_m} \sum_{k=1}^{K_m} \sum_{\gamma=1}^{\Gamma_m} \sum_{x^{r \cdot m}} e(x^{r \cdot m}|k, \gamma) w_{s^{r \cdot m}}(D_k^{\gamma}|x^{r \cdot m}) \geq 1 - \epsilon_m, \quad (10.148)$$

where $\mathbf{x}_{k,\gamma} \in U_r^m$ are codewords (each $x_{k,\gamma,i}$ is an element of $\mathcal{X}^r$) for $(W_{s^r} \circ U)_{s^r \in \mathcal{S}^r}$, and the stochastic encoder is $e(x^{r \cdot m}|k, \gamma) = \prod_{i=1}^{m} u(x_{ij}|x_{k,\gamma,i})$ for $x^{r \cdot m}$. The rate of the code is given by

$$\liminf_{m \to \infty} \frac{1}{m} \log K_m \geq \min_{q \in \mathcal{P}(\mathcal{S}^r)} I(p; W_q^r \circ U) \quad (10.149)$$

$$- \max_{s^r \in \mathcal{S}^r} I(p; V_{s^r} \circ U) + r \cdot G - \delta.$$

For every $n = r \cdot m + t$ with $r > t > 0$ one can now define a trivial extension $\hat{\mathcal{K}}_n$ of the code $\mathcal{K}_n$ such that this code is asymptotically reliable, and operates at an asymptotic rate satisfying

$$\liminf_{n \to \infty} \frac{1}{n} \log \hat{K}_n \geq \frac{1}{r} \left(\min_{q \in \mathcal{P}(\mathcal{S}^r)} I(p; W_q^r \circ U) \quad (10.150) \right.$$

$$\left. - \max_{s^r \in \mathcal{S}^r} I(p; V_{s^r} \circ U) + r \cdot G - \delta \right).$$

To see that every number $C^*(\mathfrak{W}, \mathfrak{V}) - \epsilon$ is an achievable rate, take r, U, and p such that

$$C^*(\mathfrak{W}, \mathfrak{V}) - \epsilon/2 \leq \frac{1}{r} \left(\min_{q \in \mathcal{P}(\mathcal{S}^r)} I(p; W_q^r \circ U) - \max_{s^r \in \mathcal{S}^r} I(p; V_{s^r} \circ U) \right). \quad (10.151)$$

This is possible due to Remark 10.7, where it was proven that the equality

$$C^*(\mathfrak{W}, \mathfrak{V}) = \lim_{r \to \infty} \frac{1}{r} \max_{p \in \mathcal{P}(\mathcal{X}^n)} \max_{U \in C(\mathcal{X}^n, \mathcal{X}^n)}$$

$$\left(\min_{q \in \mathcal{P}(\mathcal{S}^r)} I(p; W_q^r \circ U) - \max_{s^r \in \mathcal{S}^r} I(p; V_{s^r} \circ U) \right) \quad (10.152)$$

holds. We set $\delta = r \cdot \epsilon/4$. Then our preceding arguments imply that there is a sequence $\hat{\mathcal{K}}$ of asymptotically reliable codes at an asymptotic rate

$$\liminf_{n \to \infty} \frac{1}{n} \log \hat{K}_n \geq \frac{1}{r} \left(\min_{q \in \mathcal{P}(\mathcal{S}^r)} I(p; W_q^r \circ U) - \max_{s^r \in \mathcal{S}^r} I(p; V_{s^r} \circ U) + r \cdot G \right) - \frac{\epsilon}{4}$$

$$\geq C^*(\mathfrak{W}, \mathfrak{V}) + G - \epsilon. \quad (10.153)$$

This proves the direct part of the coding theorem.

Proof of Theorem 10.4

We prove the properties of C_S in the same order as they were stated:

1. This is clear from [5], where it was proven that symmetrizability makes it impossible to reach reliable transmission of messages.

2. The strategy of the proof is to use Lemma 10.4 with $\Gamma = 1$. The reason for this is that, by assumption, $\mathfrak{W}$ is non-symmetrizable. Now, we know from Example 10.1 that this does not imply that every $\mathfrak{W} \circ U$ is non-symmetrizable as well. More precisely, to a given $r \in \mathbb{N}$ there may exist an alphabet $\mathcal{X}^r$, a $p \in \mathcal{P}(\mathcal{X}^r)$, and a channel $U_r \in C(\mathcal{X}^r, \mathcal{X}^r)$ such that

$$\min_{q \in \mathcal{P}(\mathcal{S}^r)} I(p; W_q \circ U_r) - \max_{s^r \in \mathcal{S}^r} I(p; V_{s^r} \circ U_r) \tag{10.154}$$

$$= \max_{p' \in \mathcal{P}(\mathcal{X}^r)} \max_{U'_r \in C(\mathcal{X}^r, \mathcal{X}^r)} \min_{q \in \mathcal{P}(\mathcal{S}^r)} I(p'; W_q \circ U'_r) \tag{10.155}$$

$$- \max_{s^r \in \mathcal{S}^r} I(p'; V_{s^r} \circ U'_r)$$

$$\geq C_{S,\mathrm{ran}}^{\mathrm{mean}}(\mathfrak{W}, \mathfrak{V}) - \epsilon \tag{10.156}$$

but, additionally, $(W_{s^r} \circ U_r)_{s^r \in \mathcal{S}^r}$ is symmetrizable. We provide here two approaches to deal with this problem: First, we will use the fact that $\mathfrak{W}$ is non-symmetrizable for transmission of a small number of messages that can be read by Eve but, since backwards communication from Eve to James is forbidden, are sufficient to counter any of the allowed jamming strategies.

Second, we will consider a variant of the optimization problem (10.14) where optimization of U'_r is restricted to maps of the form $U'_r = Id \otimes U''_{2,...,r}$, and we will prove that these restricted maps are asymptotically as good as those that are derived from the original problem when it comes to calculating capacity. However, these maps have the additional property that they cannot turn a non-symmetrizable AVC into a symmetrizable one.

Now let $r \in \mathbb{N}$ be arbitrary but fixed and p, U_r as above. Let $k, l \in \mathbb{N}$ be such that $n = k + l$ and $l = \lfloor \lambda \cdot n \rfloor$, where $\lambda \in (0, 1)$ is arbitrary but fixed for the moment. Then, from [11, Lemma 5], if $\hat{K}$ satisfies the assumptions of Lemma 10.4 with L set to one based on the properties (10.68), (10.69), and (10.70) of the lemma.

So, on the grounds of Lemma 10.4 and of the results proven in [11], we see that for every $m' \in \mathbb{N}$, $r \in \mathbb{N}$, $\delta > 0$, $p \in \mathcal{P}_0^{m'}(\mathcal{X}^r)$, and $U \in C(\mathcal{X}^r, \mathcal{X}^r)$ there exists a code $\mathcal{K} = (\mathcal{K}_m)_{m=1}^\infty$ such that for every $s^{r \cdot m} \in (\mathcal{S}^r)^m = \mathcal{S}^{r \cdot m}$ we have

$$\frac{1}{K'_k} \sum_{a=1}^{K'_k} \sum_{x^k} w_{s^k}(D'_a | x'_a) \geq 1 - \epsilon_k, \tag{10.157}$$

where $\{\epsilon_k\}_{k \in \mathbb{N}} \subset [0, 1]$, $\lim_{k \to \infty} \epsilon_k = 0$ and it may be assumed that $K'_k = l^3$. In addition to that, we know from [64] that there exist codes for $(\mathfrak{W}, \mathfrak{V})$ such that

$$\min_{s^l \in \mathcal{S}^l} \sum_{a,b=1}^{\Gamma_l, K''_l} \sum_{x^l \in \mathcal{X}^l} \frac{u_l(x^l | a, b) w_{s^l}(D''_{a,b} | x''_{ab})}{\Gamma_l \cdot K''_l} \geq 1 - \delta_l, \tag{10.158}$$

where $\{\delta_l\}_{l\in\mathbb{N}} \subset [0,1]$, $\lim_{l\to\infty}\epsilon_l = 0$, $\Gamma_l = l^3$, $U_l \in C([\Gamma_l] \times [K_l''], \mathcal{X}^l)$ is stochastic precoding, and $D_{a,b} \cap D_{a,b'} = \emptyset$ whenever $b \neq b'$ ($a \in [\Gamma_l]$ is used as common randomness in [64], whereas here we will substitute the messages that were sent on the first k channel uses for it. Note that the messages on the first k channel uses are not secure against Eve). In addition,

$$\lim_{l\to\infty} \frac{1}{l} \log K_l'' = C_{S,\text{ran}}^{\text{mean}}(\mathfrak{W}, \mathfrak{V}) - \nu \tag{10.159}$$

for some arbitrarily small $\nu > 0$, and

$$\lim_{l\to\infty} \frac{1}{l} \max_{\gamma\in[\Gamma_l]} \max_{s^l\in\mathcal{S}^l} I(\mathfrak{K}_l''; \mathfrak{Z}_{s^l} | \Gamma_l = a) = 0. \tag{10.160}$$

The mutual information is evaluated on the random variables defined via

$$\mathbb{P}_{s^l}((\mathfrak{K}_l'', \mathfrak{Z}_{s^l}, \Gamma_l) = (b, z^l, a)) := \frac{1}{\Gamma_l} \frac{1}{K_l''} \sum_{x^l\in\mathcal{X}^l} u_l(x^l|a,b) v(z^l|s^l, x^l). \tag{10.161}$$

We concatenate the two codes by defining new stochastic encodings $E_n \in C([K_l''], \mathcal{X}^n)$ via

$$e_n((x^k, x^l)|b) := \sum_{a=1}^{\Gamma_l} \delta_{\mathbf{x}_a}(x^k) u_l(x^l|a,b), \tag{10.162}$$

and new decoding sets via

$$D_b := \cup_a D_a' \times D_{a,b}'' \subset \mathcal{X}^n. \tag{10.163}$$

Then $D_b \cap D_{b'} = \cup_{a,a'}(D_a \times D_{a,b} \cap D_{a'} \times D_{a',b'}) = \emptyset$. We set $K_n := K_l''$, $\alpha_n := \epsilon_k$, and $\beta_n := \delta_l$ for the l satisfying $l = \lfloor \lambda \cdot n \rfloor$ and the k satisfying $k = n - l$. Then $\lim_{n\to\infty} \alpha_n = \lim_{n\to\infty} \beta_n = 0$. As a consequence of the Innerproduct Lemma in [2], we know that for every $s^n = (s^k, s^l)$ we have

$$\frac{1}{K_n} \sum_{b=1}^{K_n} \sum_{x^n\in\mathcal{X}^n} e_n(x^n|b) w(D_b|s^n, x^n)$$

$$\geq \sum_{a,b=1}^{\Gamma_k, K_l'} \sum_{x^l} \frac{u(x^l|a,b) w(D_a'|s^k, x^k) w(D_{a,b}''|s^l, x^l)}{K_l} \tag{10.164}$$

$$\geq 1 - 2\max\{\alpha_n, \beta_n\}. \tag{10.165}$$

The fact that the messages $b \in [K_n]$ are also asymptotically secure in the sense that

$$\lim_{n\to\infty} \frac{1}{n} \max_{s^n\in^n} I(\mathfrak{K}_n; \mathfrak{Z}_{s^n}) \leq \lim_{l\to\infty} \frac{\lambda}{l} \max_{s^l\in\mathcal{S}^l} I(\mathfrak{K}_l''; \mathfrak{Z}_{s^l} | \Gamma_l) \tag{10.166}$$

$$= 0 \tag{10.167}$$

follows from the independence of the distributions of the messages b and the values a of the common randomness as described in the inequalities (10.49)–(10.55). In particular, inequality (10.49) is valid as a consequence of (10.160). The rate of the code is calculated as

$$\lim_{n \to \infty} \frac{1}{n} \log K_n = \lambda \left(C_{S,\text{ran}}^{\text{mean}}(\mathfrak{W}, \mathfrak{V}) - \nu \right). \tag{10.168}$$

Since ν can be arbitrarily close to 0 and λ can be chosen arbitrarily close to 1, we have proven the desired result.

We now explain the second approach to proving statement 2 in Theorem 10.4. Here we aim to utilize the full power of Lemma 10.4 with $\Gamma = 1$. Our starting point are the distributions p and the channels U arising from the optimization (10.14) for fixed $r \in \mathbb{N}$. Set, for every $r \in \mathbb{N}$,

$$C_r := \max_{p \in \mathcal{P}(\mathcal{X}^r)} \max_{U \in C(\mathcal{X}^r, \mathcal{X}^r)} \min_{q \in \mathcal{P}(\mathcal{S}^r)} \left(I(p; W_q \circ U) - \max_{s^r \in \mathcal{S}^r} I(p; V_{s^r} \circ U) \right). \tag{10.169}$$

Let $r \in \mathbb{N}$ and $\epsilon \geq 0$ be arbitrary but fixed. Let p and U_r be such that

$$C_r - \epsilon = \min_{q \in \mathcal{P}(\mathcal{S}^r)} I(p; W_q \circ U_r) - \max_{s^r \in \mathcal{S}^r} I(p; V_{s^r} \circ U_r). \tag{10.170}$$

Now define $\tilde{U}_{r+1}$ by

$$\tilde{u}_{r+1}((x_1, \ldots, x_{r+1}) | (x, u)) := \sum_{x' \in \mathcal{X}} u_r((x', x_2, \ldots, x_{r+1}) | u) \delta_x(x_1)$$

for all $x, x_1, \ldots, x_{r+1} \in \mathcal{X}$ and $u \in \mathcal{X}^r$. Then, with $t := r + 1$, it holds that

$$C_{r+1} \geq \min_{q \in \mathcal{P}(\mathcal{S}^t)} I(p \otimes \pi; W_q \circ U_t) - \max_{s^t \in \mathcal{S}^t} I(p \otimes \pi; V_{s^t} \circ U_t) \tag{10.171}$$

$$\geq \min_{q \in \mathcal{P}(\mathcal{S}^r)} I(p; W_q \circ U_r) - \max_{s^r \in \mathcal{S}^r} I(p; V_{s^r} \circ U_r) - \log |\mathcal{X}| \tag{10.172}$$

$$= C_r - \epsilon - \log |\mathcal{X}|, \tag{10.173}$$

where $\pi \in \mathcal{P}(\mathcal{X})$ is defined by $\pi(x) := |\mathcal{X}|^{-1}$ for all $x \in \mathcal{X}$. This latter estimate is due to the equality $I(p \otimes \pi; V_{s^{r+1}} \circ U_{r+1}) = I(p; V_{s^r} \circ U) + I(\pi; V_{s_{r+1}})$, the data processing inequality, and the fact that for arbitrary channels $S \in C(\mathcal{A} \times \mathcal{B}, \mathcal{C})$ and $T \in C(\mathcal{A}' \times \mathcal{B}', \mathcal{C}')$, as well as distributions $q \in \mathcal{S}(\mathcal{B} \times \mathcal{B}')$ with respective marginal distributions $q_B \in \mathcal{P}(\mathcal{B})$ and $q_{B'} \in \mathcal{P}(\mathcal{B}')$ and $p \in \mathcal{S}(\mathcal{A} \times \mathcal{A}')$ with respective marginal distributions $p_A \in \mathcal{P}(\mathcal{A})$ and $q_{A'} \in \mathcal{P}(\mathcal{A}')$ we have for all $(a, b, c) \in \mathcal{A} \times \mathcal{B} \times \mathcal{C}$ that

$$\sum_{a',c'} \sum_{b,b'} s(c|a,b) t(c'|a',b') p(a,a') q(b,b') = \sum_b q_B(b) p_A(a) t(c|a,b). \tag{10.174}$$

Since $\mathfrak{W}$ is non-symmetrizable, we know that $\mathfrak{W}^{\otimes r} \circ \tilde{U}_r$ is non-symmetrizable for every $r \geq 2$, for the following reason. Again let S, T be channels as above. Assume that S is symmetrizable but T is not. Then $S \otimes T$ is non-symmetrizable. This can be seen by assuming the existence of a symmetrizing map $Q \in C(\mathcal{A} \times \mathcal{A}', \mathcal{B} \times \mathcal{B}')$. The statement

$$\forall (a_1, a_2, a_1', a_2') \in \mathcal{A}^2 \times \mathcal{A}'^2 :$$

$$\sum_{b,b'} s(\cdot|a_1,b) t(\cdot|a_1',b') q(b,b'|a_2,a_2') = \sum_{b,b'} s(\cdot|a_2,b) t(\cdot|a_2',b') q(b,b'|a_1,a_1') \tag{10.175}$$

would obviously imply for any fixed choice of (a_1, a_2) that

$$\sum_{b'} t(\cdot|a_1', b') q_{B'}(b'|a_2, a_2') = \sum_{b'} t(\cdot|a_2', b') q_{B'}(b'|a_1, a_1'), \qquad (10.176)$$

regardless of the choice of $(a_1', a_2') \in \mathcal{A}' \times \mathcal{A}'$ and setting $q_{B'}(b'|a_1, a_1') := \sum_b q(b, b'|a_1, a_1')$. This would contradict the non-symmetrizability of T. Since $\tilde{U}_r = U_{r-1} \otimes Id$, we can thus conclude that $\mathfrak{W}^{\otimes r} \circ \tilde{U}_r$ is non-symmetrizable. We now proceed with the proof of Theorem 10.4.

With this approach we have evaded the problem that $\mathfrak{W}^{\otimes r} \circ U_r$ may well be symmetrizable (see our Example 10.1).

By [11, Lemma 4], non-symmetrizability of $\mathfrak{W}^{\otimes r} \circ \tilde{U}_r$ implies that it is possible to define a decoder according to [11, Definition 3], with $N = K \cdot L$ and $[N]$ replaced by $[K] \times [L]$. Since only the number of codewords and their type ever enters the proof, it makes no difference whether we enumerate them by one index taken from $[N]$ or by two indices taken from $[K] \times [L]$. This decoder is proven to work reliably in [11, Lemma 5] (even with an exponentially fast decrease of average error) if $N = K \cdot L$ satisfies the assumptions of Lemma 10.4 based on the properties (10.68), (10.69), and (10.70) of the lemma.

So, on the grounds of Lemma 10.4 and of the results proven in [11], we see that for every $m \in \mathbb{N}$, $r \in \mathbb{N}\setminus\{1\}$, $\delta > 0$, $p \in \mathcal{P}_0^m(\mathcal{X}^r)$, and $U \in C(\mathcal{X}^{r-1}, \mathcal{X}^{r-1})$, there exists a code $\mathcal{K} = (\mathcal{K}_m)_{m=1}^\infty$ such that for every $s^{r \cdot m} \in (\mathcal{S}^r)^m = \mathcal{S}^{r \cdot m}$ we have

$$\sum_{k,l=1}^{K_m, L_m} \sum_{x^{r \cdot m}} \frac{w_{s^{r \cdot m}}(D_{kl}|x^{r \cdot m}) u^{\otimes m}(x^{m \cdot r}|u_{kl})}{K_m L_m} \geq 1 - \epsilon_m,$$

where $\{\epsilon_m\}_{m \in \mathbb{N}} \subset [0, 1]$, $\lim_{m \to \infty} \epsilon_m = 0$, and

$$\liminf_{m \to \infty} \frac{\log K_m \cdot L_m}{m} \geq \min_{q \in \mathcal{P}(\mathcal{S}^r)} I(p; W_q^m \circ \tilde{U}_r) - \delta \qquad (10.177)$$

(the code we use here is defined by using the codewords $\mathbf{x}_{kly}$ together with the decoder from [11, Definition 3] defined for the AVC $\mathfrak{W}^{\otimes r} \circ \tilde{U}_r := (W_{s^r} \circ (U_{r-1} \otimes Id))_{s^r \in \mathcal{S}^r}$), and

$$\max_{s^r \in \mathcal{S}^r} I(p; V_{s^r} \circ \tilde{U}_r) + 2\delta \geq \liminf_{m \to \infty} \frac{1}{m} \log L_m \qquad (10.178)$$

$$\geq \max_{s^r \in \mathcal{S}^r} I(p; V_{s^r} \circ \tilde{U}_r) + \delta, \qquad (10.179)$$

implying that, for a sequence $(p_m)_{m \in \mathbb{N}}$ of choices for p_m converging to some p having a decomposition $p = p' \otimes \pi$ for $p' \in \mathcal{P}(\mathcal{X}^{r-1})$ being an optimal choice in the sense of (10.169), we get

$$\liminf_{m \to \infty} \frac{1}{m} \log K_m \geq \min_{q \in \mathcal{P}(\mathcal{S}^r)} I(p; W_q \circ \tilde{U}_r) - \max_{s^r \in \mathcal{S}^r} I(p; V_{s^r} \circ \tilde{U}_r) - 3\delta \qquad (10.180)$$

$$\geq C_{r-1} - \log |\mathcal{X}| - 3\delta. \qquad (10.181)$$

Also, it is clear from the last part of Lemma 10.4 [Eq. (10.71)] together with [64, Lemma 20] that the codes employed here are asymptotically secure in the strong sense:

$$\limsup_{m\to\infty} \max_{s^{r\cdot m}} I(\mathfrak{K}_m; \mathfrak{Z}_{s^{r\cdot m}}) = 0. \tag{10.182}$$

We now apply the code for the extended channel $(\mathfrak{W}^{\otimes r} \circ \tilde{U}_r, \mathfrak{V}^{\otimes r})$ to the original channel $(\mathfrak{W}, \mathfrak{V})$. Define values $t_n \in \{0, \ldots, r-1\}$ by requiring $n = r \cdot m + t_n$ for them to hold for some suitably chosen $m = m(n) \in \mathbb{N}$. This quantity satisfies $-1 + n/r \leq m(n) \leq n/r$. For every $n \in \mathbb{N}$ we then define new decoding sets by

$$\hat{D}_{kl} := D_{kl} \times \mathcal{Y}^{t_n}, \tag{10.183}$$

and new randomized encodings by setting, for some arbitrary but fixed x^{t_n},

$$E(\hat{x}^n | k) := \sum_{l=1}^{L} \frac{1}{L} u^{\otimes n}(x^{r\cdot m} | u_{kl}) \cdot \delta_{x^{t_n}}(\hat{x}^n). \tag{10.184}$$

From the choice of codewords and the decoding rule it is clear that this code is asymptotically reliable. The asymptotic number of codewords (note that $\hat{K}_n = K_{m(n)}$) calculated and normalized with respect to n is

$$\liminf_{n\to\infty} \frac{1}{n} \log \hat{K}_n = \liminf_{n\to\infty} \frac{1}{m(n)\cdot r + t_n} K_{m(n)} \tag{10.185}$$

$$\geq \liminf_{n\to\infty} \frac{1}{r} \cdot \frac{1}{m(n)+1} K_{m(n)} \tag{10.186}$$

$$= \frac{1}{r}(C_{r-1} - 3\delta - \log|\mathcal{X}|). \tag{10.187}$$

In addition to that, the code is secure: For each $n \in \mathbb{N}$, the distribution of the input codewords and Eve's outputs is

$$\mathbb{P}(\mathfrak{K}_n = k, \mathfrak{Z}_{s^n} = z^n) = \sum_{l=1}^{L} \frac{1}{L} \sum_{x^{r\cdot m}} \sum_{x^{t_n}} u^{\otimes m}(x^{r\cdot m} | u_{kl})$$

$$\cdot v^{\otimes r\cdot m}(z^{r\cdot m} | x^{r\cdot m}, s^{r\cdot m}) v^{\otimes t_n}(z^{t_n} | x^{t_n}, s^{t_n}) \tag{10.188}$$

$$= \mathbb{P}(\mathfrak{K}_n = k, \mathfrak{Z}_{s^{r\cdot m}} = z^{r\cdot m}) \cdot v^{\otimes t_n}(z^{t_n} | x^{t_n}, s^{t_n}). \tag{10.189}$$

This demonstrates that (uniformly in $s^n \in \mathcal{S}^n$, and since $\mathfrak{K}_n = \mathfrak{K}_m$ holds) we have

$$I(\mathfrak{K}_n; \mathfrak{Z}_{s^n}) = I(\mathfrak{K}_n; \mathfrak{Z}_{s^{r\cdot m}}) + 0 = I(\mathfrak{K}_m; \mathfrak{Z}_{s^{r\cdot m}}). \tag{10.190}$$

Since the right-hand side of the above equation goes to zero for n going to infinity, and since $\lim_{r\to\infty} \frac{r-1}{r} = 1$, we see that the capacity C_S is bounded below by $\lim_{r\to\infty} \frac{1}{r} C_r$. It is not an immediate consequence that this implies we can reach the capacity

$C_{S,\text{ran}}^{\text{mean}}(\mathfrak{W},\mathfrak{V}) = C^*(\mathfrak{W},\mathfrak{V})$. Fortunately, it was proven in [64] that

$$C^*(\mathfrak{W},\mathfrak{V}) = \lim_{r\to\infty} \frac{1}{r} \max_{p\in\mathcal{P}(\mathcal{X}^n)} \max_{U\in\mathcal{C}(\mathcal{X}^n,\mathcal{X}^n)} \quad (10.191)$$

$$\left(\min_{q\in\mathcal{P}(\mathcal{S}^r)} I(p; W_q \circ U) - \max_{s^r\in\mathcal{S}^r} I(p; V_{s^r} \circ U) \right)$$

holds. Thus, $\lim_{r\to\infty} \frac{1}{r} C_r = C^*(\mathfrak{W},\mathfrak{V})$. This finally implies the desired result.

Proof of Theorem 10.5

If $C_S(\mathfrak{W},\mathfrak{V}) = 0$, there is nothing to prove. Assume that $C_S(\mathfrak{W},\mathfrak{V}) > 0$. It is evident that, in this case, $\mathfrak{W}$ is not symmetrizable. The function F defined in Definition 10.12 is continuous with respect to the Hausdorff distance (proving this statement is in complete analogy to the corresponding part in the proof of Theorem 5 in [68]). Thus, if $F(\mathfrak{W}) > 0$, then there is an $\epsilon > 0$ such that for all $\mathfrak{W}'$ satisfying $d(\mathfrak{W},\mathfrak{W}') < \epsilon$ we know that $F(\mathfrak{W}') > 0$ as well. Thus, each such $\mathfrak{W}'$ is non-symmetrizable.

For some suitably chosen $\epsilon' < \epsilon$, we deduce from Theorem 10.2 that $C_{S,\text{ran}}^{\text{mean}}(\mathfrak{W}',\mathfrak{V}) > 0$ for all those $\mathfrak{W}'$ for which $d(\mathfrak{W},\mathfrak{W}') < \epsilon'$. But since Theorem 10.3 shows that $F(\mathfrak{W}') > 0 \Rightarrow C_S(\mathfrak{W}',\mathfrak{V}) = C_{S,\text{ran}}^{\text{mean}}(\mathfrak{W}',\mathfrak{V})$, this implies that

$$C_S(\mathfrak{W},\mathfrak{V}) > 0 \,\forall\, \mathfrak{W}' : d(\mathfrak{W},\mathfrak{W}') < \epsilon'. \quad (10.192)$$

From Theorem 10.4, we know that positivity of $C_S(\mathfrak{W}',\mathfrak{V})$ ensures that it equals $C_{S,\text{ran}}^{\text{mean}}(\mathfrak{W}',\mathfrak{V})$. Since the latter is continuous, the proof is completed.

Proof of Theorem 10.6

Again, we prove everything in the same order as it is listed in the theorem.

1. Let C_S be discontinuous in the point $(\mathfrak{W},\mathfrak{V})$. By Theorem 10.5, we know that this is possible only if $C_S(\mathfrak{W},\mathfrak{V}) = 0$. In addition, let $C_{S,\text{ran}}^{\text{mean}}(\mathfrak{W},\mathfrak{V}) = 0$. Then, since $C_{S,\text{ran}}^{\text{mean}}$ is continuous, it follows that for every $\epsilon > 0$ there is $\delta > 0$ such that for all $(\mathfrak{W}_\delta,\mathfrak{V})$ satisfying $d(\mathfrak{W}_\delta,\mathfrak{W}) < \delta$ we have $C_{S,\text{ran}}^{\text{mean}}(\mathfrak{W}_\delta,\mathfrak{V}) \leq \epsilon$. Since $C_{S,\text{ran}}^{\text{mean}} \geq C_S$ this would imply that C_S is continuous as well, in contradiction to the assumption. Thus $C_{S,\text{ran}}^{\text{mean}}(\mathfrak{W},\mathfrak{V}) > 0$.

Of course, this immediately implies that $\mathfrak{W}$ has to be symmetrizable, by property 2. But symmetrizability is equivalent to $F(\mathfrak{W}) = 0$. The definition of F is given in Definition 10.12 – its connection to symmetrizability is obvious from the definition. The notion of symmetrizability is explained in Eq. (10.10)). Clearly, if for all $\epsilon > 0$ and $\mathfrak{W}'$ satisfying $d(\mathfrak{W},\mathfrak{W}') < \epsilon$ we would have $F(\mathfrak{W}') = 0$, then $C_S(\mathfrak{W}',\mathfrak{V}')$ would be zero in a whole vicinity of $(\mathfrak{W},\mathfrak{V})$. Thus for all $\epsilon > 0$ there has to be at least one $\mathfrak{W}_\epsilon$ such that $d(\mathfrak{W},\mathfrak{W}_\epsilon) < \epsilon$ but $F(\mathfrak{W}_\epsilon) > 0$.

The reverse direction is basically established by using all our arguments backwards: For all $\epsilon > 0$, let there be at least one $\mathfrak{W}_\epsilon$ such that $d(\mathfrak{W},\mathfrak{W}_\epsilon) < \epsilon$ but $F(\mathfrak{W}_\epsilon) > 0$. In addition, let $F(\mathfrak{W}) = 0$ but $C_{S,\text{ran}}^{\text{mean}}(\mathfrak{W},\mathfrak{V}) > 0$. Since $C_{S,\text{ran}}^{\text{mean}}$ is continuous, there is a $\delta > 0$ such that $C_{S,\text{ran}}^{\text{mean}}(\mathfrak{W}',\mathfrak{V}') > (1/2) \cdot C_{S,\text{ran}}^{\text{mean}}(\mathfrak{W},\mathfrak{V}) =: \alpha$ whenever $d((\mathfrak{W},\mathfrak{V}),(\mathfrak{W}',\mathfrak{V}')) < \delta$.

For every $\epsilon' \leq (1/2)\min\{\epsilon,\delta\}$ we can therefore deduce the following: $C_S(\mathfrak{W}_{\epsilon'},\mathfrak{V}) = C_{S,\text{ran}}^{\text{mean}}(\mathfrak{W}_{eps'},\mathfrak{V}) \geq \alpha > 0$ (since $F(\mathfrak{W}_{\epsilon'}) > 0$), but $C_S(\mathfrak{W}_0,\mathfrak{V}) = 0$. Thus C_S is discontinuous in the point $(\mathfrak{W},\mathfrak{V})$.

2. Let C_S be discontinuous in the point $(\mathfrak{W},\mathfrak{V})$. By property 4, this implies that for all $\epsilon > 0$ there is $\mathfrak{W}_\epsilon$ such that $d(\mathfrak{W},\mathfrak{W}_\epsilon) < \epsilon$ but $F(\mathfrak{W}_\epsilon) > 0$. If $\hat{\mathfrak{V}}$ is such that $C_{S,\text{ran}}^{\text{mean}}(\mathfrak{W},\hat{\mathfrak{V}}) > 0$, then the pair $(\mathfrak{W},\hat{\mathfrak{V}})$ fulfills all the points in the second of the two equivalent formulations in statement 4, and this implies that C_S is discontinuous in the point $(\mathfrak{W},\hat{\mathfrak{V}})$.

Proof of Theorem 10.7 (Super-Activation Results)

We divide this proof into four parts, each corresponding to its counterpart in Theorem 10.7.

1. We start with the "only if" statement. Clearly, if $\mathfrak{W}_1 \otimes \mathfrak{W}_2$ is symmetrizable then $C_S(\mathfrak{W}_1 \otimes \mathfrak{W}_2, \mathfrak{V}_1 \otimes \mathfrak{V}_2) = 0$.

If, on the other hand, $\mathfrak{W}_1 \otimes \mathfrak{W}_2$ is not symmetrizable and $C_{S,\text{ran}}^{\text{mean}}(\mathfrak{W}_1 \otimes \mathfrak{W}_2, \mathfrak{V}_1 \otimes \mathfrak{V}_2) > 0$, then on account of statement 1 from Theorem 10.4, we know that $C_S(\mathfrak{W}_1 \otimes \mathfrak{W}_2, \mathfrak{V}_1 \otimes \mathfrak{V}_2) > 0$.

This proves the first part of the theorem.

2. In [30, Section VI], an explicit example of a pair $(\mathfrak{W}_i,\mathfrak{V}_i)_{i=1,2}$ was given where $\mathfrak{W}_1$ is symmetrizable but $\mathfrak{W}_2$ is not. By elementary calculus, this implies that $\mathfrak{W}_1 \otimes \mathfrak{W}_2$ is non-symmetrizable.

Since this holds, our statement 1 from Theorem 10.4 shows that the uncorrelated coding capacity of $(\mathfrak{W}_1 \otimes \mathfrak{W}_2, \mathfrak{V}_1 \otimes \mathfrak{V}_2)$ equals its randomness assisted capacity.

In [30], it was further shown that $C_{S,\text{ran}}^{\text{mean}}(\mathfrak{W}_1,\mathfrak{V}_1) > 0$ and $C_S(\mathfrak{W}_i,\mathfrak{V}_i) = 0$ $(i = 1, 2)$.

3. By assumption, $C_{S,\text{ran}}^{\text{mean}}(\mathfrak{W}_i,\mathfrak{V}_i) = 0$ $(i = 1,2)$ but $C_{S,\text{ran}}^{\text{mean}}(\mathfrak{W}_1 \otimes \mathfrak{V}_1, \mathfrak{W}_2 \otimes \mathfrak{V}_2) > 0$. The former implies $C_S(\mathfrak{W}_i,\mathfrak{V}_i) = 0$ $(i=1,2)$. If $\mathfrak{W}_1$ and $\mathfrak{W}_2$ are symmetrizable then clearly $\mathfrak{W}_1 \otimes \mathfrak{W}_2$ is symmetrizable, and by [5] the message transmission capacity of $\mathfrak{W}_1 \otimes \mathfrak{W}_2$ is zero, implying $C_S(\mathfrak{W}_1 \otimes \mathfrak{W}_2, \mathfrak{V}_1 \otimes \mathfrak{V}_2) = 0$. If, on the other hand, either $\mathfrak{W}_1$ or $\mathfrak{W}_2$ is not symmetrizable then $\mathfrak{W}_1 \otimes \mathfrak{W}_2$ is not symmetrizable, and this implies

$$C_S(\mathfrak{W}_1 \otimes \mathfrak{W}_2, \mathfrak{V}_1 \otimes \mathfrak{V}_2) = C_{S,\text{ran}}^{\text{mean}}(\mathfrak{W}_1 \otimes \mathfrak{W}_2, \mathfrak{V}_1 \otimes \mathfrak{V}_2) > 0, \qquad (10.193)$$

where the equality is due to part 1 of Theorem 10.4, and the lower bound is true by assumption.

4. We again rely on Theorem 10.4. Let both $\mathfrak{W}_1$ and $\mathfrak{W}_2$ be symmetrizable. Then $\mathfrak{W}_1 \otimes \mathfrak{W}_2$ is symmetrizable. Since by assumption $C_{S,\text{ran}}^{\text{mean}}$ shows no super-activation on the pair $(\mathfrak{W}_i,\mathfrak{V}_i)$ $(i=1,2)$, it follows that C_S also cannot show super-activation. Thus, at least one of the two AVCs has to be non-symmetrizable. Without loss of generality, let this channel be $\mathfrak{W}_1$.

If, in addition, $\mathfrak{W}_2$ is non-symmetrizable, then $C_S(\mathfrak{W}_i,\mathfrak{V}_i) = C_{S,\text{ran}}^{\text{mean}}(\mathfrak{W}_i,\mathfrak{V}_i)$ holds for $i=1,2$. Since $\mathfrak{W}_1 \otimes \mathfrak{W}_2$ is symmetrizable as well, we additionally have $C_S(\mathfrak{W}_1 \otimes$

$\mathfrak{W}_2, \mathfrak{V}_1 \otimes \mathfrak{V}_2) = C_{S,\text{ran}}^{\text{mean}}(\mathfrak{W}_1 \otimes \mathfrak{W}_2, \mathfrak{V}_1 \otimes \mathfrak{V}_2)$. But since $C_{S,\text{ran}}^{\text{mean}}$ shows no super-activation on the pair $(\mathfrak{W}_i, \mathfrak{V}_i)$ ($i = 1, 2$), this cannot be. Thus, again without loss of generality, we have $\mathfrak{W}_2$ is symmetrizable.

Since we are talking about the super-activation of C_S, it has to be that $C_S(\mathfrak{W}_i, \mathfrak{V}_i) = 0$ holds for $i = 1, 2$. But since $\mathfrak{W}_1$ is non-symmetrizable this requires that $C_{S,\text{ran}}^{\text{mean}}(\mathfrak{W}_1, \mathfrak{V}_1) = 0$ holds. If in addition we would have $C_{S,\text{ran}}^{\text{mean}}(\mathfrak{W}_2, \mathfrak{V}_2) = 0$ then C_S could not be super-activated since $C_{S,\text{ran}}^{\text{mean}}$ cannot be super-activated by assumption. Thus $C_{S,\text{ran}}^{\text{mean}}(\mathfrak{W}_2, \mathfrak{V}_2) > 0$.

This completes the proof of Theorem 10.7.

Acknowledgments

This work was supported by the German Research Foundation (DFG) under Grants NO 1129/1-1 (J.N.) and BO 1734/20-1 (H.B.), and by the BMBF via the grants 01BQ1050 and 16KIS0118 (H.B., J.N.).

Further funding (J.N.) was provided by the ERC Advanced Grant IRQUAT, the Spanish MINECO Project No. FIS2013-40627-P, and the Generalitat de Catalunya CIRIT Project No. 2014 SGR 966.

References

[1] D. Blackwell, L. Breiman, and A. J. Thomasian, "The capacities of certain channel classes under random coding," *Ann. Math. Stat.*, vol. 31, no. 3, pp. 558–567, 1960.

[2] R. Ahlswede, "Elimination of correlation in random codes for arbitrarily varying channels," *Z. Wahrscheinlichkeitstheorie verw. Gebiete*, vol. 44, pp. 159–175, 1978.

[3] I. Csiszár and J. Körner, *Information Theory: Coding Theorems for Discrete Memoryless Systems*, 2nd edn. Cambridge: Cambridge University Press, 2011.

[4] R. Ahlswede and N. Cai, "Correlated sources help transmission over an arbitrarily varying channel," *IEEE Trans. Inf. Theory*, vol. 43, no. 4, pp. 1254–1255, Jul. 1997.

[5] T. Ericson, "Exponential error bounds for random codes in the arbitrarily varying channel," *IEEE Trans. Inf. Theory*, vol. 31, no. 1, pp. 42–48, Jan. 1985.

[6] I. G. Stiglitz, "Coding for a class of unknown channels," *IEEE Trans. Inf. Theory*, vol. 12, no. 2, pp. 189–195, Apr. 1966.

[7] I. Csiszár and P. Narayan, "Capacity of the Gaussian arbitrarily varying channel," *IEEE Trans. Inf. Theory*, vol. 37, no. 1, pp. 18–26, Jan. 1991.

[8] I. Csiszár, "Arbitrarily varying channels with general alphabets and states," *IEEE Trans. Inf. Theory*, vol. 38, no. 6, pp. 1725–1742, Nov. 1992.

[9] R. L. Pickholtz, D. L. Schilling, and L. B. Milstein, "Theory of spread-spectrum communications – a tutorial," *IEEE Trans. Commun.*, vol. 30, no. 5, pp. 855–884, May 1982.

[10] I. Csiszár and P. Narayan, "Arbitrarily varying channels with constrained inputs and states," *IEEE Trans. Inf. Theory*, vol. 34, no. 1, pp. 27–34, Jan. 1988.

[11] I. Csiszár and P. Narayan, "The capacity of the arbitrarily varying channel revisited: Positivity, constraints," *IEEE Trans. Inf. Theory*, vol. 34, no. 2, pp. 181–193, Mar. 1988.

[12] D. P. Bertsekas and R. Gallager, *Data Networks*, 2nd edn. Upper Saddle River, NJ: Prentice-Hall International, 1992.
[13] J. Wolfowitz, "Simultaneous channels," *Arch. Rational Mech. Analysis*, vol. 4, no. 4, pp. 371–386, 1960.
[14] J. Wolfowitz, *Coding Theorems of Information Theory*, 3rd edn. Berlin: Springer-Verlag, 1978.
[15] D. Blackwell, L. Breiman, and A. J. Thomasian, "The capacity of a class of channels," *Ann. Math. Stat.*, vol. 30, no. 4, pp. 1229–1241, Dec. 1959.
[16] T. M. Cover, "Broadcast channels," *IEEE Trans. Inf. Theory*, vol. 18, no. 1, pp. 2–14, Jan. 1972.
[17] A. Lapidoth and I. E. Telatar, "The compound channel capacity of a class of finite-state channels," *IEEE Trans. Inf. Theory*, vol. 44, no. 3, pp. 396–400, May 1998.
[18] I. Csiszár and P. Narayan, "Capacity and decoding rules for classes of arbitrarily varying channels," *IEEE Trans. Inf. Theory*, vol. 35, no. 4, pp. 752–769, Jul. 1989.
[19] J. Kiefer and J. Wolfowitz, "Channels with arbitrarily varying channel probability functions," *Inform. Contr.*, vol. 5, pp. 44–54, Mar. 1962.
[20] R. Ahlswede, "A note on the existence of the weak capacity for channels with arbitrarily varying channel probability functions and its relation to Shannon's zero error capacity," *Ann. Math. Stat.*, vol. 41, no. 3, pp. 1027–1033, 1970.
[21] R. Ahlswede and J. Wolfowitz, "The capacity of a channel with arbitrarily varying channel probability functions and binary output alphabet," *Z. Wahrscheinlichkeitstheorie verw. Gebiete*, vol. 15, pp. 186–194, 1970.
[22] C. E. Shannon, "The zero error capacity of a noisy channel," *IRE Trans. Inf. Theory*, vol. 2, no. 3, pp. 8–19, Sep. 1956.
[23] W. Haemers, "On some problems of Lovász concerning the Shannon capacity of a graph," *IEEE Trans. Inf. Theory*, vol. 25, no. 2, pp. 231–232, Mar. 1979.
[24] W. Haemers, "An upper bound for the Shannon capacity of a graph," *Algebraic Methods in Graph Theory*, pp. 267–272, 1978.
[25] N. Alon, "The Shannon capacity of a union," *Combinatorica*, vol. 18, no. 3, pp. 301–310, Mar. 1998.
[26] L. Lovász, "On the Shannon capacity of a graph," *IEEE Trans. Inf. Theory*, vol. 25, no. 1, pp. 1–7, Jan. 1979.
[27] F. Guo and Y. Watanabe, "On graphs in which the Shannon capacity is unachievable by finite product," *IEEE Trans. Inf. Theory*, vol. 36, no. 3, pp. 622–623, May 1990.
[28] U. M. Maurer, "Secret key agreement by public discussion from common information," *IEEE Trans. Inf. Theory*, vol. 39, no. 3, pp. 733–742, May 1993.
[29] M. Wiese, "Multiple access channels with cooperating encoders," Ph.D. dissertation, Technische Universität München, Munich, Germany, 2013.
[30] H. Boche and R. F. Schaefer, "Capacity results and super-activation for wiretap channels with active wiretappers," *IEEE Trans. Inf. Forensics Security*, vol. 8, no. 9, pp. 1482–1496, Sep. 2013.
[31] C. E. Shannon, "Communication theory of secrecy systems," *Bell Syst. Tech. J.*, vol. 28, no. 4, pp. 656–715, Oct. 1949.
[32] A. D. Wyner, "The wire-tap channel," *Bell Syst. Tech. J.*, vol. 54, pp. 1355–1387, Oct. 1975.
[33] M. R. Bloch and J. N. Laneman, "On the secrecy capacity of arbitrary wiretap channels," in *Proc. 46th Annual Allerton Conf. Commun., Control, Computing*, Monticello, IL, USA, Sep. 2008, pp. 818–825.

[34] M. Bloch and J. Barros, *Physical-Layer Security: From Information Theory to Security Engineering*. Cambridge: Cambridge University Press, 2011.
[35] M. Wiese, J. Nötzel, and H. Boche, "The arbitrarily varying wiretap channel – communication under uncoordinated attacks," in *Proc. IEEE Int. Symp. Inf. Theory*, Hong Kong, Jun. 2015, pp. 2146–2150.
[36] I. Csiszár and J. Körner, "Broadcast channels with confidential messages," *IEEE Trans. Inf. Theory*, vol. 24, no. 3, pp. 339–348, May 1978.
[37] W. Kang and N. Liu, "Wiretap channel with shared key," in *Proc. IEEE Inf. Theory Workshop*, Dublin, Ireland, Aug. 2010, pp. 1–5.
[38] H. Yamamoto, "Rate–distortion theory for the Shannon cipher system," *IEEE Trans. Inf. Theory*, vol. 43, no. 3, pp. 827–835, May 1997.
[39] N. Merhav, "Shannon's secrecy system with informed receivers and its application to systematic coding for wiretapped channels," *IEEE Trans. Inf. Theory*, vol. 54, no. 6, pp. 2723–2734, Jun. 2008.
[40] R. Liu and W. Trappe, eds., *Securing Wireless Communications at the Physical Layer*. Boston, MA: Springer US, 2010.
[41] X. Zhou, L. Song, and Y. Zhang, eds., *Physical Layer Security in Wireless Communications*. London: CRC Press, 2013.
[42] Y. Liang, H. V. Poor, and S. Shamai (Shitz), "Information theoretic security," *Found. Trends Commun. Inf. Theory*, vol. 5, no. 4–5, pp. 355–580, 2009.
[43] N. Sklavos and X. Zhang, *Wireless Security and Cryptography: Specifications and Implementations*. London: CRC Press, 2007.
[44] D. Welch and S. Lathrop, "Wireless security threat taxonomy," in *Proc. IEEE Systems, Man and Cybernetics Society Information Assurance Workshop*, West Point, NY, USA, Jun. 2003, pp. 76–83.
[45] G. Fettweis *et al.*, "The tactile internet," ITU-T Tech. Watch Rep., Tech. Rep., Aug. 2014.
[46] D. N. C. Tse and P. Viswanath, *Fundamentals of Wireless Communication*. Cambridge: Cambridge University Press, 2005.
[47] S. Goel and R. Negi, "Secret communication in presence of colluding eavesdroppers," in *Proc. IEEE Military Commun. Conf.*, Atlantic City, NJ, USA, 2005, pp. 1501–1506, Vol. 3.
[48] S. Goel and R. Negi, "Guaranteeing secrecy using artificial noise," *IEEE Trans. Wireless Commun.*, vol. 7, no. 6, pp. 2180–2189, Jun. 2008.
[49] X. Li, J. Hwu, and E. Ratazzi, "Array redundancy and diversity for wireless transmissions with low probability of interception," in *Proc. IEEE Int. Conf. Acoustics, Speech, Signal Process.*, vol. 4, 2006.
[50] X. Li, J. Hwu, and E. Ratazzi, "Using antenna array redundancy and channel diversity for secure wireless transmissions," *J. Commun.*, vol. 2, no. 3, pp. 24–32, May 2007.
[51] A. Khisti and G. W. Wornell, "Secure transmission with multiple antennas I: The MISOME wiretap channel," *IEEE Trans. Inf. Theory*, vol. 56, no. 7, pp. 3088–3104, Jul. 2010.
[52] S. Shafiee, N. Liu, and S. Ulukus, "Towards the secrecy capacity of the Gaussian MIMO wire-tap channel: The 2-2-1 channel," *IEEE Trans. Inf. Theory*, vol. 55, no. 9, pp. 4033–4039, Sep. 2009.
[53] F. Oggier and B. Hassibi, "The secrecy capacity of the MIMO wiretap channel," *IEEE Trans. Inf. Theory*, vol. 57, no. 8, pp. 4961–4972, Aug. 2011.
[54] T. T. Kim and H. V. Poor, "Secure communications with insecure feedback: Breaking the high-SNR ceiling," *IEEE Trans. Inf. Theory*, vol. 56, no. 8, pp. 3700–3711, Aug. 2010.
[55] X. He, A. Khisti, and A. Yener, "MIMO multiple access channel with an arbitrarily varying

eavesdropper: Secrecy degrees of freedom," *IEEE Trans. Inf. Theory*, vol. 59, no. 8, pp. 4733–4745, Aug. 2013.

[56] X. He and A. Yener, "MIMO wiretap channels with unknown and varying eavesdropper channel states," *IEEE Trans. Inf. Theory*, vol. 60, no. 11, pp. 6844–6869, Nov. 2014.

[57] Y. Liang, G. Kramer, H. V. Poor, and S. Shamai (Shitz), "Compound wiretap channels," in *Proc. 45th Annual Allerton Conf. Commun., Control, Computing*, Monticello, IL, USA, 2007.

[58] Y. Liang, G. Kramer, H. V. Poor, and S. Shamai (Shitz), "Compound wiretap channels," *EURASIP J. Wireless Commun. Netw.*, article ID 142374, pp. 1–13, 2009.

[59] I. Bjelaković, H. Boche, and J. Sommerfeld, "Secrecy results for compound wiretap channels," *Probl. Inf. Transmission*, vol. 49, no. 1, pp. 73–98, Mar. 2013.

[60] I. Bjelaković, H. Boche, and J. Sommerfeld, *Information Theory, Combinatorics, and Search Theory*, ser. Lecture Notes in Computer Science, vol. 7777, pp. 123–144. Berlin, Heidelberg: Springer, 2013.

[61] E. MolavianJazi, "Secure communications over arbitrarily varying wiretap channels," Master's thesis, Graduate School, Notre Dame, Indiana, Dec. 2009.

[62] E. MolavianJazi, M. Bloch, and J. N. Laneman, "Arbitrary jamming can preclude secure communication," in *Proc. 47th Annual Allerton Conf. Commun., Control, Computing*, Monticello, IL, USA, Sep. 2009, pp. 1069–1075.

[63] M. Wiese and H. Boche, *Information Theory, Combinatorics, and Search Theory*, ser. Lecture Notes in Computer Science, vol. 7777, pp. 71–122. Berlin, Heidelberg: Springer, 2013.

[64] M. Wiese, J. Nötzel, and H. Boche, "A channel under simultaneous jamming and eavesdropping attack – correlated random coding capacities under strong secrecy criteria," *IEEE Trans. Inf. Theory*, vol. 62, no. 7, pp. 3844–3862, Jul. 2016.

[65] J. Nötzel, M. Wiese, and H. Boche, "The arbitrarily varying wiretap channel – secret randomness, stability and super-activation," *IEEE Trans. Inf. Theory*, vol. 62, no. 6, pp. 3504–3531, Jun. 2016.

[66] J. Körner and K. Marton, "Comparison of two noisy channels," in *Colloquia Mathematica Societatis, János Bolyai, 16, Topics in Info. Th.*, North Holland, 1977, pp. 411–424.

[67] G. Smith and J. Yard, "Quantum communication with zero-capacity channels," *Science*, vol. 321, pp. 1812–1815, 2008.

[68] H. Boche and J. Nötzel, "Positivity, discontinuity, finite resources and nonzero error for arbitrarily varying quantum channels," *J. Mathematical Physics*, vol. 55, p. 122201, 2014.

[69] H. Boche, R. F. Schaefer, and H. V. Poor, "On the continuity of the secrecy capacity of compound and arbitrarily varying wiretap channels," *IEEE Trans. Inf. Forensics Security*, vol. 12, no. 10, pp. 2531–2546, Dec. 2015.

[70] H. Boche and R. F. Schaefer, "Arbitrarily varying wiretap channels with finite coordination resources," in *Proc. IEEE Int. Conf. Commun. Workshops*, Sydney, Australia, Jun. 2014, pp. 746–751.

[71] D. P. Dubhasi and A. Panconesi, *Concentration of Measure for the Analysis of Randomized Algorithms*. Cambridge: Cambridge University Press, 2009.

[72] R. Ahlswede and A. Winter, "Strong converse for identification via quantum channels," *IEEE Trans. Inf. Theory*, vol. 48, no. 3, pp. 569–579, Mar. 2002.

[73] R. Ahlswede, "Coloring hypergraphs: A new approach to multi-user source coding II," *J. Comb. Inform. Syst. Sci.*, vol. 5, no. 3, pp. 220–268, 1980.

[74] R. Ahlswede, "Arbitrarily varying channels with states sequence known to the sender," *IEEE Trans. Inf. Theory*, vol. 32, no. 5, pp. 621–629, Sep. 1986.

11 Super-Activation as a Unique Feature of Secure Communication over Arbitrarily Varying Channels

Rafael F. Schaefer, Holger Boche, and H. Vincent Poor

The question of whether the capacity of a channel is additive or not goes back to Shannon, who asked this for the zero-error capacity function. Despite the common sense that the capacity is usually additive, there is surprisingly little known for non-trivial channels. This chapter addresses this question for the *arbitrarily varying wiretap channel* (AVWC), which models secure communication in the presence of arbitrarily varying channel (AVC) conditions. For orthogonal AVWCs it has been shown that the strongest form of non-additivity occurs: the phenomenon of super-activation. That is, there are orthogonal AVWCs, each having zero secrecy capacity, which allow for transmission with positive rate if they are used together. Subsequently, the single-user AVC is studied and it is shown that in this case, super-activation for non-secure message transmission is not possible, making it a unique feature of secure communication over AVWCs. However, the capacity for message transmission of the single-user AVC is shown to be super-additive, including a complete characterization. Super-activation was known for a long time in the area of quantum communication, where it is common opinion that this is solely a phenomenon of quantum physics and that this cannot occur for classical communication. However, the results in this chapter show that super-activation is indeed a feature of secure communication and therewith occurs in classical, non-quantum, communication as well.

11.1 Introduction

Information theory goes back to Shannon's seminal work "A Mathematical Theory of Communication" [1]. Since then it has been proven to be an indispensable concept for analyzing complex communication systems to obtain insights and optimal design criteria. Among many things, it is used to address the important issue of medium access control: How should available resources be divided among multiple users in the best possible way? This is a crucial question, especially for wireless communication systems since they are usually composed of orthogonal sub-systems such as those that arise via time division multiplexing (TDM) or orthogonal frequency division multiplexing (OFDM). Of particular interest then is to know how the capacity of the overall system depends on the orthogonal sub-systems. Common sense tells us that for such systems it should be given by the sum of the capacities of all sub-systems. This goes along with the inherent world view of the additivity of classical resources.

To this end, consider a system consisting of two orthogonal ordinary discrete memoryless channels (DMCs) W_1 and W_2. If both channels are accessed in an orthogonal way by using independent encoders and decoders for each sub-channel, the rate $C(W_1) + C(W_2)$ is clearly an achievable transmission rate for the overall system, where $C(\cdot)$ denotes the capacity of the corresponding channel. The interesting question is now: Are there gains in capacity by bonding the orthogonal resources and jointly accessing the resulting system $W_1 \otimes W_2$ by using a joint encoder and decoder? The operational definition of the capacity immediately yields

$$C(W_1 \otimes W_2) \geq C(W_1) + C(W_2), \qquad (11.1)$$

since a joint use of both channels can only increase the performance compared to a separate use of both. However, for ordinary DMCs it is known that the capacity under the average error criterion is indeed *additive*. This means the relation (11.1) is actually satisfied with equality, i.e.,

$$C(W_1 \otimes W_2) = C(W_1) + C(W_2), \qquad (11.2)$$

so that there are no gains in capacity by using joint encoding and decoding over orthogonal channels. Thus, the overall capacity of a TDM or OFDM system is indeed given by the sum of the capacities of all orthogonal sub-channels.

Although this confirms what one would usually expect for the capacity of a system consisting of orthogonal sub-channels, the question of additivity of the capacity function is in general by no means obvious or trivial to answer. This is reflected in the fact that Shannon, for example, already asked this question in 1956 for the zero-error capacity [2], which was then restated in 1979 by Lovász [3, Problem 2]. Shannon conjectured that the zero-error capacity $C_0(\cdot)$ is additive, thus possessing the same behavior as ordinary DMCs: $C_0(W_1 \otimes W_2) = C_0(W_1) + C_0(W_2)$; in a similar way as in (11.2). This was disproved by Haemers in 1979, who explicitly constructed a counterexample for which the overall capacity $C_0(W_1 \otimes W_2)$ is strictly greater than the sum of the individual capacities $C_0(W_1) + C_0(W_2)$, cf. [4]. Subsequently, it was Alon in 1998 who constructed an even stronger counterexample with an arbitrarily large discrepancy between the overall capacity and the sum of the individual capacities [5]. Thus, there are channels for which the zero-error capacity is strictly greater when encoding and decoding are done jointly instead of independently. This means that the zero-error capacity is *super-additive* and there exist channels for which "$\geq$" in (11.1) can actually be replaced by "$>$" so that

$$C_0(W_1 \otimes W_2) > C_0(W_1) + C_0(W_2)$$

holds. To date, only certain explicit examples are known which possess this property of super-additivity. A general characterization of which channels are super-additive or what further properties such channels possess remains open.

In 1970 it was Ahlswede who showed that the capacity of AVCs under the maximum error criterion includes the characterization of the zero-error capacity as a special case [6]. Thus, Shannon's zero-error capacity is closely related to AVCs, making it also worth studying this question from an AVC perspective.

11.2 Problem Motivation

Information theoretic approaches to security realize security directly at the physical layer by exploiting the noisy and imperfect nature of the underlying communication channel. This line of thinking goes back to Wyner, who introduced the so-called wiretap channel in [7]. This area of research provides a promising approach to achieve secrecy and to embed secure communication into wireless networks, and not surprisingly it has drawn considerable attention in recent years [8–12] and has also been identified by national agencies and operators of communication systems as a key technique to secure future communication systems [13–15]. Accordingly, it has become a discussion item in standardization bodies for future system design.

The open nature of the wireless medium makes wireless communication systems inherently vulnerable to eavesdropping – transmitted signals are received not only by intended receivers, but also by eavesdroppers. Accordingly, such information theoretic approaches to security are particularly relevant for these systems. Although not feasible for practical systems, a widely used assumption of many studies is perfect knowledge of all channels for all users. However, the quality of channel state information (CSI) will always be limited due to the nature of the wireless medium, but also due to practical limitations such as estimation/feedback inaccuracy. This is particularly true for the eavesdropper's channel, since malevolent eavesdroppers will not share any information about their channels. As a consequence, to make such security approaches applicable in practical systems, they must incorporate imperfect CSI assumptions. A recent overview on secure communication under channel uncertainty and adversarial attacks is given in [16].

One approach of having more realistic and practically relevant CSI assumptions is the concept of *compound channels* [17, 18]. Instead of knowing the precise channel realization, for compound channels it is assumed that all users know only that the true channel realization belongs to a known uncertainty set and that it remains constant for the whole duration of transmission. Then the *compound wiretap channel* models secure communication over compound channels, and has been investigated, for example, in [19–23].

In this chapter, we model the uncertainty in CSI by assuming arbitrarily varying channels [24–26]. Here, the actual channel realization is unknown; rather, it is only known that it is from a known uncertainty set and that it may vary in an arbitrary and unknown manner from channel use to channel use. Due to its generality, the concept of AVCs provides a very powerful framework to model not only channel uncertainty, but also scenarios with malevolent adversaries who maliciously influence or jam legitimate transmissions.

Accordingly, secure communication over AVCs is then modeled by the AVWC, which has been studied in [27–33]. In contrast to the case of perfect CSI, it makes a substantial difference whether unassisted or common randomness (CR) assisted codes are used by the transmitter and legitimate receiver. To this end, if the AVC to the legitimate receiver possesses the so-called property of symmetrizability (as defined in Definition 11.2), then the unassisted secrecy capacity is zero, while the CR-assisted secrecy capacity may be positive. A complete characterization of the unassisted secrecy capacity in terms of its CR-assisted one is given in [28] and [31]. However, a single-letter characterization of the CR-assisted secrecy capacity itself is only known for the special case of type constraint states [33], and remains open in general. Only a multi-letter description has been established [30].

Very recently, the phenomenon of *super-activation* has been observed for secure communication over AVCs [29]. This is the extreme case of non-additivity as discussed above, cf. (11.1), and it occurs when all orthogonal channels are "useless," i.e., each having zero capacity $C(\mathbf{W}_1) = C(\mathbf{W}_2) = 0$, but joint encoding and decoding yields $C(\mathbf{W}_1 \otimes \mathbf{W}_2) > 0$. Useless channels each with zero capacity can be used together to super-activate the overall system, giving it a positive capacity. Accordingly, this shows that the world view of additivity of orthogonal resources does not hold anymore; for AVWCs we may have "$0 + 0 > 0$."

The goal of this chapter is to further explore and to discuss the non-additivity of (secure) communication over AVCs. To do so, we introduce the system model in Section 11.4. Secure communication over orthogonal AVCs is then discussed in Section 11.5, and it is argued that super-activation can happen for orthogonal AVWCs. Section 11.6 then considers reliable message transmission over orthogonal AVCs, providing a complete characterization of the capacity. It is shown that super-activation cannot happen for non-secure communication over AVCs, making it a unique feature of secure communication. However, the capacity of orthogonal AVCs is shown to be super-additive, including a complete characterization of the capacity behavior. A discussion in Section 11.7 concludes the chapter.

11.3 Notation

Discrete random variables are denoted by capital letters, and their realizations and ranges by lower case and script letters; all information quantities and logarithms are taken to base 2; $\mathcal{P}(\mathcal{X})$ is the set of all probability distributions on $\mathcal{X}$; X–Y–Z denotes a Markov chain of random variables X, Y, and Z; the mutual information between the input random variable X and the output random variable Y of a channel W is denoted by $I(X;Y)$.

11.4 Arbitrarily Varying Wiretap Channel

In this section we introduce the AVWC [27–32], which models secure communication over arbitrarily varying channels [24–26]. Such channel conditions appear, for example, in fast fading environments, but also in situations in which malevolent adversaries

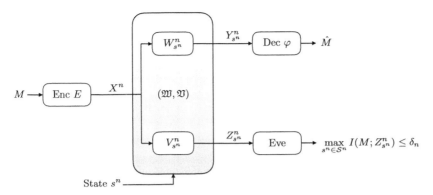

Figure 11.1 Arbitrarily varying wiretap channel $(\mathfrak{W}, \mathfrak{V})$. The confidential message M is encoded by the transmitter into the codeword $X^n = E(M)$ and transmitted over the AVC $\mathfrak{W}$ to the legitimate receiver, which has to decode its message $\hat{M} = \varphi(Y_{s^n}^n)$ for any state sequence $s^n \in \mathcal{S}^n$. At the same time, the eavesdropper eavesdrops upon the transmission via the AVC $\mathfrak{V}$ and must be kept ignorant of M by requiring $\max_{s^n \in \mathcal{S}^n} I(M; Z_{s^n}^n) \leq \delta_n$.

actively influence or jam legitimate transmissions. The corresponding system model is depicted in Fig. 11.1.

11.4.1 System Model

All communication parties are facing a channel which may vary in an unknown and arbitrary manner from channel use to channel use. This uncertainty in CSI is captured with the help of a finite state set $\mathcal{S}$. Then the channels to the legitimate receiver and the eavesdropper are given by stochastic matrices $W : \mathcal{X} \times \mathcal{S} \to \mathcal{P}(\mathcal{Y})$ and $V : \mathcal{X} \times \mathcal{S} \to \mathcal{P}(\mathcal{Z})$, with $\mathcal{X}$ the finite input alphabet and $\mathcal{Y}$ and $\mathcal{Z}$ the finite output alphabets at the legitimate receiver and eavesdropper respectively. We also interchangeably write $W_s : \mathcal{X} \to \mathcal{P}(\mathcal{Y})$ and $V_s : \mathcal{X} \to \mathcal{P}(\mathcal{Z})$ with $s \in \mathcal{S}$.

Let us first specify the channel to the legitimate receiver. For a fixed state sequence $s^n = (s_1, s_2, \ldots, s_n) \in \mathcal{S}^n$ of length n, the discrete memoryless channel is given by $W_{s^n}^n(y^n|x^n) = W^n(y^n|x^n, s^n) = \prod_{i=1}^n W(y_i|x_i, s_i)$ for all input and output sequences $x^n \in \mathcal{X}^n$ and $y^n \in \mathcal{Y}^n$.

DEFINITION 11.1   The *arbitrarily varying channel* $\mathfrak{W}$ to the legitimate receiver is defined as the family of channels for all state sequences $s^n \in \mathcal{S}^n$,

$$\mathfrak{W} = \{W_{s^n}^n : s^n \in \mathcal{S}^n\}.$$

We also need the definition of an *averaged channel*, which is given for any probability distribution $q \in \mathcal{P}(\mathcal{S})$ as

$$\overline{W}_q(y|x) = \sum_{s \in \mathcal{S}} W(y|x,s) q(s) \qquad (11.3)$$

for all $x \in \mathcal{X}$ and $y \in \mathcal{Y}$.

The concept of symmetrizability is another important property of an AVC, and is introduced next.

DEFINITION 11.2 An AVC $\mathfrak{W}$ is called *symmetrizable* if there exists a channel (stochastic matrix) $\sigma : \mathcal{X} \to \mathcal{P}(\mathcal{S})$ such that

$$\sum_{s \in \mathcal{S}} W(y|x,s)\sigma(s|x') = \sum_{s \in \mathcal{S}} W(y|x',s)\sigma(s|x) \tag{11.4}$$

holds for all $x, x' \in \mathcal{X}$ and $y \in \mathcal{Y}$.

Roughly speaking, this means that the state sequence can "simulate" a valid channel input, which makes it impossible for the receiver to decide on the correct codeword sent by the transmitter. This becomes more visible by writing the left-hand side of (11.4) as $\widetilde{W}(y|x,x') = \sum_{s \in \mathcal{S}} W(y|x,s)\sigma(s|x')$. Now, symmetrizability means that the channel $\widetilde{W}$ is symmetric in both inputs x and x', i.e., $\widetilde{W}(y|x,x') = \widetilde{W}(y|x',x)$.

The following presents a simple example of a symmetrizable AVC which first appeared in [24] and which was subsequently further discussed in [25, Example 1].

EXAMPLE 11.1 *Consider input and output alphabets of sizes $|\mathcal{X}| = 2$, $|\mathcal{Y}| = 3$, $|\mathcal{Z}| = 2$, and state alphabet of size $|\mathcal{S}| = 2$. We define the AVC $\mathfrak{W}$ as*

$$\mathfrak{W} = \{W_1, W_2\},$$

with

$$W_1 = \begin{pmatrix} 1 & 0 & 0 \\ 0 & 0 & 1 \end{pmatrix} \quad \text{and} \quad W_2 = \begin{pmatrix} 0 & 0 & 1 \\ 0 & 1 & 0 \end{pmatrix}.$$

Then we know from [25] that the AVC $\mathfrak{W}$ is symmetrizable so that its unassisted capacity is zero.

Following these lines, we can define the channel to the eavesdropper accordingly. For a fixed state sequence $s^n \in \mathcal{S}^n$ the discrete memoryless channel to the eavesdropper is given by $V_{s^n}^n(z^n|x^n) = V^n(z^n|x^n,s^n) = \prod_{i=1}^n V(z_i|x_i,s_i)$. We also set $\mathfrak{V} = \{V_{s^n}^n : s^n \in \mathcal{S}^n\}$ and $\overline{V}_q(z|x) = \sum_{s \in \mathcal{S}} V(z|x,s)q(s)$ for $q \in \mathcal{P}(\mathcal{S})$.

Combining both (marginal) AVCs to the legitimate receiver and eavesdropper finally results in the arbitrarily varying wiretap channel.

DEFINITION 11.3 The *arbitrarily varying wiretap channel* $(\mathfrak{W}, \mathfrak{V})$ is given by its marginal AVCs $\mathfrak{W}$ and $\mathfrak{V}$ with common input as

$$(\mathfrak{W}, \mathfrak{V}) = \big(\{W_{s^n}^n : s^n \in \mathcal{S}^n\}, \{V_{s^n}^n : s^n \in \mathcal{S}^n\}\big).$$

REMARK 11.1 *Here, the AVWC is defined in terms of the marginal probabilities of the channel. We want to note that it can equivalently be defined in terms of its joint probability. However, Theorem 11.1 shows that the secrecy capacity depends only on the marginal probabilities and not on the joint probability. More specifically, reliability depends on the legitimate link while secrecy solely depends on the eavesdropper link; see also [30].*

11.4.2 Code Concepts

It has been observed in [24–26] that it makes a substantial difference for communication over AVCs whether unassisted (deterministic) or more sophisticated code concepts based on *common randomness* are used. Indeed, the unassisted capacity of an AVC can be zero, while the corresponding CR-assisted capacity may be positive.

Unassisted Codes

The concept of unassisted codes refers to codes whose encoder and decoder are prespecified and fixed prior to the transmission. This is the case shown in Fig. 11.1.

DEFINITION 11.4   An *unassisted* (n, M_n)-*code* $\mathcal{C}$ consists of a stochastic encoder at the transmitter,

$$E : \mathcal{M} \to \mathcal{P}(\mathcal{X}^n), \tag{11.5}$$

i.e., a stochastic matrix, a set of messages $\mathcal{M} = \{1, \ldots, M_n\}$, and a deterministic decoder at the legitimate receiver,

$$\varphi : \mathcal{Y}^n \to \mathcal{M}. \tag{11.6}$$

REMARK 11.2   *Since encoder* (11.5) *and decoder* (11.6) *are fixed prior to the transmission of the message, they must be universally valid for all possible state sequences $s^n \in \mathcal{S}^n$ simultaneously.*

REMARK 11.3   *In the context of AVCs, unassisted codes are often referred to as deterministic codes (see, for example, [25, 26]), since the encoder and decoder are fixed and do not depend on a common random source like CR-assisted codes discussed below. But note that for unassisted codes we still allow randomness in the encoding* (11.5) *which is unknown to the legitimate receiver and eavesdropper.*

The average probability of error of such a code for a given state sequence $s^n \in \mathcal{S}^n$ is given by

$$\bar{e}_n(s^n) = \frac{1}{|\mathcal{M}|} \sum_{m \in \mathcal{M}} \sum_{x^n \in \mathcal{X}^n} \sum_{y^n : \varphi(y^n) \neq m} W^n(y^n | x^n, s^n) E(x^n | m).$$

The confidentiality of the message is ensured by requiring $\max_{s^n \in \mathcal{S}^n} I(M; Z^n_{s^n}) \leq \delta_n$ for some $\delta_n > 0$, with M the random variable uniformly distributed over the set of messages $\mathcal{M}$ and $Z^n_{s^n} = (Z_{s_1}, Z_{s_2}, \ldots, Z_{s_n})$ the channel output at the eavesdropper for state sequence $s^n \in \mathcal{S}^n$. This criterion is termed *strong secrecy* [34,35], and the reasoning is to control the total amount of information leaked to the eavesdropper. The usage of strong secrecy has some desirable practical implications; for example, for compound wiretap channels it implies that the average decoding error at the eavesdropper approaches one exponentially fast – see, for example, [28, Section 2.2] or [23, Remark 3] for further details.

This yields the following definition.

DEFINITION 11.5   A rate $R_S > 0$ is an *achievable secrecy rate* for the AVWC $(\mathfrak{W}, \mathfrak{V})$ if for all $\tau > 0$ there exists an $n(\tau) \in \mathbb{N}$, positive null sequences $\{\lambda_n\}_{n \in \mathbb{N}}$, $\{\delta_n\}_{n \in \mathbb{N}}$, and a sequence of (n, M_n)-codes $\{\mathcal{C}_n\}_{n \in \mathbb{N}}$ such that for all $n \geq n(\tau)$ we have $\frac{1}{n} \log M_n \geq R_S - \tau$,

$$\max_{s^n \in \mathcal{S}^n} \bar{e}_n(s^n) \leq \lambda_n$$

and

$$\max_{s^n \in \mathcal{S}^n} I(M; Z_{s^n}^n) \leq \delta_n.$$

The *unassisted secrecy capacity* $C_S(\mathfrak{W}, \mathfrak{V})$ of the AVWC $(\mathfrak{W}, \mathfrak{V})$ is given by the maximum of all achievable rates R_S.

While unassisted codes with prespecified encoder and decoder are suitable for many channels, they will not work for symmetrizable channels – see Definition 11.2. Thus, the unassisted capacity will be zero [26] and more sophisticated code concepts based on CR are needed.

CR-Assisted Codes
Based on a common synchronization procedure or a satellite signal, for example, it is possible for the transmitter and the receiver to generate CR. This is a powerful coordination resource and is usually modeled by a random variable Γ which takes values in a finite set $\mathcal{G}_n$ according to a certain distribution $P_\Gamma \in \mathcal{P}(\mathcal{G}_n)$. CR enables the transmitter and the receiver to choose their encoder (11.5) and decoder (11.6) according to the actual realization $\gamma \in \mathcal{G}_n$, as shown in Fig. 11.2.

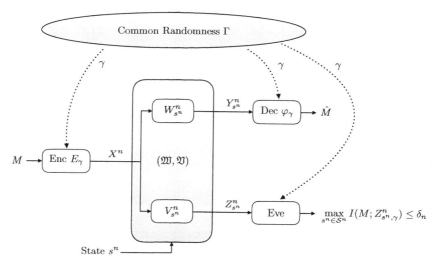

Figure 11.2   CR is available to all users, including the eavesdropper. The transmitter and receiver can adapt their encoder and decoder according to the actual CR realization $\gamma \in \mathcal{G}_n$.

DEFINITION 11.6 A *CR-assisted* $(n, M_n, \mathcal{G}_n, P_\Gamma)$-*code* $\mathcal{C}_{CR}$ is given by a family of unassisted codes

$$\{\mathcal{C}(\gamma) : \gamma \in \mathcal{G}_n\}$$

together with a random variable Γ taking values in $\mathcal{G}_n$, with $|\mathcal{G}_n| < \infty$ according to $P_\Gamma \in \mathcal{P}(\mathcal{G}_n)$.

The reliability and secrecy constraints extend to CR-assisted codes by taking the expectation over the family of codes: The mean average probability of error becomes

$$\bar{e}_{CR,n} = \max_{s^n \in \mathcal{S}^n} \frac{1}{|\mathcal{M}|} \sum_{m \in \mathcal{M}} \sum_{\gamma \in \mathcal{G}_n} \sum_{x^n \in \mathcal{X}^n} \sum_{y^n : \varphi_\gamma(y^n) \neq m} W^n(y^n | x^n, s^n) E_\gamma(x^n | m) P_\Gamma(\gamma)$$

where E_γ and φ_γ indicate that the encoder and decoder are chosen according to the CR realization $\gamma \in \mathcal{G}_n$. Accordingly, the strong secrecy criterion becomes

$$\max_{s^n \in \mathcal{S}^n} \sum_{\gamma \in \mathcal{G}_n} I(M; Z^n_{s^n, \gamma}) P_\Gamma(\gamma) \leq \delta_n, \tag{11.7}$$

where $Z^n_{s^n, \gamma}$ indicates that the observed output at the eavesdropper depends on the chosen encoder E_γ, $\gamma \in \mathcal{G}_n$.

REMARK 11.4 *Note that the secrecy criterion (11.7) can further be strengthened by requiring*

$$\max_{s^n \in \mathcal{S}^n} \max_{\gamma \in \mathcal{G}_n} I(M; Z^n_{s^n, \gamma}) \leq \delta_n,$$

i.e., the average information leakage of all codebooks in (11.7) is replaced by the information leakage of the worst codebook. Surprisingly, this strengthening does not decrease the secrecy capacity – see [30]. The stronger criterion has the advantage that it protects the message even in the scenario in which the eavesdropper is aware of the CR realization $\gamma \in \mathcal{G}_n$ and therewith of the codebook realization that is actually used by the transmitter and the receiver (cf. Fig. 11.2).

Then the definitions of a *CR-assisted achievable secrecy rate* and the *CR-assisted secrecy capacity* $C_{S,CR}(\mathfrak{W}, \mathfrak{V})$ of the AVWC $(\mathfrak{W}, \mathfrak{V})$ follow accordingly as in Definition 11.5.

11.4.3 Capacity Results

If CR is available to the transmitter and the receiver, they can coordinate their choice of encoder and decoder [27–30, 32]. To date, a single-letter characterization remains unknown (if it exists at all) and only a multi-letter description has been found.

THEOREM 11.1 ([30]) *A multi-letter description of the CR-assisted secrecy capacity* $C_{S,CR}(\mathfrak{W}, \mathfrak{V})$ *of the AVWC* $(\mathfrak{W}, \mathfrak{V})$ *is given by*

$$C_{S,CR}(\mathfrak{W}, \mathfrak{V}) = \lim_{n \to \infty} \frac{1}{n} \max_{U - X^n - (\overline{Y}^n_q, Z^n_{s^n})} \left(\min_{q \in \mathcal{P}(S)} I(U; \overline{Y}^n_q) - \max_{s^n \in \mathcal{S}^n} I(U; Z^n_{s^n}) \right),$$

with $\overline{Y}_q^n$ the random variable associated with the output of the averaged channel $\overline{W}_q^n = \sum_{s^n \in \mathcal{S}^n} q^n(s^n) W_{s^n}$, $q \in \mathcal{P}(\mathcal{S})$.

If CR is not available to the transmitter and legitimate receiver, unassisted codes must be used. The concept of symmetrizability allows the complete characterization of the corresponding unassisted secrecy capacity in terms of its CR-assisted secrecy capacity.

THEOREM 11.2 ([28, 31]) *The unassisted secrecy capacity $C_S(\mathfrak{W}, \mathfrak{V})$ of the AVWC $(\mathfrak{W}, \mathfrak{V})$ possesses the following symmetrizability properties:*

1. *If the AVC $\mathfrak{W}$ is symmetrizable, then $C_S(\mathfrak{W}, \mathfrak{V}) = 0$.*
2. *If the AVC $\mathfrak{W}$ is non-symmetrizable, then $C_S(\mathfrak{W}, \mathfrak{V}) = C_{S,CR}(\mathfrak{W}, \mathfrak{V})$.*

Similar to the single-user AVC [25, 26], the unassisted secrecy capacity displays a dichotomous behavior: The unassisted secrecy capacity $C_S(\mathfrak{W}, \mathfrak{V})$ either equals its CR-assisted secrecy capacity $C_{S,CR}(\mathfrak{W}, \mathfrak{V})$ or else is zero. Interestingly, we see from Theorem 11.2 that it is only the symmetrizability of the legitimate AVC $\mathfrak{W}$ that controls whether the unassisted secrecy capacity is zero or positive.

11.5 Super-Activation of Orthogonal AVWCs

So far, secure communication over a single AVWC has been studied. In what follows, we consider multiple orthogonal AVWCs. Then the behavior of the communication system changes drastically. In particular, it was demonstrated in [29] that two orthogonal AVWCs which are themselves "useless" in the sense of each having zero secrecy capacity can be used jointly to allow for secure transmission with a positive rate. This phenomenon of super-activation was then further studied in [31], which in particular provides a characterization of when super-activation is possible.

In the following we discuss secure communication over two orthogonal AVWCs and the phenomenon of super-activation in more detail. For this purpose, let $\mathcal{S}_i$, $\mathcal{X}_i$, $\mathcal{Y}_i$, and $\mathcal{Z}_i$, $i = 1, 2$, be finite state sets, input, and output alphabets. We define the two AVWCs $(\mathfrak{W}_1, \mathfrak{V}_1)$ and $(\mathfrak{W}_2, \mathfrak{V}_2)$ exactly as in Section 11.4.1, see Definitions 11.1 and 11.3. Then the parallel (or orthogonal) use of both AVWCs $(\mathfrak{W}_1, \mathfrak{V}_1)$ and $(\mathfrak{W}_2, \mathfrak{V}_2)$ creates a "new" combined AVWC,

$$(\widetilde{\mathfrak{W}}, \widetilde{\mathfrak{V}}) = (\mathfrak{W}_1, \mathfrak{V}_1) \otimes (\mathfrak{W}_2, \mathfrak{V}_2)$$
$$= (\mathfrak{W}_1 \otimes \mathfrak{W}_2, \mathfrak{V}_1 \otimes \mathfrak{V}_2),$$

where the notation $\otimes$ indicates the orthogonal use of $(\mathfrak{W}_1, \mathfrak{V}_1)$ and $(\mathfrak{W}_2, \mathfrak{V}_2)$. For given state sequences $s^n = (s_1^n, s_2^n) \in \mathcal{S}_1^n \times \mathcal{S}_2^n$, the discrete memoryless channel to the legitimate receiver is then specified by

$$\widetilde{W}^n(y^n | x^n, s^n) = W_{1,s_1^n}^n(y_1^n | x_1^n) W_{2,s_2^n}^n(y_2^n | x_2^n)$$
$$= W_1^n(y_1^n | x_1^n, s_1^n) W_2^n(y_2^n | x_2^n, s_2^n)$$
$$= \prod_{i=1}^n W_1(y_{1,i} | x_{1,i}, s_{1,i}) \prod_{i=1}^n W_2(y_{2,i} | x_{2,i}, s_{2,i}), \qquad (11.8)$$

with $\mathbf{x}^n = (x_1^n, x_2^n) \in \mathcal{X}_1^n \times \mathcal{X}_2^n$ and $\mathbf{y}^n = (y_1^n, y_2^n) \in \mathcal{Y}_1^n \times \mathcal{Y}_2^n$. Accordingly, the AVC $\widetilde{\mathfrak{W}}$ is then given by

$$\widetilde{\mathfrak{W}} = \{\widetilde{W}_{\mathbf{s}^n}^n : \mathbf{s}^n \in \mathcal{S}_1^n \times \mathcal{S}_2^n\}$$
$$= \{W_{1,s_1^n}^n W_{2,s_2^n}^n : s_1^n \in \mathcal{S}_1^n, s_2^n \in \mathcal{S}_2^n\}, \qquad (11.9)$$

and the AVWC $(\widetilde{\mathfrak{W}}, \widetilde{\mathfrak{V}})$ by

$$(\widetilde{\mathfrak{W}}, \widetilde{\mathfrak{V}}) = \big(\{\widetilde{W}_{\mathbf{s}^n}^n : \mathbf{s}^n \in \mathcal{S}_1^n \times \mathcal{S}_2^n\}, \{\widetilde{V}_{\mathbf{s}^n}^n : \mathbf{s}^n \in \mathcal{S}_1^n \times \mathcal{S}_2^n\}\big),$$

with $\widetilde{\mathfrak{V}}$ the AVC to the eavesdropper defined as in (11.9).

Note that a parallel use of both AVWCs $(\mathfrak{W}_1, \mathfrak{V}_1)$ and $(\mathfrak{W}_2, \mathfrak{V}_2)$ means that for each $(\mathfrak{W}_i, \mathfrak{V}_i)$ we have individual encoders $E_i : \mathcal{M}_i \to \mathcal{P}(\mathcal{X}_i^n)$ and decoders $\varphi_i : \mathcal{Y}_i^n \to \mathcal{M}_i$, $i = 1, 2$, according to Definitions 11.4 and 11.6. On the other hand, a joint use of both AVWCs results in a joint encoder $E : \mathcal{M} \to \mathcal{P}(\mathcal{X}_1^n \times \mathcal{X}_2^n)$ and a joint decoder $\varphi : \mathcal{Y}_1^n \times \mathcal{Y}_2^n \to \mathcal{M}$. Both designs of individual and joint encoders and decoders are visualized in Figs. 11.3 and 11.4.

For such orthogonal AVWCs, the phenomenon of super-activation has been analyzed and completely characterized.

THEOREM 11.3 ([31]) *Let $(\mathfrak{W}_1, \mathfrak{V}_1)$ and $(\mathfrak{W}_2, \mathfrak{V}_2)$ be two orthogonal AVWCs. Then the following properties hold:*

1. *If $C_S(\mathfrak{W}_1, \mathfrak{V}_1) = C_S(\mathfrak{W}_2, \mathfrak{V}_2) = 0$, then*

$$C_S(\mathfrak{W}_1 \otimes \mathfrak{W}_2, \mathfrak{V}_1 \otimes \mathfrak{V}_2) > 0$$

if and only if $\mathfrak{W}_1 \otimes \mathfrak{W}_2$ is non-symmetrizable and $C_{S,\mathrm{CR}}(\mathfrak{W}_1 \otimes \mathfrak{W}_2, \mathfrak{V}_1 \otimes \mathfrak{V}_2) > 0$. If $(\mathfrak{W}_1, \mathfrak{V}_1)$ and $(\mathfrak{W}_2, \mathfrak{V}_2)$ can be super-activated, then

$$C_S(\mathfrak{W}_1 \otimes \mathfrak{W}_2, \mathfrak{V}_1 \otimes \mathfrak{V}_2) = C_{S,\mathrm{CR}}(\mathfrak{W}_1 \otimes \mathfrak{W}_2, \mathfrak{V}_1 \otimes \mathfrak{V}_2).$$

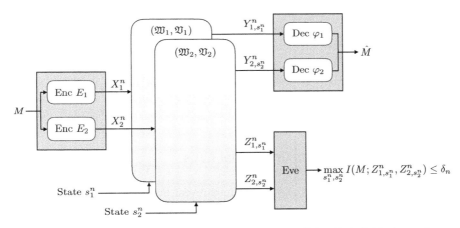

Figure 11.3 Parallel use of two orthogonal AVWCs $(\mathfrak{W}_1, \mathfrak{V}_1)$ and $(\mathfrak{W}_2, \mathfrak{V}_2)$ with individual encoders and decoders for each AVWC.

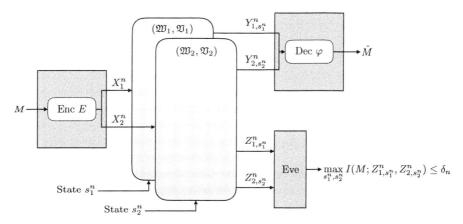

Figure 11.4 Joint use of two orthogonal AVWCs $(\mathfrak{W}_1, \mathfrak{V}_1)$ and $(\mathfrak{W}_2, \mathfrak{V}_2)$ with one joint encoder and decoder.

2. If $C_{S,CR}$ shows no super-activation for $(\mathfrak{W}_1, \mathfrak{V}_1)$ and $(\mathfrak{W}_2, \mathfrak{V}_2)$, then super-activation of C_S can only happen if $\mathfrak{W}_1$ is non-symmetrizable and $\mathfrak{W}_2$ is symmetrizable and $C_{S,CR}(\mathfrak{W}_1, \mathfrak{V}_1) = 0$ and $C_{S,CR}(\mathfrak{W}_2, \mathfrak{V}_2) > 0$. The statement is independent of the specific labeling.
3. There exist AVWCs that exhibit the behavior described by the second property.

To get some feeling as to why super-activation can happen for secure communication over AVCs, let us consider the following scenario: Assume there are two orthogonal "useless" AVWCs $(\mathfrak{W}_1, \mathfrak{V}_1)$ and $(\mathfrak{W}_2, \mathfrak{V}_2)$ each having zero unassisted secrecy capacity, i.e., $C_S(\mathfrak{W}_1, \mathfrak{V}_1) = C_S(\mathfrak{W}_2, \mathfrak{V}_2) = 0$. More specifically, we assume that one of the capacities is zero because the corresponding legitimate AVC, say $\mathfrak{W}_1$, is symmetrizable, cf. Definition 11.2, and the other capacity is zero because the eavesdropper AVC is "stronger" than the legitimate AVC such that $C(\mathfrak{V}_2) > C(\mathfrak{W}_2) > 0$. Since, in principle, the legitimate AVC $\mathfrak{W}_2$ supports a positive rate (although non-secure), it can be used to transmit information to the legitimate receiver (and eavesdropper) to generate CR. With this, the legitimate users can then use CR-assisted codes to achieve a positive CR-assisted secrecy rate. Note that we only need CR of size $|\mathcal{G}_n| = \mathcal{O}(n^2)$, i.e., sub-exponential, to achieve the CR-assisted secrecy capacity (see, e.g., [28]), the overhead is negligible, and an arbitrarily small rate is sufficient to transmit this information.

11.6 Super-Additivity of Orthogonal AVCs

The previous results and discussions reveal that the phenomenon of super-activation depends particularly on the legitimate AVC. A more detailed discussion is given in [36]. Accordingly, it is interesting to drop the eavesdropper and the security requirement for a while and study reliable message transmission over single-user AVCs in more detail.

By considering only the legitimate AVC between the transmitter and legitimate receiver, the definition of the single-user AVC follows as in Section 11.4.1. Having no eavesdropper in the system, there remains the task of reliable message transmission. The single-user AVC has been well studied, and its capacity has been established for both unassisted [25, 26] and CR-assisted [24] codes.

THEOREM 11.4 ([24]) *The CR-assisted capacity $C_{CR}(\mathfrak{W})$ of the AVC $\mathfrak{W}$ is*

$$C_{CR}(\mathfrak{W}) = \max_{P_X \in \mathcal{P}(\mathcal{X})} \inf_{q \in \mathcal{P}(\mathcal{S})} I(X; \overline{Y}_q), \tag{11.10}$$

where $\overline{Y}_q$ denotes the random variable associated with the output of the averaged channel $\overline{W}_q$, $q \in \mathcal{P}(\mathcal{S})$; cf. (11.3).

The unassisted capacity is then completely characterized in terms of its CR-assisted capacity.

THEOREM 11.5 ([25, 26]) *The unassisted capacity $C(\mathfrak{W})$ of the AVC $\mathfrak{W}$ is*

$$C(\mathfrak{W}) = \begin{cases} C_{CR}(\mathfrak{W}) & \text{if } \mathfrak{W} \text{ is non-symmetrizable} \\ 0 & \text{if } \mathfrak{W} \text{ is symmetrizable.} \end{cases}$$

The problem of reliable message transmission over orthogonal AVCs is already implicitly addressed by Shannon's question of the additivity of the zero-error capacity [2]. But it was Ahlswede in [6] who explicitly connected the zero-error question with AVCs by showing that the capacity of the AVC under the maximum error criterion includes the characterization of the zero-error capacity as a special case. Thus, it is even more surprising that reliable message transmission over orthogonal AVCs has not received much attention to date. In the following we completely characterize the behavior of the capacity of orthogonal AVCs for the average error criterion.

We start with the CR-assisted capacity, and show that it is additive.

THEOREM 11.6 ([36]) *Let $\mathfrak{W}_1$ and $\mathfrak{W}_2$ be two orthogonal AVCs. Then the CR-assisted capacity is additive, i.e.,*

$$C_{CR}(\mathfrak{W}_1 \otimes \mathfrak{W}_2) = C_{CR}(\mathfrak{W}_1) + C_{CR}(\mathfrak{W}_2). \tag{11.11}$$

This result shows that the CR-assisted capacity of two orthogonal AVCs is the sum of their CR-assisted capacities. This confirms Shannon's conviction of the additivity of the capacity. The consequence is that joint encoding and decoding for both AVCs does not yield any gains in terms of CR-assisted capacity.

Next, we move on to the unassisted capacity of two orthogonal AVCs.

PROPOSITION 11.1 ([36]) *Let $\mathfrak{W}_1$ and $\mathfrak{W}_2$ be two orthogonal AVCs. If the unassisted capacities satisfy $C(\mathfrak{W}_1) > 0$ and $C(\mathfrak{W}_2) > 0$, then the unassisted capacity is additive, i.e.,*

$$C(\mathfrak{W}_1 \otimes \mathfrak{W}_2) = C(\mathfrak{W}_1) + C(\mathfrak{W}_2). \tag{11.12}$$

PROPOSITION 11.2 ([36]) *Let $\mathfrak{W}_1$ and $\mathfrak{W}_2$ be two orthogonal AVCs. If the unassisted capacities satisfy $C(\mathfrak{W}_1) = C(\mathfrak{W}_2) = 0$, then the unassisted capacity is additive, i.e.,*

$$C(\mathfrak{W}_1 \otimes \mathfrak{W}_2) = C(\mathfrak{W}_1) + C(\mathfrak{W}_2) = 0.$$

These two results still confirm the additivity of the capacity. Specifically, the overall unassisted capacity is additive whenever the unassisted capacities are both positive or both zero. Moreover, from the latter proposition it follows immediately that super-activation is not possible for reliable message transmission over orthogonal AVCs.

COROLLARY 11.1 *Let $\mathfrak{W}_1$ and $\mathfrak{W}_2$ be two orthogonal AVCs. If the unassisted capacities satisfy $C(\mathfrak{W}_1) = C(\mathfrak{W}_2) = 0$, then super-activation is not possible for the combined AVC $\mathfrak{W}_1 \otimes \mathfrak{W}_2$.*

Finally, the following result solves the remaining case.

THEOREM 11.7 ([36]) *Let $\mathfrak{W}_1$ and $\mathfrak{W}_2$ be two orthogonal AVCs. The unassisted capacity $C(\mathfrak{W}_1 \otimes \mathfrak{W}_2)$ is super-additive, i.e.,*

$$C(\mathfrak{W}_1 \otimes \mathfrak{W}_2) > C(\mathfrak{W}_1) + C(\mathfrak{W}_2), \tag{11.13}$$

if and only if either of $\mathfrak{W}_1$ or $\mathfrak{W}_2$ is symmetrizable and has a positive CR-assisted capacity.

Without loss of generality, let $\mathfrak{W}_1$ be symmetrizable; then

$$C(\mathfrak{W}_1 \otimes \mathfrak{W}_2) = C_{\text{CR}}(\mathfrak{W}_1) + C(\mathfrak{W}_2)$$
$$> C(\mathfrak{W}_1) + C(\mathfrak{W}_2) = C(\mathfrak{W}_2).$$

Similar effects happen here as for the super-activation of AVWCs in Theorem 11.3. Intuitively, the super-additivity works as follows: $\mathfrak{W}_2$ is non-symmetrizable and has a non-zero capacity $C(\mathfrak{W}_2) > 0$. Accordingly, it can be used to transmit CR information at a negligible rate to the receiver allowing for a CR-assisted coding strategy for $\mathfrak{W}_1$. An unassisted coding strategy would result in zero capacity $C(\mathfrak{W}_1) = 0$ due to the symmetrizability of the channel. However, the transmission of CR enables the use of CR-assisted codes, bringing it to a non-zero capacity $C_{\text{CR}}(\mathfrak{W}_1) > 0$ so that the overall system shows a super-additive behavior.

Interestingly, this result shows that the capacity of reliable message transmission over orthogonal AVCs is actually super-additive under certain circumstances. This breaks with the world view of classical additivity of resources.

11.7 Discussion

For public message transmission over orthogonal AVCs, the behavior of the unassisted and CR-assisted capacity is completely understood. While the CR-assisted capacity is additive, the unassisted capacity is super-additive and consequently there are orthogonal AVCs for which joint encoding and decoding result in a higher capacity than individual encoding and decoding. However, if all orthogonal AVCs are useless, i.e., each having zero capacity, the overall capacity remains zero even for joint processing; thus, there is no super-activation possible.

The situation changes if secrecy requirements are imposed on the message transmission; the capacity behavior becomes even more involved. In this case, the phenomenon of super-activation occurs. A joint use of two completely useless AVWCs, i.e., with zero unassisted secrecy capacity, can result in a combined AVWC whose unassisted secrecy capacity is positive. From a practical point of view this has important consequences for medium access control, and in particular for resource allocation.

The possibility of super-activation for AVWCs has substantial consequences for jamming strategies of potential adversaries. Assume that there are two orthogonal AVWCs that can be super-activated, and further that for each AVWC an adversary has a suitable jamming strategy to drive the unassisted secrecy capacity to zero. Specifically, for each AVWC the adversary can choose a corresponding state sequence that symmetrizes the legitimate AVC, prohibiting any reliable communication between transmitter and legitimate receiver. Now, joint encoding and decoding allows super-activation of the combined AVWC to make the communication robust: they can now transmit at a positive secrecy rate. This means that for the adversary there is no suitable jamming strategy for the combined AVWC, although there is one for each AVWC individually. As there are no restrictions on the strategy space of the adversary, this even includes the case of a product strategy consisting of both individually working jamming strategies.

Finally, we want to address some interesting and open problems. So far, no restrictions have been considered for the state sequences. Imposing such constraints yields an AVC with state constraints whose CR-assisted and unassisted capacity is known [26, 37]. But this is only true for the single-user AVC and it would be interesting to study how orthogonal AVCs and AVWCs with state constraints behave.

The zero-error capacity is known to be super-additive and, particularly, the discrepancy between the overall capacity resulting from joint use and the sum of the individual capacities can be arbitrarily large [5]. Now it would be interesting to study whether similar unbounded discrepancies occur for AVCs as well. The results on super-activation for AVWCs show that this is indeed possible for the secrecy capacity. Here, it is an even stronger result since, in contrast to [5], the input and output alphabets can be chosen in an arbitrary way. A more detailed discussion on this can be found in [38].

Acknowledgments

R. F. Schaefer's and H. Boche's work has been motivated by questions about the wiretap channel and jamming raised by the German Federal Office for Information Security (BSI). H. Boche would like to thank Dr. R. Plaga, BSI, for motivating and fruitful discussions over recent years on this topic. The results of this work were presented at the BSI-TUM/LTI Workshop at the BSI, Bonn, Germany, May 2016.

This work was supported in part by the German Research Foundation (DFG) under Grants WY 151/2-1, BO 1734/31-1, BO 1734/33-1, and in part by the U.S. National Science Foundation under Grant CMMI-1435778.

References

[1] C. E. Shannon, "A mathematical theory of communication," *Bell Syst. Tech. J.*, vol. 27, pp. 379–423, 623–656, Jul., Oct. 1948.

[2] C. E. Shannon, "The zero error capacity of a noisy channel," *IRE Trans. Inf. Theory*, vol. 2, no. 3, pp. 8–19, Sep. 1956.

[3] L. Lovász, "On the Shannon capacity of a graph," *IEEE Trans. Inf. Theory*, vol. 25, no. 1, pp. 1–7, Jan. 1979.

[4] W. Haemers, "On some problems of Lovász concerning the Shannon capacity of a graph," *IEEE Trans. Inf. Theory*, vol. 25, no. 2, pp. 231–232, Mar. 1979.

[5] N. Alon, "The Shannon capacity of a union," *Combinatorica*, vol. 18, no. 3, pp. 301–310, Mar. 1998.

[6] R. Ahlswede, "A note on the existence of the weak capacity for channels with arbitrarily varying channel probability functions and its relation to Shannon's zero error capacity," *Ann. Math. Stat.*, vol. 41, no. 3, pp. 1027–1033, 1970.

[7] A. D. Wyner, "The wire-tap channel," *Bell Syst. Tech. J.*, vol. 54, pp. 1355–1387, Oct. 1975.

[8] Y. Liang, H. V. Poor, and S. Shamai (Shitz), "Information theoretic security," *Found. Trends Commun. Inf. Theory*, vol. 5, no. 4–5, pp. 355–580, 2009.

[9] R. Liu and W. Trappe, eds., *Securing Wireless Communications at the Physical Layer*. Boston, MA: Springer US, 2010.

[10] M. Bloch and J. Barros, *Physical-Layer Security: From Information Theory to Security Engineering*. Cambridge: Cambridge University Press, 2011.

[11] H. V. Poor and R. F. Schaefer, "Wireless physical layer security," *Proc. Natl. Acad. Sci. U.S.A.*, vol. 114, no. 1, pp. 19–26, Jan. 3, 2017.

[12] R. F. Schaefer and H. Boche, "Physical layer service integration in wireless networks – signal processing challenges," *IEEE Signal Process. Mag.*, vol. 31, no. 3, pp. 147–156, May 2014.

[13] U. Helmbrecht and R. Plaga, "New challenges for IT-security research in ICT," in *World Federation of Scientists International Seminars on Planetary Emergencies*, Erice, Italy, Aug. 2008, pp. 1–6.

[14] Deutsche Telekom AG Laboratories, "Next generation mobile networks: (R)evolution in mobile communications," *Technology Radar Edition III/2010*, Feature Paper, 2010.

[15] G. Fettweis *et al.*, "The tactile internet," ITU-T Tech. Watch Rep., Tech. Rep., Aug. 2014.

[16] R. F. Schaefer, H. Boche, and H. V. Poor, "Secure communication under channel uncertainty and adversarial attacks," *Proc. IEEE*, vol. 102, no. 10, pp. 1796–1813, Oct. 2015.

[17] D. Blackwell, L. Breiman, and A. J. Thomasian, "The capacity of a class of channels," *Ann. Math. Stat.*, vol. 30, no. 4, pp. 1229–1241, Dec. 1959.

[18] J. Wolfowitz, "Simultaneous channels," *Arch. Rational Mech. Analysis*, vol. 4, no. 4, pp. 371–386, 1960.

[19] Y. Liang, G. Kramer, H. V. Poor, and S. Shamai (Shitz), "Compound wiretap channels," *EURASIP J. Wireless Commun. Netw.*, article ID 142374, pp. 1–13, 2009.

[20] I. Bjelaković, H. Boche, and J. Sommerfeld, "Secrecy results for compound wiretap channels," *Probl. Inf. Transmission*, vol. 49, no. 1, pp. 73–98, Mar. 2013.

[21] A. Khisti, "Interference alignment for the multiantenna compound wiretap channel," *IEEE Trans. Inf. Theory*, vol. 57, no. 5, pp. 2976–2993, May 2011.

[22] E. Ekrem and S. Ulukus, "Degraded compound multi-receiver wiretap channels," *IEEE Trans. Inf. Theory*, vol. 58, no. 9, pp. 5681–5698, Sep. 2012.

[23] R. F. Schaefer and S. Loyka, "The secrecy capacity of compound MIMO Gaussian channels," *IEEE Trans. Inf. Theory*, vol. 61, no. 10, pp. 5535–5552, Dec. 2015.

[24] D. Blackwell, L. Breiman, and A. J. Thomasian, "The capacities of certain channel classes under random coding," *Ann. Math. Stat.*, vol. 31, no. 3, pp. 558–567, 1960.

[25] R. Ahlswede, "Elimination of correlation in random codes for arbitrarily varying channels," *Z. Wahrscheinlichkeitstheorie verw. Gebiete*, vol. 44, pp. 159–175, 1978.

[26] I. Csiszár and P. Narayan, "The capacity of the arbitrarily varying channel revisited: Positivity, constraints," *IEEE Trans. Inf. Theory*, vol. 34, no. 2, pp. 181–193, Mar. 1988.

[27] E. MolavianJazi, M. Bloch, and J. N. Laneman, "Arbitrary jamming can preclude secure communication," in *Proc. 47th Annual Allerton Conf. Commun., Control, Computing*, Monticello, IL, USA, Sep. 2009, pp. 1069–1075.

[28] I. Bjelaković, H. Boche, and J. Sommerfeld, "Capacity results for arbitrarily varying wiretap channels," *Lecture Notes in Computer Science*, vol. 7777, pp. 123–144, 2013.

[29] H. Boche and R. F. Schaefer, "Capacity results and super-activation for wiretap channels with active wiretappers," *IEEE Trans. Inf. Forensics Security*, vol. 8, no. 9, pp. 1482–1496, Sep. 2013.

[30] M. Wiese, J. Nötzel, and H. Boche, "A channel under simultaneous jamming and eavesdropping attack – correlated random coding capacities under strong secrecy criteria," *IEEE Trans. Inf. Theory*, vol. 62, no. 7, pp. 3844–3862, Jul. 2016.

[31] J. Nötzel, M. Wiese, and H. Boche, "The arbitrarily varying wiretap channel – secret randomness, stability and super-activation," *IEEE Trans. Inf. Theory*, vol. 62, no. 6, pp. 3504–3531, Jun. 2016.

[32] H. Boche, R. F. Schaefer, and H. V. Poor, "On the continuity of the secrecy capacity of compound and arbitrarily varying wiretap channels," *IEEE Trans. Inf. Forensics Security*, vol. 12, no. 10, pp. 2531–2546, Dec. 2015.

[33] Z. Goldfeld, P. Cuff, and H. H. Permuter, "Arbitrarily varying wiretap channels with type constrained states," Jan. 2016. [Online]. Available: http://arxiv.org/abs/1601.03660

[34] I. Csiszár, "Almost independence and secrecy capacity," *Probl. Pered. Inform.*, vol. 32, no. 1, pp. 48–57, 1996.

[35] U. M. Maurer and S. Wolf, "Information-theoretic key agreement: From weak to strong secrecy for free," *Lecture Notes in Computer Science*, vol. 1807, pp. 351–368, 2000.

[36] R. F. Schaefer, H. Boche, and H. V. Poor, "Super-activation as a unique feature of arbitrarily varying wiretap channels," in *Proc. IEEE Int. Symp. Inf. Theory*, Barcelona, Spain, Jul. 2016.

[37] I. Csiszár and P. Narayan, "Arbitrarily varying channels with constrained inputs and states," *IEEE Trans. Inf. Theory*, vol. 34, no. 1, pp. 27–34, Jan. 1988.

[38] A. Ahlswede, I. Althöfer, C. Deppe, and U. Tamm, eds., *Rudolf Ahlswede's Lectures on Information Theory 3 – Hiding Data: Selected Topics*. New York: Springer, 2016.

Part III

Secret Key Generation and Authentication

12 Multiple Secret Key Generation: Information Theoretic Models and Key Capacity Regions

Huishuai Zhang, Yingbin Liang, Lifeng Lai, and Shlomo Shamai (Shitz)

This chapter reviews recent results on simultaneously generating multiple keys over a network via public discussion. Models of various structures, including the hierarchical model, the cellular model, and the pairwise independent network (PIN) model are introduced. For each model, the design of key generation schemes is described, and characterization of the key capacity region is presented. Insights into the optimality of the schemes are provided. Future directions on the topic are discussed in the end.

12.1 Introduction

The problem of secret key generation via public discussion under the source model was initiated by [1, 2]. In the basic source-type model, two legitimate terminals observe correlated source sequences and wish to establish a common secret key by communicating with each other over a public channel, which can be accessed by eavesdroppers. The secret key is required to be kept secure from eavesdroppers. The main observation is that, due to the correlation between two source sequences, terminal $\mathcal{Y}$ can recover terminal $\mathcal{X}$'s source sequence by letting terminal $\mathcal{X}$ send a limited amount of information using the distributed source coding technique [3]. Then both terminal $\mathcal{X}$ and terminal $\mathcal{Y}$ can generate a shared secret key based on terminal $\mathcal{X}$'s source sequence. The key capacity is given by

$$C(K) = H(X) - H(X|Y) = I(X;Y), \tag{12.1}$$

which can be interpreted as the information rate in terminal $\mathcal{X}$'s source sequence subtracting the rate of information released over the public channel. The close connection between distributed source coding and secret key generation also holds for more general source-type models [4]. In particular, [4] studied a general network with multiple terminals, in which a subset of terminals need to agree on a common secret group key and the remaining terminals act as dedicated helpers. It was shown in [4] that the secret key capacity is equal to the joint entropy of all source observations subtracting the minimum amount of communications needed to enable the subset of terminals to recover all source observations. Consider an example case, in which all terminals in the set $\mathcal{M}$ wish to agree on a secret key. Then, in [4], the secret key capacity is given by

$$C(K) = H(X_\mathcal{M}) - R_{\text{CO}}, \tag{12.2}$$

where $X_\mathcal{M} = (X_i : i \in \mathcal{M})$ and R_{CO} is the minimum rate of "communication for omniscience." Here, omniscience means that all legitimate terminals recover the source observations of all terminals. More specifically, R_{CO} is characterized as

$$R_{\text{CO}} = \min_{R_\mathcal{M} \in \mathcal{R}(\mathcal{M})} \sum_{i \in \mathcal{M}} R_i, \qquad (12.3)$$

where $\mathcal{R}(\mathcal{M}) = \left\{ R_\mathcal{M} : \sum_{i \in \mathcal{B}} R_i \geq H(X_\mathcal{B}|X_{\mathcal{B}^c}), \forall \mathcal{B} \subsetneq \mathcal{M}, \mathcal{B} \neq \emptyset \right\}$

and $R_\mathcal{M} = (R_i : i \in \mathcal{M})$, $X_\mathcal{B} = (X_i, i \in \mathcal{B})$.

We note that the secret key capacity in (12.2) can be intuitively interpreted as the entire source information rate $H(X_\mathcal{M})$ minus the minimum transmission rate R_{CO} revealed to the public in order to achieve omniscience at all terminals.

Such problems of single-key generation were also studied for other scenarios in [5–9]. In fact, there are various practical scenarios in which multiple keys need to be simultaneously generated. For instance, suppose a number of terminals have different security clearance levels, and each terminal is allowed to access confidential documents up to its own clearance level. In such a case, terminals with the same clearance level should share the same key, and should be kept ignorant of higher-level keys.

Ye and Narayan [10] studied a multi-key generation model that captures the above hierarchical scenario. In the model of [10], three terminals with correlated source observations wish to agree on a common secret key, required to be secure from an eavesdropper, while two designated terminals aim to generate a private key, required to be secure from the eavesdropper and the third terminal. This *hierarchical model* has also been referred to as the secret-key–private-key model in the literature. [10] provided outer and inner bounds on the key capacity region and showed that the bounds match for one special case. Recently, [11, 12] developed a random binning and joint decoding scheme which achieves the outer bound for the other cases. Thus, the key capacity region is fully characterized. These results are presented in Section 12.2.

Another type of multi-key model is referred to as the *cellular model*, e.g., [13–17], in which a central terminal (base station) wishes to agree on independent keys respectively with a number of (mobile) terminals. These models are well motivated in cellular networks, in which mobile terminals need to share independent secret keys with the base station in order to achieve secure communication. Recent results on the problem of generating two keys in a cellular model are presented in Section 12.3, where further generalizations to models with a helper are also presented.

When the number of terminals becomes large, the analysis of the multi-key generation problem under the general source is very complex. Thus, the pairwise independent network (PIN) model, a specific and important instance of the general source model, is studied in, e.g., [18, 19]. Reference [19] established interesting connections between the problem of generating multiple keys under the PIN model and a multi-commodity flow problem in a graph, and then exploited rich tools in graph theory for the design of efficient key generation schemes. These results are reviewed in Section 12.4.

12.2 Hierarchical Model

In this section, we introduce the hierarchical model with three terminals, and review results on the key capacity region under this model.

12.2.1 Model Description

Consider a discrete memoryless source, whose outputs at each time instant are generated based on the joint distribution of random variables (X,Y,Z) with corresponding alphabets $(\mathcal{X}, \mathcal{Y}, \mathcal{Z})$. We consider a system with three terminals $(\mathcal{X}, \mathcal{Y}, \mathcal{Z})$ and an eavesdropper. Here, we use the alphabet symbols to denote the terminals. Terminal $\mathcal{X}$ observes n independent and identically distributed (i.i.d.) repetitions of X, i.e., $X^n = (X_1, \ldots, X_n)$, and terminals $\mathcal{Y}$ and $\mathcal{Z}$ observe $Y^n = (Y_1, \ldots, Y_n)$ and $Z^n = (Z_1, \ldots, Z_n)$, respectively. We assume that the eavesdropper does not have source observations, and terminals are allowed to communicate with each other over a public noiseless channel with no rate constraint. We further assume that all transmissions over the public channel are observable to all parties, including the eavesdropper. The public discussion can be interactive. Without loss of generality, we assume that terminals $(\mathcal{X}, \mathcal{Y}, \mathcal{Z})$ take turns to transmit for r rounds over $3r$ consecutive time slots. We use $3r$ random variables $F_1, \ldots, F_{3r}$ to denote these transmissions, where F_t denotes the transmission in time slot t for $1 \leq t \leq 3r$. The transmission F_t can be any function of its own observation and all previous transmissions $F_{[1,t-1]} = (F_1, \ldots, F_{t-1})$. We use $\mathbf{F} = (F_1, \ldots, F_{3r})$ to denote all transmissions in $3r$ time slots.

In this model (see Fig. 12.1), terminals $\mathcal{X}$, $\mathcal{Y}$, and $\mathcal{Z}$ wish to generate a common secret key K_S, which is required to be kept secure from the eavesdropper (that has access only to the public discussion). Furthermore, terminals $\mathcal{X}$ and $\mathcal{Y}$ wish to generate a private key K_P, which is required to be secure not only from the eavesdropper but also from terminal $\mathcal{Z}$.

We next introduce the mathematical definition of the secret key and the private key. A random variable U is said to be ϵ-recoverable from another random variable V if there

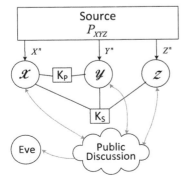

Figure 12.1 Hierarchical model.

exists a function f such that

$$\Pr\{U \neq f(V)\} < \epsilon. \tag{12.4}$$

DEFINITION 12.1 A pair (K_S, K_P) is said to be an ϵ-*hierarchical key pair* if K_S and K_P satisfy the following requirements:

- K_S is ϵ-*recoverable* at each of the three terminals with the public transmission $\mathbf{F}$, i.e., it can be ϵ-*recoverable* from $(X^n, \mathbf{F})$, $(Y^n, \mathbf{F})$, and $(Z^n, \mathbf{F})$, respectively.
- K_P is ϵ-*recoverable* at terminals $\mathcal{X}$ and $\mathcal{Y}$ with public transmission $\mathbf{F}$, i.e., it can be ϵ-*recoverable* from $(X^n, \mathbf{F})$ and $(Y^n, \mathbf{F})$, respectively.
- K_S and K_P satisfy the secrecy condition

$$\frac{1}{n} I(K_S; \mathbf{F}) < \epsilon, \tag{12.5}$$

$$\frac{1}{n} I(K_P; \mathbf{F}, Z^n) < \epsilon \tag{12.6}$$

for large enough n, where ϵ can be arbitrarily small.
- K_S and K_P satisfy the uniformity condition

$$\frac{1}{n} H(K_S) \geq \frac{1}{n} \log |\mathcal{K}_S| - \epsilon, \tag{12.7}$$

$$\frac{1}{n} H(K_P) \geq \frac{1}{n} \log |\mathcal{K}_P| - \epsilon \tag{12.8}$$

for large enough n, where $|\mathcal{K}_S|$ and $|\mathcal{K}_P|$ denote the alphabet sizes of the random variables K_S and K_P, respectively.

We note that the secrecy conditions (12.5) and (12.6) are in the weak sense, and can be strengthened to the strong sense without loss of key rates as in [20].

DEFINITION 12.2 A rate pair (R_S, R_P) is said to be an achievable rate pair if for every $\epsilon > 0$, $\delta > 0$, and for sufficiently large n, there exists an ϵ-hierarchical key pair $(K_S^{(n)}, K_P^{(n)})$ such that

$$\frac{1}{n} H(K_S^{(n)}) > R_S - \delta, \qquad \frac{1}{n} H(K_P^{(n)}) > R_P - \delta. \tag{12.9}$$

Furthermore, the key capacity region contains all achievable rate pairs (R_S, R_P).

12.2.2 Key Capacity Region

The key capacity region for the hierarchical model is presented in Theorem 12.1. For notational convenience, we define

$$R_A := I(Z; XY), \tag{12.10a}$$

$$R_B := \min\{I(X; YZ), I(Y; XZ)\}, \tag{12.10b}$$

$$R_C := \frac{1}{2}(H(X) + H(Y) + H(Z) - H(X, Y, Z)). \tag{12.10c}$$

THEOREM 12.1 *The key capacity region for the hierarchical model contains the rate pairs (R_S, R_P) satisfying the following inequalities:*

$$R_S \leq R_A, \qquad (12.11)$$

$$R_P \leq I(X;Y|Z), \qquad (12.12)$$

$$R_S + R_P \leq R_B, \qquad (12.13)$$

$$2R_S + R_P \leq 2R_C, \qquad (12.14)$$

where the constants R_A, R_B, and R_C are defined in (12.10a)–(12.10c).

The converse of Theorem 12.1 was developed in [10,21]. The bounds (12.11)–(12.13) can be intuitively understood as cut-set bounds. A cut-set bound [22, Chapter 15] means that if a cut partitions a network into two sets S and S^c of terminals, then the key rate across this cut is bounded above by the mutual information $I(X_S; X_{S^c})$. The cut-set bound can be achieved when terminals in S as well as S^c cooperate to establish a key. More specifically, the upper bound on R_S in (12.11) is due to the cut separating Z and (X, Y) for generating K_S. The upper bound on R_P in (12.12) is due to the cut separating X and Y for generating K_P, and the conditioning on Z is due to the requirement that K_P is concealed from Z. The sum rate bound (12.13) is due to the cut separating X and Y for generating both K_S and K_P simultaneously.

It can also be observed that some corner points of the region have already been shown to be achievable. First, if we choose to generate the private key K_P without considering the secret key K_S, then the model becomes the private key model, which was originally studied in [1], and later in [4] for a more general case. The outer bound on R_P is (12.12), which can be achieved by letting terminal Z reveal all its information publicly. Here, terminal Z is curious but honest, and helps to generate the private key. Second, if we choose to generate the secret key K_S without considering the private key K_P, then the model reduces to the secret key model studied in [4]. Correspondingly, the above outer bound reduces to $R_S \leq \min\{R_A, R_B, R_C\}$ based on (12.11), (12.13), and (12.14). According to [4], this bound is achievable by applying the "omniscience" scheme, which requires each terminal to recover the sources of all three terminals after the public discussion.

In order to show that the entire region in Theorem 12.1 is achievable, we first note that the region (i.e., the outer bound) takes three different structures corresponding to the following three cases: case 1, with $R_B = \min\{R_A, R_B, R_C\}$; case 2 with $R_C = \min\{R_A, R_B, R_C\}$; and case 3 with $R_A = \min\{R_A, R_B, R_C\}$.

For case 1, the outer bound (as illustrated in Fig. 12.2) is shown to be achievable in [10]. It is clear that the point B with the rate coordinates $(R_B, 0)$ is achievable by applying the "omniscience" scheme in [4]. The corner point T with the rate coordinates $(R_B - I(X;Y|Z), I(X;Y|Z))$ is achievable by the following scheme. Let Z reveal information at the rate $R_Z = \max\{H(Z|X), H(Z|Y)\}$, then both X and Y can recover Z^n correctly with high probability. Now Z^n is the information shared by three terminals, and hence the secret key K_S can be generated based on Z^n with the rate $R_S = H(Z) - R_Z = \min\{I(X;Z), I(Y;Z)\}$. Then, given Z^n, terminals X and Y can generate

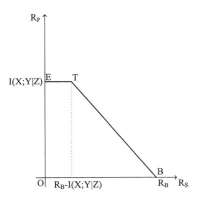

Figure 12.2   Outer bound for case 1: O–E–T–B–O.

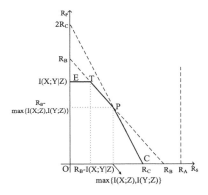

Figure 12.3   Outer bound for case 2: O–E–T–P–C–O.

a private key with the rate $R_P = I(X;Y|Z)$ if terminal $\mathcal{X}$ further reveals information at the rate $R_{\mathcal{X}} = H(X|YZ)$ to terminal $\mathcal{Y}$. Finally, the entire outer bound can be achieved by a time-sharing scheme.

The achievability of Theorem 12.1 for cases 2 and 3 is developed in [11, 12] by employing a novel random binning and joint decoding scheme. The general ideas of achievable schemes for cases 2 and 3 are as follows.

In case 2, $R_C = \min\{R_A, R_B, R_C\}$. The outer bound is plotted in Fig. 12.3 as the pentagon O–E–T–P–C–O. It has been shown in case 1 that the corner points E, T, and C are achievable. It is thus sufficient to show that the point P is achievable. Then the entire pentagon can be achieved by time sharing.

The coordinates of the point P are

$$(\max\{I(X;Z), I(Y;Z)\}, R_B - \max\{I(X;Z), I(Y;Z)\}).$$

Without loss of generality, we assume that $I(X;Z) > I(Y;Z)$ (the argument for the opposite assumption is similar), and hence $R_B = I(Y;XZ)$ and the point P becomes $(I(X;Z), I(Y;XZ) - I(X;Z))$. The secret key rate $R_S = I(X;Z)$ suggests that the highest rate that $\mathcal{Z}$ can transmit publicly is $H(Z|X)$, with which $\mathcal{X}$ recovers Z^n, but $\mathcal{Y}$ cannot recover Z^n. Then $\mathcal{X}$ must transmit some information to help $\mathcal{Y}$ to recover Z^n so that all three terminals can generate a secret key based on Z^n. Furthermore, the information transmitted by terminal $\mathcal{X}$ also helps $\mathcal{Y}$ to recover X^n so that $\mathcal{X}$ and $\mathcal{Y}$ can generate a private key. The critical part of our achievable scheme lies in that terminal $\mathcal{X}$'s transmission should help $\mathcal{Y}$ to recover Z^n without revealing more information about Z^n publicly beyond terminal $\mathcal{Z}$'s transmission. Otherwise, the secret key rate $R_S = I(X;Z)$ is not achievable. The idea is that $\mathcal{X}$ helps $\mathcal{Y}$ to improve its identifiability of Z^n without revealing information about Z^n directly.

More specifically, the achievable scheme of point P is based on random binning and joint decoding. Typical z^n sequences are randomly and independently thrown into 2^{nR_Z} bins and 2^{nR_S} sub-bins. Given a source realization Z^n, the bin index of Z^n is revealed publicly and the sub-bin index of Z^n is set to be the secret key K_S. Similarly, typical x^n sequences are randomly and independently thrown into 2^{nR_X} bins and 2^{nR_P} sub-bins. Given a source realization X^n, the bin index of X^n is revealed publicly and the sub-bin index of X^n is set to be the private key K_P. The values of the above rates are given by

$$R_Z = H(Z|X) + \epsilon, \tag{12.15}$$

$$R_S = I(X;Z) - 2\delta(\epsilon) - 2\epsilon, \tag{12.16}$$

$$R_X = H(XZ|Y) - H(Z|X), \tag{12.17}$$

$$R_P = I(XZ;Y) - I(X;Z) - 2\delta(\epsilon) - \epsilon. \tag{12.18}$$

It can be verified that with these rate settings the agreement and secrecy of the generated keys are guaranteed. More details can be found in [12].

In case 3, $R_A = \min\{R_A, R_B, R_C\}$. The outer bound is plotted in Fig. 12.4 as the hexagon O–E–T–P–Q–A–O. It has been shown in case 1 that the corner points E, T, and A are achievable. The point P can be achieved by applying the same scheme as in case 2. It is thus sufficient to show that the point Q is achievable. Then the entire hexagon can be achieved by time sharing.

The rate coordinates of the point Q are given by $(I(Z;XY), I(X;Y) - I(Z;XY))$. The secret key rate $I(Z;XY)$ suggests that the highest rate that $\mathcal{Z}$ can transmit publicly is

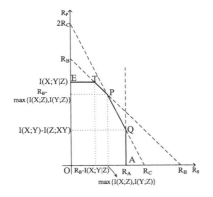

Figure 12.4 Outer bound for case 3: O–E–T–P–Q–A–O.

$H(Z|XY)$, with which neither $\mathcal{X}$ nor $\mathcal{Y}$ can recover Z^n. Then both $\mathcal{X}$ and $\mathcal{Y}$ must help each other to recover Z^n so that all three terminals can generate a secret key based on Z^n. Furthermore, terminal $\mathcal{Y}$ also helps terminal $\mathcal{X}$ to recover Y^n so that $\mathcal{X}$ and $\mathcal{Y}$ can generate a private key. The critical part lies in that $\mathcal{X}$'s and $\mathcal{Y}$'s transmissions help each other to recover Z^n without revealing more information about Z^n publicly beyond terminal $\mathcal{Z}$'s transmission. Otherwise, the secret key rate $R_S = I(Z;XY)$ is not achievable.

More specifically, the achievable scheme of point Q is also based on random binning and joint decoding. The binning and transmission steps are the same as in case 2, but with different rate settings:

$$R_\mathcal{Z} = H(Z|XY) + \epsilon + 2\delta(\epsilon), \tag{12.19}$$

$$R_S = I(Z;XY) - 2\epsilon - 4\delta(\epsilon), \tag{12.20}$$

$$R_\mathcal{X} = H(X|Y) + \epsilon, \tag{12.21}$$

$$R_P = I(X;Y) - I(Z;XY) - 2\epsilon - 2\delta(\epsilon), \tag{12.22}$$

$$R_\mathcal{Y} = H(Y|X) - 2\delta(\epsilon). \tag{12.23}$$

With these rate settings, the agreement and secrecy of the generated keys are guaranteed. More details can be found in [12].

12.3 Cellular Model

In this section, we present the results of multiple-key generation over cellular models. Such models are well motivated in cellular networks, in which mobile devices need to establish independent keys with a base station to encrypt data communications. In this section, cellular models with various secrecy constraints are introduced. In all these models, there are a central terminal (or base station) $\mathcal{X}_0$ and two cellular terminals $\mathcal{X}_1$ and $\mathcal{X}_2$, each of which observes a component of correlated source sequences. The assumption of public discussion is the same as in Section 12.2: terminals are allowed to communicate over a public noiseless channel and all transmissions are observed by all parties. An eavesdropper is assumed to have access to the public channel but does not have additional source observations. Terminal $\mathcal{X}_0$ wishes to agree on two independent keys K_1 and K_2 with terminals $\mathcal{X}_1$ and $\mathcal{X}_2$, respectively. In the rest of this section, we first introduce the models with three terminals, and then introduce models with an additional helper.

12.3.1 Two Key Generation over Three Terminals

In this subsection, we present three models of generating two independent keys K_1 and K_2 over three terminals. These models are differentiated from each other in the secrecy constraints on K_1 and K_2.

Model 1a. Two Private Key Generation

For this model, terminal $\mathcal{X}_0$ wishes to agree on two private keys K_1 and K_2 with terminals $\mathcal{X}_1$ and $\mathcal{X}_2$, respectively, where K_1 is required to be concealed from terminal $\mathcal{X}_2$ and K_2 is required to be concealed from terminal $\mathcal{X}_1$, i.e., K_1 and K_2 should satisfy the secrecy conditions

$$\frac{1}{n}I(K_1;X_2^n,\mathbf{F}) < \epsilon, \quad \frac{1}{n}I(K_2;X_1^n,\mathbf{F}) < \epsilon. \qquad (12.24)$$

This model was studied by Ye and Narayan in [13, 21], who provided outer and inner bounds on the key capacity region as given in Theorems 12.2 and 12.3.

THEOREM 12.2 *An outer bound on the key capacity region for* Model 1a *consists of rate pairs (R_1,R_2) satisfying*

$$R_1 \leq I(X_0;X_1|X_2), \quad R_2 \leq I(X_0;X_2|X_1), \qquad (12.25)$$

$$R_1 + R_2 \leq \min_U I(X_0;X_1X_2|U), \qquad (12.26)$$

where the minimum is over all random variable U that satisfy the Markov conditions U–X_1–X_0X_2 and U–X_2–X_0X_1.

If one chooses to generate K_1 with $\mathcal{X}_2$ being a helper, or alternatively to generate K_2 with $\mathcal{X}_1$ being a helper, then the model becomes the case of single-key generation which was originally studied in Theorem 3 of [1], and later in Theorem 2 of [4] for a more general case. The largest achievable rate of K_1 is $I(X_0;X_1|X_2)$ as in (12.25). The minimum in (12.26) is attainable, with U taking values in a set $\mathcal{U}$ of cardinality $|\mathcal{U}| \leq |\mathcal{Y}| \cdot |\mathcal{Z}| + 1$.

As commented in [21], one permissible choice of U in (12.26) is $U = U_{\text{mcf}(X_1,X_2)}$, where $U_{\text{mcf}(X_1,X_2)}$ is a maximal common function of X_1 and X_2. A common function of X_1 and X_2 is any random variable $U_{\text{cf}(X_1,X_2)}$ that satisfies $U_{\text{cf}(X_1,X_2)} = f(X_1) = g(X_2)$ for some deterministic mappings f and g. Furthermore, $U_{\text{mcf}(X_1,X_2)}$ is called a maximal common function of X_1 and X_2 if every other common function of X_1 and X_2 is a function of $U_{\text{mcf}(X_1,X_2)}$. It is clear that $U_{\text{mcf}(X_1,X_2)}$ satisfies the two Markov conditions in Theorem 12.2. Thus, an outer bound on R_1 and R_2 is:

$$\left\{ (R_1, R_2) : \begin{array}{l} R_1 \leq I(X_0;X_1|X_2), \quad R_2 \leq I(X_0;X_2|X_1), \\ R_1 + R_2 \leq I(X_0;X_1X_2|U_{\text{mcf}(X_1,X_2)}) \end{array} \right\}. \qquad (12.27)$$

THEOREM 12.3 *An inner bound on the key capacity region for* Model 1a *is the convex hull of the union of regions*

$$\left\{ (R_1, R_2) : \begin{array}{l} R_1 \leq \max\limits_{U:\, U-X_1-X_0X_2}[I(X_0;X_1|U) - I(X_1;X_2|U)], \\ R_2 \leq I(X_0;X_2|X_1) \end{array} \right\} \qquad (12.28)$$

and

$$\left\{ (R_1, R_2) : \begin{array}{l} R_1 \leq I(X_0;X_1|X_2), \\ R_2 \leq \max\limits_{V:\, V-X_2-X_0X_1}[I(X_0;X_2|V) - I(X_1;X_2|V)] \end{array} \right\}. \qquad (12.29)$$

The maxima in (12.28) and (12.29) are attainable with U, V taking values in sets $\mathcal{U}, \mathcal{V}$ of cardinalities $|\mathcal{U}| \leq |\mathcal{Y}| + 1$ and $|\mathcal{V}| \leq |\mathcal{Z}| + 1$.

The achievability proof of Theorem 12.3 is provided in [21]. Here, we describe the main idea. By symmetry, it suffices to show the achievability of region (12.28). Then, by a time-sharing argument, it suffices to show the achievability of rate pair

$$\left(\max_{U:\ U-X_1-X_0X_2} [I(X_0;X_1|U) - I(X_1;X_2|U)],\ I(X_0;X_2|X_1) \right).$$

The scheme to achieve this rate pair is as follows. Let terminal $\mathcal{X}_1$ transmit information such that terminal $\mathcal{X}_0$ can recover X_1^n and then K_1 can be established with rate as much as $\max_{U:\ U-X_1-X_0X_2}[I(X_0;X_1|U) - I(X_1;X_2|U)]$ due to the result on wiretap secret key capacity in [6]. Then let terminal $\mathcal{X}_2$ transmit information such that terminal $\mathcal{X}_0$ can recover X_2^n and then K_2 can be established with key rate $I(X_0;X_2|X_1)$.

Similarly, it is permissible to set $U = V = U_{\mathrm{mcf}(X_1,X_2)}$. Hence, an inner bound on R_1 and R_2 is [21]:

$$\left\{ (R_1, R_2) : \begin{array}{l} R_1 \leq I(X_0;X_1|X_2),\quad R_2 \leq I(X_0;X_2|X_1), \\ R_1 + R_2 \leq I(X_0;X_1X_2|U_{\mathrm{mcf}(X_1,X_2)}) - I(X_1;X_2|U_{\mathrm{mcf}(X_1,X_2)}) \end{array} \right\}. \quad (12.30)$$

Comparing the outer bound (12.27) with the inner bound (12.30), the gap is $I(X_1;X_2|U_{\mathrm{mcf}(X_1,X_2)})$ for the sum rate. More interpretation of these two regions can be found in [21].

Model 1b. Two Secret Key Generation

For this model, terminal $\mathcal{X}_0$ wishes to agree on two independent keys K_1 and K_2 with terminals $\mathcal{X}_1$ and $\mathcal{X}_2$, respectively. Two keys K_1 and K_2 are required to be independent and concealed from the eavesdropper, who has access to the public channel, i.e., K_1 and K_2 should satisfy the following secrecy condition:

$$\frac{1}{n} I(K_1 K_2; \mathbf{F}) < \epsilon. \quad (12.31)$$

Compared to Model 1a, Model 1b relaxes the secrecy requirement by removing the privacy constraints. This model can be viewed as a special case of the model of two secret key generation with a helper presented in Section 12.3.2, by setting the helper's source sequence to be constant. Thus we only present the key capacity region in the following theorem. Further details can be found in Section 12.3.2.

THEOREM 12.4 *The key capacity region for* Model 1b *contains rate pairs* (R_1, R_2) *satisfying*

$$R_1 \leq I(X_1; X_0, X_2), \quad (12.32)$$

$$R_2 \leq I(X_2; X_0, X_1), \quad (12.33)$$

$$R_1 + R_2 \leq I(X_0; X_1, X_2). \quad (12.34)$$

Model 1c: Secret–Private Key Generation
In this model, terminal $\mathcal{X}_0$ wishes to agree on two independent keys K_1 and K_2 with terminals $\mathcal{X}_1$ and $\mathcal{X}_2$, respectively. The two keys have asymmetric secrecy constraints, i.e., K_1 is required to be secure from an eavesdropper while K_2 is required to be secure from both the eavesdropper and terminal $\mathcal{X}_1$.

This model is well motivated in practice. For instance, a cellular system may serve two types of wireless users, e.g., one military and one civilian cellphone. Each user needs to establish a key with the base station to protect their information exchange. Moreover, communication between the military user and the base station should also be kept secure from the civilian user. This implies that the two generated keys should satisfy asymmetric secrecy constraints, which is exactly captured by this model.

This model was studied in [15], and the key capacity region is characterized by the following theorem.

THEOREM 12.5 *The key capacity region for* Model 1c *contains rate pairs* (R_1, R_2) *satisfying*

$$R_1 \leq I(X_1; X_0, X_2), \tag{12.35}$$

$$R_2 \leq I(X_0; X_2 | X_1), \tag{12.36}$$

$$R_1 + R_2 \leq I(X_0; X_1, X_2). \tag{12.37}$$

The converse of Theorem 12.5 can be argued by cut-set bounds. In particular, the upper bound on R_1 in (12.35) is due to the one cut separating $\mathcal{X}_0$ and $\mathcal{X}_1$ for generating K_1. The upper bound on R_2 in (12.36) is due to the cut separating $\mathcal{X}_0$ and $\mathcal{X}_2$ for generating K_2, and the conditioning on X_1 is due to the requirement that K_2 is concealed from $\mathcal{X}_1$. The sum rate bound (12.37) is due to the cut separating $\mathcal{X}_0$ and $(\mathcal{X}_1, \mathcal{X}_2)$ for generating two keys K_1 and K_2 simultaneously.

The achievability of Theorem 12.5 follows the unified scheme introduced in Section 12.3.2. There, a more complex model that can be viewed as Model 1c with an additional helper is studied, and thus the achievability of Theorem 12.5 is immediate by setting the helper's source sequence to be constant in the more complex model.

12.3.2 Two Key Generation Assisted by a Helper

In this subsection, we review the results on two key generation with a helper. Compared to the models studied in Section 12.3.1, the problem here becomes more challenging because the cut-set bounds become much more complex as the fourth terminal (i.e., a helper) is involved. For the three-terminal cellular model, there are three cuts for generating two keys. However, for the four-terminal cellular model (with one base station, two mobile devices, and one dedicated helper), there are six cuts for generating two keys. Furthermore, these six cuts yield eight possible cases in terms of the cut-set bound structures due to different source distributions. It is not clear at the outset whether the cut-set bound can still be achieved, because in general, a cut-set bound is less likely to be achievable as the system gets more complex. Moreover, if one tries to achieve corner points in the eight cut-set bound structures one by one, there are 16 schemes to design in total. It is not clear whether there exists a unified design of schemes to achieve the cut-set bound for all cases.

It turns out that [17] provided affirmative answers to the above two questions for models under various secrecy requirements and proposed a unified design of schemes to achieve the cut-set bound for all cases for each model. More specifically, in [17], four models (see Figs. 12.5, 12.6, 12.7, and 12.8) are considered. In all models, there are four terminals, and each terminal observes one component of a correlated vector source. Terminals $\mathcal{X}_0$ and $\mathcal{X}_1$ wish to agree on a key K_1, and terminals $\mathcal{X}_0$ and $\mathcal{X}_2$ wish to agree on another independent key K_2. Terminal $\mathcal{X}_3$ serves as a helper to assist key generation. The four terminals are allowed to communicate over a public channel, and an eavesdropper is assumed to have access to the public discussion without ambiguity. The four models differentiate from each other in secrecy constraints as we describe in the sequel.

For all of the above four models, the cut-set bound is established to be the key capacity region in [17]. Furthermore, a unified achievable strategy is constructed to achieve the corner points of the cut-set bound corresponding to all cases of source

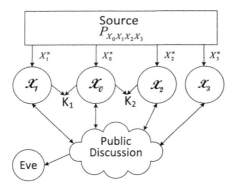

Figure 12.5 Model 2a. Two secret key generation with a trusted helper.

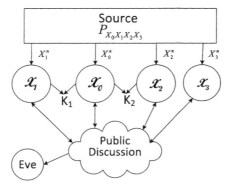

Figure 12.6 Model 2b. Two secret key generation with an untrusted helper.

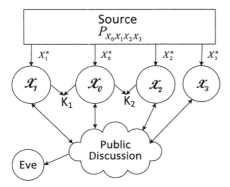

Figure 12.7 Model 2c. Secret–private key generation with a trusted helper.

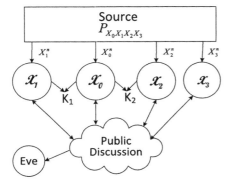

Figure 12.8 Model 2d. Secret–private key generation with an untrusted helper.

distributions. The schemes to achieve the different cases vary only in the rate at which each terminal reveals information publicly. More specifically, the achievable strategy is based on random binning and joint decoding. Given such a unified strategy, one derives the Slepian–Wolf conditions that guarantee correct key agreement and derives secrecy conditions that are sufficient to guarantee the secrecy requirements. Then, for each individual case it is sufficient to verify that the public transmission rates of the terminals satisfy the derived Slepian–Wolf conditions and secrecy conditions, which can be performed easily.

Model 2a: Two Secret Key Generation with a Trusted Helper

In this model, the two generated keys K_1 and K_2 are independent and required to be secure only from the eavesdropper, i.e.,

$$\frac{1}{n}I(K_1, K_2; \mathbf{F}) < \epsilon. \tag{12.38}$$

To assist the presentation, we introduce the following notation:

$$R_A := \min\{I(X_1, X_3; X_0, X_2), I(X_1; X_0, X_2, X_3)\}, \tag{12.39a}$$

$$R_B := \min\{I(X_0, X_1; X_2, X_3), I(X_2; X_0, X_1, X_3)\}, \quad (12.39b)$$

$$R_C := \min\{I(X_0; X_1, X_2, X_3), I(X_0, X_3; X_1, X_2)\}. \quad (12.39c)$$

The following theorem [17] characterizes the key capacity region of this model.

THEOREM 12.6 *The key capacity region for* Model 2a *contains rate pairs* (R_1, R_2) *satisfying*

$$R_1 \leq R_A, \quad (12.40)$$

$$R_2 \leq R_B, \quad (12.41)$$

$$R_1 + R_2 \leq R_C. \quad (12.42)$$

If X_3^n is independent of (X_0^n, X_1^n, X_2^n), then the helper terminal is not able to help in the key generation. In such a case, the key capacity region characterized in Theorem 12.6 reduces to Model 1b in Section 12.3.1.

Since the secrecy constraints on K_1 and K_2 are symmetric, the bounds on R_1 and R_2 are also symmetric. These bounds can be intuitively understood as cut-set bounds. The upper bound on R_1 in (12.40) is due to two cuts separating $\mathcal{X}_0$ and $\mathcal{X}_1$ for generating K_1. In fact there are two more cuts separating $\mathcal{X}_0$ and $\mathcal{X}_1$, which become redundant due to the sum rate bound (12.42). The upper bound on R_2 in (12.41) is due to two cuts separating $\mathcal{X}_0$ and $\mathcal{X}_2$ for generating K_2. Similarly, two other bounds become redundant due to the sum rate bound (12.42). The sum rate bound (12.42) is due to the two cuts separating $\mathcal{X}_0$ and $(\mathcal{X}_1, \mathcal{X}_2)$ for generating the two keys K_1 and K_2 simultaneously.

The structure of the key capacity region is illustrated in Fig. 12.9 as the pentagon O–A–P–Q–B–O. The idea of constructing an achievable scheme to achieve the key capacity region is as follows. It suffices to show the achievability of the points P and Q. Since the secrecy constraints on K_1 and K_2 are symmetric, it is sufficient to show that the corner point P is achievable, and then the achievability of the point Q follows by symmetry.

The key rate pair of the point P is given by $R_1 = R_A$ and $R_2 = R_C - R_A$. Here, $R_A < R_C$ is assumed, because otherwise the point P would collapse to the point A, which is shown to be achievable in [4]. Corresponding to different source distributions, each of R_A and R_C can take one of the two mutual information terms given in (12.39a) and (12.39c),

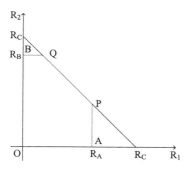

Figure 12.9 Key capacity region for two secret keys with a trusted helper.

respectively. Hence, the coordinates of the point P can take four forms, i.e., case 1 with $R_A = I(X_1;X_0X_2X_3)$ and $R_C = I(X_0;X_1X_2X_3)$, case 2 with $R_A = I(X_1;X_0X_2X_3)$ and $R_C = I(X_0X_3;X_1X_2)$, case 3 with $R_A = I(X_1X_3;X_0X_2)$ and $R_C = I(X_0;X_1X_2X_3)$, and case 4 with $R_A = I(X_1X_3;X_0X_2)$ and $R_C = I(X_0X_3;X_1X_2)$.

One constructs a unified scheme to achieve the rate point P for all cases. In the unified scheme, terminals $\mathcal{X}_1$, $\mathcal{X}_2$, and $\mathcal{X}_3$ reveal enough information publicly so that terminal $\mathcal{X}_0$ can recover X_1^n, X_2^n, and X_3^n. Then, K_1 is generated by terminals $\mathcal{X}_0$ and $\mathcal{X}_1$ based on X_1^n, and K_2 is generated by terminals $\mathcal{X}_0$ and $\mathcal{X}_2$ based on X_2^n. The schemes for the four cases are different only in the rate at which each terminal reveals information to public. Let $\tilde{R}_1$, $\tilde{R}_2$, and $\tilde{R}_3$ denote the rates at which terminals $\mathcal{X}_1$, $\mathcal{X}_2$, and $\mathcal{X}_3$ reveal information publicly, respectively. To guarantee key agreement, one only requires the Slepian–Wolf conditions given below:

$$\tilde{R}_1 > H(X_1|X_0X_2X_3), \qquad (12.43)$$

$$\tilde{R}_2 > H(X_2|X_0X_1X_3), \qquad (12.44)$$

$$\tilde{R}_3 > H(X_3|X_0X_1X_2), \qquad (12.45)$$

$$\tilde{R}_1 + \tilde{R}_2 > H(X_1X_2|X_0X_3), \qquad (12.46)$$

$$\tilde{R}_1 + \tilde{R}_3 > H(X_1X_3|X_0X_2), \qquad (12.47)$$

$$\tilde{R}_2 + \tilde{R}_3 > H(X_2X_3|X_0X_1), \qquad (12.48)$$

$$\tilde{R}_1 + \tilde{R}_2 + \tilde{R}_3 > H(X_1X_2X_3|X_0). \qquad (12.49)$$

Starting from the secrecy requirement (12.38), one can derive a set of sufficient conditions that guarantee (12.38):

$$\tilde{R}_1 + R_1 < H(X_1) - 2\delta(\epsilon), \qquad (12.50)$$

$$\tilde{R}_2 + R_2 < H(X_2|X_1) - 2\delta(\epsilon), \qquad (12.51)$$

$$\tilde{R}_1 < H(X_1|X_2) - 2\delta(\epsilon), \qquad (12.52)$$

$$\tilde{R}_3 < \min\{H(X_3|X_2), H(X_3|X_1)\} - 2\delta(\epsilon), \qquad (12.53)$$

$$\tilde{R}_1 + \tilde{R}_3 \leq H(X_1X_3|X_2) - 2\delta(\epsilon), \qquad (12.54)$$

$$\tilde{R}_2 + R_2 + \tilde{R}_3 < H(X_2X_3|X_1) - 2\delta(\epsilon). \qquad (12.55)$$

It is shown in [17] that for all four cases, the rate point P is achievable by setting $\tilde{R}_1$, $\tilde{R}_2$, and $\tilde{R}_3$ properly to satisfy (12.43)–(12.49) and (12.50)–(12.55). For example, in case 1 ($R_A = I(X_1;X_0X_2X_3)$ and $R_C = I(X_0;X_1X_2X_3)$), the rates are set as follows:

$$\tilde{R}_1 = H(X_1|X_0X_2X_3) + \epsilon, \qquad (12.56)$$

$$R_1 = I(X_1;X_0X_2X_3) - 2\delta(\epsilon) - 2\epsilon, \qquad (12.57)$$

$$\tilde{R}_2 = H(X_2X_3|X_0) - \tilde{R}_3 + \epsilon, \qquad (12.58)$$

$$R_2 = H(X_2X_3|X_1) - H(X_2X_3|X_0) - 4\delta(\epsilon) - 3\epsilon, \qquad (12.59)$$

$$\tilde{R}_3 = \min\{H(X_3|X_2), H(X_3|X_1), H(X_3|X_0)\} - 2\delta(\epsilon) - 2\epsilon. \qquad (12.60)$$

Details for other cases can be found in [17].

We next briefly present an alternative way to show the achievability of Theorem 12.6. Applying Theorem 1 of [23], a set of sufficient secrecy conditions can be obtained:

$$\tilde{R}_1 + R_1 < H(X_1), \tag{12.61}$$

$$\tilde{R}_2 + R_2 < H(X_2), \tag{12.62}$$

$$\tilde{R}_3 < H(X_3), \tag{12.63}$$

$$\tilde{R}_1 + R_1 + \tilde{R}_2 + R_2 < H(X_1 X_2), \tag{12.64}$$

$$\tilde{R}_1 + R_1 + \tilde{R}_3 < H(X_1 X_3), \tag{12.65}$$

$$\tilde{R}_2 + R_2 + \tilde{R}_3 < H(X_2 X_3), \tag{12.66}$$

$$\tilde{R}_1 + R_1 + \tilde{R}_2 + R_2 + \tilde{R}_3 < H(X_1 X_2 X_3). \tag{12.67}$$

Combine the above conditions together with Slepian–Wolf conditions (12.43)–(12.49) and apply Fourier–Motzkin elimination to eliminate $\tilde{R}_1$, $\tilde{R}_2$, and $\tilde{R}_3$. Then we obtain bounds only on R_1 and R_2, which can be shown to match the cut-set outer bounds.

Model 2b: Two Secret Key Generation with an Untrusted Helper

In this model, the two generated keys K_1 and K_2 are required to be independent and secure from both the eavesdropper and the helper terminal $\mathcal{X}_3$, i.e.,

$$\frac{1}{n} I(K_1, K_2; X_3^n, \mathbf{F}) < \epsilon. \tag{12.68}$$

The following theorem [17] characterizes the key capacity region of this model.

THEOREM 12.7 *The key capacity region for* Model 2b *contains rate pairs* (R_1, R_2) *satisfying the following inequalities:*

$$R_1 < I(X_1; X_0 X_2 | X_3), \tag{12.69}$$

$$R_2 < I(X_2; X_0 X_1 | X_3), \tag{12.70}$$

$$R_1 + R_2 < I(X_0; X_1 X_2 | X_3). \tag{12.71}$$

These bounds can also be intuitively understood as cut-set bounds as before. Comparing to Theorem 12.6, it is clear that the key capacity region for two key generation with an untrusted helper is contained in the key capacity region for two key generation with a trusted helper, because it always holds that $I(X_1; X_0 X_2 | X_3) \leq R_A$, $I(X_2; X_0 X_1 | X_3) \leq R_B$, and $I(X_0; X_1 X_2 | X_3) \leq R_C$. This is reasonable due to the additional requirement for the keys to be concealed from the untrusted helper.

The structure of the key capacity region is illustrated in Fig. 12.10 as the pentagon O–A–P–Q–B–O. In order to justify the achievability of the region, by symmetry, it is sufficient to show the achievability of the point P in Fig. 12.10. The rate pair at point P is given by $(I(X_1; X_0 X_2 | X_3), H(X_2 | X_1 X_3) - H(X_2 | X_0 X_3))$. The idea to achieve the point P follows the unified strategy described for Model 2a, i.e., public discussion first guarantees that terminal $\mathcal{X}_0$ recovers X_1^n, X_2^n, and X_3^n correctly, and then K_1 is generated by terminals $\mathcal{X}_0$ and $\mathcal{X}_1$ based on X_1^n and K_2 is generated by terminals $\mathcal{X}_0$ and $\mathcal{X}_2$ based on X_2^n. Since the generated keys should be concealed from the helper terminal $\mathcal{X}_3$, X_3^n

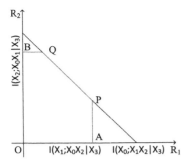

Figure 12.10 Key capacity region for two secret keys with an untrusted helper.

cannot be used as a random resource for generating the keys, although the helper can still participate in the public discussion to assist the recovery of source sequences.

More specifically, since the rate of K_1 needs to be maximized at point P, terminal $\mathcal{X}_1$ should reveal as little information as possible. Hence, terminals $\mathcal{X}_2$ and $\mathcal{X}_3$ first reveal information at the rate $\tilde{R}_2 + \tilde{R}_3 = H(X_2 X_3 | X_0)$ so that terminal $\mathcal{X}_1$ needs to release only at the rate $\tilde{R}_1 = H(X_1 | X_0 X_2 X_3)$ in order for $\mathcal{X}_0$ to recover X_1^n. Thus, the rate of K_1 can be as large as $I(X_1; X_0 X_2 | X_3)$. Since $\mathcal{X}_3$ is an untrusted helper, it can reveal information at any rate. Hence, $\tilde{R}_2$ can be set to be $H(X_2 | X_0 X_3)$. Consequently, $R_2 = H(X_2 | X_1 X_3) - H(X_2 | X_0 X_3)$. Since the generation of K_1 already uses up information contained in X_1^n, K_2 can be generated only based on information contained in X_2^n given X_1^n and X_3^n.

Model 2c: Secret–Private Key Generation with a Trusted Helper

In this model, the two generated keys K_1 and K_2 are required to be kept secure from the eavesdropper, and furthermore, the key K_2 is also required to be kept secure from terminal $\mathcal{X}_1$, i.e.,

$$\frac{1}{n}I(K_1; \mathbf{F}) < \epsilon, \quad \frac{1}{n}I(K_2; \mathbf{F}, X_1^n) < \epsilon. \tag{12.72}$$

For concise presentation, we introduce the following notation:

$$R_A := \min\{I(X_1 X_3; X_0 X_2), I(X_1; X_0 X_2 X_3)\}, \tag{12.73a}$$

$$R_B' := \min\{I(X_0; X_2 X_3 | X_1), I(X_2; X_0 X_3 | X_1)\}, \tag{12.73b}$$

$$R_C := \min\{I(X_0; X_1 X_2 X_3), I(X_0 X_3; X_1 X_2)\}. \tag{12.73c}$$

The expressions of R_A and R_C remain unchanged compared with those in Model 2a, but R_B' is different from R_B by having X_1 in the conditioning.

The following theorem [17] characterizes the key capacity region of this model.

THEOREM 12.8 *The key capacity region for* Model 2c *contains rate pairs* (R_1, R_2) *satisfying the following inequalities:*

$$R_1 \leq R_A, \tag{12.74}$$

$$R_2 \leq R'_B, \tag{12.75}$$

$$R_1 + R_2 \leq R_C. \tag{12.76}$$

If X_3^n is independent of (X_0^n, X_1^n, X_2^n), then the helper terminal is not able to help in the key generation. In such a case, the key capacity region characterized in Theorem 12.8 reduces to Model 1c in Section 12.3.1.

The bounds in Theorem 12.8 can be intuitively understood as cut-set bounds as before. The key capacity region is illustrated in Fig. 12.11 as the pentagon O–A–P–Q–B–O. The idea of constructing an achievable scheme to achieve the key capacity region is as follows. Since the secrecy requirements for the two keys are different, two achievable schemes should be separately designed to achieve the points P and Q. Corresponding to different source distributions, the rate pairs of the points P and Q can take different forms. Interestingly, the same unified strategy described for the previous models can achieve the points P and Q for all cases. Namely, terminals $\mathcal{X}_1$, $\mathcal{X}_2$, and $\mathcal{X}_3$ reveal information publicly such that terminal $\mathcal{X}_0$ can recover (X_1^n, X_2^n, X_3^n) correctly. Then K_1 is generated based on X_1^n and K_2 is generated based on X_2^n. The schemes for different cases vary only in the rate at which each terminal reveals information publicly. Here, we still use $\tilde{R}_1$, $\tilde{R}_2$, and $\tilde{R}_3$ to denote the rates at which terminals $\mathcal{X}_1$, $\mathcal{X}_2$, and $\mathcal{X}_3$ reveal information publicly, respectively.

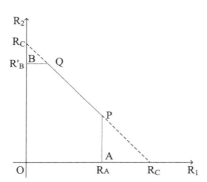

Figure 12.11 Key capacity region for secret–private key generation with a trusted helper.

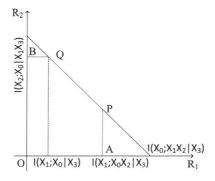

Figure 12.12 Key capacity region for secret–private key generation with an untrusted helper.

For the point P, it can be observed that its rate coordinates are exactly the same as those in Model 2a, and can be shown to be achievable by the same schemes designed for Model 2a of two secret key generation with a trusted helper. This is because both models should reach the same maximum rate R_1 of K_1 due to the same secrecy requirement for K_1. Furthermore, since both models should exhaust all random resources in X_1^n for generating K_1, K_2 should be generated from random resources independent from X_1^n even if it is not required to be concealed from terminal $\mathcal{X}_1$ in two secret key generation. Thus, the two models also have the same rate R_2 at the point P.

For the point Q, the rate coordinates are given by $(R_C - R'_B, R'_B)$. Corresponding to different source distributions, each of R_C and R'_B can take one of the two mutual information terms given in (12.75) and (12.76), respectively. Hence, the coordinates of the point Q can take four forms, i.e., case 1 with $R_C = I(X_0; X_1 X_2 X_3)$ and $R'_B = I(X_0; X_2 X_3 | X_1)$; case 2 with $R_C = I(X_0; X_1 X_2 X_3)$ and $R'_B = I(X_2; X_0 X_3 | X_1)$; case 3 with $R_C = I(X_0 X_3; X_1 X_2)$ and $R'_B = I(X_0; X_2 X_3 | X_1)$; and case 4 with $R_C = I(X_0 X_3; X_1 X_2)$ and $R'_B = I(X_2; X_0 X_3 | X_1)$. Then the achievability of point Q follows the same steps as the achievability of point P in Model 2a. Binning rates are set differently according to the different source distributions for each case, which satisfy the Slepian–Wolf conditions and the sufficient conditions that guarantee secrecy. We refer to [17] for the detailed proof.

For the model of secret–private key generation with a trusted helper, the point Q achieves the same sum rate as that for two secret key generation with a trusted helper. This is because although the rate of K_2 decreases in the secret–private key model due to the additional secrecy requirement for K_2 to be concealed from $\mathcal{X}_1$, the random resource contained in X_1^n can still be used to generate K_1 so that there is no loss in the sum rate.

Model 2d: Secret–Private Key Generation with an Untrusted Helper

In this model, the two generated keys K_1 and K_2 are required to be kept secure from both the eavesdropper and the helper terminal $\mathcal{X}_3$, and furthermore, the key K_2 is required to be kept secure from terminal $\mathcal{X}_1$, i.e.,

$$\frac{1}{n}I(K_1; \mathbf{F}, X_3^n) < \epsilon, \quad \frac{1}{n}I(K_2; \mathbf{F}, X_1^n, X_3^n) < \epsilon. \tag{12.77}$$

The following theorem [17] characterizes the key capacity region for this model.

THEOREM 12.9 *The key capacity region for* Model 2d *contains rate pairs* (R_1, R_2) *satisfying the following inequalities:*

$$R_1 < I(X_1; X_0 X_2 | X_3), \tag{12.78}$$

$$R_2 < I(X_2; X_0 | X_1 X_3), \tag{12.79}$$

$$R_1 + R_2 < I(X_0; X_1 X_2 | X_3). \tag{12.80}$$

These bounds can also be understood intuitively as cut-set bounds as before. All these bounds (12.78)–(12.80) are conditioned on X_3 because both K_1 and K_2 are required to be kept secure from terminal $\mathcal{X}_3$.

Similarly to the two secret key case, comparing to Theorem 12.8, it is clear that the key capacity region for secret–private key generation with an untrusted helper is contained in that for the secret–private key generation with a trusted helper, because it always holds that $I(X_1;X_0X_2|X_3) \leq R_A$, $I(X_2;X_0|X_1X_3) \leq R'_B$, and $I(X_0;X_1X_2|X_3) \leq R_C$. This is anticipated due to the additional requirement for the keys to be concealed from the helper.

12.4 Generating Multiple Keys under the PIN Model

In this section, we review results obtained in [19] on the generation of multiple keys for a general network consisting of m terminals indexed by $\mathcal{M} = \{1,\ldots,m\}$. In this general network, the goal is to generate T independent keys $\{K_t, t = 1,\ldots,T\}$, one for each pair of users indexed by $(t, T+t), t = 1,\ldots,T$. All the remaining users ranging $(2T+1,\ldots,m)$ serve as helpers that are neither required to recover any of the keys nor are they required to be kept ignorant of these keys.

In this network, each terminal $i \in \mathcal{M}$ observes $\tilde{X}_i^n = (\tilde{X}_{i1},\ldots,\tilde{X}_{in})$, which are n i.i.d. repetitions of $\tilde{X}_i$. As the problem under the general model is very complex, all existing results are related to a special model: the pairwise independent network (PIN) model.[1] More specifically, under the PIN model, each $\tilde{X}_i$ is of the form $\tilde{X}_i = \{X_{ij}, j \in \mathcal{M} \setminus \{i\}\}$, and the pairs $\{(X_{ij}, X_{ji}), 1 \leq i < j \leq m\}$ are mutually independent. As the result, in the PIN model,

$$P_{\tilde{X}_1,\ldots,\tilde{X}_m}(\tilde{x}_1,\ldots,\tilde{x}_m) = \prod_{1 \leq i < j \leq m} P_{X_{ij},X_{ji}}(x_{ij},x_{ji}). \tag{12.81}$$

Similar to other sections, these users are allowed to exchange information with each other using a public channel with infinite capacity, and we use $\mathbf{F}$ to denote the whole set of public discussions. Combining $\tilde{X}_t^n$ and $\mathbf{F}$, terminal t generates an estimate $\hat{K}_t$ of the key K_t for $1 \leq t \leq T$ (or K_{t-T} for $T+1 \leq t \leq 2T$), where K_t is defined on an alphabet $\mathcal{K}_t$. For any $\epsilon > 0$, the generated keys should satisfy the following requirements:

$$\Pr\{K_t = \hat{K}_t = \hat{K}_{t+T}\} \geq 1 - \epsilon, \forall t \in \{1,\ldots,T\}, \tag{12.82}$$

$$H(K_1,K_2,\ldots,K_T) = \sum_{t=1}^{T} H(K_t), \tag{12.83}$$

$$\log|\mathcal{K}_t| - H(K_t) \leq \epsilon, \forall t \in \{1,\ldots,T\}. \tag{12.84}$$

$$I(K_1,\ldots,K_T;\mathbf{F}) \leq \epsilon. \tag{12.85}$$

Here, (12.82) means that the pair $(t, t+T)$ generates the same key (e.g., Fig. 12.13), (12.83) implies that these T keys are mutually independent, (12.84) says that the keys

---

[1] The PIN model first studied in [24] is well motivated by real-life scenarios. In particular, it is particularly suitable for studying the key agreement problem in wireless networks [25–31]. This model, as a special case of the general model considered in [4], was motivated by the observation that each pair of wireless terminals can obtain correlated estimates of the channel gain between them. This pair of estimates are independent of estimates associated with the channel gains from other channels. Hence, the name PIN was used.

Figure 12.13 An example of pairwise key generation.

are close to being uniformly generated, and (12.85) implies that Eve learns a limited amount of information about the generated keys.

A rate vector $(R_1,\ldots,R_T)$ is said to be achievable if there exists a communication strategy $\mathbf{F}$ such that conditions (12.82)–(12.85) are satisfied, and

$$R_t = \frac{1}{n}H(K_t), \quad t=1,\ldots,T \qquad (12.86)$$

as $n \to \infty$. The set of all achievable rate vectors is called the capacity region. Furthermore, the maximal sum of key rates

$$C_{\text{sum}} = \sup \sum_{t=1}^{T} R_t \qquad (12.87)$$

is of interest.

For this problem, [19] proposed a low complexity scheme that is a combination of local point-to-point key generation and linear programming (LP). Furthermore, [19] showed that this low complexity scheme achieves the whole capacity region for the case of generating two keys, and is a constant fraction away from the maximum sum rate for the case of multiple keys. The key idea is to convert the key generation problem to a multi-commodity flow over a network problem, and exploit various results in graph theory [32].

12.4.1 Two Pairs Case

We first review the case of $T=2$ in which one is required to generate key K_1 for terminals $(1,3)$ and key K_2 for terminals $(2,4)$. All other terminals serve as helpers that will assist in the key generation process. They are neither required to recover the value of keys nor to be kept ignorant of the generated keys.

In the following, we describe a graph-based approach proposed in [19] that allows nodes to propagate keys over the network. There are two main steps: (1) graph construction via local key establishment, and (2) key propagation via multi-commodity flow.

ALGORITHM 12.1 Generating two keys for two pairs

Step 1: Graph construction via local key establishment
Construct an undirected graph $G_n(\mathcal{V},\mathcal{E})$ in which $\mathcal{V}$ and $\mathcal{E}$ are the set of nodes and edges of the graph, respectively. In the graph, $\mathcal{V}$ includes all the nodes in $\mathcal{M}$. For each node pair (i,j), an undirected secure link with link capacity

$e_{ij} = n(I(X_{ij}; X_{ji}) - \epsilon)$ is added. This is done by asking nodes i and j to establish a local key via the existing point-to-point key establishment protocol with the correlated observations (X_{ij}^n, X_{ji}^n) [1]. We use K_{ij} to denote the value of this local key at node i and K_{ji} to denote the value of this key at node j. For any $\epsilon_1 > 0$, there exists a scheme [1] such that $\Pr\{K_{ij} \neq K_{ji}\} \leq \epsilon_1$. In the following, instead of using both K_{ij} and K_{ji} to denote the value of the local key between (i,j), we will use K_{ij} to denote both keys with the understanding that there is a small probability that the values of local keys at (i,j) are different. We use F_{ij} to denote the public discussion information exchanged in order to establish the local key between (i,j). For any $\epsilon_2 > 0$, there exists a scheme [1,4] such that $I(K_{ij}; F_{ij}) \leq \epsilon_2$.

Step 2: Key propagation

Nodes 1 and 2 independently and randomly generate keys K_1 and K_2 from sets $\{1,\ldots,2^{nR_1}\}$ and $\{1,\ldots,2^{nR_2}\}$ using a uniform distribution. Hence, K_1 has nR_1 bits while K_2 has nR_2 bits. Node 1 then sends these nR_1 bits of information to node 3 using a secure routing approach. More specifically, let $\mathcal{P}_l^1 = (1, i_{l,2}, i_{l,3},\ldots, 3)$ be the lth route between node 1 and node 3, and Q_l^1 be the total number of hops in this route. Node 1 divides key K_1 into L_1 non-overlapping parts $(K_1^1, K_2^1,\ldots, K_{L_1}^1)$, each having length W_l bits, and sends K_l^1 through the lth route. Hence, we have a total of L_1 routes for key 1. In the qth hop of the lth route $(i_{l,q}, i_{l,q+1})$, node $i_{l,q}$ encrypts K_l^1 using W_l bits of the local key $K_{i_{l,q}, i_{l,q+1}}$. We use $K_{i_{l,q}, i_{l,q+1}}^{1,l}$ to denote this part of the local key. In this case, node $i_{l,q}$ uses the one-time pad scheme for encryption, and broadcasts $K_l^1 \oplus K_{i_{l,q}, i_{l,q+1}}^{1,l}$ over the public channel. After that, node $i_{l,q+1}$ decrypts K_l^1 using the same part of the local key $K_{i_{l,q}, i_{l,q+1}}$, namely $K_{i_{l,q}, i_{l,q+1}}^{1,l}$. Finally, the node pair $(i_{l,q}, i_{l,q+1})$ will discard $K_{i_{l,q}, i_{l,q+1}}^{1,l}$, i.e., $K_{i_{l,q}, i_{l,q+1}}^{1,l}$, which will not be used again.[2] Similarly, node 2 divides key K_2 into L_2 parts, and sends them to node 4 using a one-time pad through L_2 different secure routes, each having Q_l^2 hops.

It is clear that the above secure routing key propagation protocol converts the simultaneous key agreement problem into a multi-commodity flow problem over the graph $G_n(\mathcal{V}, \mathcal{E})$ [32]. In this equivalent multi-commodity flow problem, one has two commodities that need to be transferred from node 1 to node 3 and from node 2 to node 4 with the constraint that the total amount of flows on each link cannot exceed the flow capacity. Maximizing the achievable key rates using this approach is the same as maximizing the rates of these two flows by carefully selecting the routes and the amount of flow over each route. It is shown in [19] that, by suitable routes, this secure routing approach is optimal.

THEOREM 12.10 *The scheme in Algorithm 12.1 satisfies conditions* (12.82)–(12.85) *and achieves the whole capacity region for the case of generating two keys.*

---

[2] This will guarantee that the total amount of key information of node 1 and node 2 passing through each link will not be larger than the capacity of the corresponding link.

The proof of this theorem is available in [19]. It has two main components: (1) it develops a genie-aided outer bound on the capacity region; and (2) using the "max bi-flows min-cut" theorem in graph theory established in [33], it then shows that the scheme described in Algorithm 12.1 achieves the outer bound and hence is optimal. More details are available in [19]. One can further use the cycle flow method proposed in [33] to efficiently find the routes and the corresponding flows that achieve the capacity region. The basic idea is to recursively construct routes for each user under the rate constraints.

12.4.2 General Case

We now review the results for the general case in which one is required to generate $T > 2$ keys, one key for each pair $(t, t+T)$, $t = 1, \ldots, T$. In this general case, [19] focused on the sum of key rates. In particular, [19] generalized Algorithm 12.1 to this general case and showed that the routing-based key propagation approach can achieve a sum rate that is a constant factor away from the optimal value.

The secure routing-based key propagation scheme discussed in Section 12.4.1 can be used in this general scenario. In particular, one can again construct an undirected graph $G_n(\mathcal{V}, \mathcal{E})$ with $\mathcal{V}$ being the same as $\mathcal{M}$ and $\mathcal{E}$ being the set of edges of capacity $e_{ij} = n(I(X_{ij}; X_{ji}) - \epsilon)$. Node t, $1 \leq t \leq T$, then randomly generates a key K_t using a uniform distribution from the set $\{1, \ldots, 2^{nR_t}\}$. Node t further divides K_t into non-overlapping L_t parts $(K_1^t, \ldots, K_{L_t}^t)$, where each part is sent over a route from node t to node $t+T$. During the routing, each key part K_i^t is encrypted and decrypted using the local keys established from the pairwise correlated observations. One can show that as long as the sum of key parts flowing through each edge (i,j) is less than the edge capacity e_{ij}, there is an arbitrarily small error probability of key recovery and Eve can learn a negligible amount of information about the established keys. It is clear that this routing-based approach converts the problem into a multi-commodity flow problem in the graph $G_n(\mathcal{V}, \mathcal{E})$. Finding the maximum achievable sum of rates using this approach is equivalent to finding the maximum sum of the rates of fractional multi-commodity flows,[3] which has been extensively studied in graph theory. In particular, one can formulate an LP problem to characterize C_r. To write the largest achievable rate in a concise manner, we add special edges $(T+t,t)^s$ with infinite capacity to the set $\mathcal{E}$ that allows only commodity of type t to flow from user $T+t$ to user t. We use $\tilde{\mathcal{E}} = \mathcal{E} \cup \{(T+t,t)^s, t = 1, \ldots, T\}$ to denote this enhanced set of edges. Then, C_r is the value of the following LP problem [32]:

$$\max \quad \frac{1}{n} \sum_{t=1}^{T} f_{(T+t,t)^s}^t \tag{12.88}$$

$$\text{s.t.} \tag{12.89}$$

$$\sum_{(j,i) \in \tilde{\mathcal{E}}} f_{ji}^t - \sum_{(i,j) \in \tilde{\mathcal{E}}} f_{ij}^t = 0, \forall i \in \mathcal{V}, \forall t \in [1, \ldots, T], \tag{12.90}$$

---

[3] Although the number of bits passed through the network should be an integer, rounding an optimal fractional solution to an integer solution will not affect the rate.

$$\sum_{t=1}^{T} f_{ij}^t + \sum_{t=1}^{T} f_{ji}^t \leq e_{ij}, \forall (i,j) \in \mathcal{E}, \tag{12.91}$$

$$f_{ij}^t \geq 0, \forall (i,j) \in \tilde{\mathcal{E}}, \forall t \in [1,\ldots,T], \tag{12.92}$$

where f_{ij}^t is the amount of key information of user pair $(t, t+T)$ that passed through from i to j. (12.90) implies that the total flow of each commodity into node i is the same as the total flow out of it, and (12.91) implies that the total amount of information flow through edge (i,j) must be smaller than $e_{i,j}$. Since this is an LP problem, efficient algorithms to find the best routes and the corresponding largest achievable rate exist.

Reference [19] further developed an upper bound for the sum of key rates for any key generation protocol (not necessarily limited to the two-step approach discussed above). We will use a graph $G_n^*(\mathcal{V}, \mathcal{E})$ that is the same as $G_n(\mathcal{V}, \mathcal{E})$ constructed above with the modification that the link capacity is $e_{ij} = nI(X_{ij}; X_{ji})$. A set of edges $\mathcal{E}'$ of the graph $G_n^*(\mathcal{V}, \mathcal{E})$ is called a multicut if removing the set $\mathcal{E}'$ from the graph $G_n^*(\mathcal{V}, \mathcal{E})$ disconnects node t from $t+T$ for $t = 1,\ldots,T$. Equivalently, a set $\mathcal{E}'$ is a multicut if for all $t = 1,\ldots,T$ there is no path between nodes t and $t+T$ in the graph $G_n^*(\mathcal{V}, \mathcal{E} \setminus \mathcal{E}')$. This implies that a multicut $\mathcal{E}'$ divides the set of nodes $\mathcal{V}$ into U non-overlapping subsets $\mathcal{V}_1, \mathcal{V}_2, \ldots, \mathcal{V}_U$ such that, for all $t = 1,\ldots,T$, nodes t and node $t+T$ are in two different subsets. It is easy to see that $U \leq m$. For each node set $\mathcal{V}_u$ with $u = 1,\ldots,U$, we define a set $\mathcal{E}'_{\mathcal{V}_u} \subset \mathcal{E}'$ such that an edge $(i,j) \in \mathcal{E}'$ is in the set $\mathcal{E}'_{\mathcal{V}_u}$ if either $i \in \mathcal{V}_u$ or $j \in \mathcal{V}_u$. Clearly, each edge $(i,j) \in \mathcal{E}'$ belongs to two different $\mathcal{E}'_{\mathcal{V}_u}$s. Figure 12.14 illustrates a multicut and the associated definitions. The value of a multicut $\mathcal{E}'$ is defined as

$$C_{\mathcal{E}'} = \sum_{(i,j) \in \mathcal{E}'} e_{ij} = \sum_{(i,j) \in \mathcal{E}'} nI(X_{ij}; X_{ji}). \tag{12.93}$$

Reference [19] established the following upper bound on the sum rate of key rates for any key generation protocol.

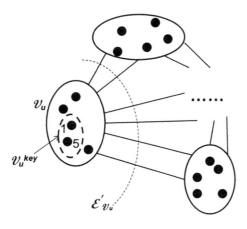

Figure 12.14 An example of a multicut and related definitions.

THEOREM 12.11 *For any key generation protocol (not necessarily limited to the two-step approach), the sum capacity is bounded above by*

$$C_{\text{sum}} \leq \frac{1}{n}\min_{\mathcal{E}'} C_{\mathcal{E}'} = \min_{\mathcal{E}'} \sum_{(i,j)\in\mathcal{E}'} I(X_{ij}; X_{ji}). \tag{12.94}$$

The following theorem [19] characterizes the relationship between the sum rate C_r achieved using the routing-based approach, which is characterized in (12.88), and that of the upper bound derived in Theorem 12.11.

THEOREM 12.12

$$C_r \geq C_{\text{sum}}/O(\log T). \tag{12.95}$$

The proof of this theorem is an application of a result in graph theory that characterizes the relationship between max sum flow and min multicut [34] of a graph.

12.5 Discussion and Future Topics

In this chapter, we have reviewed recent results on multi-key generation over source-type models. For general source models, we have summarized the characterization of the key capacity region for the hierarchical model and cellular models. For the PIN model, we have presented a simple graph-based algorithm, which is optimal for the case of generating two keys and achieves a sum rate that is a constant factor away from the optimal value.

Besides these recent results summarized in this chapter, there are many interesting and promising topics about simultaneously generating multiple keys. First, the key capacity region for the two private key model is still unknown, due to its stringent secrecy requirements. If we choose to generate one single key in the two private key model, the key capacity can be achieved by applying the scheme involving Slepian–Wolf source coding [4]. However, if we need to generate two private keys with non-zero rates simultaneously, neither of the two cellular terminals can recover each other's information, which rules out using a pure Slepian–Wolf source coding argument. In order to achieve the full key capacity region for the two private key model, a new scheme needs to be designed.

As we observe in the preceding section, cut-set bounds become very complex as the number of terminals increases. Hence, it is not clear whether our argument for a small number of terminals can be generalized to models of generating multiple keys over an arbitrary number of terminals. Thus, one future direction is to consider the hierarchical model beyond three terminals.

When the number of terminals becomes large, multiple-key generation over a general source model is very complex, and thus the PIN model is introduced. The key capacity region for generating two pairwise keys under the PIN model is established by converting the key agreement problem to a multi-commodity flow problem. One future direction is to study generating two or more group keys under the PIN model. This problem is very challenging since neither the Steiner tree packing scheme [35] nor the network coding scheme [19] achieves the general single group key capacity.

Another direction under the general source model is to study multi-key generation problems with rate constraints on the public discussion. For such a case, previous studies of the general source model with public discussion subject to finite rate constraints in [6] and of the vector Gaussian source model with public discussion subject to finite rate constraints in [36] may provide useful techniques. Furthermore, recent studies in [37,38] proposed a variant of the multi-terminal source model in which there is a newly introduced common communicator. For such a model, secret key generation under rate-limited communication was studied and the tradeoff between the communication rate and the key rate was characterized. It is interesting to generalize such studies to multi-key source models.

The model of generating multiple keys over three terminals with an eavesdropper observing source sequences is studied in [39]. An achievable region is derived by employing superposition coding and double layer binning. It is challenging to characterize the key capacity region, as even the key capacity for the corresponding single-key model is not fully understood.

Besides establishing key capacity regions for the introduced models, studying key generation problems over a general source model also gives us insights on the wiretap channel. Transmitted reliable codewords to the legitimate receiver, which are not fully decoded by the adversary, can play the role of correlated available observations to all parties in source-type models. Recall the basic source-type model [1] of two terminals with correlated observations (X, Y) agreeing on a key via a public channel with an eavesdropper observing side information Z. The forward key capacity with rate-limited public communication is given in [6] as

$$R_K \leq I(V;Y|U) - I(V;Z|U); \quad (12.96)$$

$$\text{s.t.} \quad (U,V)\text{–}X\text{–}(Y,Z) \quad \text{and} \quad \tilde{R} \geq I(U,V;X) - I(U,V;Y), \quad (12.97)$$

where R_K is an achievable secret key rate and $\tilde{R}$ is the public communication rate. Correspondingly in the wiretap channel, the transmitter and receiver share nR bits of common information (i.e., 2^{nR} equally probable codewords V'^n). Further assume that the eavesdropper observes Z'^n and hence gleans no more than $I(V'^n;Z'^n)$ information about the common information. It is then clear by setting $V = X = Y = V'^n$ and $U = \emptyset$ in (12.96) that the secret key rate of $nR - I(V'^n;Z'^n)$ bits can be achieved without public communication (since $\tilde{R} \geq 0$ if $X = Y$). With such an observation, we further understand such a rate in a more explicit form. We share $X = Y = V'^n$ at the transmitter and receiver by transmitting information over the wiretap channel with input X' and output Y' via a precoding variable V' that satisfies the Markov chain $V'\text{–}X'\text{–}(Y',Z')$, where Z' is the eavesdropper's knowledge. In such a system, one can guarantee that the transmitter conveys $nI(V';Y')$ bits reliably to the legitimate receiver, while the eavesdropper obtains no more information about those bits than $nI(V';Z')$. The latter is due to rate distortion theory, where the distortion is measured by the equivocation and the optimal source channel coding is employed. Thus, using such a scheme, we can produce a secret key (as a function of the message V'^n) of at least $nI(V';Y') - nI(V';Z')$ bits. As we know, this is indeed the capacity of the wiretap channel (thus we cannot produce a larger key rate), which implies that the rate distortion bound is tight.

We further note that the above idea of connecting the wiretap channel to a secret key generation system via codewords as correlated information between the transmitter and legitimate receiver can yield deeper insights. For example, consider the wiretap model with secrecy outside a bounded range [40, 41], in which the messages with no secrecy demands play the role of correlated signals (i.e., codewords). Such understanding motivates a new viewpoint for further generalization of the studies in [40, 41]. Furthermore, the wiretap channel with artificial noise added (see [42, 43]) can be interpreted as the key generation model with rate-limited communications, where the artificial noise serves as correlated information to generate a secret key. This is specially helpful to encrypt data in time-varying situations when the eavesdropper has instantaneously better channel state than the legitimate receiver.

Acknowledgments

The work of H. Zhang and Y. Liang was supported by a National Science Foundation CAREER Award under Grant CCF-10-26565 and by the National Science Foundation under Grant CNS-11-16932. The work of L. Lai was supported by a National Science Foundation CAREER Award under Grant CCF-13-18980 and by the National Science Foundation under Grant CNS-13-21223. The work of S. Shamai was supported by the S. and N. Grand Research Fund, and by the European Commission in the framework of the Network of Excellence in Wireless COMmunications NEWCOM#.

References

[1] R. Ahlswede and I. Csiszár, "Common randomness in information theory and cryptography – Part I: Secret sharing," *IEEE Trans. Inf. Theory*, vol. 39, no. 4, pp. 1121–1132, Jul. 1993.

[2] U. M. Maurer, "Secret key agreement by public discussion from common information," *IEEE Trans. Inf. Theory*, vol. 39, no. 3, pp. 733–742, May 1993.

[3] D. Slepian and J. Wolf, "Noiseless coding of correlated information sources," *IEEE Trans. Inf. Theory*, vol. 19, no. 4, pp. 471–480, Jul. 1973.

[4] I. Csiszár and P. Narayan, "Secrecy capacities for multiple terminals," *IEEE Trans. Inf. Theory*, vol. 50, no. 12, pp. 3047–3061, Dec. 2004.

[5] U. M. Maurer and S. Wolf, "Unconditionally secure key agreement and the intrinsic conditional information," *IEEE Trans. Inf. Theory*, vol. 45, no. 2, pp. 499–514, Feb. 1999.

[6] I. Csiszár and P. Narayan, "Common randomness and secret key generation with a helper," *IEEE Trans. Inf. Theory*, vol. 46, no. 2, pp. 344–366, Mar. 2000.

[7] U. M. Maurer and S. Wolf, "Secret-key agreement over unauthenticated public channels – Part I. Definitions and a completeness result," *IEEE Trans. Inf. Theory*, vol. 49, no. 4, pp. 822–831, Apr. 2003.

[8] U. M. Maurer and S. Wolf, "Secret-key agreement over unauthenticated public channels – Part II. The simulatability condition," *IEEE Trans. Inf. Theory*, vol. 49, no. 4, pp. 832–838, Apr. 2003.

[9] U. M. Maurer and S. Wolf, "Secret-key agreement over unauthenticated public channels – Part III. Privacy amplification," *IEEE Trans. Inf. Theory*, vol. 49, no. 4, pp. 839–851, Apr. 2003.

[10] C. Ye and P. Narayan, "The secret key–private key capacity region for three terminals," in *Proc. IEEE Int. Symp. Inf. Theory*, Adelaide, Australia, Sep. 2005, pp. 2142–2146.

[11] H. Zhang, L. Lai, Y. Liang, and H. Wang, "The secret key–private key generation over three terminals: Capacity region," in *Proc. IEEE Int. Symp. Inf. Theory*, Honolulu, HI, USA, Jul. 2014, pp. 1141–1145.

[12] H. Zhang, L. Lai, Y. Liang, and H. Wang, "The capacity region of the source-type model for secret key and private key generation," *IEEE Trans. Inf. Theory*, vol. 60, no. 10, pp. 6389–6398, Oct. 2014.

[13] C. Ye and P. Narayan, "The private key capacity region for three terminals," in *Proc. IEEE Int. Symp. Inf. Theory*, Chicago, IL, USA, Jun. 2004, p. 44.

[14] L. Lai and L. Huie, "Simultaneously generating multiple keys in many to one networks," in *Proc. IEEE Int. Symp. Inf. Theory*, Istanbul, Turkey, Jul. 2013, pp. 2394–2398.

[15] H. Zhang, Y. Liang, and L. Lai, "Key capacity region for a cellular source model," in *Proc. IEEE Inf. Theory Workshop*, Hobart, TAS, Australia, Nov. 2014, pp. 321–325.

[16] H. Zhang, Y. Liang, L. Lai, and S. Shamai (Shitz), "Two-key generation for a cellular model with a helper," in *Proc. IEEE Int. Symp. Inf. Theory*, Hong Kong, Jun. 2015, pp. 715–719.

[17] H. Zhang, Y. Liang, L. Lai, and S. Shamai (Shitz), "Multi-key generation over a cellular model with a helper," submitted to *IEEE Trans. Inf. Theory*, 2015, available at http://hzhan23.mysite.syr.edu

[18] L. Lai and S.-W. Ho, "Simultaneously generating multiple keys and multi-commodity flow in networks," in *Proc. IEEE Inf. Theory Workshop*, Lausanne, Switzerland, Sep. 2012, pp. 627–631.

[19] L. Lai and S.-W. Ho, "Key generation algorithms for pairwise independent networks based on graphical models," *IEEE Trans. Inf. Theory*, vol. 61, no. 9, pp. 4828–4837, Sep. 2015.

[20] U. M. Maurer and S. Wolf, "From weak to strong information-theoretic key agreement," in *Proc. IEEE Int. Symp. Inf. Theory*, Sorrento, Italy, Jun. 2000, p. 18.

[21] C. Ye, "Information theoretic generation of multiple secret keys," Ph.D. dissertation, University of Maryland, College Park, MD, USA, 2005.

[22] A. El Gamal and Y.-H. Kim, *Network Information Theory*. Cambridge: Cambridge University Press, 2011.

[23] M. H. Yassaee, M. R. Aref, and A. Gohari, "Achievability proof via output statistics of random binning," *IEEE Trans. Inf. Theory*, vol. 60, no. 11, pp. 6760–6786, Nov. 2014.

[24] C. Ye and A. Reznik, "Group secret key generation algorithms," in *Proc. IEEE Int. Symp. Inf. Theory*, Nice, France, Jun. 2007, pp. 2596–2600.

[25] R. Wilson, D. Tse, and R. A. Scholtz, "Channel identification: Secret sharing using reciprocity in ultrawideband channels," *IEEE Trans. Inf. Forensics Security*, vol. 2, no. 3, pp. 364–375, Sep. 2007.

[26] C. Ye, S. Mathur, A. Reznik, W. Trappe, and N. Mandayam, "Information-theoretic key generation from wireless channels," *IEEE Trans. Inf. Forensics Security*, vol. 5, no. 2, pp. 240–254, Jun. 2010.

[27] L. Lai, Y. Liang, and H. V. Poor, "A unified framework for key agreement over wireless fading channels," *IEEE Trans. Inf. Forensics Security*, vol. 7, no. 2, pp. 480–490, Apr. 2012.

[28] T.-H. Chou, A. M. Sayeed, and S. C. Draper, "Impact of channel sparsity and correlated eavesdropping on secret key generation from multipath channel randomness," in *Proc. IEEE Int. Symp. Inf. Theory*, Austin, TX, USA, Jun. 2010, pp. 2518–2522.

[29] S. Mathur, W. Trappe, N. Mandayam, C. Ye, and A. Reznik, "Radiotelephathy: Extracting a secret key from an unauthenticated wireless channel," in *Proc. ACM Int. Conf. Mobile Computing and Networking*, San Francisco, CA, USA, Sep. 2008, pp. 128–139.

[30] A. Khisti, "Interactive secret key generation over reciprocal fading channels," in *Proc. 50th Annual Allerton Conf. Commun., Control, Computing*, Monticello, IL, USA, Sep. 2012, pp. 1374–1381.

[31] M. J. Siavoshani, C. Fragouli, S. Diggavi, U. Pulleti, and K. Argyraki, "Group secret key generation over broadcast erasure channels," in *Proc. 44th Asilomar Conf. Signals, Systems, Computers*, Pacific Grove, CA, USA, Nov. 2010, pp. 719–723.

[32] R. Ahuja, T. Magnanti, and J. Orlin, *Network Flows*. Upper Saddle River, NJ: Prentice Hall, 1993.

[33] T. Hu, "Multi-commodity network flows," *Operations Research*, vol. 11, no. 3, pp. 344–360, May 1963.

[34] N. Garg, V. Vazirani, and M. Yannakakis, "Approximate max-flow min-(multi)cut theorems and their applications," *SIAM J. Comput.*, vol. 25, no. 2, pp. 235–251, Apr. 1996.

[35] S. Nitinawarat, C. Ye, A. Barg, P. Narayan, and A. Reznik, "Secret key generation for a pairwise independent network model," *IEEE Trans. Inf. Theory*, vol. 56, no. 12, pp. 6482–6489, Dec. 2010.

[36] S. Wantanabe and Y. Oohama, "Secret key agreement from vector Gaussian sources by rate limited public communication," *IEEE Trans. Inf. Forensics Security*, vol. 6, no. 3, pp. 541–550, Sep. 2011.

[37] J. Liu, P. Cuff, and S. Verdú, "Secret key generation with one communicator and a one-shot converse via hypercontractivity," in *Proc. IEEE Int. Symp. Inf. Theory*, Hong Kong, Jun. 2015, pp. 710–714.

[38] P. Cuff, J. Liu, and S. Verdú, "Secret key agreement with rate-limited communication among three nodes," in *Proc. Inf. Theory Applications Workshop*, La Jolla, CA, USA, Feb. 2015.

[39] P. Babaheidarian, S. Salimi, and M. R. Aref, "A new secret key agreement scheme in a four-terminal network," in *Proc. Canadian Workshop Inf. Theory*, Kelowna, BC, Canada, May 2011, pp. 151–154.

[40] S. Zou, Y. Liang, L. Lai, and S. Shamai (Shitz), "Degraded broadcast channel: Secrecy outside of a bounded range," in *Proc. IEEE Inf. Theory Workshop*, Jerusalem, Israel, Apr. 2015, pp. 1–5.

[41] S. Zou, Y. Liang, L. Lai, and S. Shamai (Shitz), "Rate splitting and sharing for degraded broadcast channel with secrecy outside a bounded range," in *Proc. IEEE Int. Symp. Inf. Theory*, Hong Kong, Jun. 2015, pp. 1357–1361.

[42] S. Yang, P. Piantanida, M. Kobayashi, and S. Shamai (Shitz), "On the secrecy degrees of freedom of multi-antenna wiretap channels with delayed CSIT," in *Proc. IEEE Int. Symp. Inf. Theory*, St. Petersburg, Russia, Jul. 2011, pp. 2866–2870.

[43] R. Liu, T. Liu, H. V. Poor, and S. Shamai (Shitz), "Multiple-input multiple-output Gaussian broadcast channels with confidential messages," *IEEE Trans. Inf. Theory*, vol. 56, no. 9, pp. 4215–4227, Sep. 2010.

13 Secret Key Generation for Physical Unclonable Functions

Michael Pehl, Matthias Hiller, and Georg Sigl

Secure storage of cryptographic keys is a popular application for responses generated from physical unclonable functions (PUFs). It is, however, required to correct these noisy PUF responses in order to derive the same key under all environmental conditions. This is enabled by mapping the random response pattern of the PUF to codewords of error correcting codes using so called helper data, and by proper error correction mechanisms.

This chapter maps the process of key storage with PUFs to the information theoretic model of key agreement from a compound source and shows theoretical bounds. It introduces a unified algebraic description of helper data generation schemes that is able to represent most state-of-the-art approaches. This is used together with the theoretic bounds to analyze the existing schemes. The focus here is secrecy leakage through the helper data. The new representation will allow the analysis of future schemes in an early design phase.

13.1 Introduction

Physical circuit properties such as exact run times vary for each manufactured chip. The root cause for this phenomenon is slight variations in process parameters that affect, e.g., the threshold voltages and electron mobility in the transistors of the circuit. To ensure predictable and reliable behavior of circuits, much effort is spent to mitigate the effect of such unpredictable variations. However, they turn out to be unavoidable and, moreover, the influence of these variations on the circuit properties increases with decreasing process sizes. While conventional circuits suffer from this fact, silicon-based *physical unclonable functions* take advantage of the variations: they capture randomness in the manufacturing process and transform the analog physical variations into digital numbers that can be interpreted as the outcome of a random variable. Then, the quantized result can be used for authentication in a challenge–response protocol or to embed a key into a device and only reproduce it on demand to avoid permanent storage of secret keys in non-volatile memory.

Since silicon PUFs are constructed from transistors, other standard devices, or even from standard cells, their implementation fits in seamlessly with the standard digital design flow and manufacturing process. Therefore, PUFs can be easily added to a standard integrated circuit and bridge the gap between the increasing demand for

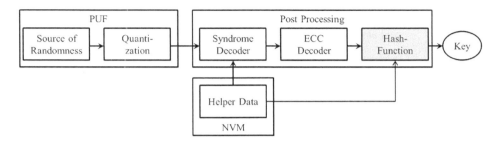

Figure 13.1 Structural setup of key recovery scenario for PUFs.

security and the restriction of a low additional cost overhead. Furthermore, PUFs can still be built in deep sub-micron technologies where standard solutions for secure key storage, e.g., secured non-volatile memory, are no longer available.

In this chapter, we focus on the key storage application of PUFs shown in Fig. 13.1, where the PUF provides a sequence of bits, called the *PUF response*, which is derived by sampling and quantizing an analog source. The analog source in such a case can be, e.g., the frequency difference of two ring oscillators (ROs) in an RO PUF [1], the difference in the threshold voltages of the transistors in an SRAM cell for an SRAM PUF [2], or the run-time difference of two equally designed paths in an arbiter PUF [3]. Once the PUF is manufactured and the variations are fixed to a specific value, the PUF response can be reproduced whenever needed. However, the analog source is influenced deterministically by environmental changes like shifts in temperature or supply voltage, by aging effects, and also randomly by noise. This leads to bit errors in the sequence provided by the PUF. The mean bit error probability is typically in the range of up to 15%–20%, so that the PUF output is too unreliable to be directly used as a key [4].

To overcome this problem and to generate sufficiently reliable keys with error probabilities in the range of 10^{-6} down to 10^{-9}, error correction is required. The building blocks in Fig. 13.1 show the different postprocessing steps that are required for secret key generation with PUFs. Error correction is enabled by mapping the initial PUF output to a codeword of the *error correcting code* (ECC) during enrollment and storing the mapping as side information, called *helper data*. However, this leads to different problems:

1. The helper data is typically stored in non-secured non-volatile memory since the PUF is used to replace the secure non-volatile memory. Therefore, the helper data must be considered as public information that is accessible to an attacker, and must not reveal any information about the secret key.
2. Non-volatile memory is an expensive resource, especially in the low-cost applications targeted by PUF designs. Therefore, the amount of helper data should be as small as possible.
3. The ECC should offer a good tradeoff between output word error probability, rate (which translates directly into number of required PUF bits), and implementation

complexity. In this chapter, we focus on efficient and secure helper data generation and assume a good ECC as a given.

In recent years, several practical algorithms have been introduced and implemented to address these issues, e.g., [5–9]. However, some of these approaches can be attacked by manipulating the helper data [10, 11] so that the optional hash function in Fig. 13.1 is required to prevent these attacks or to compensate for leakage through the helper data. If the PUF bits only have a low entropy per bit, the compression property of the hash function can also be used to compress a larger secret to a shorter key with good cryptographic properties.

Currently, the development of new error correction algorithms for PUFs is driven from a design perspective. New solutions are compared to the state of the art by analyzing practical quantities such as error probabilities, leakages, and implementation sizes.

In parallel, the information theory community studied the problem of key generation from a correlated source [12–14], which led to capacity results and optimal coding strategies based on random codes. In this chapter, we show that the information theoretic model and limits are valid for the practical problem of key generation with PUFs. We also introduce a unified algebraic description of the process of helper data generation which applies to most currently available schemes available for the syndrome decoder block in Fig. 13.1. The unified description allows the discovery of similarities and differences in these schemes and provides a tool to evaluate the leakage of the approach early, during the design of the code.

The rest of this chapter is structured as follows. In Section 13.2, we describe the notation used. This is followed by Section 13.3, showing the parallels between key agreement from a compound source and secret key generation from a PUF, which gives a deeper insight into the nature of key storage with PUFs. Section 13.4 provides an algebraic description of helper data generation for PUFs and explains different properties from this new description. The results are applied to state-of-the-art coding schemes in Section 13.5, which also provides a comparison of the different schemes. We draw some conclusions on the chapter in Section 13.6.

13.2 Notation

In this chapter, functions are denoted by small letters, e.g., $f(\cdot)$. Line vectors are denoted by capital letters, e.g., X, where $[X, Y]$ denotes the concatenation of vectors X and Y to a new line vector. Matrices are provided in bold capital letters, e.g., $\mathbf{A}$. The dimension of a vector or matrix is denoted by $\dim(\cdot)$, and evaluates, e.g., to $\dim(\mathbf{A}) = \langle a \times b \rangle$, where the first entry (a) is the number of rows and the second entry (b) is the number of columns. The rank of a matrix is computed by the $\text{rank}(\cdot)$ operation; a diagonal matrix is created from a vector X by applying $\text{diag}(X)$. A matrix $\mathbf{A}^{\times N}$ can be constructed from matrix $\mathbf{A}$ by repeating each column of $\mathbf{A}$ N times, e.g., let $\mathbf{A} = [A_1^T A_2^T]$ with columns A_1^T and A_2^T, then $\mathbf{A}^{\times 2} = [A_1^T, A_1^T, A_2^T, A_2^T]$.

The entropy of a random vector X can be evaluated to $H(X)$; the joint entropy of X and Y is given by $H(XY)$. $I(X; Y)$ denotes the mutual information between the vectors X and Y.

For some special matrices and vectors we use the following nomenclature:

- **I** is the identity matrix.
- **G** defines the generator matrix of a code, with the special case $\mathbf{G} = [\mathbf{IP}]$ for a linear code in standard form.
- **P** is the part of the generator matrix of a linear code in standard form which creates the redundancy.
- **H** is the parity check matrix of a code.

13.3 An Information Theoretical View on Key Storage with PUFs

The problem of secret key generation is well established and studied in information theory. Ahlswede and Csiszár describe in [12] the idea of establishing a common secret between two terminals $\mathcal{X}$ and $\mathcal{Y}$ from a discrete memoryless multiple source with two outputs providing correlated variables (X, Y) such that $\mathcal{X}$ has access to X and $\mathcal{Y}$ to Y. The authors give a one-way communication protocol in the same publication, and provide a corresponding model later in [15]. In [14], the initial model is extended to a source model that has an internal state. It considers an eavesdropper to control the state of the compound source and a wiretapper that receives a correlated output of the source Z as shown in Fig. 13.2(a). In the following, we discuss the similarities to the key storage scenario with PUFs in Fig. 13.2(b).

13.3.1 Source Model

In the information theoretic model, the correlated random variables X_t and Y_t, depending on the state $t \in \mathcal{T}$, are output by the compound source where terminal $\mathcal{X}$ observes X_t and terminal $\mathcal{Y}$ observes Y_t. In addition, the wiretapper has access to a degraded correlated sequence Z_t. Analogously, in the PUF scenario the manufacturing process acts as source of randomness and a specific PUF instantiation is drawn as a sample. The PUF also has an internal state t defined by the environmental conditions. It outputs a random sequence X_t during enrollment, i.e., at the point in time when the key is implanted in or read out from the PUF and the helper data is generated. Since the enrollment is carried out in a secured environment, the environmental conditions are assumed as fixed to nominal values and the t of X_t is dropped.

Later, however, during reproduction of the key, $\mathcal{Y}$ receives a noisy version of X represented by Y_t depending on t. Since $Y_t = X + E$ for an error vector E, X and Y_t can be considered as correlated random variables provided by a compound source. This represents the randomness from the manufacturing process and from noise.

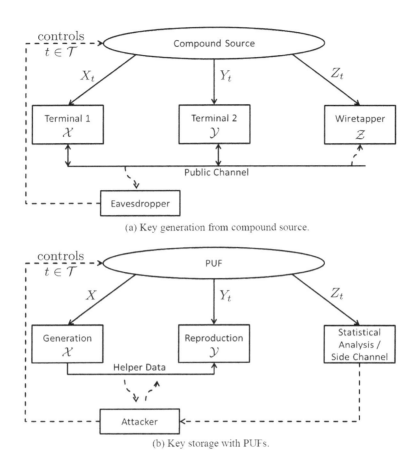

Figure 13.2 Similarity of key generation from compound source and key storage with PUFs.

13.3.2 Communication Channel

In the original model [12], the public communication is performed over a noiseless but authenticated channel, so that the attacker can read the transmitted information but cannot tamper with it. This channel is realized in the PUF scenario by an unprotected non-volatile memory. In contrast to classical information theoretic scenarios, the key storage scenario with PUFs assumes a communication not over time and space but only over time.

13.3.3 Key Agreement

Reference [14] provides the following key agreement for the compound source and one-way communication: Given X_t and Y_t as correlated outputs of the source, it is assumed that $\mathcal{X}$ computes the helper data as a function $f(X_t)$ and transmits it over a noiseless channel to Y_t. The keys are generated as $K = \varphi(X_t)$ at terminal $\mathcal{X}$ and as $L = \psi(Y_t, f(X_t))$ at terminal $\mathcal{Y}$ for an encoding function $\varphi(\cdot)$ and decoding function

$\psi(\cdot)$. A key pair (K, L) can be derived reliably if the error probability $P(K \neq L)$ is sufficiently small, i.e., $P(K \neq L) < \epsilon$ for a small $\epsilon > 0$.

In the PUF scenario, the manufacturing process acts as source. The initial response of a specific PUF instantiation is given by X and used to derive a key $K = \varphi(X)$. From this data, helper data $W = f(X)$ is computed and stored in non-volatile memory or an external server. Later, the key is reproduced from the stored helper data and the PUF output Y_t. That is, the key is computed as a function of Y_t and $f(X)$ as $L = \psi(Y_t, f(X))$.

A single protocol run is embedded into a larger scenario of the use case where it is applied. In the key generation scenario from a compound source, the key agreement is carried out multiple times to generate new session keys, for example for repeated communication between the same parties. Analogously, several PUFs are manufactured in the same batch with different pairs of output sequences for each PUF instance. Both applications have in common that the same source is used multiple times and each sequence X is only used once to generate one secret. In the PUF use case, Y_t is read multiple times and the same key is generated multiple times.

13.3.4 Rate and Capacity

Following [12] and assuming only a passive eavesdropper, and considering $\mathcal{K}$ as the key space, the rate R_{key} of a secret key generation strategy is called an *achievable key rate* if for any $\epsilon > 0$ and sufficiently large n the following four conditions hold for sequences of length n:

$$P(K \neq L) < \epsilon, \quad (13.1)$$

$$\frac{1}{n} I(W; K) < \epsilon, \quad (13.2)$$

$$\frac{1}{n} H(K) > R_{\text{key}} - \epsilon, \quad (13.3)$$

$$\frac{1}{n} \log_2(|\mathcal{K}|) < \frac{1}{n} H(K) + \epsilon. \quad (13.4)$$

In [14], an even stronger version of Eq. (13.2) is used, namely

$$I(W; K) < \epsilon. \quad (13.5)$$

Equation (13.1) requires that the error probability of the key is bounded by some small $\epsilon > 0$. In the PUF scenario, this can be seen as a design criterion required for the ECC, so that a code is selected in such a way that Eq. (13.1) is fulfilled. Equation (13.2) requires in the classical scenario that the mutual information between key K and information on the public channel $f(X)$ is bounded by ϵ. The public data in the PUF context is the helper data W stored in the non-volatile memory. In fact, many different variants have been presented to generate helper data for PUFs. Therefore, the mutual information $I(W; K)$ together with the helper data size are one of the main concerns in the field of error correction for key storage with PUFs. We discuss this problem in more detail in the next section, and also show that several approaches do not fulfill this criterion.

Next, Eq. (13.3) defines the achievable rate such that the entropy of the key has to be very close to the given rate. This quantifies the efficiency of the code. In the PUF context, this is given by the ratio between the number of key bits k_K and the number of PUF output bits n_0 used to generate the key:

$$R_{key} = \frac{k_K}{n_0}. \tag{13.6}$$

The maximal achievable rate is given by the key capacity C_{key},

$$C_{key} = \sup_{R_{key} \text{ is achievable rate}} R_{key} = \min_{t \in \mathcal{T}} I(X; Y_t), \tag{13.7}$$

which for PUFs is bounded by the mutual information between the noise-free and noisy PUF response.

The last equation required for a secret key generation scheme, Eq. (13.4), ensures that the key space is nearly uniformly distributed and all possible key candidates are selected with a similar probability. For PUFs, the base to achieve this goal is tackled in hardware by careful design of the circuit. However, a bias or correlations in the PUF design can also be counteracted by compressing multiple bits. This point will be further discussed in the next section.

The relative size of the helper data is defined as the helper data rate R_{hd},

$$R_{hd} = \frac{\log_2 |\mathcal{W}|}{n_0}. \tag{13.8}$$

For later analysis, note that using the compound source scenario, the helper data capacity, i.e., the minimum amount of public discussion, can be defined as the minimum helper data size that is required to restore Y_t from X for all states $t \in \mathcal{T}$ [14]:

$$C_{hd} = \inf_{R_{hd} \text{ is achievable rate}} R_{hd} = \max_{t \in \mathcal{T}} H(X|Y_t). \tag{13.9}$$

13.3.5 Attack Vectors

In the compound source scenario described in [14], two types of attackers are mentioned. The first one is an eavesdropper who can observe the public channel and manipulate the state of the source. The second is a wiretapper, who can observe the public channel and also has access to an output variable Z_t of the source which is correlated to X_t and Y_t.

Similar adversaries can be assumed for PUFs. The eavesdropper is able to observe the helper data of the PUF, which corresponds to the public channel, and can try to reconstruct the key from his observations. Since the attacker has physical access to the device, he can set the environmental parameters such as temperature or supply voltage which correspond to the state of the source. However, as long as the attacker is not the manufacturer of the chip, it can be assumed that he does not have access to the state of the source during enrollment, i.e., he cannot manipulate the generation of the initial key or the helper data.

Table 13.1 Overview of similarities and differences between the information theoretic model and the scenario of key storage with PUFs.

Aspect	Information theoretic model	Key storage with PUFs
Source of randomness	Compound source	PUF (manufacturing variations and noise)
Dependency of source	State dependency	Dependency on environment
Communication channel	• Public channel • Over time and space • One or two way	• Helper data in non-secured non-volatile memory • Over time • One way • Unauthenticated channel
Communication partners	Terminal 1, terminal 2	Same PUF during enrollment and key reproduction in field
Attacker Model	• Eavesdropper (access to public channel, controls state) • Wiretapper (access to public channel, knowledge on source statistics)	Attacker with capabilities • read and write helper data • control on environment knowledge on source statistics • capability to do side-channel attacks

The obvious wiretapper in the PUF scenario corresponds to side-channel attacks that observe data that is directly correlated to Y_t [16, 17]. In addition, the wiretapper can be added to conserve the i.i.d. assumption of the source for PUFs with correlated outcomes. Instead of modifying the distributions of X and Y_t, we can add Z_t such that the additional output reveals the correlation.

In addition to the eavesdropper and wiretapper mentioned in the compound source model, the attacker in the PUF scenario can also manipulate the helper data to reveal information about K [10, 11], because we cannot assume an authenticated channel in practice.

For the rest of this chapter, we target the case of a passive eavesdropper but point to other attackers where necessary. Note that in [14], the compound scenario and the wiretapper scenario are separated. In practice we have to face attackers combining all possible attacks, which is closer to the more recent work in [18].

13.3.6 Summary

In this section we have shown the similarities and differences between the information theoretic scenario of secret sharing from a compound source and key storage with PUFs, as summarized in Table 13.1. This shows a series of links between the theoretical and the practical world for our application. Keeping in mind the complete scenario, for the rest of the chapter we focus on the special case that an attacker only has access to the helper data, i.e., is a passive eavesdropper. For this case, the source coding is of special interest, since in this block the helper data is used or generated during enrollment. To analyze state-of-the-art codes, we use the mutual information $I(W;K)$ between helper data and key and the key and helper data rates, R_{key} and R_{hd} from Section 13.3.4.

13.4 Unified Algebraic Description of Secret Key and Helper Data Generation

In Section 13.3 we have shown the similarity of the information theoretical secret key generation model and the key storage scenario for PUFs. All attacker models described had in common that the attackers have access to the public data transmitted over the channel, i.e., for PUFs to the helper data stored in non-volatile memory. The mutual information $I(W;K)$ between the secret and data transmitted over the channel – or the helper data stored in non-volatile memory – is the most critical attack vector, so we will focus on this metric in the following. In Eq. (13.2), $\frac{1}{n}I(W;K) < \epsilon$ is required for a permissible secret sharing protocol, raising the question of how helper data for PUFs has to be computed to have a key storage protocol that fulfills this requirement, i.e., to be permissible in terms of mutual information.

In the key storage scenario, a source decoder is used to map the random PUF response to a codeword and thus enable error correction (see Fig. 13.1). The value of the mutual information $I(W;K)$, i.e., the information leaked by the helper data, is determined by the helper data generation for this block. Currently, most helper data generation schemes are given in an algorithmic way so that their fundamental theoretical properties are hard to compare, and it is also hard to see the root cause for potential leakages. To provide better insight into why and how much the helper data for different coding schemes leaks in the PUF scenario, we introduce a unified algebraic description for helper data generation schemes. The description holds for linear mappings as the most widely used class of schemes, with very few exceptions such as [9,10]. Surprisingly, the pointer-based approaches IBS [6] and C-IBS [7] can also be brought into an algebraic form.

For all key storage approaches with PUFs, obviously the PUF output X is one mandatory input to construct the key and also to compute the helper data. In addition, some scenarios require a random number R, which is only available during helper data generation, either as the key to be embedded or as a mask to secure the PUF output.

Furthermore, in some schemes additional reliability input information is used, e.g., to perform a selection of stable PUF output bits or to postprocess the results from helper data generation. We assume that the reliability information can be taken into account as parts of the additional pre- and postprocessing matrices $\mathbf{M}_{pre}$ and $\mathbf{M}_{post}$ defined below.

The secret S is always one output of the helper data generation scheme. The other output of the helper data generation step is the helper data W. The setup is sketched in Fig. 13.3. Note that for some approaches, the secret and the helper data are postprocessed by a hash function, $f(\cdot)$ such that $K = f(S,W)$, to counter helper data

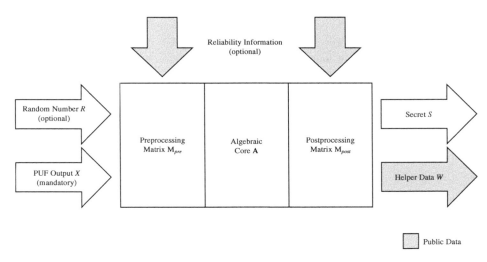

Figure 13.3 Sketch of helper data generation for PUFs with possible inputs and outputs.

manipulation [10]. Also, the secret can be hashed according to $S' = f(S)$ to mitigate leakage through the helper data and still achieve cryptographic security [5].

Here, we focus on the linear part of the scheme for helper data generation, which provides clearer insights into the properties of the underlying method.

All helper data generation schemes for PUFs that use linear codes can be brought into the generic basic form

$$[S, W] = [R, X] \, \mathbf{M}_{\text{pre}} \, \mathbf{A} \, \mathbf{M}_{\text{post}}. \tag{13.10}$$

The random number and PUF response $[R, X]$ with dimensions $\langle 1 \times k_R \rangle$ and $\langle 1 \times l_X \rangle$ are first multiplied with the preprocessing matrix $\mathbf{M}_{\text{pre}}$. The fundamental encoding operations are performed in the algebraic core $\mathbf{A}$. Then, the result is multiplied with the postprocessing matrix $\mathbf{M}_{\text{post}}$. The following analysis of the algebraic core requires isolated intermediate results. Therefore, we define the auxiliary vectors $U = [U_R, U_X]$ and $V = [V_S, V_W]$.

$$U = [R, X] \, \mathbf{M}_{\text{pre}}, \tag{13.11}$$

$$V = [R, X] \, \mathbf{M}_{\text{pre}} \mathbf{A} = U \mathbf{A}. \tag{13.12}$$

This permits us to look at the isolated algebraic core in the following.

Given U, the *algebraic core* $\mathbf{A}$ consists of a left part that computes the secret V_S and a right part that computes the helper data V_W. These two sub-matrices combined with the random number R and the PUF output X result in four sub-matrices:

$$\mathbf{A} = \begin{bmatrix} \mathbf{A}_L & \mathbf{A}_R \end{bmatrix} = \begin{bmatrix} \mathbf{A}_{\text{UL}} & \mathbf{A}_{\text{UR}} \\ \mathbf{A}_{\text{LL}} & \mathbf{A}_{\text{LR}} \end{bmatrix}. \tag{13.13}$$

Note that input U_R is multiplied with the upper half while input U_X is multiplied with the lower half of $\mathbf{A}$. Different constructions of the four submatrices of $\mathbf{A}$ and their properties will be addressed in the next section.

The pre- and postprocessing matrices $\mathbf{M}_{\text{pre}}$ and $\mathbf{M}_{\text{post}}$ can be identical for all devices, e.g., to perform a linear mapping of the PUF outputs before the actual helper data generation, or unique for each device, e.g., to consider reliability information on the PUF. We define that all operations where R and X interact take place in $\mathbf{A}$. Therefore, the pre- and postprocessing matrices have to be in the form

$$\mathbf{M} = \begin{bmatrix} \mathbf{M}_{\text{UL}} & \mathbf{0} \\ \mathbf{0} & \mathbf{M}_{\text{LR}} \end{bmatrix}, \tag{13.14}$$

with an upper-left sub-matrix $\mathbf{M}_{\text{UL}}$ and lower-right sub-matrix $\mathbf{M}_{\text{LR}}$. The remainder of the matrix is filled with zeros.

For typical scenarios, the secret is forwarded as $V_S = S$, so that in the postprocessing matrix $\mathbf{M}_{\text{UL}} = \mathbf{I}$.

The matrices in Eq. (13.10) are given *a priori* by the code or are derived from reliability information of the PUF. That is, the variable components have to be stored as additional helper data. The fixed part can be hard-wired in hardware. Thus, we assume that an attacker has full access to this public device-specific information.

The question that we try to answer below is, how much information about the secret S is leaked or can be attacked through the helper data W and variable public matrices $\mathbf{M}_{\text{pre}}, \mathbf{M}_{\text{post}}$?

13.4.1 Background for Further Analysis

The following discussion about leakage is driven by the fact that the entropy of the output is bounded by the rank of the matrices in use. Given two random sequences $X \in \mathbb{F}_2^n$ and $Y \in \mathbb{F}_2^m$ and a matrix $\mathbf{A}$ with $\dim(\mathbf{A}) = \langle n \times m \rangle$ and $q = \text{rank}(\mathbf{A})$, then

$$Y = X \cdot \mathbf{A} \tag{13.15}$$

can be interpreted as using $\mathbf{A}$ to map $X \in \mathbb{F}_2^n$ into a subspace $\mathcal{Q} \subseteq \mathbb{F}_2^m$ with $|\mathcal{Q}| = 2^q$ points. It follows that $H(Y) = q$ if all points in $\mathcal{Q}$ occur with equal probability, and $H(Y) = q - \epsilon$ otherwise.

13.4.2 Vulnerability of the Pre- and Postprocessing

The preprocessing and postprocessing matrices can contain reliability information about the PUF. The preprocessing matrix is frequently used to select reliable bits. The stored data does not contain information on the value of the bit but only on which bits are selected. Assuming no spatial correlations within the PUF bits, it does not reveal secret information directly. Since the data that defines the matrix is read without authentication, an attacker might manipulate it to find out, e.g., if two PUF bits are equal [10]. To achieve this goal, the adversary modifies the matrix and observes if the key stays the same or not.

Example 13.1 Preprocessing by dark bit masking [19] selects reliable bits from the PUF output. This can be interpreted as a preprocessing matrix that maps l_X inputs to l_U values used for key storage and helper data generation. Assume a subset of a PUF output of length three, $X = [1\ 0\ 0]$, and let the first and last bits in this sequence be sufficiently reliable. Then, the corresponding preprocessing matrix has the form

$$\mathbf{M}_{\mathrm{pre}} = \begin{bmatrix} 1 & 0 \\ 0 & 0 \\ 0 & 1 \end{bmatrix}. \quad (13.16)$$

An attacker can modify this matrix in the simplest case by switching columns one and two. Let us assume he observes that the key is no longer valid. Thus, he learned that bits one and three are different, which is equivalent to the loss of one bit of entropy in U. Even if the attacker cannot exchange columns of $\mathbf{M}_{\mathrm{pre}}$, he can still use this approach, e.g., by comparing the first and the last bits to the bit in the middle and using that the probability of this bit being one or zero is not equal to 50%. That is, he can observe if the probability of receiving the correct key is equally high for the case where the first bit is unselected and second bit is selected and for the case where the second bit is selected and third bit is unselected. If the probabilities are clearly different, then the bit values of the first and the last bits differ.

The problem of helper data manipulation is typically addressed by adding a hash function to the final outputs of the algorithm which hashes the helper data W together with the secret S to a final key. The hash function then prevents the attacker from modifying the helper data without modifying the key at the same time. More precisely, if a hash function is used, the attacker has to find a collision of the hash function subject to the corresponding modification leading to a preprocessing matrix he can use for his attack.

In certain cases, the postprocessing matrix can also be attacked. The main difference compared to using the preprocessing matrix is that the selection of a certain bit not only depends on the reliability but also on the embedded key. Theoretically, the same attacks are possible for postprocessing that can be done for preprocessing if they are not prevented by implementation. In addition, for some PUF types, an attacker can take advantage of the information in the postprocessing matrix in combination with information about the PUF.

Example 13.2 Consider a challenge–response (sometimes also called strong) PUF, which derives a secret bit from four analog values δ_i according to

$$X[i] = \mathrm{sign}\left(c_{i,1} \cdot \delta_1 + c_{i,2} \cdot \delta_2 + c_{i,3} \cdot \delta_3 + c_{i,4} \cdot \delta_4\right). \quad (13.17)$$

Each element of the challenge $C_i = [c_{i,1}, c_{i,2}, c_{i,3}, c_{i,4}]$ is derived from a binary challenge bit and can take values -1 or $+1$. The PUF responds $X[i]$ to challenge i is a logical 1 if the weighted sum is positive or a logical 0 if the weighted sum is negative. A typical example for such a PUF is the SUM PUF [20], where the δ_i are frequency differences of a ring oscillator.

Now, assume that challenges $C_0 = [-1,-1,-1,-1]$ to $C_7 = [-1,+1,+1,+1]$ are applied, that all PUF bits are reliable, and that we use a C-IBS (see Section 13.5.7) scheme. We use block length 4, i.e., the PUF response for the eight challenges is subdivided into two four-bit blocks and one secret bit is embedded into each block. If we encode in C-IBS each secret bit by two PUF bits, we know that for each block, pointers point to a sequence [01] (for a secret bit value 0) or [10] (for a secret bit value 1). Due to the fixed block length in C-IBS an attacker cannot compare bits in the first and second blocks, and the attack in the previous example is prevented.

However, through $\mathbf{M}_{\text{post}}$ the attacker has the information on which bits have been selected in each block. Assume for our example that in the first block the PUF response bit for $C_0 = [-1,-1,-1,-1]$ is the first and $C_1 = [-1,-1,-1,+1]$ is the second selected output. In the second block $C_6 = [-1,+1,+1,-1]$ is selected as the first and $C_7 = [-1,+1,+1,+1]$ as the second bit. Since the attacker knows that only sequences [01] and [10] are valid, he also knows that the responses of C_0 and C_1 as well as of C_6 and C_7 must be different.

Now the attacker can take advantage of his knowledge about the structure of the PUF: C_0 and C_1 as well as C_6 and C_7 differ only in one challenge bit, i.e., only one weight (more precisely $c_{i,4}$) in Eq. (13.17) has changed from -1 to $+1$. However, we know that this causes a bit flip and – since δ_4 is constant – that the flip is equal for both cases: either twice from 0 to 1 or twice from 1 to 0. Therefore, the adversary can conclude that the secret bits encoded in the first and second blocks are equal, and one bit of entropy has been lost.

Obviously, in this example we simplified the task for the attacker since we assumed only reliable bits and flipped the same challenge bit in both blocks. However, the attack can also be generalized for other cases and the example shows the general issue.

Note that leakage through the postprocessing matrix in the example is directly connected to the structure of the PUF. Hash functions and careful selection of challenges can help to prevent the attack up to a certain limit. However, other schemes that do not provide a link between the challenges and a change in the responses are recommended for challenge–response PUFs.

13.4.3 Leakage of the Algebraic Core

All approaches are based on an algebraic core $\mathbf{A}$ which determines the main differences and similarities of the approaches. Therefore, our further analysis focuses on this part of Eq. (13.10).

Recall that $\mathbf{A}$ maps $U = [U_R, U_X]$ to $V = [V_S, V_W]$. The goal of the following analysis is to discuss how much information about V_S is leaked by V_W.

Let $\dim(U_R) = \langle 1 \times k_U \rangle$ and $\dim(U_X) = \langle 1 \times l_U \rangle$. Furthermore, let $\dim(V_S) = \langle 1 \times k_V \rangle$ and $\dim(V_W) = \langle 1 \times l_V \rangle$, so that the dimensions of the sub-matrices in $\mathbf{A}$ are given by

$$\dim(\mathbf{A}_{UL}) = \langle k_U \times k_V \rangle \qquad \dim(\mathbf{A}_{UR}) = \langle k_U \times l_V \rangle \qquad (13.18)$$

$$\dim(\mathbf{A}_{LL}) = \langle l_U \times k_V \rangle \qquad \dim(\mathbf{A}_{LR}) = \langle l_U \times l_V \rangle. \qquad (13.19)$$

$\mathbf{A}$ maps space $\mathcal{U}$ to space $\mathcal{V} \subseteq \mathbb{F}_2^{k_V + l_V}$, and the cardinality of $\mathcal{V}$ is determined by two to the rank of $\mathbf{A}$, as discussed in Section 13.4.1. The rank is equivalent to the number of base vectors in the index set of the output space.

Let λ_i be binary coefficients. Spaces $\mathcal{V}_S$ and $\mathcal{V}_W$ have bases formed by rows $A_{L,i}$ and $A_{R,i}$ of $\mathbf{A}_L$ and $\mathbf{A}_R$, and the corresponding index sets are

$$\mathcal{I}_L = \left\{ i \in \{1, \ldots, k_U + l_U\} \;\middle|\; \sum_i \lambda_i \cdot A_{L,i} = 0 \Leftrightarrow \lambda_i = 0 \right\}, \qquad (13.20)$$

$$\mathcal{I}_R = \left\{ i \in \{1, \ldots, k_U + l_U\} \;\middle|\; \sum_i \lambda_i \cdot A_{R,i} = 0 \Leftrightarrow \lambda_i = 0 \right\}, \qquad (13.21)$$

with $|\mathcal{I}_L| \leq \operatorname{rank}(\mathbf{A}_L)$ and $|\mathcal{I}_R| \leq \operatorname{rank}(\mathbf{A}_R)$. Analogously, we get for the full algebraic core $\mathbf{A}$

$$\mathcal{I}_A = \left\{ i \in \{1, \ldots, k_U + l_U\} \;\middle|\; \sum_i \lambda_i \cdot A_i = 0 \Leftrightarrow \lambda_i = 0 \right\}, \qquad (13.22)$$

with $|\mathcal{I}_A| \leq \operatorname{rank}(\mathbf{A})$. For later interpretation, we define the rank loss Δ_L as

$$\Delta_L = \min\{k_U + l_U, k_V\} - \operatorname{rank}(\mathbf{A}_L). \qquad (13.23)$$

Accordingly, we get for $\mathbf{A}_R$ and $\mathbf{A}$,

$$\Delta_R = \min\{k_U + l_U, l_V\} - \operatorname{rank}(\mathbf{A}_R), \qquad (13.24)$$

$$\Delta = \min\{k_U + l_U, k_V + l_V\} - \operatorname{rank}(\mathbf{A}). \qquad (13.25)$$

Here, Δ, Δ_L, and Δ_R correspond to the difference between the maximum possible rank of the matrix given by the smaller dimension and the actual rank. We define the minimum rank loss of $\mathbf{A}$ as $g(k_U + l_U, k_V, l_V)$:

$$g(k_U + l_U, k_V, l_V) = \min\{k_U + l_U, k_V\} + \min\{k_U + l_U, l_V\} - \min\{k_U + l_U, k_V + l_V\}. \qquad (13.26)$$

Looking at the difference between the minimum rank loss and the actual rank loss, i.e., $g(k_U + l_U, k_V, l_V) - (\Delta_L + \Delta_R - \Delta)$, can be used later to decide which compression properties are required for a subsequent hash function.

To fulfill $I(V_S; V_W) \leq \epsilon_0$ in Eq. (13.2),

$$\exists_{\mathcal{I}_A, \mathcal{I}_R, \mathcal{I}_L} ((\mathcal{I}_A = \mathcal{I}_R \cup \mathcal{I}_L) \wedge (|\mathcal{I}_A| = |\mathcal{I}_R| + |\mathcal{I}_L| = r_R + r_L)) \qquad (13.27)$$

is a necessary and sufficient condition. To avoid leakage there must be rows in $\mathbf{A}_L$ and $\mathbf{A}_R$ that form the bases of two complementary vector spaces with dimensions r_R and r_L, and the two vector spaces together form the vector space built by a basis formed by rows from $\mathbf{A}$. That is, the indices of the rows selected for the basis from $\mathbf{A}_L$ and $\mathbf{A}_R$ are different. If there is some overlap, $\Delta - \Delta_L - \Delta_R$ is increased by one for each overlapping index, i.e., the leakage of the algebraic core is increased by one bit per such overlapping index. This value of $g(k_U + l_U, k_V, l_V) - (\Delta_L + \Delta_R - \Delta)$ can be used later to decide which compression properties are required from a subsequent hash function.

For realistic scenarios, U_R and U_X might not have full entropy. Thus, we assume $H(U_R) = k_U - \epsilon_{in,1}$ and $H(U_X) = l_U - \epsilon_{in,2}$. This causes a reduced entropy of V_S, V_W, and V. The reduction in entropy for each vector V_S, V_W, and V due to this effect is described by ϵ_L, ϵ_R, and ϵ. If for high-entropy sources $\epsilon_{in,1}$ and $\epsilon_{in,2}$ are small, we can assume that ϵ, ϵ_L, and ϵ_R are also small. Therefore, the derived value $\epsilon_0 = \epsilon - \epsilon_L - \epsilon_R$ also has a small absolute value. The entropy of V_S, V_W, and V can be given as

$$\begin{aligned}
H(V_S) &= H(U_R \mathbf{A}_L) \\
&= \text{rank}(\mathbf{A}_L) - \epsilon_L \\
&= \min\{k_U + l_U, k_V\} - \Delta_L - \epsilon_L, \\
H(V_W) &= H(U_X \mathbf{A}_R) \\
&= \text{rank}(\mathbf{A}_R) - \epsilon_R \\
&= \min\{k_U + l_U, l_V\} - \Delta_R - \epsilon_R, \\
H(V) &= H(U\mathbf{A}) \\
&= \text{rank}(\mathbf{A}) - \epsilon \\
&= \min\{k_U + l_U, k_V + l_V\} - \Delta - \epsilon.
\end{aligned} \qquad (13.28)$$

Using the entropy values and Eq. (13.28), the mutual information of V_S and V_W can be computed as

$$I(V_S; V_W) = H(V_S) + H(V_W) - H(V) \qquad (13.29)$$

$$= \text{rank}(\mathbf{A}_R) + \text{rank}(\mathbf{A}_L) - \text{rank}(\mathbf{A}) + \epsilon_0 \qquad (13.30)$$

$$= g(k_U + l_U, k_V, l_V) - (\Delta_L + \Delta_R - \Delta) + \epsilon_0. \qquad (13.31)$$

In current practical applications, U_R or U_X is always mapped to V_S, so either $\mathbf{A}_{UL} = \mathbf{I}$ or $\mathbf{A}_{LL} = \mathbf{I}$, while the other matrix on the left side of $\mathbf{A}$ is set to zero. The result is then either directly output as the secret key or compressed using a hash function. In these cases, it is sufficient that for the part of $\mathbf{A}_R$ that corresponds to the part in $\mathbf{A}_L$ which is zero, the rank is equal to the number of columns, e.g., if $\mathbf{A}_{UL} = \mathbf{0}$ then for $\mathbf{A}_{UR}$ the number of columns should be the rank of $\mathbf{A}_{UR}$. Note that Eq. (13.27) is always fulfilled if $\mathbf{A}$ has full rank.

13.5 Algebraic Core Representations of State-of-the-Art Helper Data Generation

In recent years, several different helper data generation algorithms for PUFs have been introduced. We present the state of the art in the conventional form, and represent them all in the algebraic notation discussed in the previous section. The observations from above are used to discuss leakage properties. Note that for the following approaches dark bit masking can always be used to improve the bit error probabilities at the input as soon as reliability information is available. This can be done during the preprocessing step, i.e., by multiplying the matrix $\mathbf{M}_{\text{pre}}$.

13.5.1 Fuzzy Commitment

Fuzzy commitment is one of the oldest members of this zoo of algorithms [21]. In this scheme, a random input R is used as a secret. To enable reproduction of this key word, the secret is encoded to a codeword $C = R\,\mathbf{G}$ of the code, which has to be stored to enable error correction. However, storing C in plain would leak the complete secret. To overcome this problem, it is masked with the PUF response X in the form

$$\begin{aligned} S &= R, \\ W &= R\,\mathbf{G} \oplus X. \end{aligned} \quad (13.32)$$

To transform this into the generalized description from Section 13.4, first note that we only operate on the algebraic core such that no pre- or postprocessing is required. That is, $\mathbf{M}_{\text{pre}}$ and $\mathbf{M}_{\text{post}}$ can be replaced by an identity matrix or can simply be dropped. Furthermore, R is directly mapped to S, i.e., for $\mathbf{A}_{\text{UL}} = \mathbf{I}$ and $\mathbf{A}_{\text{LL}} = \mathbf{0}$. The helper data is computed from $R\,\mathbf{G}$ XORed with X so that $\mathbf{A}_{\text{UR}} = \mathbf{G}$ is the generator matrix and $\mathbf{A}_{\text{UL}} = \mathbf{I}$ is the identity. The equation system can then be written as

$$[S\ W] = [R\ X]\begin{pmatrix} \mathbf{I} & \mathbf{G} \\ \mathbf{0} & \mathbf{I} \end{pmatrix}. \quad (13.33)$$

Obviously, (13.33) is an upper triangular matrix. Therefore, (13.27) is fulfilled, i.e., the leakage of the algorithm only depends on the entropy of the PUF output and of the random numbers, and their joint entropy. As a consequence, the secret S can directly be used as a key K if the entropy of the input data R and X is sufficiently high. If the entropy is not sufficient at the input, the compression property of a hash function can be used to reach sufficient entropy per bit in the key.

A scheme which is very similar to fuzzy commitment is the computational fuzzy extractor [22]. The difference here is that the generator matrix $\mathbf{G}$ of the code is replaced by a random matrix $\mathbf{R}$ that is chosen in advance. Note that no efficient decoding algorithms exist for the random code, so it is hard to solve the equation system if errors occur. This is exploited in the trapdoor extension [23]. The alphabet of X is extended from $\{0, 1\}$ to $\{0, 1, \otimes\}$. After readout, unreliable PUF response bits in X are marked as erasures ($\otimes$) and ignored during decoding to reduce the complexity. The helper data has a fixed size n_0 which is larger than necessary.

13.5.2 Code-Offset Fuzzy Extractor

The code-offset fuzzy extractor [5] is a direct offspring of fuzzy commitment. However, in contrast to fuzzy commitment, the secret S is derived from the PUF output X and not from a random number R. Due to leakage, a hash function is used to derive the key from this PUF output. The helper data W is computed from a codeword, which is computed from a random number R and XORed with the PUF output:

$$K = f(S),$$
$$S = X, \quad (13.34)$$
$$W = (R\,\mathbf{G}) \oplus X.$$

In Eq. (13.34), use of the PUF output as a secret can be interpreted in the algebraic description as dropping the pre- and postprocessing matrix and setting $\mathbf{A}_{UL} = \mathbf{0}$ and $\mathbf{A}_{LL} = \mathbf{I}$. The computation of the helper data then corresponds to setting the right side of $\mathbf{A}$ in the algebraic description to $\mathbf{A}_{UR} = \mathbf{G}$ and $\mathbf{A}_{LR} = \mathbf{I}$:

$$[S\ W] = [R\ X] \begin{pmatrix} \mathbf{0} & \mathbf{G} \\ \mathbf{I} & \mathbf{I} \end{pmatrix}, \quad (13.35)$$
$$K = f(S).$$

It can be seen that $\mathbf{A}_L$ and $\mathbf{A}_R$ have full rank (i.e., in Eq. (13.28), $\Delta_L = 0$ and $\Delta_R = 0$). However, they overlap such that the right part only has k_R rows that do not overlap with the index set of the left part. As a consequence, $\Delta \neq 0$, and the generated helper data leaks information about the secret. The amount of entropy leaked by the approach depends on the rank of $\mathbf{G}$. In practice, this rank is equal to the number of rows k_R, so that the rank of the $(k_R + l_X) \times (2\,l_X)$-dimensional matrix $\mathbf{A}$ can be given as $k_R + l_X$, where k_R is the length of the random number, and l_X is the length of the PUF output, $l_X > k_R$, and $\Delta = 0$. The mutual information can, therefore, be given according to Eq. (13.31) as $I(S; W) = l_X - k_R + \epsilon_0$. That is, if an attacker knows the helper data, the entropy of an l_X-bit secret is reduced to k_R. As a consequence, the hash function in 13.35 has to be designed so that the remaining k_R bits of entropy are distributed equally to the bits of a k_R-bit key.

In comparison to fuzzy commitment, where a codeword of the error correction is masked with the PUF response, here, the PUF response is masked with the codeword. This leads to different leakage. The practical security level of the key is the same for both approaches [24] if a well-designed hash function is used. However, for the code-offset fuzzy extractor it comes at the cost of an additional hash function.

13.5.3 Fuzzy Extractor with Syndrome Construction

In syndrome construction, also presented in [5], no additional random sequence is used, only the PUF output. The PUF output X is multiplied with parity check matrix $\mathbf{H}$ to create the helper data and to allow error correction. Similar to the code-offset fuzzy extractor, a hash function $K = f(S)$ is used to compress the PUF output since the helper data leaks information about the key.

Since no random number is used and all PUF bits contribute to the secret key as well as to the helper data, we set R and the corresponding upper sub-matrices to zero. Then, the unified mathematical description of the scheme can be given by

$$[S\ W] = [0\ X] \begin{pmatrix} 0 & 0 \\ \mathbf{I} & \mathbf{H}^T \end{pmatrix}, \qquad (13.36)$$
$$K = f(S).$$

The rank of $\mathbf{I}$ in Eq. (13.36) is $\text{rank}(\mathbf{I}) = l_X$. Since $X\mathbf{H}^T$ is a syndrome, the number of lines is smaller than the number of columns and the rank of $\mathbf{H}$ is equal to the size of the helper data l_W. So the mutual information between secret and helper data can be computed as $I(S; W) = l_W + \epsilon_0$, i.e., the algebraic core causes a leakage of l_W bits, which has to be compensated by the hash function in Eq. (13.36). The hash must compress the secret S to a key of length at most $k_K = l_X - l_W$ bits, thus, $l_W = l_X - k_K$. Note that this scheme requires only $l_X - k_K$ bits of helper data, which gives the lowest possible rate.

Compared to that, the previously introduced code-offset fuzzy extractor in Section 13.5.2 required l_X bits of helper data. Since, for a fixed number of PUF bits l_X and a fixed key length $k_K = k_R$, the code-offset fuzzy extractor and the syndrome construction approach both leak $l_X - k_K$ bits, this reduction in helper data size is the greatest benefit of the syndrome coding scheme.

13.5.4 Parity Construction

In the construction presented in [25], the entire PUF response X is treated as the information part of an ECC with systematic encoding, i.e., $\mathbf{G} = (\mathbf{I}\ \mathbf{P})$ with parity part $\mathbf{P}$. The redundancy is stored as helper data W. Again, no random number is used and all PUF bits contribute to the secret key as well as to the helper data, so that we set R and the corresponding matrices to zero. Then, the unified mathematical description is:

$$[S\ W] = [0\ X] \begin{pmatrix} 0 & 0 \\ \mathbf{I} & \mathbf{P} \end{pmatrix}, \qquad (13.37)$$
$$K = f(S).$$

The rank of $\mathbf{I}$ in Eq. (13.37) is the length of the secret $\text{rank}(\mathbf{I}) = k_S = l_X$; $\mathbf{P}$ can be assumed to have full rank, i.e., $\text{rank}(\mathbf{P}) = \min\{l_X, l_W\}$ is determined by the number of columns or rows of $\mathbf{P}$. The rank of the complete algebraic core is again equal to l_X. So the mutual information can be computed as $I(S; W) = \min\{l_X, l_W\} + \epsilon_0$, i.e., the algebraic core causes $\min\{l_X, l_W\}$ bits of leakage and a hash function is required to compress S down to at most $k_K = l_X - \min\{l_X, l_W\}$ bits. That is, as soon as the amount of helper data is larger than l_X, which is, in contrast to the syndrome construction, not inherently given, no secret is left to build a key.

13.5.5 Systematic Low Leakage Coding

Systematic low leakage coding (SLLC) [8] also uses codes with systematic encoding. Similar to parity construction, it does not use additional PUF bits and computes helper data by computing redundancy to correct these bits. However, similar to the fuzzy commitment scheme, it masks the redundancy with fresh PUF bits.

Therefore, we divide the PUF output X into a part $X_1^{k_S}$ (which can be seen as the part which is an analog of the random number in fuzzy commitment, with the difference that it can be reproduced), and a part $X_{k_S+1}^{l_X}$, which is an analog of the PUF bits in the fuzzy commitment scheme. Then, the approach can be described as

$$\begin{aligned} S &= X_1^{k_S}, \\ W &= \left(X_1^{k_S}\, \mathbf{P}\right) \oplus X_{k_S+1}^{l_X}. \end{aligned} \quad (13.38)$$

This corresponds to the unified algebraic form

$$[S\ W] = \begin{bmatrix} X_1^{k_S} & X_{k_S+1}^{l_X} \end{bmatrix} \begin{pmatrix} \mathbf{I} & \mathbf{P} \\ \mathbf{0} & \mathbf{I} \end{pmatrix}. \quad (13.39)$$

Obviously, the left and right parts of the algebraic core, as well as the concatenation of both, all have full rank. It follows that the mutual information $I(S; W) = \epsilon_0$, i.e., no information leaks due to the structure of the algebraic core. So, it can be concluded that this approach combines the benefit of zero leakage through the structure of the algebraic core from fuzzy commitment with the benefit of low helper data size from parity and syndrome construction.

13.5.6 Index-Based Syndrome Coding

Index-based syndrome (IBS) coding differs from the previously discussed schemes in the sense that it is a pointer-based approach that stores pointers to specific PUF bits as helper data, as shown in Algorithm 13.1. For a given random number R, a codeword $C = R\,\mathbf{G}$ is initially computed which can, but does not necessarily have to, consist of distinct information and redundancy parts. Then, the sequence X of l_X PUF bits is divided into equally long segments with length ν. In each block X_i, the PUF bit $X_i[j]$ is selected from the sequence which most likely takes the same value as the codeword bit that should be stored. The ith helper data is the index j of the selected PUF bit in the ith block. Finally, the algorithm returns the sequence of helper data.

Algorithm 13.1 IBS encoding

Input: Random number R; sequence of PUF bits X; block length ν; expectation value for each PUF bit
Output: Helper data sequence W; secret S
$S = R$
$C = R\,\mathbf{G}$
for $i = 0$ **to** $\frac{l_X}{\nu} - 1$ **do**
 $W[i] = \arg\max\limits_{j \in \{0,\ldots,\nu-1\}} \mathrm{P}(X[\nu \cdot i + j] = C[i])$
end for

For an algebraic representation, the random number R must be mapped to a codeword $C = R\,\mathbf{G}$ in the algebraic core. The index of the selected PUF bit in a block is represented by an $\frac{lx}{v}$-bit vector with the one-hot encoding below. That is, the bit which corresponds to the selected index is set to 1 and all other bits of the v-bit word are set to 0.

Assume for a first step toward the algebraic representation that there is at least one bit with expectation < 0.5 and at least one bit with expectation > 0.5 in each block of the PUF response. A single one-hot encoded element of the helper data sequence can then be computed in four steps:

1. Select codeword bit $c_i = C[i]$ and create word $\hat{C}_i = [c_i, c_i, \ldots, c_i]$ of length v where all elements are equal to $C[i]$.
2. For the current ith (with $i = 0, \ldots, \frac{lx}{v} - 1$) v-bit block $X_i = X[iv, \ldots, (i+1)v - 1]$ in the PUF response, initialize a v-bit vector $B_{a,i} = 0$.
3. Set the indices of $B_{a,i}$ that correspond to the bits with highest and lowest expectation to 1.
4. The one-hot encoded helper data for block i is now

$$W_i = \left(\hat{C}_i \oplus \bar{X}_i\right) \cdot \mathrm{diag}\left(B_{a,i}\right), \qquad (13.40)$$

where $\bar{X}_i$ is the inverted PUF response of block i.

Equation (13.40) suggests identifying bits that have the same expectation as the codeword (XOR) and selecting the most reliable out of these (multiplication). However, there is a certain chance that no bit in the block of the PUF response tends to have the value of the codeword. In this case, Eq. (13.40) would lead to a word $W_i = 0$. For example, let $X_i = [0, \ldots, 0] \Rightarrow \bar{X}_i = [1, \ldots, 1]$ and $c_i = 1$; then $W_i = [0, \ldots, 0]$. However, in the IBS algorithm the bit with the highest expectation is still considered a 1. This behavior is carried over to the algebraic description by an additional vector $B_{b,i}$, which is also initialized as an all-zero vector. If now in step 2 no bit of the PUF response has expectation < 0.5 for logical 0 (or > 0.5 for logical 1), the bit of $B_{b,i}$ which corresponds to the PUF response with the lowest (or highest) expectation is not set in $B_{a,i}$ but in $B_{b,i}$. This value should be taken, although a codeword bit with the *inverse* value of the PUF response for this PUF bit must be embedded. Thus, Eq. (13.40) must be extended to

$$W_i = \left(\hat{C}_i \oplus \bar{X}_i\right) \cdot \mathrm{diag}\left(B_{a,i}\right) \vee \left(\hat{C}_i \oplus X_i\right) \cdot \mathrm{diag}\left(B_{b,i}\right), \qquad (13.41)$$

with bitwise logical or $\vee$, which can be replaced by XOR since, by definition, $B_{a,i} \cdot B_{b,i}^T = 0$.

Built on these observations, the linear algebraic description of IBS can be given. The algebraic core operates on a random number, the PUF output, and the inverted PUF output. Thus, preprocessing is used to compute

$$[U_R | U_X] = \left[R | \bar{X}\ X\right] = [R\ X\ 1] \begin{bmatrix} \mathbf{I} & 0 & 0 \\ 0 & \mathbf{I} & \mathbf{I} \\ 0 & 1 & 0 \end{bmatrix}, \qquad (13.42)$$

where **1** is the all-one line vector which together with the scalar 1 at the input causes an inversion of the PUF bits.

In the subsequent algebraic core, the codeword $C = R\,\mathbf{G}$ is generated from the random number. Each bit of the generated codeword is reproduced v times, which can be interpreted as multiplying each bit with the generator matrix of a repetition code. To come to a closed form, a matrix $\mathbf{I}^{\times v}$ is used, which contains ones in every row i in column $i \cdot v$ to $(i+1) \cdot v - 1$ for $i = 0, \ldots, \frac{l_X}{v} - 1$. $\hat{C} = R\,\mathbf{G}\,\mathbf{I}^{\times v}$ contains each code bit v times. The result can be compared to the inverted and non-inverted PUF response as in Eq. (13.41):

$$V_a = \bar{X} \oplus \mathbf{G}\mathbf{I}^{\times v}, \qquad (13.43)$$

$$V_b = X \oplus \mathbf{G}\mathbf{I}^{\times v}. \qquad (13.44)$$

V_a and V_b contain the information of which bits of the PUF response are equal to and inverse of the code bits. To select the most reliable PUF bits in each block, the previous knowledge on which PUF bits are most reliable is used. Selection vectors $B_{a,i}$ and $B_{b,i}$ are defined for each block as previously described and converted into matrices $\mathbf{B}_{a,i} = \mathrm{diag}(B_{a,i})$ and $\mathbf{B}_{b,i} = \mathrm{diag}(B_{b,i})$. $\mathbf{B}_{a,i}$ and $\mathbf{B}_{b,i}$ are used in a matrix $\mathbf{B}$ to map V_a and V_b to the helper data:

$$W = [V_a\ V_b]\,\mathbf{B} = [V_a\ V_b] \begin{bmatrix} \mathbf{B}_{a,0} & 0 & \cdots & 0 \\ 0 & \mathbf{B}_{a,1} & \ddots & 0 \\ \vdots & \ddots & \ddots & \vdots \\ 0 & \ddots & \ddots & \mathbf{B}_{a,\frac{l_X}{v}-1} \\ \mathbf{B}_{b,0} & 0 & \cdots & 0 \\ 0 & \mathbf{B}_{b,1} & \ddots & 0 \\ \vdots & \ddots & \ddots & \vdots \\ 0 & \ddots & \ddots & \mathbf{B}_{b,\frac{l_X}{v}-1} \end{bmatrix}. \qquad (13.45)$$

Then, the algebraic representation of the IBS scheme is

$$[S\ W] = [R\ X\ 1] \underbrace{\begin{bmatrix} \mathbf{I} & 0 & 0 \\ 0 & \mathbf{I} & \mathbf{I} \\ 0 & \mathbf{1} & 0 \end{bmatrix}}_{\mathbf{M}_{\mathrm{pre}}} \underbrace{\begin{bmatrix} \mathbf{I} & \mathbf{G}\mathbf{I}^{\times v} & \mathbf{G}\mathbf{I}^{\times v} \\ 0 & \mathbf{I} & 0 \\ 0 & 0 & \mathbf{I} \end{bmatrix}}_{=\mathbf{A}} \underbrace{\begin{bmatrix} \mathbf{I} & 0 \\ 0 & \mathbf{B} \end{bmatrix}}_{\mathbf{M}_{\mathrm{post}}} \qquad (13.46)$$

$$\underbrace{\phantom{[R\ X\ 1]\begin{bmatrix} \mathbf{I} & 0 & 0 \\ 0 & \mathbf{I} & \mathbf{I} \\ 0 & \mathbf{1} & 0 \end{bmatrix}}}_{=[R\ \bar{X}\ X]}$$

$$\underbrace{\phantom{[R\ X\ 1]\begin{bmatrix} \mathbf{I} & 0 & 0 \\ 0 & \mathbf{I} & \mathbf{I} \\ 0 & \mathbf{1} & 0 \end{bmatrix}\begin{bmatrix} \mathbf{I} & \mathbf{G}\mathbf{I}^{\times v} & \mathbf{G}\mathbf{I}^{\times v} \\ 0 & \mathbf{I} & 0 \\ 0 & 0 & \mathbf{I} \end{bmatrix}}}_{=[R\ V_a\ V_b]}$$

The algebraic core is used to compute intermediate values for the helper data $V_W = [V_a \ V_b]$. The left side of the algebraic core has rank k_R, the right side has rank $2 \cdot l_X$, and the complete matrix $\mathbf{A}$ has rank $k_R + 2 \cdot l_X$. Thus, the mutual information between the intermediate values of the helper data and the secret is given by $I(S; V_W) = \epsilon_0$.[1] Note that using the $2 \cdot l_X$-bit input $[\bar{X} \ X]$ with entropy of only $l_X - \epsilon$ bits does not affect security in this case, since the same one-by-one mapping from $\bar{X}$ to V_a and from X to V_b is used and bits are later taken either from V_a or from V_b.

Since each code bit is used ν times for the computation of $\bar{X}$, an adversary who knows V_W gains up to $\nu - 1$ bits of information about the PUF per IBS block and, like in every other scheme, additional information through dependencies between bits in the codeword which do not correspond to the same block. The latter is that the bits in the codeword cannot be independent if errors should be corrected. This knowledge about the PUF is considered less critical but might be used for attacks in future. The postprocessing matrix in the algebraic description of IBS reduces the leakage about the PUF down to 0: instead of using all repeated bits of a single codeword bit, the index of the bit with expectation closest to the code bit is stored in the helper data. This step eliminates the knowledge an attacker can gain through the repetition and causes the positions of the bits, which represent the actual codeword, to be stored. Also, the values of the bits are obfuscated in this way and thus dependency between bits in the codeword cannot be exploited. However, in addition to the helper data discussed so far, reliability information about the PUF bits might be stored for soft decision decoding, which – especially in conjunction with additional information, e.g., about a bias – can cause additional leakage.

Note that $\mathbf{M}_{\text{post}}$, which needs to be known only during enrollment and is not to be stored, cannot only be used to select the most reliable bits but also to sort out bits with undesirable statistical properties, e.g., biased or strongly correlated bits. This can be used to improve the entropy of the key. Due to this possible improvement and to the selection of certain bits in the PUF response, it can be assumed that $I(S; W) < \epsilon_0$.

13.5.7 Complementary IBS

Complementary IBS (C-IBS) [7] is an extension of IBS and performs the IBS encoding η times on the same ν-bit PUF response blocks X_i – see Algorithm 13.2. In the algorithm, the iteration over i corresponds to the selected codeword bit. The inner loop over m tries to encode a 1 by a sequence 1010... and a 0 by a sequence 0101..., both of length η.

Thus, in odd iterations m the not yet selected bit in X_i is taken that has the highest probability of being equal to the current (ith) codeword bit. In even iterations, the bit

---

[1] This is similar to the fuzzy commitment scheme where the codeword is XORed with PUF bits and the result is stored as helper data: V_b can be seen as a masked codeword, where each bit of the actual codeword is again protected by a $(\nu, 1)$ repetition code.

with the highest probability of taking the inverted value of the current codeword bit is selected.

IBS and C-IBS differ in the algebraic description only in the bits that are selected after the computation in the algebraic core. Thus, as for IBS, the helper data is encoded as v-bit words with one-hot encoding in the following and Eq. (13.46) is used as a starting point, where only the postprocessing has to be modified.

In the postprocessing matrix $\mathbf{M}_{\text{post}}$, for each block of PUF bits the bits from $[V_a \ V_b]$ are selected which have highest, second highest, etc., probability of being equal to the codeword bit which should be embedded in this block. This corresponds to the min and max operations in the inner loop of the C-IBS algorithm.

Algorithm 13.2 $(l,4)$ C-IBS encoding

Input: Random number R; sequence of PUF bits X; block length v; expectation value for each PUF bit; code bits per block $\eta \leq v$
Output: Helper data sequence W; secret S
$S = R$
$C = R\,\mathbf{G}$
for $i = 0$ **to** $\frac{l_X}{v} - 1$ **do**
    $\mathcal{J}_i = \emptyset$
    for $m = 0$ **to** $\eta - 1$ **do**
        if $m \bmod 2 = 0$ **then**
$$W[i\eta + m] = \arg\max_{j \in \{0,\ldots,v-1\} \setminus \mathcal{J}_i} P(X[iv + j] = C[i])$$
        else
$$W[i\eta + m] = \arg\min_{j \in \{0,\ldots,v-1\} \setminus \mathcal{J}_i} P(X[iv + j] = C[i])$$
        end if
        $\mathcal{J}_i = \mathcal{J}_i \cup \{W[i\eta + m]\}$
    end for
end for

For C-IBS, η bits are selected per block. Similar to IBS, this is done by selection vectors denoted by $B_{a,i}^{\{0\}}$ to $B_{a,i}^{\{\frac{\eta}{2}-1\}}$ and $B_{b,i}^{\{0\}}$ to $B_{b,i}^{\{\frac{\eta}{2}-1\}}$. For $B_{a,i}^{\{0\}}$ the bits which correspond to the PUF bits in the block with highest and lowest expectation value are set to one if there is at least one bit each with expectation value $\mu \geq 0.5$ and $\mu < 0.5$ in the ith block. $B_{b,i}^{\{0\}}$ is an all-zero vector in this case. If no bit with $\mu \geq 0.5$ or $\mu < 0.5$ exists in the ith block, the bit of $B_{b,i}^{\{0\}}$ which corresponds to the PUF bit with the highest or, respectively, lowest expectation value is set while the corresponding bit of $B_{a,i}^{\{0\}}$ is zero.

For the sake of a simpler description in the following, an even number of bits per block is assumed. The procedure of generating the selection vectors is than continued for the bits with second highest probability of being one and zero, and vectors $B_{a,i}^{\{1\}}$ and $B_{b,i}^{\{1\}}$ to $B_{a,i}^{\{\frac{\eta}{2}-1\}}$ and $B_{b,i}^{\{\frac{\eta}{2}-1\}}$ are derived. This leads to $\frac{\eta}{2}$ vectors, pointing to the η bits which are most likely to be one or zero. With these selection vectors, the matrix $\mathbf{B}$ of the postprocessing matrix can be constructed as

$$\mathbf{B} = \begin{bmatrix} \mathbf{B}_{a,0}^{\{0\}} & 0 & \cdots & \cdots & \cdots & 0 \\ 0 & \mathbf{B}_{b,0}^{\{0\}} & \ddots & \ddots & \ddots & 0 \\ \vdots & \ddots & \mathbf{B}_{a,0}^{\{1\}} & \ddots & \ddots & 0 \\ \vdots & \ddots & \ddots & \mathbf{B}_{b,0}^{\{1\}} & \ddots & 0 \\ \vdots & \ddots & \ddots & \ddots & \ddots & \vdots \\ 0 & \ddots & \ddots & \ddots & \ddots & \mathbf{B}_{b,\frac{l_X}{\nu}-1}^{\{\frac{\eta}{2}-1\}} \\ \mathbf{B}_{b,0}^{\{0\}} & 0 & \cdots & \cdots & \cdots & 0 \\ 0 & \mathbf{B}_{a,0}^{\{0\}} & \ddots & \ddots & \ddots & 0 \\ \vdots & \ddots & \mathbf{B}_{b,0}^{\{1\}} & \ddots & \ddots & 0 \\ \vdots & \ddots & \ddots & \mathbf{B}_{a,0}^{\{1\}} & \ddots & 0 \\ \vdots & \ddots & \ddots & \ddots & \ddots & \vdots \\ 0 & \ddots & \ddots & \ddots & \ddots & \mathbf{B}_{a,\frac{l_X}{\nu}-1}^{\{\frac{\eta}{2}-1\}} \end{bmatrix}. \quad (13.47)$$

The complete description of C-IBS is than identical to the one in Eq. (13.46), and only has differences in the part **B** of the postprocessing matrix.

Like in the IBS scheme, the left side of the algebraic core has rank k_R, the right side has rank $2 \cdot l_X$, and **A** has rank $k_R + 2 \cdot l_X$. Thus, the mutual information between the intermediate values of the helper data and the secret is again given by $I(S; V_W) = \epsilon_0$. Also, an adversary who knows V_W gains up to $\nu - 1$ bits of information about the PUF per C-IBS block and – like in every other scheme – additional information through dependencies between bits in the codeword which do not correspond to the same block.

The encoding of each code bit by multiple bits allows a first stage of error correction during decoding. This significantly increases the reliability of the codeword, which is then input to the second stage of error correction. This comes with the price that – while $\mathbf{M}_{\text{post}}$ in the algebraic description of IBS reduces the leakage about the PUF down to zero – after postprocessing up to $\eta - 1$ bits about the PUF are leaked in the C-IBS scheme. The leaked information is which of the η bits per block are expected to take the same and the inverted value. Reliability information about the PUF bits, which is typically stored for soft decision decoding of the C-IBS codewords, strengthens this effect. The criticality of this knowledge depends on the PUF used (see Section 13.4.2).

Since in C-IBS, like in IBS, the helper data does not store the code bit itself but only the index of the bits, no information about the secret leaks and appropriate indexing can be used to cancel out bias and undesirable statistical dependencies of the PUF bits. Again, $I(S; W) < \epsilon_0$.

13.5.8 Summary of State-of-the-Art Syndrome Decoders

It has been shown that many state-of-the-art decoders can be brought into the same form, which allows comparison of the individual properties. Interestingly, this is also

Table 13.2 Theoretical comparison of algorithms for helper data generation.

Key generation scheme	$\max R_{\text{key}}$	$\min R_{\text{hd}}$	$I(S; W)$
Fuzzy commitment	C_{key}	1	ϵ_0
Code-offset	C_{key}	1	$l_X - k_R + \epsilon_0$
Syndrome construction	C_{key}	C_{hd}	$l_X - k_K + \epsilon_0$
Parity construction	$\leq C_{\text{key}}$	C_{hd}	$\min\{l_X, l_W\} + \epsilon_0$
SLLC	C_{key}	C_{hd}	ϵ_0
IBS	$< C_{\text{key}}$	$\lceil \log_2 \nu \rceil$	$< \epsilon_0$
C-IBS	$< C_{\text{key}}$	$\eta \lceil \log_2 \nu \rceil$	$< \epsilon_0$

possible for pointer-based approaches, which can be analyzed with the same methods. The methods are applicable to further schemes like differential sequence coding (DSC) [10], which are not discussed here.

Wrapping up the results of this section, Table 13.2 provides an overview of the properties of the discussed algorithms. For an optimal ECC and an ideal bias-free PUF with i.i.d. output bits, the maximum key rate and also the minimum helper data rate can approach capacity. Reaching the key capacity is possible with fuzzy commitment, code-offset, syndrome construction, and SLLC, since all these methods use all generated PUF bits and by design cannot leak the complete secret of the key. For parity construction, the leakage is not constrained and the number of key bits per PUF bit can go down to 0. Since for practical approaches typically the redundancy part must be relatively large, information about the key is leaked [26] and capacity cannot be achieved in these cases. Also, IBS and C-IBS cannot reach the maximum key rate, since they use only dedicated bits from the PUF response.

The minimum helper data rate can be reached with approaches that only require information about the syndrome or the parity in the helper data, namely, with syndrome construction, parity construction, and SLLC. Fuzzy commitment and the code-offset fuzzy extractor always store the complete codeword. Thus, the helper data size is always at least as long as the secret and thus as the key. For IBS and C-IBS, the one-hot encoded codewords from above which define the index in a ν-bit block can alternatively be stored as $\lceil \log_2 \nu \rceil$-bit words. Thus, for IBS one, and for C-IBS at least η, such words have to be stored in the helper data per key bit.

The leakage through the helper data which was derived in this section for non-ideal PUF and random number is listed in the right column of Table 13.2. To derive a secure key from the secret, a compression or even a hash function is required which ensures sufficient entropy per key bit at the cost of a larger number of bits in the secret. For a PUF and random number generator with i.i.d. and bias-free outputs, the value of ϵ_0 goes to zero. In all cases where the leakage is not bounded by or equal to ϵ_0, a hash function is mandatory to derive the key. Note that in the case of IBS and C-IBS, ϵ_0 can be reduced for non-ideal PUFs by careful selection of the used PUF bits.

13.6 Conclusions

Secret key generation with PUFs is a maturing research area with the first products in the field and widespread commercial application in sight. Several algorithms were proposed that already provide good practical results. In order to further improve the state of the art, theoretical tools are necessary to evaluate approaches and focus new research.

In this chapter, we have shown that key agreement with a compound source provides a theoretical model that closely matches the practical problem. Capacity results show what is possible and permit the evaluation of whether or not approaches can achieve this capacity with optimal codes.

The introduction of the unified algebraic notation facilitates splitting the problem into up to three algebraic parts. Analyzing the rank of the core plays a key role for the security of the schemes. Pre- and postprocessing can be used to select reliable bits to improve error correction capabilities in practice, and to reduce the leakage that is introduced via the algebraic core. A systematic analysis and improvement of these processing steps might lead to more efficient practical solutions in the future.

Acknowledgments

The first two authors contributed equally. This work was partly funded by the German Federal Ministry of Education and Research in the project SIBASE through grant number 01S13020A.

References

[1] G. E. Suh and S. Devadas, "Physical unclonable functions for device authentication and secret key generation," in *Proc. 44th ACM/IEEE Design Automation Conf.*, San Diego, CA, USA, Jun. 2007, pp. 9–14.

[2] J. Guajardo, S. S. Kumar, G. J. Schrijen, and P. Tuyls, "FPGA intrinsic PUFs and their use for IP protection," *Lecture Notes in Computer Science*, vol. 4727, pp. 63–80, 2007.

[3] B. Gassend, D. Clarke, M. van Dijk, and S. Devadas, "Delay-based circuit authentication and applications," in *Proc. ACM Symp. Applied Computing*, Melbourne, FL, USA, Mar. 2003, pp. 294–301.

[4] S. Katzenbeisser, U. Kocabaş, V. Rožić, A.-R. Sadeghi, I. Verbauwhede, and C. Wachsmann, "PUFs: Myth, fact or busted? A security evaluation of physically unclonable functions (PUFs) cast in silicon," *Lecture Notes in Computer Science*, vol. 7428, pp. 283–301, 2012.

[5] Y. Dodis, L. Reyzin, and A. Smith, "Fuzzy extractors: How to generate strong keys from biometrics and other noisy data," *Lecture Notes in Computer Science*, vol. 3027, pp. 523–540, 2004.

[6] M. Yu and S. Devadas, "Secure and robust error correction for physical unclonable functions," *IEEE Design & Test Comp.*, vol. 27, no. 1, pp. 48–65, Jan. 2010.

[7] M. Hiller, D. Merli, F. Stumpf, and G. Sigl, "Complementary IBS: Application specific error correction for PUFs," in *Proc. IEEE Int. Symp. Hardware-Oriented Security Trust*, San Francisco, CA, USA, Jun. 2012, pp. 1–6.

[8] M. Hiller, M. Yu, and M. Pehl, "Systematic low leakage coding for physical unclonable functions," in *Proc. 10th ACM Symp. Inf., Comp. Commun. Security*, Singapore, Apr. 2015, pp. 155–166.

[9] M. Yu, M. Hiller, and S. Devadas, "Maximum likelihood decoding of device-specific multi-bit symbols for reliable key generation," in *Proc. IEEE Int. Symp. Hardware-Oriented Security Trust*, Washington, DC, USA, May 2015, pp. 38–43.

[10] M. Hiller, M. Weiner, L. R. Lima, M. Birkner, and G. Sigl, "Breaking through fixed PUF block limitations with differential sequence coding and convolutional codes," in *Proc. 3rd Int. Workshop Trustworthy Embedded Devices*, Berlin, Germany, Nov. 2013, pp. 43–54.

[11] J. Delvaux, D. Gu, D. Schellekens, and I. Verbauwhede, "Helper data algorithms for PUF-based key generation: Overview and analysis," *IEEE Trans. Computer-Aided Design Integrated Circuits Systems*, vol. 34, no. 6, pp. 889–902, Jun. 2015.

[12] R. Ahlswede and I. Csiszár, "Common randomness in information theory and cryptography – Part I: Secret sharing," *IEEE Trans. Inf. Theory*, vol. 39, no. 4, pp. 1121–1132, Jul. 1993.

[13] U. M. Maurer, "Secret key agreement by public discussion from common information," *IEEE Trans. Inf. Theory*, vol. 39, no. 3, pp. 733–742, May 1993.

[14] H. Boche and R. F. Wyrembelski, "Secret key generation using compound sources – optimal key-rates and communication costs," in *Proc. 9th Int. ITG Conf. Systems, Communications and Coding*, Munich, Germany, Jan. 2013, pp. 1–6.

[15] R. Ahlswede and I. Csiszár, "Common randomness in information theory and cryptography – Part II: CR capacity," *IEEE Trans. Inf. Theory*, vol. 44, no. 1, pp. 225–240, Jan. 1998.

[16] D. Merli, D. Schuster, F. Stumpf, and G. Sigl, "Side-channel analysis of PUFs and fuzzy extractors," *Lecture Notes in Computer Science*, vol. 6740, pp. 33–47, 2011.

[17] D. Merli, J. Heyszl, B. Heinz, D. Schuster, F. Stumpf, and G. Sigl, "Localized electromagnetic analysis of RO PUFs," in *Proc. IEEE Int. Symp. Hardware-Oriented Security Trust*, Austin, TX, USA, Jun. 2013, pp. 19–24.

[18] N. Tavangaran, H. Boche, and R. F. Schaefer, "Secret-key capacity of compound source models with one-way public communication," in *Proc. IEEE Inf. Theory Workshop – Fall*, Jeju, Korea, Oct. 2015, pp. 252–256.

[19] F. Armknecht, R. Maes, A.-R. Sadeghi, B. Sunar, and P. Tuyls, "Memory leakage-resilient encryption based on physically unclonable functions," *Lecture Notes in Computer Science*, vol. 5912, pp. 685–702, 2009.

[20] M. Yu and S. Devadas, "Recombination of physical unclonable functions," in *Proc. 35th Annual GOMACTech Conf.*, Reno, NV, USA, Mar. 2010, pp. 1–4.

[21] A. Juels and M. Wattenberg, "A fuzzy commitment scheme," in *Proc. 6th ACM Conf. on Computer and Communications Security*. Singapore: ACM Press, Nov. 1999, pp. 26–36.

[22] B. Fuller, X. Meng, and L. Reyzin, "Computational fuzzy extractors," *Lecture Notes in Computer Science*, vol. 8269, pp. 174–193, 2013.

[23] C. Herder, L. Ren, M. van Dijk, M. Yu, and S. Devadas, "Trapdoor computational fuzzy extractors and stateless cryptographically-secure physical unclonable functions," *IEEE Trans. Dependable Secure Computing*, Mar. 2016.

[24] T. Ignatenko and F. M. J. Willems, "Biometric security from an information-theoretical perspective," *Found. Trends Commun. Inf. Theory*, vol. 7, no. 2–3, pp. 135–316, 2012.

[25] G. I. Davida, Y. Frankel, and B. J. Matt, "On enabling secure applications through off-line

biometric identification," in *Proc. IEEE Symp. Security Privacy*, Oakland, CA, USA, May 1998, pp. 148–157.

[26] A. Stoianov, T. Kevenar, and M. van der Veen, "Security issues of biometric encryption," in *Proc. IEEE Toronto Int. Conf. Science and Technology for Humanity*, Toronto, ON, Canada, Sep. 2009, pp. 34–39.

14 Wireless Physical-Layer Authentication for the Internet of Things

Gianluca Caparra, Marco Centenaro, Nicola Laurenti, Stefano Tomasin, and Lorenzo Vangelista

Authentication of messages in an Internet of Things (IoT) is a key security feature that may involve heavy signaling and protocol procedures, not suitable for small devices with very limited computational capabilities and energy availability. In this chapter we address the problem of message authentication in an IoT context, by using physical-layer approaches. We propose a solution based on the use of trusted *anchor* nodes that estimate the channel from the transmitting node and report them to a *concentrator* node, which takes a decision on the message authenticity. Assuming that the anchor nodes have a limited energy availability, we analyze the lifespan of the authenticating network and propose both centralized and distributed approaches to determine which anchor nodes report the information to the concentrator. The authenticating network overhead is also discussed and a tradeoff between energy efficiency and signaling traffic is found.

14.1 IoT Authentication Overview

In the near future it is expected that many devices in common use will be connected to the Internet, thus enabling enhanced features and applications, from flexible home automation to customization of body area networks. Huge security challenges must be faced in this new scenario. We will focus on the authentication problem, i.e., the problem of determining whether a message has been truly transmitted by a specific device. In other words, we want to make sure that no malicious node is transmitting messages in place of a legitimate node.

In an IoT scenario the dramatically large number of nodes calls for simple authentication techniques. As will be discussed in more detail in Section 14.2, the most popular IoT standards address the problem of authentication only with approaches based on cryptography that require complex processing procedures and the exchange (and refresh) of keys among the devices. An open-minded approach, trying to include new techniques, e.g., at the physical layer, could contribute strongly to the solution of the problem.

Therefore, here we investigate solutions that exploit the features of wireless transmissions and can integrate well other authentication procedures implemented in

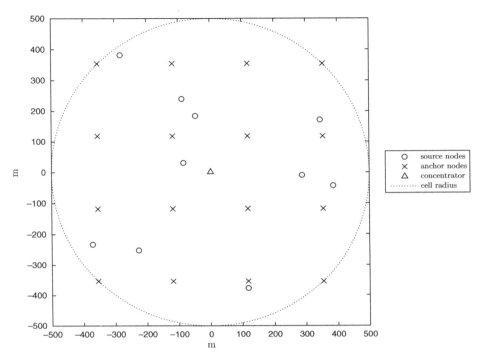

Figure 14.1 Map of a planar IoT including source nodes (circles), anchor nodes (crosses), and the concentrator node c (triangle). The dotted line indicates the coverage radius of the cell.

the higher layers. In particular, the random nature of the wireless channel depending on the spatial position of the transmitter or relatively long time periods (a few seconds) can be exploited to distinguish nodes that are placed in different locations. At the same time, we will also take advantage of the time-invariant nature of the wireless channel for static nodes over short time intervals (up to 100 milliseconds).

We consider the map of a planar IoT in Fig. 14.1, in which a specific *concentrator node* c is collecting data from many *source nodes* by wireless links. Moreover, a number of *anchor nodes* are available to cooperate with the concentrator and are considered trusted. In general, the source nodes can communicate only with anchor nodes or also with the concentrator, depending on the network topology.

In this chapter, for data communication we consider a *star* topology, in which the source nodes communicate with the concentrator node. Anchor nodes overhear the communication for authentication purposes, but may act as relays as well if needed. We assume that a good link is available between the concentrator and each anchor node. We want to remark that we could easily also fit a *tree* topology, in which source nodes communicate only with the anchor nodes and anchor nodes communicate with the concentrator, in order to fit existing IoT standards. For example, the Zigbee® standard provides a *coordinator node* (which plays the role of our concentrator), assisted by many routers (which play the role of our anchor nodes) and end devices (which play the role of our source nodes). The IEEE 802.15.4 standard provides a *first personal area network (PAN) coordinator* (which plays the role of our concentrator), assisted by many PAN coordinators (which play the role of our anchor nodes) and devices (which play the role of our source nodes).

Consider that the generic source node s is transmitting. We will assume that the first transmission between s and c is authenticated by some initial pairing procedure, e.g., authentication at high layers by previously shared secrets, or manual pairing by the user checking the consistency of transmitted and received messages. The rest of this chapter will consider how to authenticate forthcoming packets. For broadband transmission a first option is that the concentrator node c compares the estimate of channel impulse response of the newly received message with that of the initial transmission. An attacking node trying to impersonate a source node but transmitting from another location will in general have a different channel impulse response to c and therefore it will not be authenticated. The wider the bandwidth of the signal, the higher the precision of this authentication process. For narrowband transmissions, which are typical of an IoT scenario, this authentication process is highly imprecise. In this case the wireless channel between two devices is characterized by an attenuation (path loss) depending on the distance between the two devices. However, we can exploit the availability of many low-cost nodes in the IoT to implement a distributed node authentication procedure. Therefore, we will assume having anchor nodes (whose authenticity is guaranteed) spread over the area of the source nodes estimating the channel gain of ongoing transmissions. By considering the aggregate estimates of the anchor nodes, it is possible to obtain precise authentication of the messages.

A second problem that we address in this chapter is the energy balance of the anchor nodes. In fact, we envision that in the IoT scenario the anchor nodes are small devices equipped with batteries that are possibly recharged by some renewable energy source (e.g., by solar cells). Therefore, the number of transmissions that can be authenticated is limited. However, since multiple source–concentrator pairs will be communicating over the IoT area, with different distances to the various anchor nodes, we study the possibility of activating different anchor nodes according to the transmitting source, thus making the authentication process more efficient.

On the other hand, the signaling traffic required to perform the authentication procedure must be taken into account. Indeed, from the point of view of a network administrator, an intensive exchange of control messages between the anchor nodes and the concentrator containing the channel estimates to perform the authentication may put the network under strain, eventually causing the collapse of the network itself. Therefore, we study strategies that are efficient from the point of view of the signaling, and we find a tradeoff between energy efficiency and the amount of network traffic needed to perform the authentication of the data packets.

14.2 State of the Art

IoT and Wireless Sensor and Actuator Networks
The IoT concept is much broader than that of wireless sensor and actuator networks (WSAN), to which the present chapter mostly refers. As a matter of fact, the IoT is not

bound to be wireless. Consider the Industrial IoT [1], which embraces the Microgrids control, and aims to build on top of existing wired protocols. Furthermore, IoT includes gateways, information technology infrastructures, and platforms for service delivery and control (see, e.g., [2]).

It is, however, pretty common to use the two words IoT and WSAN in an interchangeable way: as the most prominent feature of the IoT is its *pervasiveness*, the easiest – and actually the first used – technology to realize it is a wireless, battery-operated one. Nevertheless, a proper setting for the concept is needed: in the most common paradigms for IoT, a WSAN is at the base of the IoT, providing the two lower layers of the communication system, i.e., the wireless physical layer and the medium access control (MAC).

Authentication in WSAN-based IoT

Authentication is a key feature specified in different flavors in almost all WSANs. However, although the basic mechanisms are specified, very often the actual authentication, i.e., how actually to check whether the message has been truly sent by the alleged node, is typically done at layers higher than the MAC by cryptographic approaches.

As a matter of fact, the IEEE 802.15.4 [3] standard provides procedures for encrypting and/or authenticating frames as well as for replay protection. The IEEE 802.15.4 standard uses the CCM* ("extension of counter mode encryption and cipher block chaining message authentication code" [4]) algorithm: as usual for the CCM [5] on which it is based, it encrypts and adds a signature (MIC – MAC integrity check) to the input data. In addition, CCM* has the possibility of only either encrypting or adding a signature to the input data without encryption. It must be pointed out that in this context, some PAN information base (PIB) attributes (in particular the keys) must be properly configured. Here is where higher-layer cooperation or some other mechanisms are needed. Actually, in some implementations, the MAC security tables (including the keys) are set at software compile time and cannot be changed dynamically at runtime.

14.2.1 Physical-Layer Authentication

Since the initial work by Shannon in the 1940s [6], the idea of implementing security at the physical layer by exploiting information theoretic properties of the channels was neglected for about 30 years until Wyner introduced the wiretap channel [7]. A significant literature on confidentiality has flourished since then, while much fewer works have appeared on other security topics (see also [8] for an overview). The authentication (as discussed in this chapter) implemented at the physical layer is one of the less explored subjects.

The existing solutions can be classified into two main categories: the *keyless* and *key-based* solutions. Schemes in which no secret key is shared among legitimate users to achieve authentication fall in the first category, while for schemes in the latter category the secret key is exploited to this end.

Keyless Authentication

When no key is used, authentication only relies on the characteristics of the channel over which the communication occurs. Such schemes require that an initial procedure (in general not based on the channel characteristics) establishes an authenticated transmission, and that the channel features are extracted in this initial trusted transmission [9]. An attacker may be passive, i.e., just trying to get its message accepted without trying to modify the characteristics of its own channel to the legitimate receiver. Otherwise, the attacker can also attempt to modify its channel by applying some signal processing procedure (e.g., precoding or beamforming) in order to deceive the legitimate receiver, which will erroneously recognize the correct channel characteristics in the received signal [10, 11]. However, the attacker often does not have access to the legitimate channel characteristics but can exploit the statistical relation of his own channel to the legitimate one in order to improve the attack effectiveness. Analysis of this kind of attack for typical wireless channels was performed in [12]. In [13], it was proved that secure authentication is possible when the legitimate transmitter has a noisy channel to the receiver whose behavior cannot be completely simulated by the attacker.

Among features that can be used to characterize the channel, and thus perform the authentication, the impulse response of a wideband channel has been considered in [14–19], the channel gain with respect to multiple antennas has been considered in [20–22], and the frequency response of orthogonal frequency division multiplexing (OFDM) systems has been used in [23]. In [24], the variations in the radio frequency (RF) chain are exploited for authentication purposes; bounds for both the error exponents of the legitimate nodes and the success exponent of the adversary are provided as a function of their fingerprint.

Key-Based Authentication

Another approach for authentication requires that a secret key is shared between the legitimate users. This approach has been investigated since early studies on information theoretic security, where coding schemes were used as key [25, 26], then better framed in a proper theory in the 1980s by Simmons [27]. The same author also obtained for a noiseless channel a lower bound on the impersonation probability [28]. Various extensions of the bound have been obtained for the case of substitution attacks [29, 30] and untrusted transmitters and receivers [31]. In [32], message authentication is interpreted as a hypothesis testing problem, thus extending the previous scenarios. All these works, however, considered transmissions without noise, although this is a key feature of communication channels, and in particular for the wireless scenario where interference can also be modeled as additional noise. For key-based authentication schemes, [33, 34] have considered the presence of noise. Suitable coding strategies for authentication over noisy channels are derived in [35]. Other important results on this topic include [36, 37], where it was shown that noise reduces the detection capabilities of the authentication scheme. More recently, however, it has been shown that by joint channel and authentication coding noise can be exploited as a protection against attacks [38]. While in most cases the key is used to encode the message, [39] proposed adding a fingerprint sequence to a multiple-input multiple-output (MIMO) transmission in order to ease authentication at the receiver.

14.3 IoT Channel-Based Authentication

Let us consider an IoT with M legitimate sources, N anchor nodes (with indices $i = 1, \ldots, N$), and one concentrator node c.[1]

In the IoT scenario, where devices transmit at a low rate, we assume that the communication channel between each source s and each anchor node i is narrowband and can be represented by a single complex coefficient. We suppose that the communication between the anchor nodes and the concentrator node c is secure, either by some additional communication feature or also because these nodes are connected to node c by a wired network. The anchor nodes are battery powered, and we assume that for each transmission a fixed amount of energy E_0 is consumed.

Let $h_i(\mathsf{s})$ be the channel gain from the generic source node s to anchor node $i = 1, \ldots, N$. $h_i(\mathsf{s})$ includes the effects of path loss, typically deterministic given the node positions, and those of shadowing and fading, modeled as random.[2] The channel power gain is

$$\mathrm{E}[|h_i(\mathsf{s})|^2] = \lambda_i, \quad i = 1, \ldots, N. \tag{14.1}$$

In the following we will assume independent fading over each source–anchor link and for each transmission, thus

$$\mathrm{E}[h_i(\mathsf{s}) h_j^*(\mathsf{s})] = 0, \quad \forall i \neq j. \tag{14.2}$$

We collect the channel gains into the N-sized row vector $\boldsymbol{h}(\mathsf{s})$, i.e.,

$$\boldsymbol{h}(\mathsf{s}) = [h_1(\mathsf{s}), \ldots, h_N(\mathsf{s})]. \tag{14.3}$$

Attacker Model

The *attacker node* a aims at transmitting a message by impersonating a legitimate source node. The attacker is able to both listen to ongoing transmissions and transmit signals. Moreover, we assume that the attacker may be equipped with multiple antennas, in order to be able to beamform signals, conveying messages with different gains to the various anchor nodes.

We indicate by $\boldsymbol{z}$ the random vector of observations available to the attacker (e.g., the channel gains from s to a and/or the channel gains from a to the N anchor nodes) that we assume to be somewhat correlated with $\boldsymbol{h}(\mathsf{s})$. In particular, if $\boldsymbol{z} = [z_1, \ldots, z_N]$ is a

---

[1] In principle we can combine the channel estimates of the concentrator with those of the anchor nodes to perform the authentication procedure. However, without loss of generality, in the rest of this chapter we will not consider the channel estimate of the concentrator.

[2] $\mathrm{E}[x]$ denotes the expectation of random variable x.

vector of N independent observations, we assume that each one is correlated only with the channel from s to a single anchor node, with

$$\frac{\mathrm{E}[z_i h_i^*(\mathrm{s})]}{\mathrm{E}[|z_i|^2]} = \rho, \quad i = 1,\ldots,N, \tag{14.4}$$

$$\mathrm{E}[z_i h_j^*(\mathrm{s})] = \mathrm{E}[z_i z_j^*] = 0, \quad \forall i \neq j. \tag{14.5}$$

Note that (14.4) establishes that the correlation coefficient is the same for all anchor nodes and (14.5) yields that only source–anchor and attacker–anchor channels relative to the same anchor are correlated. However, we will assume that the attacker knows neither vector $\boldsymbol{h}(\mathrm{s})$ nor has access to its estimate. This is reasonable since the only way to have access to this estimate is to replace the anchor nodes.[3] The only knowledge available to the attacker on channel $\boldsymbol{h}(\mathrm{s})$ is its joint statistics with the observations z.

We denote by $\boldsymbol{g} = [g_1,\ldots,g_N]$ the vector containing the forged channel gains from the attacker to c and the N anchor nodes. We also assume that both source and anchor nodes do not know of the presence of the attacker, and thus they do not have an estimate of z or $\boldsymbol{g}$ available.

14.3.1 Authentication Protocol

We assume now that all N anchor nodes cooperate for authentication purposes, although later in the chapter we will also consider the case in which a subset of anchor nodes is active in the authentication process. We observe that c is empowered by the N assisting nodes for the authentication procedure, thus we can see c as a receiver with N distributed antennas. The authentication procedure then operates in two phases.

Phase 1: By some pairing procedure we are sure that the message transmitted in this phase is originated by s. The N anchor nodes listen to the transmission and estimate the channel to s (which by reciprocity is assumed to be the same as that from s to the anchor nodes). Let $\hat{h}_i^{(0)}(\mathrm{s})$ be the channel estimate at anchor node $i = 1,\ldots,N$. This estimate is reported to c.

Phase 2: Forthcoming messages transmitted by s contain an unencrypted header that indicates the message source. Whenever the anchor nodes detect the s transmission header, they estimate the channel from the packet. Upon transmission of the kth message ($k \geq 1$) with s's header, let $\hat{h}_i^{(k)}(\mathrm{s})$ be the estimated channel at anchor node i, which is forwarded to c through an *authentication packet* over a secure channel. We collect all channel estimates relative to packet k into the row vector

$$\hat{\boldsymbol{h}}^{(k)}(\mathrm{s}) = [\hat{h}_1^{(k)}(\mathrm{s}),\ldots,\hat{h}_N^{(k)}(\mathrm{s})], \quad k \geq 0. \tag{14.6}$$

c will take a decision about the message's authenticity on the basis of the obtained channel estimates. If the actual transmitter is s, $\hat{h}_i^{(k)}(\mathrm{s})$ will be an estimate of

---

[3] It could be achieved only by placing spoofing nodes very close to each anchor node in order to estimate the same channel.

$h_i(\mathtt{s})$, $i=1,\ldots,N$. Instead, if the actual transmitter is the attacker, since it can induce any channel g to the anchor nodes, $\hat{h}_i^{(k)}(\mathtt{s})$ will be an estimate of g_i, $i=1,\ldots,N$.

The authentication performed by c on $\hat{\boldsymbol{h}}^{(k)}(\mathtt{s})$ must discern between two hypotheses:

- $\mathcal{H}_0$: packet k comes from s.
- $\mathcal{H}_1$: packet k has been transmitted by the attacker a.

The decision between the two hypotheses is taken by comparing the estimates $\hat{\boldsymbol{h}}^{(k)}(\mathtt{s})$, $k>0$ with the estimates $\hat{\boldsymbol{h}}^{(0)}(\mathtt{s})$.

In the following, we assume that the channel realization in two subsequent phases is subject to different fading (while still correlated), while sources do not move over the phases, so that the path loss for each node remains constant. Once a packet is deemed to be authentic, the estimate $\hat{\boldsymbol{h}}^{(0)}(\mathtt{s})$ is updated with the estimate $\hat{\boldsymbol{h}}^{(k)}(\mathtt{s})$ in order to track channel variations over time. Moreover, we assume additive white Gaussian noise (AWGN).

Decision Process

When the transmission is not performed by s, we expect that the channel estimates of Phase 2 significantly differ from those of Phase 1. However, even when the transmission is actually performed by s, the estimates in the two phases may differ due to channel variations, noise, and interference. Therefore, the decision process is prone to two well-known types of error: (1) *false alarms* (FAs), occurring when a legitimate packet is deemed as not being transmitted by s, and (2) *missed detections* (MDs), occurring when the impersonation attack succeeds, and the message coming from the attacker a is accepted as authentic. The quality of the detection process is determined by the probabilities of these two events, and in general a lower value of one yields a higher value of the other.

The detection procedure that, for a given FA probability, minimizes the MD probability is the likelihood ratio test (LRT). However, this approach requires knowledge of the statistics of the channel of the attacker node. Moreover, if a is able to forge the channels to the anchor nodes, the LRT technique requires knowledge of the attacking strategy, i.e., vector g. It is unrealistic to expect to have all this knowledge, and therefore LRT must be dropped in favor of generalized LRT (GLRT) [40]. In GLRT, the knowledge of g is replaced by its maximum likelihood (ML) estimate, i.e., $\hat{\boldsymbol{h}}^{(k)}(\mathtt{s})$.

The test works as follows. Let $f_{\hat{\boldsymbol{h}}^{(k)}(\mathtt{s})|\mathcal{H}_0}(\boldsymbol{a})$ be the probability distribution function (PDF) of $\hat{\boldsymbol{h}}^{(k)}(\mathtt{s})$ under hypothesis $\mathcal{H}_0$. Similarly, let $f_{\hat{\boldsymbol{h}}^{(k)}(\mathtt{s})|\mathcal{H}_1,\boldsymbol{g}}(\boldsymbol{a}|\boldsymbol{b})$ be the PDF of $\hat{\boldsymbol{h}}^{(k)}(\mathtt{s})$ under hypothesis $\mathcal{H}_1$ and given that $\boldsymbol{g}=\boldsymbol{b}$. Then the log likelihood ratio (LLR) of the estimated channel $\hat{\boldsymbol{h}}^{(k)}(\mathrm{s})$ is defined as[4]

$$\log \frac{f_{\hat{\boldsymbol{h}}^{(k)}(\mathrm{s})|\mathcal{H}_1,\boldsymbol{g}}(\hat{\boldsymbol{h}}^{(k)}(\mathrm{s})|\hat{\boldsymbol{h}}^{(k)}(\mathrm{s}))}{f_{\hat{\boldsymbol{h}}^{(k)}|\mathcal{H}_0}(\hat{\boldsymbol{h}}^{(k)}(\mathrm{s}))} \propto \frac{2}{\sigma^2}\|\hat{\boldsymbol{h}}^{(k)}(\mathrm{s}) - \hat{\boldsymbol{h}}^{(0)}(\mathrm{s})\|^2 = \Psi, \qquad (14.7)$$

where $\sigma^2 = \mathrm{E}[\|\hat{\boldsymbol{h}}^{(k)}(\mathrm{s}) - \hat{\boldsymbol{h}}^{(0)}(\mathrm{s})\|^2]$. According to the GLRT, authenticity is established by comparing the LLR (14.7), or its proportional variable Ψ, with a threshold (θ), i.e.,

$$\Psi \leq \theta : \text{decide for } \mathcal{H}_0, \qquad (14.8)$$

$$\Psi > \theta : \text{decide for } \mathcal{H}_1. \qquad (14.9)$$

We note from (14.7) that Ψ is a random variable depending both on the estimate accuracy and on the fact that the transmitting node is either s or a. In particular, conditioned on $\mathcal{H}_0$ and for any realization of $\boldsymbol{h}(\mathrm{s})$, as shown in [11], Ψ is a central chi-squared distributed random variable with $2N$ degrees of freedom, yielding the FA probability

$$P_{\mathrm{FA}} = \mathrm{P}[\Psi > \theta | \mathcal{H}_0] = 1 - F_{2N,0}(\theta), \qquad (14.10)$$

where $F_{Q,y}(x)$ is the cumulative distribution function (CDF) of a non-central chi-squared random variable with Q degrees of freedom and non-centrality parameter y. On the other hand, conditioned on $\mathcal{H}_1$, specific realizations of $\boldsymbol{h}(\mathrm{s})$, and the forged vector $\boldsymbol{g}$, Ψ is a non-central chi-squared distributed random variable with $2N$ degrees of freedom and non-centrality parameter

$$\beta = \frac{2}{\sigma^2}\|\boldsymbol{g} - \boldsymbol{h}(\mathrm{s})\|^2, \qquad (14.11)$$

yielding the MD probability, i.e., the probability that (14.8) is verified when a is transmitting,

$$P_{\mathrm{MD}}(\boldsymbol{h}(\mathrm{s}),\boldsymbol{g}) = \mathrm{P}[\Psi \leq \theta | \mathcal{H}_1, \boldsymbol{h}(\mathrm{s}), \boldsymbol{g}] = F_{2N,\beta}(\theta). \qquad (14.12)$$

We observe that the MD probability depends on the attack channel vector $\boldsymbol{g}$, which is random because it depends on the attacker observations and on its attack strategy. For instance, if Rayleigh fading is assumed and $\boldsymbol{h}(\mathrm{s})$ and $\boldsymbol{z}$ are jointly circularly symmetric complex Gaussian (CSCG) vectors, the optimal attack – both in the maximum MD probability sense of [11] and in the minimum divergence sense of [12] – is itself jointly CSCG with $\boldsymbol{h}(\mathrm{s})$ and $\boldsymbol{z}$ and can be written as

$$\boldsymbol{g} = \Xi \boldsymbol{z} = \Omega \boldsymbol{h}(\mathrm{s}) + \boldsymbol{\epsilon}, \qquad (14.13)$$

with Ξ and Ω complex matrices, and $\boldsymbol{\epsilon}$ a zero-mean CSCG vector independent of $\boldsymbol{h}(\mathrm{s})$.

Under the assumption of (14.5), both the matrices Ξ and Ω in (14.13), as well as the covariance matrices of $\boldsymbol{z}$ and $\boldsymbol{\epsilon}$, are diagonal (see [11, Appendix A] or [12, Section V]),

---

[4] We use log for the natural (base e) logarithm.

so we can write

$$g_i = \xi_i z_i = \omega_i h_i(\text{s}) + \epsilon_i, \qquad (14.14)$$

where $\omega_i = \xi_i \rho_{z_i h_i(\text{s})} \sigma_{z_i}/\sigma_{h_i(\text{s})}$ and $\sigma_{\epsilon_i}^2 = |\xi_i|^2 \sigma_{z_i}^2 (1 - |\rho_{z_i h_i(\text{s})}|^2)$, while $\sigma_{h_i(\text{s})}$ and σ_{z_i} represent the standard deviations of $h_i(\text{s})$ and z_i, respectively.

It is worth assessing the average MD probability when the optimal attack is performed and the channel $\boldsymbol{h}(\text{s})$ is Gaussian distributed with independent and identically distributed (i.i.d.) entries, i.e., $P_{\text{MD}} = \text{P}[\Psi \leq \theta | \mathcal{H}_1]$. This measure is relevant when the sequence of transmitted messages is long enough to span a significant portion of the channel fading statistics.[5] In the case of N independent observations, $\beta = 2\sum_i |(1 - \omega_i) h_i(\text{s}) + \epsilon_i|^2 / \sigma^2$ becomes the sum of N independent exponentially distributed random variables, each with mean (see also [11, Appendix A])

$$\frac{1}{\zeta_i} = \frac{2}{\sigma^2}(|1 - b_i|^2 \lambda_i + \sigma_{\epsilon_i}^2) = \frac{2}{\sigma^2}(1 - |\rho_{z_i h_i(\text{s})}|^2) \lambda_i. \qquad (14.15)$$

Then, under the simplifying assumption[6] that the ζ_i are all distinct, the average MD probability is

$$P_{\text{MD}} = 2 \sum_{i=1}^{N} \zeta_i \left(\prod_{j \neq i} \frac{1}{1 - \zeta_i/\zeta_j} \right) \left(\sum_{m=0}^{\infty} \frac{\bar{\gamma}(N+m; \theta/2)}{(2\zeta_i + 1)^{m+1}} \right), \qquad (14.16)$$

where $\bar{\gamma}(r; a) = \frac{1}{\Gamma(r)} \int_0^a x^{r-1} e^{-x} dx$ denotes the normalized lower incomplete gamma function.

Observe that, since $F_{2N,x}(\theta)$ is a decreasing function of x for every θ, and the CDF of β is a decreasing function of each λ_i once $\rho_{z_i h_i(\text{s})} = \rho$ is kept fixed, P_{MD} is itself a decreasing function of each λ_i for a fixed ρ. In other words, better legitimate channel gains yield a lower probability of confusing an attacker as a legitimate source.

14.3.2 Authentication Protocol Performance

In the following we will consider a cellular IoT (CIoT) scenario, where the IoT is deployed in the global system for mobile communications (GSM) frequency bands, which has been standardized by the third-generation partnership project (3GPP) [42]. Let us consider a CIoT for a campus, extending over a circular area with a radius of 500 m. The anchor nodes are placed on a square grid inside the circle; the number of anchor nodes is in the set $\{4, 9, 16, 25, 36\}$. We assume a unitary transmit power. The deterministic component of the wireless channel, i.e., the path loss, is computed (in dB) as

$$(\Gamma(\eta, d))_{\text{dB}} \triangleq -10 \cdot \eta \cdot \log_{10}\left(\frac{4\pi d}{\Lambda}\right), \qquad (14.17)$$

---

[5] We remark that fading is independent on each phase, and therefore MD is averaged over the fading. The case of constant fading over the phases can be addressed by a similar approach but leads to hardly tractable expressions.

[6] This assumption in only made here for the sake of obtaining a more compact expression in (14.16). In case it does not hold, the PDF of β can be derived with only a slight complication as described in [41].

where η is the path loss exponent (PLE), d is the distance between the transmitter and the receiver, and Λ is the wavelength, defined as the ratio between the speed of light and the carrier frequency f. We assume that $f = 900\,\text{MHz}$, which is the typical carrier frequency value considered in the context of CIoT. The parameter η is in the interval $[2, 3]$, which is a reasonable assumption for radio wave propagation in an urban scenario. We recall that as η increases, the propagation environment becomes harsher.

MD Probability

We now assess the performance of the proposed authentication protocol when choosing the detection threshold that ensures a FA probability $P_{\text{FA}} = 10^{-4}$.

Figure 14.2 shows (in log scale) the average (with respect to noise and channel realization) MD probability as a function of the legitimate source node position, with $N = 9$ anchor nodes and a correlation factor for the attacker node $\rho \in \{0.1, 0.5\}$. The PLE is set to $\eta = 2$. The signal-to-noise ratio (SNR), defined as the average (over fading) power ratio for a sensor–anchor distance of $250\,\text{m}$, is $15\,\text{dB}$. We observe that the positions at the center of the circle provide a lower MD probability, since the average channel gain sensed by the anchor nodes is higher than for external positions, especially as the source node moves to the circle border.

The same scenario is considered for Figs. 14.3 and 14.4, which report the complementary CDF (CCDF) of the MD probability for two different values of SNR. It can easily be seen that an increasing value of η adversely affects the performance of the proposed authentication protocol much more than an increasing number of anchor nodes N.

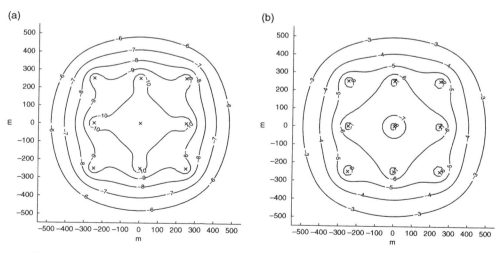

Figure 14.2 Logarithm of the MD probability as a function of legitimate source node position for (a) $\rho = 0.1$ and (b) $\rho = 0.5$. $N = 9$, SNR $= 15\,\text{dB}$ at a distance of $250\,\text{m}$.

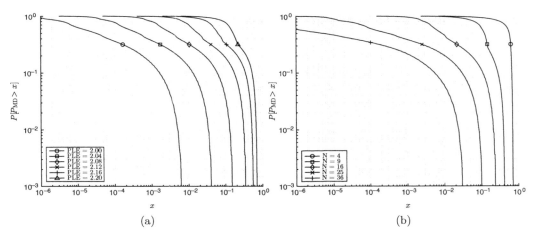

Figure 14.3 CCDF of MD probability in the case of a fixed number of anchor nodes N and fixed PLE. In (a) $N = 16$ and PLE varies, while in (b) PLE = 2.1 and N varies. SNR = 12 dB at a distance of 250 m and $\rho = 0.1$.

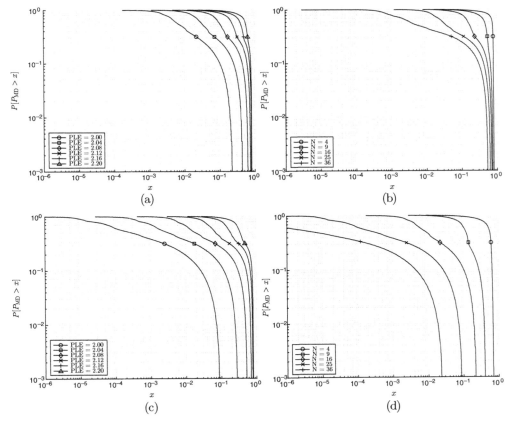

Figure 14.4 CCDF of MD probability in the case of a fixed number of anchor nodes and PLE, with SNR = 8 dB at 250 m, $\rho = 0.1$, $N = 16$ in (a), $N = 25$ in (c), PLE = 2.1 in (b), and PLE = 2.0 (d).

14.4 Centralized Anchor Node Selection

Until now, we have considered the case in which all anchor nodes provide the channel gain estimate to the concentrator node c for each data packet. It is reasonable to assume that most of the energy cost of the anchor nodes comes from their transmission of authentication packets to the concentrator node c. From (14.16), we observe that nodes having a better channel can better contribute to the authentication as they provide a lower MD probability, once the FA probability, and hence the decision threshold, is fixed. On the other hand, in a scenario in which the anchor nodes are battery-powered wireless devices with limited energy storage capabilities, it is important to optimize the use of the anchor nodes. The optimization procedure must on the one hand ensure accurate message authentication, and on the other hand reduce the power consumption of the anchor nodes.

We now proceed with some definitions that will be useful for the optimization procedure.

Configuration

We indicate a *configuration* as a set of anchor nodes that are active in authenticating a message presumably coming from source s. Clearly, with N anchor nodes 2^N configurations are possible, and each configuration can be described by an N-sized binary vector, where the i-th entry is set to one if anchor node i is active in the authentication process, and zero otherwise. Therefore, the ℓth configuration is that having a vector representation $c_\ell(s)$ being the binary representation of ℓ. For example, for $N = 4$, vector $c_8(s) = [1000]^T$ denotes the eighth configuration for the authentication of node s, where anchor node 1 is active, while anchor nodes 2, 3, and 4 are not active. For a selected configuration ℓ, the authentication procedure is carried out as described in the previous section, where the channel vector $h^{(k)}(s)$ now has length

$$L_\ell(s) = \sum_{i=1}^{N} [c_\ell(s)]_i, \qquad (14.18)$$

i.e., the Hamming weight of vector $c_\ell(s)$.

Admissible Configurations

Not all configurations allow the target FA and MD probabilities to be achieved; such configurations must be discarded in the optimization process. The configurations that provide at most the target MD and FA probabilities will be termed *admissible configurations*. In particular, configuration $c_\ell(s)$ is admissible if we can find a θ such that both the constraints on FA and MD probabilities, adapted to the configuration $c_\ell(s)$, are satisfied, that is,

$$P_{\text{FA}} = 1 - F_{2L_\ell(s),0}(\theta) \qquad (14.19)$$

and

$$P_{\text{MD}} = 2\sum_{i=1}^{N}[c_\ell(\text{s})]_i \zeta_i \left[\prod_{j \neq i} \frac{1}{(1-\zeta_i/\zeta_j)^{[c_\ell(\text{s})]_i}}\right] \sum_{m=0}^{\infty} \frac{\bar{\gamma}(L_\ell(\text{s})+m;\theta/2)}{(2\zeta_i+1)^{m+1}} \quad (14.20)$$

are not higher than their respective target values.

For source node s, let a_s be the number of admissible configurations. The total number of considered configurations is

$$A = \sum_{m=1}^{M} a_m. \quad (14.21)$$

Then we build the $N \times A$ matrix C having as columns all the configurations, i.e.,

$$C = \begin{bmatrix} c_1(1) & \cdots & c_{a_1}(1) & \cdots & c_1(M) & \cdots & c_{a_M}(M) \end{bmatrix}. \quad (14.22)$$

Anchor Network Lifespan

As target of the anchor optimization we consider the *anchor network lifespan*, i.e., the average time after which the first anchor node runs out of power. In fact, we assume that if an anchor node is not available anymore, there will be some source position for which no admissible configuration exists, and therefore no reliable authentication can be performed.

The optimization process will select, among the admissible configurations, those that will minimize the average energy consumption of the most used anchor node. Instead of selecting a single configuration and always using it, we allow for a mixed utilization of configurations, where each configuration is used for a fraction of the times in which a message coming (presumably) from source s must be authenticated. In practice, this mixed use of configurations can also be implemented by picking at random the configuration to be used according to specific probability distribution. Therefore, the optimization process aims at selecting the usage probability distribution in order to maximize the anchor network lifespan.

Let $\pi_\ell(m)$ be the probability (or the fraction times) of using the configuration $c_\ell(m)$, and let us stack the PMFs of configurations into the A-sized column vector

$$\boldsymbol{\pi} = \begin{bmatrix} \pi_1(1) & \cdots & \pi_{a_1}(1) & \cdots & \pi_1(M) & \cdots & \pi_{a_M}(M) \end{bmatrix}^{\text{T}}. \quad (14.23)$$

Define $\boldsymbol{u}$ as the N-sized column vector having as entry i the probability that anchor node i is used for authentication (*anchor utilization probability*), irrespective of the source node used. Assume that ϕ_m is the probability that source node $m \in \{1,\ldots,M\}$ is transmitting; therefore, $\sum_{m=1}^{M} \phi_m = 1$. Let us define the $A \times A$ diagonal matrix $\boldsymbol{\Phi}$ that weights the admissible configurations by the probabilities that the corresponding transmitter is active, i.e.,

$$\boldsymbol{\Phi} = \begin{bmatrix} \phi_1 \cdot \boldsymbol{I}_{a_1} & 0 & 0 & 0 \\ 0 & \phi_2 \cdot \boldsymbol{I}_{a_2} & 0 & 0 \\ 0 & 0 & \ddots & 0 \\ 0 & 0 & 0 & \phi_M \cdot \boldsymbol{I}_{a_M} \end{bmatrix}, \quad (14.24)$$

where I_n is the identity matrix of size $n \times n$. Then u can be written as

$$u = C\Phi\pi. \qquad (14.25)$$

In most cases it is reasonable to assume that each source is transmitting with the same probability $\phi_m = 1/M \; \forall m$, and in this case we have

$$u = \frac{1}{M}C\pi. \qquad (14.26)$$

14.4.1 Energy-Efficient Anchor Node Selection

We remark that the higher the anchor utilization probability u_i, the higher its power consumption. Therefore, assuming that the initial battery charge is the same for all nodes, the lifespan of the anchor network is related to the maximum anchor utilization probability. In order to maximize the network lifespan, we minimize the maximum (among all anchor nodes) utilization probability, under the constraint that only admissible configurations are used each time. The optimization problem can then be written as

$$\min_{\pi} \max_{i} u_i \qquad (14.27a)$$

subject to (14.23), (14.26), and

$$0 \leq \pi_\ell(m) \leq 1, \quad m = 1, \ldots, M, \; \ell = 1, \ldots, a_m, \qquad (14.27b)$$

$$\sum_{\ell=1}^{a_m} \pi_\ell(m) = 1, \quad m = 1, \ldots, M. \qquad (14.27c)$$

Note that constraint (14.27c) ensures that for each node to be authenticated there is always a configuration to be used when requested. We remark that the min–max problem (14.27) can be linearized by introducing the auxiliary variable t as follows:

$$\min_{\pi, t} t \qquad (14.28a)$$

subject to (14.27b), (14.27c), and

$$\frac{1}{M}C\pi \leq t\mathbf{1}_{N\times 1}, \qquad (14.28b)$$

where $\mathbf{1}_{N\times 1}$ is an N-sized column vector with entries all equal to 1.

We observe that the complexity of the problem grows with the number of nodes (both M and N). We can reduce the number of configurations in matrix C by observing that if a configuration $c_\ell(m)$ is admissible and anchor node i is not active ($[c_\ell(m)]_i = 0$), the new configuration $c_{\ell'}(m)$ obtained by activating node i ($[c_{\ell'}(m)]_i = 1$, and $[c_{\ell'}(m)]_q = [c_\ell(m)]_q$ for $q \neq i$) yields additional power consumption while still being admissible. In other words, the new configuration is worse in terms of energy consumption than the original $c_\ell(m)$, and will never be used for the optimal probabilities, i.e., on solution of (14.27) we have $\pi_{\ell'}(m) = 0$. Therefore, the new configuration can be discarded and

Table 14.1 Simulation parameters for the reference scenario.

Parameter	Value
f	900 MHz
η	2
ρ	0.5
Cell radius	500 m
SNR	30 dB (at 250 m)
N	16
M	10
Δ	0.1

not included in matrix C. In other words, matrix C contains the distinct admissible configurations containing the minimum number of anchor nodes.

The proposed authentication protocol works as follows: when a packet (presumably) coming from source node s is received by the concentrator node c, the latter draws a configuration according to the PMFs π obtained as the outcome of the optimization problem (14.27) and sends it in broadcast to the anchor nodes, triggering the participation of the selected anchors. Then, the authentication process continues as described in Section 14.3.

Example 14.1 We consider the same scenario as Section 14.3.2, but here we optimize the usage of anchor nodes according to the min–max problem (14.27). In particular, we consider the parameters reported in Table 14.1 and the IoT network deployment of Fig. 14.1.

For a single realization of the source node deployment, Fig. 14.5 shows the anchor node utilization probabilities u_i $\forall i$ after optimization. As expected, we observe that all anchor nodes are used on average with a similar probability. Moreover, if we compare it with the case in which all anchors are always used ($u_i = 1$ $\forall i$) we note a sharp decrease of the anchor node utilization probability to 0.12, thus providing 10 times the network lifespan.

14.4.2 Signaling-Efficient Anchor Selection

By solving problem (14.27) we aim to distribute the burden of the authentication procedure between all the anchor nodes, so that their utilization is as similar as possible and, therefore, the lifespan of the authenticating network is maximized. However, to achieve this objective, the optimization procedure may employ configurations involving multiple anchor nodes, i.e., configurations with a high Hamming weight, resulting in a heavy amount of signaling traffic. In this section we tackle the problem of minimizing the Hamming weight $H_\ell(\text{s})$ $\forall \ell, \text{s}$ of the admissible configurations, regardless of the overall utilization probability of the anchor nodes.

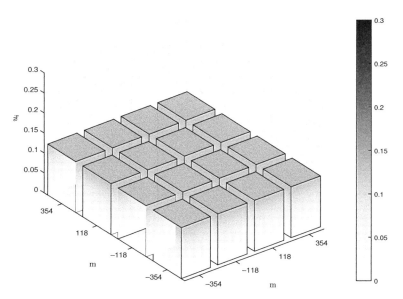

Figure 14.5 Anchor utilization probabilities u_i $\forall i$, solving the min–max problem for the IoT deployment in Fig. 14.1.

Recalling that anchor nodes having a better channel can yield a lower MD probability, for each source node we order the anchor nodes with decreasing channel gain, and we run through the ordered list adding anchor nodes to the configuration until we find a configuration that satisfies the target FA probability. With this approach, no min–max algorithm must be solved and we have a single broadcast message from the concentrator node c to indicate the selected configuration, i.e., we minimize the signaling traffic due to authentication. Note that this SNR-based anchor node selection does not properly take into account the energy consumption of the nodes. Indeed, while the minimum number of nodes to achieve authentication is used, it may occur that, e.g., if the source node distribution is not uniform, some anchor nodes are often activated while others are never activated, thus limiting the anchor network lifespan. Therefore, this technique does not maximize the network lifespan in general and is sub-optimal with respect to the solution of (14.27) from the point of view of the energy consumption.

Example 14.2 Let us consider the scenario of Section 14.3.2 with the parameters reported in Table 14.1 and deployment of Fig. 14.1. Figure 14.6 shows the anchor node utilization probabilities u_i $\forall i$ when the SNR-based anchor selection method is used. It can be seen that the utilization is not equal among the anchor nodes: some of them are never used, while others remain active most of the time to authenticate source nodes. The maximum anchor node utilization probability is about 0.27, providing a network lifespan which is 45% of the maximum lifespan obtained solving problem (14.27). Still, when compared to the case in which all anchor nodes are always used for authentication, we have four times higher network lifespan.

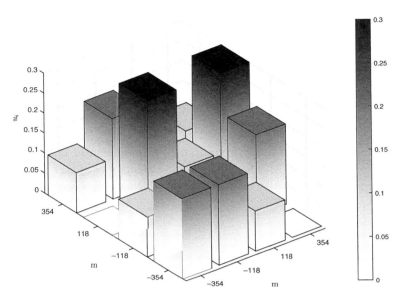

Figure 14.6 Anchor node utilization probabilities u_i $\forall i$, employing the SNR-based approach, as for the IoT deployment in Fig. 14.1.

14.4.3 A Tradeoff between Energy Efficiency and Signaling Efficiency

A possible way to reduce the amount of authentication overhead and, at the same time, to balance the anchor node utilization consists in limiting the Hamming weight $L_\ell(\mathrm{s})$ $\forall \ell, \mathrm{s}$ of the admissible configurations in matrix C. Two distinct approaches can be identified:

- put a *hard* constraint on the Hamming weight of the configurations, modifying the computation of matrix C by using an upper bound for the Hamming weight of every configuration;
- set a *soft* constraint on the Hamming weight of the configurations, modifying the min–max problem formulation in order to bound the average Hamming weight of the admissible configurations.

Min–Max Problem with Hard Constraint

This strategy consists in modifying the construction of matrix C so that it contains only admissible configurations $c_\ell(\mathrm{s})$ with Hamming weight $L_\ell(\mathrm{s}) \leq K$, where

$$K \geq \max_{\mathrm{s}} \min_{\ell} L_\ell(\mathrm{s}) \tag{14.29}$$

is the minimum feasible Hamming weight for an efficient admissible configuration, yielding a *reduced* configuration matrix C'. We remark that a particular value K_s for every source s can also be set, provided that $K_\mathrm{s} \geq \min_\ell L_\ell(\mathrm{s})$ $\forall \mathrm{s}$. The problem formulation is the following:

$$\min_{\pi} \max_{i} u_i \qquad (14.30a)$$

subject to (14.23),

$$u = C'\Phi\pi, \qquad (14.30b)$$

and (14.27b), (14.27c).

Note that the SNR-based anchor selection policy presented in Section 14.4.2 is a special case of problem (14.30), where a single configuration of anchor nodes is allowed for each source node s. Also in this case the problem can be linearized for an efficient solution.

Example 14.3 Let us consider the scenario of Table 14.1 in Section 14.3.2. Figure 14.7 shows the anchor node utilization probabilities u_i $\forall i$ when the min–max problem with hard constraints (14.30) is solved, imposing $K_s = \min_\ell L_\ell(s)$ $\forall s$. We observe that in this deployment example we achieve a similar utilization probability among the anchor nodes with respect to the original min–max strategy in (14.27). However, the maximum usage is slightly higher than the min–max.

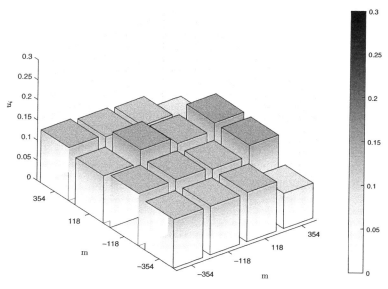

Figure 14.7 Anchor node utilization probabilities u_i $\forall i$, solving problem (14.30), as for the IoT deployment in Fig. 14.1.

Min–Max Problem with Soft Constraint

This approach consists in imposing a constraint on the *average* Hamming weight of the admissible configurations. Note that now we are shaping the structure of the vector of the admissible configuration probabilities π, rather than the intrinsic structure of matrix C, as done in problem (14.30). The proposed optimization problem is the following:

$$\min_{\pi} \max_{i} u_i \quad (14.31a)$$

subject to (14.27b), (14.27c), (14.26), and

$$\sum_{\ell=1}^{a_s} L_\ell(s) \cdot \pi_\ell(s) \leq K_s, \quad s = 1, \ldots, M. \quad (14.31b)$$

We denote with Δ the fixed lag that exceeds the minimum Hamming weight $\min_\ell L_\ell(s)$ $\forall s$. Note that if we impose that $\Delta = 0$, then the problems (14.31) and (14.30) provide the same solution.

Example 14.4 In the reference scenario of Table 14.1 of Section 14.3.2, Fig. 14.8 shows the anchor node utilization probability u_i obtained by solving the optimization problem (14.31) for a single realization of the source node deployment, with $\Delta = 0.1$. It can be seen that this method provides a flatter utilization probability among the anchor nodes with respect to the solution of problem (14.30). Moreover, note that in this case the maximum utilization probability is greater than the maximum utilization probability of problem (14.27) and lower than the utilization probability of problem (14.30).

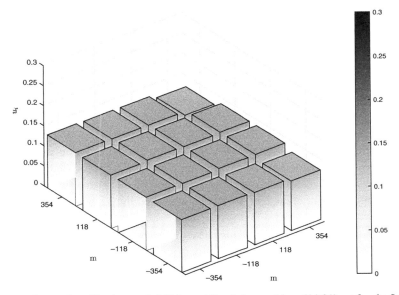

Figure 14.8 Anchor node utilization probabilities u_i $\forall i$, solving problem (14.31), as for the IoT deployment in Fig. 14.1.

14.5 Distributed Anchor Node Selection

The optimization procedures of Section 14.4 are centralized, because:

- the concentrator node c collects all the channel estimates of Phase 1 to build the matrix C;
- the concentrator node c solves the min–max problem (14.27);
- for each packet transmission, the concentrator node c sends N control messages, one per anchor node, indicating which configuration is selected.[7]

Therefore, the centralized procedure requires intense exchange of control messages from the concentrator c to the anchor nodes.

In this section, decentralized solutions that have reduced message exchange requirements are considered instead.

14.5.1 Distributed Configuration Selection

A first method of distributing the centralized procedure consists in eliminating the transmission of the control message from the concentrator node to the anchor nodes. In this solution we assume that the messages from the anchor nodes to the concentrator node can be overheard by all anchor nodes, e.g., because they are transmitted over the wireless medium or because they all go through a wired bus. To this end, we must consider a preliminary operation (performed at the end of Phase 1) in which the concentrator node c sends the admissible configurations matrix C and the optimal configurations usage probability vector π in multicast to all anchor nodes.

Then, in Phase 2, upon the transmission of each data packet, a round-robin procedure is used to select the desired configuration in a distributed fashion. In this procedure, anchor nodes each have a fixed time slot assigned in which they can provide their channel estimate to the concentrator. The first anchor node, i.e., the anchor node that owns the first slot, provides the feedback with probability u_1. The other anchor nodes, in the meantime, listen to the control channel and are able to detect whether anchor node 1 transmits or not. Anchor node 2 then selects among the admissible configurations those that match the initial behavior of node 1, and decides whether to transmit or not according to the probability of transmission conditioned on the transmission of anchor node 1. The third anchor node overhears what happens in the two previous slots and again transmits with a probability that is determined by the subset of admissible configurations that have been identified by the behavior of the two anchor nodes.

In general, to authenticate node m, at slot τ, define $\mathcal{C}_\tau$ as the set of configurations that are compatible with the transmissions performed in the previous $\tau - 1$ slots and that have a non-zero probability of being used. Moreover, let $\mathcal{R}_\tau$ be the set of configurations

---

[7] When the same control message can reach more nodes simultaneously (e.g., in a wireless broadcast scenario), the concentrator c can send a single message indicating the configuration.

in $\mathcal{C}_\tau$ in which node i is active, i.e.,

$$\mathcal{R}_\tau = \{\ell \in \mathcal{C}_\tau : [c_\ell(m)]_i = 1\}. \tag{14.32}$$

Then anchor node i will transmit with probability

$$p_{\text{tx}} = \frac{\sum_{\ell \in \mathcal{R}_\tau} \pi_\ell(m)}{\sum_{\ell \in \mathcal{C}_\tau} \pi_\ell(m)}. \tag{14.33}$$

Example 14.5 A simple example is now provided to better understand the proposed approach. Let us consider the configurations matrix C and the optimal configurations usage probability vector π in (14.34), where $N = 5$, $a_s = 3$, and the section of the matrix C relative to the source $m = 3$ is

$$C = \begin{bmatrix} \cdots & \begin{matrix} 0 & 0 & 0 \\ 1 & 0 & 0 \\ 0 & 1 & 1 \\ 0 & 1 & 0 \\ 0 & 0 & 1 \end{matrix} & \cdots \end{bmatrix}, \quad \pi = \begin{bmatrix} \vdots \\ 0.8 \\ 0.1 \\ 0.1 \\ \vdots \end{bmatrix}, \tag{14.34}$$

where clearly π contains zeros in all entries pertaining to the same source, except the three highlighted in (14.34).

When s sends the packet, anchor node 1 knows it must not participate. Therefore, slot 1 will remain empty. Then, in the next slot, anchor node 2 participates with probability 0.8. Observe that node 2 can authenticate the source by itself and hence the only considered configuration that includes node 2 is $c_1(3) = [0\,1\,0\,0\,0]^\text{T}$. Therefore, if anchor node 2 participates, no other node will participate. Otherwise, in the following slot, node 3 is required to participate (since it is active in all remaining configurations with non-zero probability) and, therefore, it sends its report to c. In slot 4, anchor node 4 collaborates with probability 0.5, and finally node 5 remains silent if node 4 transmits, otherwise it sends its report to the concentrator c.

14.5.2 Distributed SNR-Based Anchor Node Selection

The SNR-based anchor node selection described in Section 14.4.2 could be partially distributed by avoiding the reporting of the channel gains in Phase 1 when not used in the selected configuration. In particular, the proposed SNR-based distributed algorithm works as follows. In Phase 1, each anchor node estimates the channel, without immediately transmitting it to the concentrator c. Instead, anchor node i waits a time $w(|h_i^{(0)}(\text{s})|)$, which is a decreasing function of the estimated SNR; thus, for anchor nodes having a higher SNR the forwarding of the channel estimate to node c will be faster. When the anchor node c has obtained an admissible configuration, it sends a broadcast message to stop the forwarding from the anchor nodes.

Table 14.2 Simulation parameters for the performance evaluation.

Parameter	Value
f	900 MHz
η	2
ρ	0.1
Cell radius	500 m
SNR	25 dB (at 250 m)
N	9
M	20
Δ	0.1

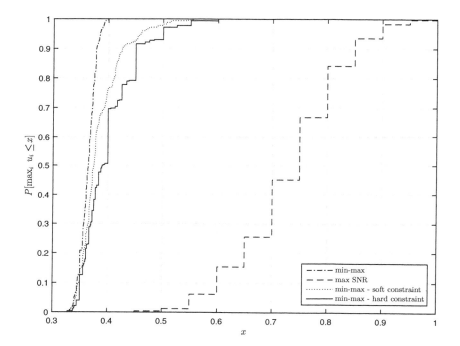

Figure 14.9 CDF of the anchor node utilization probability in the proposed authentication methods.

Now, let $F^c(x) = P[|h_i^{(0)}(\mathsf{s})| \geq x] = 1 - F(x)$ be the CCDF of the SNR over source node position statistics and fading statistics. Then we set the wait time as

$$\begin{aligned} w(|h_i^{(0)}(\mathsf{s})|) &= F^c(|h_i^{(0)}(\mathsf{s})|) \cdot T_0 \\ &= [1 - F(|h_i^{(0)}(\mathsf{s})|)] \cdot T_0, \end{aligned} \quad (14.35)$$

where T_0 is a constant chosen in order to minimize authentication packet collisions. The choice of the waiting time according to (14.35) ensures a uniform distribution of the transmissions within the interval T_0, thus minimizing the duration of the authentication procedure.

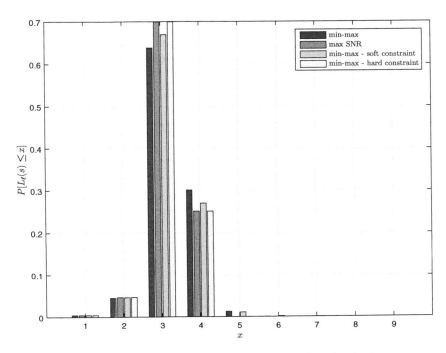

Figure 14.10 PMF of the number of anchor nodes involved in a source node authentication.

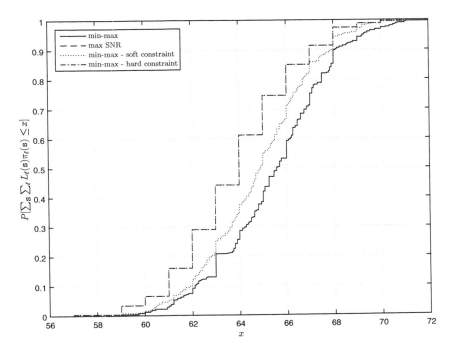

Figure 14.11 CDF of the energy consumption per authentication round.

However, the collection of channel gains in Phase 1 can take longer than the fully centralized approach, where a simple round-robin approach is used to gain all values. Moreover, possible collisions of control packets transmitted by the anchor nodes must be handled.

14.6 Performance Summary and Conclusions

In this section the performance of the three proposed approaches are compared, and pros and cons for each of them are discussed. For the comparison we use the parameters in Table 14.2. For the characterization of problems (14.30) and (14.31), we remark that only minimum Hamming weight configurations, i.e., $K_s = \min_\ell H_\ell(s)\ \forall s$, are considered.

Figure 14.9 shows the CDF of the maximum anchor node utilization probability, $\max_{i=1}^{N} u_i$. It can be seen that the min–max-based optimization problems (14.27), (14.30), and (14.31) provide a much lower utilization probability. In particular, as expected, the solution of problem (14.31) lies between the solutions of problems (14.27) and (14.31). On the other hand, the SNR algorithm is clearly the worst solution in terms of anchor node lifespan.

Figure 14.10, instead, depicts the PMF of the *number of anchor nodes involved in the authentication of a source node*. We note that the SNR-based policy and problem (14.30) have approximately the same behavior, using fewer reports from the anchor nodes with respect to problems (14.27) and (14.31), and, thus, being signaling-efficient methods. Moreover, it is clear that the intermediate approach with soft constraint (14.31) represents a tradeoff between the min–max approach of (14.27) and the intermediate approach with hard constraint (14.30).

Finally, Fig. 14.11 depicts the CDF of the *average number of reports used in a whole authentication round*, i.e., for the authentication of all the M source nodes. This metric is obtained as $\sum_{s=1}^{M} \sum_{\ell=1}^{a_s} L_\ell(s) \cdot \pi_\ell(s)$. We note that the SNR-based policy and problem (14.30) employ the minimum number of anchor nodes and, therefore, the two curves overlap. other hand, the min–max problem (14.27) tends to use more anchor nodes in performing the authentication, while the soft approach (14.31) is once again in-between.

14.6.1 Summary

Due to the pervasive nature of devices and their impact on our daily life, the future IoT needs new low-cost and efficient techniques to improve security. A key issue is the authentication of each message exchanged over the network. In this chapter we have analyzed a solution based on a protocol operating with physical layer signaling, in particular based on the authentication of messages through the characteristics of the channel over which transmission occurs. We have characterized the proposed approach in an IoT context and discussed the FA and MD probabilities. Then, assuming that anchor nodes have limited energy resources, we have proposed techniques for the maximization of the anchor node network lifespan under constraints on the FA and MD probabilities. Numerical results obtained in a typical CIoT scenario show that the proposed approach is able to achieve a low error probability, and the optimization procedure significantly increases the anchor node network lifespan compared to the case in which all anchor nodes are always used for authentication.

References

[1] II Consortium, 2015. [Online]. Available: www.iiconsortium.org/
[2] ThingWorx, 2015. [Online]. Available: www.thingworx.com/
[3] "IEEE Standard 802.15.4," IEEE, Tech. Rep., 2011.
[4] NIST, "Pub 800-38c, recommendation for block cipher modes of operation – the CCM mode for authentication and confidentiality," U.S. Department of Commerce/N.I.S.T., Tech. Rep., 2004.
[5] ANSI, "ANSI X9.63-2001, public key cryptography for the financial services industry – key agreement and key transport using elliptic curve cryptography," ANSI, Tech. Rep., 2001.
[6] C. E. Shannon, "Communication theory of secrecy systems," *Bell Syst. Tech. J.*, vol. 28, no. 4, pp. 656–715, Oct. 1949.
[7] A. D. Wyner, "The wire-tap channel," *Bell Syst. Tech. J.*, vol. 54, pp. 1355–1387, Oct. 1975.
[8] E. A. Jorswieck, S. Tomasin, and A. Sezgin, "Broadcasting into the uncertainty: Authentication and confidentiality by physical-layer processing," *Proc. IEEE*, vol. 103, no. 10, pp. 1702–1724, Oct. 2015.
[9] T. Daniels, M. Mina, and S. F. Russell, "A signal fingerprinting paradigm for general physical layer and sensor network security and assurance," in *Proc. IEEE First Int. Conf. Security and Privacy for Emerging Areas in Commun. Networks*, Athens, Greece, Sep. 2005, pp. 1–3.
[10] P. Baracca, N. Laurenti, and S. Tomasin, "Physical layer authentication over an OFDM fading wiretap channel," in *Proc. Int. ICST Conf. Performance Evaluation Methodologies and Tools*, Paris, France, 2011, pp. 648–657.
[11] P. Baracca, N. Laurenti, and S. Tomasin, "Physical layer authentication over MIMO fading wiretap channels," *IEEE Trans. Wireless Commun.*, vol. 11, no. 7, pp. 2564–2573, Jul. 2012.
[12] A. Ferrante, N. Laurenti, C. Masiero, M. Pavon, and S. Tomasin, "On the error region for channel estimation based physical layer authentication over Rayleigh fading," *IEEE Trans. Inf. Forensics Security*, vol. 10, no. 5, pp. 941–952, May 2015.
[13] S. Jiang, "Keyless authentication in a noisy model," *IEEE Trans. Inf. Forensics Security*, vol. 9, no. 6, pp. 1024–1033, Jun. 2014.
[14] L. Xiao, L. J. Greenstein, N. B. Mandayam, and W. Trappe, "Fingerprints in the ether: using the physical layer for wireless authentication," in *Proc. IEEE Int. Conf. Commun.*, Glasgow, UK, Jun. 2007, pp. 4646–4651.
[15] L. Xiao, L. J. Greenstein, N. B. Mandayam, and W. Trappe, "A physical-layer technique to enhance authentication for mobile terminals," in *Proc. IEEE Int. Conf. Commun.*, Beijing, China, May 2008, pp. 1520–1524.
[16] L. Xiao, L. J. Greenstein, N. B. Mandayam, and W. Trappe, "Channel-based spoofing detection in frequency-selective Rayleigh channels," *IEEE Trans. Wireless Commun.*, vol. 8, no. 12, pp. 5948–5956, Dec. 2009.

[17] L. Xiao, A. Reznik, W. Trappe, C. Ye, Y. Shah, and L. J. Greenstein, "PHY-authentication protocol for spoofing detection in wireless networks," in *Proc. IEEE Global Commun. Conf.*, Miami, FL, USA, Dec. 2010, pp. 1–6.

[18] F. He, W. Wang, and H. Man, "REAM: rake receiver enhanced authentication method," in *Proc. IEEE Military Commun. Conf.*, San Jose, CA, USA, Oct. 2010, pp. 2205–2210.

[19] J. Liu, A. Refaey, X. Wang, and H. Tang, "Reliability enhancement for CIR-based physical layer authentication," *Security and Communication Networks*, vol. 8, no. 4, pp. 661–671, 2015.

[20] D. B. Faria and D. R. Cheriton, "Detecting identity-based attacks in wireless networks using signalprints," in *Proc. ACM Workshop Wireless Security*, Los Angeles, CA, USA, Sep. 2006, pp. 43–52.

[21] M. Demirbas and Y. Song, "An RSSI-based scheme for sybil attack detection in wireless sensor networks," in *Proc. IEEE Int. Symp. World of Wireless, Mobile and Multimedia Networks*, Buffalo, NY, USA, Jun. 2006, p. 5.

[22] Y. Chen, W. Trappe, and R. P. Martin, "Detecting and localizing wireless spoofing attacks," in *Proc. IEEE Conf. Sensor, Mesh and Ad Hoc Commun. and Networks*, San Diego, CA, USA, Jun. 2007, pp. 193–202.

[23] F. He, H. Man, D. Kivanc, and B. McNair, "EPSON: enhanced physical security in OFDM networks," in *Proc. IEEE Int. Conf. Commun.*, Dresden, Germany, Jun. 2009, pp. 1–5.

[24] O. Gungor, C. Koksal, and H. El Gamal, "An information theoretic approach to RF fingerprinting," in *Proc. 47th Asilomar Conf. Signals, Systems, Computers*, Pacific Grove, CA, USA, Nov. 2013, pp. 61–65.

[25] E. N. Gilbert, F. J. MacWilliams, and N. J. A. Sloane, "Codes which detect deception," *Bell Syst. Tech. J.*, vol. 53, no. 3, pp. 405–424, 1974.

[26] V. Fåk, "Repeated use of codes which detect deception," *IEEE Trans. Inf. Theory*, vol. 25, pp. 233–234, Mar. 1979.

[27] G. J. Simmons, "Authentication theory/coding theory," *Lecture Notes in Computer Science*, vol. 196, pp. 411–431, 1985.

[28] G. J. Simmons, "A survey of information authentication," *Proc. IEEE*, vol. 76, no. 5, pp. 603–620, May 1988.

[29] A. Sgarro, "Information divergence bounds for authentication codes," *Lecture Notes in Computer Science*, vol. 434, pp. 93–101, 1985.

[30] R. Johannesson and A. Sgarro, "Strengthening Simmons' bound on impersonation," *IEEE Trans. Inf. Theory*, vol. 37, pp. 1182–1185, Jul. 1991.

[31] T. Johansson, "Lower bounds on the probability of deception in authentication with arbitration," *IEEE Trans. Inf. Theory*, vol. 40, pp. 1573–1585, Sep. 1994.

[32] U. M. Maurer, "Authentication theory and hypothesis testing," *IEEE Trans. Inf. Theory*, vol. 46, no. 4, pp. 1350–1356, Jul. 2000.

[33] P. L. Yu, J. S. Baras, and B. M. Sadler, "Physical-layer authentication," *IEEE Trans. Inf. Forensics Security*, vol. 3, no. 1, pp. 38–51, Mar. 2008.

[34] P. L. Yu, J. S. Baras, and B. M. Sadler, "Power allocation tradeoffs in multicarrier authentication systems," in *Proc. IEEE Sarnoff Symp.*, Princeton, NJ, USA, Mar. 2009, pp. 1–5.

[35] E. Martinian, G. W. Wornell, and B. Chen, "Authentication with distortion criteria," *IEEE Trans. Inf. Theory*, vol. 51, no. 7, pp. 2523–2542, Jul. 2005.

[36] Y. Liu and C. G. Boncelet, "The CRC-NTMAC for noisy message authentication," *IEEE Trans. Inf. Forensics Security*, vol. 1, no. 4, pp. 517–523, Dec. 2006.

[37] C. G. Boncelet, "The NTMAC for authentication of noisy messages," *IEEE Trans. Inf. Forensics Security*, vol. 1, no. 1, pp. 35–42, Mar. 2006.

[38] L. Lai, H. El Gamal, and H. V. Poor, "Authentication over noisy channels," *IEEE Trans. Inf. Theory*, vol. 55, no. 2, pp. 906–916, Feb. 2009.

[39] P. L. Yu and B. M. Sadler, "MIMO authentication via deliberate fingerprinting at the physical layer," *IEEE Trans. Inf. Forensics Security*, vol. 6, no. 3, pp. 606–615, Sep. 2011.

[40] S. Kay, *Fundamentals of Statistical Signal Processing: Detection Theory*. Upper Saddle River, NJ: Prentice Hall, 1993.

[41] H. V. Khuong and H. Y. Kong, "General expression for pdf of a sum of independent exponential random variables," *IEEE Commun. Letters*, vol. 10, no. 3, pp. 159–161, 2006.

[42] A. Biral, M. Centenaro, A. Zanella, L. Vangelista, and M. Zorzi, "The challenges of m2m massive access in wireless cellular networks," *Digital Communications and Networks*, vol. 1, no. 1, pp. 1–19, 2015.

Part IV

Data Systems and Related Applications

15 Information Theoretic Analysis of the Performance of Biometric Authentication Systems

Tanya Ignatenko and Frans M. J. Willems

In this chapter we analyze the performance of biometric authentication systems in terms of their typical performance measures, i.e., false rejection rate (FRR) and false acceptance rate (FAR). In biometric authentication systems the goal is to reliably authenticate individuals based on their biometric information. Recently, however, it was also concluded that biometric information itself has to be protected in these systems, due to privacy concerns. This gave rise to the development of biometric systems with template protection. In this work we analyze four types of biometric systems, i.e., traditional authentication systems, authentication systems with storage constraints, secret-based authentication systems, and secret-based authentication systems with privacy protection. For all these systems we present the fundamental limits on the false acceptance exponent. Moreover, for the last system we determine the tradeoff between the false acceptance exponent and the amount of information that the exchanged message leaks about the biometric sequence (privacy leakage).

15.1 Introduction

Nowadays, securing and regulating access to various systems and services relies heavily on passwords. However, password-based access control systems have a number of drawbacks. From the usability perspective, these systems are not user friendly, since users have to remember a large number of passwords. The latter, in its turn, results in weak security guarantees, since users tend to choose passwords that are easy to remember, and thus also easy to guess, as well as to reuse them in different applications. With the recent advances in biometric technologies, biometric information that uniquely characterizes individuals is a promising alternative to passwords. Biometric authentication is the process of establishing the identity of an individual using measurements of his/her biological characteristics, such as irises, fingerprints, face, etc.

The attractive property of uniqueness of biometrics also introduces privacy concerns related to their use in various access control systems. Unlike passwords, biometric data cannot be easily canceled and substituted with new biometrics, as they are unique for individuals and, moreover, individuals have limited resources of biometric information. Therefore secure storage and communication of biometric information in the corresponding access control systems becomes crucial. The corresponding biometric systems are called systems with template protection.

Biometric authentication systems with template protection are often modeled as systems for secret key generation; see, e.g., [1]. In these systems, two terminals observe enrollment and authentication biometric sequences and aim to form a common secret by exchanging a public message (helper data). To guarantee the security of the system, the shared secret key should be as large as possible and the public message should provide negligible information about the secret. This setting forms the foundation of template protection systems, e.g., the fuzzy commitment system proposed by Juels and Wattenberg [2].

The security of biometric systems with respect to access control is typically assessed in terms of the secret key rate that characterizes the secret key size. In traditional biometric systems, a FAR is used for this purpose, which characterizes the ability of an impostor to get access to the system based on some biometric sequence. In this work we extend a typical biometric secret generation system by introducing an authentication step, and show how the secret key rate relates to the FAR in authentication systems. We demonstrate that the performance of traditional biometric systems without template protection in terms of FAR is equivalent to one with template protection when it operates at maximum secret key rate (secrecy capacity). Moreover, we consider the biometric model for secret generation where the objective is not only to achieve high secret key rate but also to minimize privacy leakage, i.e., the amount of biometric information that is leaked via the public message. While the fundamental tradeoff between secret key rate and privacy leakage was studied in [3], here we extend this result to the authentication setting and argue that the secret key rate achieved in this setting is equal to the false acceptance exponent in the authentication setting. Wang et al. [4] investigated biometric systems based on linear error correcting code with a focus on the tradeoff between both the FRR and mFAR exponents. The problems addressed in this chapter are studied from an information theoretic perspective.

In this chapter we make the following modeling assumptions for our biometric data: we assume biometric sequences to be discrete and independent and identically distributed (i.i.d.). The independence of biometric sequence components can be assumed for many practical biometric systems, since signal processing steps applied to biometric measurements often involve principal component analysis, linear discriminant analysis, and other transformations (see, e.g., [5]) that result in more or less independent components of biometric sequences. In general, different components of biometric sequences may have different ranges of correlation. However, for reasons of simplicity we only discuss identically distributed biometric sequences here. Finally, discrete representation of biometrics can be obtained using quantization. Fingerprints and irises are typical examples of biometric sources that produce biometric sequences satisfying our assumptions. As a final remark, we note that the discrete assumption we employ here can be relaxed, as our results (proofs) can be easily generalized to Gaussian biometric sources.

15.1.1 Chapter Organization

We start with the traditional biometric authentication system where, during the enrollment phase, an individual presents his/her biometric. The corresponding biometric sequence X^N is observed and sent to an authenticator. At a later point in time, the legitimate person can get authenticated by presenting his/her (legitimate) biometric authentication sequence Y^N to the authenticator. Also, an impostor can try to get positively authenticated. To accomplish this, he has to present a sequence Z^N to the authenticator that results in a positive authentication decision. We assume here that the impostor can only perform an active attack (i.e., inject a sequence) before the authentication point (or matching point); see [6]. We study a system that is designed such that the FRR characterizing the reliability of authentication is negligible and the FAR is exponentially as small as possible. We show that the maximal false acceptance exponent in this setting is equal to the mutual information between enrollment and authentication biometric sequences, i.e., $I(X;Y)$.

Then we consider a setting where authentication is based on secret generation. We discuss the impostor strategy that has access to public helper data M in the biometric authentication system with template protection and argue that high secret key rates do not necessarily guarantee low FAR. Nevertheless, we can show that the largest false acceptance exponent that can be achieved in this setting equals the secret key capacity, i.e., $I(X;Y)$. We show that it can be achieved by having the secret key labels uniformly distributed. Interestingly, the largest false acceptance exponents for unprotected and protected cases are equal.

Finally, we turn to biometric authentication systems with protected templates where privacy leakage, characterized by $I(X^N;M)$, also has to be minimized. We show that for authentication there exists a tradeoff between the false acceptance exponent and privacy leakage. Moreover, we demonstrate that the corresponding achievable region is equivalent to the achievable region of secret key and privacy leakage rate pairs in the secret generation model [3].

15.2 Enrollment and Authentication Statistics

Biometric systems operate on three type of sequences: enrollment, legitimate authentication, and impostor authentication sequences. Here we describe their statistical properties.

The symbols of the enrollment sequence X^N and legitimate authentication sequence Y^N assume values in the finite alphabets $\mathcal{X}$ and $\mathcal{Y}$, respectively. We assume that the sequences are composed of N symbols. The joint probability of the sequences x^N and y^N is given by

$$\Pr\{X^N = x^N, Y^N = y^N\} = \prod_{n=1}^{N} Q(x_n, y_n), \qquad (15.1)$$

for all $x^N \in \mathcal{X}^N$, $x^N \in \mathcal{Y}^N$. Here, $\{Q(x,y), x \in \mathcal{X}, y \in \mathcal{Y}\}$ is a probability distribution, and the pairs (X_n, Y_n) for $n = 1, 2, \ldots, N$ are i.i.d. according to $Q(x,y)$.

Moreover, the impostor sequence Z^N also has length N and its symbols assume values in the alphabet $\mathcal{Y}$.

15.3 Traditional Authentication Systems

15.3.1 Scenario and Objective

We start with a basic, traditional authentication system – see Fig. 15.1. During enrollment, a biometric sequence X^N of the individual is observed and stored in the database. This sequence is also used by an authenticator. Later, when the legitimate individual would like to be authenticated, he presents a legitimate biometric sequence Y^N to the authenticator. The authenticator observes both X^N and Y^N, and decides if they come from the same individual. Also, an impostor can try to get authenticated. To accomplish this he has to present a sequence Z^N to the authenticator that results in a positive authentication decision. In this scenario, it is assumed that the impostor cannot access the enrollment sequence X^N. Therefore, this sequence has to be stored and conveyed to the authenticator in a secure manner – see Fig. 15.1.

Consider now the authentication mechanism for this scenario. The authenticator uses an authentication set $\mathcal{B}(XY) \in \mathcal{X}^N \times \mathcal{Y}^N$ and authenticates pairs $(x^N, y^N) \in \mathcal{B}(XY)$ and rejects pairs $(x^N, y^N) \notin \mathcal{B}(XY)$. We denote $\mathcal{B}(X|y^N) \triangleq \{x^N : (x^N, y^N) \in \mathcal{B}(XY)\}$. Now the false rejection rate, FRR, and maximum false acceptance rate, mFAR, are defined as

$$\text{FRR} = \Pr\{(X^N, Y^N) \notin \mathcal{B}(XY)\},$$
$$\text{mFAR} = \max_{z^N \in \mathcal{Y}^N} \Pr\{X^N \in \mathcal{B}(X|z^N)\}. \quad (15.2)$$

The system should be designed such that the FRR is negligible and the mFAR is exponentially as small as possible.

15.3.2 Achievability Definition and Result

To characterize the performance of the traditional authentication system, we first present our definition of the achievable FAR, more precisely its exponent version. Then we state our result for the traditional (basic) authentication, and present the corresponding proof.

DEFINITION 15.1 We say that the false acceptance exponent E is achievable in the basic authentication scenario if, for all $\delta > 0$ and for all N large enough, there exist

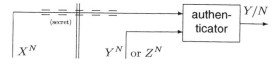

Figure 15.1 Traditional authentication system.

authenticators that achieve

$$\text{FRR} \leq \delta,$$

$$\frac{1}{N}\log_2 \frac{1}{\text{mFAR}} \geq E - \delta. \tag{15.3}$$

THEOREM 15.1 *All false acceptance exponents $E \leq I(X;Y)$ are achievable, and therefore $I(X;Y)$ is the fundamental limit in the basic authentication scenario.*

Proof The proof consists of two parts, i.e., the achievability and converse parts.

1. First we prove the converse. Consider an authentication set $\mathcal{B}(XY)$, and define C such that $C = 1$ if $(X^N, Y^N) \in \mathcal{B}(XY)$, and $C = 0$ otherwise. Now let

$$\text{FRR} = \sum_{(x^N,y^N) \notin \mathcal{B}(XY)} P(x^N, y^N) = P(C=0) \leq \delta. \tag{15.4}$$

Then

$$\text{mFAR} = \max_{y^N \in \mathcal{Y}^N} \sum_{x^N \in \mathcal{B}(X|y^N)} P(x^N)$$

$$= \max_{y^N \in \mathcal{Y}^N} \sum_{x^N \in \mathcal{B}(X|y^N)} P(x^N|C=1)P(C=1)$$

$$\geq \sum_{(x^N,y^N) \in \mathcal{B}(XY)} P(x^N|C=1)P(y^N|C=1)P(C=1)$$

$$= P(C=1) \sum_{(x^N,y^N) \in \mathcal{B}(XY)} P(x^N, y^N|C=1) \cdot \frac{P(x^N|C=1)P(y^N|C=1)}{P(x^N, y^N|C=1)}$$

$$= P(C=1) \sum_{(x^N,y^N) \in \mathcal{B}(XY)} P(x^N, y^N|C=1) \cdot 2^{-\log_2 \frac{P(x^N,y^N|C=1)}{P(x^N|C=1)P(y^N|C=1)}}$$

$$\geq P(C=1) 2^{-I(X^N;Y^N|C=1)}$$

$$\geq (1-\delta) 2^{-I(X^N;Y^N|C=1)}, \tag{15.5}$$

where in the last step we used Jensen's inequality.
Next consider

$$I(X^N;Y^N|C) \leq I(X^N;Y^N,C)$$

$$= I(X^N;Y^N) + I(X^N;C|Y^N)$$

$$\leq I(X^N;Y^N) + H(C)$$

$$\leq NI(X;Y) + 1. \tag{15.6}$$

On the other hand,

$$I(X^N;Y^N|C) = I(X^N;Y^N|C=1)P(C=1) + I(X^N;Y^N|C=0)P(C=0)$$
$$\geq -\log_2(P(C=1)2^{-I(X^N;Y^N|C=1)} + P(C=0)2^{-I(X^N;Y^N|C=0)})$$
$$= -\log_2(P(C=1)2^{-I(X^N;Y^N|C=1)} + P(C=0))$$
$$\geq -\log_2((1-\delta)2^{-I(X^N;Y^N|C=1)} + \delta), \quad (15.7)$$

where we again used Jensen's inequality in the second step. Then, combining the expressions above we obtain

$$\text{mFAR} \geq (1-\delta)2^{-I(X^N;Y^N|C=1)} \geq 2^{-N(I(X;Y)+\frac{1}{N})} - \delta, \quad (15.8)$$

and for achievable mFAR we get

$$E - \delta \leq \frac{1}{N}\log_2\frac{1}{\text{mFAR}} \leq \frac{1}{N}\log_2\frac{1}{2^{-N(I(X;Y)+\frac{1}{N})} - \delta}. \quad (15.9)$$

Now letting $\delta \downarrow 0$ and $N \to \infty$, we obtain the converse.

2. The achievability proof is based on weak typicality – for the properties of typical sets, see [7]. Fix an N and an $\varepsilon > 0$, and take an authentication set to be a set of jointly typical sequences, i.e., $\mathcal{B}(XY) = \mathcal{A}_\varepsilon^N(XY)$.

Now, from the properties of jointly typical sequences it follows that FRR $= \Pr\{(X^N, Y^N) \notin \mathcal{A}_\varepsilon^N(XY)\} \leq \varepsilon$ for $N \to \infty$. The mFAR can be upper bounded as

$$\text{mFAR} = \max_{y^N \in \mathcal{Y}^N} \sum_{x^N \in \mathcal{A}_\varepsilon^N(X|y^N)} P(x^N|y^N) \frac{P(x^N)P(y^N)}{P(x^N, y^N)}$$
$$\leq 2^{-N(I(X;Y)-3\varepsilon)} \max_{y^N \in \mathcal{Y}^N} \sum_{x^N \in \mathcal{A}_\varepsilon^N(X|y^N)} P(x^N|y^N)$$
$$\leq 2^{-N(I(X;Y)-3\varepsilon)}. \quad (15.10)$$

Now letting $N \to \infty$ for every $\varepsilon > 0$, the achievability follows.

15.3.3 Discussion

The result that we have obtained is not a surprise considering Stein's lemma (see [8]).

Observe that an impostor that has access to X^N can easily get authenticated. Therefore, in the traditional biometric authentication system the enrollment sequence has to be stored and communicated to the authenticator in a secure way, e.g., by using cryptographic techniques. However, to perform authentication, the enrollment sequence X^N that first has to be encrypted for secure communication then needs to be decrypted again, since the authenticator needs the original X^N. Thus, this basic authentication system has a number of disadvantages, i.e., it requires extra secure key management mechanisms to perform encryption and decryption, and it is not secure against insider attacks.

15.4 Rate-Constrained Authentication Systems

15.4.1 Scenario and Objective

In a rate-constrained authentication scenario, see Fig. 15.2, the legitimate individual presents his enrollment biometric sequence X^N to an encoder, who creates and stores an index $V \in \mathcal{V} = \{1, 2, \ldots, |\mathcal{V}|\}$. We write $V = e(X^N)$. The authenticator will use this index in order to make an authentication decision about presented legitimate or impostor authentication sequences Y^N or Z^N, respectively.

It is the objective in this rate-constrained setting to constrain the storage rate $(1/N)\log_2 |\mathcal{V}|$, and study the effect of this constraint on the exponential behavior of the mFAR given that the FRR is negligible.

We use the following authentication procedure here. The authenticator uses an authentication set $\mathcal{B}(VY) \in \mathcal{V} \times \mathcal{Y}^N$ and positively authenticates y^N if pairs $(V, y^N) \in \mathcal{B}(VY)$, and rejects y^N if $(V, y^N) \notin \mathcal{B}(VY)$. Again we denote $\mathcal{B}(V|y^N) \triangleq \{v : (v, y^N) \in \mathcal{B}(VY)\}$. The false rejection rate FRR and maximum false acceptance rate mFAR are now defined as

$$\text{FRR} = \Pr\{(e(X^N), Y^N) \notin \mathcal{B}(VY)\},$$
$$\text{mFAR} = \max_{z^N \in \mathcal{Y}^N} \Pr\{e(X^N) \in \mathcal{B}(V|z^N)\}. \tag{15.11}$$

15.4.2 Achievability Definition and Result

Again, we first present our definition of the achievable FAR exponents and storage rates, and then state the result for the rate-constrained authentication.

DEFINITION 15.2 The false acceptance exponent and storage rate combination (E, R) is achievable in the rate-constrained authentication scenario if, for all $\delta > 0$ and for all N large enough, there exist encoders and authenticators that achieve

$$\text{FRR} \leq \delta,$$
$$\frac{1}{N} \log_2 \frac{1}{\text{mFAR}} \geq E - \delta,$$
$$\frac{1}{N} \log_2 |\mathcal{V}| \leq R + \delta. \tag{15.12}$$

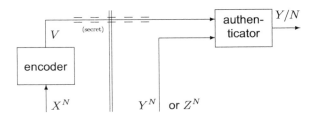

Figure 15.2 A rate-constrained authentication system.

THEOREM 15.2 *In the rate-constrained authentication scenario, any false acceptance exponent and storage rate combination (E,R) is achievable if and only if $(E,R) \in \mathcal{R}_{ER}$, where*

$$\mathcal{R}_{ER} \triangleq \{(E,R) : 0 \leq E \leq I(U;Y), \qquad (15.13)$$
$$R \geq I(U;X),$$
$$\text{for some } P(u,x,y) = Q(x,y)P(u|x)\}.$$

Proof 1. To prove the achievability part, we transform N copies of the source (X,Y) into a source (V,Y^N) with (roughly) $H(Y^N|V) \leq NH(Y|U)$ and $V \in \{1,2,\ldots,|\mathcal{V}|\}$, $|\mathcal{V}| \approx 2^{NI(U;X)}$; hence, V is a quantized version of X^N – see [9]. Then, applying the achievability part of the basic Theorem 15.1 on this new source, we obtain the result.

2. The converse part is an extension of the converse of the basic theorem. The proof for the mFAR part is similar to the basic theorem. The difference here is that we use an authentication set $\mathcal{B}(VY)$, and define C such that $C = 1$ if $(V,Y^N) \in \mathcal{B}(VY)$, and $C = 0$ otherwise; then the bounds for FRR and mFAR are obtained in terms of this set as

$$\text{FRR} = \sum_{(v,y^N) \notin \mathcal{B}(VY)} P(v,y^N) \leq \delta, \qquad (15.14)$$

$$\text{mFAR} = \max_{y^N \in \mathcal{Y}^N} \sum_{v \in \mathcal{B}(V|y^N)} P(v) \geq (1-\delta)2^{-I(V;Y^N|C=1)}. \qquad (15.15)$$

Then we can write. using a similar series of steps to the converse of the basic theorem,

$$-\log_2((1-\delta)2^{-I(V;Y^N|C=1)} + \delta) \leq I(V;Y^N|C)$$
$$\leq I(V;Y^N) + 1$$
$$= \sum_{n=1}^{N} I(V;Y_n|Y^{n-1}) + 1$$
$$\leq \sum_{n=1}^{N} I(V,Y^{n-1};Y_n) + 1$$
$$\leq \sum_{n=1}^{N} I(V,X^{n-1};Y_n) + 1$$
$$= NI(U;Y) + 1; \qquad (15.16)$$

here, however, the last two steps require some attention. In the last inequality we used the fact that $I(V,Y^{n-1};Y_n) \leq I(V,X^{n-1},Y^{n-1};Y_n) = I(V,X^{n-1};Y_n)$, since $Y^{n-1} - (V,X^{n-1}) - Y_n$. To obtain the last equality, we first define $U_n \triangleq (V,X^{n-1})$.

Then, if we take a time-sharing variable T uniform over $\{1,2,\ldots,N\}$ and independent of all other variables and set $U \triangleq (U_n, n)$, $X \triangleq X_n$ and $Y \triangleq Y_n$ for $T = n$, we obtain

$$\sum_{n=1}^{N} I(V, X^{n-1}; Y_n) = \sum_{n=1}^{N} I(U_n; Y_n)$$
$$= NI(U_T; Y_T | T)$$
$$= NI((U_T, T); Y_T)$$
$$= NI(U; Y). \tag{15.17}$$

Finally, note that U_n–X_n–Y_n and, consequently, U–X–Y. Then, combining (15.15) and (15.16), and letting $\delta \downarrow 0$ and $N \to \infty$, we obtain the converse result for mFAR.

To obtain the bound for the storage rate, we use the facts that $V = e(X^N)$ and that X^N is an i.i.d. sequence; then,

$$H(V) = I(V; X^N)$$
$$= \sum_{n=1}^{N} I(V; X_n | X^{n-1})$$
$$= \sum_{n=1}^{N} I(V, X^{n-1}; X_n)$$
$$= NI(U; X), \tag{15.18}$$

for the same $P(u,x,y) = Q(x,y)P(u|x)$ as before.

Now, for achievable R we obtain

$$R + \delta \geq \frac{1}{N} \log_2 |\mathcal{V}| \geq H(V) = I(U; X), \tag{15.19}$$

for some $P(u,x,y) = Q(x,y)P(u|x)$. This finalizes the converse.

15.4.3 Discussion

The advantage of constraining the storage rate is that we have to transmit, store, and encrypt/decrypt less data than in traditional biometric authentication. Note that taking $U \equiv X$ results in $E = I(X;Y)$ and $R = H(X)$, which is the result for the basic setting. Nevertheless, constraining the storage rate does not eliminate the need for V to be encrypted and decrypted. Therefore, rate-constrained systems have, in principle, the same security issues as the basic setting.

Note that our Theorem 15.2 is similar to but different from the result on one-sided rate-constrained hypothesis testing in [10].

15.5 Secret-Key-Based Authentication Systems

15.5.1 Scenario and Objective

The next scenario we consider is secret-based authentication. Here, during enrollment an individual presents his enrollment biometric sequence X^N to an encoder. From this enrollment sequence X^N, the encoder generates a secret S. This secret is used later on by the authenticator, which is just an equality checker here – see Fig. 15.3. The encoder also produces a helper message M that is stored in a public database.

In order to get authenticated, a legitimate individual presents his authentication biometric sequence Y^N to a decoder. The decoder produces an estimated secret $\widehat{S}_y$ using the helper message M that it can retrieve from the public database. Also, an impostor can try to get authenticated. An impostor who also has access to the helper message M can present an impostor sequence $Z^N(M)$ to the decoder that now forms the estimated secret $\widehat{S}_z$ using M. The equality checker compares the estimated secret $\widehat{S}_y$ or $\widehat{S}_z$ to the enrolled secret S, and takes a positive authentication decision about the individual if the secrets are equal, rejecting the individual otherwise.

The secret-key-based authentication system must be designed such that the mFAR is exponentially as small as possible, while we require FRR to be negligible.

15.5.2 System Building Blocks: Encoder, Decoder, and Equality Checker; FRR and mFAR

Consider now the operations performed in a secret-based biometric authentication system. In this system, the encoder is specified by the encoding function $(S, M) = e(X^N)$, where $S \in \{\phi_e, 1, 2, \ldots, |\mathcal{S}|\}$ is the generated secret and $M \in \{1, 2, \ldots, |\mathcal{M}|\}$ the public helper message. Here, ϕ_e is the secret value if the encoder could not assign a secret to the enrollment sequence X^N.

The decoder operations are given by the decoding function $\widehat{S}_y = d(M, Y^N)$, where $\widehat{S}_y \in \{\phi_d, 1, 2, \ldots, |\mathcal{S}|\}$ is the estimated secret. Now, ϕ_d is the estimated secret value if the decoder could not find a secret for the authentication sequence Y^N. Note that the decoder can operate on both legitimate sequences, Y^N, and impostor sequences, Z^N. An impostor can choose his sequence $Z^N = i(M)$, depending on the helper data M. When this impostor sequence $z^N \in \mathcal{Y}^N$ is presented to the decoder, the decoder forms the estimate $\widehat{S}_z = d(M, Z^N) = d(M, i(M))$.

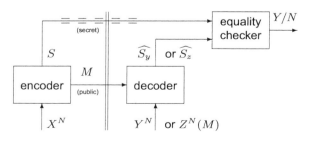

Figure 15.3 A secret-key-based authentication system.

Finally, the equality checker compares the output of the encoder, i.e., the secret S, to the output of the decoder, i.e., the estimated secret $\widehat{S}_y$ or $\widehat{S}_z$.

The FRR and mFAR in a secret-based biometric authentication system are now defined as follows:

$$\text{FRR} \triangleq \Pr\{\widehat{S}_y \neq S\},$$

$$\text{mFAR} \triangleq \max_{i(M)} \Pr\{\widehat{S}_z = S\}. \quad (15.20)$$

15.5.3 Definition of Achievability and Statement of Result

Now we give a formal definition of the achievable false acceptance exponents in secret-key-based biometric authentication systems, and present the corresponding result on the fundamental limit for these systems. The proof of this result is postponed to Section 15.6.

DEFINITION 15.3 *The false acceptance exponent E is achievable in the secret-key-based authentication scenario if, for all $\delta > 0$ and all N large enough, there exist encoders and decoders such that*

$$\text{FRR} \leq \delta, \quad (15.21)$$

$$\frac{1}{N} \log_2 \frac{1}{\text{mFAR}} \geq E - \delta. \quad (15.22)$$

THEOREM 15.3 *For a secret-key-based biometric authentication system, the maximum achievable false acceptance exponent E is equal to $I(X;Y)$; this is also the fundamental limit in the secret-key-based authentication scenario.*

15.5.4 Relation to Ahlswede–Csiszár Secret Generation

Observe that secret-key-based biometric authentication is very similar to the Ahlswede–Csiszár secret generation setting [1]. In the Ahlswede–Csiszár setting, an encoder and decoder observe two correlated sequences X^N and Y^N. It is the objective of both the encoder and decoder to form a common secret by interchanging a public helper message. Here, both the enrolled, S, and estimated, $\widehat{S}_y$, secrets assume values in the same alphabet $\{1, 2, \ldots, |\mathcal{S}|\}$. In this system the secret must be reliably recoverable by the decoder, it should be as large as possible and uniform, and the helper message should

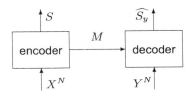

Figure 15.4 Secret key generation.

be uninformative about the secret. These requirements lead to the following definition of achievability.

DEFINITION 15.4 In the Ahlswede–Csiszár setting, the secrecy rate R is achievable if, for all $\delta > 0$ and all N large enough, there exist encoders and decoders such that

$$\Pr\{\widehat{S}_y \neq S\} \leq \delta,$$
$$\frac{1}{N}H(S) + \delta \geq \frac{1}{N}\log_2 |\mathcal{S}| \geq R - \delta,$$
$$\frac{1}{N}I(S;M) \leq \delta. \tag{15.23}$$

The fundamental limit for the secrecy rate was found in [1] and is stated in the following theorem.

THEOREM 15.4 (Ahlswede–Csiszár, 1993) *For a secret generation system the maximum achievable secrecy rate, also called the secrecy capacity, is given by $C_s = I(X;Y)$.*

Note that in the Ahlswede–Csiszár setting, the security of the system is defined in terms of the secret key rate, and the above theorem only gives the statement for the FRR. The question is, what happens if we take into account an impostor and the fact that the impostor has access to the helper data M?

To see the consequences of the introduction of the impostor on the system security in terms of mFAR, consider the following important example.

Example 15.1 Consider two distributions $P(m,s)$ realized by an encoder that satisfies Theorem 15.4. These distributions satisfy the achievability constraints

$$\frac{1}{N}I(S;M) \leq \delta,$$
$$\frac{1}{N}H(S) + \delta \geq \frac{1}{N}\log_2 |\mathcal{S}| \geq R - \delta. \tag{15.24}$$

First consider a distribution $P(s|m)$ as in Fig. 15.5.

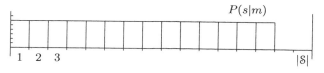

Figure 15.5 A conditional distribution with low mFAR.

We define this distribution such that for each m, $P(s|m) = 1/(\alpha|\mathcal{S}|)$ or $P(s|m) = 0$, for some $\alpha \leq 1$. Observe then that an impostor can achieve

$$\begin{aligned}
\log_2 \frac{1}{\text{mFAR}} &= \log_2(\alpha|\mathcal{S}|) \\
&= H(S|M) \\
&= H(S) - I(S;M) \\
&\geq N(R - 2\delta) - N\delta \\
&= N(R - 3\delta),
\end{aligned} \qquad (15.25)$$

and therefore

$$\frac{1}{N} \log_2 \frac{1}{\text{mFAR}} \geq R - 3\delta. \qquad (15.26)$$

Thus for this distribution we can achieve low mFAR.

Next, focus on another distribution $P(s|m)$, as in Fig. 15.6.

This distribution is defined such that for each m, $P(s|m) = 1 - \beta$ for a single s, and $P(s|m) = \beta/(|\mathcal{S}| - 1)$ for all the other s. Then we can write

$$\begin{aligned}
H(S|M) &= H(S) - I(S;M) \\
&\geq H(S) - N\delta \\
&= (H(S) + N\delta) - 2N\delta;
\end{aligned} \qquad (15.27)$$

on the other hand, we have

$$\begin{aligned}
H(S|M) &= h(\beta) + \beta \log_2(|\mathcal{S}| - 1) \\
&\leq 1 + \beta \log_2 |\mathcal{S}| \\
&\leq 1 + \beta(H(S) + N\delta).
\end{aligned} \qquad (15.28)$$

Combining these two expressions, we obtain

$$(1 - \beta)(H(S) + N\delta) \leq 1 + 2N\delta. \qquad (15.29)$$

Observe that the best impostor strategy is to take the most probable secret s, then he can achieve

$$\text{mFAR} = (1 - \beta) \leq \frac{1 + 2N\delta}{H(S) + N\delta} \leq \frac{3\delta}{R - \delta}. \qquad (15.30)$$

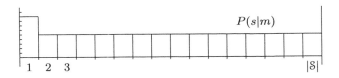

Figure 15.6   A conditional distribution with large mFAR.

Hence, for large enough N and for a maximum *a posteriori* impostor, we see that we cannot guarantee a low false acceptance exponent, since

$$\frac{1}{N}\log_2\frac{1}{\mathrm{mFAR}} \geq \frac{1}{N}\log_2\frac{R-\delta}{3\delta}. \qquad (15.31)$$

This example demonstrates that in the Ahlswede–Csiszár setting, a small $I(S;M)$ does not guarantee an exponentially small mFAR for all impostors. Nevertheless, we can still guarantee $\mathrm{mFAR} \approx 2^{-NC_s} = 2^{-NI(X;Y)}$ in secret-based authentication systems for all impostors, and show that $I(X;Y)$ is a fundamental limit for the false acceptance exponent, just as it is the fundamental limit for secret key rate – see Theorem 15.3. As seen from the example for the first distribution, it is important to uniformly distribute the secret key labels. The proof of this result is given in Section 15.6.

15.5.5 Discussion

We investigated a so-called secret-key-based authentication system. Observe that the secret S should be stored and communicated to the equality checker securely using cryptographic techniques. It is crucial that in this setting we do not have to decrypt the secret S in order to perform authentication. It is possible to apply a one-way function on both the enrollment secret S and the estimated secret $\widehat{S}_y$ or $\widehat{S}_z$ and check equality in the encrypted domain. This technique is also used for traditional password authentication.

It is also of interest that the same false acceptance exponent for the basic setting is also achievable with secret-based authentication.

The investigated scheme is an extension of the Ahlswede–Csiszár [1] model in which no impostor is present. The relation between mutual information and reliability was investigated by Ho [11] and in Ho and Verdu [12]. Strengthening the secrecy constraint from weak to strong does not lead to Theorem 15.3.

15.6 Proof of Theorem 15.3

15.6.1 Achievability Proof for Theorem 15.3

First, note that in our achievability proof we must demonstrate that there exist encoders and decoders that (1) achieve the FRR constraint (15.21), and that (2) guarantee that, for all impostor strategies, the mFAR constraint (15.22) is met for $E = I(X;Y)$.

FRR, M-Labeling

First, we show that there exists a Slepian–Wolf (SW) code for reconstruction of $\widehat{X^N}$ by the decoder – see Fig. 15.7. This code defines the M-labeling.

Fix $\varepsilon > 0$ and N, and consider the set of jointly ε-typical sequences $\mathcal{A}_\varepsilon^N(XY)$, based on the joint distribution $\{Q(x,y), x \in \mathcal{X}, y \in \mathcal{Y}\}$ of the biometric source.

> **Random labeling:** Assign a label m that is *uniformly chosen* from $\{1, 2, \ldots, \mathcal{M}\}$ to each $x^N \in \mathcal{X}^N$. Denote this label by $m(x^N)$.

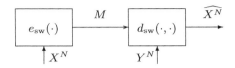

Figure 15.7 Slepian–Wolf coding.

Encoding: Upon observing x^N, the encoder determines $m(x^N)$ based on the random codebook. This label is then sent to the decoder.

Decoding: The decoder observes an authentication sequence y^N and chooses the unique $\widehat{x^N}$ such that $m(\widehat{x^N}) = m(x^N)$ and $(\widehat{x^N}, y^N) \in \mathcal{A}_\varepsilon^N(XY)$. If such an estimate $\widehat{x^N}$ cannot be found, the decoder declares an error. It is not important what value $\widehat{x^N}$ gets in that case.

Error probability: Now consider the error probability averaged over the ensemble of labeling. Using the properties of typical sequences, we obtain

$$\Pr\{\widehat{X^N} \neq X^N\}$$

$$= \Pr\left\{((X^N, Y^N) \notin \mathcal{A}_\varepsilon^N) \cup \left(\bigcup_{\tilde{x}^N \neq X^N, (\tilde{x}^N, Y^N) \in \mathcal{A}_\varepsilon^N} M(\tilde{x}^N) = M(X^N)\right)\right\}$$

$$\leq \Pr\{(X^N, Y^N) \notin \mathcal{A}_\varepsilon^N\} + \Pr\left\{\bigcup_{\tilde{x}^N \neq X^N, (\tilde{x}^N, Y^N) \in \mathcal{A}_\varepsilon^N} M(\tilde{x}^N) = M(X^N)\right\}$$

$$\leq \varepsilon + \sum_{x^N, y^N} P(x^N, y^N) \cdot \sum_{\tilde{x}^N \neq x^N, (\tilde{x}^N, y^N) \in \mathcal{A}_\varepsilon^N} \Pr\{M(\tilde{x}^N) = M(x^N)\}$$

$$\leq \varepsilon + \sum_{x^N, y^N} P(x^N, y^N) |\mathcal{A}_\varepsilon^N(X|y^N)| \frac{1}{|\mathcal{M}|}$$

$$\leq \varepsilon + 2^{N(H(X|Y)+2\varepsilon)} \frac{1}{2^{N(H(X|Y)+3\varepsilon)}} = \varepsilon + 2^{-N\varepsilon} \leq 2\varepsilon \qquad (15.32)$$

for N large enough, if we take

$$|\mathcal{M}| = 2^{N(H(X|Y)+3\varepsilon)}. \qquad (15.33)$$

Thus there exists an M-labeling with

$$\Pr\{\widehat{X^N} \neq X^N\} \leq 2\varepsilon. \qquad (15.34)$$

S-labeling and the decoder strategy: Next, consider an S-labeling that can be used by the encoder during enrollment. The encoder can use any labeling that satisfies $s(x^N) : \mathcal{X}^N \to \{1, 2, \ldots, |\mathcal{S}|\}$ for $x^N \in \mathcal{A}_\varepsilon^N(X)$, and $s(x^N) = \phi_e$ for $x^N \notin \mathcal{A}_\varepsilon^N(X)$. Then the decoder can estimate the secret as $s(\widehat{x^N})$, where $\widehat{x^N}$ is the output of the SW decoder if this decoder did not declare an error, and as ϕ_d if an error was declared by the SW decoder.

As a result, if no error occurred, the SW encoder input x^N equals the SW decoder output $\widehat{x^N} \in \mathcal{A}_\varepsilon^N(X)$. This implies that, for a legitimate individual, our encoder and decoder guarantee that

$$\text{FRR} = \Pr\{\widehat{S}_y \neq S\} \leq \Pr\{\widehat{X^N} \neq X^N\} \leq 2\varepsilon. \tag{15.35}$$

FAR, S-Labeling

Next, fix a Slepian–Wolf code constructed before, and define, for all $m \in \mathcal{M}$, the sets of typical sequences

$$\mathcal{A}(m) \triangleq \{x^N \in \mathcal{A}_\varepsilon^N(X) \text{ for which } m(x^N) = m\}. \tag{15.36}$$

Impostor strategy: Now consider a helper message $m \in \mathcal{M}$. An impostor, knowing the helper message m, tries to pick a sequence z^N such that the resulting estimated secret $\widehat{S}_z$ equals the secret key S of the individual he claims to be. Given m, the impostor can decide for the most promising secret key $\widehat{S}_z$ and then choose the sequence z^N that results, together with m, in this most promising key. Note that for a given m, the impostor need only consider secrets $\widehat{S}_z$ that result from typical sequences, i.e., from $x^N \in \mathcal{A}(m)$. Other sequences cannot be valid outputs of the SW decoder.

S-labeling: Now consider a specific S-labeling that satisfies the properties of S-labeling used by our encoder. For each m, we distribute all sequences $x^N \in \mathcal{A}(m)$ roughly uniformly over s-labels. All non-typical sequences get label ϕ_e. The number of typical sequences with label m is upper bounded by

$$\Pr\{M = m\}/2^{-N(H(X)+\varepsilon)}. \tag{15.37}$$

Distributing these typical sequences over all s-labels uniformly leads to at most

$$\lceil \Pr\{M = m\}/(2^{-N(H(X)+\varepsilon)}|\mathcal{S}|) \rceil \tag{15.38}$$

typical sequences having a certain secret label. The joint probability that m occurs and an impostor, knowing m, chooses the correct secret, is therefore upper bounded by

$$\left\lceil \frac{\Pr\{M = m\}}{2^{-N(H(X)+\varepsilon)}|\mathcal{S}|} \right\rceil \cdot 2^{-N(H(X)-\varepsilon)}. \tag{15.39}$$

FAR: An upper bound for the mFAR follows if we carry out the summation over all m, i.e.,:

$$\text{mFAR} \leq \sum_{m=1}^{|\mathcal{M}|} \left\lceil \frac{\Pr\{M = m\}}{2^{-N(H(X)+\varepsilon)}|\mathcal{S}|} \right\rceil \cdot 2^{-N(H(X)-\varepsilon)}$$

$$\leq \sum_{m=1}^{|\mathcal{M}|} \left(\frac{\Pr\{M = m\}}{2^{-N(H(X)+\varepsilon)}|\mathcal{S}|} + 1 \right) 2^{-N(H(X)-\varepsilon)}$$

$$= \sum_{m=1}^{|\mathcal{M}|} \frac{\Pr\{M=m\}}{2^{-N(H(X)+\varepsilon)}|\mathcal{S}|} 2^{-N(H(X)-\varepsilon)} + \sum_{m=1}^{|\mathcal{M}|} 2^{-N(H(X)-\varepsilon)}$$

$$= 2^{-N(I(X;Y)-4\varepsilon)} + 2^{-N(I(X;Y)-4\varepsilon)}$$

$$\leq 2^{-N(I(X;Y)-4\varepsilon-\frac{1}{N})} \tag{15.40}$$

for large enough N, for all impostors, if we take the number of s-labels $|\mathcal{S}| = 2^{N(I(X;Y)-2\varepsilon)}$.

Combining the upper bound (15.35) on the FRR and the upper bound (15.40) on the mFAR results in the achievability of $E = I(X;Y)$.

15.6.2 Converse for Theorem 15.3

We will show that for all encoders and decoders that achieve the FRR constraint (15.21), there is at least one impostor that does not satisfy the mFAR constraint (15.22) for $E > I(X;Y)$.

Definition of Set $\mathcal{C}(m)$ and C-Function

First, consider an encoder and decoder achieving the FRR constraint (15.21). We define the set of secrets $\mathcal{C}(m)$ that can be reconstructed from a helper label m as

$$\mathcal{C}(m) \triangleq \{s : \text{there exists } y^N \text{ such that } d(m, y^N) = s\}. \tag{15.41}$$

Moreover, let $C(\cdot,\cdot)$ be a function of s and m such that $C(s,m) = 1$ for $s \in \mathcal{C}(m)$, and $C(s,m) = 0$ otherwise. Now,

$$\delta \geq \Pr\{\widehat{S}_y \neq S\} \geq \sum_m \Pr\{M=m, S \notin \mathcal{C}(m)\} = P(C=0), \tag{15.42}$$

since $S \notin \mathcal{C}(M)$ will always lead to an error.

A Conditional Maximum A Posteriori Impostor Strategy

Knowing m, an optimal impostor's strategy is to choose a target secret $\widehat{s}_z \in \mathcal{C}(m)$ with maximum conditional probability, i.e.,

$$\widehat{s}_z(m) = \arg \max_{s \in \mathcal{C}(m)} P(s|m). \tag{15.43}$$

Since the target secret can be realized, this impostor achieves

$$\text{FAR} = \sum_m P(m) P(C=1|m) \max_{s \in \mathcal{C}(m)} \frac{P(s|m)}{P(C=1|m)}$$

$$= \sum_m P(m) P(C=1|m) \max_{s \in \mathcal{C}(m)} \frac{P(s, C=1|m)}{P(C=1|m)}$$

$$= \sum_m P(m, C=1) \max_s P(s|m, C=1). \tag{15.44}$$

Consider next the relation between conditional entropy and FAR.

$$
\begin{aligned}
H(S|M,C=1) &= \sum_m P(m|C=1) \sum_s P(s|m,C=1) \log_2 \frac{1}{P(s|m,C=1)} \\
&\geq \sum_m P(m|C=1) \sum_s P(s|m,C=1) \log_2 \frac{1}{\max_s P(s|m,C=1)} \\
&= \sum_m P(m|C=1) \log_2 \frac{1}{\max_s P(s|m,C=1)} \\
&\geq \log_2 \frac{1}{\sum_m P(m|C=1) \max_s P(s|m,C=1)} \\
&= \log_2 \frac{P(C=1)}{\text{FAR}} \geq \log_2 \frac{P(C=1)}{\text{mFAR}},
\end{aligned}
\tag{15.45}
$$

where to get the last inequality we used Jensen's inequality; see also [12].
Now can write

$$
\begin{aligned}
P(C=1)H(S|M,C=1) &\leq H(S|M,C) \\
&\leq H(S|M) \\
&\leq I(S;Y^N|M) + F \\
&\leq H(Y^N) - H(Y^N|M,S,X^N) + F \\
&= H(Y^N) - H(Y^N|X^N) + F \\
&= NI(X;Y) + F,
\end{aligned}
\tag{15.46}
$$

where $F = 1 + \Pr\{\widehat{S}_y \neq S\} \log_2 |\mathcal{X}|^N \leq 1 + \delta \log_2 |\mathcal{X}|^N$ is the Fano term.

Completing the Proof
Combining (15.45) and (15.46), we get

$$
\begin{aligned}
P(C=1) \log_2 \frac{P(C=1)}{\text{mFAR}} &\leq P(C=1)H(S|M,C=1) \\
&\leq NI(X;Y) + 1 + \delta \log_2 |\mathcal{X}|^N.
\end{aligned}
\tag{15.47}
$$

Then, for an achievable exponent E, noting that $P(C=1) = 1 - P(C=0) \geq 1 - \delta$, we have

$$
\begin{aligned}
(1-\delta)N(E-\delta) &\leq P(C=1) \log_2 P(C=1) + \log_2 \frac{1}{\text{mFAR}} \\
&\leq NI(X;Y) + 1 + \delta \log_2 |\mathcal{X}|^N.
\end{aligned}
\tag{15.48}
$$

If we now let $\delta \downarrow 0$ and $N \to \infty$, we get that for achievable false acceptance exponents $E \leq I(X;Y)$. This finalizes the converse.

15.7 Privacy Leakage in Secret-Based Systems

Up to now, we have only been concerned with the security in biometric authentication systems. However, in biometric systems, privacy leakage, characterizing the amount of information leaked about biometric data in the system, is another important parameter.

We define the privacy leakage as the mutual information $I(X^N;M)$ between the enrollment biometric sequence X^N and the public helper message M. For our code that demonstrates the achievability of $E = I(X;Y)$, we can write that

$$I(X^N;M) \leq H(M) \leq \log_2 |\mathcal{M}| = N(H(X|Y) + 3\varepsilon). \tag{15.49}$$

Clearly, in biometric systems the goal is to maximize the false acceptance exponent and minimize the privacy leakage. Therefore we would like to find out what the tradeoff is between the false acceptance exponent and the privacy leakage rate. In order to investigate this, consider again the secret-key-based authentication system in Fig. 15.3 and the following definition of achievability.

DEFINITION 15.5 The false acceptance exponent / privacy leakage rate combination (E,L) is achievable in secret-key-based authentication systems if, for all $\delta > 0$ and all N large enough, there exist encoders and decoders such that

$$\text{FRR} \leq \delta,$$
$$\frac{1}{N} \log_2 \frac{1}{\text{mFAR}} \geq E - \delta,$$
$$\frac{1}{N} I(X^N;M) \leq L + \delta. \tag{15.50}$$

The fundamental tradeoff between the false acceptance exponent and privacy leakage rate is given by the following theorem.

THEOREM 15.5 *For a secret-key-based authentication system, the region $\mathcal{R}_{EL}$ of achievable false acceptance exponent and privacy leakage combinations is given by*

$$\mathcal{R}_{EL} \triangleq \{(E,L) : 0 \leq E \leq I(U;Y),$$
$$L \geq I(U;X) - I(U;Y),$$
$$\text{for some } P(u,x,y) = Q(x,y)P(u|x)\}. \tag{15.51}$$

Note that the result found here coincides with the corresponding result in [3] and also in [13], if we replace the secret key rate with the false acceptance exponent. There is also a connection to Csiszár and Narayan [14], who considered the secret key rate vs. helper data rate tradeoff.

15.8 Proof of Theorem 15.5

The proof of Theorem 15.5 again consists of two parts: achievability and converse. The main idea of the achievability part is again (as in the rate-constrained scenario) to transform N copies of the biometric source (X,Y) into a source (V,Y^N), where V is a quantized version of X^N. The converse is a mix of the converses in Section 15.6.2 and that for Theorem 1 in [3].

15.8.1 Achievability Part for Theorem 15.5

Before beginning the achievability proof, we first define a modified typical set and present its properties. Fix an auxiliary alphabet $\mathcal{U}$ and conditional probabilities $\{P(u|x), u \in \mathcal{U}, x \in \mathcal{X}\}$, and let $0 < \varepsilon < 1$.

DEFINITION 15.6 Assuming that transition probability matrix $P(u|x)$ determines the joint probability distribution $\{P(u,x,y) = Q(x,y)P(u|x), u \in \mathcal{U}, x \in \mathcal{X}, y \in \mathcal{Y}\}$, we define

$$\mathcal{B}_\varepsilon^N(UX) \triangleq \{(u^N, x^N) :$$
$$\Pr\{Y^N \in \mathcal{A}_\varepsilon^N(Y|u^N, x^N) | (U^N, X^N) = (u^N, x^N)\} \geq 1 - \epsilon\}, \quad (15.52)$$

where Y^N is the output sequence of a "memoryless channel" $Q(y|x) = Q(x,y)/\sum_y Q(x,y)$, and the input was a sequence x^N.

PROPERTY 15.1 If $(u^N, x^N) \in \mathcal{B}_\varepsilon^N(UX)$, then also $(u^N, x^N) \in \mathcal{A}_\varepsilon^N(UX)$.

PROPERTY 15.2 Let (U^N, X^N) be i.i.d. according to $P(u,x)$, then, for all N large enough,

$$\Pr\{(U^N, X^N) \in \mathcal{B}_\varepsilon^N(UX)\} \geq 1 - \epsilon. \quad (15.53)$$

For proofs of these properties, see, e.g., [3].

We can now proceed with the achievability part.

Random coding: For each index $v \in \{1, 2, \ldots, |\mathcal{V}|\}$ generate an auxiliary sequence $u^N(v)$ at random according to $P(u) = \sum_{x,y} Q(x,y)P(u|x)$.
Quantization: Upon observing x^N, let V be the smallest value of v such that $(u^N(v), x^N) \in \mathcal{B}_\varepsilon^N(UX)$. If no such v is found, we set $V = \phi_q$.
Events: Let X^N and Y^N be the observed biometric source sequences, and V be the index determined by the quantizer. Define, for $v = 1, 2, \ldots, |\mathcal{V}|$, the events

$$E_v \triangleq \{(u^N(v), X^N) \in \mathcal{B}_\varepsilon^N(UX)\}. \quad (15.54)$$

Error probability: As in [15, p. 454], we can write for the error probability of a quantizer averaged over the ensemble of auxiliary sequences:

$$\Pr\left\{\bigcap_v E_v^c\right\} = \sum_{x^N \in \mathcal{X}^N} Q(x^N) \prod_v \left(1 - \sum_{u^N \in \mathcal{B}_\varepsilon^N(U|x^N)} P(u^N)\right)$$

$$\stackrel{(a)}{\leq} \sum_{x^N \in \mathcal{X}^N} Q(x^N)(1 - 2^{-N(I(U;X)+3\varepsilon)} \cdot \sum_{u^N \in \mathcal{B}_\varepsilon^N(U|x^N)} P(u^N|x^N))^{|\mathcal{V}|}$$

$$\overset{(b)}{\leq} \sum_{x^N \in \mathcal{X}^N} Q(x^N)(1 - \sum_{u^N \in \mathcal{B}_\varepsilon^N(U|x^N)} P(u^N|x^N)$$

$$+ \exp(-|\mathcal{V}|2^{-N(I(U;X)+3\varepsilon)}))$$

$$\leq \sum_{(u^N,x^N) \notin \mathcal{B}_\varepsilon^N(UX)} P(u^N,x^N) + \sum_{x^N \in \mathcal{X}^N} Q(x^N)\exp(-2^{N\varepsilon})$$

$$\overset{(c)}{\leq} 2\varepsilon, \tag{15.55}$$

for N large enough, if $|\mathcal{V}| = 2^{N(I(U;X)+4\varepsilon)}$. Here, (a) follows from the fact that for $(u^N, x^N) \in \mathcal{B}_\varepsilon^N(UX)$, using Property 15.1, we get

$$P(u^N) = P(u^N|x^N)\frac{Q(x^N)P(u^N)}{P(x^N,u^N)}$$

$$\geq P(u^N|x^N)\frac{2^{-N(H(X)+\varepsilon)}2^{-N(H(U)+\varepsilon)}}{2^{-N(H(U,X)-\varepsilon)}}$$

$$= P(u^N|x^N)2^{-N(I(U;X)+3\varepsilon)}, \tag{15.56}$$

(b) from the inequality $(1-\alpha\beta)^K \leq 1 - \alpha + \exp(-K\beta)$, which holds for $0 \leq \alpha, \beta \leq 1$ and $K > 0$, and (c) from Property 15.2.

We have shown that in the ensemble of auxiliary sequences, for N large enough, there exists a set of auxiliary sequences achieving

$$\Pr\{V = \phi_q\} \leq 2\varepsilon. \tag{15.57}$$

We concentrate on such a set of auxiliary sequences (a quantizer).

Now we have that a sequence x^N for which there is v such that $(u^N(v), x^N) \in \mathcal{B}_\varepsilon^N(UX)$ occurs with probability greater than or equal to $1 - 2\varepsilon$. Then the authentication sequence y^N is also in $\mathcal{A}_\varepsilon^N(Y|u^N(v), x^N)$, and consequently in $\mathcal{A}_\varepsilon^N(Y|u^N(v))$ with probability greater than or equal to $1 - \varepsilon$. Moreover, note that $|\mathcal{A}_\varepsilon^N(Y|u^N(v))| \leq 2^{N(H(Y|U)+2\varepsilon)}$.

Thus we can write

$$H(Y^N|V) \leq (2\varepsilon + (1-2\varepsilon)\varepsilon)\log_2|\mathcal{Y}|^N + (1-2\varepsilon)(1-\varepsilon)\log_2 2^{N(H(Y|U)+2\varepsilon)}$$

$$\leq N((1-3\varepsilon+2\varepsilon^2)(H(Y|U)+2\varepsilon) + (3\varepsilon-2\varepsilon^2)\log_2|\mathcal{Y}|); \tag{15.58}$$

letting $\varepsilon \downarrow 0$ and $N \to \infty$, we can get $H(Y^N|V)/N$ arbitrarily close to $H(Y|U)$, and thus also $I(V;Y^N)/N = H(Y) - H(Y^N|V)/N$ arbitrary close to $I(U;Y)$.

Moreover, we can also write

$$\frac{1}{N}H(V|Y^N) = \frac{1}{N}(H(V) + H(Y^N|V) - H(Y^N))$$

$$\leq I(U;X) + 4\epsilon + \frac{1}{N}H(Y^N|V) - H(Y), \tag{15.59}$$

and then letting $\delta \downarrow 0$ and $N \to \infty$, we can make $\frac{1}{N}H(V|Y^N)/N$ arbitrary close to $I(U;X) - I(U;Y)$.

Now we can apply the achievability proof for Theorem 15.3 to demonstrate the achievability of the false acceptance exponent. Furthermore, the achievability of the privacy leakage rate follows from (15.49) and the result for $\frac{1}{N}H(V|Y^N)$. This leads to the achievability of the combination

$$(E,L) = (I(U;Y), I(U;X) - I(U;Y)). \tag{15.60}$$

15.8.2 Converse for Theorem 15.5

For the converse, we only consider the main steps. First we bound

$$\begin{aligned}
H(S|M) &\leq I(S;Y^N|M) + H(S|Y^N,M) \\
&\leq I(S;Y^N|M) + H(S|\widehat{S}_y) \\
&\leq I(S,M;Y^N) + F \\
&= \sum_{n=1}^{N} I(S,M;Y_n|Y^{n-1}) + F \\
&= \sum_{n=1}^{N} I(S,M,Y^{n-1};Y_n) + F \\
&\leq \sum_{n=1}^{N} I(S,M,Y^{n-1},X^{n-1};Y_n) + F \\
&= \sum_{n=1}^{N} I(S,M,X^{n-1};Y_n) + F \\
&= NI(U;Y) + F, \tag{15.61}
\end{aligned}$$

where $F \triangleq 1 + \delta \log |\mathcal{X}|^N$ is a Fano term again. Here we used Y^{n-1}–(S,M,X^{n-1})–Y_n in the last but one equality, and to obtain the last step, we defined $U_n \triangleq (S,M,X^{n-1})$ and used the time-sharing argument as before. Here it holds again that U–X–Y. This expression can now be plugged into the FAR part of the converse for Theorem 15.3.

Now we continue with the privacy leakage:

$$\begin{aligned}
I(X^N;M) &= H(M) - H(M|X^N) \\
&\geq H(M|Y^N) - H(S,M|X^N) \\
&= H(S,M|Y^N) - H(S|M,Y^N,\widehat{S}_y) - H(S,M|X^N) \\
&\geq H(S,M|Y^N) - H(S|\widehat{S}_y) - H(S,M|X^N) \\
&\geq H(S,M|Y^N) - F - H(S,M|X^N) \\
&= I(S,M;X^N) - I(S,M;Y^N) - F \\
&= \sum_{n=1}^{N} I(S,M;X_n|X^{n-1}) - \sum_{n=1,N} I(S,M;Y_n|Y^{n-1}) - F
\end{aligned}$$

$$= \sum_{n=1}^{N} I(S,M,X^{n-1};X_n) - \sum_{n=1}^{N} I(S,M,Y^{n-1};Y_n) - F$$

$$\geq \sum_{n=1}^{N} I(S,M,X^{n-1};X_n) - \sum_{n=1,N} I(S,M,X^{n-1};Y_n) - F$$

$$\geq NI(U;X) - NI(U;Y) - F, \tag{15.62}$$

for the same $P(u,x,y) = Q(x,y)P(u|x)$ as before.

Now the converse follows if we let $\epsilon \downarrow 0$ and $N \to \infty$.

15.9 Conclusions and Final Remarks

In this chapter we have analyzed a number of biometric authentication systems starting with traditional biometric systems and ending up with biometric systems with template protection. We have extended the secret key generation biometric scenario by considering false acceptance rates. We found the fundamental limits and tradeoffs for the false acceptance exponents. These exponents are equal to the known secret key rates. When no privacy leakage is taken into account, the maximum false acceptance in the secret-based scenario coincides with that in the traditional authentication system. When the goal is also to minimize privacy leakage, the false acceptance exponent has to go down. Just like in [3] and [13], where there is a tradeoff between the secret key and the privacy leakage rates, here there is a tradeoff between the false acceptance exponent and the privacy leakage rate.

The cardinalities of auxiliary alphabets are easy to obtain, using the Fenchel–Eggleston strengthening of the Carathéodory lemma (see [16]), but are not considered here. Also, the secret key rate as a parameter is not considered here. It is, however, clear that the secret key rate and the false acceptance exponent are related. It is obvious that the false acceptance exponent cannot exceed the secret key rate. Finally, note that limiting the privacy leakage rate is similar to bounding the rate of message M.

Acknowledgments

Partly supported by NWO via grant agreement 632.012.001.

References

[1] R. Ahlswede and I. Csiszár, "Common randomness in information theory and cryptography – Part I: Secret sharing," *IEEE Trans. Inf. Theory*, vol. 39, no. 4, pp. 1121–1132, Jul. 1993.

[2] A. Juels and M. Wattenberg, "A fuzzy commitment scheme," in *Proc. 6th ACM Conf. on Computer and Communications Security*, Singapore: ACM Press, Nov. 1999, pp. 26–36.

[3] T. Ignatenko and F. M. J. Willems, "Biometric systems: Privacy and secrecy aspects," *IEEE Trans. Inf. Forensics Security*, vol. 4, no. 4, pp. 956–973, Dec. 2009.

[4] S. Wang, S. Rane, S. C. Draper, and P. Ishwar, "An information-theoretic analysis of revocability and reusability in secure biometrics," in *Proc. Inf. Theory Applications Workshop*, La Jolla, CA, USA, Feb. 2011, pp. 1–10.

[5] J. Wayman, A. Jain, and D. Maltoni, Eds., *Biometric Systems: Technology, Design and Performance Evaluation*. London: Springer-Verlag, 2005.

[6] N. Ratha, J. Connell, and R. Bolle, "Enhancing security and privacy in biometrics-based authentication systems," *IBM Syst. J.*, vol. 40, no. 3, pp. 614–634, 2001.

[7] T. M. Cover and J. A. Thomas, *Elements of Information Theory*, 2nd edn. Chichester: Wiley & Sons, 2006.

[8] I. Csiszár and J. Körner, *Information Theory – Coding Theorems for Discrete Memoryless Systems*, 1st edn. Waltham, MA: Academic Press, 1981.

[9] R. Ahlswede and J. Körner, "Source coding with side information and a converse for degraded broadcast channels," *IEEE Trans. Inf. Theory*, vol. 21, no. 6, pp. 629–637, Nov. 1975.

[10] R. Ahlswede and I. Csiszár, "Hypothesis testing with communication constraints," *IEEE Trans. Inf. Theory*, vol. 32, no. 4, pp. 533–542, Jul. 1986.

[11] S.-W. Ho, "On the interplay between Shannon's information measures and reliability criteria," in *Proc. IEEE Int. Symp. Inf. Theory*, Seoul, Korea, Jun. 2009, pp. 154–158.

[12] S.-W. Ho and S. Verdú, "On the interplay between conditional entropy and error probability," *IEEE Trans. Inf. Theory*, vol. 56, no. 12, pp. 5930–5942, Dec. 2010.

[13] L. Lai, S.-W. Ho, and H. V. Poor, "Privacy–security trade-offs in biometric security systems – Part I: Single use case," *IEEE Trans. Inf. Forensics Security*, vol. 6, no. 1, pp. 122–139, Mar. 2011.

[14] I. Csiszár and P. Narayan, "Common randomness and secret key generation with a helper," *IEEE Trans. Inf. Theory*, vol. 46, no. 2, pp. 344–366, Mar. 2000.

[15] R. G. Gallager, *Information Theory and Reliable Communication*. Chichester: Wiley & Sons, 1968.

[16] A. D. Wyner and J. Ziv, "The rate–distortion function for source coding with side information at the decoder," *IEEE Trans. Inf. Theory*, vol. 22, no. 1, pp. 1–10, Jan. 1976.

16 Joint Privacy and Security of Multiple Biometric Systems

Adina Goldberg and Stark C. Draper

This paper explores the design of biometric authentication in the context of a single user that has enrolled in multiple (distinct) authentication systems. The compromise of some subset of these systems will generally impact both the privacy of the user's biometric information and the security of the balance of the systems. In this work we consider how to design the systems jointly to minimize losses in privacy and security in the case of such compromise. It turns out that there is a tension between the two objectives, resulting in a privacy/security tradeoff. We introduce worst-case privacy and security measures, and consider the tradeoff between them, in the context of the "secure sketch" architecture. Secure sketch systems are based on error correction codes, and the considerations of joint design that we pose result in a novel code design problem. We first study the design problem algebraically and identify an equivalence with a type of subspace packing problem. While the packing problem fully characterizes the design space, it does not yield an explicit characterization. We then turn to a "fixed-basis" subspace of the general design space. We map a relaxed version of the fixed-basis design problem to a linear program which, after exploiting much symmetry, leads to an explicit tradeoff between security and privacy. While we show that fixed-basis designs are restrictive in terms of the achievable privacy/security tradeoffs, they have the advantage of being easily mapped to existing codes (e.g., low-density parity check codes), and thence to immediate deployment. Finally, we conjecture that the achievable privacy/security tradeoff of fixed-basis designs is characterized by an extremely simple analytic expression, one that matches our numerical results.

16.1 Introduction

The goal of an *authentication system* is to ensure that only legitimate individuals gain access to a secured resource or area. Increasingly popular are methods of authentication that use biometric data – unique information present in a person's physical attributes. An example of such a *biometric system* is a laptop-mounted fingerprint scanner, or an iris scanner at an airport.

As biometric data is bound to a particular person, is a finite resource, and cannot be renewed, there are a number of distinct considerations that apply to biometric authentication, in contrast to, say, password-based authentication. While security – the ability to recognize a legitimate user and reject an illegitimate user – is central to authentication, maintaining privacy of the underlying biometric data is also of great importance. Thus, paralleling the rise in popularity of biometrics has been an increase in our theoretical understanding of privacy and security in such systems.

In this chapter we focus on the less-studied area of *multiple* biometric systems. In this setting, a person uses the same biometric data to enroll in distinct systems. We pose a version of this problem where some subset of the systems may be compromised and ask how one should jointly design the systems to maintain security and privacy. We connect our design problem to a novel problem of error correction coding. We both formulate a general design problem (which appears to be computationally intractable) and solve a restricted, relaxed version that leads to a quantifiable tradeoff between security and privacy.

An outline of the chapter is as follows. First, in the spirit of a general introduction that a book chapter can provide, we spend the rest of this section discussing the design requirements of biometric systems generally, the tradeoff between privacy and security in the design of multiple systems, and related work. Then, in Section 16.2, we formalize the problem setting and define our measures of security and privacy. In Section 16.3 we introduce the design space in which we will work. Our first contribution here is to connect our problem formulation to the geometry of fixed-dimensional vector spaces, the interdependencies of which can be represented graphically. We show that the Grassmann graph does not encode sufficient structure to characterize our problem, while the subspace Hasse diagram does. While the intent of this section is to characterize the design space, the resultant characterization does not provide us a tractable objective with which to work. Thus, in Section 16.4 we study a particular "fixed-basis" case that looks at only a subgraph of the Hasse diagram. A relaxed version of the fixed-basis problem (which will be increasingly tight as block length gets large) is amenable to characterization as a linear program. Solving this linear program yields designs that demonstrate a natural tradeoff between security and privacy. We conclude the chapter in Section 16.5.

16.1.1 Biometric System Design Requirements

Biometric authentication takes place in two stages: enrollment and authentication. At enrollment a version of the user's biometric data is stored in the system. At authentication a second copy of the user's biometric is extracted for comparison with the data stored at enrollment. As biometric measurements are subject to noise, e.g., resulting from changes in makeup, skin conditions, lighting, etc., biometric authentication must be robust to some level of noise in the data. Such noise is an aspect that differentiates biometric-based from password-based authentication. We now list a number of requirements in the design and characterization of biometric systems.

> **Accuracy:** The system should allow access to a legitimate user, i.e., it should have a low false rejection rate (FRR). Conversely, the system should not allow access to illegitimate individuals, i.e., it should have a low false acceptance rate (FAR).

Efficiency: Efficiency refers to both storage and speed. The system should require a reasonably small amount of storage. Ideally, authentication should happen in no more than a matter of seconds.

The next two properties are of particular relevance in our setting, where some subset of systems have been compromised. A system is compromised when an attacker gains access to the enrollment data stored on that system.

Privacy: In the case of system compromise, the information that the attacker learns about the user's underlying biometric data should be minimal.

Security: In the case of system compromise, the enrollment data acquired by an attacker should minimally improve the attacker's ability to gain access to uncompromised systems.

Biometric enrollment can be characterized by a map from the space of processed biometric data to the space of stored data. When we design a biometric system we select such a map along with some protocol for authentication. We refer to the input to the map as a *feature vector*, which is the user's biometric data after preprocessing. We refer to the output of the map as a *template*.

Jointly Designed Systems

A large body of work discusses the above criteria in the context of single biometric systems; some of this work is referenced in Section 16.1.3. When designing multiple systems that will be accessed by the same user, many of the results developed for single systems still apply. For example, each system can be independently designed for accuracy and efficiency. From a privacy and security perspective, however, the systems must be jointly designed as enrollment data is correlated across systems. Leaking data from one system can therefore compromise the privacy and security of another. This is the motivation to consider the joint properties of biometric systems. Jointly designing multiple biometric systems to be optimally private and/or secure is still a largely open problem. Note that this problem is not the same as designing multi-factor biometric systems, i.e., systems that use a combination of various biometrics/techniques/algorithms to improve authentication performance.

16.1.2 Security and Privacy Leakage

The goal of this chapter is to characterize the tradeoff between security and privacy leakage when selecting the maps that characterize a set of biometric authentication systems, and to find choices of those maps that guarantee both high security and low privacy leakage. Note that throughout the chapter, we assume that the map characterizing each system is publicly available information. Only the feature vectors and templates are private.

In order to understand this problem, we develop some intuition regarding what we mean by security and privacy leakage. In this chapter, each of these terms is a measure of information. We define the *security* of a system as the amount of additional information (about a user's feature vector) that an intelligent attacker would need to learn in order to gain access to the system (which is proportional to the number of guesses that the attacker would need to make to gain access). The lower the security of a system, the easier it is for an attacker to gain access to the protected resource. Note that if an attacker discovers the templates stored in several systems, that may decrease the security of some uncompromised systems, as the discovered templates may provide partial information about the other systems to the attacker. In the worst case, the "non-leaked" (or uncompromised) systems could have zero security. When we discuss security, we are only interested in the security of non-leaked systems. To illustrate: When two systems use exactly the same subvector of the feature vector to form their templates, and one system is compromised, the other system has zero security since there is nothing left unknown about that second system to the attacker. However, if the two systems use disjoint and independent subvectors, and one system is compromised, the other system will remain fully secure.

We define the *privacy leakage* of a set of systems as the amount of information about a user's feature vector that would be known to an attacker in the case that the attacker learned the templates stored in each system. We will see that the greater the distinction between systems, the more aggregate information is stored in any fixed number of systems. When two systems use disjoint and independent subvectors of the feature vector to form their templates, compromise of both will result in twice as much privacy leakage as when the two systems use the same subvector.

The above examples of disjoint and identical subvectors illustrate the fact that security and privacy are at odds with each other. Using disjoint subvectors guarantees high security, but also risks high privacy leakage. Using identical subvectors guarantees low privacy leakage, but provides no security. There is a fundamental tension between the two measures. This tension forces the designer to consider both measures at once, as opposed to considering each measure independently.

These definitions of security and privacy leakage depend on a specific set of compromised systems. A designer of multiple biometric systems will not know in advance which (if any) systems will be compromised, but still needs to measure these quantities when choosing between various design options. Thus, we need a way to predict or guarantee the privacy leakage and security of any set of systems. We mention two natural ways to do this:

Worst case: First, we assume an intelligent attacker, who can compromise a maximum number, L, of systems. The intelligent attacker will always attack the set of systems that will yield the most useful information. In this case, the most appropriate measures are worst-case measures. We will define the *worst-case joint security* to be the minimum security of any non-compromised system with respect to any compromised set of L systems, and we define the *worst-case joint privacy leakage* to be the maximum privacy leakage of any L systems.

Expected: In scenarios wherein the attacker is not intelligent, or the maps characterizing the systems are not public, or systems are compromised at random, we can assert a probability distribution across all possible leakage scenarios. We can then use that distribution to find the *expected joint security* and the *expected joint privacy leakage*.

In both cases, the joint privacy/security measures are a function of L, the number of systems compromised. In this chapter we focus on worst-case measures.

Now that we have an idea of what we mean by privacy leakage and security, we develop a sense of how these measures are related. The high-level relationship between the two boils down to how much similarity there is within a set of systems. As alluded to earlier, similarity between systems is helpful from a privacy perspective, but undesirable from a security perspective. We now construct a simple example in order to illustrate the tradeoff involved.

EXAMPLE 16.1 *Let's say you own and operate the world's largest and most successful rubber duck factory. Your method of rubber duck manufacturing is top secret, and you don't trust your employees. In order to preserve the secret of your success, you design the factory so that five automatic aspects of the manufacturing process occur in five separated and secured areas that only you can access via fingerprint scans.*

Once your fingerprint is scanned, it is processed into a 5000-bit binary string. You don't want any of the scanners to store the entire string for authentication purposes, because you are concerned about your privacy. After all, if one of your more devious employees was able to take apart a scanner and find your entire fingerprint stored inside, they could steal your identity and impersonate you in a variety of situations. That would be dreadful. You therefore decide that each scanner will only store a part of your fingerprint. You decide that 1000 bits[1] will be enough to authenticate you.

Now you must determine which 1000 bits to store in each of the five scanners. At first you think it would be best to store different bits in each scanner. You like this idea because it provides you with a high level of security. Even in the unlikely case that an employee fools the first scanner by using a fingerprint matching yours in the first 1000 bits, the employee would not be able to fool any of the other scanners with this same false fingerprint. Stealing only one company secret wouldn't be enough to start a rival rubber duck factory.

After some thinking, however, you're ashamed to realize you were almost naïve enough to distribute your entire fingerprint of bits among the scanners. This would be almost as bad as storing all 5000 bits in one scanner. A perseverant employee, intent on impersonating you, could methodically take apart each scanner, revealing the information inside, and reconstruct your entire fingerprint! Your design would be a privacy nightmare.

You wish you could avoid this privacy issue by storing the same 1000 bits in each scanner, but it's all too clear that with that design, stealing one company secret would put an intelligent employee in the position to run off with all five. This is when you begin to understand the complex nature of the tradeoff between privacy leakage and security. It might be even more complex than manufacturing rubber ducks ...

---

[1] Let's assume that each bit contains the same amount of information about your fingerprint, and there is no correlation between bits. That is to say, the bit sequence is sampled from a uniform independent and identically distributed binary random process. Thus, any 1000 bits will contain the same amount of information.

16.1.3 Related Work

Biometric authentication is a topical and quickly growing area of research that draws on a variety of disciplines, including computer science, biology, and information theory [1]. The applications of this research range from healthcare to robotics to network security [2] and more.

In particular, privacy and security of single biometric systems is explored by many authors. In [3], Ratha et al. discuss the advantages and disadvantages (from a privacy and security standpoint) of using biometrics for authentication, highlighting where biometric systems' vulnerabilities lie, and emphasizing the importance of privacy in the design of biometric systems. The book by Campisi [4] is a compilation of a variety of research focusing on privacy and security of single biometric systems. In particular, the chapter in [4] by Ignatenko and Willems [5] has close ties with the work presented here. In their chapter, Ignatenko and Willems find a region of achievable privacy–security pairs, but only for a single system, and they assume a Gaussian biometric. In contrast, we look at multiple systems and assume an i.i.d. Bernoulli(0.5) feature vector. In [6], Wang et al. provide a framework for studying the privacy–security tradeoff in a biometric system, and perform analysis for single systems. They extend their framework to multiple systems, and pose the problem of determining an analogous tradeoff for multiple systems. This chapter tackles that problem.

When dealing with multiple systems as opposed to a single system, it's necessary to consider how systems affect each other, in addition to considering how each system behaves in isolation. The problem of balancing privacy and security in multiple biometric systems has already been approached from an information theoretic angle, by Lai et al. in [7]. Their measures of privacy and security are based on mutual information, and their constructions are information theoretic, based on random coding. Conversely, our work approaches the problem from a deterministic code design perspective, using privacy and security measures based on the number of bits discovered by an attacker in a worst-case scenario.

Our problem framing has interesting similarities to a few other areas of research. One thread of work that is highly relevant to a special case of our problem is the work of Koetter and Kschischang in [8], followed by the work of Khaleghi et al. on subspace codes in [9]. In the context of our problem, these works provide a full analysis of converse and achievability bounds on designs when we restrict ourselves to the setting wherein only one system is compromised, i.e., $L = 1$. As we will see, each codeword in [9] will correspond to one system in our setting. For $L = 1$, privacy leakage will turn out to be constant and the security of one system with respect to another is simply the injection distance between them.

Another similar area is distributed storage, as explored in [10, 11]. The goal in distributed storage is to store redundant information on a set of distributed servers to meet two competing objectives. On the one hand, it is desirable to have high storage efficiency, as measured by the total number of bits stored. On the other hand, the system should be efficiently reparable in the face of losing some subset of the servers, in that the bandwidth required to replace or regenerate the information on a lost server should be minimal. We have already seen that the subspace codes of [8], used in a particular type of network coding, correspond to a special case of our design problem. Hence, it is natural that distributed storage, which relies on network coding as well, should, in some way, be related to our problem. In both our problem and distributed storage, there are two competing measures which loosely relate to the amount of redundant information stored. In each case, one measure responds well to having a lot of redundancy (privacy leakage or bandwidth efficiency), and one responds well to having little redundancy (security or storage efficiency). Additionally, in both problems, the way the redundancy is structured is of crucial importance. However, one difference between the two problems is that we want to avoid reconstruction of the user's biometric, whereas distributed storage looks to facilitate source reconstruction. Another difference is that in the distributed storage setting, server losses are modeled as random erasures, whereas in our setting, systems are strategically targeted by an intelligent attacker.

A third related area is that of secret sharing, introduced by Shamir in [12]. In secure secret sharing, the goal is to store a secret in a distributed way among a set of participants, such that it is impossible for a subset of fewer than t participants to obtain any information about the value of the secret. This idea is loosely linked to the idea of privacy in our problem, in that the goal is to distribute biometric data among the systems so that if an attacker breaks into a small number of them, the attacker will have tremendous difficulty reconstructing the user's biometric. The requirements of the secret sharing problem are quite different, however. Secure secret sharing requires zero information to be knowable by any set of fewer than t participants. It relies on randomness to do this. In contrast, the information stored in our systems is non-random, and even compromising a single system may provide information to an attacker.

16.2 Problem Formulation

In this section we formalize the operation of the biometric systems and quantify the central measures of security and privacy that we study. We first introduce the secure sketch architecture. This is the type of biometric system we study. We then discuss the design requirements introduced in Section 16.1: accuracy and efficiency, as well as privacy and security. When we introduce the latter two, we formalize how the attack model (compromise of any L systems) affects these measures. Finally, we argue for studying the worst-case measures (as opposed to expected measures), which results in our joint measures of security and privacy. These joint measures are the central object of study in the rest of the chapter.

Secure Sketch Systems

In this chapter we focus on the *secure sketch* system architecture. We follow the definition of secure sketch systems detailed in [6].

When a user enrolls in a secure sketch system, the user's biometric data is preprocessed into a binary feature vector, A, of length n. We model A as an independent and identically distributed (i.i.d.) Bernoulli(0.5) random process. The preprocessing is an important step. We do not discuss the preprocessing in this chapter other than to state our assumption that it produces a uniformly distributed feature vector. This is necessary for the validity of the authentication procedure. (We note, however, that the authentication procedure can be modified to accommodate other statistics.) For a discussion of feature extraction algorithms that are designed to yield such statistics (approximately) for fingerprint biometrics, see [13].

Next, A is mapped by the $m \times n$ *parity check* matrix H to the length-m binary vector, $\mathbf{s} = HA$, referred to as the *syndrome* of A. This $\mathbf{s}$ is the template that is stored by the system. Operations are in the finite field $\mathbb{F}_2$.

At authentication, new biometric data is extracted and preprocessed to produce a binary vector D. This vector is mapped to syndrome $\mathbf{s}' = HD$. The system then estimates the biometric noise vector $A \oplus D$ as

$$\hat{W} = \underset{W:HW=\mathbf{s}'\oplus\mathbf{s}}{\arg\min} \; d(W), \qquad (16.1)$$

where $d(W)$ is the Hamming weight of W. If $d(\hat{W})$ is below some threshold θ, i.e., $d(\hat{W}) < \theta$, then authentication succeeds. Otherwise, the user is rejected. The underlying model is that $d(A \oplus D)$ will be small if D comes from a legitimate user and will be large (roughly $n/2$) if it comes from an illegitimate user. Further, if the ratio m/n is set correctly, $\hat{W}$ will equal $A \oplus D$ with high probability. The parameter θ should be set proportional to the expected noise in the reading of a legitimate user's biometric. Both enrollment and authentication are depicted in Fig. 16.1.

Note that under the assumptions we've made, secure sketch systems are based on linear error correcting codes in that biometric data can be thought of as codewords. Authentication is then equivalent to syndrome decoding over the binary symmetric channel.

The linear map H, used to send biometric data (the feature vector) to syndromes (the templates), is the only thing we choose when designing a system.

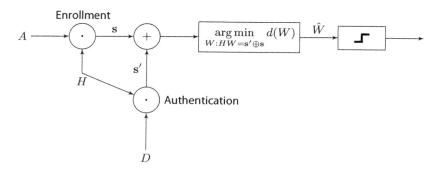

Figure 16.1 Block diagram of a keyless secure sketch system, modeled on [6].

Accuracy and Efficiency

Since the feature vector A is a length-n binary vector, for some $n \in \mathbb{N}$, the linear maps we choose must have domain $\mathbb{F}_2^n$. The co-domain of the linear maps is $\mathbb{F}_2^m$, where m is the length of the desired syndrome, and $m < n$. The choice of m, and the choice of the threshold θ, depends on the noise model that relates A to D for a legitimate user. The matrix H defines an error correcting code of rate $1 - \frac{m}{n}$, and $A \oplus D$ can be thought of as channel noise. If the ratio $\frac{m}{n}$ is selected so that the code can correct the channel noise and θ is suitably selected, then one can attain a low FRR and FAR. The specifics of this analysis are well known and are, e.g., discussed in detail in [6].

In this chapter we do not concentrate on the selection of $\frac{m}{n}$, we simply assume it is set to provide low FRR and low FAR. Instead, we focus on how to select a set of u binary parity check matrices, each of dimension $m \times n$, to optimize security and privacy leakage. The ith such matrix is denoted H_i. The matrix H_i characterizes the linear map that takes feature vectors to syndromes for the ith system. Henceforth, when we refer to a *design*, it can be thought of as a tuple of u matrices, $\mathcal{H} = (H_1, \ldots, H_u)$. The ith system needs to store H_i, which is public information, and the user's syndrome, $\mathbf{s}_i \in \mathbb{F}_2^m$, which is kept secret.

Privacy and Security Measures

Now that we have formalized our definition of a design, we formalize our measures of privacy and security. These definitions were first presented in [6], wherein additional justification for the measures as well as some analysis is provided. As we have already discussed, our measures of privacy and security will be contingent on the set of systems that are compromised. Say that the set of compromised systems is $\mathcal{S} \subseteq \{1, \ldots, u\}$, where $|\mathcal{S}| = L$. We denote[2] the *privacy leakage* that has occurred by $r(\mathcal{S})$. To quantify $r(\mathcal{S})$ we calculate the number of bits of information about a user's feature vector that can be gleaned from the syndromes learned. Since $\mathbf{s}_i = H_i A$, each coordinate in the syndrome provides the attacker one linear equation constraining the feature vector A. When L systems are compromised, the attacker acquires a system of mL (possibly dependent) binary linear equations in n unknowns. The number of bits an attacker can learn is exactly the number of linearly independent equations the attacker has learned. If we construct an $mL \times n$ matrix H by vertically concatenating all L matrices H_i such that $i \in \mathcal{S}$, we can write

$$r(\mathcal{S}) = \text{rank}(H).$$

Note that $r(\mathcal{S}) \in \mathbb{Z}_{\geq 0}$.

Similarly, when L systems have been compromised we can measure the (residual) *security* of the jth system, denoted $t_j(\mathcal{S})$, by calculating how many more bits of information an attacker would need in order to gain access to the jth system. This corresponds to the amount of information needed to learn the syndrome stored in the

---

[2] We note that $r(\mathcal{S})$ is also a function of $\mathcal{H}$; $\mathcal{S}$ only contains the indices of the compromised systems and we need $\mathcal{H}$ to calculate privacy leakage. However, since we discuss only one $\mathcal{H}$ at a time, we suppress this dependence for notational compactness.

uncompromised system. If the attacker were to learn the new syndrome in addition to the L syndromes he already knows, he would know $r(\mathcal{S} \cup \{j\})$ bits. Currently, he knows $r(\mathcal{S})$ bits. Therefore,

$$t_j(\mathcal{S}) = r(\mathcal{S} \cup \{j\}) - r(\mathcal{S}).$$

Note that if $j \in \mathcal{S}$, security is zero. Note also that $t_j(\mathcal{S}) \in \mathbb{Z}_{\geq 0}$.

One way to think about this measure of security is as follows. Knowing the syndromes in $\mathcal{S}$, the attacker can construct a set of possible syndromes that may be stored in the jth system. If the size of this set is denoted M, then the attacker needs to learn close to $\log_2(M)$ additional bits to specify which syndrome among the M is the correct one. It is easy to verify that $\log_2(M) = r(\mathcal{S} \cup \{j\}) - r(\mathcal{S})$.

We will later normalize the privacy leakage and security measures by block length (feature vector length) n. We delay this normalization in order to preserve the intuitive correspondence between our measures and learned bits.

Choice of Worst-Case Measures

As our measures of privacy leakage and security depend on the number L of systems that get compromised, we need to decide how to control this variable for our analysis. Our options are to fix L, bound L, or (as mentioned previously) take an expectation over L. We choose to fix L. This is equivalent to setting an upper bound on the number of compromised systems. An alternative is to assert a probabilistic model on L and then to calculate privacy leakage as an expected value, and to do the same for security. We don't explore that possibility in this chapter, but such an analysis would be a direct extension of our results.

Joint Worst-Case Measures of Privacy and Security

Finally, we now introduce joint measures that quantify privacy leakage and residual security in the event of an attack. As mentioned, since an intelligent attacker would target the most valuable information, we use worst-case measures of privacy leakage and security. First we define *worst-case security* as

$$t(\mathcal{S}) = \min_{j \notin \mathcal{S}} t_j(\mathcal{S}).$$

This measures the minimum number of bits an attacker would need in order to gain access to some other system.

We next respectively define *worst-case joint privacy leakage* and *worst-case joint security* of a design as follows:

$$R_L = \max_{\substack{\mathcal{S} \subseteq \{1,\ldots,u\} \\ |\mathcal{S}|=L}} r(\mathcal{S}),$$

$$T_L = \min_{\substack{\mathcal{S} \subseteq \{1,\ldots,u\} \\ |\mathcal{S}|=L}} t(\mathcal{S}).$$

When we write the (joint) privacy leakage or (joint) security of a design, it is the above measures to which we are referring. We refer to the tuple (R_L, T_L) as a *privacy–security pair*. Good performance is characterized by high security and low privacy leakage. Since we are working in two dimensions, performance of designs cannot necessarily be compared unless either privacy leakage or security is fixed, which gives a partial order on design performance.

The worst-case measures we study are intended as a guarantee. If a design corresponds to a particular privacy–security pair, then no possible scenario can cause a higher level of privacy leakage or a lower level of security. It is, however, not necessarily the case that there exists a leakage scenario which results in both limits being met. Accordingly, note that the worst-case leaked set S maximizing $r(S)$ can be different from the one minimizing $t(S)$.

16.3 Design Space

In this section we seek to understand the space of possible designs. Given the secure sketch system and the joint privacy/security measure defined in the last section, there remain a number of parameters that characterize the design space. These are n, the feature vector length; m, the syndrome vector length; u, the number of systems to be designed; and L, the maximum number of compromised systems.

We assume that the total number of systems u and the number of compromised systems L are set by the application. Earlier we noted that the ratio $\frac{m}{n}$ must be set to manage FRR and FAR. Initially, we consider m and n to be fixed integers. Later, in Section 16.4, we study a relaxed design problem, the results of which can be approximated arbitrarily well as n is allowed to grow large, with m growing proportionally, to maintain the proper ratio.

We call an ordered pair (r,t) an *achievable privacy–security pair* for some fixed problem parameters n, m, u, L if there exists a design with $R_L = r$ and $T_L = t$. Our goal, given parameters n, m, u, L, is to characterize the region, contained in $\mathbb{Z}^2$, of achievable privacy–security pairs. In addition, we want to find designs that maximize T_L for a fixed R_L, and designs that minimize R_L for a fixed T_L.

To understand better the privacy and security of various designs we next provide a number of examples. Recall that a design is a tuple, $\mathcal{H}$, of $m \times n$ matrices, $(H_1, \ldots, H_u)$, each with entries in $\mathbb{F}_2$. We require each matrix to be full rank, i.e., to have rank m, since $m < n$.

EXAMPLE 16.2 *Let feature vector length $n = 4$, syndrome length $m = 2$, number of systems $u = 3$, and maximum number of compromised systems $L = 1$. Since $L = 1$, the cardinality $|S| = 1$ and so $R_L = m = 2$. Accordingly, for this example the only interesting design problem is to maximize security.*

> **Zero bits of security:** If we choose all matrices to be the same, then regardless of which system is compromised, all other systems match the compromised system, so $t_j(S) = 0$ for any S such that $|S| = 1$ and for any $j \notin S$. In this case $T_L = 0$.
> **One bit of security:** To get $T_L = 1$, we need to ensure that $t_j(S) \geq 1$ for all choices of S and $j \notin S$. We can do this by making sure that each pair, H_i, H_j, contains at least

three linearly independent rows. An example of a design that achieves $T_L = 1$ is as follows:

$$H_1 = \begin{bmatrix} 1 & 0 & 0 & 0 \\ 0 & 1 & 0 & 0 \end{bmatrix}, \quad H_2 = \begin{bmatrix} 0 & 0 & 1 & 0 \\ 0 & 0 & 0 & 1 \end{bmatrix}, \quad H_3 = \begin{bmatrix} 1 & 0 & 0 & 0 \\ 0 & 0 & 0 & 1 \end{bmatrix}.$$

One can verify that $t_1(\{2\}) = 2$ and $t_3(\{1\}) = 1$.

Two bits of security: To get $T_L = 2$, we need to ensure that $t_j(\mathcal{S}) \geq 2$ for all allowable $j, \mathcal{S}$. Note that we cannot have $t_j(\mathcal{S}) > 2$, since that would imply $r(\mathcal{S} \cup \{j\}) = t_j(\mathcal{S}) + r(\mathcal{S}) = t_j(\mathcal{S}) + 2 > 4$, but $r(\mathcal{S} \cup \{j\})$ is the rank of a 4×4 matrix. Thus, we need to choose a design that always has $t_j(\mathcal{S}) = 2$ for $j \notin \mathcal{S}$. Any two H_i matrices taken together must have their rows constitute a basis for $\mathbb{F}_2^4$. One design that achieves $T_L = 2$ is:

$$H_1 = \begin{bmatrix} 1 & 0 & 0 & 0 \\ 0 & 1 & 0 & 0 \end{bmatrix}, \quad H_2 = \begin{bmatrix} 0 & 0 & 1 & 0 \\ 0 & 0 & 0 & 1 \end{bmatrix}, \quad H_3 = \begin{bmatrix} 0 & 1 & 0 & 1 \\ 1 & 0 & 1 & 0 \end{bmatrix}.$$

Adding systems: Although we have shown that $T_L > 2$ is not possible with three systems, we have not examined whether three is the most systems we can have with $T_L = 2$. It turns out that with $L = 1$, $n = 4$, and $m = 2$, we can find a design achieving $T_L = 2$ for all $u \leq 5$, and we cannot find a design for $u > 5$. Consider adding

$$H_4 = \begin{bmatrix} 1 & 1 & 0 & 1 \\ 1 & 0 & 1 & 1 \end{bmatrix}, \quad H_5 = \begin{bmatrix} 0 & 1 & 1 & 1 \\ 1 & 1 & 1 & 0 \end{bmatrix}.$$

It is not difficult to verify that we still have $T_L = 2$. However, if we were to have six or more systems, the rowspaces of some pair of H_i would have to intersect non-trivially, as each rowspace contains three non-zero vectors, but there are only $2^4 - 1 = 15 = 5 \cdot 3$ non-zero vectors in $\mathbb{F}_2^4$. Those vectors are already contained in the union of the rowspaces of $H_1, \ldots, H_5$. The intersecting pair of rowspaces necessarily introduced by a sixth system would lead to some $t_j(\mathcal{S}) < 2$ and thus $T_L < 2$.

Example 16.2 hints that there are fundamental bounds on R_L and T_L that characterize the region of achievable privacy–security pairs. We have already seen via Example 16.2 that we must always have

$$0 \leq T_L \leq m \leq R_L \leq n.$$

The proof of this fact follows from the definitions of $r(\mathcal{S})$ and $t_j(\mathcal{S})$. Note that we also have $R_L \leq mL$. If $L = 1$, this means that we must have $R_L = m$, as in the above example. In the case of $L = 1$, the region of achievable privacy–security pairs is therefore one-dimensional, making $L = 1$ a fairly simple case to analyze, but non-representative of the general problem. As mentioned in Section 16.1.3, bounds on T_L for the $L = 1$ case can be determined from the work of Koetter and Kschischang in [8] and Khaleghi et al. in [9].

16.3.1 Geometric Intuition

It is difficult to gain intuition for how different designs behave under privacy and security measures by looking at our current matrix representations. It will be helpful rather to consider the nullspaces[3] of these matrices as opposed to the matrices themselves. This yields a more geometric problem. If we let $V_i = \text{null}(H_i)$, and refer to a design as $\mathcal{V} = (V_1, \ldots, V_u)$, then for $\mathcal{S} \subseteq \{1, \ldots, u\}$, we can rewrite privacy leakage as

$$r(\mathcal{S}) = n - \dim\left(\bigcap_{i \in \mathcal{S}} V_i\right),$$

and security as

$$t_j(\mathcal{S}) = \dim\left(\bigcap_{i \in \mathcal{S}} V_i\right) - \dim\left(V_j \cap \bigcap_{i \in \mathcal{S}} V_i\right).$$

These definitions agree with our earlier matrix-oriented definitions of privacy leakage and security. The definitions of R_L and T_L remain the same as before.

Note that $\dim(V_i) = n - m =: k$ for all i. Now that we can think of our systems as k-dimensional subspaces of a finite-dimensional vector space over a finite field, there are many mathematical results that provide insight into the problem. Subspaces are widely studied mathematical objects. We would like to form analogies between our problem and more abstract characterizations of subspaces. In order to do this, we introduce two graphs that depict the relationship between the subspaces of a vector space over a given finite field. Even if our problems (framed in terms of subspaces) are unsolved or have high complexity, establishing the connections will allow us to understand much more about our problem.

The Grassmann Graph

Before we make these connections, we need a few definitions. The first object that comes up in the study of equidimensional subspaces is the *Grassmann graph*. Denoted $\mathcal{G}_q(n, k)$, the Grassmann graph has a vertex set that consists of all k-dimensional subspaces of $\mathbb{F}_q^n$. An edge exists between a pair of subspaces U and V if $\dim(U \cap V) = k - 1$.

To develop some intuition for this graph, think about traveling along an edge between two vertices, U and V. This journey begins with a basis for U, removes one basis vector to obtain a basis for $U \cap V$, and adds a new basis vector, arriving at a basis for V. Accordingly, the shortest path from U to V is the least number of basis vector swaps one would need in order to take a basis for U and transform it into a basis for V. This distance is known as the *injection distance*, and can be shown to be a metric (see, e.g., [14]).

Note that while we have framed our problem in terms of binary vector spaces, the following discussion applies to vector spaces over any finite field $\mathbb{F}_q$. For simplicity we stick with $q = 2$.

---
[3] We could just as easily have chosen the rowspaces of the H_i as our subspace representation. The two choices can be thought of as duals of each other.

We can view our problem as the problem of selecting u nodes in $\mathcal{G}_q(n,k)$ where $k = n - m$, in order to have low R_L and high T_L. Ideally, we would want to compute our privacy leakage and security measures directly from the graph itself. Doing this would translate our entire problem to a problem about the (well-studied) Grassmann graph. For $L = 1$, we can accomplish this translation. We have that

$$R_1 = n - k = m,$$
$$T_1 = \min_{U,V} d(U,V),$$

where $U \neq V$ and $d(U,V)$ is the shortest distance in the graph between nodes U and V. To see why this is the expression for T_1, note that for any U, V in $\mathcal{G}_q(n,k)$, $d(U,V) + \dim(U \cap V) = k$ [9]. The intuition behind this equality comes from thinking of performing basis vector swaps to get from U to V. U has a total of k basis vectors. There is no need to swap out $\dim(U \cap V)$ of them, because those vectors can also form part of a basis for V. The other $k - \dim(U \cap V)$ vectors do need to be swapped. Therefore, $k - \dim(U \cap V)$ is the expression for $d(U,V)$.

In the case of $L = 1$, we want to select u nodes in the graph that maximize the minimum pairwise distance. This problem is shown in [15] to be NP-complete for general graphs, and is referred to as the discrete p-dispersion problem. Given that the Grassmann graph has plenty of structure, there may be an efficient algorithm to find such a subgraph. Some of these are explored in [8,9] in the context of code construction for the subspace coding problem.

In the case that $L > 1$, we need to look for ways to uncover more subtle structure. To compute T_2, we need to know the dimension of the intersection of three subspaces. The Grassmann graph only stores information about pairwise intersections (or unions) of subspaces. The dimension of the intersection (or union) of three subspaces cannot be directly computed from the Grassmann graph. For this purpose, we need to turn to the Hasse diagram.

The Hasse Diagram

The *subspace Hasse diagram*, denoted $\mathcal{H}_q(n)$, and illustrated in Fig. 16.2 for $q = 2$ and $n = 3$, is defined as follows. First, $\mathcal{H}_q(n)$ is a graph, the nodes of which represent all subspaces of $\mathbb{F}_q^n$. An edge exists between two nodes U and V if $U \subseteq V$ and $\dim(U) = \dim(V) - 1$. This graph is a lattice of $n+1$ levels, where the level of each node, starting from zero, is equal to the dimension of the subspace represented by that node.

As was the case for the Grassmann graph, traversing the edges in the Hasse diagram can be understood by manipulating bases. Traveling from any node, U, to another (not necessarily neighboring) node, V, can be thought of as follows. Start with a basis for U. Traveling up (to a higher level) on an edge corresponds to adding one basis vector to the basis. Traveling down an edge corresponds to removing one basis vector. Therefore, if P is a path from U to V, $N_{\text{up}}(P)$ is the number of edges traveled up on P, and $N_{\text{down}}(P)$ is the number of edges traveled down in P, then $\dim(V) = \dim(U) + N_{\text{up}}(P) - N_{\text{down}}(P)$.

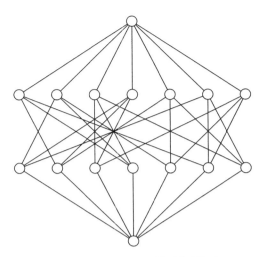

Figure 16.2 The subspace Hasse diagram $\mathcal{H}_2(3)$. The bottom node is the trivial subspace, $\{0\}$. The nodes on the first (second) level are all the one- (two-)dimensional subspaces of $\mathbb{F}_2^3$. The topmost node is $\mathbb{F}_2^3$ itself. For example, if the leftmost node on the second level is the subspace spanned by $(1,0,0)$ and $(0,1,0)$, then its children (the second, fourth, and sixth nodes on the first level) are the subspaces spanned by $(0,1,0)$, $(1,0,0)$, and $(1,1,0)$.

Recall that the sum of two subspaces U and V of some vector space is the unique minimal-dimension subspace that contains both U and V. That is, a subspace $W = U + V$ if and only if

1. $U \subseteq W$ and $V \subseteq W$;
2. if W' also satisfies condition (1), then $\dim(W) \leq \dim(W')$.

Similarly, the intersection of two subspaces U and V is the unique maximal-dimension subspace that is contained in both U and V.

With this observation, we can find sums and intersections of subspaces using the Hasse diagram. To check if $U \subseteq W$, we simply need to check if W is an ancestor of U, i.e., if there is a path from U to W that only increases in level. Thus, the sum of two or more subspaces corresponds to the lowest common ancestor of those subspaces in the Hasse diagram. Similarly, the intersection of two or more subspaces corresponds to their highest-level common descendant. Knowing this, it is simple to find the sum or intersection of arbitrary subspaces in the Hasse diagram. Then, to find the dimension of this sum or intersection, one need only check the level of the resultant node.

The reason the Grassmann graph does not contain sufficient information to encode our problem structure is that it does not contain information about intersections or unions of more than two subspaces. However, such information is embedded in the Hasse diagram.

The following two results, first published in [16], relate our problem to the subspace Hasse diagram. The results follow from our observations about finding sums and intersections of subspaces, and from the definitions of R_L and T_L.

THEOREM 16.1 *Given a design, $\mathcal{V} = (V_1, \ldots, V_u)$ and $r \in \mathbb{Z}$, the following are equivalent:*

1. $R_L \leq r$.
2. *Every L nodes from $\mathcal{V}$ have a common descendant at level $n - r$.*

THEOREM 16.2 *Given a design, $\mathcal{V} = (V_1, \ldots, V_u)$, and $t \in \mathbb{Z}$, the following are equivalent:*

1. $T_L \geq t$.
2. *If the highest common descendant of a set of L nodes from $\mathcal{V}$ is on level ℓ, then those L nodes have no common descendant with any other nodes in $\mathcal{V}$ above level $\ell - t$.*

These two theorems allow us to identify good designs without performing algebraic calculations. Instead, we can trace paths on the Hasse diagram to form designs. Given properties of those paths, we can make statements about privacy leakage and security without computing ranks of matrices or dimensions of intersections.

Using the relationship between our problem and the Grassmann graph and Hasse diagram, we may be able to translate statements from existing graph theory and linear algebraic research into the language of privacy and security. For reference, papers focusing on the Grassmann graph and its properties include [17–19]. Additionally, there are many existing results regarding finding subgraphs of general graphs with certain distance properties [20].

16.3.2 Scaling Complexity

We will now discuss the complexity of the design problem, in terms of both the size of the design space and the time to compute privacy and security measures. We will look for ways to simplify the design space and/or the measures, even if it requires some loss of generality. If we can solve our problem for a special case, it may give us more insight into how to tackle the general problem.

Our problem comes down to selecting fixed-size subgraphs of the Hasse diagram with certain properties. Doing this by generating each subgraph and comparing them would require generating $(N_g)^u$ different designs (if a design can contain multiple copies of the same subspace), where

$$N_g = \begin{bmatrix} n \\ m \end{bmatrix}_q = \frac{[n]_q!}{[m]_q![n-m]_q!}$$

is the *Gaussian binomial coefficient* counting the number of m-dimensional subspaces of $\mathbb{F}_q^n$, and $[n]_q! = \prod_{i=1}^{n}(q^i - 1)$ is the *q-factorial* of n. Without allowing repetition, we would still need to search through $\binom{N_g}{u}$ designs. To get an idea of how fast N_g grows, we can fix $n = 2m$, and see that in that case,

$$N_g = \prod_{i=1}^{m} \frac{q^{m+i} - 1}{q^i - 1}.$$

For large m, the -1 has little effect, so most of the m factors are approximately q^m. As $m \to \infty$, N_g grows like q^{m^2}. Practical values of n in the hundreds or thousands make a brute force search for designs impracticable.

On top of the fact that there are too many designs to search through, it is also computationally non-trivial to calculate the worst-case privacy leakage and security of arbitrary designs. For each design, using a brute force search, we would need to find the ranks of fairly large matrices. To be exact, it would require computing the rank of $\binom{u}{L}$ matrices, each of dimension $mL \times n$, and $\binom{u}{L+1}$ matrices, each of dimension $m(L+1) \times n$. The time complexity of finding the rank of a matrix over a finite field is on the order of the product of the matrix dimensions [21].

16.3.3 Equivalent Designs

In this section we restrict our design space to a smaller space of designs that is easier to search and for which it is simpler to compute privacy and security. One way to restrict our design space is by dealing with equivalence classes of designs rather than with designs themselves. Then, when searching for optimal designs, we can merely consider a single representative from each equivalence class. We want an equivalence relation, $\equiv$, on designs to be such that if $\mathcal{V} \equiv \mathcal{V}'$ then $(R_L, T_L) = (R'_L, T'_L)$. That way, as long as we consider one representative from each equivalence class, we will find all possible (R_L, T_L) pairs. A restriction of this sort would not limit the performance of designs.

The most naïve and general type of equivalence relation satisfying our requirement is *performance equivalence*, where $\mathcal{V} \equiv_P \mathcal{V}'$ if and only if $(R_L, T_L) = (R'_L, T'_L)$. Performance equivalence induces the largest possible equivalence classes (alternatively, the least number of equivalence classes) we can use without missing some (R_L, T_L) pairs. However, performance equivalence is practically useless in that privacy and security still need to be computed for every design in order to determine equivalence classes in the first place.

The following insight allows us to construct a second type of design equivalence. Viewing a design as a collection of subspaces, we know that our measures depend on relationships between the constituent subspaces, but not on the absolute orientation of those subspaces. Consider taking a design and applying an invertible transformation to the basis vectors in the constituent subspaces. The new design will have the same measures as the original. If two designs are related by an invertible transformation of this sort, we will refer to them as *transformationally equivalent*, which we denote $\equiv_T$. If we are able to invertibly transform a design that we have already analyzed, each design obtained in this way is redundant and can therefore be discarded. Unfortunately, it is immensely computationally intensive to find or store all invertible transformations, as there are $\prod_{k=0}^{n-1}(q^n - q^k)$ such transformations. On the bright side, this means that there are very many redundant designs, which indicates that there is likely deeper structure to this problem that we have yet to uncover.

A third equivalence relation can be formulated in terms of the graphical structure of the relevant Hasse diagram. We define the *representative subgraph* of a design to be the subgraph of the Hasse diagram containing all nodes of the Hasse diagram corresponding to the subspaces in the design, all ancestors and descendants of these nodes, and all edges between the mentioned nodes. Two designs with the same parameters are then said to be *graphically equivalent* ($\equiv_G$) if their representative subgraphs are isomorphic.[4] Essentially, two designs are equivalent if one is a structure-preserving relabeling of the other. Unfortunately, even determining the existence of a subgraph that is isomorphic to a specified graph is an NP-hard problem. Finding an entire class of isomorphic subgraphs thus appears intractable.

None of the introduced types of equivalence appear to be practical in simplifying our search for the best designs. However, they do provide insight into the underlying structure of the problem, hinting at symmetry in potential solutions.

We next introduce another way to restrict our choice of designs that will be of great practical use. This restriction is not based on equivalence classes, and hence will not produce generally optimal designs. However, it will greatly simplify the computation of privacy and security. This simplification will come from reducing our rank-dependent measures to simple counting measures. With this simplification, we will be able to solve our problem on this restricted set!

16.4 The Fixed-Basis Case

We now restrict our problem to choosing designs only comprised of a particular type of subspace: those that are the nullspace of a matrix, H_i, the rows of which are selected from a fixed basis for $\mathbb{F}_q^n$, denoted $(\mathbf{b}_1, \ldots, \mathbf{b}_n)$. That is to say, the H_i matrices can no longer be arbitrary full-rank matrices, but their constituent rows must be selected from $\{\mathbf{b}_1, \ldots, \mathbf{b}_n\}$. For any set of basis vectors $\mathbf{b}_i$ there are only $\binom{n}{m}$ such designs. We will refer to our new class of designs as *fixed-basis designs*.

Note that the particular choice of basis does not affect our analysis. This follows from transformational equivalence. However, the freedom to choose any basis is appealing from a practical perspective. It allows us to choose a basis related to a tractable code, which addresses efficiency issues concerning implementation, as well as FRR and FAR requirements per [6]. For example, each $\mathbf{b}_i$ can be chosen to be low density, meaning the codes selected will be low-density parity check (LDPC) codes. This implies that the authentication problem (16.1) can be solved using belief-propagation-type algorithms.

As fixed-basis designs are a strict subset of all possible designs, better designs will generally exist. However, fixed-basis designs are simpler to represent and to analyze than designs composed of arbitrary subspaces.

16.4.1 Impact of the Restriction

Each system is now represented by a full-rank matrix H_i with distinct vectors $\mathbf{b}_\ell$ for rows. Thanks to this restriction, instead of this representation, we can represent a system

---

[4] For this isomorphism, we take the edges on the Hasse diagram to be directed from top to bottom, i.e., from higher levels to lower ones.

as an indicator vector, $\mathbf{h}_i$, where $\mathbf{h}_i(\ell) = 1$ if and only if $\mathbf{b}_\ell$ is a row in H_i. Thus, an entire design, $(H_1, \ldots, H_u)$, can be represented as a collection of u binary vectors of length n, which takes at most $u \times n$ bits. In contrast, in the general case we needed $u \times n \times m$ bits (or more if the field is not $\mathbb{F}_2$). Note that to represent designs even more efficiently, we can use the fact that each $\mathbf{h}_i$ must have Hamming weight $\text{wt}(\mathbf{h}_i) = m$ (m non-zero entries). Thus, there are only $\binom{n}{m} \ll 2^n$ possible values[5] for each $\mathbf{h}_i$. This information could be used to store designs with a total of $u \times \lceil \log_2(\binom{n}{m}) \rceil$ bits.

Not only does representation of designs become easier in this restricted space but, more importantly, privacy leakage and security become simple to compute. Our basic privacy leakage and security measures become, respectively,

$$r(\mathcal{S}) = \text{wt}\left(\bigvee_{i \in \mathcal{S}} \mathbf{h}_i\right), \quad t_j(\mathcal{S}) = \text{wt}\left(\mathbf{h}_j \wedge \neg \left(\bigvee_{i \in \mathcal{S}} \mathbf{h}_i\right)\right),$$

where $\neg$ is logical negation, $\wedge$ is the logical AND operator, and $\bigvee$ is the logical OR operator. For fixed-basis designs, the privacy and security measures become counting measures. For privacy leakage we count how many $\mathbf{b}_\ell$ are rows in at least one H_i in the set of leaked systems, $\mathcal{S}$. For security we count how many $\mathbf{b}_\ell$ are rows in H_j but not in any H_i in $\mathcal{S}$. We no longer need to find the ranks of arbitrary large matrices.

Before we work more closely with fixed-basis designs, we again note that working in this design space does generally restrict the performance of designs. We illustrate this using Example 16.2. Note that the design used to obtain one bit of security is a fixed-basis design, using the standard basis, $(\mathbf{e}_1, \ldots, \mathbf{e}_n)$. We cannot achieve the security level of two bits with a fixed-basis design; indeed, the design used to achieve two bits of security is a more general design, as the matrix H_3 contains rows that are not elements of the standard basis. Furthermore, it is not possible to achieve $T_L = 2$ using only $\mathbf{e}_i$ for rows. The proof is that there are only four choices for $\mathbf{e}_i$. Each matrix must have two distinct rows in order to be full rank. If $T_L = 2$, H_2 must have two rows that are not in H_1. Therefore, together, H_1 and H_2 will contain all $\mathbf{e}_i$ as rows. To construct H_3, an $\mathbf{e}_i$ used previously must be reused. If we reuse two rows of either matrix we get $T_L = 0$. If, instead, we take one from each matrix we get $T_L = 1$. However, we assumed $T_L = 2$, so we arrive at a contradiction.

Many more examples can be found demonstrating that fixed-basis designs cannot always attain the best (R_L, T_L) pairs, and in fact rarely can. However, fixed-basis designs are simpler to find, and can be practical to implement (e.g., if each $\mathbf{b}_\ell$ is low density). We now attempt to find the best designs within the class of fixed-basis designs, and to characterize the region of achievable privacy–security pairs for the fixed-basis case.

16.4.2 Optimization of Fixed-Basis Designs

Both finding the best designs and characterizing the tradeoff between privacy leakage and security can be tackled by studying the following simple-looking optimization

---
[5] Unless $m = \frac{n}{2}$, in which case the two quantities grow similarly. We still have $\binom{n}{m} < 2^n$.

problem. This problem maximizes security subject to an upper bound on privacy leakage.

PROBLEM 16.1 (Naive optimization)   Pick $\mathbf{h}_1, \ldots, \mathbf{h}_u \in \mathbb{F}_2^n$ to solve

$$\text{maximize} \quad T_L$$
$$\text{s.t.} \quad R_L \leq R',$$
$$\text{wt}(\mathbf{h}_i) = m, \quad i \in \{1, \ldots, u\}.$$

The problem is written in a form that is simple to understand, but is not a standard problem that we can immediately find a method to solve, either analytically or computationally. The main issue is that this problem is discrete; it optimizes over binary vectors. One standard way to ease such intractability is to relax the integer constraints, and to optimize over $[0,1]^n$. In the following section we build up to such a relaxation. Although the results of the relaxation can only provide loose bounds on the fixed-basis design problem for small n, for sufficiently large n the results will be quite close to the results of Problem 16.1.

Relaxation of Integer Constraints

To form a tractable relaxation we make two key observations.

1. The privacy leakage and security measures depend only on the amount of overlap between sets of systems.
2. All $\mathbf{b}_\ell$ are interchangeable from a privacy and security perspective.

The second point means that, given a design $\mathcal{D}$, if one relabels the vectors $\mathbf{b}_\ell$ to induce a new design $\mathcal{D}'$, then $\mathcal{D}$ and $\mathcal{D}'$ will be equivalent in that they will have the same R_L and T_L for all values of L. A proof is omitted, as this is quite easy to verify. In fact, this statement is a particular case of the concept that transformationally equivalent designs have the same R_L and T_L.

With these two observations in mind, we propose a representation for the $\mathbf{h}_i$ vectors. Take u copies of the interval $\mathbf{I} = [0, 1)$. The ith copy, $\mathbf{I}_i$, represents $\mathbf{h}_i$. For each value of i, take the corresponding copy of $[0, 1)$, and divide it into n equal subintervals, some of which are "colored in" and some of which are left blank. If $\mathbf{h}_i(\ell) = 1$, then the ℓth subinterval, $[\frac{\ell-1}{n}, \frac{\ell}{n})$, is colored in; otherwise, it is not colored in. Each subinterval has measure $\frac{1}{n}$. Each system is required to have m colored subintervals or, equivalently, the measure of the union of the colored intervals needs to be $\frac{m}{n}$.

Note that each subinterval has u copies, one in each interval $\mathbf{I}_i$. We now use the coloring of a given subinterval across all copies to determine a subset $\mathcal{P}$ of the u systems, as follows. Fix a subinterval. If the copy of that subinterval is colored in $\mathbf{I}_i$, then the ith system is an element of $\mathcal{P}$. We have now associated each subinterval to a subset of $\{1, \ldots, u\}$.

To perform the relaxation, we will fix $\frac{m}{n} = \mu$ and let $n \to \infty$. If we were to use the above representation of a design, we would need to keep track of $u \times n$ subintervals, which becomes impossible as $n \to \infty$. Note that permuting the subintervals (using the same permutation for each system in the design) is equivalent to relabeling the $\mathbf{b}_\ell$ vectors, which is permissible by observation 2 above.

Our next step is to order the subintervals to make it possible to keep track of them with a finite amount of memory. We first fix a total ordering of the subsets of $\{1,\ldots,u\}$. Order the subsets by increasing cardinality. When two cardinalities are equal, assign an arbitrary order. By the above correspondence between subintervals and subsets, this total ordering on subsets induces an ordering on the subintervals of $\mathbf{I}$. Note, however, that several subintervals will correspond to the same subset of $\{1,\ldots,u\}$, and therefore some subintervals are equal in the ordering. Rearrange the subintervals with respect to this ordering and merge any equal subintervals.

Now, there are only at most 2^u subintervals, as there are 2^u subsets of $\{1,\ldots,u\}$. These subintervals partition the unit interval, $\mathbf{I}$. Thus, we need only keep track of the lengths of these 2^u consecutive new intervals, and adjust their lengths in order to find the best design. We represent these lengths by a real vector $\mathbf{x} = (x_1,\ldots,x_{2^u})$, constrained so that $x_j \geq 0$ and $\sum_j x_j = 1$. We note that to correspond to a real design, each x_j must be an integer multiple of $\frac{1}{n}$ (as the corresponding interval was created by taking the union of several disjoint subintervals of length $\frac{1}{n}$). When we relax the problem we will remove this restriction.

We are almost ready to move on to the relaxation. First we consider how to calculate privacy leakage and security of a design when it is represented by a real vector $\mathbf{x}$ cataloging interval lengths. Recall that privacy leakage (of a fixed-basis design) is calculated by counting the number of $\mathbf{b}_\ell$ belonging to some system in a set $\mathcal{S}$. Accordingly, we need only count the intervals that are colored in for some system in $\mathcal{S}$. Thus, if we let $\mathcal{P}_j$ be the subset corresponding to x_j, then

$$\frac{r(\mathcal{S})}{n} = \left(\sum_{j=1}^{2^u} x_j \cdot \mathbf{1}[\mathcal{P}_j \cap \mathcal{S} \neq \emptyset] \right),$$

and, as usual,

$$t_j(\mathcal{S}) = r(\{j\} \cup \mathcal{S}) - r(\mathcal{S}).$$

In order to remove the dependence on n, we will work with the normalized measures $\bar{r}(\mathcal{S}) = \frac{r(\mathcal{S})}{n}$ and $\bar{t}_j(\mathcal{S}) = \frac{t_j(\mathcal{S})}{n} = \bar{r}(\{j\} \cup \mathcal{S}) - \bar{r}(\mathcal{S})$ instead. Normalized versions of R_L and T_L are then defined accordingly, and respectively denoted $\bar{R}_L$ and $\bar{T}_L$.

Now, the only dependence left on n is in the set of possible values for the x_i. As $n \to \infty$, real-valued x_i can be arbitrarily closely achieved by actual designs. We shall thus allow the x_i to take on any real values. The results will be increasingly accurate for large n. Even for small n they provide a converse bound on the privacy–security tradeoff for fixed-basis designs. What we have now is the following optimization:

PROBLEM 16.2 (Relaxed optimization)   Pick $\mathbf{x} \in \mathbb{R}^{2^u}$ to solve

$$\text{maximize} \quad \overline{T}_L$$
$$\text{s.t.} \quad \overline{R}_L \leq R,$$
$$\overline{r}(\{i\}) = \mu, \quad i \in \{1,\ldots,u\}$$
$$x_j \geq 0, \quad j \in \{1,\ldots,2^u\}$$
$$\mathbf{1}^T \mathbf{x} = 1,$$

where $R = \frac{R'}{n}$ in order to match Problem 16.1. The first equality constraint simply requires that each system has measure μ.

Evaluation of $\overline{T}_L$ and $\overline{R}_L$ only requires knowledge of $\mathbf{x}$ and the order and membership of the subsets of $\{1,\ldots,u\}$. The latter information is stored in a matrix.

Now we have an optimization problem over the reals. It turns out that we can rewrite this problem as a linear program (LP). The last two constraints are affine constraints, so we don't need to modify them. However, we need to rewrite the objective and the other constraints in terms of $\mathbf{x}$ in a way that clearly shows them to be affine. After introducing a new variable, t, the following LP is equivalent to Problem 16.2. For a proof, see [22].

PROBLEM 16.3 (Relaxed LP)

$$\underset{\mathbf{x} \in \mathbb{R}^{2^u},\, t \in \mathbb{R}}{\text{maximize}} \quad t$$
$$\text{s.t.} \quad T\mathbf{x} \geq t \cdot \mathbf{1},$$
$$P\mathbf{x} \leq R \cdot \mathbf{1},$$
$$\mathbf{x} \geq \mathbf{0},$$
$$\mathbf{1}^T \mathbf{x} = 1,$$
$$B\mathbf{x} = \mu \cdot \mathbf{1},$$

where the matrices T, P, and B are binary. Formally, the matrices are defined as follows. For each of T, P, and B, the ith column corresponds to the ith subinterval. T is a $\binom{u}{L}(u-L) \times 2^u$ matrix where each row corresponds to a distinct pair $(\mathcal{S},j)$ such that $|\mathcal{S}| = L$ and $j \notin \mathcal{S}$. The ith entry in that row is equal to 1 if the ith subinterval is part of the jth system but not any system in $\mathcal{S}$. P is a $\binom{u}{L} \times 2^u$ matrix where each row corresponds to a distinct set $\mathcal{S}$ of L systems. The ith entry in that row is equal to 1 if the ith subinterval is part of some system in $\mathcal{S}$. Finally, B is a $u \times 2^u$ matrix where each row corresponds to a system, j. The ith entry in that row is equal to 1 if the ith subinterval is part of the jth system.

The rows of the matrices can be thought of as linear functionals on $\mathbb{R}^{2^u}$. Each row of T maps $\mathbf{x}$ to a security value, $t_j(\mathcal{S})$, for some j and $\mathcal{S}$. These security values are each bounded below by t. Each row of P maps $\mathbf{x}$ to a privacy leakage value, $r(\mathcal{S})$, for some $\mathcal{S}$. These privacy leakage values are each bounded above by R. Each row of B maps $\mathbf{x}$ to the measure of that system (represented as a union of subintervals). These measures are each fixed equal to μ.

EXAMPLE 16.3 *In the case where $u=4$ and $L=2$, we get the following matrices for B, P, and T:*

$$B = \begin{bmatrix} 0 & 0 & 0 & 0 & 1 & 0 & 0 & 0 & 1 & 1 & 1 & 0 & 1 & 1 & 1 & 1 \\ 0 & 0 & 0 & 1 & 0 & 0 & 1 & 1 & 0 & 0 & 1 & 1 & 0 & 1 & 1 & 1 \\ 0 & 0 & 1 & 0 & 0 & 1 & 0 & 1 & 0 & 1 & 0 & 1 & 1 & 0 & 1 & 1 \\ 0 & 1 & 0 & 0 & 0 & 1 & 1 & 0 & 1 & 0 & 0 & 1 & 1 & 1 & 0 & 1 \end{bmatrix},$$

$$P = \begin{bmatrix} 0 & 1 & 1 & 0 & 0 & 1 & 1 & 1 & 1 & 1 & 0 & 1 & 1 & 1 & 1 & 1 \\ 0 & 1 & 0 & 1 & 0 & 1 & 1 & 1 & 1 & 0 & 1 & 1 & 1 & 1 & 1 & 1 \\ 0 & 0 & 1 & 1 & 0 & 1 & 1 & 1 & 0 & 1 & 1 & 1 & 1 & 1 & 1 & 1 \\ 0 & 1 & 0 & 0 & 1 & 1 & 1 & 0 & 1 & 1 & 1 & 1 & 1 & 1 & 1 & 1 \\ 0 & 0 & 1 & 0 & 1 & 1 & 0 & 1 & 1 & 1 & 1 & 1 & 1 & 1 & 1 & 1 \\ 0 & 0 & 0 & 1 & 1 & 0 & 1 & 1 & 1 & 1 & 1 & 1 & 1 & 1 & 1 & 1 \end{bmatrix},$$

$$T = \begin{bmatrix} 0 & 0 & 0 & 0 & 1 & 0 & 0 & 0 & 0 & 0 & 1 & 0 & 0 & 0 & 0 & 0 \\ 0 & 0 & 0 & 1 & 0 & 0 & 0 & 0 & 0 & 0 & 1 & 0 & 0 & 0 & 0 & 0 \\ 0 & 0 & 0 & 0 & 1 & 0 & 0 & 0 & 0 & 1 & 0 & 0 & 0 & 0 & 0 & 0 \\ 0 & 0 & 1 & 0 & 0 & 0 & 0 & 0 & 0 & 1 & 0 & 0 & 0 & 0 & 0 & 0 \\ 0 & 0 & 0 & 0 & 1 & 0 & 0 & 0 & 1 & 0 & 0 & 0 & 0 & 0 & 0 & 0 \\ 0 & 1 & 0 & 0 & 0 & 0 & 0 & 0 & 1 & 0 & 0 & 0 & 0 & 0 & 0 & 0 \\ 0 & 0 & 0 & 1 & 0 & 0 & 0 & 1 & 0 & 0 & 0 & 0 & 0 & 0 & 0 & 0 \\ 0 & 0 & 1 & 0 & 0 & 0 & 0 & 1 & 0 & 0 & 0 & 0 & 0 & 0 & 0 & 0 \\ 0 & 0 & 0 & 1 & 0 & 0 & 1 & 0 & 0 & 0 & 0 & 0 & 0 & 0 & 0 & 0 \\ 0 & 1 & 0 & 0 & 0 & 0 & 1 & 0 & 0 & 0 & 0 & 0 & 0 & 0 & 0 & 0 \\ 0 & 0 & 1 & 0 & 0 & 1 & 0 & 0 & 0 & 0 & 0 & 0 & 0 & 0 & 0 & 0 \\ 0 & 1 & 0 & 0 & 0 & 1 & 0 & 0 & 0 & 0 & 0 & 0 & 0 & 0 & 0 & 0 \end{bmatrix}.$$

Now we've arrived at a tractable problem formulation. However, our linear program is far from illuminating. Simply formulating our problem as an LP doesn't yield much insight into the structure of the problem. Additionally, this LP is difficult to run for reasonable values of u and L. The number of variables and constraints is exponential in u, and the amount of memory (time) required to store (compute) T, P, and B is exponential in u.

Simplifications via Symmetry

In this section, we exploit the symmetry of the linear program posed in Problem 16.3 to formulate an equivalent, but much simpler, LP. In the same way that the b_ℓ vectors can be reordered, systems within a design can be reordered as well. Intuitively, two sets, $\mathcal{P}_i$ and $\mathcal{P}_j$, comprised of the same number of systems, should be treated interchangeably if their objectives match. For example, if one pair of systems is penalized for security more than another pair, then one could reorder the pairs and come up with a new security cost for the LP. However, reordering the systems should not have any effect on the security of the design. That is to say, for any i,j with $|\mathcal{P}_i| = |\mathcal{P}_j|$, there should be no cost (loss in security) if we force $x_i = x_j$. We now demonstrate that there is indeed no effect on our optimal objective function from forcing these equalities.

We define a *block-symmetric* solution to be a solution **x** such that if $|\mathcal{P}_i| = |\mathcal{P}_j|$ then $x_i = x_j$. We can vastly simplify Problem 16.3 by using the fact that if Problem 16.3 has a solution, then it admits a block-symmetric optimal solution. This result is proved in [22], where Problem 16.3 is shown to be equivalent to the simpler Problem 16.4. The idea behind the proof is that, given a solution to Problem 16.3, reordering the constituent systems in all possible ways produces a family of solutions. The mean of such a family is always block symmetric. For a linear program, a convex combination of solutions is also a solution. Therefore, the block-symmetric vector we obtain by taking the mean of the family is optimal.

We now impose our new block-symmetry condition on Problem 16.3 to obtain Problem 16.4. In Problem 16.4, the set of feasible solutions is reduced, but the security level obtained is not decreased. Thus, in studying the privacy–security tradeoff, it is preferable to use the following LP, which optimizes over $\mathbf{y} \in \mathbb{R}^{u+1}$.

PROBLEM 16.4 (LP: Reduced variables)

$$\begin{aligned}
&\underset{\mathbf{y}\in\mathbb{R}^{u+1},\, t\in\mathbb{R}}{\text{maximize}} \quad t \\
&\text{s.t.} \quad T A \mathbf{y} \geq t \cdot \mathbf{1}, \\
&\qquad\quad P A \mathbf{y} \leq R \cdot \mathbf{1}, \\
&\qquad\quad \mathbf{y} \geq \mathbf{0}, \\
&\qquad\quad \mathbf{1}^T A \mathbf{y} = 1, \\
&\qquad\quad B A \mathbf{y} = \mu \cdot \mathbf{1}.
\end{aligned}$$

In this LP, A is the $2^u \times (u+1)$ matrix describing the linear map that takes $\mathbf{y} \in \mathbb{R}^{u+1}$ to **x**, such that $x_j = y_k$, where k is the cardinality of the subset corresponding to x_j, i.e., $|\mathcal{S}_j| = k$.

EXAMPLE 16.4 When $u = 4$, as in Example 16.3, we have that

$$A^T = \begin{bmatrix} 1 & 0 & 0 & 0 & 0 & 0 & 0 & 0 & 0 & 0 & 0 & 0 & 0 & 0 & 0 & 0 \\ 0 & 1 & 1 & 1 & 1 & 0 & 0 & 0 & 0 & 0 & 0 & 0 & 0 & 0 & 0 & 0 \\ 0 & 0 & 0 & 0 & 0 & 1 & 1 & 1 & 1 & 1 & 1 & 0 & 0 & 0 & 0 & 0 \\ 0 & 0 & 0 & 0 & 0 & 0 & 0 & 0 & 0 & 0 & 0 & 1 & 1 & 1 & 1 & 0 \\ 0 & 0 & 0 & 0 & 0 & 0 & 0 & 0 & 0 & 0 & 0 & 0 & 0 & 0 & 0 & 1 \end{bmatrix}.$$

We now have only $u + 2$ variables, but we still have as many constraints as in Problem 16.3. The final step we make to simplify our problem is to observe that many constraints are redundant and can be removed. The reason for the redundancy derives again from problem symmetry. For example, consider BA. Every row in BA corresponds to a particular system. It indicates how many intervals of each weight are occupied by that system. Due to the symmetry imposed by requiring solutions to be block symmetric, it turns out that all rows of BA are equal to each other. The proof of this fact and the analogous proofs for T and P can be found in [22]. After removing the redundant constraints, we end up with the following LP, which is equivalent to Problem 16.4.

PROBLEM 16.5 (LP: Reduced constraints)

$$\begin{aligned}
\underset{\mathbf{y} \in \mathbb{R}^{u+1}}{\text{maximize}} \quad & \mathbf{g}^T \mathbf{y} \\
\text{s.t.} \quad & \mathbf{c}^T \mathbf{y} \leq R, \\
& \mathbf{y} \geq \mathbf{0}, \\
& \mathbf{f}_u^T \mathbf{y} = 1, \\
& [0 \ \mathbf{f}_{u-1}^T] \mathbf{y} = \mu,
\end{aligned}$$

where $\mathbf{g} = [0 \ \mathbf{f}_{u-L-1}^T \ \mathbf{0}^T]^T$, $\mathbf{f}_n = (\binom{n}{0}, \binom{n}{1}, \binom{n}{2}, \ldots, \binom{n}{n})^T$, and $\mathbf{c} = (c_0, \ldots, c_u)^T$, with

$$c_i = \sum_{k=k^-}^{k^+} \binom{L}{k} \binom{u-L}{i-k},$$

where $k^- = \max(1, i - (u - L))$ and $k^+ = \min(L, i)$.

Solving this LP is much more computationally manageable. This LP has only $u + 1$ variables and three non-trivial constraints. The remaining $u + 1$ constraints simply restrict $\mathbf{y}$ to the positive orthant. This means that both the number of variables and the number of constraints are linear in u, which we recall is the number of systems (subspaces). This is a major improvement in the complexity of the problem.

16.4.3 Resulting Privacy/Security Tradeoff

Now that we have been able to write down our fixed-basis design problem as a compact linear program, we run the LP for various values of R, μ, u, L and plot the results. Fixing μ, u, L and varying $R \in [\mu, 1]$, we plot $\overline{T}_L$ vs. $\overline{R}_L$. In other words, we plot $\mathbf{g}^T \mathbf{y}^*$ vs. R, where $\mathbf{y}^*$ is an optimal solution to Problem 16.5. Sample results are plotted in Figure 16.3. The curve we obtain encloses a convex region, $\mathcal{C}$, along with the lines $\overline{T}_L = 0$ and $\overline{R}_L = 1$. This region depends on μ, u, L. This means that for the fixed-basis design space, any achievable privacy–security pair, $(\overline{R}_L, \overline{T}_L)$, must be in $\mathcal{C}$. Specifically, it must be the case that $\overline{T}_L \leq \mathbf{g}^T \mathbf{y}^*$, where $\mathbf{y}^*$ optimizes the corresponding LP with $R = \overline{R}_L$. This follows from the fact that we are working with a relaxation of the original problem. Indeed, the $(\overline{R}_L, \overline{T}_L)$ tradeoff curve obtained from the LP provides a converse bound on the possible privacy–security pairs of actual fixed-basis designs. This means that it is impossible to increase security or decrease privacy leakage at every point on the obtained curves.

Since we are working with a continuum of R and μ values, it is impossible that every point in the region can be achieved by an actual fixed-basis design. However, it turns out that the set of achievable points is dense in the region as $n \to \infty$. For any point (r, t) in the region, and any $\varepsilon_1 > 0$, $\varepsilon_2 > 0$, there exists some $n \in \mathbb{N}$, $m \in \mathbb{N}$, with $|\frac{m}{n} - \mu| < \varepsilon_1$, for which $\|(\overline{R}_L, \overline{T}_L) - (r, t)\| < \varepsilon_2$, where $\overline{R}_L$ and $\overline{T}_L$ correspond to an actual fixed-basis design with parameters n, m, u, L.

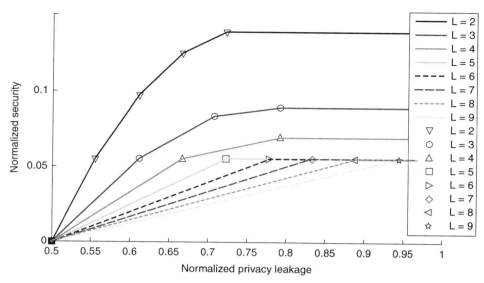

Figure 16.3 The area under each curve is the region of achievable privacy–security pairs for fixed-basis designs with $u = 10$ and $\mu = 0.5$. The lines were generated by running the LP of Problem 16.5. The superimposed points were generated independently, following Conjecture 16.1.

Note that for our problem there are many types of equivalence between designs that allow us to simplify the problem and still find optimal designs. Equivalent designs perform the same in that they have the same worst-case privacy and security measures. For our purposes, all equidimensional subspaces are interchangeable, as long as the intersections between subspaces are fixed. However, there are considerations from the implementation of secure sketch systems that do prefer certain subspaces/matrices to others. After obtaining an actual design that is close to the upper bound given by the optimization, we may then want to use its equivalence class of designs in order to find a design which is better suited to encoding syndromes. For example, we may prefer our matrices H_i to be low density. As mentioned earlier, this would allow us to apply modern techniques of error correction to solve the authentication problem in (16.1).

16.4.4 Observed Form of Optimal Solutions

Upon computing solutions $\mathbf{y}^*$ to Problem 16.5, we observe an interesting "leapfrogging" pattern. Define a *leapfrogging solution* to be one that has at most three non-zero entries, the first two of which are consecutive, y_k and y_{k+1}, and the third of which is the final entry y_u. Leapfrogging refers to the fact that, as k increases, the consecutive non-zero entries appear to leapfrog over each other.

CONJECTURE 16.1 *There exists an optimal solution to Problem 16.5 which is a leapfrogging solution. Furthermore, for $k < \lceil \frac{u}{L} - 1 \rceil$ and $R \in [1 - \alpha_k, 1 - \alpha_{k+1})$, where*

$$\alpha_k = (1 - \mu) \frac{\binom{u-L}{k}}{\binom{u-1}{k}},$$

the non-zero entries are y_k, y_{k+1}, and y_u. For $k = \lceil \frac{u}{L} - 1 \rceil$, and for $R \in [1 - \alpha_k, 1]$, y_k and y_u are the only possibly non-zero entries.

If Conjecture 16.1 holds, we can force all but three entries of **y** to be zero. This constraint can be added to Problem 16.5 to make an almost trivial LP with three variables and two equality constraints. This means we can write y_{k+1} and y_u in terms of y_k, which yields a linear optimization problem in one bounded variable. An analytic solution reduces to comparing the various bounds on y_k and identifying the tightest ones. If Conjecture 16.1 holds, we therefore have a closed-form expression for the case of fixed-basis designs. In Figure 16.3, the expressions thus derived exactly match the results of the numerical optimization.

16.5 Conclusions

The goal of this chapter was to explore privacy and security in biometric authentication from the angle of multiple-system design and enrollment. We drew on tools from algebra and graph theory to understand the design space. While the space is quite large, insight into the joint structure of subspaces yielded substantial understanding. To attain an explicit characterization we restricted ourselves to designs that used a fixed basis, a design problem that we could translate into a high-dimensional linear program. We then simplified the linear program to a tractable size by exploiting problem symmetry. We presented numerical results and a conjecture regarding the form of the tradeoff attained.

The subspace packing problems discussed in the chapter may have applications in areas beyond biometrics. It is a particular type of packing problem where there is a benefit to packing tightly, allowing some overlap, but also a benefit to each packed object to have space to itself. In other words, the problem would be to achieve a sort of global density with local sparsity. It is possible that some methods used and insights gained from our problem could be generalized to apply to other packing problems of this type. We also see that, even within the particular packing problem we study, many open questions yet remain.

Acknowledgments

The authors would like to thank Professor Frank Kschischang for his early discussions with us regarding the subspace interpretation of our problem, Steven Rayan for aiding our investigation into the Grassmann graph, and Melkior Ornik for his comments on various drafts of this work.

This work was supported by an Ontario Graduate Scholarship (OGS), a Natural Sciences and Engineering Research Council of Canada (NSERC) Discovery Grant, and an NSERC Canada Graduate Scholarship (Master's) (CGSM).

References

[1] A. Jain, P. Flynn, and A. Ross, *Handbook of Biometrics*. New York: Springer, 2008.

[2] P. Reid, *Biometrics and Network Security*. Upper Saddle River, NJ: Prentice Hall, 2003.

[3] N. Ratha, J. Connell, and R. Bolle, "Enhancing security and privacy in biometrics-based authentication systems," *IBM Syst. J.*, vol. 40, no. 3, pp. 614–634, 2001.

[4] P. Campisi, *Security and Privacy in Biometrics*. New York: Springer, 2013.

[5] T. Ignatenko and F. M. J. Willems, *Privacy Leakage in Binary Biometric Systems: From Gaussian to Binary Data*. London: Springer, 2013, pp. 105–122.

[6] Y. Wang, S. Rane, S. C. Draper, and P. Ishwar, "A theoretical analysis of authentication, privacy and reusability across secure biometric systems," *IEEE Trans. Inf. Forensics Security*, vol. 7, no. 6, pp. 1825–1840, Dec. 2012.

[7] L. Lai, S.-W. Ho, and H. V. Poor, "Privacy–security trade-offs in biometric security systems – Part II: Multiple use case," *IEEE Trans. Inf. Forensics Security*, vol. 6, no. 1, pp. 140–151, Mar. 2011.

[8] R. Koetter and F. Kschischang, "Coding for errors and erasures in random network coding," *IEEE Trans. Inf. Theory*, vol. 54, no. 8, pp. 3579–3591, Aug. 2008.

[9] A. Khaleghi, D. Silva, and F. Kschischang, *Subspace Codes*, ser. Lecture Notes in Computer Science, vol. 5921, pp. 1–21. Berlin, Heidelberg: Springer, 2009.

[10] A. Dimakis, P. Godfrey, Y. Wu, M. Wainwright, and K. Ramchandran, "Network coding for distributed storage systems," *IEEE Trans. Inf. Theory*, vol. 56, no. 9, pp. 4539–4551, Sep. 2010.

[11] P. Sobe and K. Peter, "Comparison of redundancy schemes for distributed storage systems," in *IEEE Int. Symp. Network Computing and Apps.*, Cambridge, MA, USA, Jul. 2006, pp. 196–203.

[12] A. Shamir, "How to share a secret," *Commun. ACM*, vol. 22, no. 11, pp. 612–613, Nov. 1979.

[13] Y. Sutcu, S. Rane, J. S. Yedidia, S. Draper, and A. Vetro, "Feature extraction for a Slepian–Wolf biometric system using LDPC codes," in *Proc. IEEE Int. Symp. Inf. Theory*, Toronto, ON, Canada, Jul. 2008, pp. 2297–2301.

[14] D. Silva and F. Kschischang, "On metrics for error correction in network coding," *IEEE Trans. Inf. Theory*, vol. 55, no. 12, pp. 5479–5490, Dec. 2009.

[15] E. Erkut, "The discrete p-dispersion problem," *European J. Operational Research*, vol. 46, no. 1, pp. 48–60, May 1990.

[16] A. Goldberg and S. C. Draper, "The privacy/security tradeoff across jointly designed linear authentication systems," in *Proc. 52nd Annual Allerton Conf. Commun., Control, Computing*, Monticello, IL, USA, Sep. 2014, pp. 1279–1286.

[17] K. Metsch, "A characterization of Grassmann graphs," *Eur. J. Combinatorics*, vol. 16, no. 6, pp. 639–644, Nov. 1995.

[18] J. Kosiorek, A. Matras, and M. Pankov, "Distance preserving mappings of Grassmann graphs," *Beitr. Algebra Geom.*, vol. 49, no. 1, pp. 233–242, Jan. 2008.

[19] R. Bailey and K. Meagher, "On the metric dimension of Grassmann graphs," *Discrete Math. & Theoretical Computer Science*, no. 4, pp. 97–104, Jan. 2011.

[20] D. Djokovic, "Distance-preserving subgraphs of hypercubes," *J. Comb. Theory B*, vol. 14, no. 3, pp. 263–267, Jun. 1973.

[21] H. Cheung, T. Kwok, and L. Lau, "Fast matrix rank algorithms and applications," in *ACM Symp. Theory Computing*, New York, NY, USA, May 2012, pp. 549–562.

[22] A. Goldberg, "The privacy/security tradeoff for multiple secure sketch biometric authentication systems," Master's thesis, Univ. of Toronto, 2015.

17 Information Theoretic Approaches to Privacy-Preserving Information Access and Dissemination

Giulia Fanti and Kannan Ramchandran

The Internet has revolutionized the process of obtaining and sharing information. However, this functionality comes hand-in-hand with pervasive, often subtle, privacy violations. In this chapter, we discuss algorithmic approaches for accessing and disseminating information while protecting the privacy of network users. In particular, we focus on mechanisms that offer *information theoretic* privacy guarantees; that is, these mechanisms offer protection even against computationally unbounded adversaries. We provide a high-level overview of various classes of privacy-preserving algorithms, with concrete intuition-building examples.

17.1 Introduction

People today generate, share, and consume information with unprecedented efficiency. The Internet has enabled realities that seemed impossible a few decades ago: instantaneous global message dissemination, large-scale data storage at minimal cost, and lightning-fast information search and retrieval, to name a few. However, these advances come hand-in-hand with unique privacy threats, which were largely ignored in the push to develop seamless user experiences. Namely, the very information that users willingly consume and produce on the Internet can be deeply revealing about the users themselves. More problematically, leakage of this information poses a significant privacy risk, which can have serious societal repercussions, both psychological and physical [1,2]. In this chapter, we explore several privacy challenges and offer solutions associated with two key aspects of information flow: data dissemination and data access.

When people disseminate information – for instance, by posting a message on a social network – they can inadvertently reveal sensitive information about themselves. Sometimes, this information might be inherent to the content a user posts. For example, if Alice posts on Twitter that she is going to the cinema at 8 pm, then she is implicitly telling her contacts that she will not be home at 8 pm; in the worst case, this could facilitate a robbery. Such content-based data leakage is inevitable, and must be managed by educating users about the privacy implications of posting personal information. However, sometimes information dissemination can also cause unintentional leakage of *metadata*, or data that is not directly related to content. For example, if Alice posts a picture of her cats to a social network at 3 pm, and the social network includes her GPS location when it propagates the message to her contacts, then Alice's contacts could learn her location at 3 pm. Metadata leakage typically occurs because of content propagation protocols rather than decisions made by users; it is therefore in the purview of protocol and algorithm designers to prevent or mitigate such leakage. This realization has prompted research on metadata-aware messaging protocols.

Another class of privacy leaks can result when users retrieve information from online repositories. When people query a search engine or a database, the contents of those queries may be sensitive; if so, users may be targeted for extra monitoring or even censorship. For instance, a user who types "how to build a bomb" into a search engine could trigger additional monitoring, either by the search engine itself, or by an external law enforcement agency with access to search engine records. Even if users' queries are not particularly sensitive, when aggregated over time they can enable an observer to build detailed, intimate profiles of people. Indeed, the leaked Snowden documents revealed details of a program called XKeyscore, by which U.S. government agents can allegedly peruse the emails, chats, and browsing histories of millions of people, without any special authorization [3]. Intuitively, this problem might seem fundamental: a user must reveal her query to a server in order to search a database stored on that server. Surprisingly, this is not the case; a recent body of research has investigated algorithms for searching and accessing a database without revealing the nature of users' queries to the server storing the database.

A rich body of work has emerged on exploring technological solutions to privacy concerns in both of these areas; our goal in this chapter is to give an overview of these approaches. Solutions in this space generally take the form of information-flow algorithms that prevent adversarial parties from learning about a user through the information she shares or consumes. For example, can we build a search engine that does not learn the contents of users' queries? Can we build social networks that do not leak users' metadata? Figure 17.1 gives a high-level overview of the kinds of questions discussed in this chapter, which focuses on providing contextual ties between approaches, complemented by representative, intuition-building examples. In line with the theme of this book, we emphasize solutions that offer *information theoretic* privacy guarantees. Information theoretic privacy guarantees are secure even against computationally unbounded adversaries, whereas *computational* guarantees are secure against computationally bounded adversaries.

In practice, most privacy-enhancing tools offer computational guarantees (if they offer any theoretical guarantees at all). In part, this is because information theoretic guarantees typically come with restrictive assumptions on the adversary's capabilities; for instance, a common assumption in information theoretically private tools is that protocols should be executed by multiple, non-colluding parties. As long as all the parties do not collude, perfect privacy is achieved; otherwise,

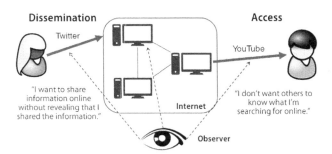

Figure 17.1 High-level view of the problem space. Privacy guarantees must be strong enough to withstand a global adversary monitoring network traffic, as well as potentially malicious service providers.

privacy is completely lost.[1] In today's technological landscape, it may be more realistic to assume that strong adversaries (e.g., government surveillance agencies) are omnipresent with bounded computational capabilities; this lends itself to computational guarantees.

However, there may be strong technical reasons for considering information theoretic models of privacy as well. One reason is that bandwidth usage is steadily increasing, and high-bandwidth content dissemination is often managed by distributed or decentralized architectures, as seen in peer-to-peer networks (e.g., Napster, BitTorrent). Incidentally, such distributed architectures are also well suited to information theoretic privacy-preserving mechanisms; they provide a natural supply of distributed nodes that could satisfy the anti-collusion assumptions made by several information theoretic privacy primitives. A second reason for considering information theoretic privacy mechanisms is that if quantum computing were to emerge in a meaningful way, then modern cryptosystems could be rendered vulnerable. The computational capabilities afforded by quantum computing may enable adversaries to compromise computational privacy or security guarantees [4]. Information theoretic secrecy is an appealing alternative because it does not depend on the adversary's computational power; it is robust to computationally unbounded adversaries.

17.2 Information Dissemination

Sharing content with one's peers is a central facet of most existing social medial platforms, such as Facebook and Twitter. People often use such platforms to post sensitive content, including controversial political or social opinions. Unfortunately, controversial opinions on these platforms can also lead to serious repercussions for authors of such messages. For example, people have been imprisoned for posting politically or socially objectionable content on social media [5]. Socially, message

---

[1] This is an oversimplification. We will see examples of information theoretic tools in which subsets of the participants may collude without losing privacy. However, such robustness often comes at the expense of more complicated algorithms.

authors may find themselves the target of cyberbullying, which can have serious psychological repercussions, especially on younger users [6]. These challenges have prompted significant research on *anonymous information dissemination*, which is the topic of this section. An anonymous messaging system enables users to disseminate messages without revealing authorship information, even to the service provider itself. For the sake of concreteness, we will assume that the information being disseminated is text-based messages; this assumption is not fundamental, but it removes challenges associated with disseminating large files or latency-sensitive content.

Within this problem domain, two specific problems have been primarily studied in the literature: anonymous broadcast messaging and anonymous point-to-point messaging. Anonymous broadcast messaging allows an individual to spread a message to as many people as possible without revealing the source of the message – even to a powerful global adversary. For instance, a truly anonymous version of Twitter would qualify as anonymous broadcast messaging. Anonymous point-to-point messaging enables an individual (Alice) to communicate with a contact (Bob) without Bob learning that he is talking to Alice. Pay phones are a real-life example of anonymous point-to-point messaging channel, since the recipient of a call cannot explicitly associate the phone number with the caller's name. While information theoretic notions of privacy have primarily been discussed in the context of anonymous broadcast messaging, we explore both problem areas in this section.

Figure 17.2 gives an overview of the classes of problems (and solutions) we consider. In the category of broadcast messaging, we consider two main classes of work: those that relate to dining cryptographer networks (colloquially, DC nets), and those that instead take a statistical approach to guaranteeing privacy. DC nets were first presented in the 1980s, and they have received a great deal of attention from the applied cryptography community; we describe these in greater detail in Section 17.2.1. They offer very strong privacy guarantees, but they are brittle to misbehaving participants. Statistical approaches emerged more recently, with the goal of considering specific, somewhat weaker, adversarial models; these concessions allow algorithms to be significantly more efficient. In the area of anonymous point-to-point messaging, we discuss the vast body of literature on onion routing, which is the dominant tool in this area.

17.2.1 Anonymous Broadcast Messaging

Anonymous broadcast messaging is typically modeled as follows: there is a fixed set of users, one of whom wants to spread a message to the other users without being detected as the message source. Users can communicate with one another, but only according to the constraints of an underlying communication network. Therefore, the setup is typically modeled as a graph, in which nodes represent users, and edges represent connectivity between users. If two users do not share an edge in this graph, they cannot communicate directly with one another. For example, if every user can talk to every other user in a network, we represent the network as a complete graph. The design of anonymous broadcasting algorithms depends in large part on the adversary's

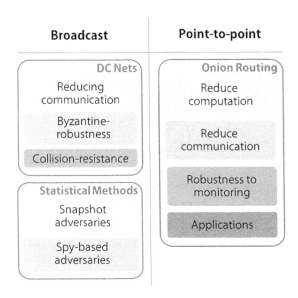

Figure 17.2 Research subtopics within anonymous information dissemination.

capabilities. Differences in threat model and assumptions on the underlying contact network have spawned a few different directions of research in this space.

Dining Cryptographer Networks

Anonymous broadcast messaging was first studied by Chaum in his seminal work on the dining cryptographers problem [7]. The dining cryptographers problem considers a scenario in which a group of cryptographers goes out to dinner, all seated around a table. After the meal, they wish to determine if one of them paid the bill, but for the sake of propriety, nobody should learn *who* paid the bill. Note that this problem setup is equivalent to that of a single user anonymously broadcasting a single bit, while the rest of the network participants stay silent. Chaum's solution proceeds as follows: each adjacent pair of cryptographers at the table flips a two-sided coin; the outcome of this coin flip is visible only to the pair. Each cryptographer therefore sees the outcome of two coin flips, one shared with the neighbor on the left and one shared with the neighbor on the right. If the two coin flips have the same outcome, the cryptographer reports his true status to the table (i.e., if he paid for the meal, he reports "True," otherwise he reports "False"). If the two coin flips are different, he reports the opposite of the truth. Given the values reported by all the cryptographers, the group determines the true answer by computing the logical exclusive-or (XOR) of all the reported statuses from all the cryptographers. If one of them paid, this value will evaluate to "True;" otherwise, "False." Figure 17.3 illustrates an example of this, in which Alice paid the bill. The black bits represent the shared pairwise coin flips, and the gray bits represent the bits broadcast by each user. She therefore reports the XOR of 1 with the XOR of her two neighboring coin flips.

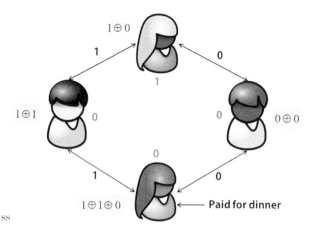

Figure 17.3 Example of the dining cryptographers solution from [7]. The black bits are secret coin flips visible only to the adjacent pair of cryptographers. The gray boxes illustrate the computation executed by each party, and the gray bits are the final value reported by each individual. Notice that in this case, since one party did pay the bill, the XOR of the gray bits evaluates to 1.

This solution, commonly referred to as a dining cryptographer network, or a DC net, enables a single individual to broadcast a single bit in such a way that the true author of the bit remains anonymous with information theoretic guarantees. Notice that each cryptographer observes only his own two shared secrets and the publicly announced bits of each other participant. The precise anonymity guarantees in [7] are as follows: Suppose we build a graph in which each participant is a vertex, and each secret bit shared between two participants constitutes an edge. In the specific instance we described, this graph would be a ring graph over all the cryptographers. Now suppose some subset of participants are colluding, and reveal their shared secrets to one another. For each compromised secret bit, we remove that edge from the graph. What remains is a set of honest nodes and the shared secrets between them. Within each connected component of this pruned graph, the adversary cannot learn anything other than the overall parity of the bits reported by nodes in the component. For example, suppose the north and west players in Fig. 17.3 are colluding. The remaining connected component is between the south and east players. Then, the colluding nodes can only learn that either the south or the east player payed the bill; they learn no information about *which* player paid. More details can be found in [7].

The dining cryptographers problem implicitly assumes the use of a complete connectivity graph, since each cryptographer shares his one-bit response with every other participant. Notably, in order for one participant to broadcast a single bit anonymously, each non-communicating participant must also broadcast one bit to the entire network. In addition, each user must have shared, secret randomness with two other users in the network. This introduces scalability challenges, which have impeded the development of large-scale systems for DC-net-based anonymous broadcasting. The dining cryptographers' adversarial model and problem setup were subsequently adopted in a number of works, with the primary goal of making DC nets more scalable and robust to stronger adversaries. For example, the "dining cryptographers in the disco" problem proposes schemes that have stronger guarantees on unconditional untraceability, even when the adversary is able to partially hijack communications and when there are Byzantine participants [8]. In a similar vein, Golle and Juels propose a DC net construction that is robust to Byzantine participants, or participants who choose not to follow the specified protocol, while retaining the attractive non-interactive property of the original DC net [9].

Research on DC nets also moved beyond the construction of primitives to actual system building. For example, Herbivore is a system that uses DC nets for anonymous point-to-point communication [10]. DC nets are designed for sender untraceability, but Herbivore harnesses the fact that by virtue of being a broadcast tool, they also provide receiver untraceability. Therefore, it divides the population into small anonymizing cliques. When Alice wants to send a message to Bob, she first broadcasts the message to her anonymizing clique, which then gets routed to Bob's anonymizing clique. Herbivore managed to scale by reducing the size of each DC net; however, this limits the anonymity that can be achieved. Another system built atop DC nets is Dissent [11, 12], which provides stronger accountability against misbehaving parties by preventing a malicious party from flooding the network with spam messages or preventing legitimate nodes from transmitting. It also considers the case where the communication load is asymmetric; for instance, one party wishes to transmit a large file, while the other participants do not wish to transmit anything. The strength and relative scalability of Dissent stems from its use of a shuffle protocol, which guarantees each party one transmit slot during each round of operation. This prevents malicious parties from corrupting other parties' transmissions. However, the original Dissent project scaled to 40 nodes, and even the follow-up work on improving scalability ran on 600 nodes. While DC nets provide strong, provable anonymity guarantees, current constructions cannot yet scale to the numbers of users seen in popular social networks.

Statistical Rumor Source Obfuscation
Whereas DC nets are tailored to provide anonymity even in the face of a strong, almost omnipresent, adversary, other work in this space instead considers statistical approaches to anonymous broadcasting in which the adversary has restricted capabilities. Work in this space typically assumes an incomplete underlying contact network. For example, this contact network might represent a social graph, implying that users of a service are only allowed to propagate content to "friends" in the social graph. Alternatively, if the network is location based, edges in the underlying contact network might represent users whose physical proximity is within some threshold. Imposing a non-trivial underlying connectivity network can facilitate social filtering to mitigate spam; in a fully connected network, it is easy to flood the network with bogus or unwanted content, which severely limits the utility of the service.

Compared to DC nets, statistical rumor source obfuscation techniques protect the author's anonymity against a weaker adversary, which uses metadata (e.g., timing or message-spreading patterns over the network) to infer the node that authored a particular message. Thus far, one main adversarial model has been considered in this research area: the *snapshot* adversary. A snapshot adversary observes the following: at time $t = 0$, the message starts to spread over the network using a fixed spreading protocol, known to the adversary. A trivial example of a spreading protocol that provides no anonymity guarantees is deterministic spreading, in which at each time step, each node passes the message to its *uninfected neighbors*, i.e., neighbors without the message. At some time $t = T$, the adversary learns which users in the network have received the message, and which have not. The adversary does not observe the timestamp T exactly, but depending on the spreading protocol, the adversary may be able to infer this information. The adversary also knows the structure of the underlying contact network and the specifications of the spreading protocol (i.e., it knows the spreading rules, but it cannot observe realizations of randomness observed by individual users). Given this information, the adversary attempts to infer the author of the message. In the trivial deterministic spreading protocol mentioned previously, the true source is easily identified at the "center" of the infected subgraph; the goal is to spread the message in such a way that makes this inference provably difficult.

The first work to adopt this adversarial model was [13], which proposes a spreading protocol called adaptive diffusion. Adaptive diffusion is designed to hide the message source from a snapshot adversary when the underlying network is a regular tree. For this special case, adaptive diffusion achieves the best possible hiding, in that the true source is equally likely to be any one of the infected nodes captured in the adversary's snapshot. The tree-structured assumption on the underlying graph is clearly stylized, since social networks, and even location-based networks, are unlikely to ever be regular trees. However, in practice, each message will spread over a tree that is embedded in the underlying contact network, so the assumption has some grounding in reality.

We briefly describe the basic adaptive diffusion algorithm on an infinite line graph, and explain why this approach hides the source perfectly. The intuition from this toy example extends to regular trees as well, with some modifications. Suppose the underlying contact network is an infinite line graph, G. A node wishes to spread a message anonymously. At each even time step t, adaptive diffusion ensures that the infected subgraph G_t – that is, the subgraph of G in which each node has received the message by time t – is always a balanced tree of depth $t/2$. Since we are considering a line graph, G_t is always a line segment with an odd number of nodes (at even time steps). The root of this tree, the central node, is called the *virtual source*. For example, Fig. 17.4 illustrates that at time $t = 2$, the infected subgraph is a tree of depth 1. In this figure, the solid-background node is the true source, and the node with a black dot inside is the virtual source; the virtual source changes over time, but the true source is fixed. Similarly, at time $t = 4$, the infected subgraph has depth 2. At $t = 0$, the true source is also the virtual source.

The idea behind adaptive diffusion is that at each even time step, the virtual source decides whether to pass the virtual source token to a different node, or keep it; this choice is probabilistic. However, at $t = 0$, the source *always* passes the virtual source. Hence, the first layer in Fig. 17.4 shows the source choosing to pass the message left or right with equal probability. In subsequent (even) time steps, if the virtual source decides to remain the virtual source, then every leaf in the infected subgraph passes the message

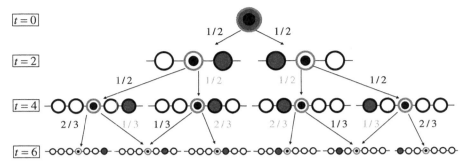

Figure 17.4 Adaptive diffusion generates an infected subgraph that is symmetric about a node at each even time step. This figure illustrates possible infection patterns on a line, with the associated probabilities. The filled-in node is the true source; nodes with a black dot inside are the virtual source at each time step. At each even time step $t \geq 2$, the virtual source must choose whether to keep the virtual source token or pass it. If the token is kept, the infected subgraph expands symmetrically in every direction, indicated by light-colored probabilities; otherwise, the subgraph expands asymmetrically, indicated by black probabilities.

to all its neighbors. In the subsequent (odd) time step, nothing happens. For example, the two inner subgraphs in Fig. 17.4 at $t = 4$ result from the virtual source keeping the token. If the virtual source instead decides to pass the virtual source token, then it must first choose one of its neighbors, excluding the previous virtual source, to be the new virtual source.[2] Since the underlying contact network is a tree (specifically a line in this example), the virtual source can only move farther away from the true source. Then, in the current (even) time step and the subsequent odd time step, the infected subgraph expands asymmetrically. That is, every leaf that is closer to the new virtual source than the old one infects its neighbors. In Fig. 17.4, this is what happens between time steps $t = 2$ and $t = 4$ to obtain the outer two subgraphs. Doing so, the infected subgraph remains balanced with the correct depth.

By carefully choosing the probability of keeping or passing the virtual source, adaptive diffusion hides the message among all nodes except the virtual source. For example, in Fig. 17.4, at $t = 4$, the adversary would observe a graph with five nodes; since there are four possible graphs, each should occur with probability $1/4$. The exact form of the probability of passing the virtual source can be derived by solving the parameters of a time-varying Markov chain that depends on the virtual source's distance from the true source and the time step. Intuitively, if the virtual source has already moved many times compared to the current time (e.g., if the virtual source moved five times in five time steps), then the virtual source is more likely to again pass the virtual source, and vice versa. This is useful because there are combinatorially many ways to generate graphs with the source as an internal node, but only one way to generate a graph with the source at a leaf of the graph; by giving more weight to the latter, we can

---

[2] The fact that the virtual source is never passed "backwards" is not fundamental – this is simply how adaptive diffusion is defined.

balance the overall likelihoods of the source ending up at different locations in the final graph.

For general trees, the probabilities of passing or keeping the virtual source are precisely tuned such that the probability of the true source being at hop distance h_0 from the virtual source at any time T is the same as the fraction of nodes at depth h_0 in a tree of depth $T/2$. On a line, the number of nodes at any given distance from the root is always two, but over d-regular trees for $d > 2$, the source should be more likely to appear at a leaf of the tree since there are (comparatively) more leaves. Adaptive diffusion therefore gives perfect obfuscation over regular trees, meaning that the source is perfectly hidden among the infected nodes.

Although the anonymity of adaptive diffusion was not originally described in information theoretic terms, it obfuscates the source in a way that the mutual information between the adversary's observed information and the true source is small. Critically, it protects authorship information even against a computationally unbounded adversary. Luo *et al.* have considered a similar problem in a game-theoretic formulation, where they develop optimal strategies for the adversary and the protocol, and present a Nash equilibirum [14].

17.2.2 Anonymous Point-to-Point Messaging

Consider a communication network modeled by a graph, in which nodes represent end hosts and edges represent connectivity between nodes. For instance, if nodes are communicating over the Internet (i.e., any node can communicate with any other node), then this graph is complete. Two of these nodes, Alice and Bob, wish to communicate. The goal is to enable such point-to-point communication, while ensuring that an adversary that has corrupted some subset of network nodes should be unable to link Alice with Bob. The adversary may or may not know the structure of the underlying connectivity graph, but it knows the strategy being used by Alice and Bob to communicate.

Anonymous point-to-point messaging has been studied extensively; some of the tools to emerge from this body of work have even seen widespread adoption by the public – Tor is a prominent example. Work in this space overwhelmingly makes use of proxies, or network nodes that forward traffic in order to mask the origin. Building on the idea of proxies, most tools for anonymous point-to-point communication use one main privacy primitive: onion routing. Onion routing refers to the layered encryption of packets in such a way that each successive hop in the chain of routing proxies can only learn the identity of the previous and following node in that chain. This approach distributes the relevant addressing information in such a way that no individual proxy can associate the source of a packet with the destination. Onion routing was used for years before a formal analysis of its security properties emerged. It provides computational privacy guarantees against a limited adversary that has corrupted a subset of onion routers that does *not* include both the first and last onion routers. However, Syverson *et al.* demonstrate that with constant probability, anonymity is lost, even against non-global adversaries [15]. Similarly, Feigenbaum *et al.* model the problem using IO automata and show that as long as the cryptographic assumptions hold, an adversary can only determine the source of an onion-routing circuit if it controls the first hop of that circuit [16].

17.3 Information Access

Web search and browser histories are extremely sensitive. Given an individual's search history, an observer can infer intimate details about the user's life. For example, in 2006, AOL famously released "anonymized" search histories; the data consisted of 20 million web search queries, in which the underlying user name was replaced with a numeric identifier [17]. These records were almost immediately deanonymized – in some cases on national media outlets – and served as a cautionary tale to companies who might be inclined to release anonymized company datasets. However, the privacy risks associated with search histories are not limited to public releases of company data, as in the AOL case. Users' search histories may be visible to a number of *third* parties, including cellular service providers [18], advertisers [19], hackers, or government agencies [3]. Clearly, the very act of accessing information online can result in significant privacy violations. Nonetheless, people have a right to access information without fear of being monitored. A large body of work has therefore studied methods for searching and retrieving database contents without revealing users' queries to the server.

In discussing information access, we assume a client–server architecture: a client wants to retrieve information from a database stored on a server. Accessing information in a privacy-preserving fashion can mean different things: a user could either try to mask the nature of the content she is accessing, or she could try to mask her identity. Both approaches have received significant attention in the literature, but information theoretic techniques have primarily been studied in the context of hiding the content of one's queries.

From a privacy and usability perspective, there are tradeoffs between masking one's identity and masking one's query contents. It is generally more computationally efficient to hide one's identity; Tor is the canonical example of a point-to-point, low-latency communication service in which the identity of the sender is masked from the receiver [20]. As discussed in Section 17.2.2, Tor routes encrypted content through a chain of relay nodes prior to delivering the content; this prevents any node in the network from learning who was communicating with whom, with computational guarantees. More generally, the approach of hiding one's identity falls under the category of anonymous point-to-point communication (Section 17.2.2). However, for certain queries, like retrieving driving directions from home to a museum, anonymity overlays like Tor may not hide the user's identity due to sensitive search contents. As such, it is important to develop tools that hide the *content* of users' queries, while still returning correct results in an efficient manner. This class of algorithms will be the focus of the rest of this section.

Figure 17.5 illustrates the research landscape in the area of privacy-preserving data access, as we will discuss it. We consider two main classes of problems:

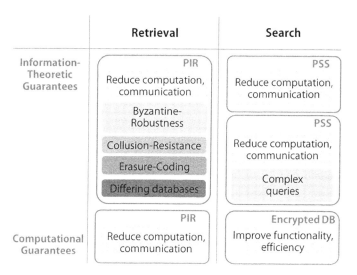

Figure 17.5 Research landscape for privacy-preserving information access. We focus mostly on information theoretic techniques (darker outline), along with related subproblems. However, some of the most widely known work in this area provides computational guarantees (lighter outline). The boxes represent different classes of privacy algorithms; there are two boxes for PIR because computational and information theoretic PIR are algorithmically different.

privacy-preserving retrieval and search. Privacy-preserving retrieval consists of accessing an indexed element of a database without revealing *which* element the client is accessing. This is commonly referred to as the private information retrieval (PIR) problem. We give an overview of the many variants of PIR that have been studied, primarily of the information theoretic variety. The privacy-preserving *search* problem consists of retrieving elements of a database that satisfy user-specified search criteria, without revealing these criteria (or the results) to the server. Generally speaking, the databases in question could either be encrypted or plaintext. If the database is plaintext, the problem is known as private streaming search (PSS). Most PSS algorithms offer computational guarantees, though there has been some work in the information theoretic space as well. There has also been significant work on implementing search algorithms over encrypted data. Although this area does not provide information theoretic guarantees, it is an important class of privacy-preserving algorithms that is gaining traction in practice.

17.3.1 Adversarial Models

Most work in this space focuses on threats posed by adversarial servers processing users' requests. The adversary's goal is to use any information provided by the user to discover the contents of her query. Adversaries are typically assumed to be "honest but curious" (i.e., they follow protocol, but they also try to learn the client's information), though some researchers have also considered malicious adversaries (i.e., the adversaries disobey protocol however they please). The exact information available to the adversary depends in large part on the problem setup. Since there are multiple problem formulations in the literature, we specify each adversary's capabilities as we present the relevant work. Note that a number of *third* parties – external government agencies, for instance – might be interested in uncovering a user's information access patterns. Particularly in the domain of information theoretic solutions, most work has not considered the non-trivial threats posed by powerful *third* parties. However, this may be an important direction for future work.

17.3.2 Private Information Retrieval

PIR is a technique that allows a client to retrieve a specific indexed record from a database without revealing *which* record to the server. A trivial PIR scheme results from sending the full database to the client, who then extracts the desired record. This solution is problematic because in web-scale applications, client devices are unlikely to have the storage or bandwidth to support such an operation. Most PIR research is therefore centered around finding efficient schemes, in terms of communication, computation, or both. Typically, it is possible to reduce the communication costs at the expense of computation, or vice versa; balancing these costs is critical in practical applications [21–23].

There are many PIR schemes in the literature, and they fall into two main categories: those that provide computational security and those that provide information theoretic security. Computationally secure schemes are safe against any adversary restricted to polynomial-time computations; they rely on cryptographic assumptions for security and require only one server to store the database. Information theoretically secure schemes are safe even against computationally unbounded adversaries, but they typically require multiple non-colluding servers to store duplicate copies of the database. In practice, computational PIR can be as much as three orders of magnitude slower than trivial database transfer [22]; however, computational PIR has the significant advantage of needing only one server. Indeed, Chor *et al.* demonstrated that among single-server information theoretic PIR schemes, trivial database download is optimal [24]; we therefore focus on information theoretic multi-server schemes. That being said, some recent work has tried to improve the practicality of PIR by combining information theoretic techniques with computational ones [25]. We do not address this work in detail, but it has been considered in various applications of PIR.

We start by introducing the basic two-server PIR scheme that launched this area of research [24]. We then discuss extensions, including more efficient constructions, collusion resistance, and Byzantine tolerance. Finally, we discuss applications of PIR.

Basic PIR Scheme

The first information theoretic PIR scheme in the literature was proposed by Chor *et al.* [24]. In this scheme, two servers store identical copies of a database $f = [f_1 \ldots f_n]^T$, where f_i denotes the ith record. Suppose the client wishes to retrieve the wth record, f_w. Each record can be of arbitrary length, but for simplicity, we can assume each database element is a single bit, 0 or 1. The user's request can be represented by the indicator

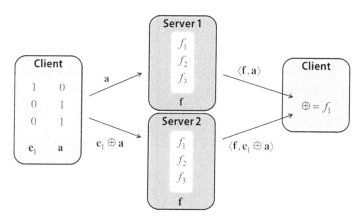

Figure 17.6 Two-server PIR scheme [24]. In this example, the client wants the first record, f_1. Each server computes the bitwise sum of a client-specified subset of database records. Since the two subsets differ only at the desired index, the binary sum of each server's results gives precisely f_1.

vector $e_w \in \{0,1\}^n$, which is 1 at index w and 0 elsewhere. If the user sent this indicator vector to the server in plaintext, it would be obvious which record is desired. The user therefore disguises her query by generating a random string $a \in \{0,1\}^n$, each entry drawn i.i.d. from a Bernoulli(1/2) distribution. The queries sent to servers 1 and 2 are $a \oplus e_w$ and a, respectively. Notice that each of these vectors is indistinguishable from random bits. This masking is equivalent to encrypting the desired query with a one-time pad, which offers perfect secrecy.

Upon receiving the query vectors, each server computes the inner product of its received query vector with the database f using bitwise addition (XOR) and returns a single-bit result. That is, the servers return $\langle f, a \rangle$ and $\langle f, a \oplus e_w \rangle$. The user XORs the results from the two servers to get f_w (see Fig. 17.6). In an honest-but-curious adversarial model, this PIR scheme is information theoretically private.

The asymptotic communication and computation costs of this scheme are both $O(n)$, but the constants are quite small since the client need only communicate a single bit per database record (of which there are n), and the server computes a simple XOR on database bits. In contrast, computationally secure schemes often rely on complex operations like modular arithmetic, which can increase runtimes by several orders of magnitude [22]. It is straightforward to lower the communication costs in the presented PIR scheme. Organizing the database into an ℓ-dimensional cube, where $\ell \geq 2$ is the number of servers, Chor et al. reduce the communication to $O(\sqrt[\ell]{n})$, or $O(\sqrt{n})$ in our two-server example [24]. In practice, these improvements can be significant.

Efficiency

Arguably the most popular research subproblem under the PIR umbrella is improving efficiency. Typically, this means reducing the communication cost (uplink, downlink, or both), but researchers have also considered methods for improving computational complexity, storage overhead, and repair efficiency for nodes that fail. While this area presents interesting theoretical problems, most of the schemes proposed in this space have not demonstrated practical gains over the basic scheme from [24]. For this reason, we give a brief overview without delving into technical details.

Communication complexity appears to depend on the number of servers. Until recently, the lowest communication complexity for a two-server scheme was $\Omega(n^{1/3})$, where n is the number of elements in the database [24]. In 2008, Yekhanin proposed the first sub-polynomial-communication PIR scheme for three servers [26], and Efremko extended the result to arbitrary numbers of servers $\ell \geq 3$ [27]. Dvir and Gopi made a breakthrough by proposing a *two*-server scheme with communication cost $n^{O(\sqrt{\log \log n}/\log n)}$, which exploits matching vector codes to achieve these gains [28]. However, if the primary concern is *downlink* communication, Shah et al. demonstrated that, information theoretically, any PIR scheme needs to incur at least one bit of communication overhead on the downlink compared to a non-private query [29]. They also showed that this one-extra-bit bound is achievable with a constructive PIR scheme that uses erasure-coded storage across servers, but with the need for an exponentially large number of servers. Fazeli et al. further demonstrated that PIR over coded databases can be used to reduce the storage overhead without affecting the communication complexity of the scheme [30].

To improve the online computational complexity, there exist PIR schemes that reduce the overall computation as a function of the number of servers. For instance, Beimel et al. propose a similar scheme that uses precomputation to achieve $O(n^{1/(2\ell-1)})$ asymptotic communication and $O(n/\log^{2\ell-2} n)$ computation, at the expense of $O(n^{1+\epsilon})$ storage for some $\epsilon > 0$ [31]. This implies that by increasing the number of server nodes in a network, we can reduce the asymptotic communication *and* computation costs.

Many of the schemes proposed in this subsection give practically negligible computational savings unless the database is large; given storage limitations on server nodes that are not operated by corporate entities, the resulting memory overhead or computation may be impractical. This tends to be a common theme in PIR research: most proposed efficiency improvements are asymptotic, and the leading constants can be large. Given how inefficient PIR is compared to non-private search, it remains to be seen if PIR over large databases will eventually become practical enough to reap these benefits.

Collusion Resistance

In the multi-server PIR scheme from [24], privacy is lost if servers collude. This is a problematic assumption because servers cannot inherently be trusted. Researchers have therefore studied PIR schemes that offer t-collusion-resistance – if no more than t of the ℓ servers collude, information theoretic security is guaranteed [32–34]. In this subsection, we explain a simplified version of the collusion-resistant scheme in [33], which uses concepts closely related to multi-party computation [35, 36].

The collusion-resistant PIR scheme in [33] allows up to t out of ℓ servers to collude without losing privacy. As we present it here, the scheme assumes honest-but-curious servers who obey the protocol but may collude in order to uncover the client's query. In reality, the PIR scheme in [33] and the follow-up work in [37] are able to handle malicious adversaries who do not follow protocol, as we will discuss in the following subsection. At a high level, the scheme in [33] proceeds as follows: The client designs a random polynomial $r(X)$ of degree t over a finite field $GF(2^\gamma)$ for some positive integer γ, such that $r(0) = f_w$ (the desired record). The client instructs each server to evaluate $r(X)$ at a distinct $X \neq 0$. The client then interpolates these evaluations to recover $r(0)$. Since $t+1$ points are required to interpolate a t-degree polynomial, up to t servers can collude without compromising privacy. Under some minor modifications, this collusion-resistant PIR scheme is also optimal in the number of servers needed and can deal with Byzantine or malicious servers.

More specifically, in the two-server case, the client transmitted a and $a \oplus e_w$ to servers 1 and 2, respectively. These queries can be interpreted as two characteristic points from a linear vector-valued polynomial $q(X) = aX + e_w$. Namely, when $X = 1$, we get $q(1) = a + e_w$ and $q'(1) = a$. We use $r(X)$ (or $r'(X)$) to denote a server's reply to the PIR query $q(X)$ (or $q'(X)$, respectively). The client's goal is to determine $r(0)$, the output for the query $q(0) = e_w$. It is straightforward to show that $r(X)$ is a linear function of X and $r(0) = f_w$, so we need only two distinct input queries to interpolate the value of $r(0)$. In the two-server example, $r(0)$ is interpolated by taking $\hat{r}(0) = r(1) + r'(1)$. This two-server query is not immune to server collusion.

Now suppose we wish to resist collusion between any set of at most t servers in the system – for concreteness, let $t = 2$. We can achieve this by querying $\ell \geq t+1$ servers; for our concrete example, let $\ell = 3$. Instead of the linear query from the two-server case, we now submit query vectors that are *quadratic* in X, i.e., of degree t. Note that this scheme requires a finite field $GF(2^\gamma)$ containing at least $t+1$ elements. So our example polynomial query is $q(X) = aX^2 + bX + e_w$, where $a,b \in \{0,\ldots,2^\gamma - 1\}^n$ are element-wise randomly drawn vectors, and e_w is a vector of zeros with a one at index w. So if the database has two elements, this is equivalent to drawing $n = 2$ independent univariate t-degree polynomials; all the n polynomials have a zero constant term except for the polynomial corresponding to the desired record index w. Concretely, suppose we want record $w = 1$. Then we draw two random quadratic polynomials (one per database record),

$$q(X) = \begin{bmatrix} q_1(X) \\ q_2(X) \end{bmatrix} = \begin{bmatrix} a_1 X^2 + b_1 X + 1 \\ a_2 X^2 + b_2 X + 0 \end{bmatrix}, \quad (17.1)$$

where a_i, b_i, q_i denote the ith elements of vectors a, b, q, respectively. The client evaluates these n polynomials (one for each database element) at ℓ distinct values of X – one for each server – and compounds the results into query vectors. For instance, the client might send $q(1)$ to server 1, $q(2)$ to server 2, and $q(3)$ to server 3. Since the queries are quadratic in X, at least three servers must collude to learn the desired record. Upon receiving such a query, each server projects the database records onto the query, and returns a single value as a result:

$$r(X) = q(X)^\mathsf{T} f$$
$$= f_1(a_1 X^2 + b_1 X + 1) + f_2(a_2 X^2 + b_2 X)$$
$$= (f_1 a_1 + f_2 a_2) X^2 + (f_1 b_1 + f_2 b_2) X + f_1.$$

Observe that $r(X)$ is quadratic in X, and $r(0) = f_w$. The client therefore needs three distinct function points to interpolate $r(0)$, or $t+1$ points in general.[3] We write

$$\underbrace{\begin{bmatrix} r(1) \\ r(2) \\ r(3) \end{bmatrix}}_{r} = \underbrace{\begin{bmatrix} 1^2 & 1^1 & 1^0 \\ 2^2 & 2^1 & 2^0 \\ 3^2 & 3^1 & 3^0 \end{bmatrix}}_{V} \times \begin{bmatrix} a^\mathsf{T} \\ b^\mathsf{T} \\ e_w^\mathsf{T} \end{bmatrix} \times \begin{bmatrix} f_0 \\ \vdots \\ f_n \end{bmatrix},$$

which is equivalent to $r = V \times [\cdots r(0)]^\mathsf{T}$. The interpolated $\hat{r}(0) = f_w$ can be computed using the bottom row of V^{-1} in the above expression (V is full rank). In our example, we get $\hat{r}(0) = [0\ 0\ 1] \cdot V^{-1}r = r(3) + 3r(2) + 3r(1)$. This approach can be extended to accommodate arbitrary collusion resistance parameters [33]. The key observation is that regardless of the number of servers, the replies from servers can be expressed as the product of a full-rank matrix V and a column vector, one of whose entries is the desired result, i.e., $r(0)$. So the desired record can be found by combining observed results in a linear combination, as in Eq. (17.3.2), which is fully determined by inverting V. These operations require a finite field with at least $t+1$ elements.

The described approach has an asymptotic communication cost of $\ell(n+L)$, where all operations are over the finite field $GF(2^\gamma)$ with $2^\gamma \geq \ell + 1$, and L is the size of a single database record in bits. Meanwhile, the computation cost is linear in the size of the database, but the fact that it only requires a single XOR for each bit in the database can provide significant savings with respect to computational PIR schemes.

Byzantine Robustness

Another significant problem that arises in PIR is dealing with Byzantine servers, or servers that do not follow protocol in order to thwart the client's ability to query the database correctly. Examples of Byzantine behavior include returning incorrect responses or not answering at all, in order to prevent the client from receiving the desired content. Questions of interest include designing schemes that return the correct answer even in the face of Byzantine servers, as well as identifying *which* servers are Byzantine. The setup is as follows: Consider ℓ servers, of which k respond (where $k \leq \ell$), and v respond incorrectly; the system should further be able to withstand up to t servers colluding, as before. This setup is called t-private v-Byzantine robust k-out-of-ℓ PIR.

Several researchers have considered the Byzantine robustness problem, starting with Biemel and Stahl [38, 39]. Their results hold when $v \leq t < k/3$; under these conditions, it is guaranteed that the protocol will return a unique, correct result. They showed that if there exists (1) a perfect hash family $H_{\ell,k}$ of size $w_{\ell,k}$, and (2) a k-out-of-k PIR protocol with communication complexity $\mathrm{PIR}_k(n)$ per server, then there also exists a k-out-of-ℓ PIR protocol with communication complexity $w_{\ell,k}\mathrm{PIR}_k(n)$ per server. This transformation can also be applied to accommodate schemes with t-collusion-resistance.

---

[3] Notice that in this particular example, we could actually decode with only two replies, since $r(X)$ is linear in f_1 and f_2. However, this special case arises because the example database only contains two elements.

Goldberg subsequently showed that if the protocol is allowed to return more than one result, the number of tolerable misbehaving servers can be increased, such that $t < k$ and $v < k - \lfloor\sqrt{kt}\rfloor$ [33]. Finally, Devet *et al.* improved the Byzantine robustness to $v < k - t - 1$, which is the theoretically largest possible number of Byzantine servers tolerated [37]. This advancement came by only changing the client-side processing of the scheme proposed in [33]. The key observation is that instead of simply interpolating a polynomial (as discussed in regards to collusion resistance), error correcting codes can be used to interpolate polynomials with noisy measurements. In particular, Reed–Solomon codes can correct both missing and corrupted symbols, and have a natural interpretation in terms of polynomial interpolation [40].

The well-known Berlekamp–Welch decoder can correctly decode a Reed–Solomon codeword when there are up to $v < (k - t)/2$ errors [41], but Devet *et al.* exploit list decoding, which allows the decoder to correct up to $v < k - \sqrt{kt}$ errors. List decoding of Reed–Solomon codes was first proposed by Guruswami and Sudan [42], albeit in the special case where the decoder interpolates a *single* polynomial with errors. Subsequent work [43, 44] considered the case in which *multiple* polynomials are being interpolated simultaneously. The key innovation of Devet *et al.* was adapting an existing heuristic list decoding scheme by Cohn and Heninger [45] for multi-polynomial reconstruction to fit the PIR setting. This multi-polynomial list decoding is currently the state of the art in Byzantine-resistant PIR [37].

Applications of PIR

A majority of existing PIR research has focused on the PIR mechanism itself – improving efficiency or extending its functionality. However, there has also been some research on systems that rely on PIR for practical data retrieval purposes. These systems have not been deployed, but are indicative of growing interest in PIR as a practical privacy tool. We briefly discuss a few of these applications.

Perhaps the best-known system to rely on PIR is PIR-Tor [46]. PIR-Tor is a system for accessing onion router information without revealing which Tor node is being accessed. The key observation is that P2P systems are often proposed for privacy-preserving systems, but they can present serious scalability challenges. On the other hand, centralized (or semi-centralized) systems scale better, but pose privacy threats because they present adversaries with a single point of failure. PIR-Tor proposes a client–server architecture for Tor, in which clients retrieve information about only a few onion routers at a time, using PIR. This approach prevents untrusted directory servers from learning about the onion routers a client plans to use. The authors consider both computational and information theoretic PIR for this application; they find that while computational PIR provides an order of magnitude improvement in communication cost over trivial database transfer, information theoretic PIR provides two orders of magnitude improvement.

A related system is DP5, which is a privacy-preserving distributed peer presence indicator [47]. Users who participate in social networks often wish to be informed when their contacts are online. The classical way to achieve this is to have users register with a central server whenever they come online. The server then passes this information along to each user's online contacts. This architecture requires the central server to know who is in contact with whom; such information is very sensitive, and there is a risk of server records being subpoenaed by the government [48]. DP5 is designed to enable such a presence service in a completely distributed privacy-preserving fashion. It does so by setting up a central registration server along with multiple PIR lookup servers. The idea is that when Alice joins the network, she registers herself with an encrypted message that only her friends can decrypt (each user shares randomness with her friends or contacts). When Bob wishes to check if Alice is online, he submits a PIR query to the lookup servers for her information. As in PIR-Tor, DP5 uses only information theoretic PIR due mainly to the significant speedups it offers.

A conceptually different application of PIR is Riposte, a system for anonymously posting messages on a public bulletin board [49]. Riposte uses a clever inversion of PIR, which enables the client to *write* a record to a database privately, instead of *reading* a record privately. Specifically, there is assumed to be a fixed set of servers that store the database, and a large number of clients. Each row of the database is a slot in which a client can write a message. When a client wants to post a message, she splits her message into secret shares [35], and writes each share to the same row of different servers' databases. To disguise which row is being overwritten, the client adds noise to the message shares in such a way that (1) the noise makes it appear as if the client is writing to *every* row, and (2) the noise cancels out across servers. In particular, recall that in the two-server PIR scheme from [24], the query vectors sent to each server looked like random bits, but their sum canceled to give the desired indicator vector. In Riposte, the same approach holds, except the "query vectors" are directly combined with the existing database contents. After enough clients have written messages, the servers combine their contents, which allows users' messages to be decoded and cancels out the noise. Since the servers do not collude prior to that, they cannot learn which row of the database was overwritten in each transaction with a client. The paper also describes methods for avoiding practical issues like collisions (clients writing to the same row) and improving bandwidth efficiency. However, the underlying idea is equivalent to information theoretic PIR.

17.3.3 Private Streaming Search

Although PIR has received a great deal of attention in the literature, it is somewhat limiting, since the user must know exactly which record she wishes to retrieve in the database. Private streaming search (PSS) relaxes this assumption. Instead of retrieving an indexed element of a database, it allows users to *search* databases using keyword-based queries. Specifically, PSS allows a user to retrieve all documents in a database that contain user-queried keywords, without revealing those keywords to the server. If a user queries more than one keyword at the same time, existing PSS schemes return all documents that contain *at least one* of the desired keywords.

Between PIR and PSS, PSS is arguably the more useful privacy primitive. It offers the user greater flexibility and more closely emulates existing methods of retrieving information from databases. However, work in this space has overwhelmingly relied on partially homomorphic encryption, which gives computational privacy guarantees. We introduce PSS and discuss the (comparatively smaller) body of work on information theoretic PSS.

Most PSS schemes rely on additively homomorphic encryption, which is a type of cryptosystem that enables computation *in the encrypted domain*. Let $\mathcal{E}(x)$ denote the encryption of x in such a cryptosystem. Informally, an additively homomorphic cryptosystem has the following property:

$$\mathcal{E}(x)\mathcal{E}(y) = \mathcal{E}(x+y).$$

The Paillier cryptosystem is probably the best-known additively homomorphic cryptosystem [50], and was used in the seminal paper on PSS by Ostrovsky and Skeith [51]. We briefly describe a simplified version of the most practical scheme from this paper, assuming the client only searches for one keyword at a time. The setup is as follows (with an example illustrated in Fig. 17.7):

1. There is an ordered dictionary of keywords D. The client chooses one of these for her query.
2. The client encrypts a vector of length $|D|$, with each entry corresponding to a keyword in the dictionary. The entry is $\mathcal{E}(1)$ if the corresponding dictionary keyword is desired, and $\mathcal{E}(0)$ otherwise. This encrypted vector gets sent to the server.[4]
3. Suppose for now that the database (stored on the server) has one document, which contains a subset of keywords in the dictionary D. For each dictionary keyword in the document, the server extracts the corresponding entry from the query vector, which is either $\mathcal{E}(1)$ or $\mathcal{E}(0)$; these entries are all multiplied together. Since the client requested only one keyword, there is at most one instance of $\mathcal{E}(1)$ in this product; so by the properties of homomorphic encryption, this product is either $\mathcal{E}(0)$ if the desired keyword is not present, or $\mathcal{E}(1)$ if it is. The server sends this encrypted product back to the client.
4. The client decrypts the reply. If the result is a zero, then the desired keyword was not present, and vice versa.

If there are multiple documents in the database, the server can place the results in a buffer in such a way that the client can recover all the results either deterministically or with high probability. Indeed, reducing the communication cost was the focus of subsequent work (at least in part) [52–55].

Most of the work on PSS gives computational privacy guarantees. Nonetheless, Olumofin and Goldberg studied the possibility of executing SQL queries on plaintext databases in a privacy-preserving way by adopting an approach similar to that of information theoretic PIR [56]. They proposed to store duplicate copies of the database to accommodate information theoretic PIR, and they convert basic SQL queries into a series of PIR queries over a database that is re-indexed to accommodate private queries. For instance, the server could use an inverted structure, where a database of

---

[4] Note that the Paillier cryptosystem is randomized, so two independent instances of $\mathcal{E}(0)$ look different with high probability.

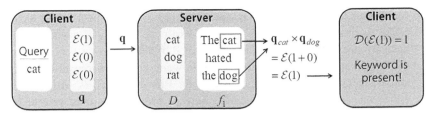

Figure 17.7 Example of PSS using homomorphic encryption, as described in [51]. Here, the client wishes to learn if the next document in the server's data stream (f_1) contains the keyword "cat." She generates an encrypted indicator vector for her query q, indexed by the keyword dictionary. The server uses properties of homomorphic encryption to generate a reply that informs the client whether the queried keyword is present or not.

documents is indexed by the keywords contained in each document. Using such an indexing scheme, PSS becomes trivial – the client need only submit a PIR query for the index that corresponds to the desired keyword(s) in the dictionary. However, processing more complicated queries, like conjunctive keyword queries, can increase the problem dimension, as well as the associated communication and computational complexity. This is a large part of the reason that PSS is not seeing practical adoptions: existing schemes are either limited in functionality, or they are too inefficient to be practical.

17.3.4 Encrypted Databases

We previously assumed that the database in question is public. However, as companies increasingly migrate storage and computation to the cloud, they may wish to mask their database contents from cloud service providers while still being able to search and query database contents normally. As such, the database must be encrypted. The question then becomes how to execute privacy-preserving database queries on *encrypted* databases. This area of work is similar in goals to PIR and PSS, but the underlying techniques are quite different.

Much work has gone toward designing efficient cryptographic primitives that can handle complex queries on encrypted data, such as equality, conjunctive, subset, and range queries on encrypted data. This area has received a great deal of attention, primarily from the applied cryptography community [57–60]. We do not present technical details since solutions inherently give computational guarantees, and the body of literature is also expansive. However, it is worth noting that this area has received a great deal of attention in both theoretical and practical circles. Applications like CryptDB [61] and Mylar [62] that rely on encrypted database searches have even seen limited adoption in industry.

17.4 Conclusions

There are many interesting and important open problems in this space. For instance, in the push toward distributed privacy-preserving search tools, it is important to account for the challenges that have traditionally faced peer-to-peer (P2P) systems as they relate to privacy. These include issues like peer churn and asynchrony, which have been studied extensively in the context of P2P networks, but very little in the context of privacy-preserving algorithms. Early work considers the effects of asynchrony and node failures [29, 63] specifically on PIR, but there is still much to be done. Other questions of interest include how to process complex private queries in an information-theoretic setting, and more traditional questions on how to reduce the complexity costs of retrieval and search to manageable levels. In the realm of anonymous messaging, there are interesting questions related to considering different adversarial models, as well as protecting the identities of relays who forward controversial content.

Privacy-preserving tools for information access and dissemination have been studied for decades. In practice, the information theoretic perspective on these problems has largely been overshadowed by tools that provide computational guarantees. This is partially due to practical constraints, since computationally secure tools often make fewer assumptions about the non-collusion of participating parties in a protocol. However, taking an information theoretic approach to privacy can result in sizeable efficiency gains over traditional cryptographic tools; in some problem areas, these gains have resulted in several orders of magnitude lower communication and/or computational complexity. Consequently, there are compelling reasons to study information theoretic approaches in the context of other privacy-preserving information flow problems. Indeed, a common criticism of privacy-preserving algorithms is that they are simply too inefficient to be practical. Moving forward, this is a solid reason to consider distributed algorithms with information theoretic guarantees, which can be significantly more efficient than traditional cryptographic tools, while simultaneously providing a stronger notion of privacy. We are optimistic that this area of research has the potential to produce privacy-preserving algorithms that are practical enough to change the way people access and share information.

Acknowledgments

Supported by NSF Grant number 1409135 and MURI CHASE grant number 556016.

References

[1] M. R. Lepper and D. Greene, "Turning play into work: Effects of adult surveillance and extrinsic rewards on children's intrinsic motivation," *J. Personality and Social Psychology*, vol. 31, no. 3, p. 479, Mar. 1975.

[2] B. J. Alge, "Effects of computer surveillance on perceptions of privacy and procedural justice," *J. Applied Psychology*, vol. 86, no. 4, p. 797, Aug. 2001.

[3] G. Greenwald. "XKeyscore: NSA tool collects 'nearly everything a user does on the internet'," *The Guardian*, Jul. 2014. [Online]. Available: www.theguardian.com/world/2013/jul/31/nsa-top-secret-program-online-data

[4] A. Nordrum. "Quantum computer comes closer to cracking RSA encryption," IEEE Spectrum, Mar. 2016. [Online]. Available: http://spectrum.ieee.org/tech-talk/computing/hardware/encryptionbusting-quantum-computer-practices-factoring-in-scalable-fiveatom-experiment

[5] A. Shontell. "7 people who were arrested because of something they wrote on Facebook," Jul. 2013. [Online]. Available: www.businessinsider.com/people-arrested-for-facebook-posts-2013-7

[6] P. K. Smith, J. Mahdavi, M. Carvalho, S. Fisher, S. Russell, and N. Tippett, "Cyberbullying: Its nature and impact in secondary school pupils," *J. Child Psychology and Psychiatry*, vol. 49, no. 4, pp. 376–385, Apr. 2008.

[7] D. Chaum, "The dining cryptographers problem: Unconditional sender and recipient untraceability," *J. Cryptology*, vol. 1, no. 1, pp. 65–75, Jan. 1988.

[8] M. Waidner and B. Pfitzmann, "The dining cryptographers in the disco: Unconditional sender and recipient untraceability with computationally secure serviceability," *Lecture Notes in Computer Science*, vol. 434, p. 690, 1989.

[9] P. Golle and A. Juels, "Dining cryptographers revisited," *Lecture Notes in Computer Science*, vol. 3027, pp. 456–473, 2004.

[10] S. Goel, M. Robson, M. Polte, and E. Sirer, "Herbivore: A scalable and efficient protocol for anonymous communication," Cornell University, Tech. Rep., 2003.

[11] H. Corrigan-Gibbs and B. Ford, "Dissent: Accountable anonymous group messaging," in *Proc. 17th ACM Conf. Computer and Communications Security*, Chicago, IL, USA, Oct. 2010, pp. 340–350.

[12] D. I. Wolinsky, H. Corrigan-Gibbs, B. Ford, and A. Johnson, "Dissent in numbers: Making strong anonymity scale," in *Proc. 10th USENIX Conf. Operating Systems Design and Implementation*, Hollywood, CA, USA, Oct. 2012, pp. 179–182.

[13] G. Fanti, P. Kairouz, S. Oh, and P. Viswanath, "Spy vs. spy: Rumor source obfuscation," in *Proc. 2015 ACM SIGMETRICS Int. Conf. Measurement and Modeling of Computer Systems*, Portland, OR, USA, Jun. 2015, pp. 271–284.

[14] W. Luo, W. P. Tay, and M. Leng, "Rumor spreading and source identification: A hide and seek game," in *Proc. 2015 IEEE/ACM Int. Conf. Advances in Social Networks Analysis and Mining*, Paris, France, Aug. 2015, pp. 186–193.

[15] P. Syverson, G. Tsudik, M. Reed, and C. Landwehr, "Towards an analysis of onion routing security," in *Proc. Int. Workshop Designing Privacy Enhancing Technologies: Design Issues in Anonymity and Unobservability*, Berkeley, CA, USA, Jul. 2001, pp. 96–114.

[16] J. Feigenbaum, A. Johnson, and P. Syverson, "A model of onion routing with provable anonymity," *Lecture Notes in Computer Science*, vol. 4886, pp. 57–71, 2007.

[17] M. Barbaro and T. Zeller. "A face is exposed for AOL searcher no. 4417749," *New York Times*, Aug. 2006. [Online]. Available: www.nytimes.com/2006/08/09/technology/09aol.html

[18] D. McCullagh, "Verizon draws fire for monitoring app usage, browsing habits," *CNET*, Oct. 2012. [Online]. Available: www.cnet.com/news/verizon-draws-fire-for-monitoring-app-usage-browsing-habits/

[19] K. Hill. "Facebook will use your browsing and app history for ads (despite saying it wouldn't 3 years ago)," *Forbes*, Jun. 2014. [Online]. Available: www.forbes.com/sites/kashmirhill/2014/06/13/facebook-web-app-tracking-for-ads/

[20] R. Dingledine, N. Mathewson, and P. Syverson, "Tor: The second-generation onion router," in *Proc. 13th Conf. USENIX Security Symposium – Volume 13*, San Diego, CA, USA, Aug. 2004, pp. 21–21.

[21] C. Gentry and Z. Ramzan, "Single-database private information retrieval with constant communication rate," *Lecture Notes in Computer Science*, vol. 3580, pp. 803–815, 2005.

[22] F. Olumofin and I. Goldberg, "Revisiting the computational practicality of private information retrieval," in *Financial Cryptography and Data Security*, ser. Lecture Notes in Computer Science, G. Danezis, Ed., Berlin, Heidelberg, 2012, vol. 7035, pp. 158–172.

[23] S. Papadopoulos, S. Bakiras, and D. Papadias, "pCloud: A distributed system for practical PIR," *IEEE Trans. Dependable Secure Computing*, vol. 9, no. 1, pp. 115–127, Jan. 2012.

[24] B. Chor, O. Goldreich, E. Kushilevitz, and M. Sudan, "Private information retrieval," in *Proc. 36th Annual Symposium Foundations of Computer Science*, Milwaukee, WI, USA, Oct. 1995, pp. 41–50.

[25] C. Devet and I. Goldberg, "The best of both worlds: Combining information-theoretic and computational PIR for communication efficiency," *Lecture Notes in Computer Science*, vol. 8555, pp. 63–82. 2014.

[26] S. Yekhanin, "Towards 3-query locally decodable codes of subexponential length," *J. ACM*, vol. 55, no. 1, pp. 1:1–1:16, Feb. 2008.

[27] K. Efremenko, "3-query locally decodable codes of subexponential length," *SIAM J. Comput.*, vol. 41, no. 6, pp. 1694–1703, 2012.

[28] Z. Dvir and S. Gopi, "2-server PIR with sub-polynomial communication," *J. ACM*, vol. 63, no. 4, pp. 39:1–39:15, Sep. 2016.

[29] N. B. Shah, K. V. Rashmi, and K. Ramchandran, "One extra bit of download ensures perfectly private information retrieval," in *Proc. IEEE Int. Symp. Inf. Theory*, Honolulu, HI, USA, Jul. 2014, pp. 856–860.

[30] A. Fazeli, A. Vardy, and E. Yaakobi, "PIR with low storage overhead: Coding instead of replication," May 2015. [Online]. Available: http://arxiv.org/abs/1505.06241

[31] A. Beimel, Y. Ishai, and T. Malkin, "Reducing the servers' computation in private information retrieval: PIR with preprocessing," *J. Cryptology*, vol. 17, no. 2, pp. 125–151, Mar. 2004.

[32] A. Beimel, Y. Ishai, and E. Kushilevitz, "General constructions for information-theoretic private information retrieval," *J. Computer and System Sciences*, vol. 71, no. 2, pp. 213–247, Aug. 2005.

[33] I. Goldberg, "Improving the robustness of private information retrieval," in *IEEE Symp. Security Privacy*, Oakland, CA, USA, May 2007, pp. 131–148.

[34] E. Yang, J. Xu, and K. Bennett, "Private information retrieval in the presence of malicious failures," in *Proc. 26th Annual Int. Comp. Software Appl. Conf.*, Oxford, UK, Aug. 2002, pp. 805–810.

[35] A. Shamir, "How to share a secret," *Commun. ACM*, vol. 22, no. 11, pp. 612–613, Nov. 1979.

[36] M. Ben-Or, S. Goldwasser, and A. Wigderson, "Completeness theorems for non-cryptographic fault-tolerant distributed computation," in *Proc. Twentieth Annual ACM Symp. Theory of Comput.*, Chicago, IL, USA, 1988, pp. 1–10.

[37] C. Devet, I. Goldberg, and N. Heninger, "Optimally robust private information retrieval," in *Proc. 21st USENIX Conf. Security Symp.*, Bellevue, WA, USA, Aug. 2012, p. 13.

[38] A. Beimel and Y. Stahl, "Robust information-theoretic private information retrieval," *Lecture Notes in Computer Science*, vol. 2576, pp. 326–341, 2003.

[39] A. Beimel and Y. Stahl, "Robust information-theoretic private information retrieval," *J. Cryptology*, vol. 20, no. 3, pp. 295–321, 2007.

[40] I. S. Reed and G. Solomon, "Polynomial codes over certain finite fields," *J. Society Industrial and Applied Mathematics*, vol. 8, no. 2, pp. 300–304, Jun. 1960.

[41] L. R. Welch and E. R. Berlekamp, "Error correction for algebraic block codes," Dec. 1986, US Patent 4633470.

[42] V. Guruswami and M. Sudan, "Improved decoding of Reed–Solomon and algebraic–geometric codes," in *Proc. 39th Annual Symp. Found. Computer Science*, Palo Alto, CA, USA, Nov. 1998, pp. 28–37.

[43] F. Parvares and A. Vardy, "Correcting errors beyond the Guruswami–Sudan radius in polynomial time," in *Proc. 46th Annual IEEE Symp. Found. Computer Science*, Pittsburgh, PA, USA, Nov. 2005, pp. 285–294.

[44] V. Guruswami and A. Rudra, "Explicit codes achieving list decoding capacity: Error-correction with optimal redundancy," *IEEE Trans. Inf. Theory*, vol. 54, no. 1, pp. 135–150, Jan. 2008.

[45] H. Cohn and N. Heninger, "Approximate common divisors via lattices," *The Open Book Series*, vol. 1, no. 1, pp. 271–293, 2013.

[46] P. Mittal, F. Olumofin, C. Troncoso, N. Borisov, and I. Goldberg, "PIR-Tor: Scalable anonymous communication using private information retrieval," in *Proc. 20th USENIX Conf. Security*, San Francisco, CA, USA, Aug. 2011, pp. 31–31.

[47] N. Borisov, G. Danezis, and I. Goldberg, "DP5: A private presence service," *Proc. Privacy Enhancing Technol.*, vol. 2015, no. 2, pp. 1–21, Jun. 2015.

[48] D. Rushe, "Lavabit founder refused FBI order to hand over email encryption keys," *The Guardian*, Oct. 2013. [Online]. Available: www.theguardian.com/world/2013/oct/03/lavabit-ladar-levison-fbi-encryption-keys-snowden

[49] H. Corrigan-Gibbs, D. Boneh, and D. Mazières, "Riposte: An anonymous messaging system handling millions of users," in *IEEE Symp. Security and Privacy*, San Jose, CA, USA, May 2015, pp. 321–338.

[50] P. Paillier, "Public-Key Cryptosystems Based on Composite Degree Residuosity Classes," *Lecture Notes in Computer Science*, vol. 1592, pp. 223–238, 1999.

[51] R. Ostrovsky and W. E. Skeith, "Private searching on streaming data," *Lecture Notes in Computer Science*, vol. 3621, pp. 223–240, 2005,

[52] J. Bethencourt, D. Song, and B. Waters, "New constructions and practical applications for private stream searching," in *Proc. IEEE Symp. Security Privacy*, Oakland, CA, USA, May 2006, pp. 134–139.

[53] J. Bethencourt, D. Song, and B. Waters, "New techniques for private stream searching," *ACM Trans. Inf. System Security*, vol. 12, no. 3, p. 16, Jan. 2009.

[54] G. Danezis and C. Diaz, "Space-efficient private search with applications to rateless codes," *Lecture Notes in Computer Science*, vol. 4886, pp. 148–162, 2007.

[55] M. Finiasz and K. Ramchandran, "Private stream search at the same communication cost as a regular search: Role of LDPC codes," in *Proc. IEEE Int. Symp. Inf. Theory*, Cambridge, MA, USA, Jul. 2012, pp. 2556–2560.

[56] F. Olumofin and I. Goldberg, "Privacy-preserving queries over relational databases," *Lecture Notes in Computer Science*, vol. 6205, pp. 75–92, 2010.

[57] M. Bellare, A. Boldyreva, and A. O'Neill, "Deterministic and efficiently searchable encryption," *Lecture Notes in Computer Science*, vol. 4622, pp. 535–552, 2007.

[58] J. Bethencourt, H. Chan, A. Perrig, E. Shi, and D. Song, "Anonymous multi-attribute encryption with range query and conditional decryption," in *Proc. IEEE Symp. Security Privacy*, Oakland, CA, USA, May 2006.

[59] D. Boneh and B. Waters, "Conjunctive, subset, and range queries on encrypted data," *Lecture Notes in Computer Science*, vol. 4392, pp. 535–554, 2007.

[60] A. Boldyreva, N. Chenette, and A. O'Neill, "Order-preserving encryption revisited: Improved security analysis and alternative solutions," *Lecture Notes in Computer Science*, vol. 6841, pp. 578–595, 2011.

[61] R. A. Popa, C. Redfield, N. Zeldovich, and H. Balakrishnan, "CryptDB: Protecting confidentiality with encrypted query processing," in *Proc. 23rd ACM Symp. Operating Systems Principles*, Cascais, Portugal, Oct. 2011, pp. 85–100.

[62] R. A. Popa, E. Stark, J. Helfer, S. Valdez, N. Zeldovich, M. F. Kaashoek, and H. Balakrishnan, "Building web applications on top of encrypted data using mylar," in *Proc. 11th USENIX Conf. Networked Systems Design Implementation*, Seattle, WA, USA, Apr. 2014, pp. 157–172.

[63] G. Fanti and K. Ramchandran, "Efficient private information retrieval over unsynchronized databases," in *Proc. 52nd Annual Allerton Conf. Commun., Control, Computing*, Monticello, IL, USA, Sep. 2014, pp. 1229–1239.

18 Privacy in the Smart Grid: Information, Control, and Games

H. Vincent Poor

The deployment of advanced cyber infrastructure in the electricity grid, leading to the so-called smart grid, introduces concerns about the privacy of the parties interconnected by the grid. In this chapter, the roles of information, control, and game theories in quantifying and exploring such privacy issues are examined, primarily in the context of smart grid communications. Absolute privacy is generally not desirable in such applications, but rather the tradeoff between the privacy of data and its usefulness, or utility, is of concern. An information theoretic setting is used to characterize the optimal such tradeoff in a general context, and this setting is then applied to two problems arising in smart grid: smart meter privacy, in which control theory also plays a role when storage is introduced into the system; and competitive privacy, in which each of a group of multiple grid actors seeks to optimize its individual privacy–utility tradeoff while interacting with the other actors in information exchange, a setting that can be examined using game theory.

18.1 Introduction

The hierarchical, centralized, electricity grid is being transformed into a cyber-physical system. This transformation involves the introduction of sensors, meters, controls, and communication networks [1], in order to allow for greater end-point participation in the management of the grid. In particular, these developments will allow for demand-side participation in grid management, better integration of renewables and storage into the grid, etc., in order to create a more efficient, sustainable energy infrastructure. A key issue that arises when cyber capability is introduced into a system is privacy. This chapter will address this issue, first in a general setting, and then in the context of some particular problems arising in smart grid.

In considering this problem, we will see that information theory, control theory, and game theory all have roles to play in helping to understand the issue of privacy in the smart grid. In particular, we will first describe a general formalism for privacy, based on an information theoretic formulation. Then we will discuss the application of this formalism in two different settings arising in the smart grid. The first of these is smart meter privacy, where control theory will also enter the toolset as we consider the control of storage to help protect privacy. As a second application, we will describe another related problem, namely competitive privacy, in which privacy becomes a transactional

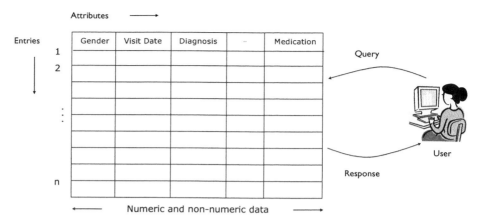

Figure 18.1 A database model.

part of a grid system involving market competitors. And, as the word "competitive" suggests, this setting is one in which game theory has a central role to play.

18.2 Information: A General Formalism

The introduction of a cyber layer into the electricity grid means that the grid becomes a source of electronic data. Some of this data will be sensitive or private [2–5]. For example, a major part of the instrumentation of the electricity grid is the introduction of smart meters, which collect and report the electricity usage at customer premises in almost real time. So, smart meters are generating data about individual consumers that can potentially reveal private information about those individuals or their households. Another aspect of the cyber layer is the widespread deployment of sensors, such as phasor measurement units (PMUs), that generate data about what is happening in the grid.

Of course, the reason why this data exists is because it is useful. But it could also leak sensitive information that should be kept private. An extreme way to keep such data private would be to simply store it somewhere in a vault and not let it be accessed by anyone. But the utility of data depends on its accessibility to those who need it to make pricing decisions, control the grid, etc. So there is a fundamental dichotomy, or tradeoff, here between the privacy and utility of data. This is true of essentially any data source, not just of smart grid data. A basic question in designing and operating information systems is, how can we characterize this fundamental tradeoff? One way of addressing this question is to examine the privacy–utility tradeoff in an information theoretic setting [6]. This approach can be described straightforwardly in the context of a database, and we will begin with that context, keeping in mind, though, that the same principles can be applied to other problems, as we will see later.

A database can be thought of as a table or matrix. The rows of the matrix correspond to different entities or entries whose data is contained in the database. So an entry might be an individual or an event, or something along those lines. The columns correspond to attributes of, i.e., data about, the entries (see Fig. 18.1). The data can consist of real numbers, categorical data, etc. So, for example, in a medical database an entry might be a patient or a patient visit, and the attributes would be things like demographic data, visit date, test results, diagnosis, treatment, and perhaps outcomes. And, of course, the way a database is used is that a user of the database sends a query to the database, and the database returns a response.

In modeling a database statistically, it is reasonable (at least to first order) to think of the rows as being independent and identically distributed (i.i.d.), since each row corresponds to a separate entity. But the columns will generally be correlated or dependent since they are related to the same entry. For example, in the example of a medical database, certainly demographics relate to test results and diagnoses, and so forth, and also to the treatment and response. So as a statistical model we have a set of n i.i.d. vectors (the entries), with the components of the vectors themselves (the attributes) being dependent.

To consider the privacy–utility tradeoff in this setting, we can divide the attributes into two types, public (or revealed) and private (or hidden). The public ones can be revealed without any concerns, while the private ones should be kept hidden to the extent possible. And, there could be overlap between the two types. For example, a financial database might contain 16-digit credit card numbers, which should be kept private in full, while the last four digits might be revealed for various purposes without concern. But even if there is no overlap, revealing only the public data can leak information about the private data because of the dependence between various attributes. So there will be a tradeoff due to this dependence, even without overlap between public and private variables.

Given the above statistical model, how can we characterize the tradeoff between utility and privacy? We can first consider utility. Clearly, if we are going to protect aspects of the database, any response to a query will not reveal the entire database. This means that there will be distortion between the true database and the database as revealed to users of the database. So, we can measure utility in terms of this distortion introduced in the public variables as they are revealed to a user of the database. This will be an inverse measure of utility since low distortion means higher utility. Similarly, we can measure privacy in a time-honored way in information theory, in terms of the entropy of the private variables conditioned on information revealed to a user of the database.[1] This conditional entropy is called the *equivocation*. This is a direct measure of privacy: higher equivocation means greater privacy. And, if the equivocation were to equal the unconditioned entropy of the private variables, then there would be no privacy leakage at all. (Note that we could also use other measures of information leakage instead of equivocation, as we will see in some of the examples below.)

Now we have a clearly defined mathematical problem. We have an i.i.d. set of vectors, we have distortion of some of the elements of these vectors, and we have equivocation of others. In order to characterize the tradeoff between these two quantities, we would

---

[1] For other ways of characterizing privacy in databases, see, e.g., [7–11].

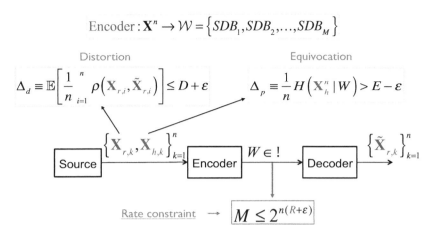

Figure 18.2 A communication model.

like to find all the possible distortion–equivocation pairs for that particular statistical model. And, once we have determined that region, the boundary of it will give us the efficient frontier in the privacy–utility tradeoff.

So, the next question that arises is, how can we solve this mathematical problem? An appealing way of doing that is to convert this problem into a traditional communications problem. Recall that the database, in this model, is a big random matrix. We can think about the way we reveal it to a user as first quantizing it into a finite set of databases, and then revealing the quantized version to the user. This quantization introduces distortion, but it also introduces a level of privacy. We can call the quantized databases *sanitized* databases, a term that comes from the computer science literature on database privacy. They are sanitized because some information has been removed from them.

Once we think of releasing the database in terms of quantization, we can examine the problem at hand as a communication problem, as shown in Fig. 18.2. We have a source, which is the database. It has two types of elements, the revealed and the hidden. We quantize the source into one of the set of sanitized databases before sending it to a receiver, which is the user of the database. The user of the database serves as a decoder, trying to reconstruct the revealed variables from the quantized database.

We measure the utility of that reconstruction in terms of distortion, which we quantify based on a distortion metric ρ and looking at the average expected distortion per entry. We would like this quantity to be less than some amount D, and to use asymptotic information theoretic arguments, we add a small amount ϵ, which will go to zero as n goes to infinity. We measure privacy in terms of equivocation, which is just the conditional entropy of the hidden variables given the sanitized database. Again, we consider the average per entry, and want that to be greater than some amount E, minus an ϵ that will go to zero as n goes to infinity. What we would like to find is the range of possible D–E pairs as $n \to \infty$ and $\epsilon \to 0$.

Now, it turns out that it is easier in general to find the set of achievable D–E pairs if we further introduce a third variable, the rate R, where R refers to the number of

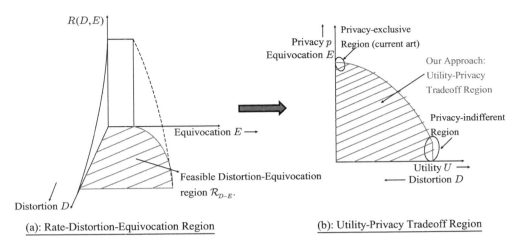

Figure 18.3 Privacy–utility tradeoff region. (Used with permission from [6].)

bits per entry that are released per query. Bounding R is equivalent to bounding M, the size of the set of sanitized databases. In this database problem, we are not particularly interested in this rate, unless we are thinking about the query being communicated over a capacity-limited line, but we introduce rate here for another reason. Now that we have three things, R, D, and E, this seems like a harder problem to solve. But in fact, now we have an easier problem, because we can now relate this to a classical problem in information theory, namely the tradeoff between rate and distortion, which is the classical rate–distortion problem. So what we have is a rate–distortion problem with an equivocation constraint. This is also a somewhat classical problem, studied by Yamamoto in the 1980s [12]. Yamamoto's work involved a simpler model than this, but he did consider the rate–distortion tradeoff with an equivocation constraint. So there is some existing machinery that we can apply to solve this problem now.

The general nature of this problem is illustrated in Fig. 18.3. In Fig. 18.3(a), we have three axes – rate, equivocation, and distortion. For a given achievable distortion–equivocation pair there is a minimal rate at and above which this pair can be achieved. This creates a surface of minimal rates, and the rate–distortion–equivocation region is everything that lies above this surface.

The projection of this surface on the rate-distortion plane is the classical rate–distortion function, whereas the projection down into the equivocation–distortion plane is what we are interested in here. So once we can solve this problem, we can look at this projection to determine the privacy–utility tradeoff region, which is shown in Fig. 18.3(b). (Of course, utility is going in the opposite direction from distortion.) Note that the outer boundary of this region is the "efficient frontier" of this tradeoff. This boundary does not tell us what to do, but it does tell how much distortion we have to accept if we want to achieve a given level of privacy. Some specific examples are found in [6], and others arising in the smart grid context will be discussed below.

To summarize this general formalism, we looked at a database as an example of an information source. We divided the attributes into public and private ones, and this led to an equivocation–distortion characterization of the privacy–utility tradeoff. We added a rate parameter to this model, and that led to a rate–distortion problem with an equivocation constraint, which put it in the context of classical, or at least established, information theoretic analysis.

It should be noted that there are other things that can be considered in this context. For example, multiple queries: if a user queries the database multiple times, it changes the tradeoff. This is a problem of successive disclosure, which can also be analyzed. We can also consider multiple sources with overlapping entries, which is a significant practical problem in databases; i.e., multiple databases can be combined to get at private information that could not be obtained from a single database alone. This issue can be examined by considering the above problem with side information, where the side information comes from a secondary database. This also changes the tradeoff region. These issues are discussed in [6] (see also [13]). We can also look at other measures of privacy and utility, so there is a lot more that can be done with this model. Now, however, we turn to the application of these ideas to some problems arising in the smart grid.

18.3 Control: Smart Meter Privacy

In this section and the next, we turn to some applications of the ideas of the previous section in the smart grid. We begin, in this section, with the problem of smart meter privacy.

As noted above, a smart meter is basically an electricity meter that measures and reports electricity usage at a home or other premises on an almost real-time basis. Typically, smart meters report every 15 minutes or so, but for the purposes of discussion we will assume this reporting is on a real-time basis. This data is useful as it permits more efficient management of electricity usage than the more traditional ways of measuring electricity usage. On the other hand, such data can also leak information about what is happening inside the home. Figure 18.4 is a cartoon, but if you look at a real smart meter trace, it looks similar to this, albeit a little noisier. It can be seen from this trace that different appliances in the home have distinctive signatures. So what is happening inside a smart-metered house can be inferred from the smart meter trace. So, there is a privacy issue with divulging electricity usage on this time scale. A number of works have investigated privacy leakage through smart meters [14–29].

Clearly, there is a privacy–utility tradeoff here, and we can apply the general ideas of the preceding section to help understand it. We will look at two approaches in this section. The first is a source coding approach in which the smart meter reporting is modified; this approach is similar to the one considered in the preceding section, in which the revealed source is modified so as to provide the desired tradeoff. The second approach is to introduce a control into the system in the form of a controllable storage device, via which the actual usage from the power company can be controlled to hide information about usage within the home.

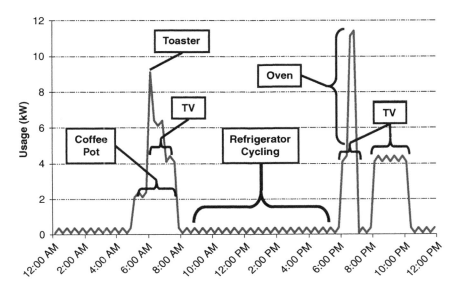

Figure 18.4 Cartoon of a smart meter trace.

In order to consider the first approach, we follow [30], and model the smart meter trace as a stationary hidden state Gauss–Markov process. The hidden state is either continuous or intermittent, where the continuous state is active when continuous loads such as air conditioners and refrigerators are dominant in the trace. In the intermittent state, things like a coffee pot, toaster, tea kettle, etc. are on, and this is going to be more informative about what is happening in the home than the continuous loads are. As noted above, we will enable the tradeoff between utility and privacy by encoding of the meter readings before they are transmitted to the power company. As in the preceding section, we will measure utility in terms of the distortion of the meter measurements after encoding; we will specifically consider the mean square error as a distortion measure. We measure privacy in terms of information leakage about the intermittent state.

Of course, this problem differs from the database problem of the previous section, notably in that the measurements are a time series and are not i.i.d., and that the private attribute corresponds to a hidden state here. Nevertheless, similar analysis can be applied to show that the optimal tradeoff between utility and privacy here is provided by a classical technique from information theory, namely *reverse water-filling*, which provides a rate-minimizing source code for Gaussian sources. This solution can be illustrated via Fig. 18.5, which represents the power spectral density of the smart meter trace. There is a "water level," ϕ, such that those spectral components whose power spectral density is below ϕ are suppressed from the usage reported by the meter, while the remaining components are reported undistorted. The intuition here is that the intermittent events, even though they are locally of high power, are actually low power overall because of their low duty cycle. So we are essentially suppressing those intermittent events when we suppress the low-power parts of the spectrum. On the other hand, the higher-power components come from things like air conditioning and

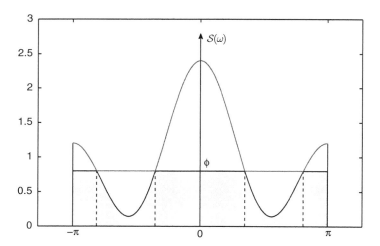

Figure 18.5 Reverse water-filling.

refrigerators which are in the continuous state parts of the signal. The efficient frontier of the privacy–utility tradeoff is traversed if we move the water level up and down. So a lower water level means more utility, less privacy; and vice versa.

Source coding of the trace is one approach to the smart meter privacy problem, but another approach is to use active control of the power drawn from the main grid to effect a privacy–utility tradeoff. To do so, we now assume that we have, in addition to a smart meter and loads, two other things, as illustrated in Fig. 18.6. We have an energy harvester, such as a wind turbine or solar array, and we have a battery in which we can store energy. The charging and discharging of the battery can be controlled, and this provides a mechanism for providing a certain level of privacy. Of course, the energy harvester also provides a certain level of privacy – if it provided all the energy needed by the loads instantaneously, this would result in full privacy from the main grid. But, active control comes by way of the energy storage.

In order to develop some understanding for the roles of energy harvesting and storage, we can consider a very simple setting from [31]. For a given discrete time instant i, we have the energy demand of the premises, which we denote by X_i. We assume this demand is i.i.d. binary – zero or one. This, of course, is not realistic in practice, but it will allow us to gain some insight. We further have the harvested energy, Z_i at time i, which we will also assume to be i.i.d. binary, and independent of the demand. We have a battery state, which we assume to be bounded by one initially, although we consider larger amounts of storage below. Finally, Y_i denotes the energy taken from the utility provider; the meter will read and report Y_i, which is the point at which information privacy leakage happens.

To control privacy leakage, there is an energy management unit, which is essentially a stochastic control; it takes the current load, the current harvested energy, and the battery state from the last time instant, and maps them to a new output load and a new battery

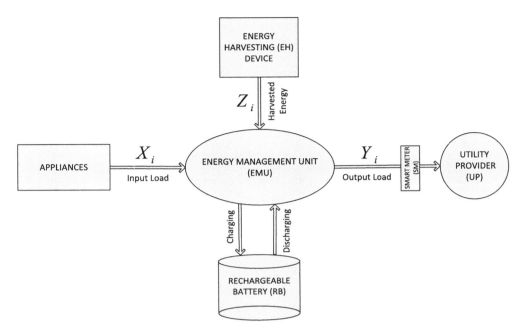

Figure 18.6 A system with energy harvesting and storage. (Used with permission from [31].)

state. That is, the energy management unit implements a stochastic mapping:

$$(X_i, Z_i, B_{i-1}) \rightarrow (Y_i, B_i). \tag{18.1}$$

And of course there are constraints, because we cannot draw more energy than exists in the whole system:

$$X_i \leq Z_i + (B_i - B_{i-1}) + Y_i. \tag{18.2}$$

As an (inverse) measure of privacy, we consider the information leakage rate as quantified by the average mutual information between the load and what is revealed at the meter, i.e.,

$$\lim_{n \to \infty} \frac{1}{n} I(X^n; Y^n), \tag{18.3}$$

where n is the number of time periods considered, $X^n = X_1, \ldots, X_n$, $Y^n = Y_1, \ldots, Y_n$, and $I(\cdot; \cdot)$ denotes mutual information. Due to the stochastic nature of the load and energy harvesting, and the limit on storage, energy may be wasted. So, as an (inverse) measure of utility, we can consider the average wasted energy:

$$\lim_{n \to \infty} \frac{1}{n} \sum_{i=1}^{n} (Z_i + Y_i - X_i). \tag{18.4}$$

To implement a stochastic control for the energy management unit, we want to choose a probability distribution on the amount of energy we draw from the utility provider, or main grid. Since the drivers are binary and i.i.d., the system evolves according to a

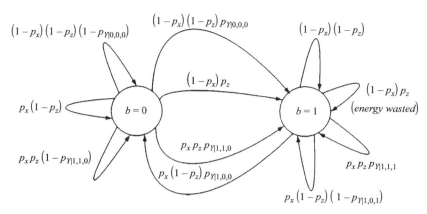

Figure 18.7 Markov chain representation of the stochastic control problem. (Used with permission from [31].)

Markov chain, as shown in Fig. 18.7. The transition probabilities here will determine the various battery states, and in turn the privacy–utility tradeoff. Because we have assumed that the independent stochastic elements are i.i.d. binary, determining the optimal control is a reasonably manageable problem. Unfortunately, even this simple problem cannot be solved analytically because the battery introduces memory that makes it very hard to get closed-form expressions. However, it can be solved numerically, because of the small dimension of the state space, by enumerating through the space of transition probabilities that we can choose until we find the efficient frontier. (Details of how the computations can be carried out are found in [31]; e.g., mutual informations are estimated using a technique from [32].) The results of such a search are shown in Fig. 18.8 for the case in which both the energy harvester and the load take on the values zero and one with equal probabilities of one-half. It is very clear from this figure that the choice of control strategy has a very strong effect on the quality of the tradeoff.

We can examine more closely the role of storage in the tradeoff, and the interaction between storage and energy harvesting, by considering the privacy and utility measures, with and without storage, as functions of the energy harvesting rate. These are shown in Fig. 18.9. In this analysis, the load is i.i.d. equiprobable binary, while the rate of i.i.d. binary energy harvesting varies along the horizontal axis. The top figure shows the minimum information leakage rate, meaning at the efficient frontier, and the bottom figure shows the wasted energy rate, again at the efficient frontier, both versus the energy harvesting rate as it ranges from zero to one.

In both cases, as the energy harvesting rate approaches one (i.e., energy is being harvested all the time), the need for storage disappears and no energy is drawn from the main grid (and hence no privacy is leaked), but energy is wasted half the time. With lower energy harvesting rates, the effects of storage are much greater, as one would expect, since the battery can smooth the load as perceived by the main grid. Without energy harvesting, no energy is wasted with or without a battery (here we do not allow wasting of grid energy), so the strongest effects of storage on utility are observed for the middle range of energy harvesting rates. These figures would change, of course, for different duty cycles of the load, but the intuition would be similar.

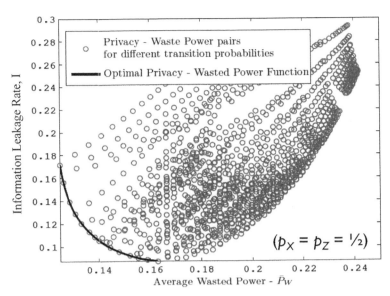

Figure 18.8 Privacy–utility tradeoff with energy harvesting and storage. (Used with permission from [31].)

We can also look at what happens without energy harvesting to examine the effects of battery capacity on the tradeoff. This analysis[2] is shown in Fig. 18.10. The left-hand figure shows the minimum information leakage rate without energy harvesting versus battery capacity, again with i.i.d. equiprobable binary load. As we would expect, as the battery capacity increases, we leak less and less information, because we can smooth out the draw from the main grid by storing energy in the battery and using that to serve the load when required. In the figure on the right, we see the tradeoff of information leakage rate versus wasted power with varying battery capacity. (In this case, we allow power from the main grid to be wasted if it helps the tradeoff.) This curve was produced numerically as in Fig. 18.8, although here we show only the efficient frontier. And, as expected, we see that as we increase the battery capacity, it gets easier to tradeoff between privacy and utility.

To summarize this section, we have examined the privacy–utility tradeoff for smart metering. We considered two approaches: source coding at the meter, and control based on storage and energy harvesting. In the first case, the optimal tradeoff comes from a classical source coding technique: reverse water-filling. In the second case, we see that local storage and energy sources give us additional flexibility in controlling the tradeoff. Other formulations using control have also been considered for this problem. For example, in [34], utility is measured in terms of the cost of electricity, while privacy is measured in terms of the variance of the load from the average load. So if the average load is always the same, privacy is complete. An adaptive controller for a continuous state is introduced, and the control adapts not only to the load, but also to price variations that the grid introduces based on meter data.

---

[2] See also [33] for an alternative analysis of this situation.

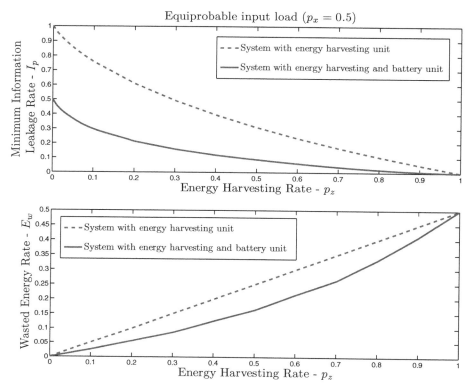

Figure 18.9 With and without energy storage. (Used with permission from [31].)

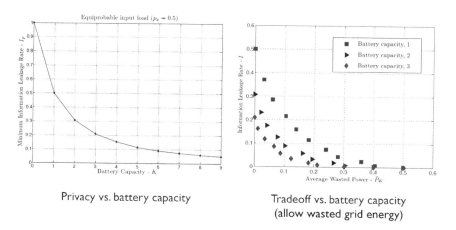

Privacy vs. battery capacity

Tradeoff vs. battery capacity (allow wasted grid energy)

Figure 18.10 Without energy harvesting. (Used with permission from [31].)

18.4 Games: Competitive Privacy

In this section, we will briefly discuss a variation on the privacy–utility tradeoff in which there are multiple interacting parties, each of which is interested in its own tradeoff,

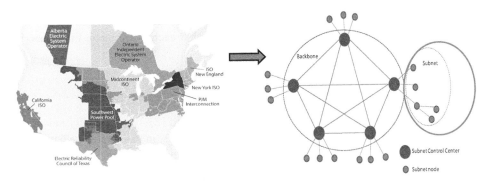

Figure 18.11 Competitive privacy. (Map reproduced with permission from the ISO/RTO Council.)

while interactions among the parties affect these individual tradeoffs. This problem is called *competitive privacy*.

By way of motivation, we can consider the North American grid, which is managed by the so-called regional transmission organizations, or RTOs, which, among other things, predict the loads in their regions of responsibility and decide what resources are needed to meet those loads (see Fig. 18.11). A supporting task for this process is estimating the state of the grid. Since the portion of a grid for which a given RTO is responsible is interconnected with those of others, more reliable state estimates can be obtained if the RTOs exchange information with one another. On the other hand, the grid is served by many private companies and involves many private entities, so there may be economic competitiveness or other reasons for keeping such measurements private. So this gives rise to a privacy–utility tradeoff, but in a competitive setting.

To develop some insight into such situations, we can consider a simple model in which each RTO's state is summarized in one real number. So the mth RTO for $m = 1, \ldots, M$ has a state X_m, and each RTO measures some linear combination of those states, plus noise, denoted by Y_k for the kth RTO. So, we have a linear measurement model

$$Y_k = \sum_{m=1}^{M} H_{k,m} X_m + Z_k, \quad k = 1, 2, \ldots, M, \tag{18.5}$$

where the $H_{k,m}$ are the coefficients in the linear model and $\{Z_k\}$ represents measurement noise.

Exchange of measurements among RTOs will lead to inevitable leakage of state information, although all RTOs already have some knowledge of all states due to the linear model. We can examine this problem in the privacy–utility tradeoff framework by defining utility and privacy measures for each RTO. A natural utility measure for RTO k is the mean square error occurred in estimating its own state, X_k, while privacy for RTO k can be measured in terms of leakage of information about its own state to other RTOs.

It is illuminating to look specifically at the case of two RTOs, which we will refer to as *agents* in the following. We assume that each agent has n i.i.d. observations of the model (18.5), i.e.,

$$Y_{1,i} = X_{1,i} + \alpha X_{2,i} + Z_{1,i}, i = 1, \ldots, n,$$
$$Y_{2,i} = \beta X_{1,i} + X_{2,i} + Z_{2,i}, i = 1, \ldots, n, \quad (18.6)$$

where we assume that $\{X_{1,i}\}$, $\{X_{2,i}\}$, $\{Z_{1,i}\}$, and $\{Z_{2,i}\}$ are mutually independent, with the $X_{j,i}$ i.i.d. $\mathcal{N}(0,1)$ and the $Z_{j,i}$ i.i.d. $\mathcal{N}(0, \sigma_j)$. Thus, in this model we have four parameters: the two "channel gains," α and β, and the two noise variances, σ_1^2 and σ_2^2.

Within this model, we wish to consider the tradeoff between utility and privacy leakage. Each agent j measures its utility as follows:

$$D_j = \frac{1}{n} \sum_{i=1}^{n} E\left[\left(X_{j,i} - \hat{X}_{j,i}\right)^2\right], \quad (18.7)$$

where $\hat{X}_{j,i}$ is agent j's estimate of $X_{j,i}$, its own state at time i. Furthermore, each agent j measures its privacy leakage as follows:

$$L_j = \frac{1}{n} I\left(X_j^n; J_j, Y_{3-j}^n\right), \quad (18.8)$$

where, again, $I(\cdot;\cdot)$ denotes mutual information, $X_j^n = X_{j,1}, \ldots, X_{j,n}$ (and similarly for Y_{3-j}^n), and J_j denotes all information transferred from agent j to its counterpart agent $3-j$.

As a first step in understanding this problem, we can state the following result from [35] that characterizes optimal information exchange between the agents:

THEOREM 18.1 *Wyner–Ziv coding maximizes privacy (i.e., minimizes L_1 and L_2) for a fixed level of utility at each agent (i.e., fixed D_1 and D_2).*

Recall that Wyner–Ziv coding [36, 37] refers to optimal distributed source coding of correlated sources, and thus that it minimizes rate of information transfer (i.e., privacy leakage) for fixed levels of distortion in (18.6) should not come as a surprise.

So the optimal way to exchange information, if one wants to do so, is through Wyner–Ziv coding. That is, this is the most efficient way of trading information in terms of providing the most reduction in mean square distortion for a given level of privacy leakage. Or, alternatively, it maximizes privacy for any desired fixed distortion levels. In other words, wherever the efficient frontier lies, Wyner–Ziv coding will be the way to implement information exchange.

This result does not solve the problem, however, because as one can see from Fig. 18.12 the leakage of one agent depends on the distortion of its counterpart, not on its own distortion. So it is not clear how the agents should behave; if an agent gives information to its counterpart, it only helps the counterpart unless there is a *quid pro quo* transfer of information from the counterpart. This suggests that this problem can be

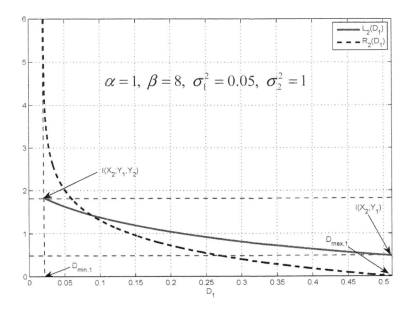

Figure 18.12 Rate and privacy leakage vs. distortion.

profitably viewed in the context of game theory, an approach explored in [38], which is summarized in the following paragraphs.

To impose a game theoretic framework on this problem, we need to establish a set of actions and a payoff function for each agent. Of course, the actions of an agent involve the transfer of information to its counterpart. Such actions can be characterized straightforwardly if we observe that, in the linear Gaussian model of (18.5), the rates of information transfer and privacy leakage of one agent as functions of the corresponding distortion experienced by the counterpart agent are both monotonically decreasing, as further illustrated in Fig. 18.12. So we can in fact think about the action of a given agent in terms of how much distortion it inflicts on its counterpart agent; i.e., an action a_j of agent j can be specified by D_{3-j}. We will assume that there is a maximal level of distortion for each agent, say $\bar{D}_j$ for agent j, and that each agent must release at least enough information to its counterpart so that the distortion achieved by the counterpart is at most this maximal level. (Such a requirement might be imposed by utility regulators, for example.)

A reasonable payoff for agent j, which of course depends on both its own action and that of its counterpart, is the following:

$$u_j(a_j, a_{3-j}) = -w_j L(a_j) + w'_j \log\left(\frac{\bar{D}_j}{a_{3-j}}\right), \quad (18.9)$$

where $L(a_j)$ is agent j's privacy leakage (18.8) due to action a_j, which is penalized in the payoff; $\log\left(\frac{\bar{D}_j}{a_{3-j}}\right)$ is a logarithmic payoff for distortion that is lower than the maximum $\bar{D}_j$; and w_j and w'_j are weights that balance the importance of these two components.

The payoff (18.9) leads to a classical prisoner's dilemma, in which there is no incentive for either agent to share any information beyond what minimal amount is necessary to achieve the maximal allowed distortion at its counterpart. That is, the only Nash equilibrium of this game is achieved at $(a_1, a_2) = (\bar{D}_2, \bar{D}_1)$ [38].

Clearly, other incentives are necessary if greater information exchange is desired. One type of incentive is pricing. For example, the payoff of (18.9) could be replaced by

$$\tilde{u}_j(a_j, a_{3-j}) = u_j(a_j, a_{3-j}) + p_j \log\left(\frac{\bar{D}_{3-j}}{a_j}\right), \quad (18.10)$$

where p_j is a price paid to agent j for improving its counterpart's distortion. With this payoff, essentially any desired behavior can be incentivized by choosing appropriate values of p_1 and p_2, as one might expect. Other non-trivial cooperative behaviors can also be induced through a common payoff function (i.e., a *common goal* game), such as

$$u_{sys}(a_1, a_2) = -L(a_1) - L(a_2) + \frac{q}{2} \log\left(\frac{\bar{D}_1 + \bar{D}_2}{a_1 + a_2}\right), \quad (18.11)$$

where q is weighting factor. (See [38] for details of the resulting equilibria.) This particular payoff gives rise to a so-called *potential game* [39]. Again, such incentives could be imposed by utility regulators, or in the case of pricing by a market structure.

Another way to provide incentives is to encourage *quid pro quo* behavior by considering a multi-stage game, in which the game is repeated over multiple time periods. So, what an agent does for its counterpart today might affect what the counterpart does tomorrow. We can examine this possibility by looking at a T-stage game with $T > 1$, in which the payoff is given by

$$\sum_{t=1}^{T} \rho^{t-1} u_j\left(a_j^{(t)}, a_{3-j}^{(t)}\right), \quad (18.12)$$

where $a_j^{(t)}$ and $a_{3-j}^{(t)}$ are the actions taken by the two parties at time t, and ρ is a discount or "forgetting" factor. Again, however, with finite T this is a prisoner's dilemma problem, in which the only Nash equilibrium[3] is $\left(a_1^{(t)}, a_2^{(t)}\right) = (\bar{D}_2, \bar{D}_1), \forall t$. On the other hand, if T is infinite, i.e., when the game will be played indefinitely, there are non-trivial Nash equilibria. In particular, for sufficiently large $\rho < 1$, any pair of strategies $\left(a_1^{(t)}, a_2^{(t)}\right) = (D_2^*, D_1^*), \forall t$ for which

$$u_j(D_j^*, D_{3-j}^*) > u_j(\bar{D}_j, \bar{D}_{3-j}), \, j = 1, 2, \quad (18.13)$$

is a (subgame perfect) Nash equilibrium. Figure 18.13 illustrates the range of ρ for which various action pairs are equilibria in a specific example.

In summary, we introduced a new element into the privacy–utility tradeoff problem, namely the presence of multiple competing parties. Using a simple two-agent model with Gaussian states and measurements, we saw that Wyner–Ziv coding provides the

---

[3] In this multi-party game the equilibrium of interest is a subgame perfect equilibrium, which essentially is an equilibrium for every subset of time periods [40].

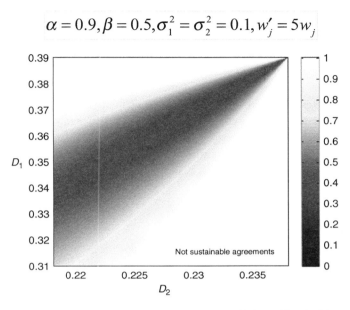

Figure 18.13 Minimal discount factors for sustaining an equilibrium. (Used with permission from [38].)

optimal means of information exchange among the parties. This result tells us how to exchange information, but not how much to exchange. To determine how much information to exchange, game theory is useful. In this context, we examined several types of games, single-stage games (with and without pricing, common goal) and multi-stage games with finite and infinite horizons, seeing that the range of equilibria varies considerably among these game types. In particular, the one-shot and finite multi-stage games without pricing are basically prisoner's dilemma problems, while the common goal and infinite multi-stage games yield more interesting behavior.

18.5 Conclusion

In this chapter, we have examined applications of information, control, and game theories in quantifying and exploring privacy issues. We have used primarily the context of smart grid communications, although most of the concepts discussed here can be applied more broadly. We began by motivating the conceptual framework by noting that absolute privacy is generally not desirable and instead is usually to be traded off with some type of utility. Then we used an information theoretic framework to characterize the optimal frontier of such tradeoffs in a database setting. Next, using smart metering as a context, we examined a more complex problem in which observations form a time series, and applied two approaches to tradeoff privacy with utility: a source coding approach, in which meter measurements are encoded to implement the tradeoff; and a control-based approach, in which battery storage is controlled to modulate the meter output and effect the tradeoff. Finally, we introduced the competitive privacy problem, in which multiple interacting parties have individual tradeoffs that can be affected by each other's actions. Here, game theory was useful in studying the behaviors that can arise in such situations.

The purpose of this chapter was to illustrate the usefulness of information, control, and game theories in examining the fundamentals of privacy in smart grid and other problems. Although practical design was not a goal, in some cases basic practical techniques from data transmission and compression arise as potentially useful solutions. The ideas described provide a basis for further exploration of privacy issues in the smart grid, and also for examining similar issues in other applications. Examples might include social networks, sensor networks, and other cyber-physical systems.

Acknowledgments

This work was supported in part by the U.S. National Science Foundation under Grant CMMI-1435778.

References

[1] E. Hossain, Z. Han, and H. V. Poor, eds., *Smart Grid Communications and Networking*. Cambridge: Cambridge University Press, 2012.
[2] P. McDaniel and S. McLaughlin, "Security and privacy challenges in the smart grid," *IEEE Security and Privacy*, vol. 7, no. 3, pp. 75–77, May 2009.
[3] E. L. Quinn, "Privacy and the new energy infrastructure," *Social Science Research Network*, Feb. 2009.
[4] G. Kalogridis, R. Cepeda, S. Z. Denic, T. Lewis, and C. Efthymiou, "Elecprivacy: Evaluating the privacy protection of electricity management algorithms," *IEEE Trans. Smart Grid*, vol. 2, no. 4, pp. 750–758, Dec. 2011.
[5] F. G. Marmol, C. Sorge, O. Ugus, and G. M. Perez, "Do not snoop my habits: Preserving privacy in the smart grid," *IEEE Commun. Mag.*, vol. 50, no. 5, pp. 166–172, May 2012.
[6] L. Sankar, S. R. Rajagopalan, and H. V. Poor, "Utility–privacy tradeoffs in databases: An information-theoretic approach," *IEEE Trans. Inf. Forensics Security*, vol. 8, no. 6, pp. 838–852, Jun. 2013.
[7] T. E. Raghunathan, J. P. Reiter, and D. B. Rubin, "Multiple imputation for statistical disclosure limitation," *IEEE Trans. Inf. Theory*, vol. 43, no. 6, pp. 1877–1894, Jun. 1997.
[8] L. Sweeney, "*k*-anonymity: A model for protecting privacy," *Int. J. Uncertain. Fuzziness Knowl.-Based Syst.*, vol. 10, no. 5, pp. 557–570, Oct. 2002.
[9] A. Machanavajjhala, D. Kifer, J. Gehrke, and M. Venkitasubramaniam, "*l*-diversity: Privacy beyond *k*-anonymity," *ACM Trans. Knowl. Discov. Data*, vol. 1, no. 1, p. 24, Mar. 2007.
[10] H. Kargupta, S. Datta, Q. Wang, and K. Sivakumar, "Random data perturbation techniques and privacy preserving datamining," *J. Knowl. Inf. Systems*, vol. 7, no. 4, pp. 387–414, May 2005.
[11] C. Dwork and A. Roth, "The algorithmic foundations of differential privacy," *Foundations and Trends in Computer Science*, vol. 9, no. 3–4, pp. 211–407, 2014.

[12] H. Yamamoto, "A source coding problem for sources with additional outputs to keep secret from the receiver or wiretappers," *IEEE Trans. Inf. Theory*, vol. 29, no. 6, pp. 918–923, Nov. 1983.

[13] R. Tandon, L. Sankar, and H. V. Poor, "Discriminatory lossy source coding: Side information privacy," *IEEE Trans. Inf. Theory*, vol. 59, no. 9, pp. 5665–5677, Sep. 2013.

[14] F. Sultanem, "Using appliance signatures for monitoring residential loads at meter panel level," *IEEE Trans. Power Delivery*, vol. 6, no. 4, pp. 1380–1385, Oct. 1991.

[15] A. Molina-Markham, P. Shenoy, K. Fu, E. Cecchet, and D. Irwin, "Private memoirs of a smart meter," in *Proc. 2nd ACM Workshop Embedded Sensing Systems Energy-Efficiency Buildings*, Zurich, Switzerland, Nov. 2010, pp. 61–66.

[16] G. W. Hart, "Nonintrusive appliance load monitoring," *Proc. IEEE*, vol. 80, no. 12, pp. 1870–1891, Dec. 1992.

[17] H. Y. Lam, G. S. K. Fung, and W. K. Lee, "A novel method to construct taxonomy of electrical appliances based on load signatures," *IEEE Trans. Consumer Electronics*, vol. 53, no. 2, pp. 653–660, May 2007.

[18] G. Kalogridis and S. Z. Denic, "Data mining and privacy of personal behaviour types in smart grid," in *Proc. IEEE Int. Conf. Data Mining*, Vancouver, Canada, Dec. 2011, pp. 636–642.

[19] A. Predunzi, "A neuron nets based procedure for identifying domestic appliances pattern-of-use from energy recordings at meter panel," in *Proc. IEEE Power Eng. Soc. Winter Meeting*, New York, NY, USA, Jan. 2002, pp. 941–946, vol. 2.

[20] U. Greveler, P. Glösekotter, B. Justus, and D. Loehr, "Multimedia content identification through smart meter power usage profiles," in *Proc. Int. Conf. Inf, Knowl. Eng.*, Las Vegas, NV, USA, Jul. 2012.

[21] M. Backes and S. Meiser, "Differentially private smart metering with battery recharging," *IACR Cryptology ePrint Archive*, vol. 2012, no. 2, p. 183, 2012.

[22] T. D. Nicol and D. M. Nicol, "Combating unauthorized load signal analysis with targeted event masking," in *Proc. 45th Hawaii Int. Conf. System Science*, Grand Wailea, HI, USA, Jan. 2012, pp. 2037–2043.

[23] C. Efthymiou and G. Kalogridis, "Smart grid privacy via anonymization of smart metering data," in *Proc. 1st IEEE Int. Conf. Smart Grid Commun.*, Gaithersburg, MD, USA, Oct. 2010, pp. 238–243.

[24] C. S. J.-M. Bohli and O. Ugus, "A privacy model for smart metering," in *Proc. IEEE Int. Conf. Commun.*, Capetown, South Africa, May 2010, pp. 1–5.

[25] F. D. Garcia and B. Jacobs, "Privacy-friendly energy-metering via homomorphic encryption," *Lecture Notes in Computer Science*, vol. 6710, pp. 226–238, 2011.

[26] E. N. Y. Kim and M. Srivastava, "Cooperative state estimation for preserving privacy of user behaviors in smart grid," in *Proce. IEEE Int. Conf. Smart Grid Commun.*, Brussels, Belgium, Oct. 2011, pp. 178–183.

[27] S. McLaughlin, P. McDaniel, and W. Aiello, "Protecting consumer privacy from electric load monitoring," in *Proc. 18th ACM Conf. Computer Commun. Security*, Chicago, IL, USA, Oct. 2011, pp. 87–98.

[28] H. Li, R. Mao, L. Lai, and R. Qiu, "Compressed meter reading for delay-sensitive and secure load report in smart grid," in *Proc. 1st IEEE Int. Conf. Smart Grid Commun.*, Gaithersburg, MD, USA, Oct. 2010, pp. 114–119.

[29] G. Kalogridis, C. Efthymiou, S. Denic, T. A. Lewis, and R. Cepeda, "Privacy for smart meters: Towards undetectable appliance load signatures," in *Proc. 1st IEEE Int. Conf. Smart Grid Commun.*, Gaithersburg, MD, USA, Oct. 2010, pp. 232–237.

[30] L. Sankar, S. R. Rajagopalan, S. Mohajer, and H. V. Poor, "Smart meter privacy: A theoretical framework," *IEEE Trans. Smart Grid*, vol. 4, no. 2, pp. 837–846, Jun. 2013.

[31] O. Tan, D. Gündüz, and H. V. Poor, "Increasing smart meter privacy through energy harvesting and storage devices," *IEEE J. Sel. Areas Commun.*, vol. 31, no. 7, pp. 1331–1341, Jul. 2013.

[32] D. M. Arnold, H. A. Loeliger, P. O. Vontobel, A. Kavcic, and W. Zeng, "Simulation-based computation of information rates for channels with memory," *IEEE Trans. Inf. Theory*, vol. 52, no. 8, pp. 3498–3508, Aug. 2006.

[33] D. Varodayan and A. Khisti, "Smart meter privacy using a rechargeable battery: Minimizing the rate of information leakage," in *Proc. IEEE Int. Conf. Acoustics, Speech, Signal Process.*, Prague, Czech Republic, May 2011, pp. 1932–1935.

[34] L. Yang, X. Chen, J. Zhang, and H. V. Poor, "Cost-effective and privacy-preserving energy management for smart meters," *IEEE Trans. Smart Grid*, vol. 6, no. 1, pp. 486–495, Jan. 2015.

[35] L. Sankar, S. K. Kar, R. Tandon, and H. V. Poor, "Competitive privacy in the smart grid: An information-theoretic approach," in *Proc. IEEE Int. Conf. Smart Grid Commun.*, Brussels, Belgium, Oct. 2011, pp. 220–225.

[36] A. D. Wyner and J. Ziv, "The rate–distortion function for source coding with side information at the decoder," *IEEE Trans. Inf. Theory*, vol. 22, no. 1, pp. 1–10, Jan. 1976.

[37] T. Berger, *The Information Theory Approach to Communications*. Berlin, Heidelberg: Springer-Verlag, 1977, pp. 170–231.

[38] E. V. Belmega, L. Sankar, and H. V. Poor, "Enabling data exchange in interactive state estimation under privacy constraints," *IEEE J. Sel. Topics Signal Process.*, vol. 9, no. 7, pp. 1285–1297, 2015.

[39] D. Monderer and L. S. Shapley, "Potential games," *Games and Economic Behavior*, vol. 14, no. 1, pp. 124–143, May 1996.

[40] D. Fudenberg and J. Tirole, *Game Theory*. Cambridge, MA: MIT Press, 1991.

19 Security in Distributed Storage Systems

Salim El Rouayheb, Sreechakra Goparaju, and Kannan Ramchandran

The proliferation of cloud applications in recent years has made securing the underlying distributed storage systems (DSSs) an important problem. This chapter focuses on the fundamental limits of information theoretic security in DSSs, with an emphasis on constructing efficient codes for achieving these limits. The challenge of studying security in DSSs results from their dynamic behavior characterized by nodes frequently leaving and joining the DSS. This dynamic behavior distinguishes them from static models typically studied in the area of information theoretic security. This chapter introduces the theoretical model of DSSs and summarizes the main results in the area. Three types of adversaries are studied: passive, active omniscient, and active limited-knowledge adversaries. For each type of attack, upper bounds on the secrecy or resiliency capacity are given, as well as capacity-achieving secure codes for certain regimes. Moreover, open problems are highlighted and discussed.

19.1 Introduction

Distributed storage systems have witnessed a rapid growth in recent years, driven by the advent of cloud applications. DSSs are used in data centers [1–6] and peer-to-peer (p2p) networks [7–10] to store large amounts of data and make it available online anywhere and anytime. The sheer volume of this data makes DSSs an obvious and lucrative target for malicious attacks, which can range from stealing private information (credit cards, fingerprints, etc.) to corrupting sensitive data [11–17]. This chapter focuses on the fundamental limits of *information theoretic security* [18–20] in DSSs, with an emphasis on constructing efficient codes for achieving these limits.

DSSs are typically formed of inexpensive and unreliable storage nodes that fail frequently, due to hardware failures [21], software failures and updates [22], and peer churning in p2p DSSs [7,9], causing temporary or even permanent data loss. Failures in DSSs are described by practitioners as being "the norm rather than the exception" [1]. To guarantee data reliability, the data is stored redundantly in DSSs. Moreover, when a node fails, or leaves the DSS, it is replaced by another node called a newcomer or replacement node, in what is referred to as the *repair process* [23]. This dynamic behavior of DSSs, in which nodes are frequently leaving and joining the system, distinguishes them from other communication systems in the literature on information theoretic security [24–27], which are static in general. To illustrate the challenge of achieving security in this dynamic setting, we give the example in Fig. 19.1 of an $(n,k,d) = (4,2,2)$ DSS. The parameter $n = 4$ represents the total number of nodes, say of one unit storage capacity each, and $k = 2$ is the number of nodes contacted by a user to retrieve a stored file. Therefore, this DSS can tolerate $n - k = 2$ simultaneous failures. A newcomer node, added to the system during the repair of a failure, contacts $d = 2$ other *helper nodes* to download its data (d is referred to as the *repair degree*). Figure 19.1 shows the failure and repair of node 1. Using a maximum distance separable (MDS) code, such as a Reed–Solomon code, the user can store a file of size two units in the DSS, which is also the information theoretical optimal size. Now, suppose that we want to protect the system against an eavesdropper that can observe a single node in the DSS, unknown to us. If the DSS is static, that is, does not experience failures and repairs, one can securely store a file $\mathcal{F}$ of one unit by "mixing" the information symbol u with a randomly generated unit-sized key r, independent of u, using the code depicted in the figure. This code can be regarded as a secret sharing scheme [24], a coset code for the wiretap channel II [25], or as a secure network code for the combination network [26, 27]. The code allows a user contacting any two nodes to decode the information symbol u (and key r), and leaks no information to the eavesdropper. A security violation occurs, however, when a node fails and is replaced by a new one. The newcomer, node 5, has to download data from two helper nodes to regenerate the lost data. If the new node is already compromised, this will reveal all the downloaded data to the eavesdropper. For instance, the figure depicts the case when node 1 fails and the coded data chunk $u + r$ is lost. The new replacement node downloads the two data chunks r and $u + 2r$ to recover the lost packet $u + r$. However, the eavesdropper can now decode the file $\mathcal{F}$. Therefore, even if we start with a perfectly secure code, the repair process can result in leaking information about the stored file to the eavesdropper.

Beyond the example above, this chapter studies information theoretic security in general dynamic DSSs under different models of attack, namely passive eavesdropping and active malicious attacks. In practice, security is typically achieved using cryptographic primitives to encrypt the data and protect it against illegitimate third parties [28, 29]. Cryptographic methods rely on the assumption that the adversary has limited computational power and cannot solve the underlying mathematical problems, such as factoring and finding discrete logarithms, which are assumed to be hard. This chapter focuses on information theoretic security, which characterizes the fundamental ability of a DSS to provide data security independently of cryptographic methods. Information theoretic security provides unconditional data security and does not assume any limitation on the computational power of the adversary. Examples of real-world DSSs that implement concepts from information theoretic security include POTSHARDS [30], which is a DSS used for long-term data storage. Over a long period of time, any encryption protocol will be broken. Therefore, the data needs to be decrypted and then reencrypted periodically with the latest cryptographic algorithm, which can be very costly. Thus, information theoretic security may be more suitable for long-term data storage. Another example is Cleversafe [31, 32], a cloud storage company that relies on dispersing the data geographically to achieve security. While we mention these practical systems, the focus of this chapter will be on theoretical results.

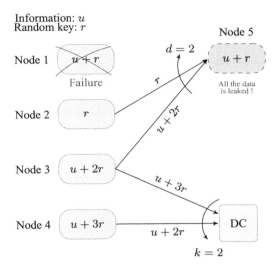

Figure 19.1 An example of how repairing a DSS can compromise the system security. The original DSS formed of nodes 1, ..., 4 is secured against a single compromised node using a secret sharing scheme or a coset code. However, repairing failed nodes can break the security of the system. For instance, consider the case when node 1 fails and is replaced by node 5, which is already compromised by a passive eavesdropper. The eavesdropper can observe all the data downloaded by node 5 and therefore decode the stored file $\mathcal{F}$.

19.1.1 Related Literature

The past few years have witnessed significant theoretical research on the information theoretic limits and code constructions for data reliability in DSS, while minimizing the cost of system repair in terms of communication cost (repair bandwidth), data reads, number of helper nodes (locality), etc. (see, e.g., [33–56] and the references within). Recently discovered codes, such as locally repairable codes [41, 44], are now implemented in practical systems [22].

Information theoretic security in DSSs was first studied by Pawar et al. in [57–59], in which they defined the *secrecy capacity* (sometimes also referred to as the secure capacity) of a DSS, gave upper bounds on it, and constructed achievable schemes in certain regimes. These bounds were shown to be achievable in more regimes in [57–62]. In [63, 64], it was shown that these bounds may not be tight in general. Additional secure code constructions were given in [65, 66], and secure codes with locality properties were studied in [67, 68].

The problem of information theoretic security in DSSs can be related to the problem of security in networks that use network coding [26, 27, 69–77]. However, the network coding literature focuses on compromised edges instead of nodes. Finding the secrecy capacity for general networks with compromised nodes is an NP-hard problem for a passive eavesdropper [78], and non-linear codes may be needed to achieve capacity [79, 80]. However, the network of DSSs have a specific structure that can be leveraged to make progress towards characterizing the fundamental limits on security in DSSs.

19.1.2 Organization

Due to space restrictions, this chapter is not meant to be comprehensive. Rather, it will introduce the theoretical model of DSSs and focus, among others, on the main results of the authors in this area. For instance, results on DSS security with locality constraints [67, 68] will not be discussed. However, pointers to the omitted results will be given throughout the chapter. The remainder of the chapter will be organized as follows. In Section 19.2, we describe the DSS model, the different adversary types and set up the notation. In Section 19.3, we discuss the results related to data secrecy against passive eavesdropping. In Sections 19.4 and 19.5, we discuss data security against active omniscient and limited-knowledge adversaries, respectively. We conclude in Section 19.6.

19.2 Model and Notation

19.2.1 System Model

A DSS is a dynamically changing system that consists of a set $\mathcal{A}$ of n *active* storage nodes. The initial nodes are indexed from 1 to n. Each node has a storage capacity equal to α symbols[1] chosen from a finite alphabet, typically a finite field $GF(q)$ of size q. A data file $\mathcal{F}$ of M symbols is stored on the DSS. We assume that the M symbols are i.i.d. and drawn uniformly at random from the finite alphabet. DSSs are characterized by frequent node failures [1] that result in a temporary or permanent loss of the data on the failed nodes.[2] To guarantee data availability, data is stored redundantly on the DSS in order to satisfy the following properties:

PROPERTY 19.1 (File reconstruction)   A legitimate user, also called a data collector (DC), contacting any k, $k < n$, active nodes and downloading their data should be able to reconstruct the original file $\mathcal{F}$. Therefore, the DSS can tolerate $n - k$ simultaneous failures.

PROPERTY 19.2 (Node repair)   We focus on single node failures[3] since they are the most common in practice [22]. When a node fails, a new replacement node contacts d active nodes (where $k \leq d < n$), called *helper* nodes, and downloads β symbols from each. We refer to the total amount of data communicated during repair as the *total repair bandwidth* and denote it by $\gamma = d\beta \geq \alpha$. The new node stores α symbols that are a function of the γ repair symbols, such that the new set of n active nodes continue to satisfy the *file reconstruction* and *node repair* properties. (Notice that the repair property is recursive to ensure data availability as the DSS evolves under failures and repairs.)

---

[1] We will adopt a homogeneous model for DSSs in which all storage nodes have the same storage capacity and the same amount of information is downloaded from all the helper nodes (symmetric repair). This is the model that is widely adopted in the literature. For results on heterogeneous DSSs, we refer the reader to [81–83], and the references within.
[2] We suppose that once a node fails, it ceases to belong to the set $\mathcal{A}$ of active nodes.
[3] Repair from multiple node failures has been studied in [84, 85].

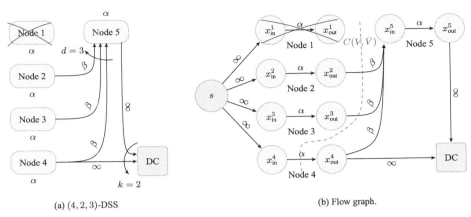

Figure 19.2 (a) An example of a failure repair in a (4,2,3)-DSS. Node 1 fails and is replaced by node 5, which recovers the lost data by contacting the helper nodes 2, 3, and 4. A data collector (DC) contacts nodes 4 and 5 to retrieve the file. We assume that the DC has infinite-capacity edges that allow him to download the data on the nodes. (b) The flow graph representation of the DSS in (a) – see Section 19.2.4.

We refer to a DSS satisfying the file reconstruction and node repair properties as an (n,k,d)-DSS. Figure 19.2(a) shows a (4,2,3)-DSS with the repair of node 1.

The literature distinguishes between two types of node repair: (1) *functional repair*, in which the reconstructed data can differ from the original lost data as long as it retains the file reconstruction and node repair properties of the DSS; and (2) *exact repair*, in which the data stored on the new node is required to be the same as the original data stored on the failed node. Typically, exact repair is desired in practice, while functional repair is more amenable to theoretical analysis. Dimakis et al. [23] showed that when storing a file of size M, there is a fundamental tradeoff between the node storage capacity α and the repair bandwidth $\gamma = d\beta$ for functional repair given by

$$M \leq C_{\mathrm{f}} \triangleq \sum_{i=1}^{k} \min\{\alpha, (d-i+1)\beta\}. \tag{19.1}$$

We refer to C_{f} as the functional repair capacity of the DSS, which is achievable using random network coding [23, 86].

For exact repair, such a characterization is still open in general [87–93]. However, the capacity C_{f} is achievable under exact repair for the two extremal points of the tradeoff: (1) the *minimum bandwidth regenerating* (MBR) point, $\alpha = d\beta$, achieves the lowest repair bandwidth in the tradeoff, and (2) the *minimum storage regenerating* (MSR) point, $\beta = \alpha/(d-k+1)$, minimizes the storage per node α; the corresponding codes, known as MSR codes, are in fact MDS with an optimal repair bandwidth. Explicit capacity-achieving code constructions corresponding to these points are given in [38,45] for MBR codes and [34, 39, 46–51] for MSR codes.

19.2.2 Security and Adversary Model

We want to ensure the security of the data against an adversary who compromises a number of nodes in the DSS. We are interested in achieving information theoretic security that does not make any assumption on the computational power of the adversary. We focus on three types of adversary:[4]

1. *Passive adversary*, or eavesdropper (Eve), who observes only the data on the compromised nodes without changing it. We assume that Eve has access to a subset $\mathcal{E}$ of ℓ nodes in the DSS (at some point of time during[5] its evolution). Eve has access to both the information (the α symbols) stored on a node and the repair information (the γ repair symbols) that is downloaded to obtain the stored information.[6] This is noteworthy as Eve may infer information about the data stored on the helper nodes, despite the fact that they may not be directly under its control.
2. *Active omniscient adversary*, who has complete knowledge of the stored file $\mathcal{F}$ and the data stored on each node. The adversary controls b nodes in the DSS. The compromised nodes can send corrupted data to the user during file reconstruction, or to other nodes when they are helping in their repair.
3. *Active limited-knowledge adversary*, whose knowledge about the stored file $\mathcal{F}$ is limited to what it observes on the b compromised nodes. The adversary can corrupt the data stored on the compromised nodes and the data transmitted during node repair or file reconstruction.

We assume that the adversary under these three models knows the secure coding scheme implemented in the DSS.

19.2.3 Secrecy Capacity and Notation

We are interested in characterizing the secrecy capacity of an (n,k,d)-DSS, which is the maximum amount of information that can be stored securely in the DSS while satisfying the reconstruction and repair properties given the storage and repair bandwidth constraints α and β.

Let U denote the random variable corresponding to the information in the stored file $\mathcal{F}$. Let W_i denote the random variable corresponding to the data stored on node i (before being possibly corrupted by an adversary). For any set $\mathcal{B}$, denote $\{W_i : i \in \mathcal{B}\}$ by $W_\mathcal{B}$. Let $\mathcal{A}$ be the set of active nodes at a given instant in the evolution of the DSS. $\mathcal{A}$ consists of n nodes if there are no failures, and $n-1$ nodes when a failure occurs and before it is repaired. The reconstruction property is given by

$$H(U|W_\mathcal{B}) = 0, \forall \mathcal{B} \subseteq \mathcal{A} \text{ s.t. } |\mathcal{B}| = k. \tag{19.2}$$

---

[4] Other types of adversaries have been considered in the literature. See, for example, [94, 95].
[5] This means that if Eve has access to a node as well as its replacement node, it is counted as having access to two nodes.
[6] This assumption is sometimes referred to as a Type II scenario in the literature, as opposed to the weaker Type I scenario where Eve has access only to the information W_i stored on node i. Our analysis can be easily adapted to the weaker Type I model or a mixed scenario.

Let node $i' \in \mathcal{A}$ fail and let its replacement node be indexed by i. (Typically, we assume that the first failure is replaced by node $n+1$, the next failure is replaced by node $n+2$, and so on.) Notice that the new set of active nodes now contains node i, but not node i'. Let $\mathcal{D} \subseteq \mathcal{A}\setminus\{i\}$, $|\mathcal{D}| = d$, denote the set of helper nodes. We denote by $S^i_j(\mathcal{D})$ the random variable corresponding to the data sent by a helper node $j \in \mathcal{D}$ to the replacement node i, and by $S^i(\mathcal{D})$ the total repair data $\{S^i_j(\mathcal{D}) : j \in \mathcal{D}\}$ downloaded by node i. We drop the notation $\mathcal{D}$ when $\mathcal{D}$ is clear from the context, or when a statement is true for all $\mathcal{D} \subseteq \mathcal{A}\setminus\{i\}$. For any set $\mathcal{B} \subset \{1,2,\ldots\}$, denote $\{S^j_i : j \in \mathcal{B}\}$ by $S^{\mathcal{B}}_i$, $\{S^i : i \in \mathcal{B}\}$ by $S^{\mathcal{B}}$, and $\{S^j_i : i \in \mathcal{B}\}$ by $S^j_{\mathcal{B}}$. The *storage and bandwidth constraints* can now be written as

$$H(W_i) \leq \alpha, \quad H(S^i_j) \leq \beta, \quad \forall\, i,j \in \{1,2,\ldots\},\, i \neq j. \tag{19.3}$$

The *repair constraint* can be written as

$$H(W_i | S^i(\mathcal{D})) = 0, \quad \forall\, i \in \{1,2,\ldots\},\ \text{and}\ \mathcal{D} \subseteq \mathcal{A}\setminus\{i\}\ \text{s.t.}\ |\mathcal{D}| = d, \tag{19.4}$$

such that the new set of active nodes $\mathcal{A}$ satisfies (19.2).

To simplify the notation for exact repair, we assume that the set of active nodes is fixed[7] at $\mathcal{A} = [n]$, and that if node i fails, it is replaced by a replacement node with the same index i. The *exact repair constraint* can therefore be written as

$$H(W_i | S^i(\mathcal{D})) = 0, \quad \forall\, i \in [n],\ \text{and}\ \mathcal{D} \subseteq [n]\setminus\{i\}\ \text{s.t.}\ |\mathcal{D}| = d. \tag{19.5}$$

We are interested in achieving *perfect secrecy*[8] in the presence of an eavesdropper. That is, Eve should not be able to infer any information about the stored file given its observation. The perfect secrecy condition is given by

$$H\left(U \,\middle|\, S^{\mathcal{E}}\right) = H(U), \quad \forall\, \mathcal{E} \subset \mathcal{A},\ \text{s.t.}\ |\mathcal{E}| = \ell. \tag{19.6}$$

DEFINITION 19.1 (Secrecy capacity) Consider an (n,k,d)-DSS with an eavesdropper compromising ℓ nodes. The secrecy capacity for functional repair $C_{s,f}(\alpha,\beta)$ is defined as the maximum amount of information that can be stored reliably in the DSS while guaranteeing perfect secrecy, i.e.,

$$C_{s,f}(\alpha,\beta) \triangleq \sup_{\substack{(19.2),(19.3),\\ (19.4),(19.6)\ \text{hold}}} H(U). \tag{19.7}$$

Similarly, we define the secrecy capacity for exact repair $C_{s,e}(\alpha,\beta)$ as

$$C_{s,e}(\alpha,\beta) \triangleq \sup_{\substack{(19.2),(19.3),\\ (19.5),(19.6)\ \text{hold}}} H(U). \tag{19.8}$$

DEFINITION 19.2 (Resiliency capacity) The resiliency capacity for functional repair $C_{r,f}(\alpha,\beta)$ is defined as the maximum amount of information that can be stored in the

---

[7] Notation: $[n]$ represents $\{1,\ldots,n\}$, and $[m:n]$ represents $\{m, m+1,\ldots,n\}$.
[8] Weaker requirements on security were studied in [96,97].

DSS under functional repair, while guaranteeing that the reconstruction property holds even in the presence of b corrupted nodes. We can write

$$C_{r,f}(\alpha,\beta) \triangleq \sup_{\substack{(19.2),(19.3), \\ (19.4) \text{ hold}}} H(U), \qquad (19.9)$$

when b nodes are corrupted by an active adversary. Similarly, we define the resiliency capacity for exact repair $C_{r,e}(\alpha,\beta)$ as

$$C_{r,e}(\alpha,\beta) \triangleq \sup_{\substack{(19.2),(19.3), \\ (19.5) \text{ hold}}} H(U). \qquad (19.10)$$

Note that the functional repair capacity is an upper bound on the exact repair capacity, i.e., $C_{s,e}(\alpha,\beta) \leq C_{s,f}(\alpha,\beta)$, and $C_{r,e}(\alpha,\beta) \leq C_{r,f}(\alpha,\beta)$.

19.2.4 Flow Graph Representation

The flow graph $\mathcal{G}$ of a DSS introduced in [23] is an essential tool for proving bounds on the secrecy and resiliency capacity under functional repair and therefore under exact repair as well. It models the DSS by an information network that captures the evolution of the DSS in time with all the possible failure and repair patterns under functional repair – see Fig. 19.2(b). The flow graph consists of a source node s that holds the file $\mathcal{F}$. Each storage node i is represented by the input vertex x_{in}^i and the output vertex x_{out}^i connected by a directed edge (x_{in}^i, x_{out}^i) having capacity α. The source s is connected to all the input vertices $x_{in}^i, i \in [n]$, by edges of infinite capacity. When node j fails, it is replaced by a new node $n+j$. The input vertex x_{in}^{n+j} of node $n+j$ is connected to the output nodes of all the helper nodes through incoming edges each of capacity β. A data collector is represented by a node connected to k output vertices by incoming edges of infinite capacity. The flow graph is an infinite graph that represents all the possible node failures, all possible helper nodes during repair, and all possible data collectors.

Let $\mathcal{V}$ be the vertex set of the flow graph $\mathcal{G}$. We define a *cut* $C(V, \bar{V})$ as a partition of $\mathcal{V}$ into two subsets V and $\bar{V} = \mathcal{V} \setminus V$ separating the source s from a data collector DC, such that $s \in V$ and DC $\in \bar{V}$. The value of a cut $C(V, \bar{V})$ is defined as the sum of the capacities of the edges that have one vertex in V and the other in $\bar{V}$ and go from V to $\bar{V}$. The tradeoff expression in (19.1) given by C_f was obtained in [23] by showing that it is the min-cut between the source and any DC in the flow graph $\mathcal{G}$.

19.3 Secrecy against Passive Eavesdropping

We focus in this section on the problem of storing a file securely in the presence of a passive eavesdropper. We first prove a general upper bound on the secrecy capacity in Theorem 19.1 for general storage and bandwidth constraints (19.3). The rest of the section will be devoted to studying the achievability (or non-achievability) of this upper bound for two important regimes: (1) $\alpha \geq d\beta$, sometimes referred to as the bandwidth-limited regime, and (2) $\beta = \alpha/(d-k+1)$. Note that these regimes correspond to the MBR and the MSR points defined in Section 19.2.1.

19.3.1 Upper Bound on the Secrecy Capacity

We start with establishing a general upper bound [58] on the secrecy capacity for a DSS when a passive eavesdropper Eve has access to some fixed number of nodes in the DSS.

THEOREM 19.1 ([58]) *The secrecy capacity of an (n,k,d)-DSS in the presence of an eavesdropper that has access to $\ell < k$ nodes is bounded above by*

$$C_{s,e}(\alpha,\beta) \leq C_{s,f}(\alpha,\beta) \leq \sum_{i=\ell+1}^{k} \min\{(d-i+1)\beta, \alpha\}, \quad (19.11)$$

where the storage and bandwidth constraints are as given in (19.3). For $\ell \geq k$, the secrecy capacity $C_{s,e}(\alpha,\beta) = C_{s,f}(\alpha,\beta) = 0$.

Proof Consider an (n,k,d)-DSS and a passive eavesdropper Eve that has access to some ℓ nodes in the system. We give the proof for the exact repair case, i.e., we prove the bound on $C_{s,e}$. The corresponding result for the secrecy capacity for functional repair $C_{s,f}$ is given in [58] and follows a similar argument, but by carefully considering the order in which the nodes fail and are replaced.

As in Section 19.2, let U denote the random variable corresponding to the information in the stored file $\mathcal{F}$. Let $[k] = \mathcal{B} \subseteq [n]$, and let $\mathcal{E} = [\ell]$. Then, we have the following:

$$H(U) = H\left(U \mid S^{\mathcal{E}}\right) - H(U \mid W_{\mathcal{B}}) \quad (19.12)$$

$$\leq H(U \mid W_{\mathcal{E}}) - H(U \mid W_{\mathcal{B}}) \quad (19.13)$$

$$= H(U \mid W_{\mathcal{E}}) - H\left(U \mid W_{\mathcal{E}}, W_{\mathcal{B} \setminus \mathcal{E}}\right) \quad (19.14)$$

$$= I\left(U; W_{\mathcal{B} \setminus \mathcal{E}} \mid W_{\mathcal{E}}\right) \quad (19.15)$$

$$\leq H\left(W_{\mathcal{B} \setminus \mathcal{E}} \mid W_{\mathcal{E}}\right) = \sum_{i=\ell+1}^{k} H\left(W_i \mid W_{[i-1]}\right), \quad (19.16)$$

where (19.12) follows from the perfect secrecy condition (19.6) and the reconstruction property (19.2), (19.13) follows from the exact repair constraint (19.5), and the remaining equalities and inequalities follow from standard definitions of entropy and mutual information. For $i \in [k]$, we have $H\left(W_i \mid W_{[i-1]}\right) \leq H(W_i) \leq \alpha$, and

$$H\left(W_i \mid W_{[i-1]}\right) \leq H\left(W_i \mid W_{[i-1]}\right) + H\left(S^i_{[i+1:d+1]} \mid W_{[i]}\right)$$

$$= H\left(W_i, S^i_{[i+1:d+1]} \mid W_{[i-1]}\right)$$

$$= H\left(S^i_{[i+1:d+1]} \mid W_{[i-1]}\right) + H\left(W_i \mid S^i_{[i+1:d+1]}, W_{[i-1]}\right)$$

$$= H\left(S^i_{[i+1:d+1]} \mid W_{[i-1]}\right) \quad (19.17)$$

$$\leq \sum_{j=i+1}^{d+1} H\left(S^i_j\right) \leq (d-i+1)\beta, \quad (19.18)$$

where (19.17) follows from exact repair constraint (19.5), and the rest from standard information theoretic relations. Note that the proof of $H(W_i|W_{[i-1]}) \leq (d-i+1)\beta$ can be found in [45]. Thus, we have

$$H\left(W_i \big| W_{[i-1]}\right) \leq \min\{\alpha, (d-i+1)\beta\}. \tag{19.19}$$

From (19.16) and (19.19), and from the definition (19.8), we obtain

$$C_{\text{s,e}}(\alpha, \beta) \leq \sum_{i=\ell+1}^{k} \min\{(d-i+1)\beta, \alpha\}. \tag{19.20}$$

19.3.2 Achievability of the Secrecy Capacity

The upper bound (19.11) on the exact repair secrecy capacity $C_{\text{s,e}}$ in Theorem 19.1 is achievable when $\alpha \geq d\beta$. The achievability scheme was first given for $d = n - 1$ in [58], and later generalized for all d in [60], using the family of *product-matrix* (PM) codes introduced in [38]. PM codes allow exact repair and achieve the functional repair non-secure capacity C_{f} for the MBR and MSR points.

Product-Matrix Codes

We briefly explain the PM code construction, known as minimum bandwidth regenerating (MBR) PM codes [38]. Without loss of generality, suppose $\beta = 1$ and $\alpha = d\beta = d$. The maximum non-secure file size [23] that can be stored in the DSS is given by substituting $\alpha = d\beta = d$ in (19.1):

$$C_{\text{mbr}} = \sum_{i=1}^{k} \min\{\alpha, (d-i+1)\beta\} = kd - \binom{k}{2}. \tag{19.21}$$

Suppose the file $\mathcal{F}$ is formed of C_{mbr} symbols over $GF(q)$, $q > n$. These symbols are arranged into two matrices S and T. S is a $k \times k$ *symmetric* matrix and T is a $k \times (d-k)$ matrix. The encoding consists of an $n \times d$ Vandermonde matrix Ψ that multiplies the $d \times d$ matrix $Z = \begin{bmatrix} S & T \\ T^{\text{t}} & 0 \end{bmatrix}$, where the superscript "t" refers to the matrix transpose. Each node i, $i \in [n]$, stores the ith row of $C = \Psi Z$.

- **Reconstruction:** A data collector contacting any k nodes downloads their data and decodes the file.
- **Repair:** Let ψ_i^{t}, $i \in [n]$, denote the ith row of Ψ that encodes the data stored on node i. When node i is helping the repair of node f, $f \neq i$, it will send the projection of its data onto ψ_f^{t}, i.e., $\psi_i^{\text{t}} Z \psi_f$, and the newcomer can repair the lost data.

We illustrate the reconstruction and repair properties of PM codes using Example 19.1. We refer the interested reader to [38] for more details and a proof of correctness.

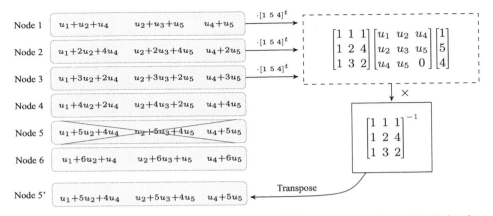

Figure 19.3 The ith row of $C = \Psi Z$ is stored on node i for $i = 1, \ldots, 6$. The exact repair of node 5 is depicted.

Example 19.1 [*Non-secure PM code*] Consider a $(6, 2, 3)$-DSS. Let $\beta = 1$ and $\alpha = d\beta = 3$; then, according to (19.21), $C_{\mathrm{mbr}} = 5$. We choose $q = 7 > n$, and let $(u_1, \ldots, u_5)$ denote the file symbols in $GF(7)$. Hence,

$$\Psi = \begin{bmatrix} 1 & 1 & 1 \\ 1 & 2 & 4 \\ 1 & 3 & 2 \\ 1 & 4 & 2 \\ 1 & 5 & 4 \\ 1 & 6 & 1 \end{bmatrix} \quad \text{and} \quad Z = \begin{bmatrix} u_1 & u_2 & u_4 \\ u_2 & u_3 & u_5 \\ u_4 & u_5 & 0 \end{bmatrix}. \tag{19.22}$$

The data stored on the nodes is shown in Fig. 19.3.

Repair: The exact repair of node i, $i \in [5]$, follows the steps shown in Fig. 19.3 (see [38] for details).

Reconstruction: Suppose a DC contacts nodes 1 and 2 to reconstruct the file. It downloads

$$\begin{bmatrix} 1 & 1 & 1 \\ 1 & 2 & 4 \end{bmatrix} \begin{bmatrix} u_1 & u_2 & u_4 \\ u_2 & u_3 & u_5 \\ u_4 & u_5 & 0 \end{bmatrix}.$$

Using the equations $u_4 + u_5$ and $u_4 + 2u_5$, the DC can decode u_4 and u_5. Then, by subtracting u_4 and u_5 from the remaining equations, the DC can decode u_i, $i \in \{1, 2, 3\}$, and hence reconstruct the file.

Achievability of the Secrecy Capacity

We now show how MBR PM codes can be appropriated [60] to achieve the secrecy capacity for the regime $\alpha \geq d\beta$.

THEOREM 19.2 (Section III, [60]) *The secrecy capacity of an (n,k,d)-DSS in the presence of an eavesdropper that has access to $\ell < k$ nodes, when $\alpha \geq d\beta$, is given by*

$$C_{s,e}(\alpha,\beta) = \sum_{i=\ell+1}^{k}(d-i+1)\beta = \left(kd - \binom{k}{2}\right)\beta - \left(\ell d - \binom{\ell}{2}\right)\beta, \quad (19.23)$$

where the storage and bandwidth constraints are as given in (19.3).

Proof It suffices to present a code construction that achieves the secrecy capacity for the case $\beta = 1$; codes for larger values of β can be obtained by concatenating multiple copies of the code for $\beta = 1$. For the regime $\alpha \geq d\beta$, we always choose to operate at the point $\alpha = d\beta = d$.

Construction
We construct an MBR PM code as in Section 19.3.2 above, but replace the

$$R = C_{\text{mbr}} - C_{s,e} = \ell d - \binom{\ell}{2} \quad (19.24)$$

message symbols [see (19.21) and (19.23)] in the first ℓ rows (and therefore, ℓ columns) of the symmetric matrix Z by R random symbols (keys), chosen i.i.d. over $GF(q)$, $q > n$, and independently of the message symbols. We denote this symbol set by $\mathcal{R} = \{r_i : i \in [R]\}$. For example, consider the $(6,2,3)$-DSS setting of Example 19.1 with $\ell = 1$ compromised node. Then the message matrix Z becomes

$$Z = \begin{bmatrix} r_1 & r_2 & r_3 \\ r_2 & u_1 & u_2 \\ r_3 & u_2 & 0 \end{bmatrix},$$

and the encoding and the repair procedures remain the same.

Proof of Secrecy
It now remains to be shown that for the resulting code, an eavesdropper that has access to the data in any $\ell < k$ nodes $W_{\mathcal{E}}$, where $\mathcal{E} \subset [n]$, $|\mathcal{E}| = \ell$, obtains no information about the message (symbol set) U. Note that $S^{\mathcal{E}} = W_{\mathcal{E}}$ in this regime, i.e., a node stores all the data downloaded during repair. This requires three steps:

1. Given the message U, the eavesdropper can recover the random symbol set $\mathcal{R}$, that is,

$$H(\mathcal{R}|W_{\mathcal{E}}, U) = 0, \quad (19.25)$$

where $W_{\mathcal{E}}$ is the same as the rows of $C = \Psi Z$ indexed by $\mathcal{E}$. Let us denote the corresponding matrices by $C_{\mathcal{E}}$ and $\Psi_{\mathcal{E}}$, respectively. It can be shown that this is equivalent to replacing the message set in Z with zeros, thus effectively storing only the random symbols in Z; call this modified Z $\tilde{Z}$. Notice then that $\Psi \tilde{Z}$ acts as an MBR PM code for an (n, ℓ, d)-DSS, and therefore, from the reconstruction property of a PM code, $\Psi_{\mathcal{E}} \tilde{Z}$ enables the recovery of the R random symbols.

2. The data accessed by an eavesdropper is a function of at most R symbols:

$$H(W_{\mathcal{E}}) \leq R. \qquad (19.26)$$

Without loss of generality, let $\mathcal{E} = [\ell]$. The proof then follows from the fact that $H(W_{[\ell]}) = \sum_{i=1}^{\ell} H(W_i | W_{[i-1]})$, and that $H(W_i | W_{[i-1]}) \leq (d-i+1)$ – see (19.18).

3. The eavesdropper has no information about U, because

$$I(U; W_{\mathcal{E}}) = H(W_{\mathcal{E}}) - H(W_{\mathcal{E}} | U)$$

$$\leq R - H(W_{\mathcal{E}} | U) \qquad (19.27)$$

$$= R - H(W_{\mathcal{E}} | U) + H(W_{\mathcal{E}} | U, \mathcal{R}) \qquad (19.28)$$

$$= R - I(W_{\mathcal{E}}; \mathcal{R} | U)$$

$$= R - (H(\mathcal{R} | U) - H(\mathcal{R} | U, W_{\mathcal{E}}))$$

$$= R - H(\mathcal{R} | U) \qquad (19.29)$$

$$= 0, \qquad (19.30)$$

where (19.27) follows from Step 2, (19.28) because $W_{\mathcal{E}}$ is a function of $\mathcal{R}$ and U, (19.29) from Step 1, and (19.30) because the random symbol set $\mathcal{R}$ is chosen independently and uniformly over $GF(q)$. (All quantities are in multiples of $\log_2 q$ bits, as the underlying field is $GF(q)$.)

19.3.3 Secrecy via Separation Schemes

The code that achieves secrecy capacity for the point $\alpha = d\beta$ (and in general, the regime $\alpha \geq d\beta$) is an instance of a so-called *separation scheme*. A separation scheme is defined as a non-secure coding scheme for a point (α, β) that can be modified to a secure coding scheme by replacing the message symbols of the non-secure scheme with a function of the message symbols for the secure scheme and possibly some random symbols (keys) that are independent of the message symbols; in other words, a secure separation-based coding scheme consists of a non-secure coding scheme attached to a *separate* precoding stage; for example, see Fig. 19.4.

Separation schemes are motivated toward prioritizing optimal storage in a non-secure scenario, and then optimizing the secure scenario in the eventuality of an attack by a passive eavesdropper. To that end, consider the stricter storage (α) and bandwidth (β) constraints that replace inequalities with equalities in (19.3):

$$H(W_i) = \alpha, \quad H(S_j^i) = \beta \quad \forall i,j \in [n], \ i \neq j, \qquad (19.31)$$

and define the corresponding secrecy capacity.

DEFINITION 19.3 (Separation secrecy capacity) Consider an (n,k,d)-DSS with an eavesdropper compromising ℓ nodes. The secrecy capacity for exact repair $C_{s,\text{sep}}(\alpha, \beta)$ is defined as the maximum amount of information that can be stored reliably in the DSS

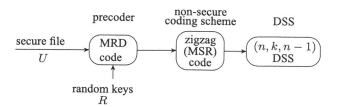

Figure 19.4 Separation secrecy capacity-achieving scheme for an $(n,k,n-1)$ linear MSR code.

while guaranteeing perfect secrecy and satisfying the stricter storage and bandwidth constraints (19.31):

$$C_{s,\text{sep}}(\alpha,\beta) \triangleq \sup_{\substack{(19.2),(19.31),\\(19.5),(19.6) \text{ hold}}} H(U). \qquad (19.32)$$

We saw in Section 19.3.2 that a separation scheme achieves the secrecy capacity for $\alpha = d\beta$. This is significant because the corresponding non-secure coding scheme (in this case, a PM code) achieves the non-secure capacity for the MBR point.

A more significant non-secure point of operation, due to its widespread use in practical DSSs, corresponds to MDS codes, which are the most space efficient (i.e., minimize α) for a specified worst-case number $(n-k)$ of simultaneous node failures. As previously mentioned, MDS codes that achieve optimal repair bandwidth β are also known as minimum storage regenerating (MSR) codes and correspond to $\beta = \alpha/(d-k+1)$. A natural question then is: can a separation scheme using MSR codes achieve the secrecy capacity upper bound in Theorem 19.1? Example 19.2 below answers this question negatively by comparing such a separation scheme and a coding scheme corresponding to $\beta = \alpha/d$ (a lower β).

Example 19.2 [*Separation is not optimal*] Consider a $(4,2,3)$-DSS, $\ell = 1$, and the MSR point $(\alpha,\beta) = (1,1/2)$. Figure 19.5(a) presents a separation scheme that achieves the optimal possible file size (see Section 19.3.4) when a separation scheme is used with a $(4,2,3)$ MSR code: $H(U) = 1/2$. Figure 19.5(b) then shows that in fact, a larger secure file size $(H(U) = 2/3)$ is achievable by reducing the repair bandwidth β to $1/3$.

In other words, using a separation scheme (that may achieve the corresponding non-secure capacity) does not necessarily lead to a capacity-achieving secrecy scheme. Example 19.2 shows the value of sometimes reducing either α or β to achieve a higher secured file size. Tandon et al. [64] gave tighter upper bounds on the secrecy capacity for certain system parameters. For instance, they showed that the exact repair secrecy capacity of an $(n,n-1,n-1)$-DSS with $\ell = n-2$ is bounded above by $C_{s,e} \leq \min\left(\dfrac{\alpha}{n-1}, \beta\right)$, and can be achieved by using secure MBR codes. An additional upper bound can be found in [64].

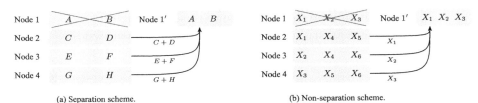

Figure 19.5 Comparison between different encodings for a secure $(4,2,3)$-DSS with storage and bandwidth constraints $\alpha \leq 1$ and $\beta \leq 1/2$. (a) Separation scheme: We use an optimal non-secure $(4,2,3)$ regenerating (MSR) code for $(\alpha, \beta) = (1, 1/2)$ of capacity 2 and store three random keys r_1, r_2, r_3 of size $1/2$ each and one message u of size $1/2$. The encoding follows the MSR PM code [38], each node i stores the ith row of the matrix $C = \begin{bmatrix} 1 & 1 & 1 \\ 2 & 4 & 8 \\ 3 & 9 & 1 \\ 4 & 3 & 12 \end{bmatrix} \begin{bmatrix} 0 & r_1 \\ r_2 & r_3 \\ r_3 & u \end{bmatrix}$, e.g., $A = r_2 + r_3$ and $B = r_1 + r_3 + u$. (b) Non-separation scheme: We construct a scheme with a lower repair bandwidth than the given constraint: $\beta = 1/3$. We use an $[n,k] = [6,5]$ MDS code to encode X_i, $i \in [6]$, (each of size $1/3$ units) and use a repetition code [42, 52] to form a $(4,2,3)$ regenerating (MBR) code. The message (u_1, u_2) of size $2/3$ is encoded with three keys r_1, r_2, r_3 as follows: $X_1 = r_1, X_2 = r_2, X_3 = r_3, X_4 = u_1 + r_1 + r_2 + r_3, X_5 = u_2 + r_1 + 2r_2 + 3r_3$, and $X_6 = u_1 + u_2 + r_1 + r_2 + 2r_3$. This example shows that a separation scheme is not always optimal and the bound in Theorem 19.1 is not tight [64].

Due to the inherent practicality of separation schemes and MDS codes, we next study the separation secrecy capacity for the corresponding operating point: $\beta = \alpha/(d-k+1)$. In particular, we consider linear coding schemes due to their tractability and practical importance.

19.3.4 Separation Secrecy Capacity of Linear Optimal Repair MDS Codes

We start by describing a linear optimal repair MDS code[9] (MSR code). Let $n = k + p$, and without loss of generality let $[k]$ be the systematic nodes, and $[k+1 : k+p]$ be the parity nodes. A linear scheme assumes linear encoding of parity nodes as well as linearity in the repair process. In other words, W_i, $i \in [k]$ denotes the systematic data of size equal to the file size $k\alpha$, and the parity data is given by

$$W_{k+i} = \sum_{j=1}^{k} A_{i,j} W_j, \qquad (19.33)$$

where $A_{i,j}$ is an $\alpha \times \alpha$ coding matrix over $GF(q)$ corresponding to the parity node $i \in [p]$ and the systematic node $j \in [k]$. The coding matrices $A_{i,j}$ can be shown to be non-singular using the MDS property of the code. Overloading the notation, for the repair of node i, a helper node $j \neq i$ transmits a vector of length β given by $S_j^i W_j$, where S_j^i is a $\beta \times \alpha$ repair matrix over $GF(q)$. This vector may be interpreted as a projection of W_j onto a subspace of dimension β. We overload the notation a little more, and use S_j^i interchangeably to denote both the matrix and the subspace spanning its rows. The remaining definitions follow naturally as in Section 19.2.

---

[9] This section follows the results in [63].

Upper Bound on the Separation Secrecy Capacity

We first give an upper bound on $C_{s,\text{sep}}(\alpha,\beta)$ for the so-called MSR point: $\beta = \alpha/(d-k+1)$.

THEOREM 19.3 ([63]) *The separation secrecy capacity of an (n,k,d)-DSS in the presence of an eavesdropper that has access to $\ell < k$ nodes, when $\beta = \alpha/(d-k+1)$, is given by*

$$C_{s,\text{sep},\text{lin}}\left(\alpha, \frac{\alpha}{d-k+1}\right) \leq (k-\ell)\left(1 - \frac{1}{d-k+1}\right)^{\ell} \alpha, \qquad (19.34)$$

where the subscript lin *indicates that the separation secrecy capacity is restricted over linear coding schemes.*

The upper bound relies on the geometry of repair subspaces corresponding to the repair of different systematic nodes by a single node.

THEOREM 19.4 ([63]) *Consider an (n,k,d)-DSS storing a linear MSR code, where each node has a storage capacity α. Then, for $d = n-1$, $\mathcal{B} \subseteq [k]\setminus i$, for $i \in [k]$, we have*

$$\dim\left(\sum_{j \in \mathcal{B}} S_i^j\right) \geq \left(1 - \left(\frac{n-k-1}{n-k}\right)^{|\mathcal{B}|}\right) \alpha, \qquad (19.35)$$

where the summation refers to the sum of subspaces, and dim *denotes the dimension of a subspace.*

The proof is given for $n-k = p = 2$ (two parity nodes), and easily extends [53, 54] to $p > 2$. For $p = 2$, the repair subspaces are of dimension $\beta = \alpha/2$ (by definition of the MSR point). It can be shown that the optimal repair of systematic nodes in an MSR code requires interference alignment [55] and leads to the following *subspace conditions* for the repair of node $j \in [k]$ (e.g., [98]):

$$\begin{aligned}
S_{k+1}^j A_{1,i} &\simeq S_{k+2}^j A_{2,i} \simeq S_i^j, \\
S_{k+1}^j A_{1,j} + S_{k+2}^j A_{2,j} &\simeq GF(q)^{\alpha},
\end{aligned} \qquad (19.36)$$

for all $i \in [k]\setminus\{j\}$, and $\simeq$ denotes an equality of subspaces. We prove the theorem using induction over $|\mathcal{B}|$.

Base Case
For $|\mathcal{B}| = 1$, $\dim(S_i^j) = \beta = \alpha/2$.

Inductive Step
Suppose the result is true for $|\mathcal{B}| = m - 1$. Without loss of generality, let $\mathcal{B} = [m]$. For $[k] \ni i \notin [m]$, we have

$$\dim\left(\sum_{j=1}^{m} S_i^j\right) = \dim\left(\sum_{j=1}^{m} S_{k+1}^j A_{1,i}\right) \qquad (19.37)$$

$$= \dim\left(\sum_{j=1}^{m} S_{k+1}^j\right) \qquad (19.38)$$

$$= \dim\left(\sum_{j=1}^{m} S_{k+1}^j A_{1,m}\right) \qquad (19.39)$$

$$\geq \dim\left(\sum_{j=1}^{m-1} S_{k+1}^j A_{1,m} \cap S_{k+2}^m A_{2,m}\right) + \dim\left(S_{k+1}^m A_{1,m}\right), \qquad (19.40)$$

where (19.37) follows from (19.36), and (19.38) and (19.39) follow from distributivity and the invertibility of the matrices $A_{1,i}$ and $A_{1,m}$. For (19.40), the two subspaces whose dimensions are considered intersect only in the zero vector [see (19.36)], and are both contained in the subspace $(\sum_{j=1}^{m} S_{k+1}^j A_{1,m})$; hence so is their *direct* sum.

Using the identity for arbitrary subspaces S_a and S_b that $\dim(S_a + S_b) + \dim(S_a \cap S_b) = \dim(S_a) + \dim(S_b)$, and the fact that the subspaces $S_{k+2}^m A_{2,m}$ and $S_{k+1}^m A_{1,m}$ have dimension $\alpha/2$, we obtain from (19.40) that

$$\dim\left(\sum_{j=1}^{m} S_i^j\right) \geq \dim\left(\sum_{j=1}^{m-1} S_{k+1}^j A_{1,m}\right) + \alpha \qquad (19.41)$$

$$- \dim\left(\sum_{j=1}^{m-1} S_{k+1}^j A_{1,m} + S_{k+2}^m A_{2,m}\right).$$

The third term on the right-hand side in inequality (19.41) equals the term on the left-hand side, because

$$\dim\left(\sum_{j=1}^{m-1} S_{k+1}^j A_{1,m} + S_{k+2}^m A_{2,m}\right) = \dim\left(\sum_{j=1}^{m-1} S_{k+1}^j A_{1,m} A_{2,m}^{-1} + S_{k+2}^m\right)$$

$$= \dim\left(\sum_{j=1}^{m-1} S_{k+2}^j + S_{k+2}^m\right), \qquad (19.42)$$

which equals $\dim(\sum_{j=1}^{m} S_i^j)$. Here, $i \in [k] \setminus [m]$ and the steps follow from similar reasons to (19.37)–(19.40). Also, similarly,

$$\dim\left(\sum_{j=1}^{m-1} S_{k+1}^j A_{1,m}\right) = \dim\left(\sum_{j=1}^{m-1} S_m^j\right). \qquad (19.43)$$

Using the induction hypothesis and (19.41)–(19.43), we have

$$2\dim\left(\sum_{j=1}^{m} S_i^j\right) \geq \dim\left(\sum_{j=1}^{m-1} S_m^j\right) + \alpha \geq \left(1 - \frac{1}{2^{m-1}}\right)\alpha + \alpha,$$

which completes the inductive step.

We are now ready to prove the upper bound on $C_{s,\text{sep}}$ for linear MSR codes.

Proof of Theorem 19.3 First we prove the theorem for $d = n - 1$. The proof almost completely follows the one in [67], except for the final part, which uses Theorem 19.4. Let $[k]$ be the systematic nodes. Let $\mathcal{E} = [\ell]$ be the eavesdropped nodes and let $\mathcal{N} = [\ell+1:k]$ be the set of $k - \ell$ systematic nodes not in $\mathcal{E}$. In order to store a file U of entropy $H(U)$ securely in the DSS, we have

$$H(U) = H\left(U \big| S^{\mathcal{E}}\right) - H\left(U \big| S^{\mathcal{E}}, W_{\mathcal{N}}\right) \quad (19.44)$$

$$= I\left(U; W_{\mathcal{N}} \big| S^{\mathcal{E}}\right)$$

$$\leq H\left(W_{\mathcal{N}} \big| S^{\mathcal{E}}\right)$$

$$\leq \sum_{i \in \mathcal{N}} H\left(W_i \big| S_i^{\mathcal{E}}\right)$$

$$= \sum_{i \in \mathcal{N}} \left(H\left(W_i, S_i^{\mathcal{E}}\right) - H\left(S_i^{\mathcal{E}}\right)\right)$$

$$= \sum_{i \in \mathcal{N}} \left(H(W_i) - H\left(S_i^{\mathcal{E}}\right)\right) \quad (19.45)$$

$$= \sum_{i \in \mathcal{N}} \left(H(W_i) - \dim\left(\sum_{j \in \mathcal{E}} S_i^j\right)\right) \quad (19.46)$$

$$\leq (k - \ell)\left(1 - \frac{1}{n-k}\right)^{\ell} \alpha, \quad (19.47)$$

where (19.44) follows from (19.2) and (19.6), (19.45) because S_i^m is a function of W_i for any $m \neq i$, (19.46) from the linearity of the MDS code being used, and (19.47) from Theorem 19.4.

For $d < n - 1$, we focus on the first $d + 1$ nodes, and can view them as an $(n' = d+1, k, d = n' - 1)$-DSS. The proof then follows along the same lines for this restricted DSS.

Achievability of the Separation Secrecy Capacity

The question of whether the upper bound on the separation secrecy capacity is tight is in general wide open. The following theorem, however, shows that the bound is tight for linear MSR codes, when $d = n - 1$, and when optimal bandwidth repair and the compromising of nodes is restricted only to (the usually more critical) systematic nodes in the DSS.

THEOREM 19.5 ([63]) *The separation secrecy capacity of an $(n,k,n-1)$-DSS in the presence of an eavesdropper that has access to $\ell < k$ nodes, for a linear MDS code such that any systematic node can be repaired using a repair bandwidth of $\beta = \alpha/(n-k)$, and the eavesdropping is restricted to systematic nodes, is given by*

$$C_{\text{s,sep,lin,sys}}\left(\alpha, \frac{\alpha}{n-k}\right) = (k-\ell)\left(1 - \frac{1}{n-k}\right)^{\ell} \alpha, \quad (19.48)$$

where sys *indicates that the capacity is for a relaxed version of MSR codes with optimal repair only for systematic nodes.*

Proof Sketch The proof follows from [67, Theorem 10]. The achievable code construction is a separation scheme that uses an (n,k) zigzag code [40] preceded by a maximum rank distance code (MRD) such as a Gabidulin code [99]; see Fig. 19.4.

19.3.5 Universally Secure Optimal Repair MDS Codes

The ideas behind a separation scheme extend beyond just separating the non-secure and secure regimes. One might ask whether a non-secure coding scheme that leads to an optimal code construction for the case of one eavesdropped node ($\ell = 1$) remains optimal for the case of two eavesdropped nodes ($\ell = 2$). In general, we have the following definition:

DEFINITION 19.4 (Universal security) A non-secure DSS coding scheme is said to exhibit *universal security* if, for an (n,k,d)-DSS, and for some storage and bandwidth constraints (19.31), there exists a separation scheme for any $\ell \in [0:k-1]$ that uses the non-secure scheme and achieves the separation secrecy capacity.

In fact, the separation scheme that achieves the secrecy capacity in Section 19.3.2, and the one that achieves the separation secrecy capacity for linear MSR codes in Section 19.3.4 satisfies this property. The only difference is that the message file U is precoded differently, using an appropriate number of random keys, as in Fig. 19.4.

However, it is not always clear if a separation scheme is universal. For instance, it might be possible that a scheme that does well with $\ell = 1$ achieves a lower secure file size for $\ell = 2$, and vice versa. In other words, there might be a *pareto-optimal* ℓ-dimensional region where the dimensions correspond to the secure file size achievable for different values of $\ell \in [0:k-1]$. In [100], it is shown that the linear MSR code that universally achieves the linear separation secrecy capacity $C_{\text{s,sep,lin,sys}}$ also achieves a pareto-optimal point corresponding to secrecy capacity $C_{\text{s,sep,sys}}$ for all (and not just linear) MSR codes. Whether there exist codes that achieve other pareto-optimal points is still an open question.

19.4 Security against an Omniscient Adversary

In this section, we study security against an active omniscient adversary. We prove an upper bound on the resiliency capacity of an (n,k,d)-DSS in the presence of an omniscient adversary and construct explicit codes achieving this capacity when $\alpha \geq (d-b)\beta$ or $\alpha \leq (d-k+1)\beta$.

19.4.1 Upper Bound on the Resiliency Capacity

We start with a Singleton-type upper bound on the resiliency capacity of a DSS in the presence of an active omniscient adversary [58, 59].

THEOREM 19.6 ([58, 59]) *The resiliency capacity of a DSS in the presence of an omniscient adversary controlling $b < k/2$ nodes is bounded above by*

$$C_{r,e}(\alpha,\beta) \leq C_{r,f}(\alpha,\beta) \leq \sum_{i=2b+1}^{k} \min\{(d-i+1)\beta, \alpha\}. \quad (19.49)$$

If $b \geq k/2$ then $C_{r,f}(\alpha,\beta) = C_{r,e}(\alpha,\beta) = 0$.

Proof We prove the upper bound on $C_{r,f}(\alpha,\beta)$, i.e., for functional repair. Then, the upper bound on $C_{e,f}(\alpha,\beta)$ follows immediately.

Consider an (n,k,d)-DSS and an omniscient adversary controlling b nodes. Assume the DSS experiences the consecutive failure of nodes $1,\ldots,k$, which are replaced respectively by nodes $n+1,\ldots,n+k$. The flow graph in Fig. 19.6 depicts the resulting repair process. Consider also a data collector that connects to nodes $n+1,\ldots,n+k$ to reconstruct the file. Consider the cut $C(V,\bar{V})$, shown in Fig. 19.6, in which $\bar{V} = \{x_{\text{out}}^{n+1},\ldots,x_{\text{out}}^{n+2b},x_{\text{in}}^{n+2b+1},\ldots,x_{\text{in}}^{n+k},DC\}$.

To get an upper bound for the capacity, we need to quantify the amount of information that the data collector can recover without error. To that end, we partition the edges crossing the cut $C(V,\bar{V})$ into three disjoint sets of edges. We denote by E_1 the set of outgoing edges from the first b nodes x_{in}^j, $j = n+1,\ldots,n+b$; E_2 the set of outgoing edges from the nodes x_{in}^j, $j = n+b+1,\ldots,n+2b$; and E_3 the set of incoming edges to nodes x_{in}^j, $j = n+2b+1,\ldots,n+k$. For each message (file) m at the source, we denote by $X_{E_i}(m)$, $i = 1,2,3$, the symbols corresponding to message m transmitted on the edges of E_i.

For the data collector to reconstruct the stored file without errors, it is necessary to have $X_{E_3}(m_1) \neq X_{E_3}(m_2)$ for any two distinct messages m_1 and m_2. We prove this claim by contradiction. Suppose there exist two different messages $m_1 \neq m_2$ such that $X_{E_3}(m_1) = X_{E_3}(m_2)$, and that the data collector observes $X_{E_1}(m_1)$, $X_{E_2}(m_2)$, and $X_{E_3}(m_1) = X_{E_3}(m_2)$ when contacting nodes $n+1,\ldots,n+k$. Then, assuming all the messages to be equally likely, the data collector will make a decoding error with probability at least $1/2$. This is true since the data collector does not know which b nodes are actually compromised and will not be able to distinguish between the following two cases:

- The true message is m_2 and nodes $n+1,\ldots,n+b$ are controlled by the adversary, who changed the transmitted symbols on the edges in the set E_1 from $X_{E_1}(m_2)$ to $X_{E_1}(m_1)$.

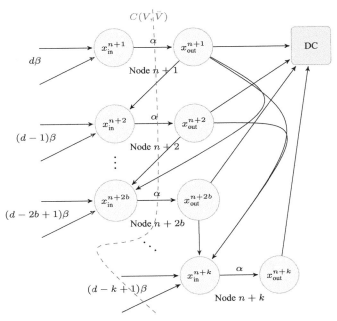

Figure 19.6 Part of the flow graph corresponding to a DSS when nodes $1,\ldots,k$ fail successively and are replaced by nodes $n+1,\ldots,n+k$. A data collector connects to nodes $n+1,\ldots,n+k$ to reconstruct the file.

- The true message is m_1 and nodes $n+b+1,\ldots,n+2b$ are controlled by the adversary, who changed the transmitted symbols on the edges in the set E_2 from $X_{E_2}(m_1)$ to $X_{E_2}(m_2)$.

Therefore, the amount of information that the data collector can reconstruct without errors is the total capacity of the links in E_3, i.e.,

$$C_{\mathrm{r,f}}(\alpha,\beta) \leq \sum_{i=2b+1}^{k} (d-i+1)\beta,$$

when $b < k/2$. If $b \geq k/2$, E_3 is empty and $C_{\mathrm{r,f}}(\alpha,\beta) = 0$.

Similar upper bounds can be obtained for each cut

$$\bar{V}_j = \{x_{\mathrm{out}}^{n+1},\ldots,x_{\mathrm{out}}^{n+2b+j}, x_{\mathrm{in}}^{n+2b+j+1},\ldots,x_{\mathrm{in}}^{n+k}, DC\}, \; j=0,\ldots,k-2b.$$

The upper bound in (19.49) is obtained by taking the minimum of all these upper bounds.

19.4.2 Achievability of the Resiliency Capacity Upper Bound

Theorem 19.7 states the regimes for which we know that the upper bound in (19.49) is achievable. The achievability was shown in [58, 59] whenever $\alpha \geq db$ and $d = n-1$, and in the remaining stated cases in [62].

THEOREM 19.7 ([58,59,62]) *Consider an (n,k,d)-DSS with an omniscient adversary controlling $b < k/2$ nodes. The resiliency capacity of the DSS under exact repair is given by*

$$C_{r,e}(\alpha,\beta) = \sum_{i=2b+1}^{k} (d-i+1)\beta \text{ for } \alpha \geq (d-2b)\beta, \quad (19.50)$$

and

$$C_{r,e}(\alpha,\beta) = (k-2b)\alpha \text{ for } \beta \geq \alpha/(d-k+1) \text{ and } d \geq 2k-2. \quad (19.51)$$

If $b \geq k/2$ then $C_{r,e}(\alpha,\beta) = 0$.

The proof of Theorem 19.7, given in [62], is constructive and uses the family of PM codes [38] introduced in the previous section. The restriction on d, $d \geq 2k-2$, in (19.51) is due to the non-existence of PM codes for smaller values of d in that regime. Beyond the cases in Theorem 19.7, the expression of the resiliency capacity of a DSS remains an open problem.

Proof of Theorem 19.7

We start with an example that illustrates the use of PM codes to achieve security.

Example 19.3 [*Secure PM codes*] Consider a $(6,4,5)$-DSS with $\alpha = d = 5$, $\beta = 1$, and $b = 1$ compromised node. To achieve the resiliency capacity $C_{r,e} = 5$, as given by (19.50), we use a non-secure $(n,d',k') = (6,2,3)$ MBR PM code (see Fig. 19.3). However, a new node contacts $d = d' + 2b = 5$ nodes upon repair, and a DC contacts $k = k' + 2b = 4$ nodes upon reconstruction. The extra information downloaded during repair and file reconstruction will be used as redundancy to correct possible errors introduced by the adversary. The secure repair process is depicted in Fig. 19.7. The data

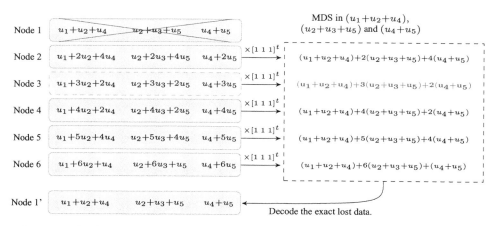

Figure 19.7 Secure repair of node 1 using PM codes in the presence of an adversary controlling $b=1$ node, node 3 in this example.

downloaded during repair forms an MDS code with minimum distance $d_{\min} = 2b+1 = 3$. Therefore, the newcomer can always correct up to b errors and recover the exact lost data. A DC will also be able to reconstruct the stored file with no errors as explained in the general proof.

Proof of Theorem 19.7 Due to space constraints, we will prove only (19.50), which uses MBR PM codes (see Section 19.3.2) [38]. The achievability of (19.51) follows similar steps but uses the so-called minimum storage regenerating (MSR) PM codes and can also be found in [62]. As illustrated in the example above, the secure scheme uses a non-secure $(n,k',d') = (n, k-2b, d-2b)$ MBR PM that stores only $\alpha' = d' \le \alpha = d$ symbols on each node.

Secure Repair
When a node f fails, the newcomer contacts $d = d' + 2b$ helper nodes h_i, $i = 1,\ldots,d$, and downloads the symbol $\psi_{h_i}^t Z \psi_f$ from each helper node h_i. Let $m_f = Z\psi_f$ and $\Psi_{\text{rep}} = [\ \psi_{h_1}\ \ldots\ \psi_{h_d}\ ]^t$ the $d \times d'$ Vandermonde matrix. Therefore, the d symbols downloaded during repair, which can be written as $\Psi_{\text{rep}} m_f$, can be seen as the encoding of m_f using a (d,d') MDS code. The minimum distance of this MDS code is $d_{\min} = d - d' + 1 = 2b + 1$, which can correct $\lfloor \frac{d_{\min}-1}{2} \rfloor = b$ possible errors introduced by the adversary [101]. Thus, we can decode the vector m_f with no errors and regenerate the lost data by taking its transpose $m_f^t = \psi_f^t Z^t = \psi_f^t Z$, since Z is symmetric.

Secure Reconstruction
Here, we think of the data stored on each node as one symbol from $GF(q)^{d'}$. Because the encoding matrix is Vandermonde, the k symbols downloaded by the DC can be seen as the encodings of a (k,k') MDS code of $d_{\min} = k - k' + 1 = 2b + 1$. Therefore, the DC can correct b possibly erroneous symbols stored on the compromised nodes and recover k' symbols with no errors. Then, the DC implements the reconstruction algorithm of the underlying non-secure PM code to reconstruct the file.

The proposed scheme achieves the resiliency capacity because, by (19.21), the maximum file size that can be stored is $\mathcal{M}^s = \sum_{i=1}^{k'} \min\{\alpha', (d'-i+1)\beta\} = C_{r,e}$.

The characterization of the resiliency capacity of a DSS in the presence of an omniscient adversary remains open in the regimes not covered by Theorem 19.7. Non-secure regenerating codes, other than PM codes, exist in some of these regimes [34, 40, 56]. However, they cannot be readily used for security following the method of the previous proof, either because they are restricted to $d = n - 1$ or because the data downloaded during repair is not MDS-coded data.

19.5 Security against a Limited-Knowledge Adversary

We continue in this section the study of security against an active adversary. However, we assume here that the adversary has limited knowledge. The limited-knowledge adversary can observe only the data on the nodes under his control. Therefore, its effect on the system may be less damaging than an omniscient adversary, but more powerful than random errors.

19.5.1 Resiliency Capacity

The next theorem gives an upper bound on the resiliency capacity in the presence of a limited-knowledge adversary [58, 59].

THEOREM 19.8 ([58, 59]) *The resiliency capacity of a DSS in the presence of a limited-knowledge adversary controlling $b < k/2$ nodes is bounded above by*

$$C_{r,e}(\alpha,\beta) \leq C_{r,f}(\alpha,\beta) \leq \sum_{i=b+1}^{k} \min\{\alpha, (d-i+1)\beta\}. \tag{19.52}$$

The proof of Theorem 19.8 can be found in [58]. The idea of the proof is that the adversary can always change the data on the nodes under its control to the all-zero vector (or any fixed vector), irrespective of the file stored, effectively deleting the data on these nodes. This explains the loss of the data on b nodes in (19.52) when compared to (19.49).

The upper bound in (19.52) is achievable for $\alpha \geq (d-b)\beta$. The achievability in this regime was first proven in [58] for $d = n - 1$, and later generalized for any d in [102].

THEOREM 19.9 ([58, 102]) *Consider an (n,k,d)-DSS with a limited-knowledge adversary controlling $b < k/2$ nodes. The resiliency capacity of the DSS under exact repair is given by*

$$C_{r,e}(\alpha,\beta) = \sum_{i=b+1}^{k} (d-i+1)\beta \text{ for } \alpha \geq (d-b)\beta, \tag{19.53}$$

and $C_{r,e} = 0$ if $b \geq k/2$.

Beyond the regime covered in Theorem 19.9, the resiliency capacity of a DSS in the presence of a limited-knowledge adversary remains an open problem.

19.5.2 Secure Scheme Example

The proof of Theorem 19.9 is constructive [58, 102] and has two main ingredients: the *correlation hashing* scheme and the *modified representation of MBR PM* code. The hashing scheme is used to determine with high probability the b compromised nodes, which are then discarded, or "erased," from the DSS. The PM code guarantees file reconstruction and node repair after the b erasures.

Modified Representation of PM Codes

Let X_{ij}, $i,j \leq n$, $i \neq j$, be the symbol sent by node i to node j when node j is being repaired. In PM codes, $X_{ij} = \psi_i^t Z \psi_j$, and $X_{ij} = X_{ji}$ because Z is symmetric. In the modified representation, we store on node i any α different symbols among

Table 19.1 Modified representation of PM codes in a (6, 2, 3)-DSS. The data between braces is not stored, but can be computed from the stored data on the corresponding node.

Node 1	X_{12}	X_{13}	X_{14}	$\{X_{15}, X_{16}\}$
Node 2	X_{21}	X_{23}	X_{24}	$\{X_{25}, X_{26}\}$
Node 3	X_{31}	X_{32}	X_{34}	$\{X_{35}, X_{36}\}$
Node 4	X_{41}	X_{42}	X_{43}	$\{X_{45}, X_{46}\}$
Node 5	X_{51}	X_{52}	X_{53}	$\{X_{54}, X_{56}\}$
Node 6	X_{61}	X_{62}	X_{63}	$\{X_{64}, X_{65}\}$

$X_{i1}, \ldots, X_{i,j-1}, X_{i,j+1}, \ldots, X_{in}$. Without loss of generality, we store the first $\alpha = 4$ symbols. Table 19.1 shows the modified representation of the (6, 2, 3) PM code given in Example 19.1. For instance, node 1 stores $X_{12} = u_1 + 3u_2 + 2u_3 + 5u_4 + 6u_5$, $X_{13} = u_1 + 4u_2 + 4u_3 + 3u_4 + 5u_5$, and $X_{14} = u_1 + 5u_2 + 4u_3 + 3u_4 + 6u_5$. In general, the data stored on node i, say W'_i, can be written as $W'_i = W_i L$, with L being an invertible matrix, hence the equivalence of the two representations. Therefore, the exact repair and file reconstruction properties are directly inherited from the original PM codes. For each node i, any of the symbols X_{ij} can be computed from any α other symbols (a property of the Vandermonde matrix). For example, in the (6, 2, 3) PM code, if node 1 is being repaired by contacting helper nodes 2, 3, and 4, then it downloads $X_{21} = X_{12}$, $X_{31} = X_{31}$, $X_{41} = X_{14}$ from each respectively. However, if it contacts, say, 2, 3, and 5, then it downloads X_{21}, X_{31}, X_{51} from each respectively. Then, it computes X_{14} using the downloaded data and stores X_{12}, X_{13}, X_{14}.

Correlation Hash [58]

We will assume in the remainder of this section that the underlying finite field is $GF(q^v)$. We abuse notation and look at $X_{ij} \in GF(q^v)$ as a vector in $GF(q)^v$, i.e., $X_{ij} = (X_{ij}^1, X_{ij}^2, \ldots, X_{ij}^v)$, where $X_{ij}^s \in GF(q)$. The hash of two different symbols X_{ij} and X_{lk} is given by the dot product $X_{ij} \cdot X_{lk}^t = \sum_{s=1}^{v} X_{ij}^s X_{lk}^s \in GF(q)$. The collection of hashes is stored on a trusted server that cannot be corrupted by the adversary.

Example 19.4 Consider the problem of securing an $(n, k, d) = (6, 3, 4)$-DSS, with $\alpha = 3$ and $\beta = 1$, against an active limited-knowledge adversary controlling $b = 1$ node. Suppose we have a file of size $C_{r,e} = 5$ symbols, as given in (19.53). We encode the file using the modified representation of an $(n, k' = k - b, d' = d - b) = (6, 2, 3)$ MBR PM code (see Example 19.1), as given in Table 19.1. Moreover, we store the hashes of any two distinct symbols X_{ij} and X_{lk} on the trusted server.

Secure Repair

Suppose that node 1 fails and a newcomer contacts $d = 4$ nodes to repair it, say nodes 2, 3, 4, and 5. From each helper node j, the newcomer downloads $X_{j1} = X_{1j}$. The newcomer also contacts the trusted server to download the corresponding hashes. The

Table 19.2 An example of the repair comparison table when node 2 is compromised. A ✓ indicates that the computed hashes of the downloaded data match the downloaded hashes from the trusted server, while an × marks a hash mismatch. A blank space means that this dot product is not computed.

	X_{21}	X_{31}	X_{41}	X_{51}
X_{21}		×	×	×
X_{31}	×		✓	✓
X_{41}	×	✓		✓
X_{51}	×	✓	✓	

newcomer computes the correlation hashes of the downloaded data and compares them to the corresponding hashes downloaded from the trusted server. Then, it constructs the symmetric comparison table, similar to the example shown in Table 19.2, in which we have assumed that node 2 is compromised.

The newcomer node looks for a column with two ×'s or more (more than $b = 1$ ×'s) to identify the compromised node. Note that a column with one ×, say in the entry corresponding to X_{21} and X_{31}, is confusing and can mean either node 2 or node 3 is compromised. In the example of Table 19.2, the newcomer can identify node 2 as a compromised node and will regenerate the lost data using the symbols received from the other helper nodes. This is possible because the underlying PM code has $d' = d - b = 2$.

The example in Table 19.2 assumes the ideal case in which all the corrupted symbols result in a hash mismatch. However, the adversary can always attempt to corrupt the data in a way to match the downloaded hashes, i.e., create a hash collision. For example, it will be able to fool the new node if it manages to change at least two ×'s into ✓'s in the column of X_{21}. However, its observation consisting of only $(X_{21}, X_{23}, X_{24}, X_{25})$ here is uncorrelated (linearly independent) with the other downloaded packets X_{31}, X_{41}, X_{51} (which follows from the construction of PM codes); its best strategy cannot be better than introducing random errors which could be caught with high probability (see the proof of Theorem 19.9).

Secure Reconstruction

Suppose a DC contacts $k = 3$ nodes, say nodes 1, 2, and 3, downloads all their, possibly corrupted, stored symbols, and computes their correlation hashes. Moreover, the DC contacts the trusted server and downloads all the corresponding hashes. The DC constructs a $k \times k = 3 \times 3$ block table, where each $(d - b) \times (d - b) = 3 \times 3$ block represents a contacted node, as depicted in Table 19.3. We define a *mismatched* block to be a block containing at least one ×. A non-compromised node can have at most $b = 1$ mismatched block. Hence, the DC declares that a node is compromised if its row or column contains more than one mismatched block. Once the DC detects the compromised node, it discards its data and reconstructs the file from the symbols on the remaining nodes. This is possible because the underlying PM code has $k' = k - b$. Here

Table 19.3 An example of the reconstruction block table formed by a DC contacting nodes 1, 2, and 3 where node 2 is compromised.

		node 1			node 2			node 3		
		X_{12}	X_{13}	X_{14}	X_{21}	X_{23}	X_{24}	X_{31}	X_{32}	X_{34}
node 1	X_{12}				✓	✓	✓	✓	✓	✓
	X_{13}				×	×	×	✓	✓	✓
	X_{14}				×	×	×	✓	✓	✓
node 2	X_{21}	✓	×	×				×	✓	×
	X_{23}	✓	×	×				×	✓	×
	X_{24}	✓	×	×				×	✓	×
node 3	X_{31}	✓	✓	✓	×	×	×			
	X_{32}	✓	✓	✓	✓	✓	✓			
	X_{34}	✓	✓	✓	×	×	×			

too, we have implicitly used the fact that a symbol corrupted by the adversary will result in a hash mismatch. This is true with high probability as detailed in the proof.

Error Analysis

We will show that the probability of failing to detect a compromised node during repair or reconstruction is bounded above by $1/q$ and therefore can be made as small as desired by increasing q. Suppose the adversary wants to flip an × to a ✓ in a comparison table, say the × corresponding to X_{21} and X_{31} in Table 19.2. It can introduce an error e to X_{21} to change it to $X_{21} + e$ and will succeed in flipping the entry in the table if e is orthogonal to X_{31}. Assuming the symbols of the original file are i.i.d. and uniformly distributed over $GF(q)^v$, for a fixed error $e \in GF(q)^v$, there are q^{v-1} possible values of X_{31} that are orthogonal to e. Hence, the probability of e being orthogonal to one symbol given the adversary's observation is

$$P(e \perp X_{31}) = \frac{q^{v-1}}{q^v} = \frac{1}{q}.$$

Evidently, if the adversary were omniscient it would know the value of X_{31} and could choose a value of e that is orthogonal to it. Note that the adversary needs to flip many entries in the comparison table to fool the secure repair and/or the reconstruction algorithms described earlier. These events are not independent and we get an upper bound for the probability of error, i.e., adversary being undetected, of $1/q$. However, this may be a loose bound.

19.5.3 Proof of Theorem 19.9

We show that the code construction described in Example 19.4, which is based on PM codes and correlation hashes, can be generalized to achieve the resiliency capacity in (19.53). To secure an (n,k,d)-DSS against an adversary controlling b nodes, the construction uses an $(n, k' = k - b, d' = d - b)$ MBR PM code with storage per node $\alpha = (d-b)\beta$. These codes can store a file of size at most $\mathcal{M} = \sum_{i=1}^{k-b} \min\{\alpha, (d-b-i+1)\beta\}$ symbols, as given by (19.21). A simple change of variables shows that $\mathcal{M} = C_{r,e}$, when $\alpha \geq (d-b)\beta$. We prove Theorem 19.9 by showing that these $\mathcal{M}$ symbols can be secured by following the secure repair and secure decoding rules illustrated in the previous example.

Secure Repair

If a node f fails, the new node contacts d helper nodes $h_1, h_2, \ldots, h_d$ for repair. Each of the helper nodes sends $X_{h_i f}$ to the new node. In addition, the new node downloads all the hashes $X_{h_i f} \cdot X_{h_j f}$ from the trusted server and forms a repair comparison table similar to Table 19.2. Without loss of generality, suppose that among the helper nodes exactly b nodes are compromised.[10] Also, suppose the ideal case that every transmitted packet corrupted by the adversary results in an × in the comparison table. An uncorrupted node will contain at most b ×'s in its column. Therefore, every helper node corresponding to more than b ×'s must be a compromised node.

SECURE REPAIR RULE: *The repaired node decides that any helper node corresponding to a column (or a row) in the comparison table with more than b ×'s is a compromised node.*

The adversary can try to avoid being detected by causing errors that lead to switching at least $d-b$ ×'s into ✓'s in columns corresponding to nodes under its control. However, since $b < k/2$ the adversary knows zero information about the symbols sent by the helper nodes that are not under its control (a property of the underlying $(n, k-b, d-b)$ PM code). Therefore, the error analysis detailed in Section 19.5.2 holds here and the probability of the adversary succeeding, i.e., causing undetected erroneous repair data, is bounded above by $1/q$.

Secure Reconstruction

A DC contacts k nodes and downloads all their stored data ($\alpha = d-b$ symbols from each node). In addition, it contacts the trusted server and downloads all the hashes corresponding to the downloaded symbols. Without loss of generality, suppose that b nodes among those which were contacted are compromised. Then, the DC forms the reconstruction block table between the trusted hashes and computed hashes as done in Table 19.3.

Each block contains the hashes cross-checking the symbols downloaded from two different nodes. A mismatched block indicates that one of these two nodes is sending corrupted packets and therefore is compromised. Hereafter, a node having more than $b+1$ mismatched blocks has to be compromised. Otherwise, the $b+1$ other nodes corresponding to the mismatched blocks are all compromised, which contradicts our initial assumption that there are at most b compromised nodes. Therefore, the user implements the following rule:

SECURE RECONSTRUCTION RULE: *The DC decides that any node having more than b mismatched blocks in its row or column is a compromised node.*

---

[10] The analysis stays the same even if fewer than b helper nodes happen to be compromised.

Treating the b compromised nodes as erasures, the user recovers the data using the $k-b$ other contacted nodes. Moreover, it reports the identity of the compromised nodes. If $b < k/2$, then on each uncompromised node there is at least one symbol that the adversary does not have information about, and the error analysis in Section 19.5.2 holds again here. Note that if $b \geq k/2$, then the limited-knowledge adversary can decode the whole file and is actually omniscient. It follows from Theorem 19.6 that the resiliency capacity here is $C_{r,e} = 0$.

Hash Storage Overhead

The total number of stored hashes is $\binom{\theta}{2}$ symbols in $GF(q)$, where $\theta = \binom{n}{2}$ is the total number of X_{ij} symbols. The ratio of the hash size to the stored data size is $\frac{\binom{\theta}{2}}{n\alpha\nu} = \mathcal{O}(\frac{1}{\nu})$, which can be made arbitrarily small by increasing the packet length ν.

19.6 Conclusion and Open Problems

In this chapter, we presented the main results on information theoretic security for distributed storage systems. We focused on three different adversarial models considered in the literature: passive eavesdropper, active omniscient adversary, and active limited-knowledge adversary. We gave upper bounds on the secrecy or resiliency capacity for each adversary model for a general set of constraints on node storage capacity α and repair bandwidth β. Moreover, we presented explicit capacity-achieving codes for some specific regimes, namely:

1. For the passive eavesdropper case, we presented capacity-achieving codes in the regime $\alpha \geq d\beta$. We also presented an upper bound for secrecy capacity in the regime $\beta = \alpha/(d-k+1)$ when the encoding is restricted to linear MDS codes, and showed the corresponding separation-based achievability schemes when $d = n-1$.
2. For the active omniscient adversary, we presented capacity-achieving codes for the regimes $\alpha \leq (d-k+1)\beta$ and $\alpha \geq d\beta$.
3. For the active limited-knowledge adversary, we presented capacity-achieving codes for the regime $\alpha \geq d\beta$.

In general, the upper bounds presented here are known not to be always achievable, and proving tighter general bounds remains an open question. In the passive eavesdropper case, even for the practically significant regime $\beta = \alpha/(d-k+1)$ the secrecy capacity is unknown when the encoding is non-linear, or when $d < n-1$, or when the repair of all (and not just the systematic) nodes is considered. Furthermore, the research direction of universally secure codes is wide open. On the other hand, less is known about the active adversary cases beyond the results presented here. The achievability of resiliency capacity for the regimes that have not been presented in this chapter is still an open problem. It should also be noted that all the presented secure codes are for the case of exact repair and it is an open problem whether codes with functional repair can outperform them.

References

[1] S. Ghemawat, H. Gobioff, and S.-T. Leung, "The Google file system," in *Proc. 19th ACM Symp. Operating Systems Principles*, Bolton Landing, NY, USA, Oct. 2003, pp. 29–43.

[2] G. DeCandia, D. Hastorun, M. Jampani, G. Kakulapati, A. Lakshman, A. Pilchin, S. Sivasubramanian, P. Vosshall, and W. Vogels, "Dynamo: Amazon's highly available key–value store," in *Proc. 21st ACM Special Interest Group on Operating System*, Stevenson, WA, USA, Oct. 2007, pp. 205–220.

[3] J. Li, *Efficient and Reliable Storage Solutions*. Redmond, WA: Microsoft Corporation, 2013.

[4] A. Lakshman and P. Malik, "Cassandra: A decentralized structured storage system," *SIGOPS Oper. Syst. Rev.*, vol. 44, no. 2, pp. 35–40, Apr. 2010.

[5] D. Beaver, S. Kumar, H. C. Li, J. Sobel, and P. Vajgel, "Finding a needle in haystack: Facebook's photo storage," in *Proc. 9th USENIX Conf. Networked Systems Design Implementation*, Vancouver, BC, Canada, Oct. 2010, pp. 47–60.

[6] B. Calder, J. Wang, A. Ogus, N. Nilakantan, A. Skjolsvold, S. McKelvie, Y. Xu, S. Srivastav, J. Wu, and H. Simitci, "Windows Azure storage: A highly available cloud storage service with strong consistency," in *Proc. 23rd ACM Symp. Operating Systems Principles*, Cascais, Portugal, Oct. 2011, pp. 143–157.

[7] J. Kubiatowicz, D. Bindel, Y. Chen, S. Czerwinski, P. Eaton, D. Geels, R. Gummadi, S. Shea, H. Weatherspoon, W. Weimer, C. Wells, and B. Zhao, "Oceanstore: An architecture for global-scale persistent storage," in *Proc. 9th Int. Conf. Architectural Support for Programming Languages and Operating Systems*, Cambridge, MA, USA, Nov. 2000, pp. 190–201.

[8] S. Rhea, P. Eaton, D. Geels, H. Weatherspoon, B. Zhao, and J. Kubiatowicz, "Pond: The Oceanstore prototype," in *Proc. 2nd USENIX Conf. File Storage Technologies*, San Francisco, CA, USA, Mar. 2003, pp. 1–14.

[9] R. Bhagwan, K. Tati, Y.-C. Cheng, S. Savage, and G. M. Voelker, "Total recall: System support for automated availability management," in *Proc. 1st USENIX Conf. Networked Systems Design Implementation*, San Francisco, CA, USA, Mar. 2004, p. 25.

[10] A. Ha, "P2P startup Space Monkey raises 2.25m led by Google Ventures and Venture51," Jul. 2012. [Online]. Available: http://techcrunch.com/2012/07/11/space-monkey-seed-round/

[11] J. McGregor, "The top 5 most brutal cyber attacks of 2014 so far," Jul. 2014. [Online]. Available: www.forbes.com/sites/jaymcgregor/2014/07/28/the-top-5-most-brutal-cyber-attacks-of-2014-so-far/

[12] G. Smith, "Home depot admits 56 million payment cards at risk after cyber attack," Sep. 2014. [Online]. Available: www.huffingtonpost.com/2014/09/18/home-depot-hack-n-5845378.html

[13] J. Silver-Greenberg, M. Goldstein, and N. Perlroth, "Was government hack our final warning?" Oct. 2014. [Online]. Available: http://dealbook.nytimes.com/2014/10/02/jpmorgan-discovers-further-cyber-security-issues/

[14] J. H. Davis, "Hacking of government computers exposed 21.5 million people," Jul. 2015. [Online]. Available: www.nytimes.com/2015/07/10/us/office-of-personnel-management-hackers-got-data-of-millions.html

[15] T. Gara, "An expensive hack attack: Target's 148 million dollars breach," Aug. 2014. [Online]. Available: http://blogs.wsj.com/corporate-intelligence/2014/08/05/an-expensive-hack-attack-targets-148-million-breach/

[16] J. Schreirer, "Sony estimates 171 million dollars loss from PSN hack," May 2011. [Online]. Available: www.wired.com/2011/05/sony-psn-hack-losses/
[17] Ponemon Institute LLC, "2014 cost of data breach study: United States," May 2014.
[18] Y. Liang, H. V. Poor, and S. Shamai (Shitz), "Information theoretic security," *Found. Trends Commun. Inf. Theory*, vol. 5, no. 4–5, pp. 355–580, 2009.
[19] I. Csiszár and J. Körner, *Information Theory: Coding Theorems for Discrete Memoryless Systems*, 2nd edn. Cambridge: Cambridge University Press, 2011.
[20] A. El Gamal and Y.-H. Kim, *Network Information Theory*. Cambridge: Cambridge University Press, 2011.
[21] B. Schroeder and G. A. Gibson, "Disk failure in the real world: What does an MTTF of 1,000,000 hours mean to you?" in *Proc. 5th USENIX Conf. File Storage Technologies*, San Jose, CA, USA, Feb. 2007, pp. 1–16.
[22] C. Huang, H. Simitci, Y. Xu, A. Ogus, B. Calder, P. Gopalan, J. Li, and S. Yekhanin, "Erasure coding in Windows Azure storage," in *Proc. USENIX Annual Technical Conf.*, Boston, MA, USA, Jun. 2012, p. 2.
[23] A. Dimakis, P. Godfrey, Y. Wu, M. Wainwright, and K. Ramchandran, "Network coding for distributed storage systems," *IEEE Trans. Inf. Theory*, vol. 56, no. 9, pp. 4539–4551, Sep. 2010.
[24] A. Shamir, "How to share a secret," *Commun. ACM*, vol. 22, no. 11, pp. 612–613, Nov. 1979.
[25] L. H. Ozarow and A. D. Wyner, "Wire-tap channel II," *AT&T Bell Lab. Tech. J.*, vol. 63, no. 10, pp. 2135–2157, Dec. 1984.
[26] N. Cai and R. W. Yeung, "Secure network coding on a wiretap network," *IEEE Trans. Inf. Theory*, vol. 57, no. 1, pp. 424–435, Jan. 2011.
[27] S. El Rouayheb, E. Soljanin, and A. Sprintson, "Secure network coding for wiretap networks of type II," *IEEE Trans. Inf. Theory*, vol. 58, no. 3, pp. 1361–1371, Mar. 2012.
[28] A. J. Menezes, P. C. Van Oorschot, and S. A. Vanstone, *Handbook of Applied Cryptography*. London: CRC Press, 1996.
[29] J. Katz and Y. Lindell, *Introduction to Modern Cryptography*. London: CRC Press, 2014.
[30] M. Storer, K. Greenan, E. Miller, and K. Voruganti, "POTSHARDS: Secure long-term storage without encryption," in *Proc. USENIX Annual Technical Conf.*, Santa Clara, CA, USA, Jun. 2007, pp. 11:1–11:14.
[31] J. K. Resch and J. S. Plank, "AONT-RS blending security and performance in dispersed storage systems," in *Proc. 9th USENIX Conf. File Storage Technol.*, San Jose, CA, USA, Feb. 2011, p. 14.
[32] www.cleversafe.com/.
[33] A. G. Dimakis, K. Ramchandran, Y. Wu, and C. Suh, "A survey on network codes for distributed storage," *Proc. IEEE*, vol. 99, no. 3, pp. 476–489, Mar. 2011.
[34] D. Papailiopoulos, A. Dimakis, and V. Cadambe, "Repair optimal erasure codes through Hadamard designs," in *Proc. 49th Annual Allerton Conf. Commun., Control, Computing*, Monticello, IL, USA, Sep. 2011, pp. 1382–1389.
[35] B. Sasidharan and P. V. Kumar, "High-rate regenerating codes through layering," in *Proc. IEEE Int. Symp. Inf. Theory*, Istanbul, Turkey, Jul. 2013, pp. 1611–1615.
[36] M. Sathiamoorthy, M. Asteris, D. Papailiopoulos, A. G. Dimakis, R. Vadali, S. Chen, and D. Borthakur, "XORing elephants: Novel erasure codes for big data," in *Proc. VLDB Endowment*, vol. 6, no. 5, Mar. 2013, pp. 325–336.
[37] N. B. Shah, K. V. Rashmi, P. V. Kumar, and K. Ramchandran, "Explicit codes minimizing

repair bandwidth for distributed storage," in *Proc. IEEE Inf. Theory Workshop*, Cairo, Egypt, Jan. 2010, pp. 1–5.

[38] K. V. Rashmi, N. B. Shah, and P. V. Kumar, "Optimal exact-regenerating codes for distributed storage at the MSR and MBR points via a product-matrix construction," *IEEE Trans. Inf. Theory*, vol. 57, no. 8, pp. 5227–5239, Aug. 2011.

[39] Z. Wang, I. Tamo, and J. Bruck, "On codes for optimal rebuilding access," in *Proc. 49th Annual Allerton Conf. Commun., Control, Computing*, Monticello, IL, USA, Sep. 2011, pp. 1374–1381.

[40] I. Tamo, Z. Wang, and J. Bruck, "Zigzag codes: MDS array codes with optimal rebuilding," *IEEE Trans. Inf. Theory*, vol. 59, no. 3, pp. 1597–1616, Mar. 2013.

[41] I. Tamo and A. Barg, "A family of optimal locally recoverable codes," *IEEE Trans. Inf. Theory*, vol. 60, no. 8, pp. 4661–4676, Aug. 2014.

[42] S. El Rouayheb and K. Ramchandran, "Fractional repetition codes for repair in distributed storage systems," in *Proc. 48th Annual Allerton Conf. Commun., Control, Computing*, Monticello, IL, USA, Sep. 2010, pp. 1510–1517.

[43] S. Pawar, N. Noorshams, S. El Rouayheb, and K. Ramchandran, "DRESS codes for the storage cloud: Simple randomized constructions," in *Proc. IEEE Int. Symp. Inf. Theory*, St. Petersburg, Russia, Jul. 2011, pp. 2338–2342.

[44] P. Gopalan, C. Huang, H. Simitci, and S. Yekhanin, "On the locality of codeword symbols," *IEEE Trans. Inf. Theory*, vol. 58, no. 11, pp. 6925–6934, Nov. 2012.

[45] N. B. Shah, K. V.Rashmi, P. V. Kumar, and K. Ramchandran, "Distributed storage codes with repair-by-transfer and nonachievability of interior points on the storage–bandwidth tradeoff," *IEEE Trans. Inf. Theory*, vol. 58, no. 3, Mar. 2012, pp. 1837–1852.

[46] D. Cullina, A. G. Dimakis, and T. Ho, "Searching for minimum storage regenerating codes," Oct. 2009. [Online]. Available: http://arxiv.org/abs/0910.2245

[47] Y. Wu, "A construction of systematic MDS codes with minimum repair bandwidth," *IEEE Trans. Inf. Theory*, vol. 57, no. 6, pp. 3738–3741, Jun. 2011.

[48] V. R. Cadambe, S. A. Jafar, H. Maleki, K. Ramchandran, and C. Suh, "Asymptotic interference alignment for optimal repair of MDS codes in distributed storage," *IEEE Trans. Inf. Theory*, vol. 59, pp. 2974–2987, May 2013.

[49] V. R. Cadambe, C. Huang, J. Li, and S. Mehrotra, "Polynomial length MDS codes with optimal repair in distributed storage," in *Proc. 45th Asilomar Conf. Signals, Systems, Computers*, Pacific Grove, CA, USA, Nov. 2011, pp. 1850–1854.

[50] A. S. Rawat, O. O. Koyluoglu, and S. Vishwanath, "Progress on high-rate MSR codes: Enabling arbitrary number of helper nodes," Jan. 2016. [Online]. Available: http://arxiv.org/abs/1601.06362

[51] S. Goparaju, A. Fazeli, and A. Vardy, "Minimum storage regenerating codes for all parameters," Feb. 2016. [Online]. Available: http://arxiv.org/abs/1602.04496

[52] K. V. Rashmi, N. B. Shah, P. V. Kumar, and K. Ramchandran, "Explicit construction of optimal exact regenerating codes for distributed storage," in *Proc. 47th Annual Allerton Conf. Commun., Control, Computing*, Monticello, IL, USA, Sep. 2009, pp. 1243–1249.

[53] S. Goparaju, "Erasure codes for optimal node repairs in distributed storage systems," Ph.D. dissertation, Princeton University, 2014.

[54] S. Goparaju, I. Tamo, and R. Calderbank, "An improved sub-packetization bound for minimum storage regenerating codes," *IEEE Trans. Inf. Theory*, vol. 60, no. 5, pp. 2770–2779, May 2014.

[55] N. B. Shah, K. V. Rashmi, P. V. Kumar, and K. Ramchandran, "Interference alignment in regenerating codes for distributed storage: Necessity and code constructions," *IEEE Trans. Inf. Theory*, vol. 58, no. 4, pp. 2134–2158, Apr. 2012.

[56] V. R. Cadambe, C. Huang, S. A. Jafar, and J. Li, "Optimal repair of MDS codes in distributed storage via subspace interference alignment," Jun. 2011. [Online]. Available: http://arxiv.org/abs/1106.1250

[57] S. Pawar, S. El Rouayheb, and K. Ramchandran, "On secure distributed data storage under repair dynamics," in *Proc. IEEE Int. Symp. Inf. Theory*, Austin, TX, USA, Jun. 2010, pp. 2543–2547.

[58] S. Pawar, S. El Rouayheb, and K. Ramchandran, "Securing dynamic distributed storage systems against eavesdropping and adversarial attacks," *IEEE Trans. Inf. Theory*, vol. 58, no. 3, pp. 6734–6753, Mar. 2012.

[59] S. Pawar, S. El Rouayheb, and K. Ramchandran, "Securing dynamic distributed storage systems from malicious nodes," in *Proc. IEEE Int. Symp. Inf. Theory*, St. Petersburg, Russia, Jul. 2011, pp. 1452–1456.

[60] N. B. Shah, K. V. Rashmi, and P. V. Kumar, "Information-theoretically secure regenerating codes for distributed storage," in *Proc. IEEE Global Commun. Conf.*, Houston, TX, USA, Dec. 2011, pp. 1–5.

[61] N. B. Shah, K. V. Rashmi, and K. Ramchandran, "Secure network coding for distributed secret sharing with low communication cost," in *Proc. IEEE Int. Symp. Inf. Theory*, Istanbul, Turkey, Jul. 2013, pp. 2404–2408.

[62] K. V. Rashmi, N. B. Shah, K. Ramchandran, and P. V. Kumar, "Regenerating codes for errors and erasures in distributed storage," in *Proc. IEEE Int. Symp. Inf. Theory*, Cambridge, MA, USA, Jul. 2012, pp. 1202–1206.

[63] S. Goparaju, S. El Rouayheb, R. Calderbank, and H. V. Poor, "Data secrecy in distributed storage systems under exact repair," in *Proc. IEEE Int. Symp. Network Coding*, Calgary, Canada, Jun. 2013, pp. 1–6.

[64] R. Tandon, S. Amuru, T. Clancy, and R. Buehrer, "Towards optimal secure distributed storage systems with exact repair," *IEEE Trans. Inf. Theory*, vol. 62, no. 6, pp. 3477–3492, Jun. 2016.

[65] O. Kosut, "Polytope codes for distributed storage in the presence of an active omniscient adversary," in *Proc. IEEE Int. Symp. Inf. Theory*, Istanbul, Turkey, Jul. 2013, pp. 897–901.

[66] N. Silberstein, A. S. Rawat, and S. Vishwanath, "Error resilience in distributed storage via rank-metric codes," in *Proc. 50th Annual Allerton Conf. Commun., Control, Computing*, Monticello, IL, USA, Sep. 2012, pp. 1150–1157.

[67] A. S. Rawat, O. O. Koyluoglu, N. Silberstein, and S. Vishwanath, "Optimal locally repairable and secure codes for distributed storage systems," *IEEE Trans. Inf. Theory*, vol. 60, no. 1, pp. 212–236, Jan. 2014.

[68] A. Agarwal and A. Mazumdar, "Security in locally repairable storage," in *Proc. IEEE Inf. Theory Workshop*, Jerusalem, Israel, May 2015, pp. 1–5.

[69] R. W. Yeung, S.-Y. R. Li, N. Cai, and Z. Zhang, "Network coding theory part I: Single sources / part II: Multiple sources," *Found. Trends Commun. Inf. Theory*, vol. 2, no. 4–5, pp. 241–381, 2005.

[70] R. W. Yeung and N. Cai, "Network error correction, I: Basic concepts and upper bounds," *Commun. Inf. Systems*, vol. 6, no. 1, pp. 19–35, 2006.

[71] N. Cai and R. W. Yeung, "Network error correction, II: Lower bounds," *Commun. Inf. & Systems*, vol. 6, no. 1, pp. 37–54, 2006.

[72] N. Cai and R. W. Yeung, "Secure network coding," in *Proc. IEEE Int. Symp. Inf. Theory*, Lausanne, Switzerland, Jun. 2002, p. 323.

[73] S. Jaggi, M. Langberg, S. Katti, T. Ho, D. Katabi, M. Medard, and M. Effros, "Resilient network coding in the presence of byzantine adversaries," *IEEE Trans. Inf. Theory*, vol. 54, no. 6, pp. 2596–2603, Jun. 2008.

[74] D. Silva, F. R. Kschischang, and R. Koetter, "A rank-metric approach to error control in random network coding," *IEEE Trans. Inf. Theory*, vol. 54, no. 9, pp. 3951–3967, Sep. 2008.

[75] R. Koetter and F. Kschischang, "Coding for errors and erasures in random network coding," *IEEE Trans. Inf. Theory*, vol. 54, no. 8, pp. 3579–3591, Aug. 2008.

[76] C. Fragouli and E. Soljanin, "Network coding fundamentals," *Found. Trends Network.*, vol. 2, no. 1, pp. 1–133, 2007.

[77] T. Ho and D. Lun, *Network Coding: An Introduction*. Cambridge: Cambridge University Press, 2008.

[78] T. Cui, T. Ho, and J. Kliewer, "On secure network coding with nonuniform or restricted wiretap sets," *IEEE Trans. Inf. Theory*, vol. 59, no. 1, pp. 166–176, Jan. 2013.

[79] O. Kosut, L. Tong, and D. N. C. Tse, "Nonlinear network coding is necessary to combat general byzantine attacks," in *Proc. 47th Annual Allerton Conf. Commun., Control, Computing*, Monticello, IL, USA, Oct. 2009, pp. 593–599.

[80] O. Kosut, L. Tong, and D. N. C. Tse, "Polytope codes against adversaries in networks," *IEEE Trans. Inf. Theory*, vol. 60, no. 6, pp. 3308–3344, Jun. 2014.

[81] T. Ernvall, S. El Rouayheb, C. Hollanti, and H. V. Poor, "Capacity and security of heterogeneous distributed storage systems," *IEEE J. Sel. Areas Commun.*, vol. 31, no. 12, pp. 2701–2709, Dec. 2013.

[82] J. Pernas, C. Yuen, B. Gastón, and J. Pujol, "Non-homogeneous two-rack model for distributed storage systems," in *Proc. IEEE Int. Symp. Inf. Theory*, Istanbul, Turkey, Jul. 2013, pp. 1237–1241.

[83] Q. Yu, K. W. Shum, and C. W. Sung, "Tradeoff between storage cost and repair cost in heterogeneous distributed storage systems," *Trans. Emerging Telecommun. Technol.*, vol. 26, no. 10, pp. 1201–1211, Oct. 2015.

[84] K. W. Shum and Y. Hu, "Cooperative regenerating codes," *IEEE Trans. Inf. Theory*, vol. 59, no. 11, pp. 7229–7258, Nov. 2013.

[85] O. O. Koyluoglu, A. S. Rawat, and S. Vishwanath, "Secure cooperative regenerating codes for distributed storage systems," *IEEE Trans. Inf. Theory*, vol. 60, no. 9, pp. 5228–5244, Sep. 2014.

[86] T. Ho, M. Médard, R. Koetter, D. R. Karger, M. Effros, J. Shi, and B. Leong, "A random linear network coding approach to multicast," *IEEE Trans. Inf. Theory*, vol. 52, no. 10, pp. 4413–4430, Oct. 2006.

[87] S. Goparaju, S. El Rouayheb, and R. Calderbank, "New codes and inner bounds for exact repair in distributed storage systems," Feb. 2014. [Online]. Available: http://arxiv.org/abs/1402.2343

[88] C. Tian, "Rate region of the $(4,3,3)$ exact-repair regenerating codes," in *Proc. IEEE Int. Symp. Inf. Theory*, Istanbul, Turkey, Jul. 2013, pp. 1426–1430.

[89] B. Sasidharan, K. Senthoor, and P. V. Kumar, "An improved outer bound on the storage-repair-bandwidth tradeoff of exact-repair regenerating codes," in *Proc. IEEE Int. Symp. Inf. Theory*, Honolulu, HI, USA, Jul. 2014, pp. 2430–2434.

[90] C. Tian, V. Aggarwal, and V. A. Vaishampayan, "Exact-repair regenerating codes via layered erasure correction and block designs," in *Proc. IEEE Int. Symp. Inf. Theory*,

Istanbul, Turkey, Jul. 2013, pp. 1431–1435.

[91] M. Elyasi, S. Mohajer, and R. Tandon, "Linear exact repair rate region of $(k+1,k,k)$ distributed storage systems: A new approach," in *Proc. IEEE Int. Symp. Inf. Theory*, Hong Kong, Jun. 2015, pp. 2061–2065.

[92] S. Mohajer and R. Tandon, "New bounds on the (n, k, d) storage systems with exact repair," in *Proc. IEEE Int. Symp. Inf. Theory*, Hong Kong, Jun. 2015, pp. 2056–2060.

[93] T. Ernvall, "Codes between MBR and MSR points with exact repair property," *IEEE Trans. Inf. Theory*, vol. 60, no. 11, pp. 6993–7005, Nov. 2014.

[94] T. K. Dikaliotis, A. G. Dimakis, and T. Ho, "Security in distributed storage systems by communicating a logarithmic number of bits," in *Proc. IEEE Int. Symp. Inf. Theory*, Austin, TX, USA, Jun. 2010, pp. 1948–1952.

[95] J. Harshan and F. Oggier, "On algebraic manipulation detection codes from linear codes and their application to storage systems," in *Proc. IEEE Inf. Theory Workshop*, Jeju, Korea, Oct. 2015, pp. 64–68.

[96] S. H. Dau, W. Song, and C. Yuen, "On block security of regenerating codes at the MBR point for distributed storage systems," in *Proc. IEEE Int. Symp. Inf. Theory*, Honolulu, HI, USA, Jun. 2014, pp. 1967–1971.

[97] S. Kadhe and A. Sprintson, "Weakly secure regenerating codes for distributed storage," in *Proc. IEEE Int. Symp. Network Coding*, Aalborg, Denmark, Jun. 2014, pp. 1–6.

[98] I. Tamo, Z. Wang, and J. Bruck, "Access vs. bandwidth in codes for storage," in *Proc. IEEE Int. Symp. Inf. Theory*, Cambridge, MA, USA, Jul. 2012, pp. 1187–1191.

[99] E. M. Gabidulin, "Theory of codes with maximum rank distance," *Probl. Peredachi Inf.*, vol. 21, no. 1, pp. 3–16, 1985.

[100] S. Goparaju, S. El Rouayheb, and R. Calderbank, "Can linear minimum storage regenerating codes be universally secure?" in *Proc. 49th Asilomar Conf. Signals, Systems, Computers*, Pacific Grove, CA, USA, Nov. 2015, pp. 549–553.

[101] F. J. MacWilliams and N. J. A. Sloane, *The Theory of Error Correcting Codes*. New York: Elsevier, 1977, vol. 16.

[102] R. Bitar and S. El Rouayheb, "Securing data against limited-knowledge adversaries in distributed storage systems," in *Proc. IEEE Int. Symp. Inf. Theory*, Hong Kong, Jun. 2015, pp. 2847–2851.

Index

achievability, 82
active limited-knowledge adversary, 524
active omniscient adversary, 524
additivity, 265, 278
Ahlswede–Csiszár secret generation, *see* secret key generation
algebraic core, 371, 374
   dimension, 375
   index set, 375
   rank, 375
   rank loss, 375
alternating CSIT, 214
   DN/ND, 223
   PD/DP, 221
anchor node, 391
anonymous broadcast messaging, 476
arbitrarily varying channel, 259, 315, 317, 324, 325
   CR-assisted capacity, 325
   unassisted capacity, 325
arbitrarily varying wiretap channel, 259, 267, 274, 315, 316, 318, 321, 322
   CR-assisted secrecy capacity, 321
   orthogonal, 322
   unassisted secrecy capacity, 322
attack vectors, 368, 372
attacks
   active attacks, 258, 259
   passive attacks, 258, 259
authentication, 390, 396, 424
   channel-based, 395
   key-based, 394
   keyless, 394
   physical-layer authentication, 393
AVC, *see* arbitrarily varying channel
average wasted energy, 507
AVWC, *see* arbitrarily varying wiretap channel

backtracking line search, 133
barrier method, 132
battery capacity, 509
beamforming, 145
Berger–Tung coding, 91
Bernoulli source, 84, 101
bin index, 82, 91

binary entropy function, 84, 101
binary erasure fading wiretap channel, 232
binary jamming, 70
binary symmetric channel, 84, 101
binning, 82, 91, 94, 99
biometric authentication, 445, 446, 450, 451
biometric authentication system, 421, 424
BSC, *see* binary symmetric channel

capacity, 263
   uncorrelated coding, 260, 263, 264, 308
   zero-error, 314
   secret common randomness assisted, 277
   with public side information, 268, 277
   zero error, 264
causal disclosure, 59, 68
CCDF, 234
cellular model, 333, 334, 340, 343, 357
channel, 184
   K-user, 195
   orthogonal, 270, 285
   broadcast, 184
   interference, 184
   multi-access, 184
   multi-antenna, 184, 189
   relay, 184
   X-channel, 187
channel enhancement, 232
channel resolvability, 165, 166, 170
channel state information, 184
   alternating, 195
   causal, 185
   mixed, 186
   no, 190
   perfect, 185
channel state information at transmitter, 203, 231
   alternating CSIT, 214
   delayed CSIT, 203
   homogeneous CSIT, 203
   hybrid CSIT, 203
   no CSIT, 203
   perfect CSIT, 203

Index

code, 274–277, 319, 321
    CR-assisted code, 275–277, 284, 290, 291, 296, 298, 321
    private/public, 275
    shared randomness assisted code, 274, 275, 296
    unassisted code, 319
    uncorrelated code, 260, 262, 274, 276, 286, 296
coding scheme at helper, 93
    decode and reencode, 93, 95, 96
    forwarding, 93, 94, 96
common goal game, 514
common randomness, 262, 265, 267, 268, 270, 271, 277–279, 286, 296, 300, 303, 320
competitive privacy, 511
    game theoretic formulation, 513
    pricing, 514
compound channel, 120, 263, 276, 291
compound source, 365
compound wiretap channel, 267, 279, 291
concentrator node, 391
continuity, 282
    capacity, 270, 278, 279, 282, 284, 307
cooperative jamming, 141, 148, 149, 152–154, 157–159, 161, 162, 178
countably infinite support, 27, 41
coupling, 234
covariance matrix, 112, 118
    full-rank, 110, 113, 115, 118
    optimal, 110, 112, 114, 116, 119, 122, 128–131
    rank of, 113
    rank-1, 113, 114
CR, *see* common randomness
CSI, *see* channel state information
CSIT, *see* channel state information at transmitter
cut-set bound, 337, 343, 344, 346, 348, 350, 351, 357
cyber-physical system, 499

dark bit masking, 373, 377
data processing inequality, 96
database
    multiple databases, 504
    multiple queries, 504
database attributes
    public (revealed) and private (hidden), 501
database model, 500
decode, 186
    interference alignment, 185
    noise injection, 186
    zero-forcing, 186
degraded, 233
degrees of freedom, 181, 196
    secure, 204
    generalized, 189, 196
    secure, 184
derandomization, 303

deterministic cipher, 22
dining cryptographer networks, 477
discrete p-dispersion problem, 458
discrete memoryless source, 55
distortion, 84, 87, 88
    Hamming, 61, 84, 101
    logarithmic loss, 73, 87, 88, 90, 94, 95
    measure of utility, 501
    quadratic, 88, 90
distributed storage system, 522
    file reconstruction, 522
    node repair, 522
        exact repair, 523
        functional repair, 523

eavesdropper, 79, 258, 259, 524
    isotropic, 120–123, 125
    negligible, 125
    omnidirectional, 110, 127
    weak, 110, 117–119
elimination of correlation, 173
encryption, 95
end-user privacy, 97, 98
energy efficiency, 404, 407
energy harvesting and storage, 506
energy harvesting rate, 508
energy management unit, 506
enrollment, 424
enrollment biometric sequence, 423, 427, 430, 439
entropy, 78, 195, 376, 501
    conditional, 78, 98
    relative, 273
entropy power inequality, 88
EPI, *see* entropy power inequality
EPS, *see* error-free perfect secrecy system
equivocation
    measure of privacy, 501
    rate, 73, 78, 97
erasure probability, 84, 101
error
    error-free, 22
    average error, 262, 275, 279, 289, 296, 298, 305
    maximal error, 264
    zero error, 264
error correcting code, 363, 452, 453, 470
error-free perfect secrecy system, 25

false acceptance exponent, 421, 422, 424, 428, 431, 439
    maximum achievable, 425, 431
    achievable, 424, 427, 431, 439
false acceptance rate, 421, 422, 424, 427
false alarm, 397, 398, 402
false rejection rate, 421, 424, 427, 431
FAR, *see* false acceptance rate
feedback, 187

fixed-basis design, 462
Fourier–Motzkin elimination, 348
FRR, *see* false rejection rate

game theory, 513
    payoff function, 513
Gaussian source, 88, 90
generalized likelihood ratio test, 397
global maximization, 131
GLRT, *see* generalized likelihood ratio test
graphical equivalence, 461
Grassmann graph, 457

hash function, 370, 373
Hasse diagram, 458
Hausdorff distance, 273
helper data, 363, 370
    capacity, 368
    leakage, 364, 370
    rate, 368, 386
helper data generation
    code-offset fuzzy extractor, 378, 386
    comparison, 385
    complementary IBS (C-IBS), 383, 386
    fuzzy commitment, 377, 386
    index-based syndrome coding (IBS), 380, 386
    parity construction, 379, 386
    syndrome construction, 378, 386
    systematic low leakage coding (SLLC), 379, 386
helper message, 430, 439
Henchman problem, 62, 70
hierarchical model, 333–337, 357
high SNR, 115–118, 122, 123, 125, 126
homogeneous CSIT, 203, 205
    DD, 203, 206, 208, 213, 216, 224
    NN, 203, 206, 216, 217
    PP, 203, 205, 216, 217
hybrid CSIT, 203, 211
    DN, 211, 212
    PD, 211
    PN, 211
hypothesis testing, 16

impostor authentication sequence, 423, 424, 427, 430
impostor strategy, 433, 434, 436, 437
information density, 165, 168
information diagrams, 38
information leakage rate, 78, 81
information theoretic security, 181
initial key requirement, 24
Internet of Things, 390
    cellular IoT, 399
IoT, *see* Internet of Things
isotropic signaling, 110, 131

jammer, 258, 259
joint privacy leakage, 454
joint security, 454

key consumption
    excess, 32
    expected, 24, 31
    minimal expected, 37–44
    minimizing, 30
key regeneration, 57
KKT conditions, 110, 112, 114, 117, 132, 133, 136, 137
Kullback–Leibler divergence, 87, 273

layered coding, 82, 83, 99
layered signaling, 232
legitimate authentication sequence, 423, 424, 427, 430
linear codes, 371
linear measurement model, 511
local statistical equivalence property, 213
locally repairable codes, 521
lossless compression, 53
low SNR, 119, 120, 123, 125, 126

MBR, *see* minimum bandwidth regenerating
MIMO, 109, 110, 116, 122, 128, 131
minimax optimization, 133
minimum bandwidth regenerating, 523
minimum storage regenerating, 523
MISO, 110, 113
missed detection, 397–400
MSR, *see* minimum storage regenerating
multi-stage game, 514
    infinite window, 514
multiple antennas, 232
multiple biometric systems, 447, 450
multiple key capacity region, 333, 334, 336, 337, 341–344, 346, 348–352, 357, 358
mutual information, 78, 81, 273, 370, 375, 376, 507

Nakagami-*m* fading, 232
Nash equilibrium, 514
network lifespan, 403–406, 414
networked secure source coding, 77–103
    under a reconstruction privacy, 96
    with a helper, 85
        one-sided, 86, 90
        two-sided, 89, 91
    with an intermediate node, 92
Newton method, 132–135
number of channel uses, 24
    minimal number, 45

omniscience, omniscience scheme, 334, 337
one-norm distance, 272
one-time pad, 29, 45, 53

partition code, 39, 40
passive adversary, 524
phasor measurement units, 500
physical layer security, 77, 183, 266
physical unclonable function
    analogy to source model, 370
    arbiter PUF, 363
    ring oscillator (RO) PUF, 363
    SRAM PUF, 363
PIN model, 333, 334, 352, 357
PM codes, *see* product-matrix codes
PMUs, *see* phasor measurement units
postprocessing matrix, 370, 371, 383, 385
preprocessing matrix, 370, 371, 381, 385
primal/dual method, 133
prisoner's dilemma, 514
privacy, 445, 447, 449, 450, 453–455, 457
privacy leakage, 421, 422, 439, 501, 504, 505, 511, 512
privacy leakage rate, 439
    achievable, 439
privacy protection, 421, 422
privacy–security pair, 454
privacy–utility tradeoff, 500, 511, 512
    efficient frontier, 502, 503, 506, 508, 509, 512
    information theoretic formulation, 503
private information retrieval, 485
private key, 334, 335, 337–341, 349, 351, 352, 357
private streaming search, 491
product-matrix codes, 528
    non-secure PM codes, 528
    secure PM codes, 540
public communication, 366
public helper, 78, 92
PUF, *see* physical unclonable function

quantization, 440

random binning, 334, 338–340, 345
randomized encoding, 260, 276, 306
rate splitting, 94
rate–distortion theory, 59, 68
rate–distortion–equivocation region, 99, 100, 503
rate–distortion–leakage region, 82, 86, 88, 89, 94, 95
rate-constrained authentication, 427
real interference alignment, 153
    complex-field extension, 155
reconstruction, 100
    causal, 100
    memoryless, 100
regional transmission organizations, 511

relaxation, 464, 465
reliability information, 370
residual secret randomness, 24, 31
resiliency capacity, 525
reverse water-filling, 506

same marginal property, 232
sanitized database, 502
secrecy capacity, 111–114, 116, 117, 121, 122, 127, 135, 183, 265, 277, 432, 525
    bound, 119, 121
    uncorrelated, 267, 268, 277, 279, 291
secrecy criterion, 276
    effective secrecy, 4, 15
    maximum strong secrecy, 276
    mean secrecy, 276, 277
    mean strong secrecy, 276
    perfect secrecy, 21, 53
    strong secrecy, 3, 143, 148, 159, 160, 173, 175, 267, 276, 277, 286, 300
    weak secrecy, 3, 143, 149, 204
secret common randomness, 268, 277, 279
secret key, 333–335, 337–340, 342, 345, 348, 351, 358, 359
    achievable rate, 367, 368
    agreement, 366
    capacity, 368, 386
    generation, 93, 95, 422, 430, 431
    rate, 368, 386, 422, 432
secret-based authentication with privacy protection, 439
secret-based biometric authentication, 430, 431
secret-based biometric authentication with privacy protection, 439
secure sketch, 451, 452
secure triangular source coding, 92
security, 445, 447, 449, 450, 453–455, 457
sensor networks, 516
separation scheme, 531
Shannon cipher system, 52
Shannon's additivity problem, 265
Shannon's fundamental bound for perfect secrecy, 21–22
shared randomness, 260, 265, 277, 284
side information, 78, 79, 81, 83, 85, 86, 89, 91–93, 95–97, 100
    causal, 100
    coded, 85, 86, 89, 91
    common, 91, 93, 95
    degraded, 83, 95, 96
    pattern, 92
signaling efficiency, 405
Slepian–Wolf coding, 91, 434
Slepian–Wolf condition, 345, 347, 348, 351, 357

smart grid
  definition, 499
  motivation, 499
smart meter, 504
  Markov chain model, 508
  privacy concerns, 504
smart meter privacy
  active control approach, 506
  source coding approach, 505
smart meters, 500
social networks, 516
source node, 391
spreading of signals, 261
statistical equivalence property, 210
stealth, 4, 16
stochastic control, 506
stochastic decoder, 53, 98
stochastic encoder, 53
stochastic order, 232
  convex order, 232
  increasing convex order, 232
  usual stochastic order, 232
storage rate, 427, 428
  achievable, 427
subspace codes, 450
super-activation, 278, 283, 285, 286, 308, 316, 323
super-additivity, 314, 326
superposition coding, *see* layered coding
symmetrizability, 278, 280, 283, 286, 292, 294, 302, 304, 307
syndrome, 451

template protection, *see* privacy protection
topology, 181, 192
  fixed, 192
total variation distance, 272
transformational equivalence, 461
typical sequences, 426, 434
typical set, 288
typicality, 288

unified algebraic description, 370
unknown and varying eavesdropper channel, 173

variable length coding, 56
variational distance, 165, 166, 170, 173

water-filling, 110, 115, 116, 123, 128, 130
WF, *see* water-filling
wireless sensor network, 392
wiretap channel, 183, 259, 265
  MIMO, 109–113, 116, 121, 127–131, 141–143, 145–150, 159–161, 178
worst-case measures, 454
Wyner–Ziv coding, 82, 83, 94, 512

zero-forcing, 110, 129
ZF, *see* zero-forcing